Study Guide and Solutions Manual

for

McMurry and Simanek's

Fundamentals of Organic Chemistry

Sixth Edition

Susan McMurry

Australia • Brazil • Canada • Mexico • Singapore • Spain • United Kingdom • United States

Printed in the United States of America
1 2 3 4 5 6 7 10 09 08 07 06

Printer: Thomson/West

0-495-01932-1
Cover image: Peter Barrett/Masterfile

Thomson Higher Education
10 Davis Drive
Belmont, CA 94002-3098
USA

Preface

If you're in a typical organic chemistry course, you attend lecture and take notes, read the text, work some problems, and take tests. For many of you, performing this ritual is all that's necessary to succeed in organic chemistry. Many others of you, however, follow all the correct steps but still feel bewildered by the course. In such a situation, supplementary material is needed.

This study guide has two functions. First, it gives an overview of the course, both in individual chapters and in appendices. Second, it furnishes solutions to the problems presented in the text. The first of these functions might be described as "the big picture," and the second as "the details." Understanding both the big picture and the details is necessary if organic chemistry is to be more than the memorization of assorted unrelated facts.

How to use this book

This *Study Guide and Solutions Manual* can't perform miracles if you don't read *Fundamentals of Organic Chemistry*. In all cases, step one is to go to class, take notes, and read the text. Then, turn to the Study Guide.

Study the Chapter Outline. The outline should help you to see how the topics in the chapter are related. Many people are able to learn the facts in each chapter, but are unable to recognize the underlying principles.The outline will clarify both the relationships between reactions and how these reactions are related to larger concepts.

Solve the problems in the text. At first, don't use the *Solutions Manual* to help you with the problems. After completing the problems, consult the *Solutions Manual* to see if your answer is correct and if your method of solution is logical and systematic. If you're totally confused by a problem, carefully read the solution; then try to solve similar problems on your own. If possible, use molecular models to make the three-dimensional nature of organic compounds more clear.

Before a test or final exam, consult the appendices in the *Study Guide*. Many of these appendices summarize or tabulate information that has been presented over several chapters. Especially helpful before an exam are the following sections: Glossary, Reagents Used in Organic Synthesis, Summary of Functional Group Preparations, and Summary of General Reaction Mechanisms. Other appendices present interesting chemical facts and tables.

Review quizzes are located at the end of the book. Each quiz covers material from a few chapters and allows you check your knowledge of more than one chapter at a time.

For most people, understanding organic chemistry takes a long time—sometimes longer than the duration of an organic chemistry course. I hope that the combination of *Fundamentals of Organic Chemistry* plus this *Study Guide and Solutions Manual* makes the study of organic chemistry easier and more rewarding for you.

Acknowledgments

I would like to thank the following people for their assistance with this book: John McMurry (of course), Eric Simanek, Debra Boehmler, John Scheffer, Rodney Boyer, Joseph Hornback and Ellen Bitter.

Contents

Solutions to Problems:

Appendices:

Chapter 1 – Structure and Bonding; Acids and Bases

Chapter Outline

I. Atomic structure (Sections 1.1–1.2).
  A. Introduction to atomic structure (Section 1.1).
    1. Atoms consist of a dense, positively charged nucleus surrounded by negatively charged electrons.
      a. The nucleus is made up of positively charged protons and uncharged neutrons.
      b. The nucleus contains most of the mass of the atom.
      c. Electrons move about the nucleus at a distance of about 10^{-10} m.
    2. The atomic number (Z) gives the number of protons in the nucleus.
    3. The mass number (A) gives the total number of protons and neutrons.
       Isotopes of an element have the same number of protons and electrons but differ in the number of neutrons.
    4. All atoms of a given element have the same value of Z.
  B. Orbitals.
    1. The distribution of electrons in an atom can be described by a wave equation.
      a. The solution to a wave equation is an orbital, represented by Ψ.
      b. Ψ^2 predicts the volume of space in which an electron is likely to be found.
    2. An atom's electrons are organized into shells.
      a. The shells differ in the numbers and kinds of orbitals they have.
      b. Electrons in different orbitals have different energies.
      c. Each orbital can hold two electrons.
      d. The two lowest-energy electrons are in the 1*s* orbital.
    3. There are four different kinds of orbitals (*s*, *p*, *d*, *f*).
      a. The *s* orbitals are spherical.
      b. The *p* orbitals are dumbbell-shaped.
      c. Four of the five *d* orbitals are cloverleaf-shaped.
  C. Electronic configuration of atoms (Section 1.2).
    1. The ground-state electron configuration of an atom is a listing of the orbitals occupied by the electrons of the atom.
    2. Rules for predicting the ground-state electron configuration of an atom:
      a. Orbitals with the lowest energy levels are filled first.
         The order of filling is 1*s*, 2*s*, 2*p*, 3*s*, 3*p*, 4*s*, 3*d*.
      b. Only two electrons can occupy each orbital, and they must be of opposite spin.
      c. If two or more orbitals have the same energy, one electron occupies each orbital until all are half-full (Hund's rule). Only then does a second electron occupy one of the orbitals.
         All of the electrons in a half-filled shell have the same spin.

II. Chemical bonds (Sections 1.3–1.5).
  A. Development of chemical bonding theory (Section 1.3).
    1. Kekulé and Couper proposed that carbon has four "affinity units" – carbon is tetravalent.
    2. Other scientists suggested that carbon can form double bonds, triple bonds and rings.
    3. Van't Hoff proposed that the 4 atoms to which carbon forms bonds sit at the corners of a regular tetrahedron.
    4. In a drawing of a tetrahedral carbon, a wedged line represents a bond pointing toward the viewer, and a dashed line points behind the plane of the page.

B. Chemical bonds (Section 1.4).
  1. Atoms bond together because the resulting compound is more stable than the individual atoms.
    a. Atoms tend to achieve the electron configuration of the nearest noble gas.
    b. Atoms either lose electrons (groups 1A, 2A) or gain electrons (group 7A) to form ionic compounds.
    c. Atoms in the middle of the periodic table share electrons by forming covalent bonds.
  2. The number of covalent bonds formed by an atom depends on the number of electrons it has and on the number it needs to achieve an octet.
  3. Covalent bonds can be represented two ways.
    a. In Lewis structures, bonds are represented as pairs of dots.
    b. In line-bond structures, bonds are represented as lines drawn between two atoms.
  4. Valence electrons not used for bonding are called lone-pair electrons.
     Lone-pair electrons are represented as dots.

C. Formation of covalent bonds (Section 1.5).
  1. Covalent bonds are formed by the overlap of two atomic orbitals, each of which contains one electron. The two electrons have opposite spins.
  2. Bond strength is the measure of the amount of energy needed to break a bond.
  3. Bond length is the optimum distance between nuclei.
  4. Every bond has a characteristic bond length and bond strength.

III. Hybridization (Sections 1.6–1.8).

A. sp^3 Orbitals (Sections 1.6, 1.7).
  1. Structure of methane (Section 1.6).
    a. When carbon forms 4 bonds with hydrogen, one 2*s* orbital and three 2*p* orbitals combine to form four equivalent atomic orbitals (sp^3 hybrid orbitals).
    b. These orbitals are tetrahedrally oriented.
    c. Because these orbitals are unsymmetrical, they can form stronger bonds than unhybridized orbitals can.
    d. These bonds have a specific geometry and a bond angle of 109.5°.
  2. Ethane has the same type of hybridization as occurs in methane (Section 1.7).

B. Double and triple bonds (Section 1.8).
  1. Properties of double and triple bonds.
    a. These bonds are more reactive than single bonds.
    b. However, they are not as strong as two or three single bonds.
    c. Double bonds are flat. Triple bonds are linear.
  2. Double bonds.
    a. If one carbon 2*s* orbital combines with two carbon 2*p* orbitals, three hybrid sp^2 orbitals are formed, and one *p* orbital remains unchanged.
    b. The three sp^2 orbitals lie in a plane at angles of 120°, and the *p* orbital is perpendicular to them.
    c. Two different types of bond form between two carbons.
      i. A σ bond forms from the overlap of two sp^2 orbitals.
      ii. A π bond forms by sideways overlap of two *p* orbitals.
      iii. This combination is known as a carbon-carbon double bond.
  3. Triple bonds.
    a. If one carbon 2*s* orbital combines with one carbon 2*p* orbital, two hybrid *sp* orbitals are formed, and two *p* orbitals are unchanged.
    b. The two *sp* orbitals are 180° apart, and the two *p* orbitals are perpendicular to them and to each other.
    c. Two different types of bonds form.
      i. A σ bond forms from the overlap of two *sp* orbitals.

ii. Two π bonds form by sideways overlap of four p orbitals.
iii. This combination is known as a carbon-carbon triple bond.

IV. Polar covalent bonds: electronegativity (Section 1.9).
A. Although some bonds are totally ionic and some are totally covalent, most chemical bonds are polar covalent bonds.
In these bonds, electrons are attracted to one atom more than to the other atom.
B. Bond polarity is due to differences in electronegativity (EN).
Elements on the right side of the periodic table are more electronegative than elements on the left side.
C. The difference in EN between two elements can be used to predict the polarity of a bond.
1. If ΔEN < 0.4, a bond is nonpolar covalent.
2. If ΔEN is between 0.4 and 2.0, a bond is polar covalent.
3. If ΔEN > 2.0, a bond is ionic.
4. The symbols $\delta+$ and $\delta-$ are used to indicate partial charges.
5. A crossed arrow is used to indicate bond polarity.
D. An inductive effect is an atom's ability to polarize a bond.

V. Acids and bases (Sections 1.10 - 1.11).
A. Brønsted-Lowry definition (Section 1.10).
1. A Brønsted-Lowry acid donates an H^+ ion; a Brønsted-Lowry base accepts H^+.
2. The product that results when a base gains H^+ is the conjugate acid of the base; the product that results when an acid loses H^+ is the conjugate base of the acid.
B. Acid and base strength.
1. A strong acid reacts almost completely with water.
2. The strength of an acid in water is indicated by K_a, the acidity constant.
3. Strong acids have large acidity constants, and weaker acids have smaller acidity constants.
4. The pK_a is normally used to express acid strength.
a. $pK_a = -\log K_a$
b. A strong acid has a small pK_a, and a weak acid has a large pK_a.
c. The conjugate base of a strong acid is a weak base, and the conjugate base of a weak acid is a strong base.
5. Predicting acid–base reactions from pK_a.
a. An acid with a low pK_a reacts with the conjugate base of an acid with a high pK_a
b. In other words, the products of an acid–base reaction are more stable than the reactants, and the reaction favors formation of the weaker acid/base pair.
6. Organic acids and organic bases.
a. There are two main types of organic acids:
i. Acids that contain hydrogen bonded to oxygen.
ii. Acids that have hydrogen bonded to the carbon next to a C=O group.
b. The main type of organic base contains a nitrogen atom with a lone electron pair.
C. Lewis acids and bases (Section 1.11).
1. A Lewis acid accepts an electron pair.
a. A Lewis acid may have either a vacant low-energy orbital or a polar bond to hydrogen.
b. Examples include metal cations, halogen acids, Group 3 compounds and transition-metal compounds.
2. A Lewis base has a pair of nonbonding electrons.
a. Most oxygen- and nitrogen-containing organic compounds are Lewis bases.
b. Many organic Lewis bases have more than one basic site.
3. A curved arrow shows the movement of electrons from a Lewis base to a Lewis acid.

Solutions to Problems

1.1 **Strategy:** The elements of the periodic table are organized into groups that are based on the number of outer-shell electrons each element has. For example, an element in group 1A has one outer-shell electron, and an element in group 5A has five outer-shell electrons. To find the number of outer-shell electrons for a given element, use the periodic table to locate its group.

Solution:
(a) Potassium is a member of group 1A and thus has one outer-shell electron.
(b) Calcium (group 2A) has two outer-shell electrons.
(c) Aluminum (group 3A) has three outer-shell electrons.

1.2 **Strategy:** (a) To find the ground-state electron configuration of an element, first locate its atomic number. For boron, the atomic number is 5; boron thus has 5 protons and 5 electrons. Next, assign the electrons to the proper energy levels, starting with the lowest level. Fill each level *completely* before assigning electrons to a higher energy level.

Solution:

	2*p*	↑ — —
Boron	2*s*	↑↓
	1*s*	↑↓

Remember that only two electrons can occupy the same orbital, and that they must be of opposite spin.

A different way to represent the ground-state electron configuration is to simply write down the occupied orbitals and to indicate the number of electrons in each orbital. For example, the electron configuration for boron is $1s^2\ 2s^2\ 2p^1$.

(b) Let's consider an element with many electrons. Phosphorus, with an atomic number of 15, has 15 electrons. Assigning these to energy levels:

	3*p*	↑ ↑ ↑
	3*s*	↑↓
Phosphorus	2*p*	↑↓ ↑↓ ↑↓
	2*s*	↑↓
	1*s*	↑↓

Notice that the $3p$ electrons are all in different orbitals; we must place one electron into each orbital of the same energy level until all orbitals are half-filled.

The more concise way to represent ground-state electron configuration for phosphorus: $1s^2\ 2s^2\ 2p^6\ 3s^2\ 3p^3$.

(c) Oxygen (atomic number 8)

$2p$ ↑↓ ↑ ↑

$2s$ ↑↓

$1s$ ↑↓

$1s^2\ 2s^2\ 2p^4$

(d) Argon (atomic number 18)

$3p$ ↑↓ ↑↓ ↑↓

$3s$ ↑↓

$2p$ ↑↓ ↑↓ ↑↓

$2s$ ↑↓

$1s$ ↑↓

$1s^2\ 2s^2\ 2p^6\ 3s^2\ 3p^6$

1.3 **Strategy:** A solid line represents a bond lying in the plane of the page, a wedged bond represents a bond pointing out of the plane of the page toward the viewer, and a dashed bond represents a bond pointing behind the plane of the page.

Solution:

H, H, H, C, Cl

Chloromethane

1.4

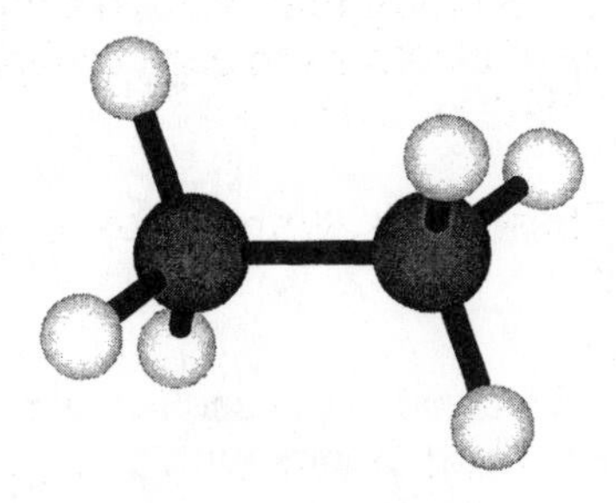

H, H, H, C, C, H, H, H

Ethane

1.5 **Strategy:** Identify the group of the central element to predict the number of covalent bonds the element can form.

Solution: (a) Carbon (group 4A) has four electrons in its valence shell and forms four bonds to achieve the noble-gas configuration of neon. Thus, a likely formula is CCl_4.

Element	*Group*	*Likely Formula*
(b) Al	3A	AlH_3
(c) C	4A	CH_2Cl_2
(d) Si	4A	SiF_4

1.6 **Strategy:** Follow these three steps for drawing the Lewis structure of a molecule.

(1) Determine the number of valence, or outer-shell, electrons for each atom in the molecule. For chloroform, we know that carbon has four valence electrons, hydrogen has one valence electron, and each chlorine has seven valence electrons.

·C· (4 dots)	4 x 1 = 4
H·	1 x 1 = 1
:Cl· (7 dots)	7 x 3 = 21
	26 total valence electrons

(2) Next, use two electrons for each single bond.

```
      H
      ••
Cl : C : Cl
      ••
      Cl
```

(3) Finally, use the remaining electrons to achieve an noble gas configuration for all atoms.

	Molecule	*Lewis structure*	*Line-bond structure*
(a)	$CHCl_3$	H, :Cl:C:Cl:, :Cl: (each Cl with 8 electrons)	Cl—C—Cl with H above C and Cl below C
(b)	H_2S 8 valence electrons	H:S: (two lone pairs on S), H below S	H—S with H below S
(c)	CH_3NH_2 14 valence electrons	H H above; H:C:N:H; H below C (one lone pair on N)	H—C—N—H with H above C and N, H below C

1.7 Each of the two carbons has 4 valence electrons. Two electrons are used to form the carbon-carbon bond, and the 6 electrons that remain can form bonds with a maximum of 6 hydrogens. Thus, the formula C_2H_7 is not possible.

1.8

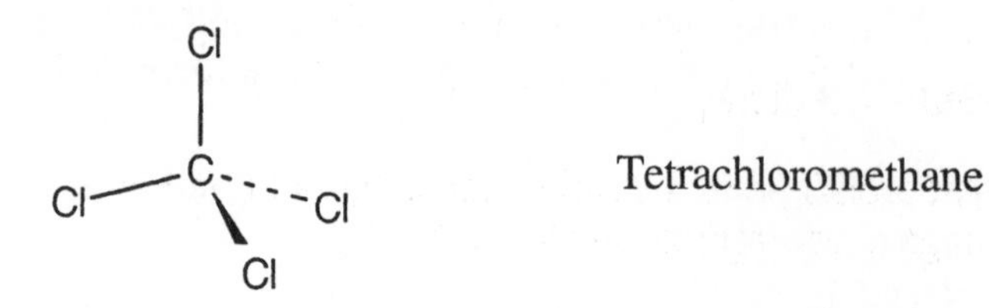

1.9 An electron in the second shell is farther from the nucleus than an electron in the first shell. Thus, a bond formed by the overlap of an sp^3 orbital of carbon (second shell) and a 1*s* orbital of hydrogen (first shell) is longer than a bond formed from two 1*s* orbitals (H–H bond).

1.10

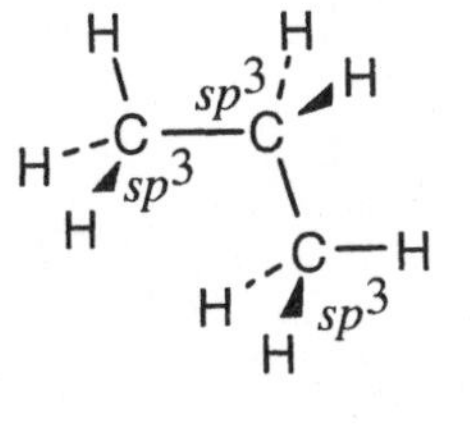

Propane

The geometry around all carbon atoms is tetrahedral, and all bond angles are approximately 109°.

1.11 The atoms of acetaldehyde contribute 18 valence electrons. Ten electrons are used in the 5 single bonds, 4 electrons are involved in the carbon-oxygen double bond, and 4 electrons form the 2 lone pairs of electrons on oxygen.

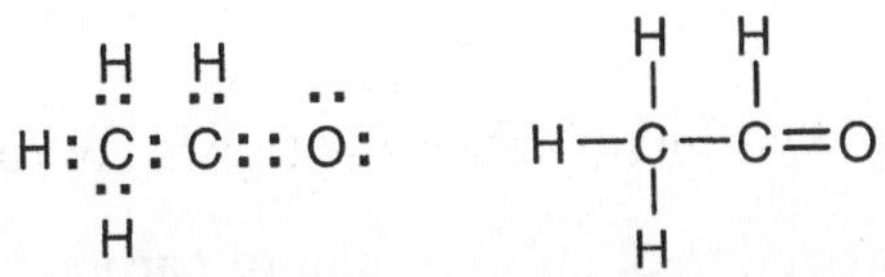

Acetaldehyde

1.12

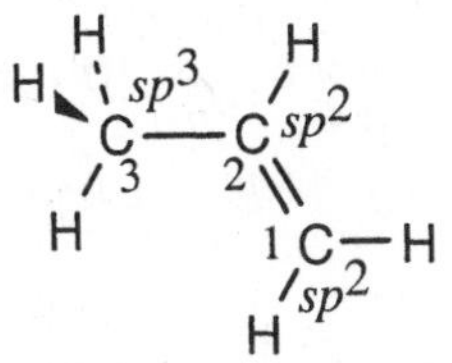

Propene

The C3–H bonds are σ bonds formed by overlap of an sp^3 orbital of carbon 3 with an *s* orbital of hydrogen. Bond angles at C3 are approximately 109°.

The C2–H and C1–H bonds are σ bonds formed by overlap of an sp^2 orbital of carbon with an *s* orbital of hydrogen.

The C2–C3 bond is a σ bond formed by overlap of an sp^3 orbital of carbon 3 with an sp^2 orbital of carbon 2.

There are two C1–C2 bonds. One is a σ bond formed by overlap of an sp^2 orbital of carbon 1 with an sp^2 orbital of carbon 2. The other bond is a π bond formed by overlap of a 2*p* orbital of carbon 1 with a 2*p* orbital of carbon 2. All four atoms connected to the carbon–carbon double bond lie in the same plane, and all bond angles between these atoms are 120°.

1.13

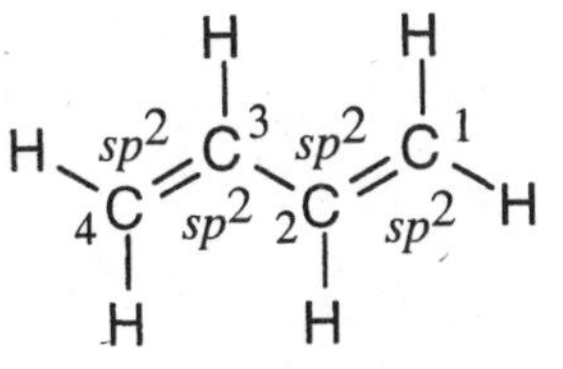

Buta-1,3-diene

All atoms lie in the same plane, and all bond angles are approximately 120°.

1.14

Aspirin

All carbons are sp^2 hybridized, with the exception of the indicated carbon. All oxygen atoms have two lone pairs of electrons.

1.15

Propyne

The C3–H bonds are σ bonds formed by overlap of an sp^3 orbital of carbon 3 with an *s* orbital of hydrogen. Bond angles at C3 are approximately 109°.

The C1–H bond is a σ bond formed by overlap of an *sp* orbital of carbon 1 with an *s* orbital of hydrogen.

The C2–C3 bond is a σ bond formed by overlap of an *sp* orbital of carbon 2 with an sp^3 orbital of carbon 3.

There are three C1–C2 bonds. One is a σ bond formed by overlap of an *sp* orbital of carbon 1 with an *sp* orbital of carbon 2. The other two bonds are π bonds formed by overlap of two 2*p* orbitals of carbon 1 with two 2*p* orbitals of carbon 2.

The three carbon atoms of propyne lie on a straight line, with a bond angle of 180°. The $H–C_1–C_2$ bond angle is also 180°.

1.16 Use Figure 1.12 to answer this problem. The larger the EN, the more electronegative the element.

	More electronegative	*Less electronegative*
(a)	H (2.1)	Li (1.0)
(b)	Br (2.8)	Be (1.6)
(c)	Cl (3.0)	I (2.5)

1.17 As in Problem 1.16, use Figure 1.12. Remember that the arrow points toward the more electronegative atom in the bond.

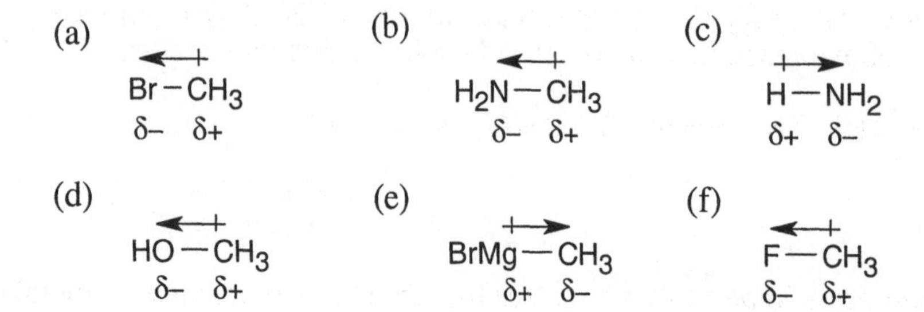

1.18 **Strategy:** Use Figure 1.12 to locate the EN of each element. The larger the difference in EN, the more ionic the bond.

Solution:

CCl_4	$MgCl_2$	$TiCl_3$	Cl_2O
Cl : EN = 3.0	Cl : EN = 3.0	Cl : EN = 3.0	Cl : EN = 3.0
C : EN = 2.5	Mg : EN = 1.2	Ti : EN = 1.5	O : EN = 3.5
Δ EN = 0.5	Δ EN = 1.8	Δ EN = 1.5	Δ EN = 0.5

Least ionic —> Most ionic

CCl_4 and Cl_2O, $TiCl_3$, $MgCl_2$

1.19 In an electrostatic potential map, the color red indicates the areas of a molecule that are most electron-rich. Because the region around the oxygen atom of methyl alcohol is red, oxygen is the more electronegative atom in the C–O bond, and the bond is polarized in the indicated direction.

H, O: δ–, δ+, C, H, H, H

Methyl alcohol

1.20 (a) On your calculator, enter 3.75, and give it a negative sign, using the +/- key. Use INV, then LOG (or use ANTILOG), and convert the answer to scientific notation. K_a of formic acid is 1.8 x 10^{-4}, and K_a of picric acid is 0.42.
(b) Since a lower pK_a value indicates a stronger acid, picric acid is stronger than formic acid.

1.21 HO–H is a stronger acid than H_2N–H. Since H_2N^- is a stronger base than HO^-, the conjugate acid of H_2N^- (H_2N–H) is a weaker acid than the conjugate acid of HO^- (HO–H).

1.22

(a)

$$H{-}CN + CH_3CO_2^-\,Na^+ \xrightarrow{?} Na^+\,{}^-CN + CH_3CO_2H$$

pK_a = 9.3 — Weaker acid; pK_a = 4.7 — Stronger acid

Remember that the lower the pK_a, the stronger the acid. Thus CH_3CO_2H, not HCN, is the stronger acid, and the above reaction will not take place in the direction written.

(b)

$$CH_3CH_2O{-}H + Na^+\,{}^-CN \xrightarrow{?} CH_3CH_2O^-\,Na^+ + HCN$$

pK_a = 16 — Weaker acid; pK_a = 9.3 — Stronger acid

Using the same reasoning as in part (a), we can see that the above reaction will not take place.

1.23 A Lewis base has a nonbonding electron pair to share. A Lewis acid has a vacant orbital to accept an electron pair. Look for a lone electron pair when identifying a Lewis base.

Lewis acids: $MgBr_2$, $B(CH_3)_3$, $^+CH_3$

Lewis bases: $CH_3\ddot{N}HCH_3$, $CH_3\ddot{P}(CH_3)CH_3$

Both: $CH_3CH_2\ddot{O}-H$

1.24 **Strategy:** Locate the electron pair(s) of the Lewis base and draw a curved arrow from the electron pair to the Lewis acid. The electron pair moves from the atom at the tail of the arrow (Lewis base) to the atom at the point of the arrow (Lewis acid).

Solution: (Note: electron dots have been omitted from Cl^- to reduce clutter.)

(a)

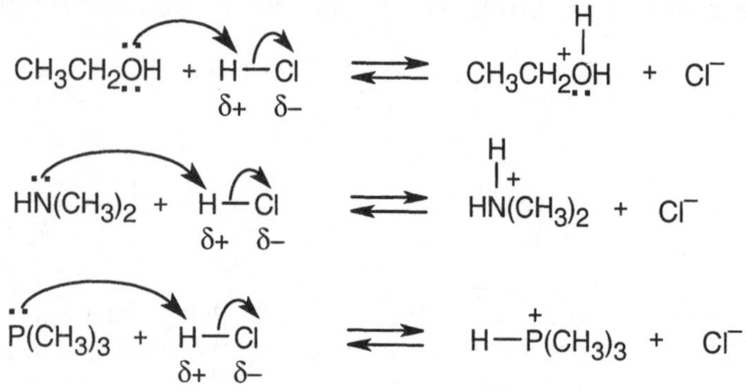

$$CH_3CH_2\ddot{O}H + \overset{\delta+}{H}-\overset{\delta-}{Cl} \rightleftharpoons CH_3CH_2\overset{+}{O}H(H) + Cl^-$$

$$H\ddot{N}(CH_3)_2 + \overset{\delta+}{H}-\overset{\delta-}{Cl} \rightleftharpoons H\overset{+}{N}(H)(CH_3)_2 + Cl^-$$

$$\ddot{P}(CH_3)_3 + \overset{\delta+}{H}-\overset{\delta-}{Cl} \rightleftharpoons H-\overset{+}{P}(CH_3)_3 + Cl^-$$

(b)

$$H\ddot{O}:^- + {}^+CH_3 \rightleftharpoons H\ddot{O}-CH_3$$

$$H\ddot{O}:^- + B(CH_3)_3 \rightleftharpoons H\ddot{O}-\overset{-}{B}(CH_3)_3$$

$$H\ddot{O}:^- + MgBr_2 \rightleftharpoons H\ddot{O}-\overset{-}{Mg}Br_2$$

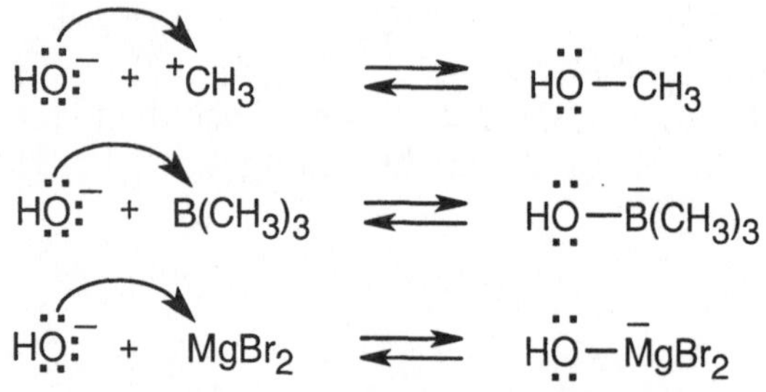

1.25 According to the electrostatic potential map, the starred hydrogen is most acidic, and the starred nitrogen is most basic (electron-rich).

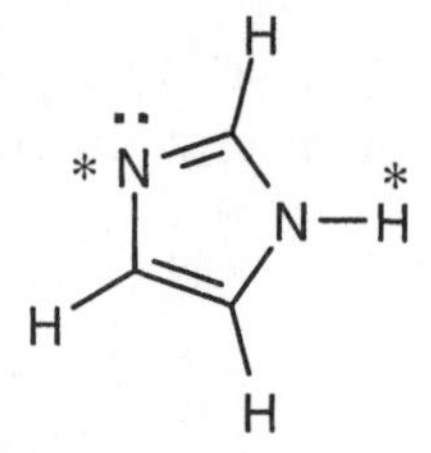

Imidazole

Visualizing Chemistry

1.26

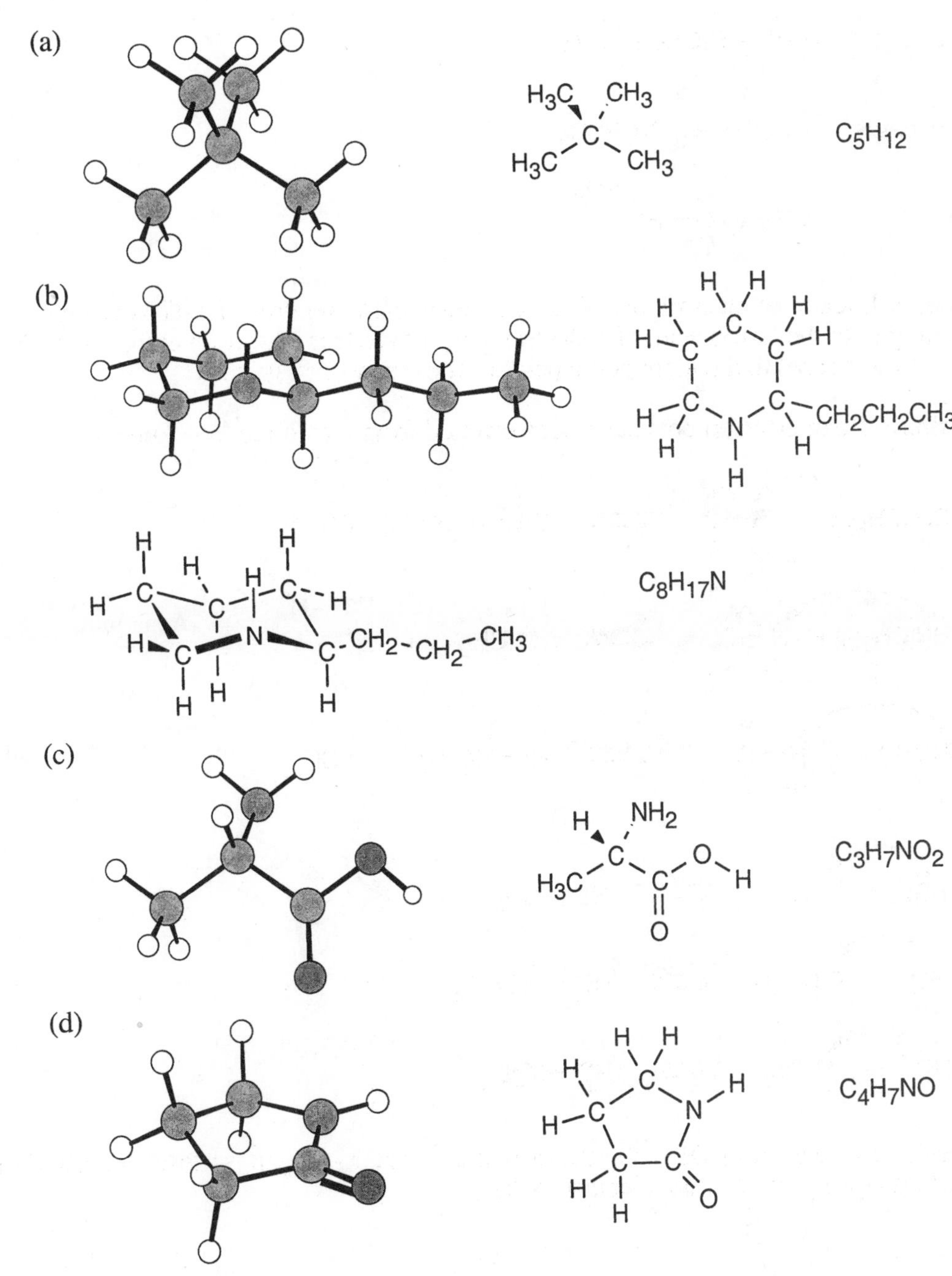

1.27

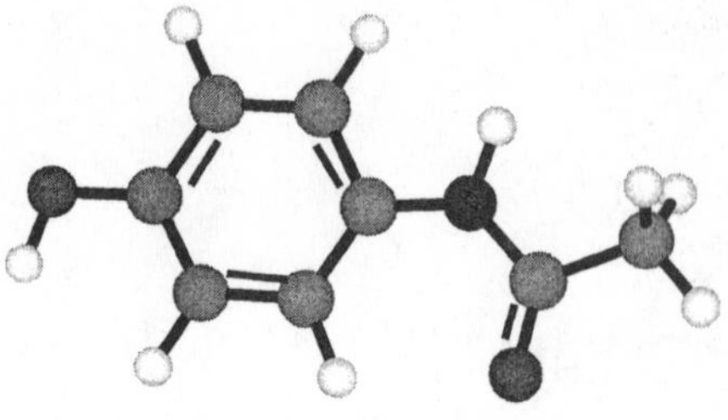

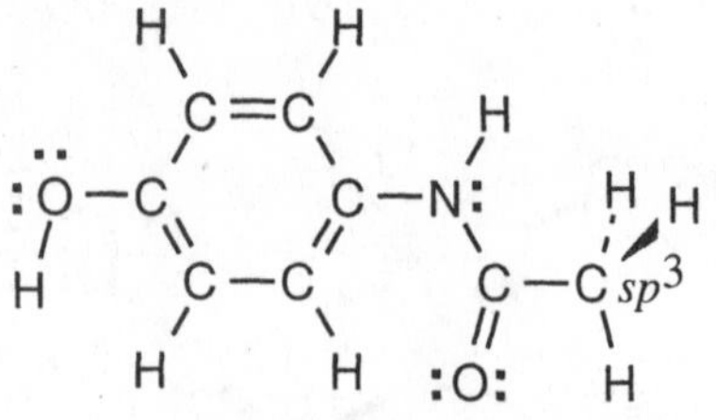

Acetaminophen

All carbons are sp^2 hybridized, except for the carbon indicated as sp^3. The two oxygen atoms and the nitrogen atom have lone pair electrons, as shown.

1.28

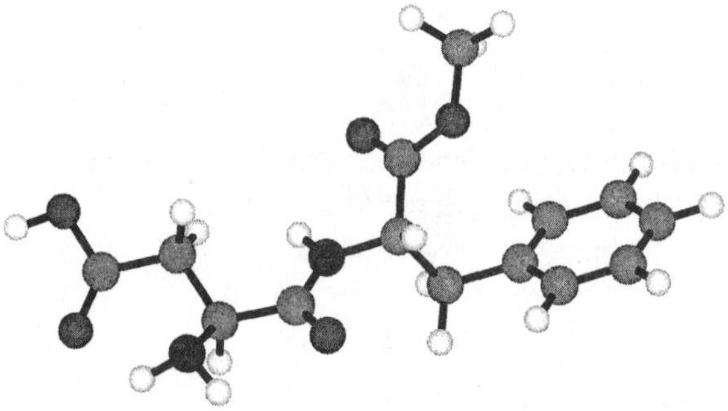

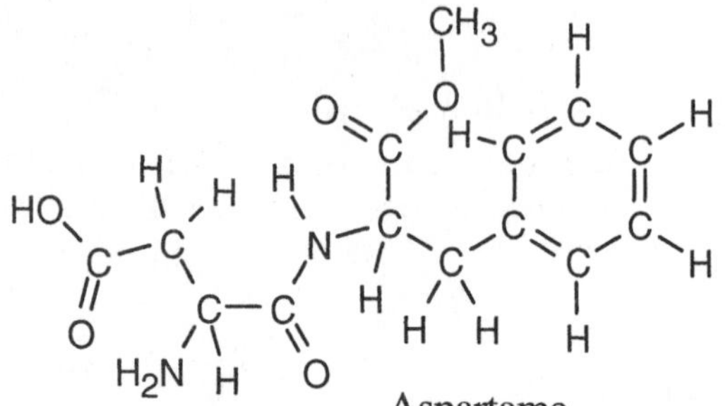

Aspartame

1.29 Electrostatic potential maps show that the most electron-rich atom in (a) is oxygen, not nitrogen. In (b), nitrogen is electron-rich and is thus more basic than the nitrogen in (a). In (a), the hydrogen atoms bonded to nitrogen are more acidic (electron-poor) than all other hydrogens in both molecules.

(a) more acidic

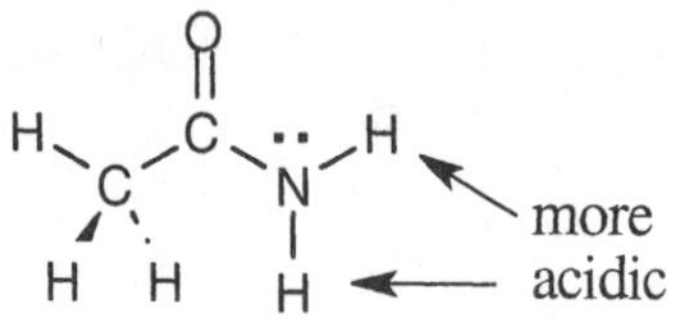

Acetamide

(b) more basic

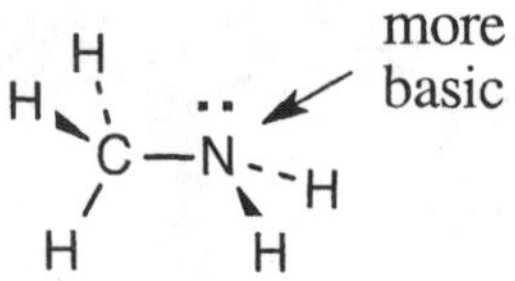

Methylamine

Additional Problems

Electron configuration

1.30 See Problem 1.1 if you need help.

Element	*Group*	*Number of outer shell electrons*
(a) Oxygen	6A	6
(b) Magnesium	2A	2
(c) Fluorine	7A	7

1.31

Element	*Atomic Number*	*Ground-state electron configuration*
(a) Li	3	$1s^2\ 2s^1$
(b) Na	11	$1s^2\ 2s^2\ 2p^6\ 3s^1$
(c) Al	13	$1s^2\ 2s^2\ 2p^6\ 3s^2\ 3p^1$
(d) S	16	$1s^2\ 2s^2\ 2p^6\ 3s^2\ 3p^4$

Lewis Structures

1.32 (a) $AlCl_3$ (b) CF_2Cl_2 (c) NI_3

1.33

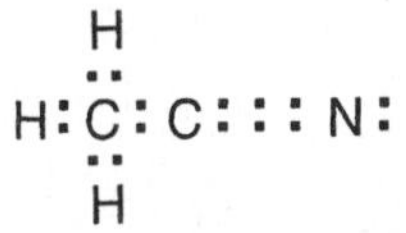

H:C:C:::N: (with H above and below the first C) Acetonitrile

In acetonitrile, nitrogen has 8 electrons in its valence shell. Six are used in the carbon–nitrogen triple bond, and two are a nonbonding electron pair.

1.34

(a) $CH_3-\ddot{O}-CH_3$

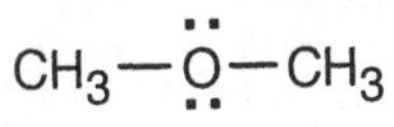

(b) $CH_3-\ddot{N}-H$ (with H below N)

(c) H–C–Cl: (with H above C and :Cl: below C)

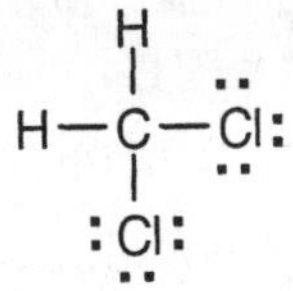

(d) $H_3C-C(=\ddot{O})-CH_3$

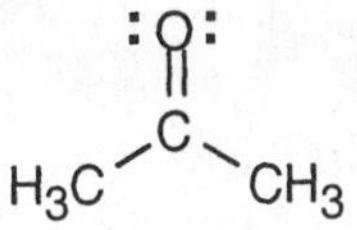

1.35

(a) $CH_3\ddot{S}H$

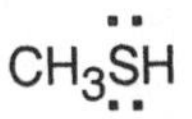

(b) $H_3C-\ddot{N}-CH_3$ (with CH_3 below N)

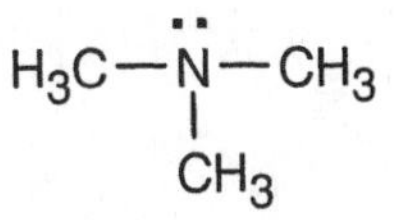

(c) $CH_3CH_2\ddot{Br}$:

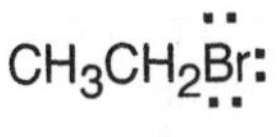

(d) $CH_3C(=\ddot{O})-\ddot{O}H$

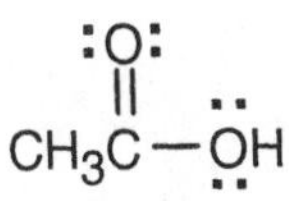

(e) $CH_3C(=\ddot{O})-\ddot{Cl}$:

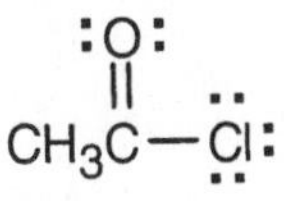

Structural Formulas

1.36

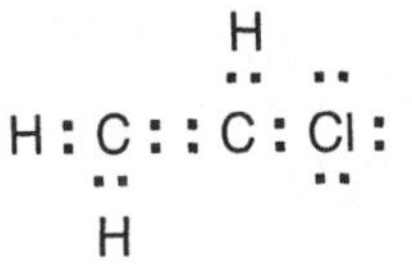

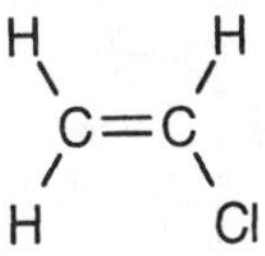

H:C::C:Cl: (with H above the second C and H below the first C)

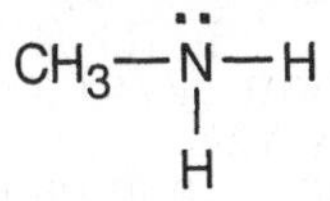

$H_2C=CHCl$ (H, H on one carbon; H, Cl on the other) Vinyl chloride

1.37

```
    H   H   H   H              H   H   H
    |   |   |   |              |   |   |
H — C — C — C — C — H      H — C — C — C — H
    |   |   |   |              |   |   |
    H   H   H   H              H   |   H
                               H — C — H
                                   |
                                   H
```

The two structures differ in the connection of the carbon atoms.

1.38 To solve a problem of this sort, draw all possible structures that have the correct number of bonds. You must systematically consider all possible arrangements of atoms, including those that have branches, rings and multiple bonds.

(a)

```
    H   H   H
    |   |   |
H — C — C — C — H
    |   |   |
    H   H   H
```

(b)

```
    H   H   H                  H   H   H
    |   |   |                  |   |   |
H — C — C — C — Br   and   H — C — C — C — H
    |   |   |                  |   |   |
    H   H   H                  H   Br  H
```

These are the only two possible structures for compounds with the formula C_3H_7Br.

(c)

```
    H   H   H                   H     H
    |   |   |                    \   /
H — C — C = C — H     and         C — C
    |                          H /  \ /  \ H
    H                               C
                                   / \
                                  H   H
```

(d)

```
    H       H                  H   H
    |       |                  |   |
H — C — O — C — H    and   H — C — C — O — H
    |       |                  |   |
    H       H                  H   H
```

1.39

```
        Cl                      H           O — H
       /                         \         /
Cl — C                       H — C ——— C
       \ Cl                      /       \ H
        H                       H         H
```

Chloroform   Ethanol

1.40

(a) Ethyl methyl ether

(b) Butane

(c) Cyclohexene

Electronegativity

1.41 Ionic bonds: BeF_2 (Δ EN = 2.4)
Covalent bonds: SiH_4 (Δ EN = 0.3), CBr_4 (Δ EN = 0.3)

1.42

(a) Br—Br nonpolar

(b) CH_3Cl: C ($\delta+$)—Cl ($\delta-$)

(c) H ($\delta+$)—F ($\delta-$)

(d) CH_3CH_2OH: C ($\delta+$)—O ($\delta-$); O—H ($\delta+$)

Molecules (b)–(d) contain polar covalent bonds. Carbon–hydrogen bonds are only slightly polar.

1.43

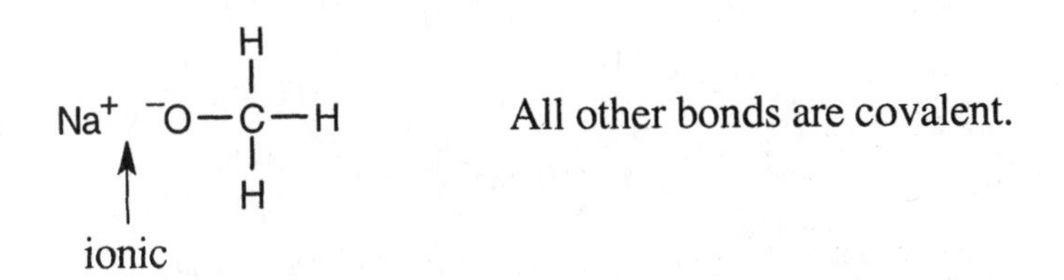

All other bonds are covalent.

1.44 Use Figure 1.12 if you need help. The most electronegative element is starred.

(a) $CH_2\overset{*}{F}Cl$ (b) $\overset{*}{F}CH_2CH_2CH_2Br$ (c) $H\overset{*}{O}CH_2CH_2NH_2$ (d) $CH_3\overset{*}{O}CH_2Li$

1.45

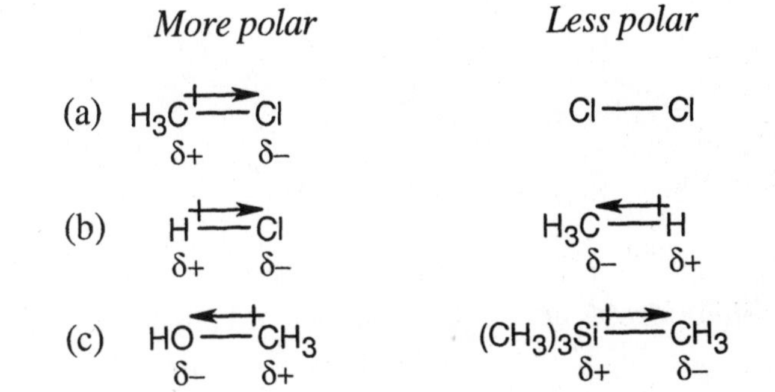

1.46 Carbon is most positive when it is bonded to the most electronegative atom.

Most negative carbon ———————> *Most positive carbon.*

CH_3Li, $CH_3–CH_3$, $CH_3–I$, $CH_3–NH_2$, $CH_3–OH$, $CH_3–F$

1.47

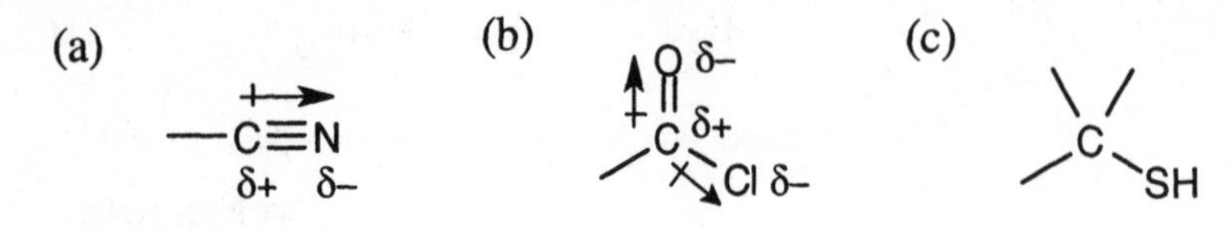

nitrile acid chloride thiol

The carbon-sulfur bond is not polar because carbon and sulfur have the same electronegativity.

Hybridization

1.48 Refer to Problem 1.33 for the structure of acetonitrile. The $H_3C–$ carbon is sp^3 hybridized, and the –CN carbon is sp hybridized.

1.49

(a) Butane (all four carbons sp^3)

(b) But-1-ene (sp^3, sp^3, sp^2, sp^2)

(c)

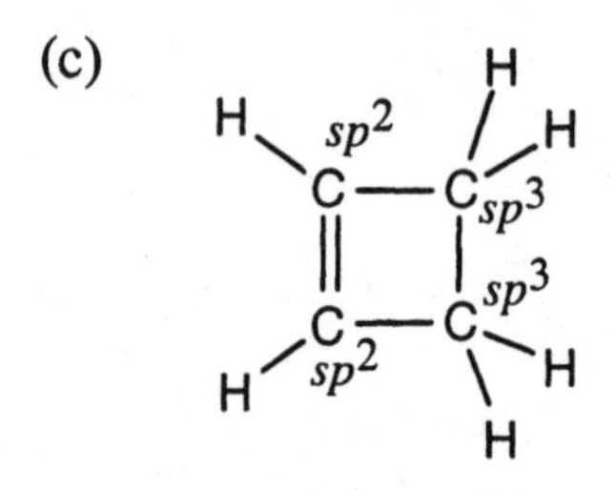

Cyclobutene

(d) But-1-en-3-yne (sp, sp, sp^2, sp^2)

1.50

Benzene

All carbon atoms of benzene are sp^2-hybridized, and all bond angles of benzene are 120°. Benzene is a planar molecule.

1.51

(a) $CH_3CH_2CH{=}CH_2$ (b) $H_2C{=}CHCH{=}CH_2$ (c) $H_2C{=}CHC{\equiv}CH$

Acid-Base Chemistry

1.52

$$CH_3COCH_3 + Na^+ \ {}^-{:}NH_2 \longrightarrow CH_3COCH_2{:}^- Na^+ + NH_3$$

$pK_a \approx 19$ stronger acid; $pK_a \approx 36$ weaker acid

The above reaction will take place as written because acetone is a stronger acid than ammonia.

1.53 Lewis acids: $AlBr_3$, HF

Lewis bases: $CH_3CH_2\ddot{N}H_2$, $CH_3\ddot{S}CH_3$

1.54 The reaction between methanol and bicarbonate does not take place in the indicated direction because methanol ($pK_a = 15.5$) is a weaker acid than H_2CO_3 ($pK_a = 6.4$).

1.55

(a) $CH_3\ddot{O}H$ (base) + H^+ (acid) $\longrightarrow$ $CH_3\overset{+}{O}H_2$

(b) $CH_3\ddot{O}H$ (acid) + ${}^-\ddot{N}H_2$ (base) $\longrightarrow$ $CH_3\ddot{O}{:}^-$ + $\ddot{N}H_3$

(c) CH_3CHO (:O:=C(H₃C)H) (base) + $ZnCl_2$ (acid) $\longrightarrow$ $CH_3CH{=}\overset{+}{O}{-}\overset{-}{Zn}Cl_2$

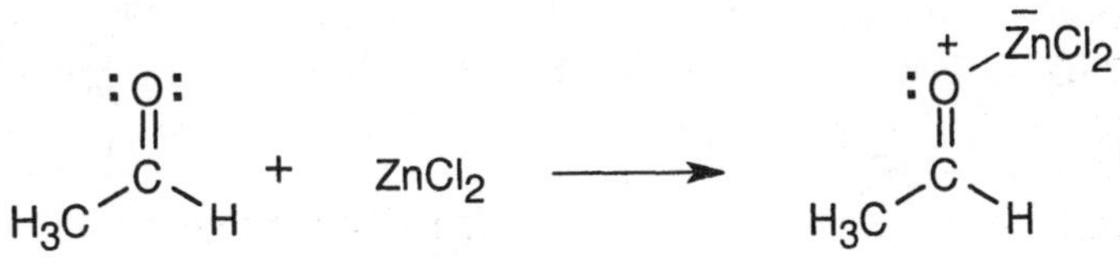

1.56 The substances with the largest values of pK_a are the least acidic.

Least acidic ⟶ *Most acidic*

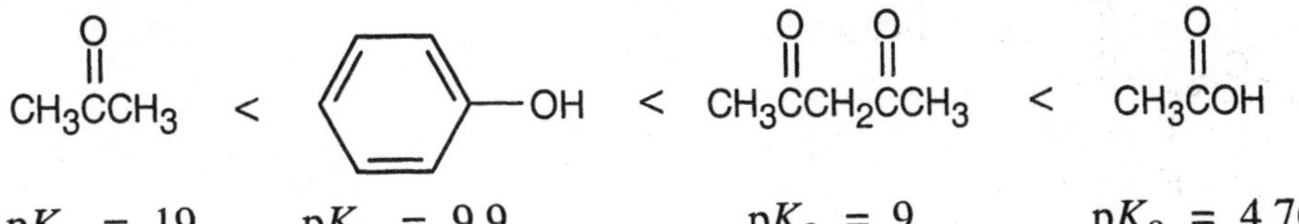

CH_3COCH_3 < C_6H_5OH < $CH_3COCH_2COCH_3$ < CH_3COOH

$pK_a = 19$ — $pK_a = 9.9$ — $pK_a = 9$ — $pK_a = 4.76$

1.57 To react completely with NaOH, an acid must have a pK_a considerably lower than the pK_a of H_2O. Thus, all substances in the previous problem except acetone react completely with NaOH.

1.58 The stronger the acid (lower pK_a), the weaker its conjugate base. Since NH_4^+ is a stronger acid than $CH_3NH_3^+$, NH_3 is a weaker base than CH_3NH_2.

Integrated Problems

1.59 (a) The 4 valence electrons of carbon can form bonds with a maximum of 4 hydrogens. Thus, it is not possible for the compound CH_5 to exist.

(b) If you try to draw a molecule with the formula C_2H_6N, you will see that it is impossible for both carbons and nitrogen to have a complete octet of electrons. Therefore, C_2H_6N is unlikely to exist.

(c) A compound with the formula $C_3H_5Br_2$ doesn't have filled outer shells for all atoms and is thus unlikely to exist.

1.60

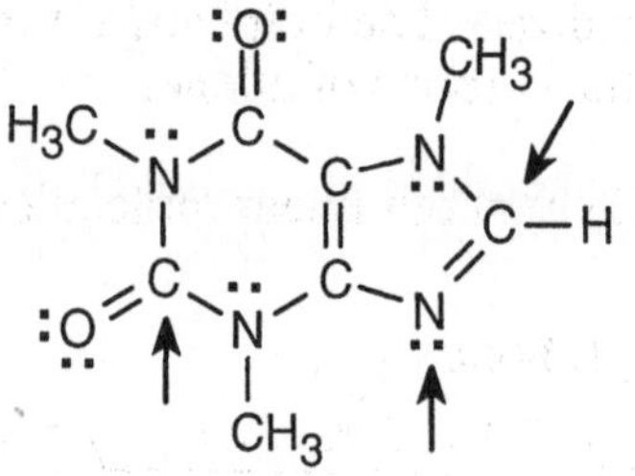

All of the indicated atoms are sp^2-hybridized.

1.61

Ammonium ion

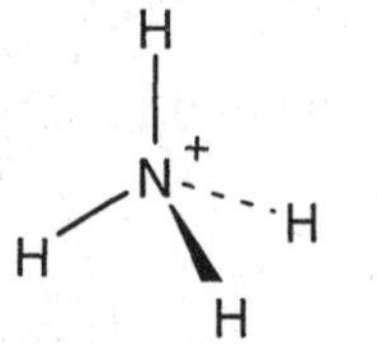

The nitrogen atom of the tetrahedral ammonium ion is sp^3-hybridized because, like the carbon atom of methane, nitrogen forms bonds to four different hydrogen atoms.

1.62

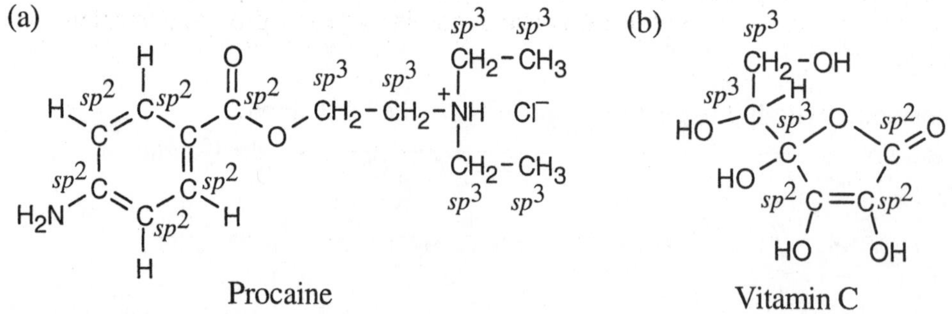

(a) Procaine

(b) Vitamin C

1.63 In a compound containing a carbon-carbon triple bond, atoms bonded to the *sp*-hybridized carbons must lie in a straight line. It is not possible to form a five-membered ring if four carbons must have a linear relationship.

1.64

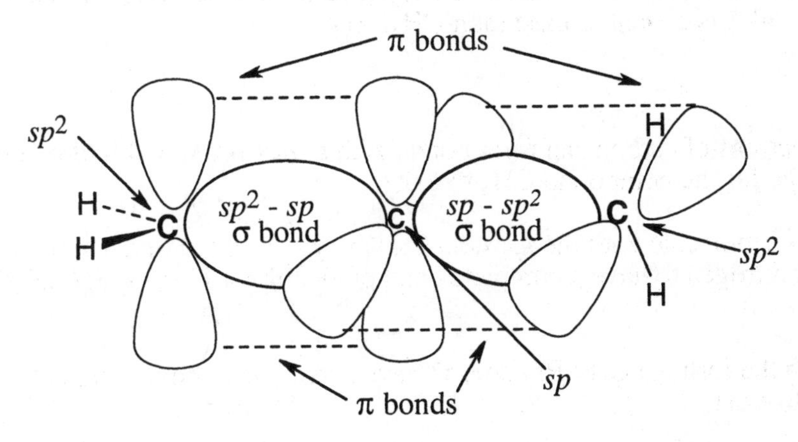

The central carbon of allene forms two σ bonds and two π bonds. The central carbon is *sp*-hybridized, and the two terminal carbons are *sp*^{2}-hybridized. The bond angle formed by the three carbons is 180°, indicating linear geometry for the carbons of allene.

1.65 The carbon atom of CO_2 is *sp*-hybridized. Allene and CO_2 are both linear molecules.

$:\ddot{O}::C::\ddot{O}:$

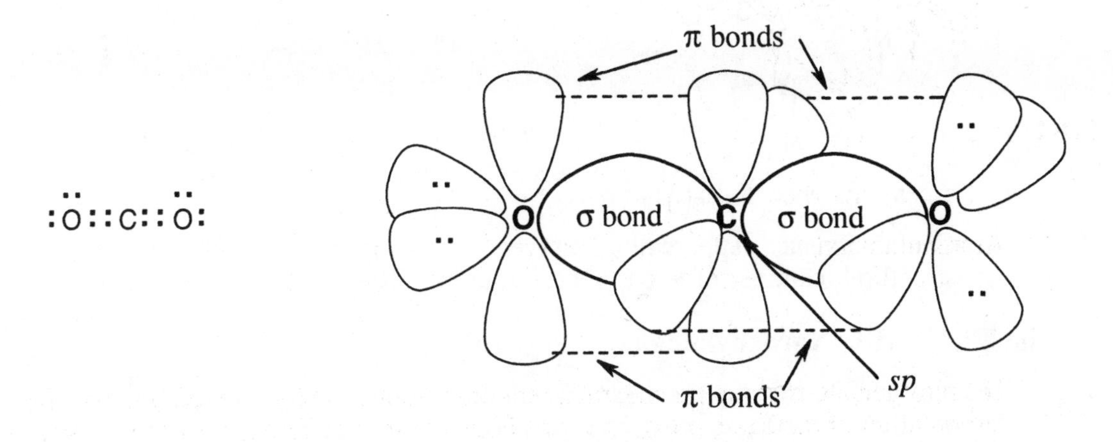

1.66 The carbon atom, which has six valence shell electrons, is *sp*^{2}-hybridized. A carbocation is planar and is *isoelectronic* with (has the same number of electrons as) a trivalent boron compound.

In the Medicine Cabinet

1.67

(a)

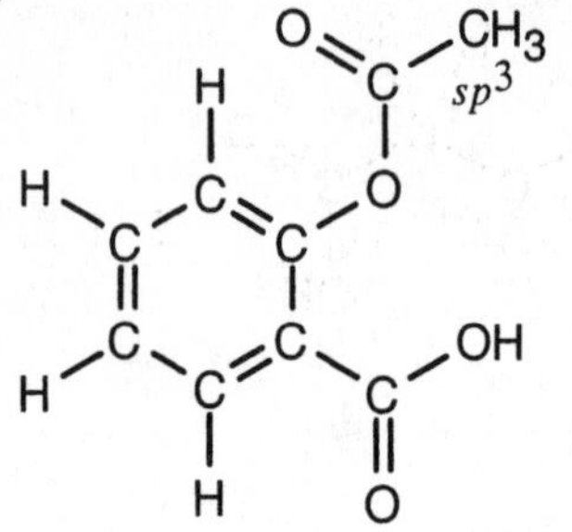

O=C–CH3 sp3
H
H C C O
C C
H C C C OH
C
H O

Acetylsalicylic acid – one sp^3 carbon

(b)

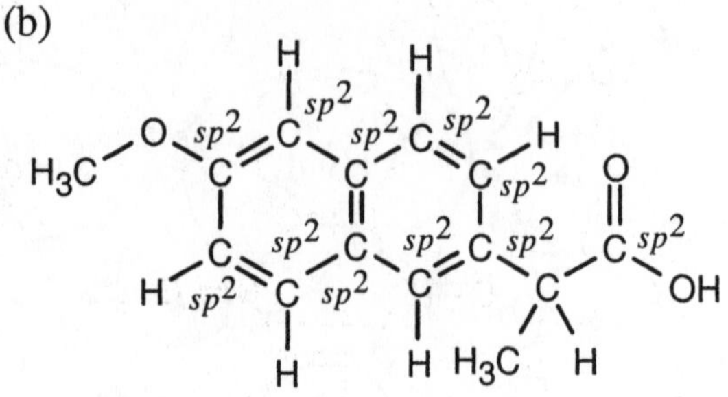

H H
sp2 sp2 sp2 sp2 H
H3C O C C C C
C sp2 O
sp2 C sp2 C sp2 C sp2 C sp2
H sp2 C sp2 C C OH
H H H3C H

Naproxen – eleven sp^2 carbons

(c)

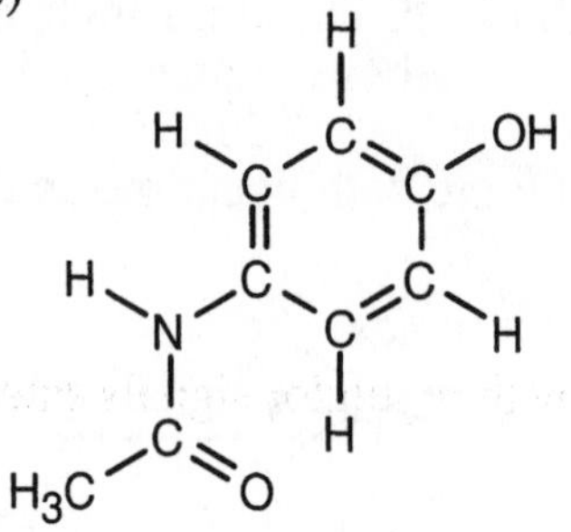

H
H C OH
C C
H C C
N C H
H
C=O
H3C

Acetaminophen $C_8H_9NO_2$

(d) Cyclooxygenase, apparently, must require compounds that interact with it to have the structural features common to these four NSAIDs.

In the Field with Agrochemicals

1.68

(a)

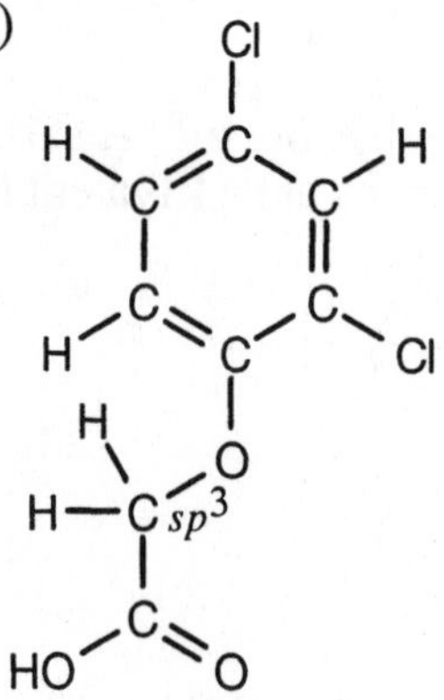

Cl
H C H
C C
H C C Cl
H O
H–C sp3
C
HO O

2,4-D
one sp^3 carbon

(b)

O
C sp2 H
HO C–H
H
N
H C–H
O=P–OH
OH

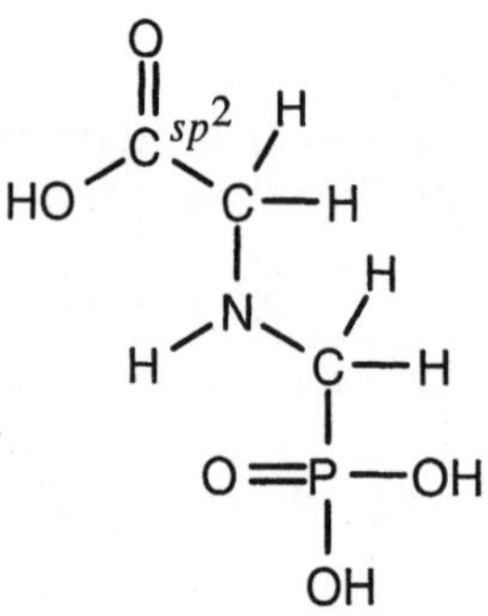

Round Up
one sp^2 carbon

(c)

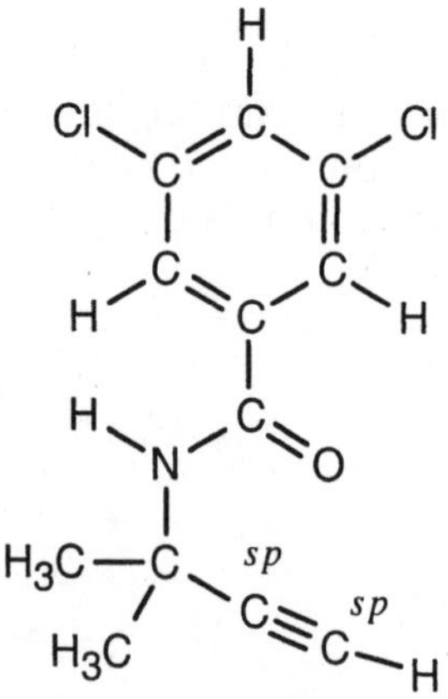

H
Cl C Cl
C C
H C C H
H N C=O
H3C–C sp
C≡C–H sp
H3C

Pronamide
two sp carbons

(d)

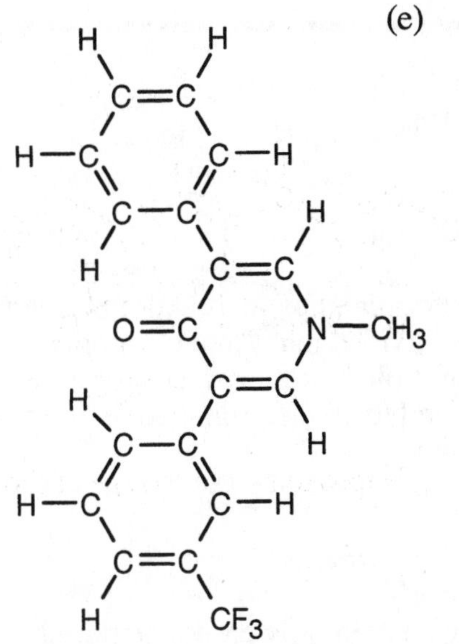

```
        H       H
         \     /
          C=C
         /     \
  H—C           C—H
     \\         //
      C—C          H
     /     \      /
    H        C=C
            /      \
       O=C          N—CH3
            \      /
             C=C
   H        /      \
    \      /        H
      C—C
     //    \\
  H—C          C—H
     \         /
      C=C
     /     \
    H        CF3
```

Fluridone $C_{19}H_{14}F_3NO$

(e)

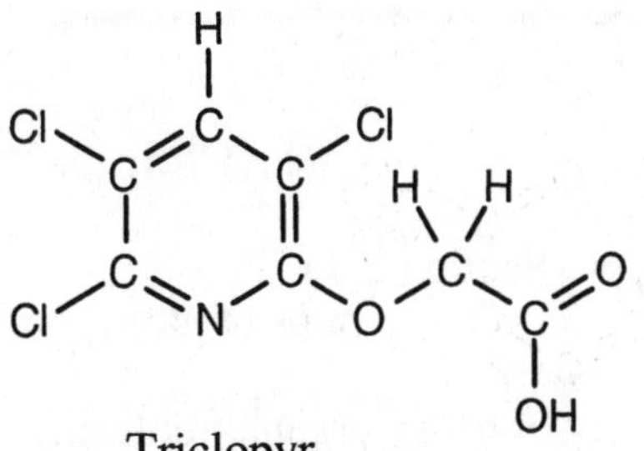

```
             H
             |
      Cl    C     Cl
        \ //  \  /        H  H
         C      C          \ /
         |      ||          C
         C     C          /   \
        / \\  / \  O    C=O
      Cl     N             |
                           OH
```

Triclopyr

(e) You would expect that Trichlopyr would disrupt growth regulator signals since it resembles 2,4-D most closely.

Chapter 2 – The Nature of Organic Molecules

Chapter Outline

I. Functional groups (Section 2.1).
   A. Functional groups are groups of atoms within a molecule that have a characteristic chemical behavior.
   B. The chemistry of every organic molecule is determined by its functional groups.
   C. Functional groups described in this text can be divided into three categories:
      1. Functional groups with carbon-carbon multiple bonds.
      2. Groups in which carbon forms a single bond to an electronegative atom.
      3. Groups with a carbon-oxygen double bond.

II. Alkanes and alkyl groups (Sections 2.2–2.5).
   A. Alkanes and alkane isomers (Section 2.2).
      1. Alkanes are formed by overlap of carbon sp^3 orbitals.
      2. Alkanes are described as saturated hydrocarbons.
         a. They are hydrocarbons because they contain only carbon and hydrogen.
         b. They are saturated because all bonds are single bonds.
         c. The general formula for alkanes is C_nH_{2n+2}.
      3. For alkanes with four or more carbons, the carbons can be connected in more than one way.
         a. If the carbons are in a row, the alkane is a straight-chain alkane.
         b. If the carbon chain has a branch, the alkane is a branched-chain alkane.
      4. Alkanes with the same molecular formula are isomers.
         a. Isomers whose atoms are connected differently are constitutional isomers.
            Constitutional isomers are always different compounds with different properties but with the same molecular formula.
         b. A given alkane can be drawn in many ways.
      5. Straight-chain alkanes are named according to the number of carbons in their chain.
   B. Alkyl groups.
      1. An alkyl group is the partial structure that results from the removal of a hydrogen atom from an alkane.
         a. Alkyl groups are named by replacing the -ane of an alkane by -yl.
         b. *n*-Alkyl groups are formed by removal of an end hydrogen of a straight-chain alkane.
         c. Branched-chain alkyl groups are formed by removal of a hydrogen atom from an internal carbon.
            The prefixes *sec*- and *tert*- refer to the degree of substitution at the branching carbon atom.
      2. There are four possible degrees of alkyl substitution for carbon.
         a. A primary carbon is bonded to one other carbon.
         b. A secondary carbon is bonded to two other carbons.
         c. A tertiary carbon is bonded to three other carbons.
         d. A quaternary carbon is bonded to four other carbons.
         e. The symbol **R** refers to the rest of the molecule.
   C. Naming branched-chain alkanes (Section 2.3).
      1. The system of nomenclature used in this book is the IUPAC system.
         In this system, a chemical name has a prefix, a parent and a suffix.
         i. The parent shows the number of carbons in the principal chain.
         ii. The suffix identifies the functional group family.
         iii. The prefix shows the location of functional groups.

2. Naming an alkane:
   a. Find the parent hydrocarbon.
      i. Find the longest continuous chain of carbons, and use its name as the parent name.
      ii. If two chains have the same number of carbons, choose the one with more branch points.
   b. Number the atoms in the parent chain.
      i. Start numbering at the end nearer the first branch point.
      ii. If branching occurs an equal distance from both ends, begin numbering at the end nearer the second branch point.
   c. Identify and number the substituents.
      i. Give each substituent a number that corresponds to its position on the parent chain.
      ii. Two substituents on the same carbon receive the same number.
   d. Write the name as a single word.
      i. Use hyphens to separate prefixes and commas to separate numbers.
      ii. Use the prefixes, *di-*, *tri-*, *tetra-* if necessary, but don't use them for alphabetizing.

D. Properties of alkanes (Section 2.4).
   1. Alkanes are prepared by fractional distillation of petroleum.
   2. Alkanes are chemically inert to most laboratory reagents.
      Alkanes react with O_2 and Cl_2.
   3. Catalytic cracking is a process in which higher boiling hydrocarbons are "cracked" into smaller hydrocarbons suitable for use in gasoline.
      The octane number of a fuel describes the performance of a fuel.

E. Conformations of ethane (Section 2.5).
   1. Rotation is possible around carbon-carbon single bonds
      The different arrangements of atoms that result from rotation are conformers.
   2. Conformations can be represented in two ways.
      a. Sawhorse representations of alkanes view the C–C bond from an oblique angle.
      b. Newman projections represent the two carbon atoms by a circle.
   3. There is a slight barrier to rotation around a C–C bond.
      a. In a staggered conformation, hydrogens are as far apart as possible.
      b. In an eclipsed conformation, hydrogens are as close together as possible.
      c. For most compounds, a staggered conformation is more stable than an eclipsed conformation.

III. Drawing chemical structures (Section 2.6).
   Skeletal structures are a shorthand way of drawing organic compounds.
   1. Carbon atoms are assumed to be at the intersection of two lines.
   2. Hydrogen atoms bonded to carbon aren't shown.
   3. All other atoms are shown.

IV. Cycloalkanes (Sections 2.7–2.11).
   A. Naming substituted cycloalkanes (Section 2.7).
      1. Count the number of carbon atoms in the ring, and add the prefix *cyclo-* to the name of the corresponding alkane.
      2. Start at a point of attachment, and number around the ring.
         For two or more substituents, begin numbering at the group with alphabetical priority, and give the other substituents the lowest possible number.
   B. Cis-trans isomerism in cycloalkanes (Section 2.8).
      1. Unlike open-chain alkanes, cycloalkanes have much less rotational freedom.
         a. Very small rings are rigid.
         b. Large rings have more rotational freedom.

2. Cycloalkanes have a "top" side and a "bottom" side.
   a. If two substituents are on the same side of a ring, the ring is cis-disubstituted.
   b. If two substituents are on opposite sides of a ring, the ring is trans-disubstituted.
3. Substituents in the two types of disubstituted cycloalkanes are connected in the same order but differ in spatial orientation.
   a. These cycloalkenes are stereoisomers that are known as cis–trans isomers.
   b. Cis-trans isomers are stable compounds that can't be interconverted.

C. Conformations of cycloalkanes (Sections 2.9–2.11).
1. Ring conformations (Section 2.9).
   a. Cyclobutane and cyclopentane are slightly puckered.
   b. Chair cyclohexane.
      i. The chair conformation of cyclohexane is strain-free.
      ii. In a standard drawing of cyclohexane, the lower bond is in front.
2. Axial and equatorial bonds in cyclohexane (Section 2.10).
   a. There are two kinds of positions on a cyclohexane ring.
      i. Six axial hydrogens are perpendicular to the plane of the ring.
      ii. Six equatorial hydrogens are roughly in the plane of the ring.
   b. Each carbon has one axial hydrogen and one equatorial hydrogen.
   c. Each side of the ring has alternating axial and equatorial hydrogens.
   d. All hydrogens on the same side of the ring are cis.
3. Conformational mobility of cyclohexanes (Section 2.11).
   a. Different chair conformations of cyclohexanes interconvert by a ring-flip.
   b. After a ring-flip, an axial bond becomes an equatorial bond, and vice versa.
   c. Both conformations aren't equally stable at room temperature.
      In methylcyclohexane, 95% of molecules have the methyl group in the equatorial position.

Solutions to Problems

2.1

(a)

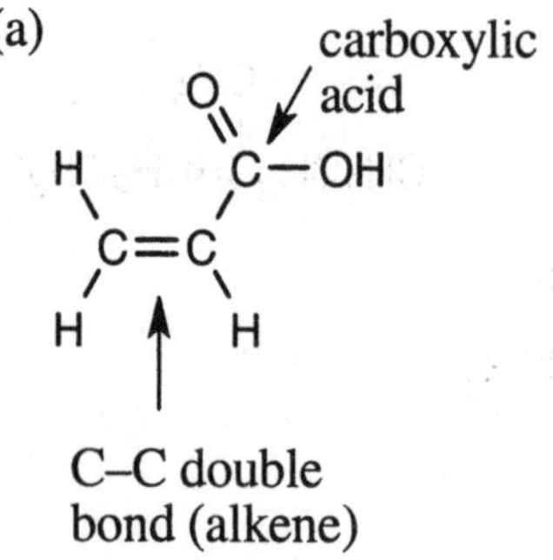

carboxylic acid

C–C double bond (alkene)

Acrylic acid

(b)

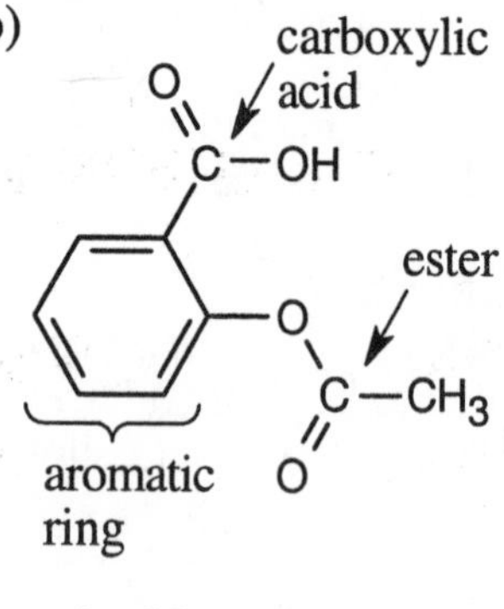

carboxylic acid

ester

aromatic ring

Aspirin

(c)

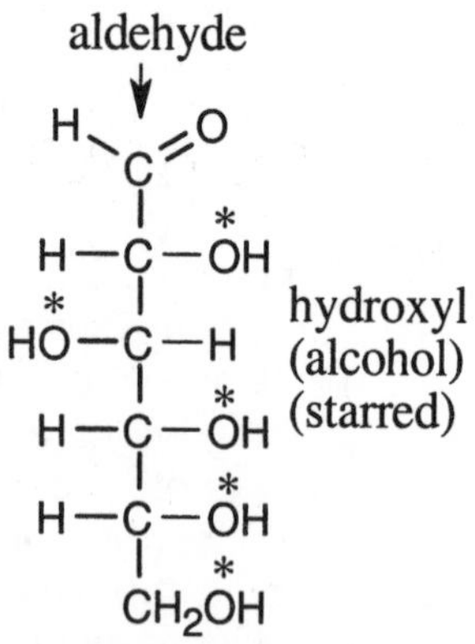

aldehyde

hydroxyl (alcohol) (starred)

Glucose

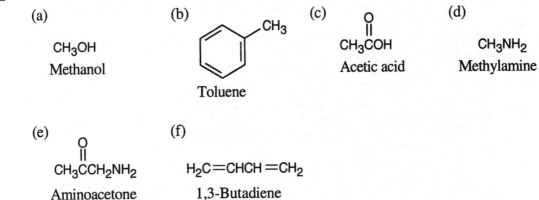

2.2

(a) CH_3OH
Methanol

(b) 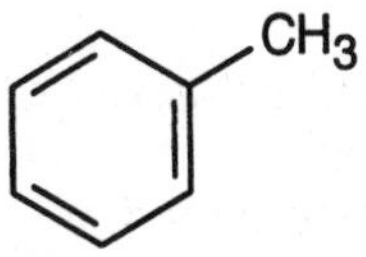
CH3
Toluene

(c) $CH_3\overset{O}{\overset{\|}{C}}OH$
Acetic acid

(d) CH_3NH_2
Methylamine

(e) $CH_3\overset{O}{\overset{\|}{C}}CH_2NH_2$
Aminoacetone

(f) $H_2C{=}CHCH{=}CH_2$
1,3-Butadiene

Many other compounds containing these functional groups can be drawn.

2.3

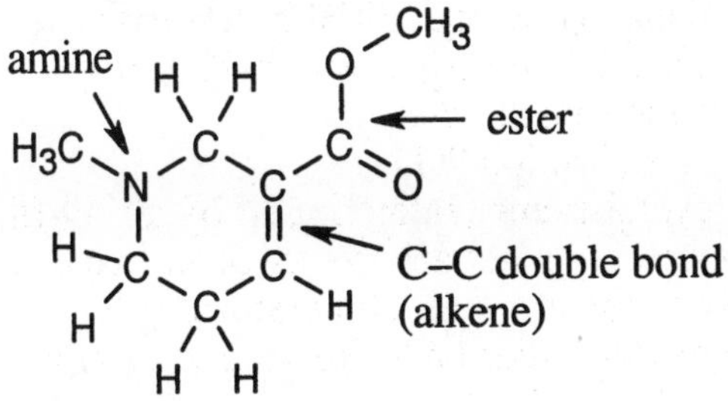
amine
ester
C–C double bond
(alkene)
Arecoline

2.4 We know that carbon forms four bonds and hydrogen forms one bond. Thus, if you draw all possible six-carbon skeletons and add hydrogens so that all carbons have four bonds, you will arrive at the following formulas.

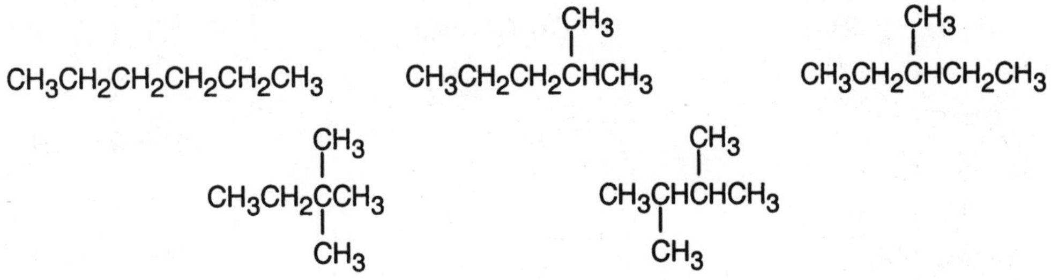

2.5 (a) There are 18 isomers with the formula C_8H_{18}.

Octane:

$CH_3CH_2CH_2CH_2CH_2CH_2CH_2CH_3$

Heptanes:

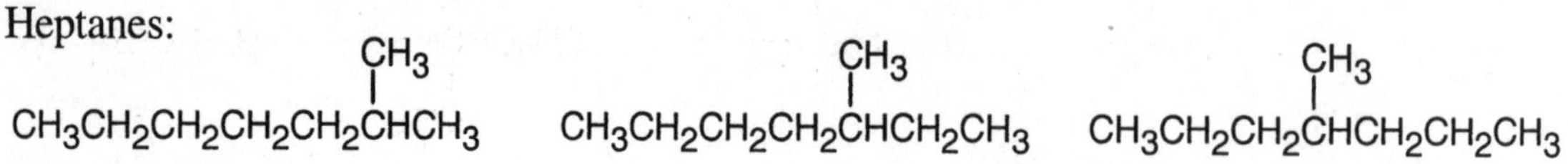

CH_3
$CH_3CH_2CH_2CH_2CH_2CHCH_3$

CH_3
$CH_3CH_2CH_2CH_2CHCH_2CH_3$

CH_3
$CH_3CH_2CH_2CHCH_2CH_2CH_3$

Hexanes:

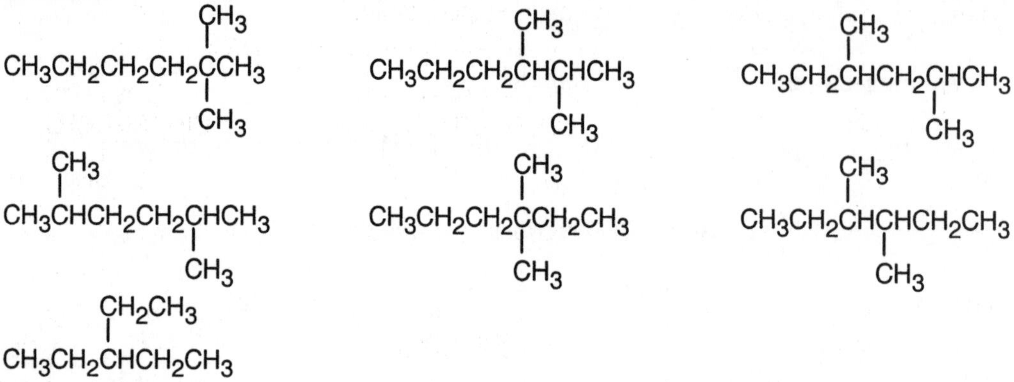

CH_3
$CH_3CH_2CH_2CH_2CCH_3$
CH_3

CH_3
$CH_3CH_2CH_2CHCHCH_3$
CH_3

CH_3
$CH_3CH_2CHCH_2CHCH_3$
CH_3

CH_3
$CH_3CHCH_2CH_2CHCH_3$
CH_3

CH_3
$CH_3CH_2CH_2CCH_2CH_3$
CH_3

CH_3
$CH_3CH_2CHCHCH_2CH_3$
CH_3

CH_2CH_3
$CH_3CH_2CHCH_2CH_3$

Pentanes:

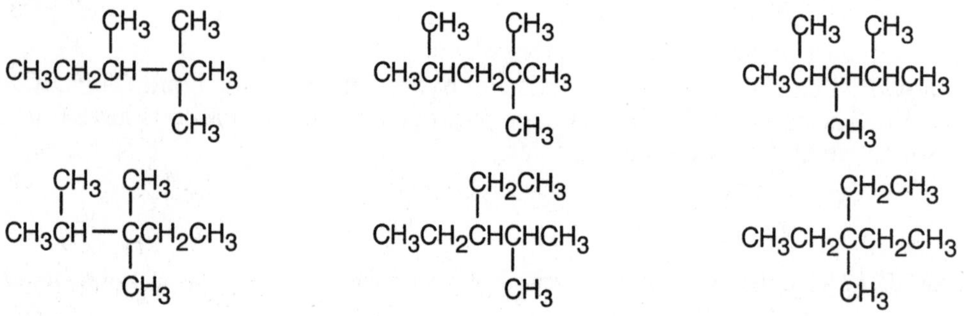

CH_3 CH_3
$CH_3CH_2CH—CCH_3$
CH_3

CH_3 CH_3
$CH_3CHCH_2CCH_3$
CH_3

CH_3 CH_3
$CH_3CHCHCHCH_3$
CH_3

CH_3 CH_3
$CH_3CH—CCH_2CH_3$
CH_3

CH_2CH_3
$CH_3CH_2CHCHCH_3$
CH_3

CH_2CH_3
$CH_3CH_2CCH_2CH_3$
CH_3

Butane:

CH_3 CH_3
$CH_3C—CCH_3$
CH_3 CH_3

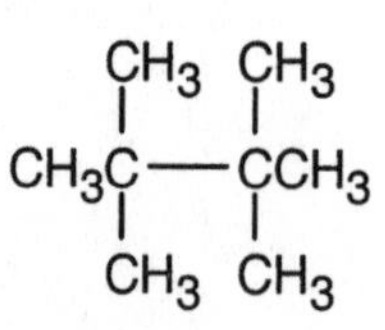

(b) Many isomers of the formula $C_4H_8O_2$ containing different functional groups can be drawn. Here are three examples:

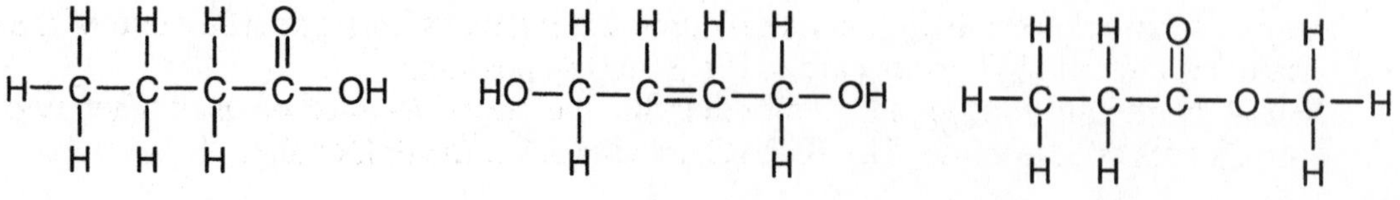

H H H O
H—C—C—C—C—OH
H H H

H H H H
HO—C—C=C—C—OH
H H

H H O H
H—C—C—C—O—C—H
H H H

2.6

$CH_3CH_2CH_2CH_2CH_2-$ $\quad$ $CH_3CH_2CH_2CH(CH_3)-$ $\quad$ $CH_3CH_2CH(CH_2CH_3)-$ $\quad$ $CH_3CH_2CH(CH_3)CH_2-$

$CH_3CH(CH_3)CH_2CH_2-$ $\quad$ $CH_3CH_2C(CH_3)_2-$ $\quad$ $CH_3CH(CH_3)CH(CH_3)-$ $\quad$ $CH_3C(CH_3)_2CH_2-$

2.7 Many other answers to these problems are acceptable.

(a) $CH_3CH(CH_3)CH(CH_3)CH_3$ — the two CH carbons are labelled *t*

(b) $CH_3CH_2CH(CH(CH_3)CH_3)CH_2CH_3$ — the CH_3CHCH_3 group is boxed

(c) $CH_3CH_2C(CH_3)_2CH_3$ — the CH_2 carbon is labelled *s*; an arrow labelled *q* points to the central C

p = primary; *s* = secondary; *t* = tertiary; *q* = quaternary

2.8

(a) $CH_3CH(CH_3)CH_2CH_2CH_3$ — labels along the chain: *p t s s p*; *p* on the branch CH_3

(b) $CH_3CH_2CH(CH(CH_3)CH_3)CH_2CH_3$ — labels along the chain: *p s t s p*; *p t p* on the branch CH_3CHCH_3

(c) $CH_3CH(CH_3)CH_2-C(CH_3)_2-CH_3$ — labels along the chain: *p t s q p*; *p* on each CH_3 branch

p = primary; *s* = secondary; *t* = tertiary; *q* = quaternary

2.9

(a)

$CH_3CH_2CH_2CH_2CH_3$

Pentane

$CH_3CH_2CH(CH_3)CH_3$

2-Methylbutane

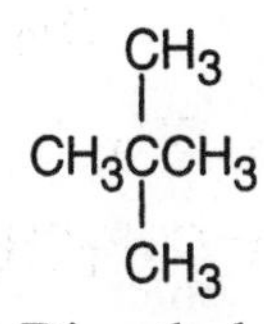

2,2-Dimethylpropane

(b)

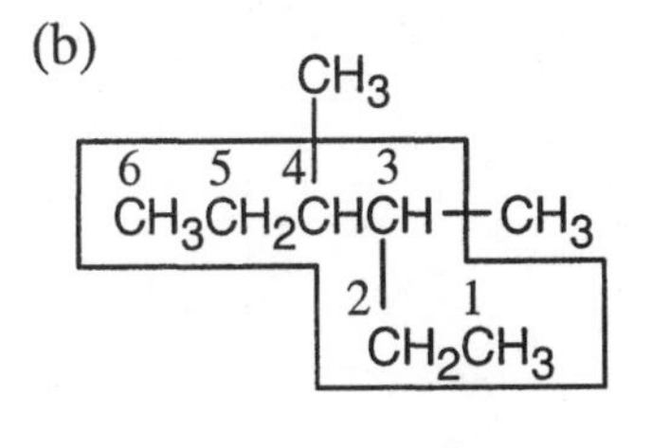

3,4-Dimethylhexane

Step 1: Find the longest continuous carbon chain and use it as the parent name. In (b), the longest chain is a hexane (boxed).
Step 2: Identify the substituents. In (b), both substituents are methyl groups.
Step 3: Number the substituents. In (b), numbering can start from either end of the carbon chain, and the methyl groups are in the 3- and 4- positions.
Step 4: Name the compound. Remember that the prefix *di-* must be used when two substituents are the same. The IUPAC name is 3,4-dimethylhexane.

(c)

$$\begin{array}{c} CH_3\ \ CH_3 \\ |\quad\ \ | \\ CH_3CHCH_2CHCH_3 \end{array}$$

2,4-Dimethylpentane

(d)

$$\begin{array}{c} CH_3\quad\ CH_2CH_3 \\ |\qquad\ \ | \\ CH_3CCH_2CH_2CHCH_3 \\ | \\ CH_3 \end{array}$$

2,2,5-Trimethylheptane

2.10 First, draw the carbon structure of the parent hydrocarbon. In this example, it is nonane.

C—C—C—C—C—C—C—C—C

Then, add the substituents—methyl groups at C3 and C4.

$$\begin{array}{c} CH_3 \\ | \\ \underset{9}{C}-\underset{8}{C}-\underset{7}{C}-\underset{6}{C}-\underset{5}{C}-\underset{4}{C}-\underset{3}{C}-\underset{2}{C}-\underset{1}{C} \\ | \\ CH_3 \end{array}$$

Finally, add hydrogens to complete the structure.

(a)

$$\begin{array}{c} CH_3 \\ | \\ CH_3CH_2CH_2CH_2CH_2CHCHCH_2CH_3 \\ | \\ CH_3 \end{array}$$

3,4-Dimethylnonane

(b)

$$\begin{array}{c} CH_3 \\ | \\ CH_3CH_2CH_2C—CHCH_2CH_3 \\ |\qquad | \\ CH_3\ \ CH_2CH_3 \end{array}$$

3-Ethyl-4,4-dimethylheptane

(c)

$$\begin{array}{c} CH_2CH_2CH_3 \\ | \\ CH_3CH_2CH_2CH_2CHCH_2C(CH_3)_3 \end{array}$$

2,2-Dimethyl-4-propyloctane

(d)

$$\begin{array}{c} CH_3\ \ CH_3 \\ |\quad\ \ | \\ CH_3CHCH_2CCH_3 \\ | \\ CH_3 \end{array}$$

2,2,4-Trimethylpentane

2.11

$$\begin{array}{c} CH_3\ CH_3 \\ |\quad\ | \\ CH_3CH_2C—CHCHCH_2CH_3 \\ |\qquad | \\ CH_3\quad CH_3 \end{array}$$

3,3,4,5-Tetramethylheptane

2.12

Most stable conformation (staggered)

Least stable conformation (eclipsed)

2.13

Staggered butane | Eclipsed butane

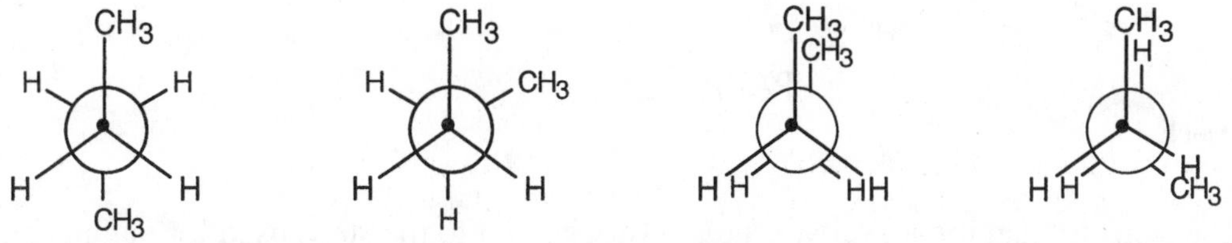

2.14 The first staggered conformation of butane (pictured above) is the most stable, because the relatively large methyl groups are as far apart as possible.

2.15

(a) Pyridine C_5H_5N

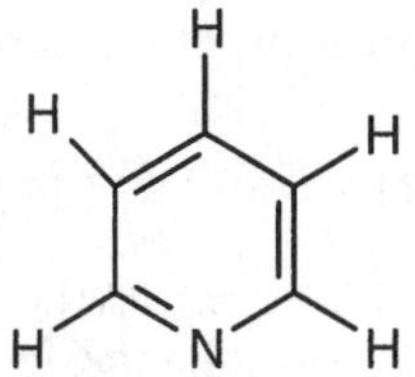

One hydrogen is bonded to each carbon.

(b) Cyclohexanone $C_6H_{10}O$

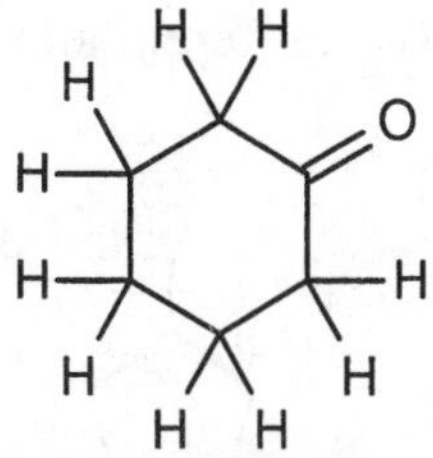

Two hydrogens are bonded to each carbon (except for the carbon bonded to oxygen)

(c) Indole C_8H_7N

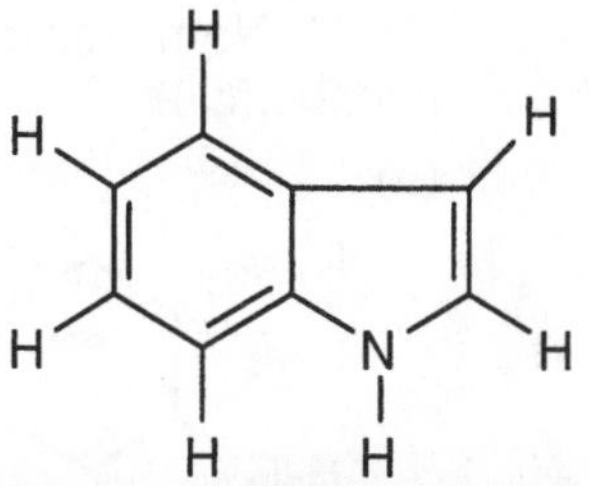

One hydrogen is bonded to each carbon (except for the two carbons at the junction of the two rings)

Although a particular shorthand structure has only one molecular formula, a particular molecular formula may represent many structures.

2.16 Several skeletal structures can satisfy each molecular formula.

(a) C_4H_8

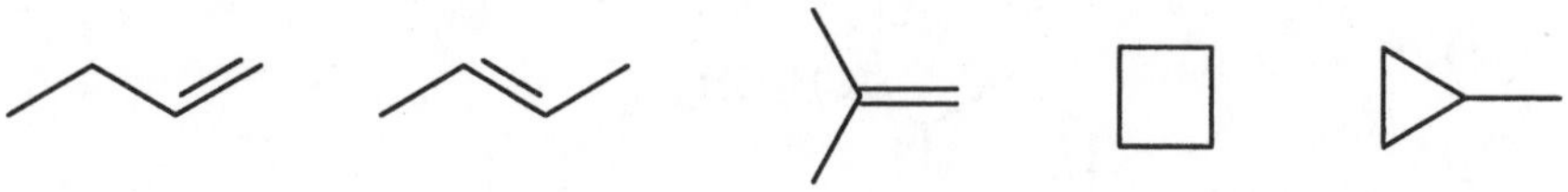

(b) C_3H_6O

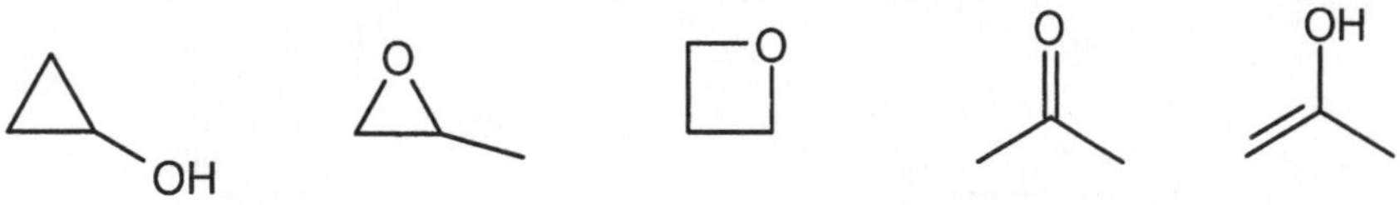

(c) C_4H_9Cl

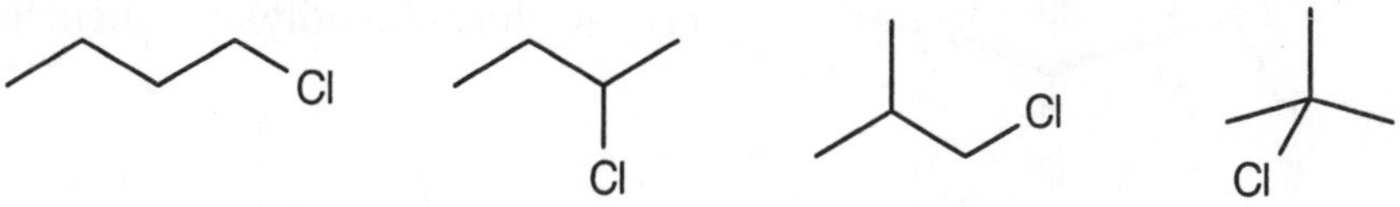

2.17

H_2N OH

PABA

2.18 The steps for naming a cycloalkane are very similar to the steps used for naming an open-chain hydrocarbon.
Step 1: Name the parent cycloalkane. In (a), the parent is cyclohexane.
Step 2: Identify the substituents. In (a), both substituents are methyl groups.
Step 3: Number the substituents so that the second substituent receives the lowest possible number. In (a), the substituents are in the 1- and 4- positions.
Step 4: Name the compound. If two different alkyl groups are present, cite them alphabetically.

(a) 1,4-Dimethylcyclohexane

(b) 1-Ethyl-3-methylcyclopentane

(c) Isopropylcyclobutaane

2.19

(a) 1-*tert*-Butyl-2-methylcyclopentane

(b) 1,1-Dimethylcyclobutane

(c) 1-Ethyl-4-isopropylcyclohexane

2.20

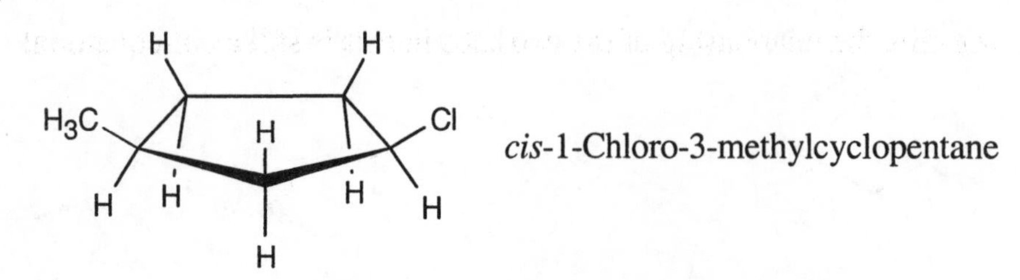

cis-1-Chloro-3-methylcyclopentane

2.21

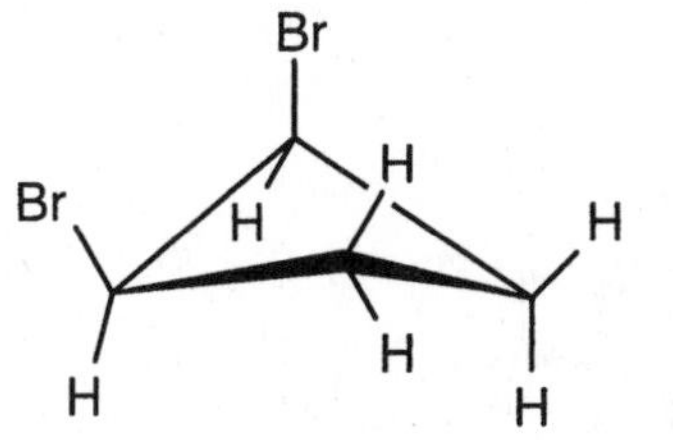

cis-1,2-dibromocyclobutane

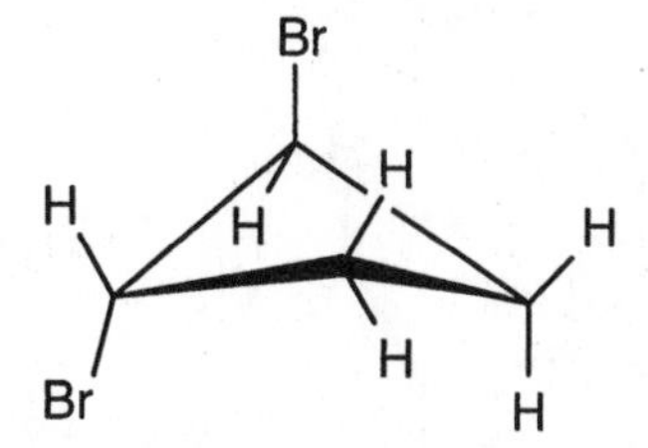

trans-1,2-dibromocyclobutane

2.22

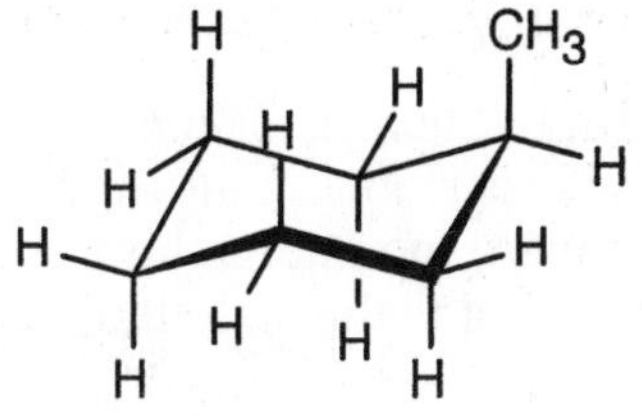

axial methyl group

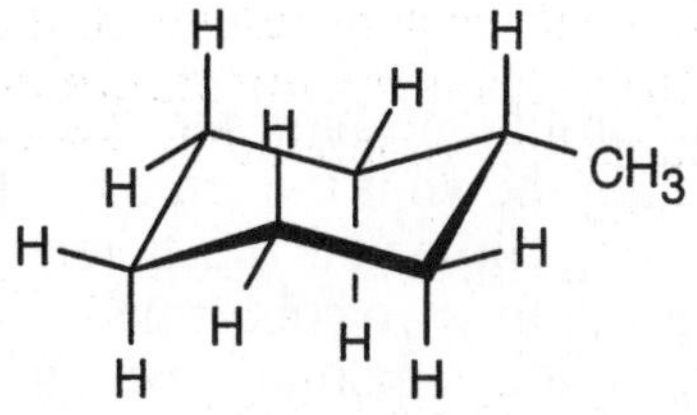

equatorial methyl group

2.23

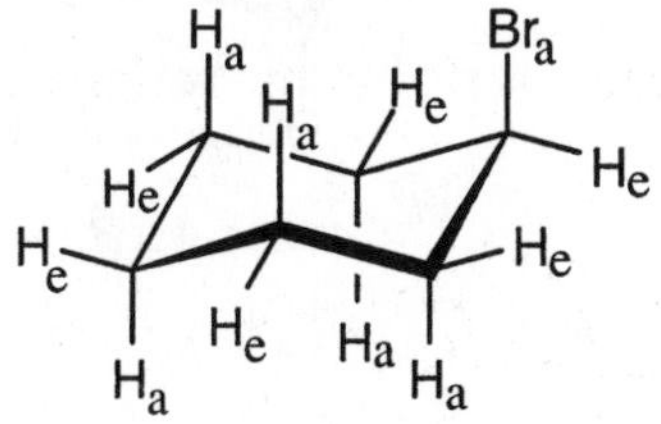

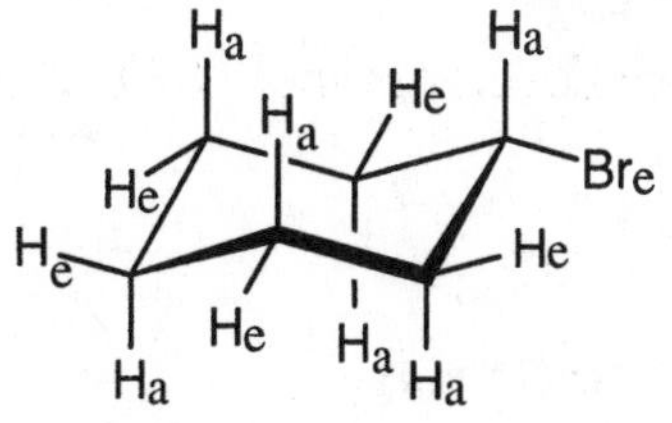

The conformation having bromine in the equatorial position is more stable.

2.24 Make a model of *cis*-1,2-dichlorocyclohexane. Notice that both cis substituents are on the same side of the ring and that two adjacent cis substituents have an axial-equatorial relationship. Now, perform a ring-flip on the cyclohexane.

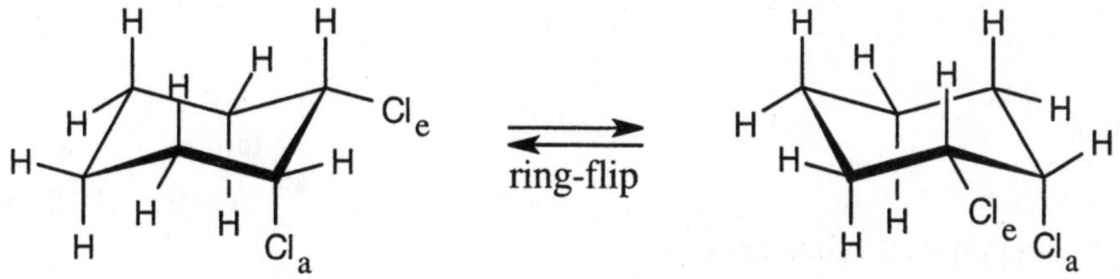

ring-flip

After the ring-flip, the relationship of the two substituents is still axial-equatorial.

2.25 For a *trans*-1,2-disubstituted cyclohexane, two adjacent substituents must be either both axial or both equatorial.

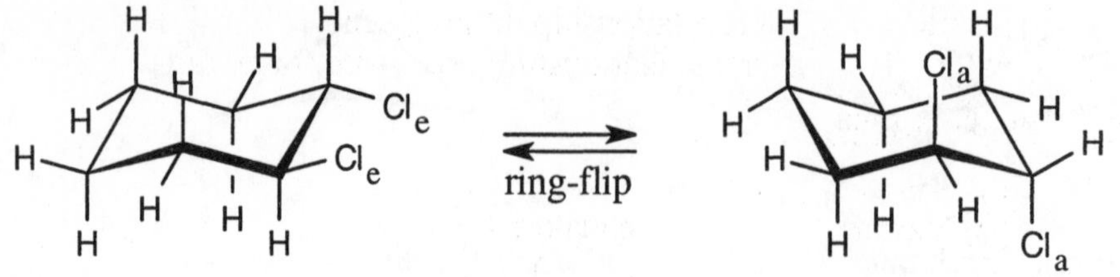

ring-flip

A ring-flip converts two adjacent axial substituents into equatorial substituents, and *vice versa.*

2.26 The three substituents have the orientations shown in the first structure. To decide if the conformation shown is the more stable conformation or the less stable conformation, perform a ring-flip on the illustrated conformation, and compare the interactions among substituents of the two conformers. The illustrated conformation is the less stable chair form because it has more interactions.

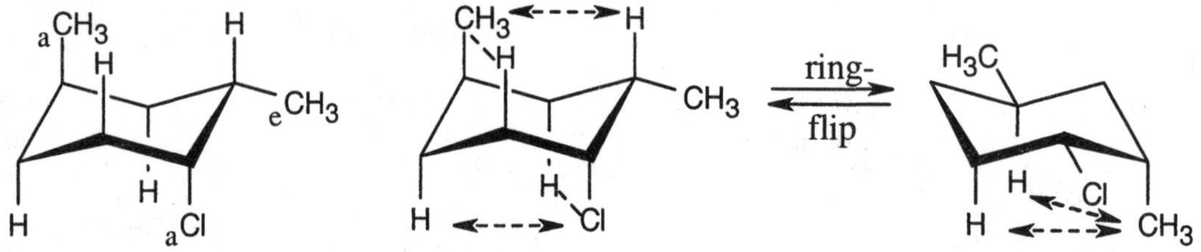

ring-flip

four interactions

two interactions

Visualizing Chemistry

2.27

(a) 3,3,5-Trimethylheptane

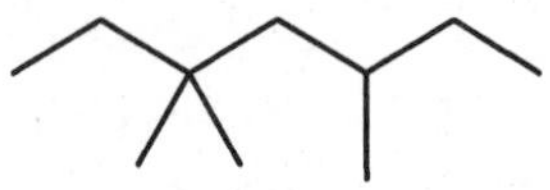

(b) *trans*-1-Ethyl-3-methylcyclopentane

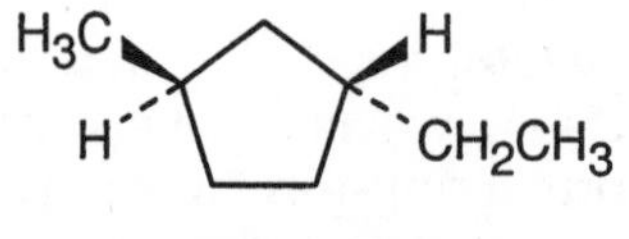

2.28

(a) $C_6H_5CH_2CH(NH_2)COOH$ — aromatic ring, amine, carboxylic acid

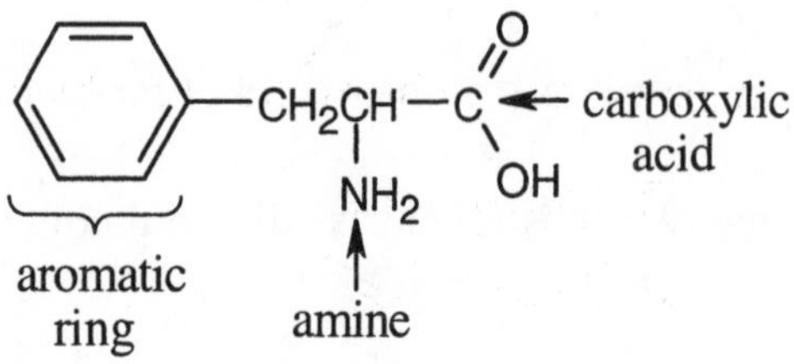

Phenylalanine $C_9H_{11}NO_2$

(b) aromatic ring, amide, amine

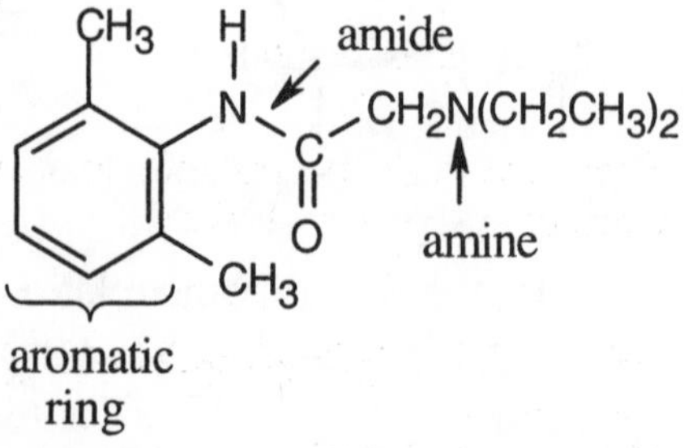

Lidocaine $C_{14}H_{22}N_2O$

2.29 The green substituent is axial, and the red and blue substituents are equatorial.

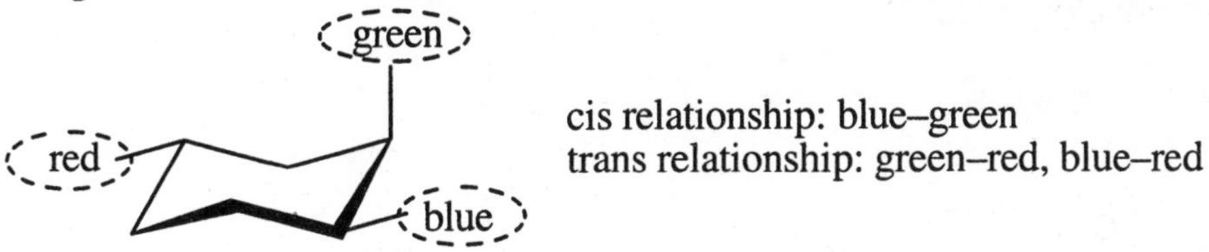

2.30

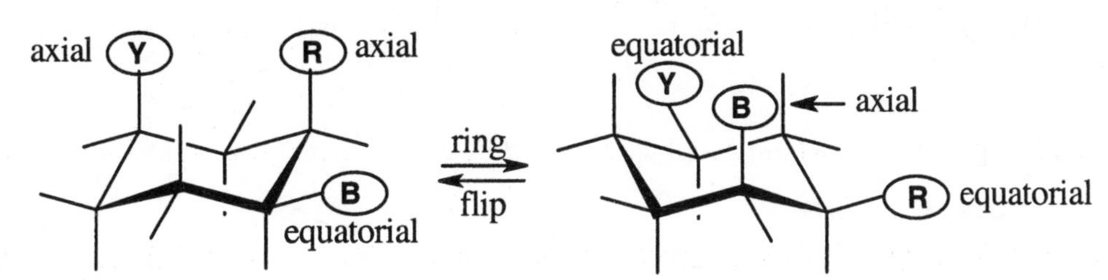

Additional Problems

Functional Groups and Isomerism

2.31

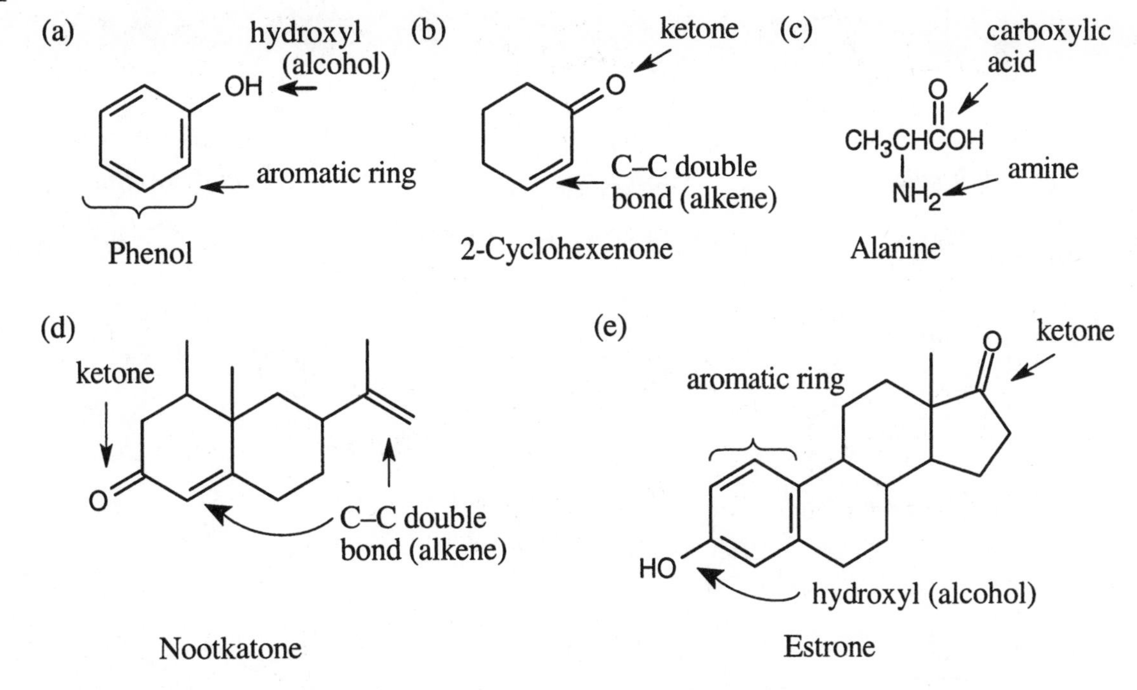

2.32 There are many acceptable answers to each part of this problem and of Problem 2.33. Correct answers include:

Skeletal structure

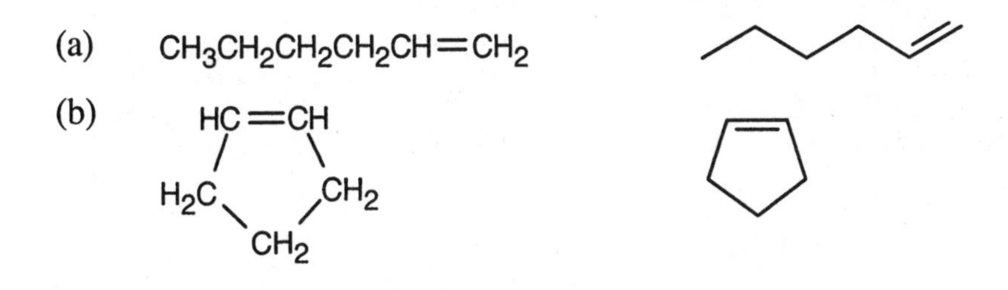

(c) 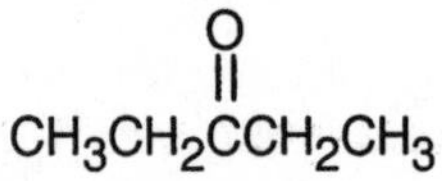 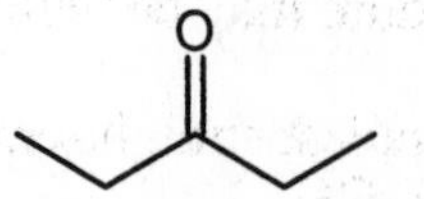 $CH_3CH_2\overset{O}{\overset{\|}{C}}CH_2CH_3$

(d) 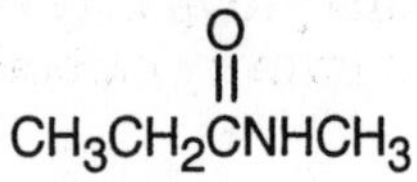 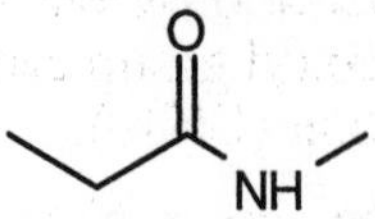 $CH_3CH_2\overset{O}{\overset{\|}{C}}NHCH_3$

(e) 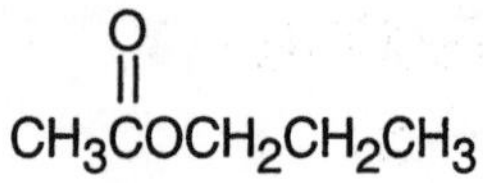 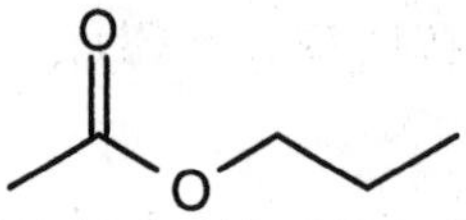 $CH_3\overset{O}{\overset{\|}{C}}OCH_2CH_2CH_3$

(f) 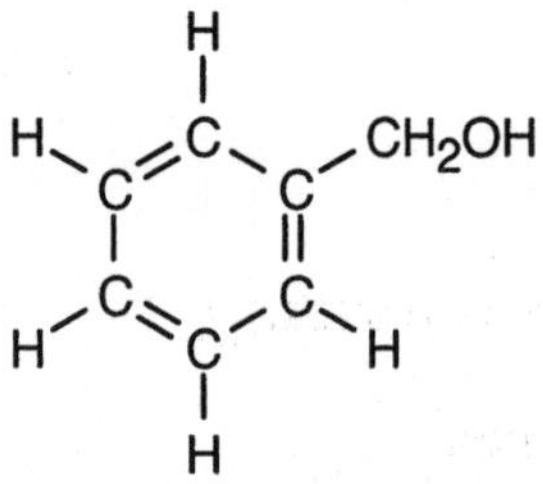 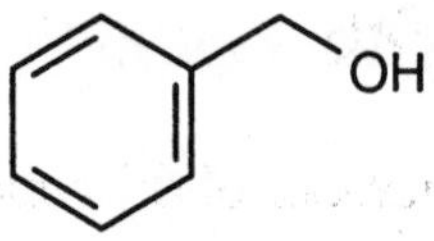 $C_6H_5CH_2OH$ (ring drawn with H on each ring carbon; skeletal: OH)

2.33

	Formula	*Structure*	*Skeletal Structure*
(a)	C_7H_{14}	$CH_3CH_2CH_2CH_2CH_2CH{=}CH_2$	
(b)	C_3H_4	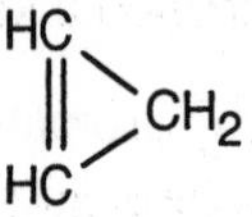 HC=HC, CH2 (cyclopropene)	
(c)	C_4H_8O	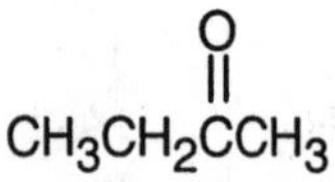 $CH_3CH_2\overset{O}{\overset{\|}{C}}CH_3$	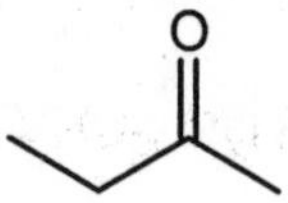
(d)	C_5H_9N	$CH_3CH_2CH_2CH_2C{\equiv}N$	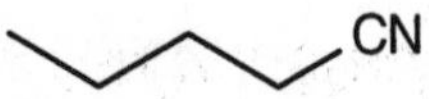 CN
(e)	C_5H_8	$CH_3CH{=}CHCH{=}CH_2$	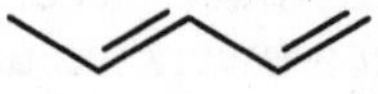
(f)	$C_4H_6O_2$	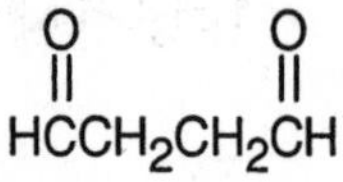 $H\overset{O}{\overset{\|}{C}}CH_2CH_2\overset{O}{\overset{\|}{C}}H$	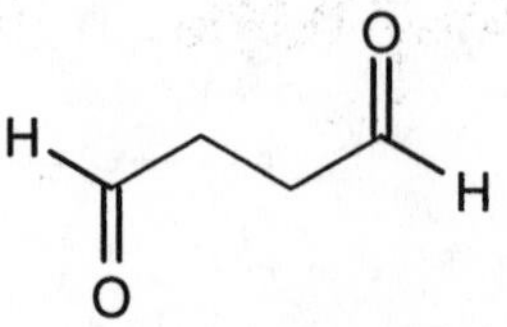 H, O, O, H

2.34 (a) To solve this problem, you must check all possibilities in a systematic way. The following procedure may be helpful.

1. Draw the simplest straight-chain parent alkane. (In this problem, it is $CH_3CH_2CH_2CH_3$.)
2. Find the number of different sites to which a functional group may be attached. (Here, the hydroxyl group can be attached to either a primary carbon (CH_3–) or to a secondary carbon (CH_2–).
3. At each different site, replace an –H with an –OH, and draw the structure.

$CH_3CH_2CH_2CH_2{-}OH$ and $CH_3CH_2CHCH_3$ (OH on C-2)

4. Draw the simplest branched alkane.

CH_3CHCH_3 (CH_3 on C-2)

5. Locate the number of different sites. (There are two sites.)
6. Replace an –H with an –OH, and draw the isomer.

$CH_3CHCH_2{-}OH$ (CH_3 on C-2) and CH_3CCH_3 (OH and CH_3 on C-2)

7. Repeat steps 4-6 with the next simplest branched alkane. In this problem, we have already drawn all isomers.

(b) There are 17 isomers of $C_5H_{13}N$. Nitrogen can be bonded to one, two or three alkyl groups.

$CH_3CH_2CH_2CH_2CH_2NH_2$ — $CH_3CH_2CH_2CHCH_3$ (NH_2 on C-2) — $CH_3CH_2CHCH_2CH_3$ (NH_2 on C-3)

$CH_3CH_2CHCH_2NH_2$ (CH_3 on C-2) — $CH_3CH_2CCH_3$ (NH_2 and CH_3 on C-2) — $CH_3CHCHCH_3$ (NH_2 on C-2, CH_3 on C-3) — $H_2NCH_2CH_2CHCH_3$ (CH_3 on C-3)

$CH_3CCH_2NH_2$ (two CH_3 on C-2) — $CH_3CH_2CH_2CH_2NHCH_3$ — $CH_3CH_2CHNHCH_3$ (CH_3 on CH)

CH_3 | $CH_3CHCH_2NHCH_3$ CH_3 | CH_3CNHCH_3 | CH_3 $CH_3CH_2CH_2NHCH_2CH_3$ CH_3 | $CH_3CHNHCH_2CH_3$

CH_3 | $CH_3CH_2CH_2NCH_3$ CH_3 | CH_3CHNCH_3 | CH_3 CH_3 | $CH_3CH_2NCH_2CH_3$

(c) There are 3 ketone isomers with the formula $C_5H_{10}O$.

O ‖ $CH_3CH_2CH_2CCH_3$ O ‖ $CH_3CH_2CCH_2CH_3$ O ‖ $CH_3CHCCH_2CH_3$ | CH_3

(d) There are 4 isomeric aldehydes with the formula $C_5H_{10}O$. Remember that the aldehyde functional group can occur only at the end of a chain.

O ‖ $CH_3CH_2CH_2CH_2CH$ O ‖ CH_3CHCH_2CH | CH_3 O ‖ CH_3CH_2CHCH | CH_3 H_3C O | ‖ CH_3C-CH | CH_3

(e) There are 3 ethers with the formula $C_4H_{10}O$.

$CH_3CH_2OCH_2CH_3$ $CH_3OCH_2CH_2CH_3$ CH_3 | CH_3OCHCH_3

(f) There are 4 esters with the formula $C_4H_8O_2$.

O ‖ $CH_3CH_2COCH_3$ O ‖ $CH_3COCH_2CH_3$ O ‖ $HCOCH_2CH_2CH_3$ O CH_3 ‖ | $HCOCHCH_3$

2.35

$CH_3CH_2CH_2CH_2CH_2Br$ Br | $CH_3CH_2CH_2CHCH_3$ Br | $CH_3CH_2CHCH_2CH_3$

2.36

CH_3 CH_3 | | $CH_3CHCH_2CH_2CHCH_2Cl$ CH_3 CH_3 | | $CH_3CHCH_2CH_2CCH_3$ | Cl CH_3 CH_3 | | $CH_3CHCH_2CHCHCH_3$ | Cl

2.37

$CH_3CH_2CH_2OH$ CH_3CHCH_3 | OH $CH_3OCH_2CH_3$

Three isomers have the formula C_3H_8O.

2.38

(a) 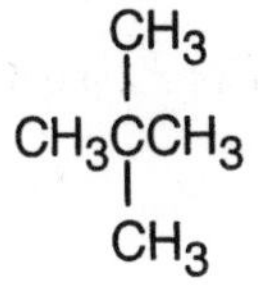 CH_3CCH_3 with two CH_3 groups on the central C (one above, one below)

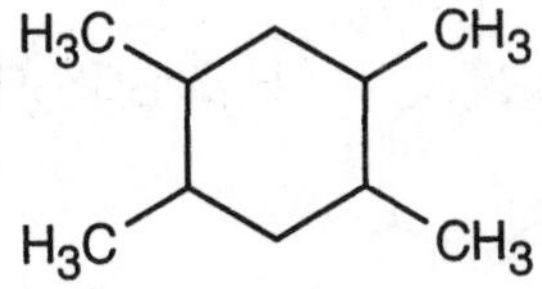

(b) H_3C, CH_3, H_3C, CH_3 (cyclohexane ring)

(c) CH_3 above $CHCH_3$ (on benzene ring)

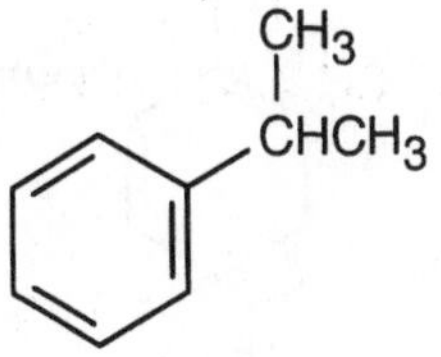

(d) 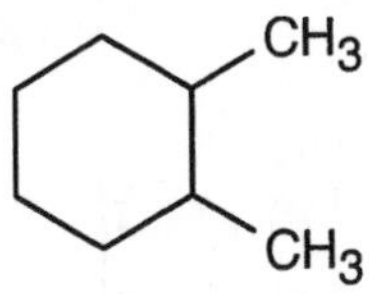 CH_3, CH_3 (cyclohexane ring)

(e) 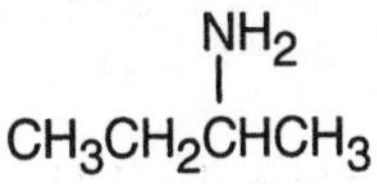 $CH_3CH_2CHCH_3$ with NH_2 on the third C

2.39

(a) 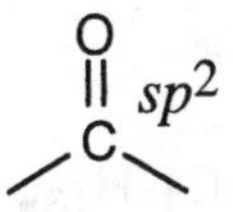 O, C, sp^2 — ketone

(b) 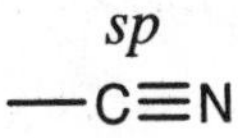 sp, —C≡N — nitrile

(c) sp^3, C—O—C 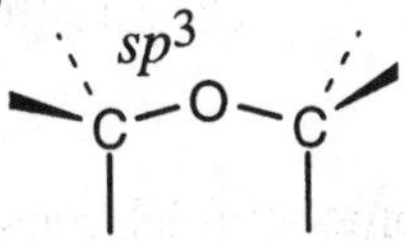 — ether

(d) 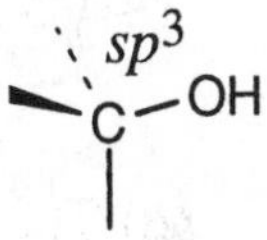 sp^3, C—OH — alcohol

Nomenclature and Representing Chemical Structures

2.40

(a)

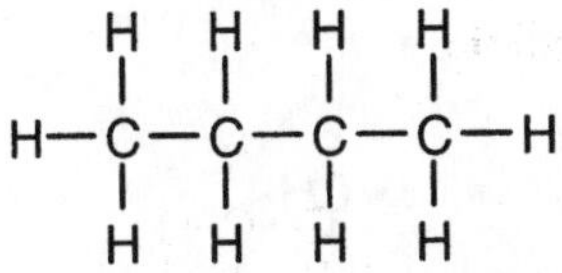

```
 H  H  H  H
 |  |  |  |
H-C--C--C--C-H
 |  |  |  |
 H  H  H  H
```

same

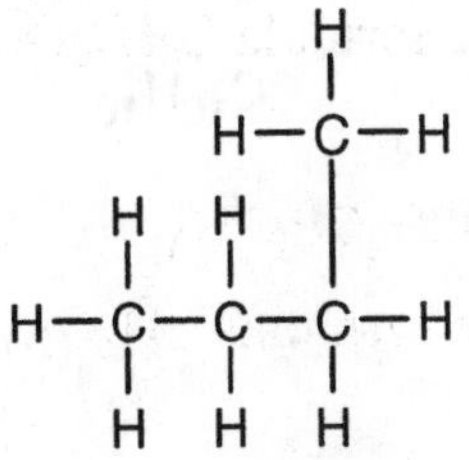

```
          H
          |
        H-C-H
 H  H     |
 |  |     |
H-C--C----C-H
 |  |     |
 H  H     H
```

same

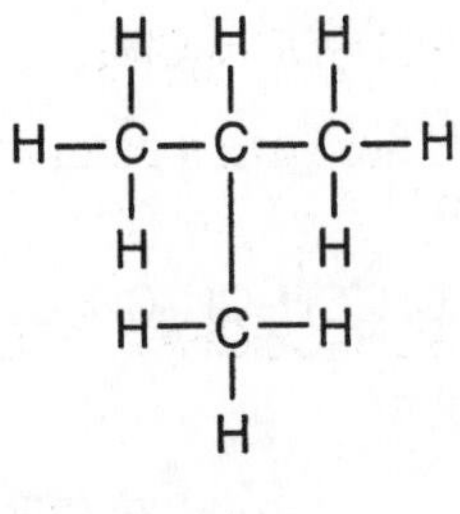

```
 H  H  H
 |  |  |
H-C--C--C-H
 |  |  |
 H  |  H
  H-C-H
    |
    H
```

(b)

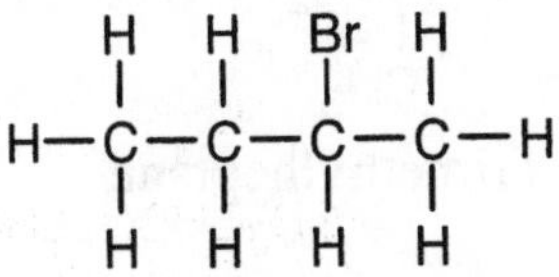

```
 H  H  Br H
 |  |  |  |
H-C--C--C--C-H
 |  |  |  |
 H  H  H  H
```

same

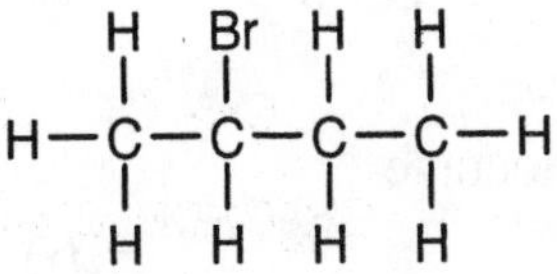

```
 H  Br H  H
 |  |  |  |
H-C--C--C--C-H
 |  |  |  |
 H  H  H  H
```

same

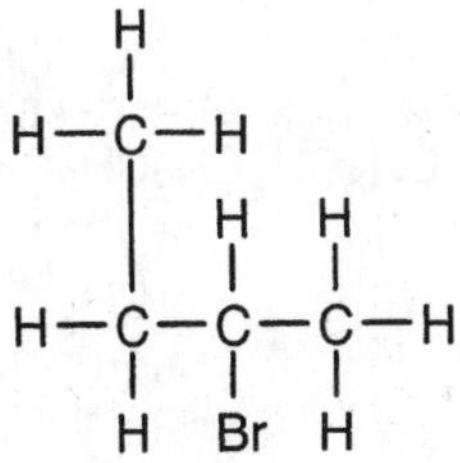

```
    H
    |
  H-C-H
    |  H  H
    |  |  |
  H-C--C--C-H
    |  |  |
    H  Br H
```

same

(c)

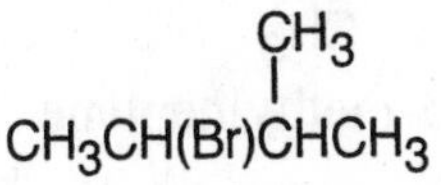

$CH_3CH(Br)CHCH_3$ with CH_3 on the fourth C

same

$CH_3CHCH(Br)CH_3$ with CH_3 on the second C

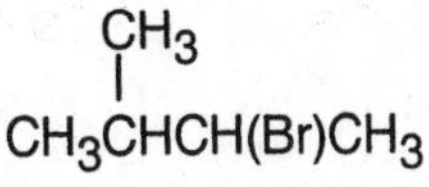

same

$(CH_3)_2CHCH(Br)CH_2CH_3$

(d)

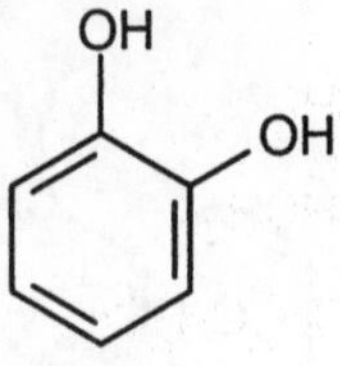

OH OH

OH OH

HO HO

same 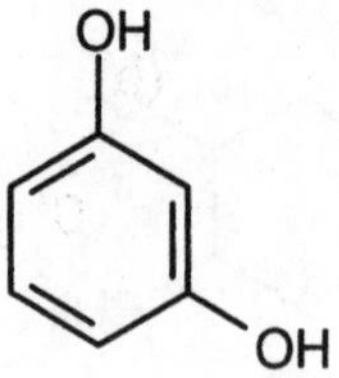 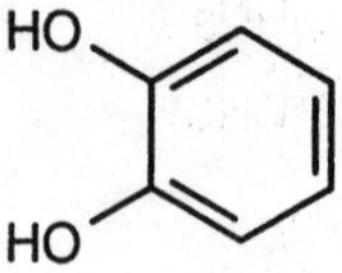 same

2.41

(a) C_8H_{16}

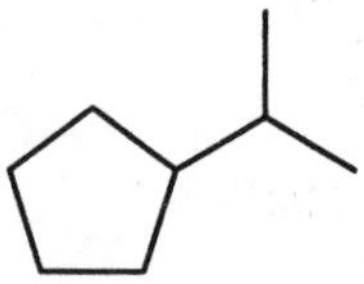

(b) $C_{10}H_{16}$

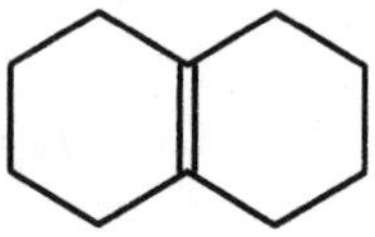

(c) $C_{13}H_{16}O$

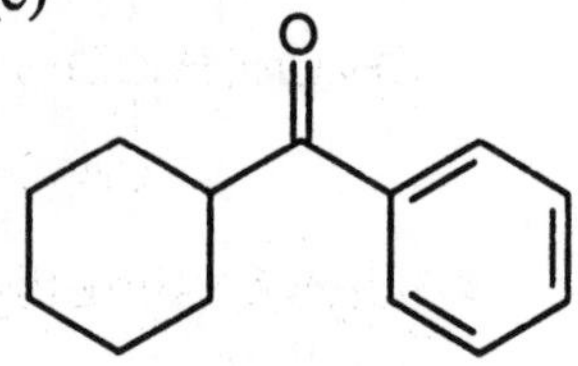

2.42

(a) C_8H_{16}

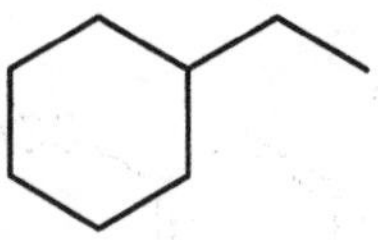

(b) $C_{10}H_{16}$

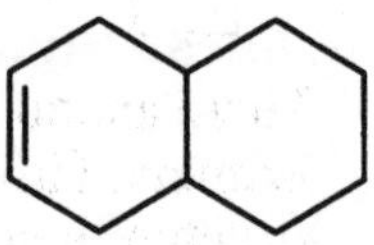

(c) O C H $C_{13}H_{16}O$

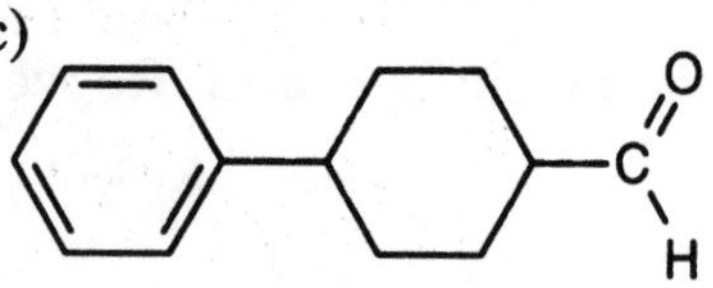

2.43

(a) CH_3 $CH_3CH_2CH_2CH_2CH_2CHCH_3$

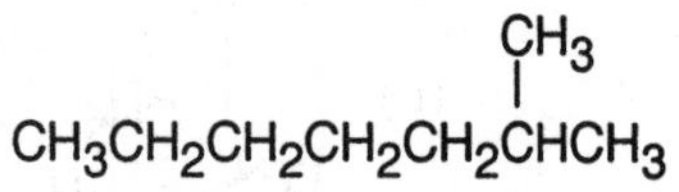

2-Methylheptane

(b) CH_3 $CH_3CH_2CHCH_2CHCH_3$ CH_2CH_3

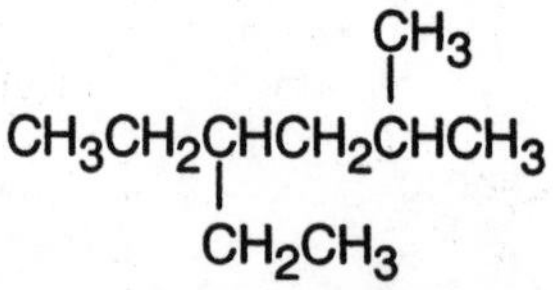

4-Ethyl-2-methylhexane

(c) CH_3 CH_3 $CH_3CH_2CH_2CH_2C$—$CHCH_2CH_3$ CH_2CH_3

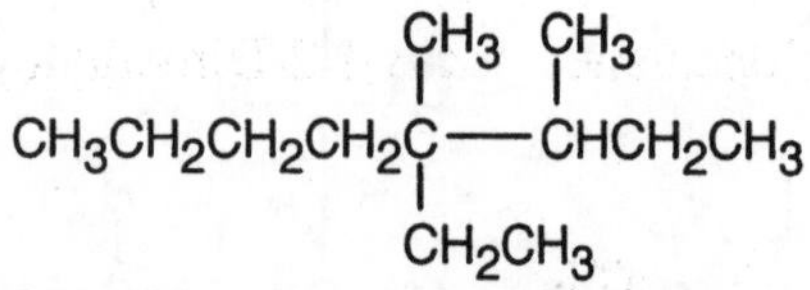

4-Ethyl-3,4-dimethyloctane

(d) CH_3 CH_3 $CH_3CH_2CH_2CCH_2CHCH_3$ CH_3

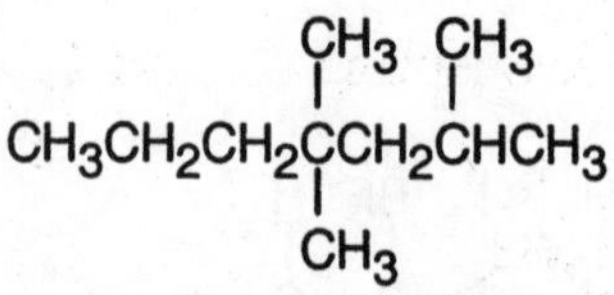

2,4,4-Trimethylheptane

(e) CH_3 CH_3

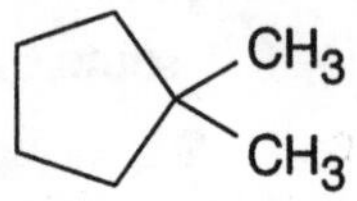

1,1-Dimethylcyclopentane

(f) CH_3CHCH_3 $CH_3CH_2CH_2CHCHCH_2CH_3$ CH_3

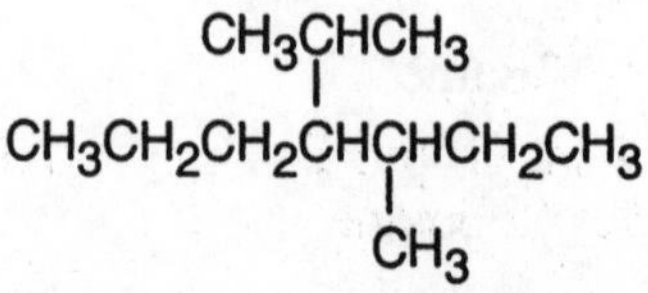

4-Isopropyl-3-methylheptane

2.44

(a)

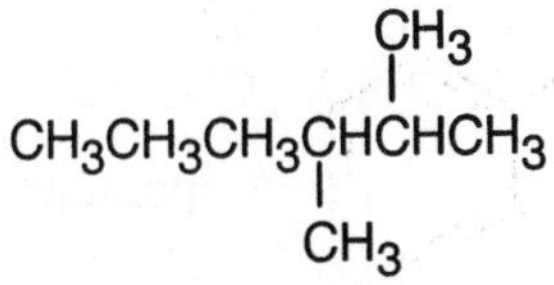

$$CH_3CH_3CH_3\underset{\displaystyle CH_3}{\underset{|}{C}}H\overset{\displaystyle CH_3}{\overset{|}{C}}HCH_3$$

2,3-Dimethylhexane

(b)

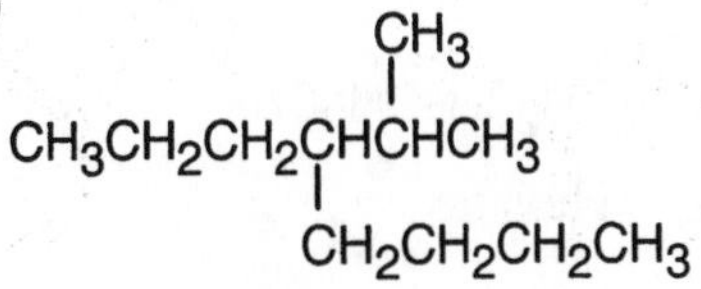

$$CH_3CH_2CH_2\underset{\displaystyle CH_2CH_2CH_2CH_3}{\underset{|}{C}}H\overset{\displaystyle CH_3}{\overset{|}{C}}HCH_3$$

4-Isopropyloctane

(c)

$$CH_3\overset{\displaystyle CH_3}{\overset{|}{C}}HCH_2\overset{\displaystyle CH_2CH_3}{\overset{|}{\underset{\displaystyle CH_2CH_3}{\underset{|}{C}}}}CH_3$$

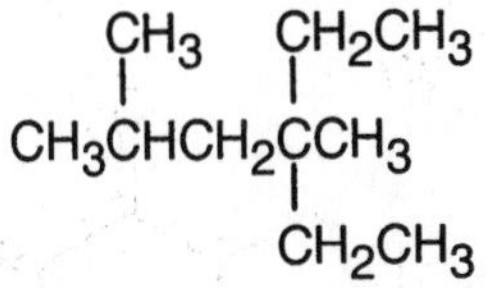

4-Ethyl-2,4-dimethylhexane

(d)

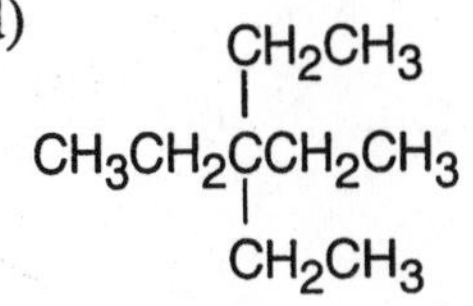

$$CH_3CH_2\overset{\displaystyle CH_2CH_3}{\overset{|}{\underset{\displaystyle CH_2CH_3}{\underset{|}{C}}}}CH_2CH_3$$

3,3-Diethylpentane

2.45

(a) $CH_3CH_2CH_2CH_2CH_2Br$

(b) (cyclobutyl)CH_2OCH_3

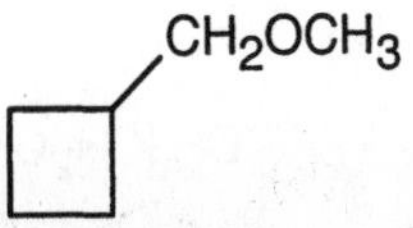

(c)

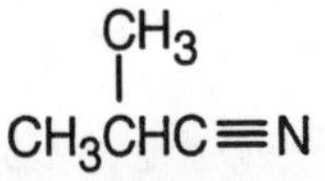

$$CH_3\overset{\displaystyle CH_3}{\overset{|}{C}}HC{\equiv}N$$

(d) (cyclopentyl)CH_2OH

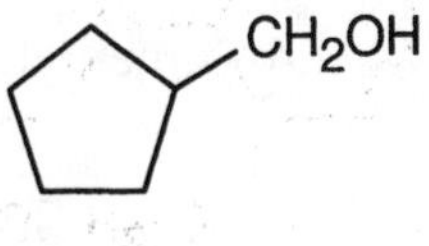

(e) There are no aldehyde isomers. However, the structure below is a ketone isomer.

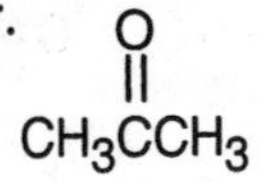

$$CH_3\overset{\displaystyle O}{\overset{\|}{C}}CH_3$$

(f) (benzene ring with COOH and CH_3 on adjacent carbons)

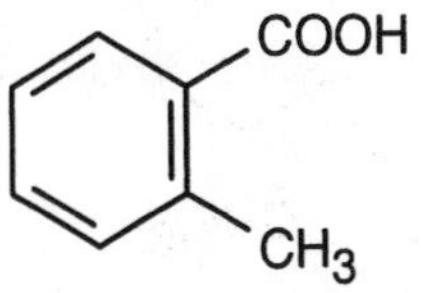

2.46

(a) (cycloheptane ring with CH_3)

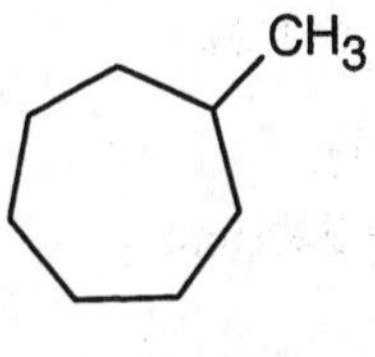

Methylcycloheptane

(b) (cyclopentane ring with H, H_3C and H, CH_3)

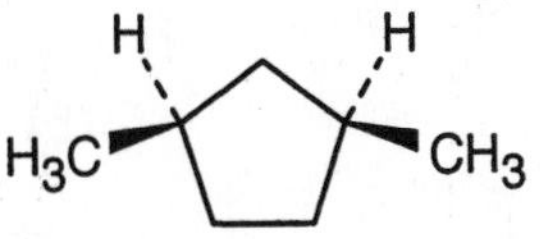

cis-1,3-Dimethylcyclopentane

(c) (cyclohexane ring with CH_3, H and H, CH_3)

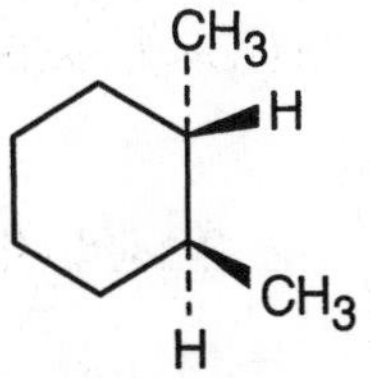

trans-1,2-Dimethylcyclohexane

(d) (cyclobutane ring with isopropyl and CH_3)

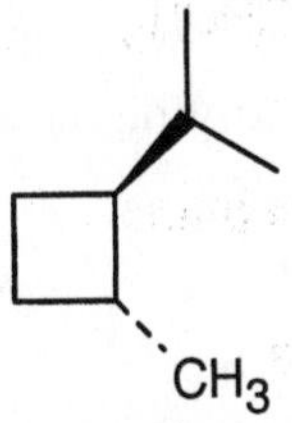

trans-1-Isopropyl-2-methylcyclobutane

(e) (cyclohexane ring with H_3C and CH_3, CH_3)

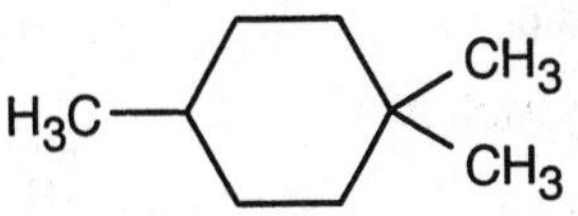

1,1,4-Trimethylcyclohexane

2.47

$CH_3CH_2CH_2CH_2CH_2CH_3$
Hexane

$$\begin{array}{c} \quad\quad\quad\quad CH_3 \\ \quad\quad\quad\quad | \\ CH_3CH_2CH_2CHCH_3 \end{array}$$
2-Methylpentane

$$\begin{array}{c} CH_3 \\ | \\ CH_3CH_2CHCH_2CH_3 \end{array}$$
3-Methylpentane

$$\begin{array}{c} \quad\quad\quad CH_3 \\ \quad\quad\quad | \\ CH_3CH_2CCH_3 \\ \quad\quad\quad | \\ \quad\quad\quad CH_3 \end{array}$$
2,2-Dimethylbutane

$$\begin{array}{c} \quad\quad\quad CH_3 \\ \quad\quad\quad | \\ CH_3CHCHCH_3 \\ | \\ CH_3 \end{array}$$
2,3-Dimethylbutane

2.48

$CH_3CH_2CH_2CH_2CH_2CH_2CH_3$
Heptane

$$\begin{array}{c} \quad\quad\quad\quad\quad CH_3 \\ \quad\quad\quad\quad\quad | \\ CH_3CH_2CH_2CH_2CHCH_3 \end{array}$$
2-Methylhexane

$$\begin{array}{c} \quad\quad CH_3 \\ \quad\quad | \\ CH_3CH_2CH_2CHCH_2CH_3 \end{array}$$
3-Methylhexane

$$\begin{array}{c} \quad\quad\quad\quad CH_3 \\ \quad\quad\quad\quad | \\ CH_3CH_2CH_2CCH_3 \\ \quad\quad\quad\quad | \\ \quad\quad\quad\quad CH_3 \end{array}$$
2,2-Dimethylpentane

$$\begin{array}{c} \quad\quad\quad\quad CH_3 \\ \quad\quad\quad\quad | \\ CH_3CH_2CHCHCH_3 \\ | \\ CH_3 \end{array}$$
2,3-Dimethylpentane

$$\begin{array}{c} \quad\quad\quad\quad\quad CH_3 \\ \quad\quad\quad\quad\quad | \\ CH_3CHCH_2CHCH_3 \\ | \\ CH_3 \end{array}$$
2,4-Dimethylpentane

$$\begin{array}{c} CH_3 \\ | \\ CH_3CH_2CCH_2CH_3 \\ | \\ CH_3 \end{array}$$
3,3-Dimethylpentane

$$\begin{array}{c} CH_3CH_2CHCH_2CH_3 \\ | \\ CH_2CH_3 \end{array}$$
3-Ethylpentane

$$\begin{array}{c} CH_3 \quad CH_3 \\ | \quad\quad\;\; | \\ CH_3CH - CCH_3 \\ \quad\quad\quad\quad | \\ \quad\quad\quad\quad CH_3 \end{array}$$
2,2,3-Trimethylbutane

2.49

(a)

$$\begin{array}{c} \quad\quad\quad\quad\quad\quad\quad\quad CH_3 \\ \quad\quad\quad\quad\quad\quad\quad\quad | \\ CH_3CH_2CH_2CH_2CH_2CH_2CCH_3 \\ \quad\quad\quad\quad\quad\quad\quad\quad | \\ \quad\quad\quad\quad\quad\quad\quad\quad CH_3 \end{array}$$
2,2-Dimethyloctane

(b)

$$\begin{array}{c} CH_3 \quad CH_2CH_3 \\ | \quad\quad\quad | \\ CH_3CCH_2CCH_2CH_3 \\ | \quad\quad\quad | \\ CH_3 \quad CH_2CH_3 \end{array}$$
4,4-Diethyl-2,2-dimethylhexane

(c)

CH_3
CH_3
CH_3

1,1,2-Trimethylcyclohexane

2.50 *Structure and Correct Name* | *Error*

(a)

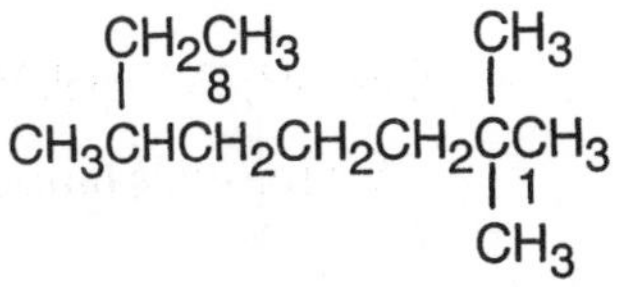

$$\begin{array}{l} \quad CH_2CH_3 \qquad\quad CH_3 \\ \quad | \;\; {}^{8} \qquad\qquad\quad | \\ CH_3CHCH_2CH_2CH_2CCH_3 \\ \qquad\qquad\qquad\qquad\quad |\,{}_1 \\ \qquad\qquad\qquad\qquad\; CH_3 \end{array}$$

Correct name: 2,2,6-Trimethyloctane

The longest chain is an octane.

(b)

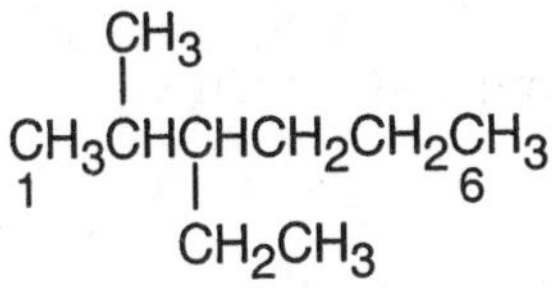

$$\begin{array}{l} \quad CH_3 \\ \quad | \\ CH_3CHCHCH_2CH_2CH_3 \\ {}_1 \qquad\; | \qquad\qquad\quad {}_6 \\ \qquad CH_2CH_3 \end{array}$$

Correct name: 3-Ethyl-2-methylhexane

The longest chain is a hexane. Numbering should start from the opposite end of the carbon chain, nearer the first branch.

(c)

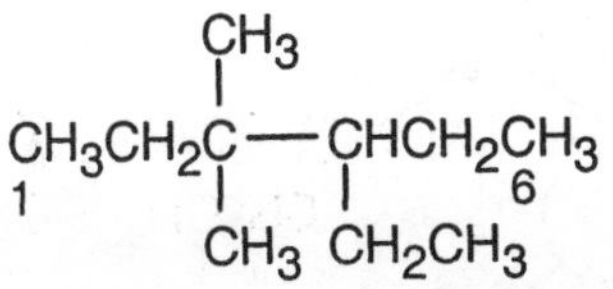

$$\begin{array}{l} \qquad\quad CH_3 \\ \qquad\quad | \\ CH_3CH_2C-CHCH_2CH_3 \\ {}_1 \qquad\quad | \quad\;\; | \qquad\quad {}_6 \\ \qquad\; CH_3 \; CH_2CH_3 \end{array}$$

Correct name: 4-Ethyl-3,3-dimethylhexane

Numbering should start from the opposite end of the carbon chain.

(d)

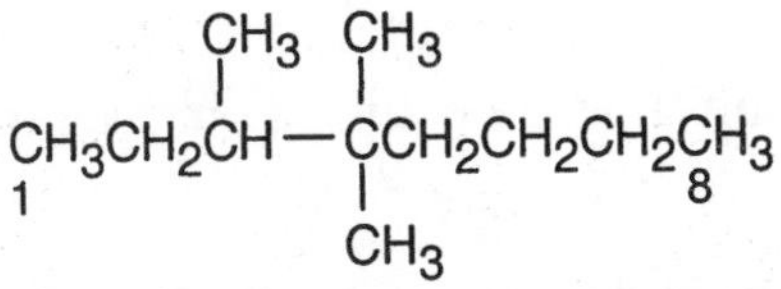

$$\begin{array}{l} \qquad\quad CH_3 \;\; CH_3 \\ \qquad\quad | \qquad\; | \\ CH_3CH_2CH-CCH_2CH_2CH_2CH_3 \\ {}_1 \qquad\qquad\qquad | \qquad\qquad\qquad {}_8 \\ \qquad\qquad\qquad CH_3 \end{array}$$

Correct name: 3,4,4-Trimethyloctane

Numbering should start from the opposite end of the carbon chain.

2.51 (a) It is stated that biacetyl contains no rings or carbon-carbon double bonds. However, it is obvious from the formula for biacetyl that some sort of multiple bond must be present. The structure for biacetyl contains two carbon-oxygen double bonds.

(b) Ethyleneimine contains a three-membered ring.

(c) Glycerol contains no multiple bonds or rings.

(a)

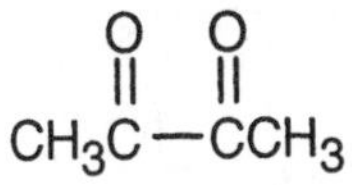

$$\begin{array}{l} \qquad\; O \quad O \\ \qquad\; \| \quad\; \| \\ CH_3C-CCH_3 \end{array}$$

Biacetyl

(b)

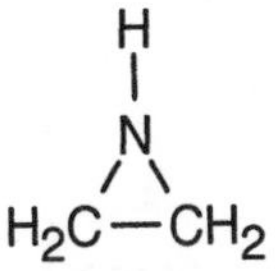

$$\begin{array}{c} H \\ | \\ N \\ / \;\; \backslash \\ H_2C-CH_2 \end{array}$$

Ethyleneimine

(c)

$$\begin{array}{l} \qquad\quad OH \\ \qquad\quad | \\ HOCH_2CHCH_2OH \end{array}$$

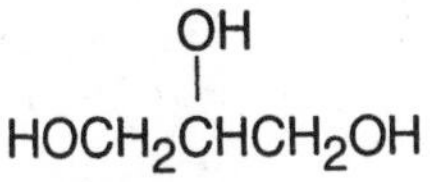

Glycerol

Conformational Analysis and Cis/Trans Isomerism

2.52

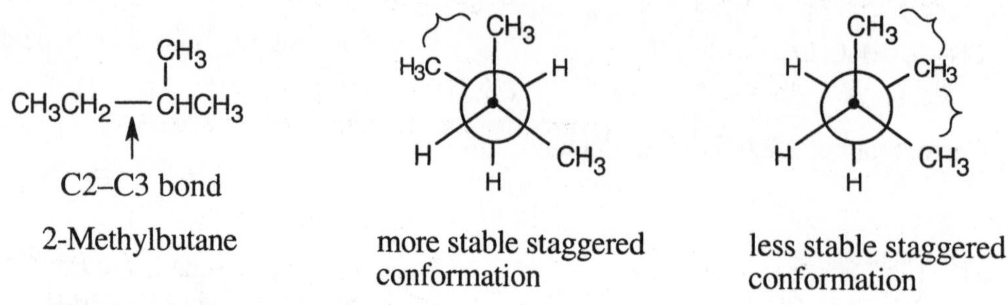

Brackets indicate steric interactions

In the less stable isomer, there are two interactions between methyl groups that are 60° apart. Only one of these interactions occurs in the more stable isomer.

2.53

CH3 H — H CH3 — HCH3

lower energy eclipsed conformation

CH3 CH3 — H H — HCH3 — strain worse here

higher energy eclipsed conformation

In the higher energy isomer, the two methyl groups that are eclipsed produce more strain than occurs in the other isomer.

2.54 Since *cis*-1-*tert*-butyl-4-methylcyclohexane exists in the conformation shown, a *tert*-butyl group must be much larger than a methyl group.

2.55 The strain energy of the highest energy conformation of bromoethane is 15.0 kJ/mol (3.6 kcal/mol). Since this includes two H–H eclipsing interactions of 3.8 kJ/mol (0.9 kcal/mol) each, the value of an H–Br interaction is 15.0 – 2(3.8) = 7.4 kJ/mol (1.8 kcal/mol).

2.56

(a)

cis-1,3-Dibromocyclohexane

trans-1,4-Dibromocyclohexane

constitutional isomers

(b)

$CH_3CH_2CH_2CH(CH_3)CH(CH_3)CH_3$ — 2,3-Dimethylhexane

$CH_3CH(CH_3)CH_2CH_2CH(CH_3)CH_3$ — 2,5,5-Trimethylpentane (correct name: 2,5-Dimethylhexane)

} constitutional isomers

(c)

Cl Cl (cyclopentane) — Cl Cl (cyclopentane)

} identical

2.57

Br Br H H — H Br Br H

cis-1,3-Dibromocyclopentane *trans*-1,3-Dibromocyclopentane

Many other constitutional isomers can also be drawn.

2.58

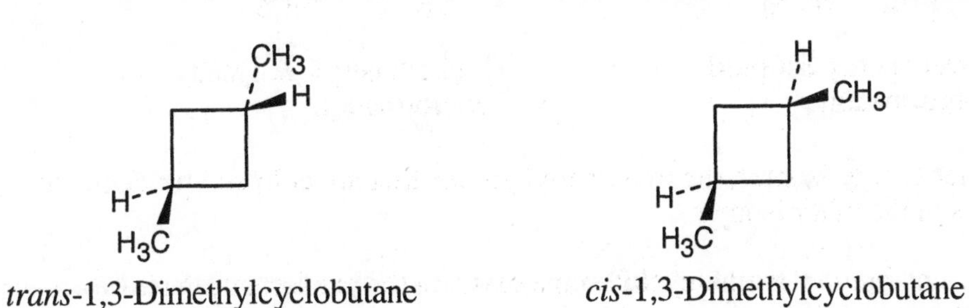

trans-1,3-Dimethylcyclobutane *cis*-1,3-Dimethylcyclobutane

2.59

H H H H H CH3 H CH3 H H H H

trans-1,2-Dimethylcyclohexane

The methyl groups are equatorial in the more stable chair conformation of *trans*-1,2-dimethylcyclohexane.

2.60

cis-1,2-Dimethylcyclohexane

There are two chair conformations of *cis*-1,2-dimethylcyclohexane that are of equal stability. In each conformation, one methyl group is axial and one is equatorial. Both cis conformations are less stable than the more stable conformation of *trans*-1,2-dimethylcyclohexane because of steric strain caused by the axial methyl group.

2.61

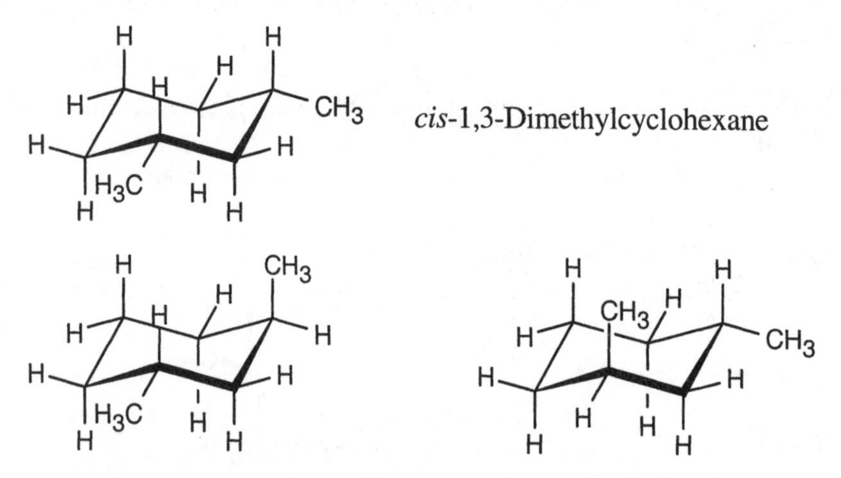

cis-1,3-Dimethylcyclohexane

trans-1,3-Dimethylcyclohexane

The lowest energy conformations of both 1,3-dimethylcyclohexanes are drawn. *cis*-1,3-Dimethylhexane is the more stable isomer because both methyl groups are equatorial in the most stable conformation. For *trans*-1,3-dimethylhexane, one methyl group must always be in the higher energy axial orientation. (A high energy diaxial conformation of *cis*-1,3-dimethylcyclohexane can also be drawn.)

Octane Ratings

2.62 A fuel with an octane rating of 87 performs like a fuel that contains 87% isooctane and 13% heptane.

2.63 According to the octane rating system, a fuel sample with a 35:65 ratio of isooctane to heptane would have an octane rating of 35.

2.64

Fuel	R	M	(R + M)/2
Pentane	62	62	62
Hexane	25	26	25.5
Heptane	0	0	0
2,3-Dimethylpentane	93	96	94.5

Integrated Problems

2.65 (a) Because malic acid has two $-CO_2H$ groups, the formula for the rest of the molecule is C_2H_4O. Possible structures for malic acid are:

$HO_2C-CH(CH_2OH)-CO_2H$	$HO_2C-CH(OH)CH_2-CO_2H$	$HO_2C-C(OH)(CH_3)-CO_2H$
primary alcohol	secondary alcohol	tertiary alcohol
$HO_2C-CH_2OCH_2-CO_2H$	$HO_2C-OCH_2CH_2-CO_2H$	$HO_2C-OCH(CH_3)-CO_2H$

(b) Because only one of these compounds (the second one) is also a secondary alcohol, it must be malic acid.

2.66

(a) $H_2C(CH_2Br)_2 \xrightarrow{\text{2 Na}}$ cyclopropane (H_2C, CH_2, CH_2 ring) + 2 NaBr

(b) $C(CH_2Br)_4 \xrightarrow{\text{4 Na}}$ spiropentane (H_2C, H_2C, CH_2, CH_2 around central C) + 4 NaBr

The two rings are perpendicular in order to keep the geometry of the central carbon as close to tetrahedral as possible.

(Structure: spiropentane with the two rings perpendicular; labels H_2C, H_2C, C, CH_2, CH_2)

2.67 Since the methyl group of *N*-methylpiperidine prefers an equatorial conformation, the steric requirements of a methyl group must be greater than those of an electron lone pair.

2.68

(Chair structure with labels H, $HOCH_2$, H, O, HO, HO, OH, H, OH, H, H)

Glucose

2.69

Two cis-trans isomers of 1,3,5-trimethylcyclohexane are possible. In one isomer (**A**), all methyl groups are cis; in **B**, one methyl group is trans to the other two.

2.70

Conformation **A** of *cis*-1-chloro-3-methylcyclohexane has no 1,3-diaxial interactions and is the more stable conformation. Steric strain in **B** is due to one $CH_3 \leftrightarrow H$ interaction, one $Cl \leftrightarrow H$ interaction and one $CH_3 \leftrightarrow Cl$ interaction.

2.71

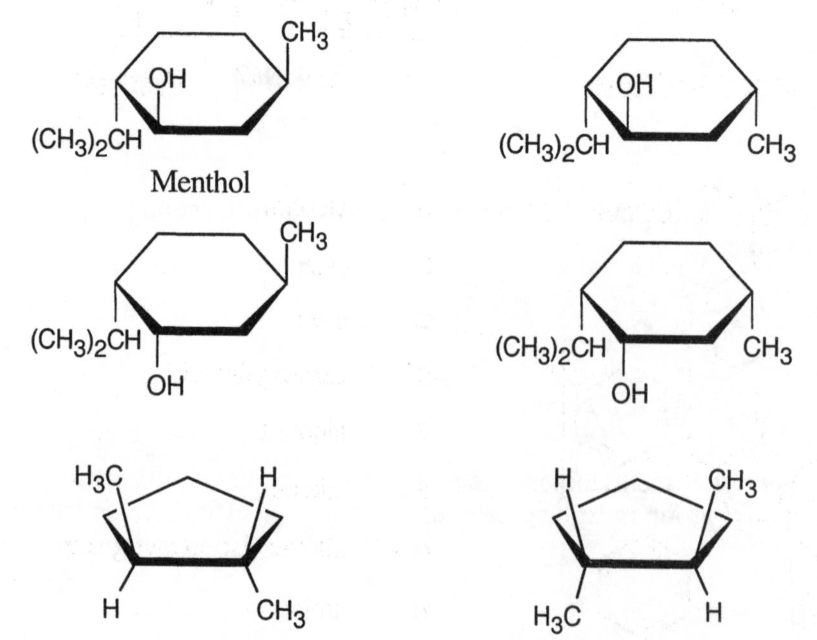

The two *trans*-1,2-dimethylcyclopentanes are mirror images.

In the Medicine Cabinet

2.73

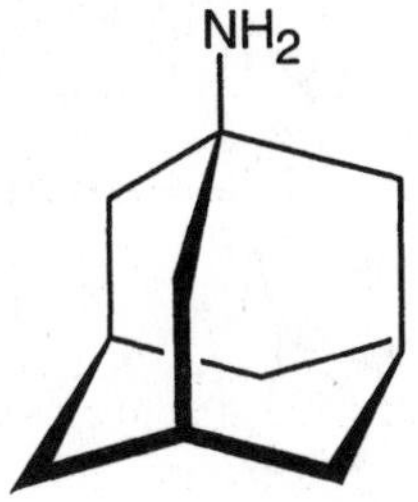

NH$_2$

Amantadine

2.74

(a), (d)

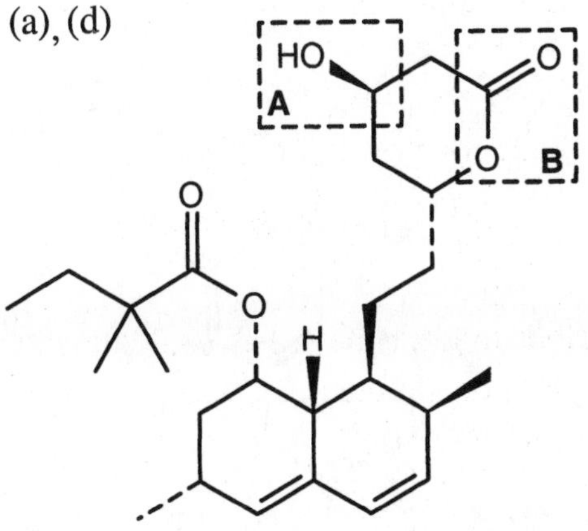

HO A O O B O

Zocor $C_{25}H_{38}O_5$

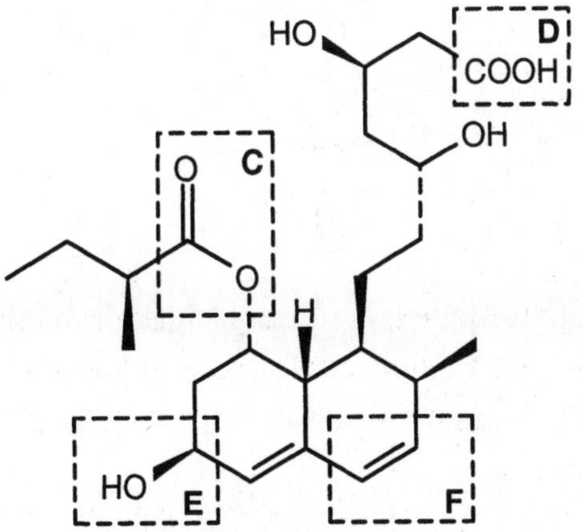

HO D COOH OH O C O H HO E F

Pravachol $C_{23}H_{36}O_7$

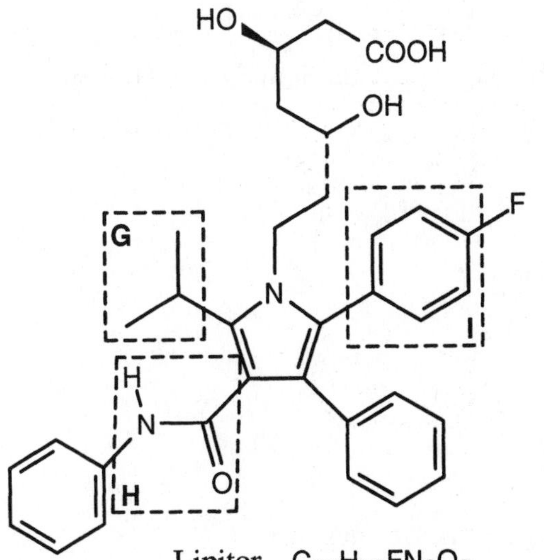

HO COOH OH F G N I H N H O

Lipitor $C_{33}H_{35}FN_2O_5$

A	alcohol
B	ester
C	ester
D	carboxylic acid
E	alcohol
F	alkene
G	alkane (isopropyl group)
H	amide
I	arene (aromatic ring)

(b) **C** and **E** have a trans relationship.

(c) Groups **G**, **H**, and **I** are bonded to sp^2 carbons of a planar ring and lie in the same plane.

2.75

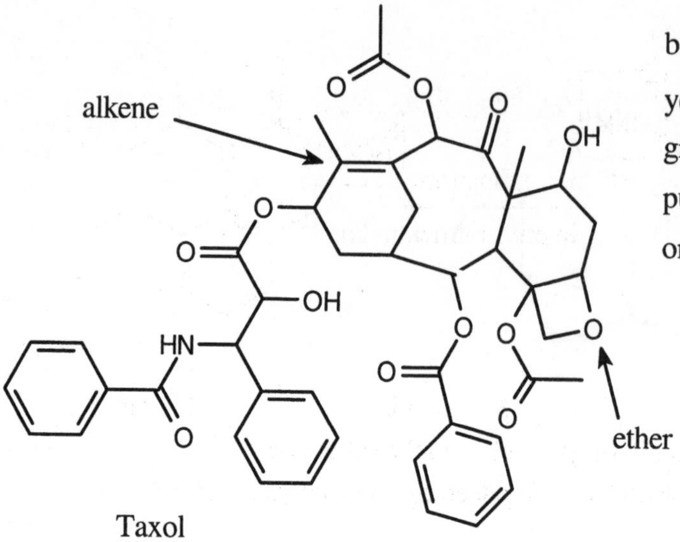

Taxol

blue = ester
yellow = aromatic ring (arene)
green = alcohol
purple = amide
orange = ketone

2.76

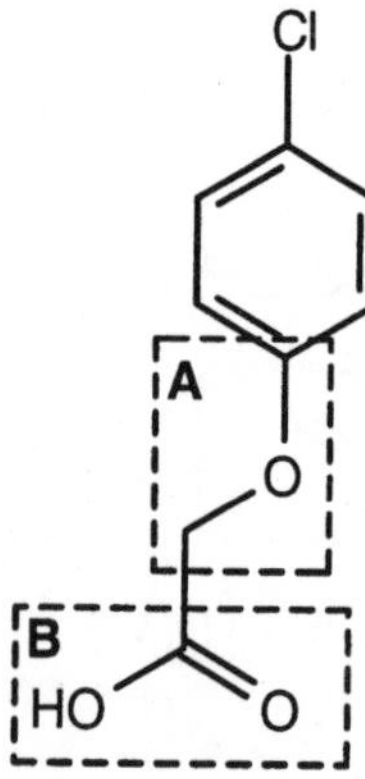

2,4-D ($C_8H_6Cl_2O_3$)

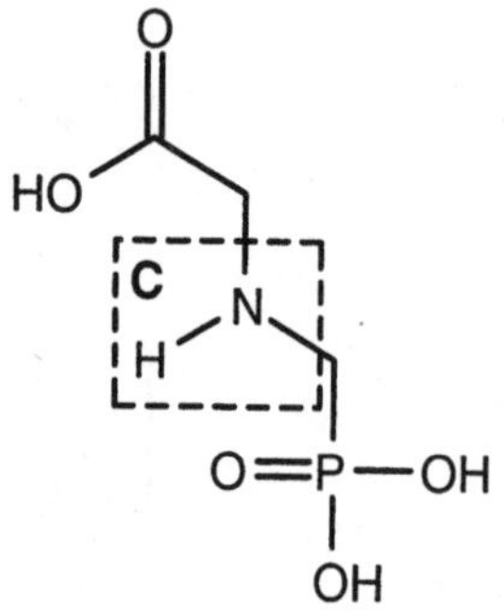

Glyphosate ($C_3H_8NO_5P$)

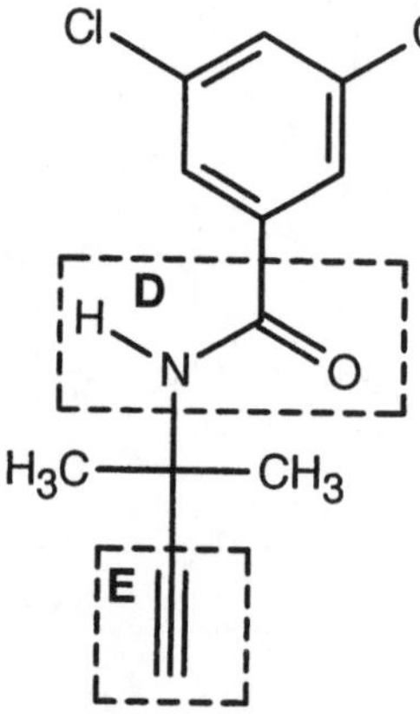

Pronamide ($C_{12}H_{11}Cl_2NO$)

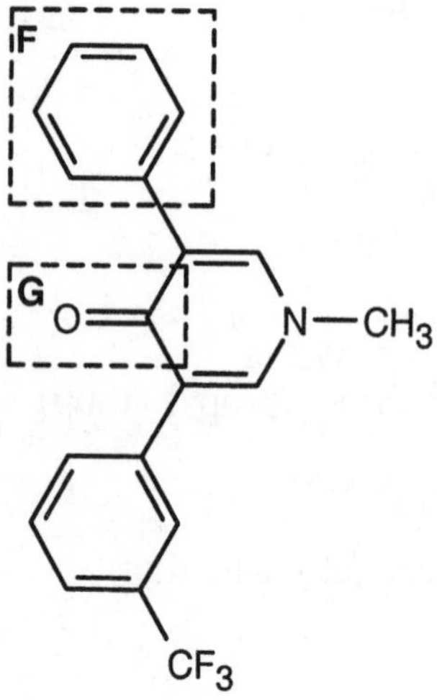

Fluridone ($C_{19}H_{14}F_3NO$)

A ether

B carboxylic acid

C amine

D amide

E alkyne

F arene

G ketone

2.77

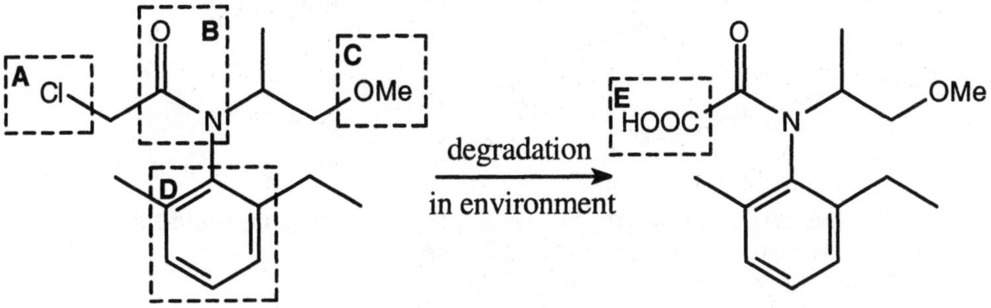

A	chloride	**B**	amide	**C**	ether
D	arene (aromatic ring)	**E**	carboxylic acid		

After degradation, the chloride (**A**) is replaced by a carboxylic acid.

Chapter 3 – The Nature of Organic Reactions: Alkenes

Chapter Outline

I. Alkenes (Sections 3.1–3.4).
  A. Naming alkenes (Section 3.1).
    1. Find the longest chain containing the double bond, and name it.
    2. Number the carbon atoms in the chain, giving the double bond the lowest number.
    3. Number the substituents and write the name.
      a. Name the substituents alphabetically.
      b. Indicate the position of the double bond.
      c. Use the suffixes -diene, -triene, etc. if more than one double bond is present.
    4. Cycloalkenes are named in a similar way, but the double bond is placed between C1 and C2, and the substituents receive the lowest possible numbers.
  B. Electronic structure of alkenes (Section 3.2).
    1. Carbon atoms in a double bond are sp^2-hybridized.
    2. The two carbons in a double bond form one σ bond and one π bond.
    3. Free rotation doesn't occur around double bonds.
    4. 268 kJ/mol of energy is required to break a π bond.
  C. Isomerism in alkenes (Sections 3.3–3.4).
    1. Cis-trans isomerism (Section 3.3).
      a. A disubstituted alkene can have substituents either on the same side of the double bond (cis) or on opposite sides (trans).
      b. These isomers don't interconvert because free rotation about a double bond isn't possible.
      c. Cis-trans isomerism doesn't occur if one carbon in the double bond is bonded to identical substituents.
      d. Cis alkenes are less stable than their trans isomers.
    2. *E,Z* isomerism (Section 3.4).
      a. The *E,Z* system is used to describe the arrangement of substituents around a double bond that can't be described by the cis-trans system.
      b. Sequence rules for *E,Z* isomers:
        i. For each double bond carbon, rank the substituents by atomic number. An atom with a high atomic number receives a higher priority than an atom with a lower atomic number.
        ii. If a decision can't be reached, look at the second or third atom until a difference is found.
        iii. Multiple-bonded atoms are equivalent to the same number of single-bonded atoms.
      c. In a *Z* isomer, the two groups of higher priority are on the same side of the alkene double bond.
      d. In an *E* isomer, the two groups of higher priority are on opposite sides of the alkene double bond.

II. Organic reactions (Sections 3.5–3.10).
  A. Kinds of organic reactions (Section 3.5).
    1. Addition reactions occur when two reactants add to form one product, with no atoms left over.
    2. Elimination reactions occur when a single reactant splits into two products.

3. Substitution reactions occur when two reactants exchange parts to yield two new products.
4. Rearrangement reactions occur when a single product undergoes a rearrangement of bonds to yield an isomeric product.

B. How reactions occur (Sections 3.6–3.8).
1. Mechanisms (Section 3.6).
   a. A reaction mechanism describes the bonds broken and formed in a chemical reaction, and accounts for all reactants and products.
   b. The electrons involved in a reaction are usually π electrons or electron lone pairs.
   c. In polar reactions, electrons move in pairs.
      i. Bond breaking in polar reactions is described as heterolytic because both electrons remain with one atom.
      ii. Bond formation is heterogenic because both electrons in the new bond come from one reactant.
   d. Radical reactions involve the movement of single electrons.
      i. Bond breaking is described as homolytic because one electron remains with each fragment.
      ii. Bond formation is homogenic if one electron in a covalent bond comes from each reactant.
   e. Movement of electrons is shown by curved arrows.
      i. A double-headed arrow indicates movement of an electron pair.
      ii. A fishhook arrow represents a single electron.
2. Mechanisms of polar reactions (Section 3.7).
   a. Polar reactions occur as a result of partial positive and negative charges within molecules.
   b. These charge differences are usually due to electronegativity differences between atoms.
   c. In polar reactions, electron-rich sites in one molecule react with electron-poor sites in another molecule.
   d. The reacting species:
      i. A nucleophile is a compound with an electron-rich atom.
      ii. An electrophile is a compound with an electron-poor atom.
      Some molecules can behave as both nucleophiles and as electrophiles
      iii. Many polar reactions can be explained in terms of acid-base reactions.
3. An example of a polar reaction: addition of HCl to ethylene (Section 3.8).
   a. This reaction is known as an electrophilic addition.
   b. The π electrons in ethylene behave as a nucleophile.
   c. The reaction begins by the addition of the π electrons of the double bond to H^+.
   d. The resulting intermediate carbocation reacts with Cl^- to form chloroethane.
   e. In each step of a mechanism, both atoms and charge are conserved.

C. Describing a reaction (Sections 3.9–3.10).
1. Energy diagrams and transition states (Section 3.9).
   a. Reaction energy diagrams show the energy changes that occur during a reaction.
      The vertical axis represents energy changes, and the horizontal axis represents the progress of a reaction.
   b. The transition state is the highest-energy species in the reaction.
      It is possible for a reaction to have more than one transition state.

c. The difference in energy between the reactants and the transition state is the energy of activation E_{act}.
Values of E_{act} range from 40 – 125 kJ/mol.
d. After reaching the transition state, a molecule can go on to form products or can revert to starting material.
e. In a reaction of at least two steps, an intermediate is the species that lies at the energy minimum between two transition states.
f. Even though an intermediate lies at an energy minimum between two transition states, it is a high-energy species and usually can't be isolated.

2. Energetics and catalysis (Section 3.10).
a. Reactions occur spontaneously if the energy of the products is less than the energy of the reactants.
b. The rate of a reaction can't be predicted by the difference in energy between reactants and products.
c. A catalyst can change the rate of a reaction but can't change the energy difference between reactant and product.

Solutions to Problems

3.1

(a)

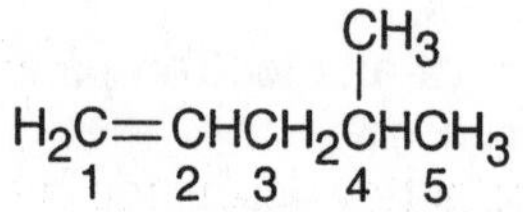

$$H_2C{=}CHCH_2CH(CH_3)CH_3$$

(numbering: C1 = H_2C, C2 = CH, C3 = CH_2, C4 = CH bearing CH_3, C5 = CH_3)

1. Find the longest carbon chain containing the double bond, and name the parent compound. In this problem, the longest chain contains five carbons, and the compound is a *pentene*.

2. Number the carbon atoms, giving to the double bond the lowest possible number.

3. Identify the substituents. In this example, there is a methyl group at C4.

4. Name the compound. The name is 4-methylpent-1-ene.

(b) $CH_3CH_2CH{=}CHCH_2CH_2CH_3$
Hept-3-ene

(c) $H_2C{=}CHCH_2CH_2CH{=}CHCH_3$
Hepta-1,5-diene

(d) $CH_3CH_2CH{=}CHCH(CH_3)_2$
2-Methylhex-3-ene

3.2

(a) 1,2-Dimethylcyclohexene

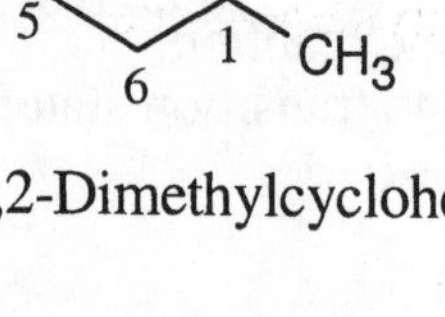

(b) 4,4-Dimethylcycloheptene

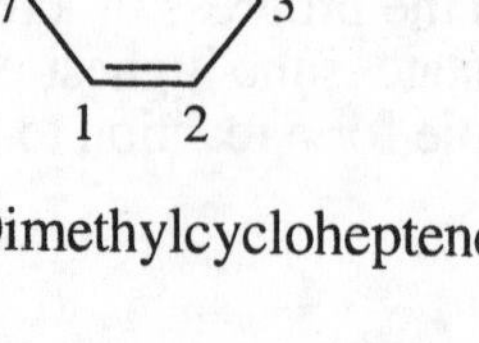

(c) 3-Isopropylcyclopentene

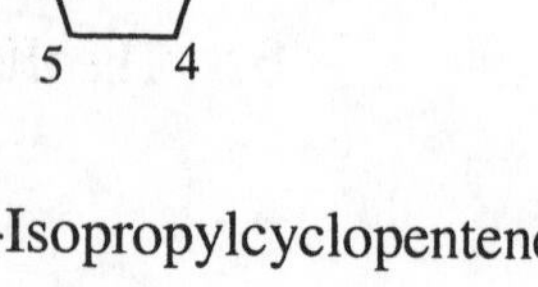

3.3

(a) $CH_3CH_2CH_2CH_2C(CH_3){=}CH_2$

2-Methylhex-1-ene

(b) $(CH_3)_3CCH{=}CHCH_3$

4,4-Dimethylpent-2-ene

(c) $H_2C{=}CHCH_2CH_2C(CH_3){=}CH_2$

2-Methylhexa-1,5-diene

(d) $CH_3CH_2CH_2CH{=}C(CH_2CH_3)C(CH_3)_3$

3-Ethyl-2,2-dimethylhept-3-ene

3.4 Compounds (c) and (d) can exist as pairs of cis–trans isomers.

(c)

cis: $(Cl)(H)C{=}C(H)(Cl)$ — H, H on the same side; Cl, Cl on the same side

trans: $(Cl)(H)C{=}C(Cl)(H)$ — H and Cl on the same side

(d)

cis: $(CH_3CH_2)(H)C{=}C(H)(CH_3)$ — H, H on the same side; CH_3CH_2, CH_3 on the same side

trans: $(CH_3CH_2)(H)C{=}C(CH_3)(H)$ — H and CH_3 on the same side

Compounds (e) and (f) can exist as pairs of isomers that are better described by *E*, *Z* designations, which are explained in Section 3.4.

(e) $(CH_3CH_2)(H)C{=}C(CH_3)(Br)$ — CH_3CH_2 and CH_3 on the same side *and* $(CH_3CH_2)(H)C{=}C(Br)(CH_3)$ — CH_3CH_2 and Br on the same side

(f) $(CH_3CH_2CH_2)(H)C{=}C(CH_2CH_3)(CH_3)$ — $CH_3CH_2CH_2$ and CH_2CH_3 on the same side *and* $(CH_3CH_2CH_2)(H)C{=}C(CH_3)(CH_2CH_3)$ — $CH_3CH_2CH_2$ and CH_3 on the same side

3.5

(a) $CH_3CH(CH_3){-}CH(CH_3)(H)C{=}C(H)CH_3$ — H, H on the same side

cis-4,5-Dimethylhex-2-ene

(b) $CH_3CH(CH_3)CH_2(H)C{=}C(H)CH_2CH_3$ — H, H on opposite sides

trans-6-Methylhept-3-ene

3.6 Review the sequence rules of Section 3.4. In summary:

Rule 1: A high-atomic-number atom has priority over a low-atomic-number atom.
Rule 2: If a decision can't be reached by Rule 1, look at the second, third or fourth atom out until the first difference is found and a decision can be made.
Rule 3: Multiple-bonded atoms are equivalent to the same number of single bonded atoms.

High	*Low*	*Rule*	*High*	*Low*	*Rule*
(a) –Br	–H	1	(b) –Br	–Cl	1
(c) $–CH_2CH_3$	$–CH_3$	2	(d) –OH	$–NH_2$	1
(e) $–CH_2OH$	$–CH_3$	2	(f) –CH=O	$–CH_2OH$	3

3.7

High	*Low*	*Rule*
$–C(=O)–O–CH_3$ (atoms 1 2 3)	$–C(=O)–O–H$ (atoms 1 2 3)	2

In this problem, priority is assigned by considering the *third* atom away from the substituent.

3.8

(a)

High CH_3O (upper left), Cl High (upper right); C=C; Low H (lower left), CH_3 Low (lower right) — *Z*

First, consider substituents on the left-hand carbon. $CH_3O–$ ranks higher than H– by Sequence Rule 1. On the right side, –Cl ranks higher than $–CH_3$. The isomer has *Z* configuration because the higher ranking substituents are on the same side of the double bond.

(b)

High H_3C (upper left), $COCH_3$ (C=O) Low (upper right); C=C; Low H (lower left), OCH_3 High (lower right) — *E*

3.9

Low H_3C (upper left), CH_2CH_3 Low (upper right); C=C; High CH_3CHCH_2 (with CH_3 on the CH) (lower left), $CH_2CH_2NH_2$ High (lower right) — *Z*

3.10

(a) $CH_3Br + KOH \longrightarrow CH_3OH + KBr$ substitution

(b) $CH_3CH_2OH \longrightarrow H_2C{=}CH_2 + H_2O$ elimination

(c) $H_2C{=}CH_2 + H_2 \longrightarrow CH_3CH_3$ addition

3.11 Use Figure 1.12 to identify the more electronegative element, and draw a δ– next to it. Draw a δ+ next to the less electronegative element.

(a) 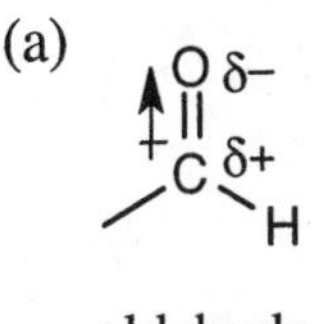

aldehyde

(b) 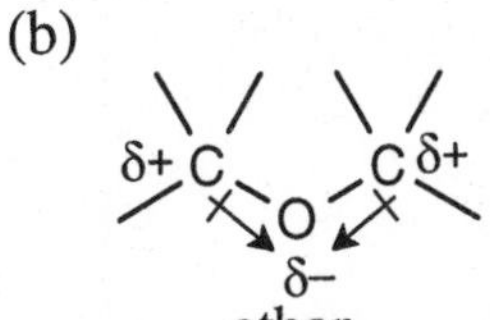

ether

(c) 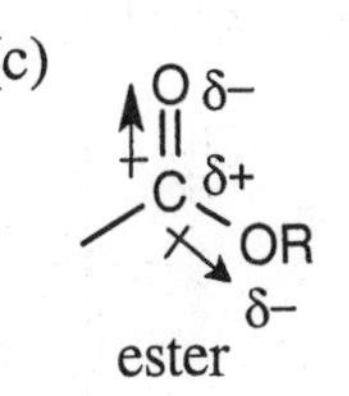

ester

(d) δ– C δ+ MgBr

alkylmagnesium bromide

3.12 *Electrophile or nucleophile?* *Reason*

Electrophile or nucleophile?	*Reason*
(a) NH_4^+ is an electrophile.	Cations (electron-poor) are electrophiles.
(b) CN^- is a nucleophile.	Anions (electron-rich) are nucleophiles.
(c) Br^+ is an electrophile.	Cations (electron-poor) are nucleophiles.
(d) CH_3NH_2 is a nucleophile.	Compounds with lone-pair electrons are usually nucleophiles.
(e) HC≡CH is a nucleophile.	The electron-rich triple bond is nucleophilic.

3.13

```
 ..     ..
:F : B :F:
 ..  ..  ..
    :F:
     ..
```

Boron trifluoride

The Lewis structure for boron trifluoride shows that boron lacks a complete electron octet and is electron-poor. Boron trifluoride is thus likely to be an electrophile.

3.14

$$(H_3C)_2C{=}CH_2 \xrightarrow{H^+\ Cl^-} \left[(H_3C)_2\overset{+}{C}{-}CH_3\right] \xrightarrow{:\ddot{Cl}:^-} (H_3C)_2C(Cl){-}CH_3$$

carbocation

3.15

$$CH_3CH_2CH{=}CHCH_3 \xrightarrow{HCl} \left[\begin{array}{c} CH_3CH_2CH_2\overset{+}{C}HCH_3 \\ \textit{and} \\ CH_3CH_2\overset{+}{C}HCH_2CH_3 \end{array}\right] \longrightarrow \begin{array}{c} CH_3CH_2CH_2CH(Cl)CH_3 \\ \textit{and} \\ CH_3CH_2CH(Cl)CH_2CH_3 \end{array}$$

3.16 A reaction with E_{act} = 60 kJ/mol is faster than a reaction with E_{act} = 80 kJ/mol. (Remember: the larger the E_{act}, the slower the reaction).

3.17

(a) One-step reaction that releases energy: (b) One-step reaction that absorbs energy:

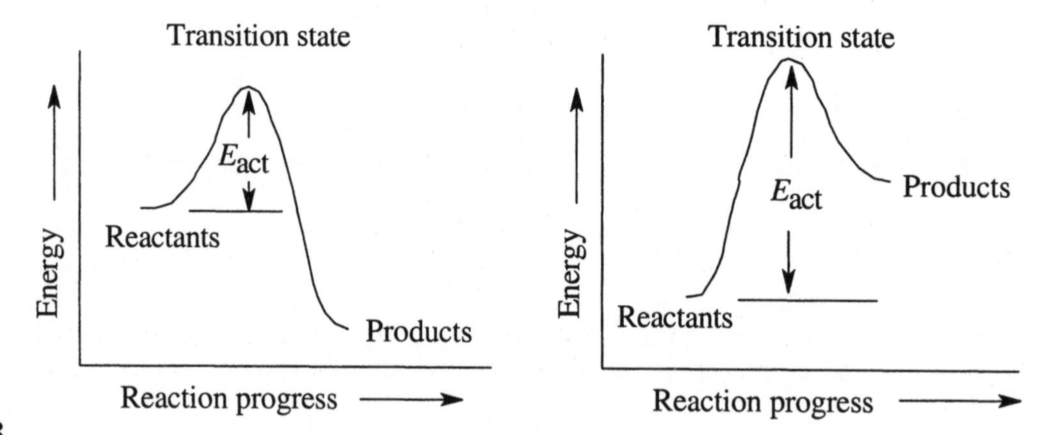

3.18

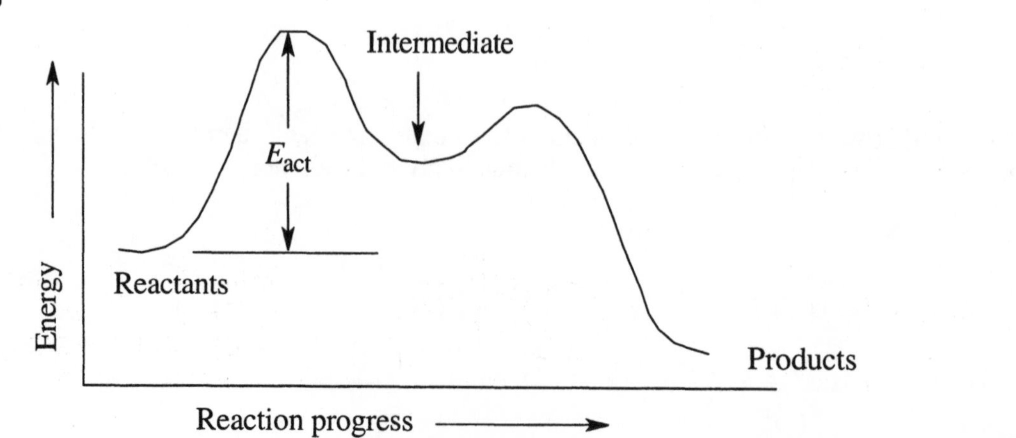

Visualizing Chemistry

3.19

(a) 2,4,5-Trimethylhex-2-ene

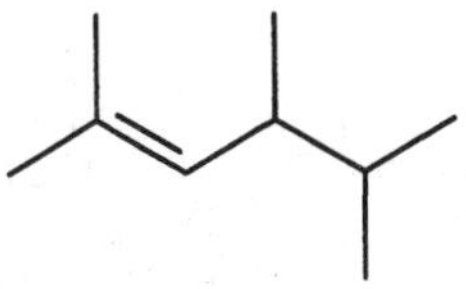

(b) 1-Ethyl-3,3-dimethylcyclohexene

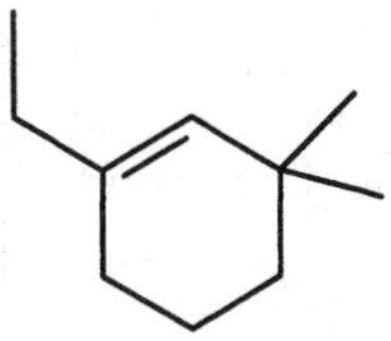

3.20

(a) High Cl, Low, Low, C=O High (H) — *E*

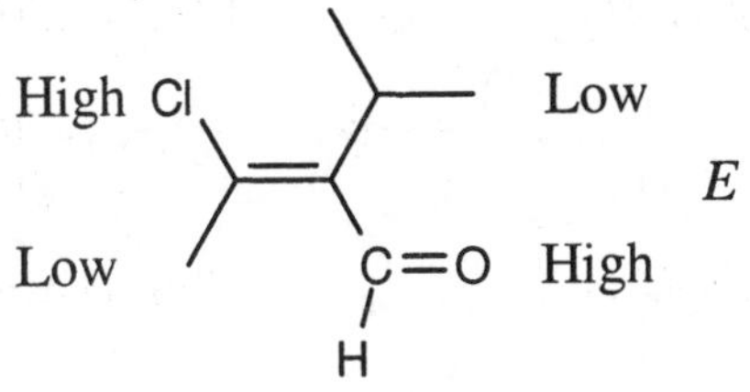

(b) High, OCH_3 ← High, Low →, Low, C=O, OH — *Z*

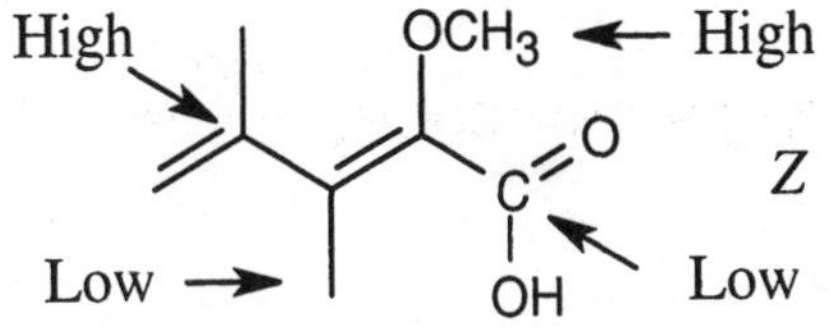

3.21

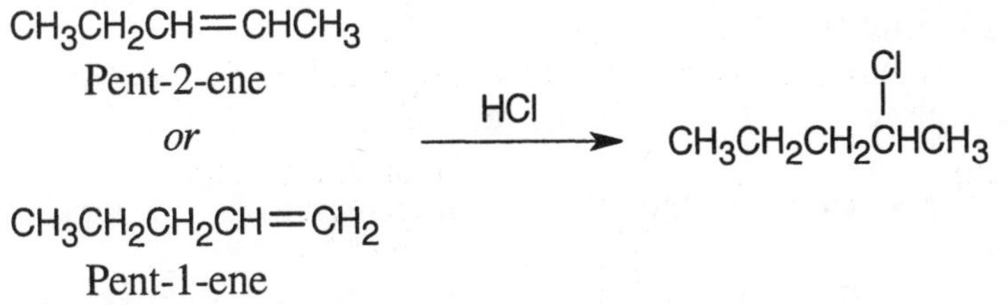

$CH_3CH_2CH{=}CHCH_3$
Pent-2-ene
or
$CH_3CH_2CH_2CH{=}CH_2$
Pent-1-ene

$\xrightarrow{HCl}$ $CH_3CH_2CH_2CHClCH_3$

3.22

1-methylcyclopentene ($-CH_3$) *or* methylenecyclopentane ($=CH_2$) $\xrightarrow{HCl}$ [methylcyclopentyl cation ($+$, $-CH_3$)] ⟶ 1-chloro-1-methylcyclopentane ($-CH_3$)

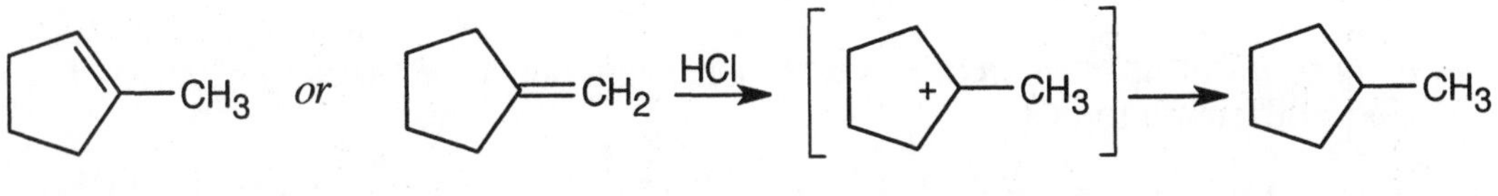

3.23 The electrostatic potential map shows that the formaldehyde oxygen is electron-rich, and the carbon-oxygen bond is polarized. The carbon atom is thus relatively electron-poor and is likely to be electrophilic.

Additional Problems

Functional Groups

3.24–3.25

(a) $CH_3CH_2C{\equiv}N$ — δ+ on C, δ– on N; nitrile

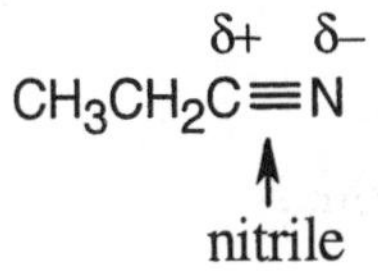

(b) cyclopentyl–O–CH_3 — δ+ on ring C, δ– on O, δ+ on CH_3; ether

(c) $CH_3CCH_2C{-}OCH_3$ with two C=O — ketone, ester; δ– on each O, δ+ on each carbonyl C, δ– on OCH_3 oxygen

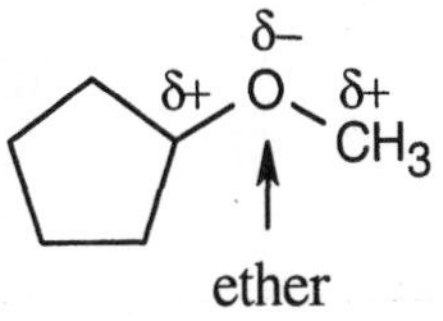

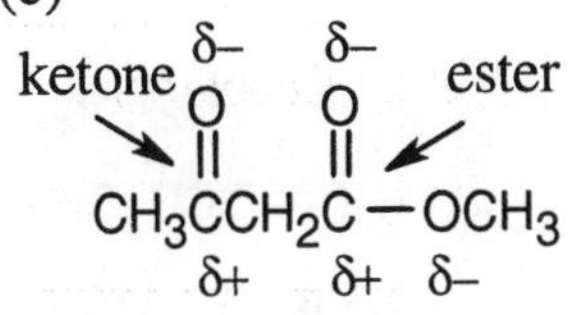

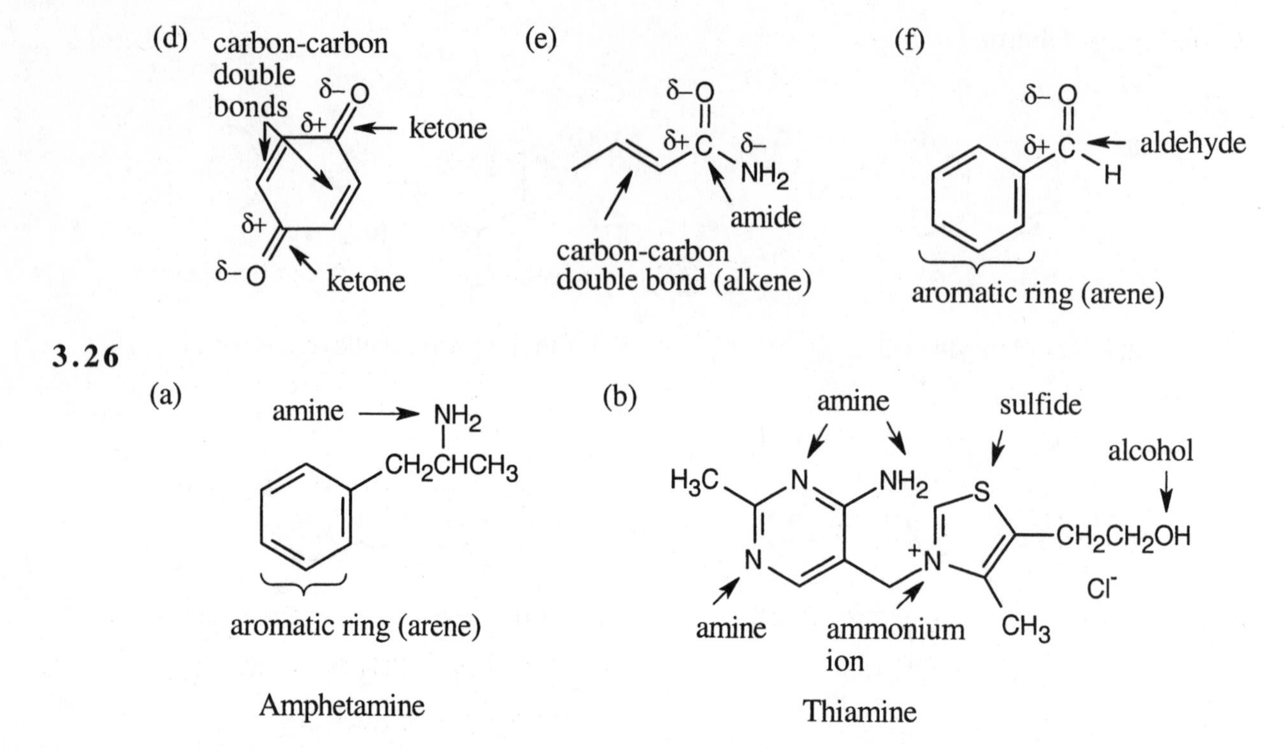

Reactions; Nucleophiles and Electrophiles

3.27 An *addition reaction* takes place when two reactants form a single product.

An *elimination reaction* takes place when one reactant splits apart to give two products.

A *substitution reaction* occurs when two reactants exchange parts to yield two different products.

A *rearrangement reaction* occurs when a reactant undergoes a reorganization of bonds to give a different product.

3.28 Nucleophiles: Cl^-, $N(CH_3)_3$, CN^-
Electrophiles: Hg^{2+}, CH_3^+

Nomenclature

3.29

(a)

$$\begin{array}{c} \qquad\quad CH_3 \\ \qquad\quad | \\ CH_3CH{=}CHCHCH_2CH_3 \end{array}$$

4-Methylhex-2-ene

(b)

$$\begin{array}{c} \qquad\qquad CH_2CH_2CH_3 \\ \qquad | \\ CH_3CH{=}CHCHCH_2CH_2CH_3 \end{array}$$

4-Propylhept-2-ene

(c)

$$\begin{array}{c} CH_2CH_3 \\ | \\ H_2C{=}CCH_2CH_3 \end{array}$$

2-Ethylbut-1-ene

(d)

$$H_2C{=}C{=}CHCH_3$$

Buta-1,2-diene

3.30

(a) 3-Methylcyclohexene

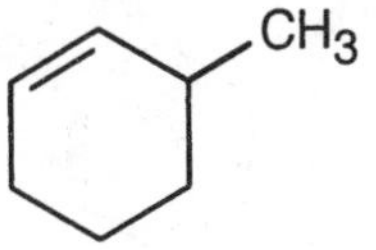

(b) 2,3-Dimethylcyclopentene

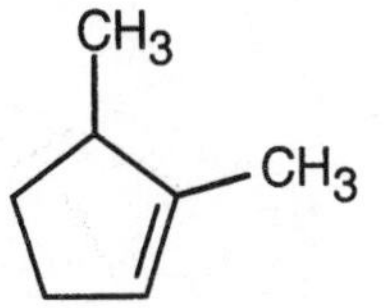

(c) Ethylcyclobutadiene

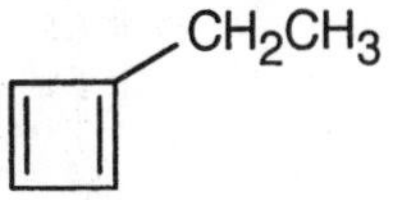

(d) 1,2-Dimethylcyclohexa-1,4-diene

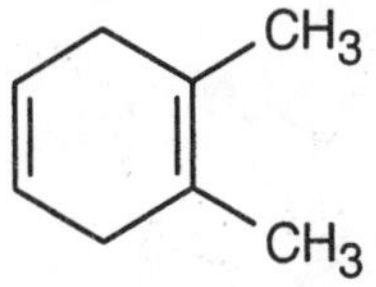

3.31

(a)

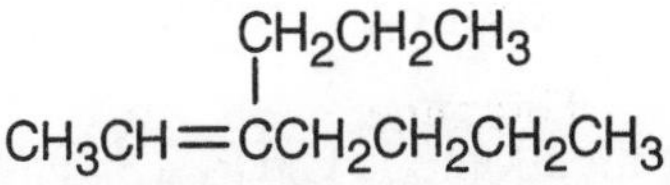

$$CH_3CH{=}C(CH_2CH_2CH_3)CH_2CH_2CH_2CH_3$$

3-Propylhept-2-ene

(b)

$$CH_3C(CH_3){=}CHCH(CH_3)CH_2CH_3$$

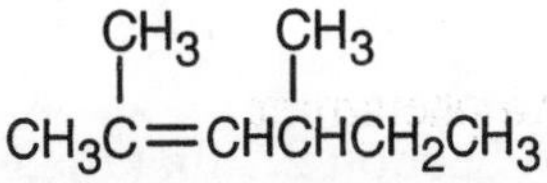

2,4-Dimethylhex-2-ene

(c)

$$CH_3CH_2CH{=}CHCH_2CH_2CH{=}CH_2$$

Octa-1,5-diene

(d)

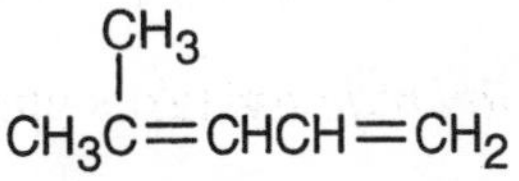

$$CH_3C(CH_3){=}CHCH{=}CH_2$$

4-Methylpenta-1,3-diene

(e)

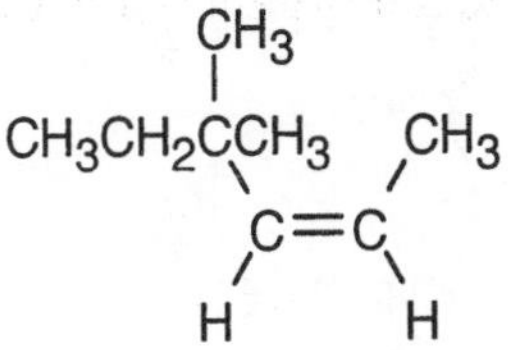

$CH_3CH_2C(CH_3)_2$ and H on one carbon, CH_3 and H on the other: $CH_3CH_2C(CH_3)_2CH{=}CHCH_3$

cis-4,4-Dimethylhex-2-ene

(f)

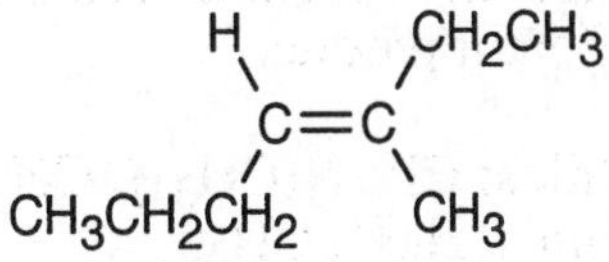

H and $CH_3CH_2CH_2$ on one carbon, CH_2CH_3 and CH_3 on the other: $CH_3CH_2CH_2CH{=}C(CH_3)CH_2CH_3$

(*E*)-3-Methylhept-3-ene

3.32

(a) *cis*-4,5-Dimethylcyclohexene

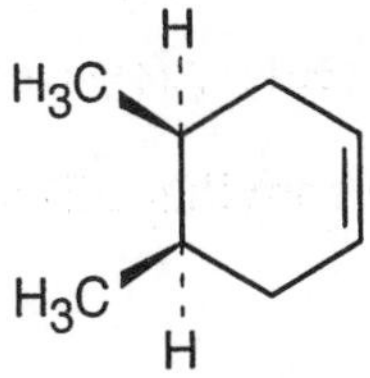

(b) 3,3,4,4-Tetramethylcyclobutene

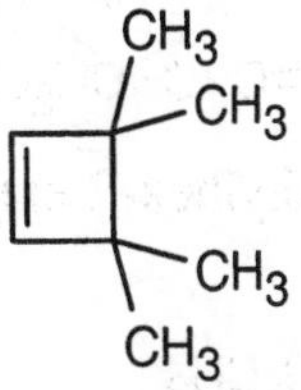

3.33

(a)

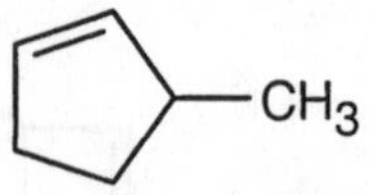

CH_3 (on cyclopentene ring)

3-Methylcyclopentene

(The double bond of a cycloalkene occurs between C1 and C2)

(b)

$CH_3CH_2CH_2CH{=}CHCH_3$

Hex-2-ene

(The methyl group is part of the carbon chain.)

(c)

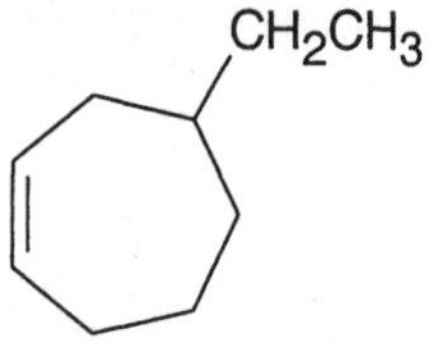

CH_2CH_3

4-Ethylcycloheptene

(The ethyl group receives the lowest possible number.)

(d)

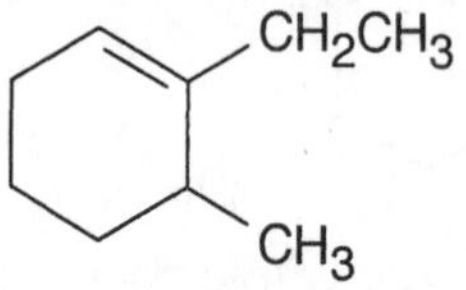

CH_2CH_3

CH_3

2-Ethyl-3-methylcyclohexene

(Substituents should be listed alphabetically.)

3.34 Recall from Section 2.2 that the general formula for an alkane is C_nH_{2n+2}. Thus, a nine-carbon alkane would contain 20 hydrogens. Since a pair of hydrogens is removed for each double bond introduced, the compound C_9H_{14} contains three double bonds because it has 3 fewer pairs of hydrogens than an alkane.

Cis and Trans Isomers

3.35

(a) No cis-trans isomerism.

(b)

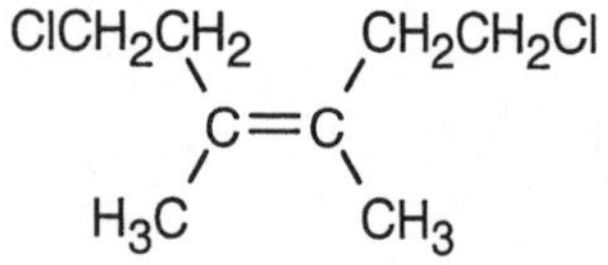

$ClCH_2CH_2$ CH_2CH_2Cl
$C{=}C$
H_3C CH_3

Z

cis

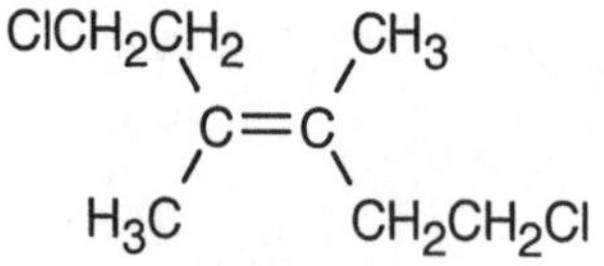

$ClCH_2CH_2$ CH_3
$C{=}C$
H_3C CH_2CH_2Cl

E

trans

(c)

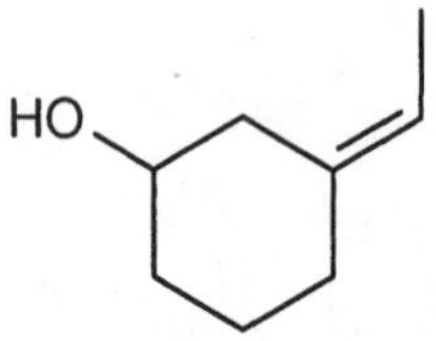

HO

Z

cis

HO

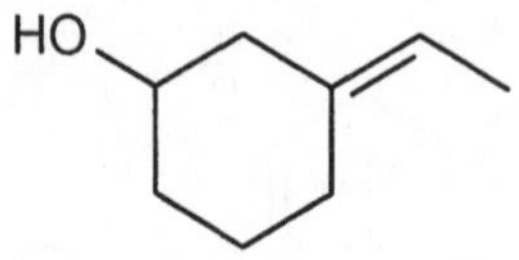

E

trans

3.36

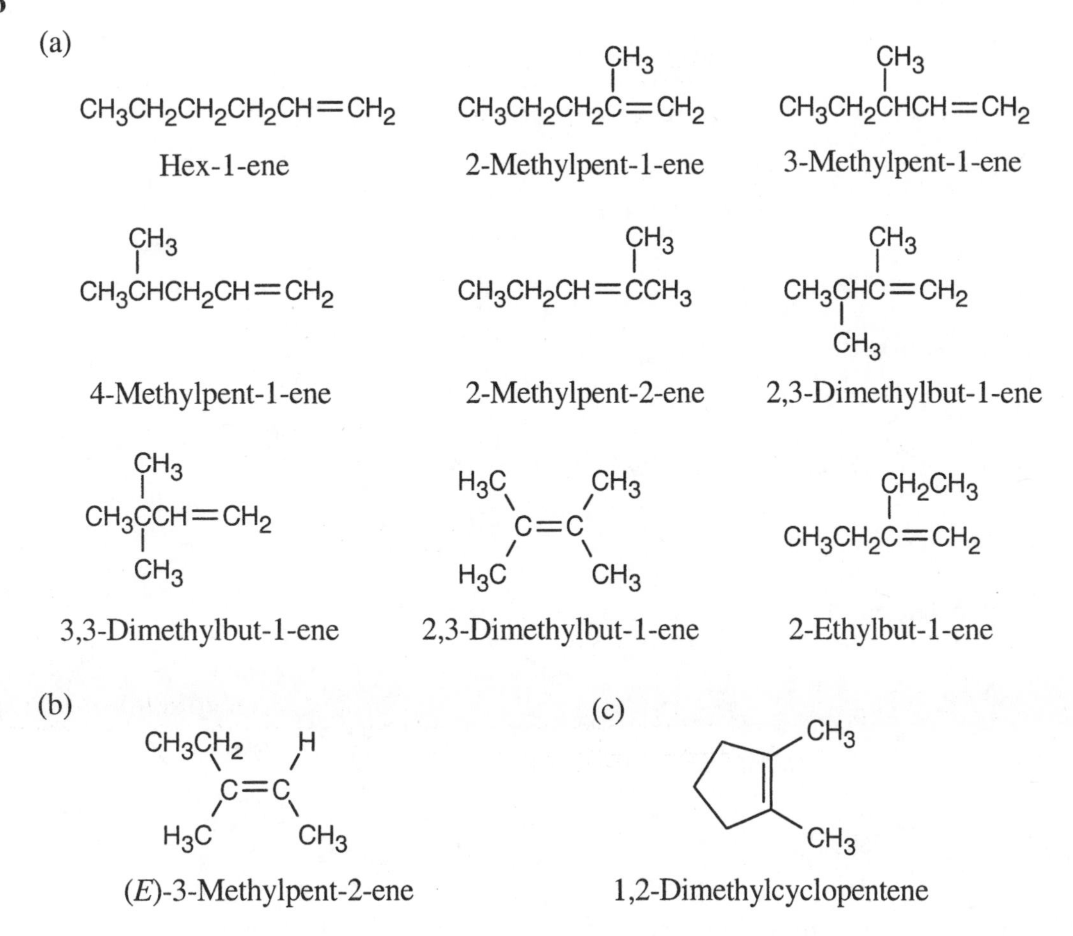

3.37

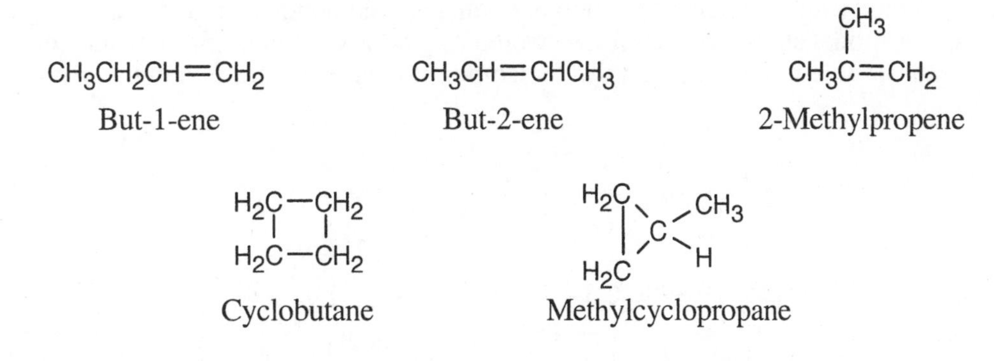

3.38 Of the above structures, only but-2-ene shows cis–trans isomerism.

cis-But-2-ene

trans-But-2-ene

Double Bond Isomers

3.39

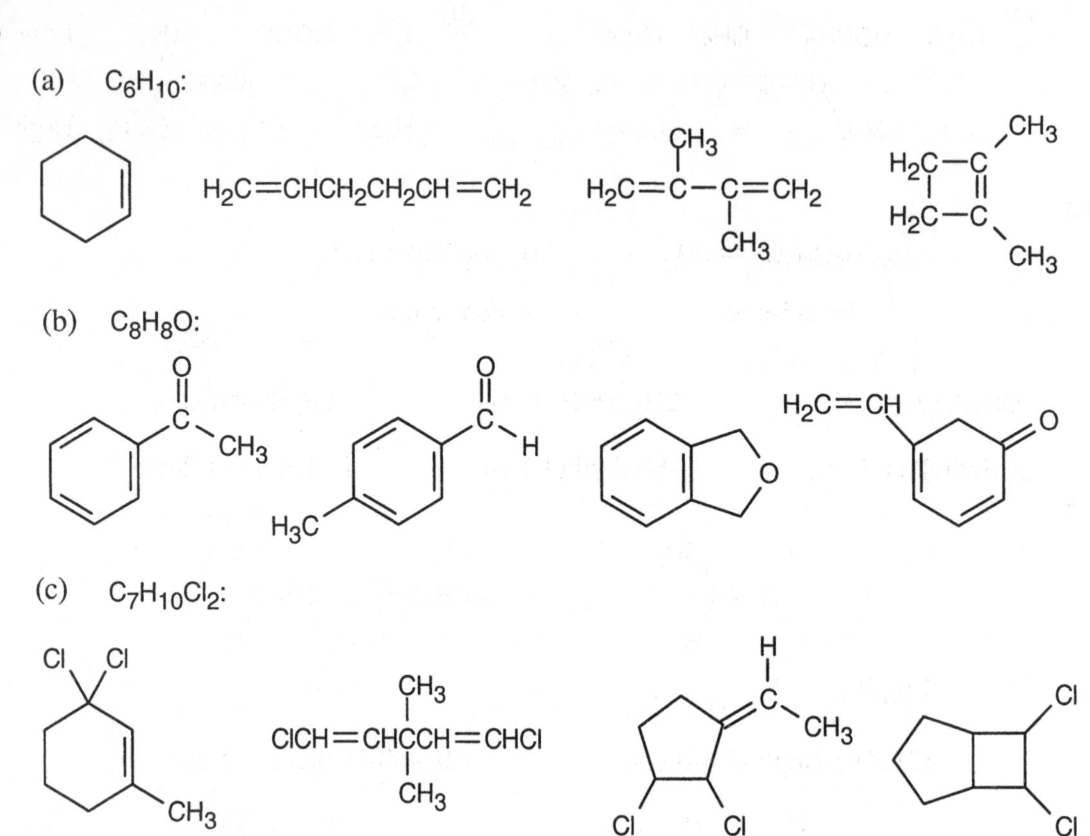

Many other structures corresponding to these formulas can be drawn.

3.40 A six membered ring is too small to contain a trans double bond without causing severe ring strain. A model shows that a ten-membered ring is flexible enough to include either a cis or a trans double bond, although the cis isomer has less ring strain than the trans isomer.

3.41

	High	*Low*		*High*	*Low*
(a)	$—CH_2CH_2CH_3$	$—CH_2CH_3$	(b)	$—C(=O)CH_3$	$—CH_2OH$
(c)	—(cyclopentyl)	$—CH(CH_3)_2$	(d)	$—CH_2CH(CH_3)CH_2Cl$	$—CH_2CH_2CH_2Br$

3.42 *Highest priority* ———> *Lowest priority*

(a) –I, –Br, $–CH_3$, –H
(b) $–OCH_3$, –OH, $–CO_2H$, –H
(c) $–CO_2H$, –CHO, $–CH_2OH$, $–CH_3$
(d) $–CH=CH_2$, $–CH(CH_3)_2$, $–CH_2CH_3$, $–CH_3$

3.43

(a) High $HOCH_2$, CH_3 High; Low H_3C, H Low — $(HOCH_2)(H_3C)C{=}C(CH_3)(H)$ — *Z*

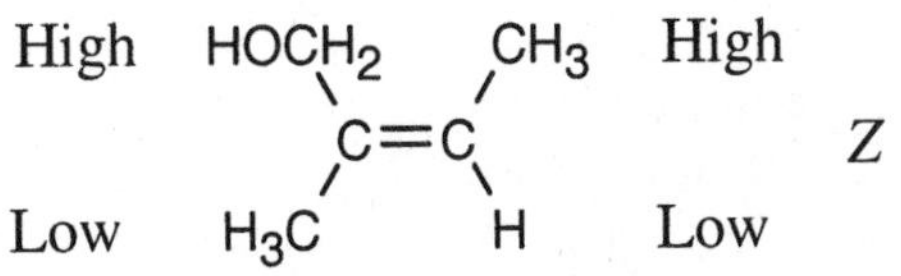

(b) Low HOOC, H Low; High Cl, OCH_3 High — $(HOOC)(Cl)C{=}C(H)(OCH_3)$ — *Z*

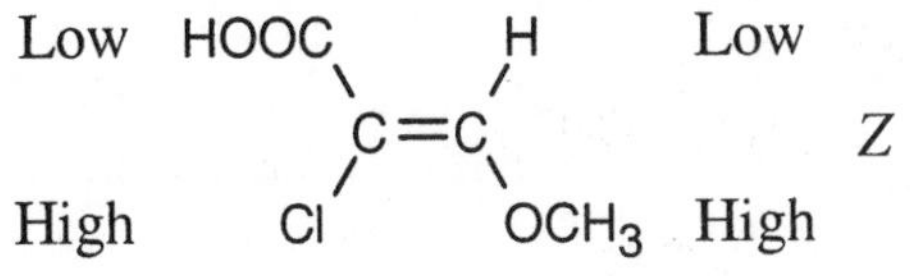

3.44

$CH_3CH_2CH_2CH{=}CH_2$ Pent-1-ene

$CH_3CH_2CH{=}CHCH_3$ Pent-2-ene

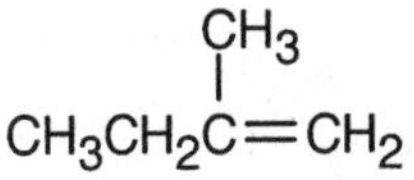

$CH_3CH_2C(CH_3){=}CH_2$ 2-Methylbut-1-ene

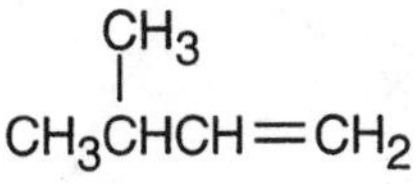

$CH_3CH(CH_3)CH{=}CH_2$ 3-Methylbut-1-ene

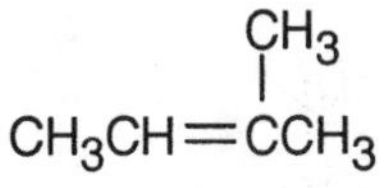

$CH_3CH{=}C(CH_3)CH_3$ 2-Methylbut-2-ene

3.45

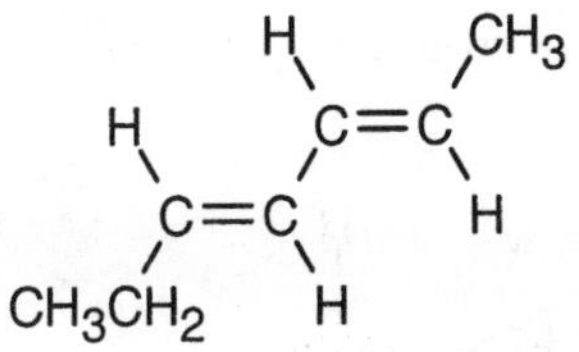

(2*E*,4*E*)-Hepta-2,4-diene

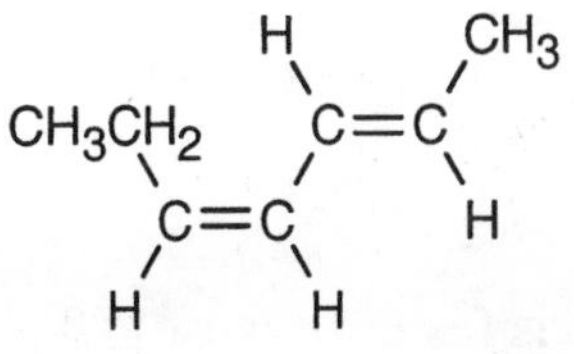

(2*E*,4*Z*)-Hepta-2,4-diene

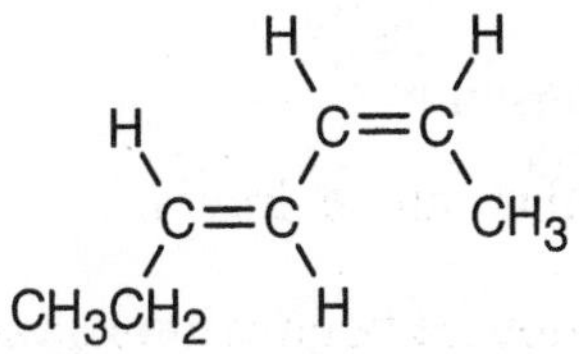

(2*Z*,4*E*)-Hepta-2,4-diene

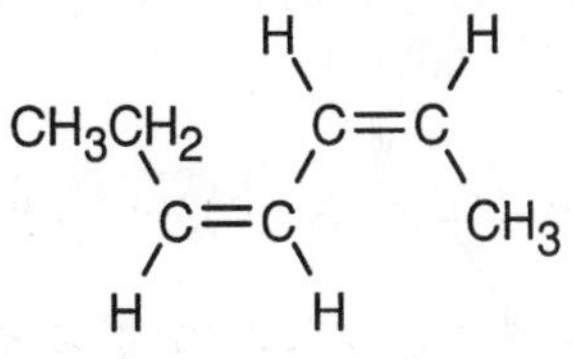

(2*Z*,4*Z*)-Hepta-2,4-diene

3.46

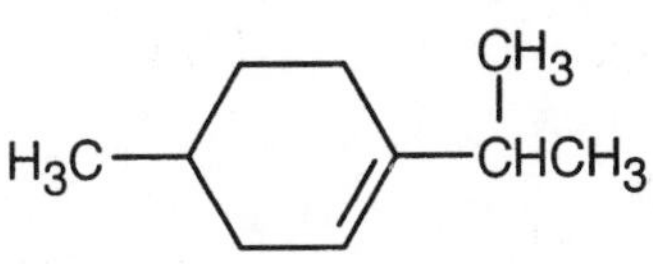

H_3C–(cyclohexene)–$CH(CH_3)CH_3$ Menthene (1-Isopropyl-4-methylcyclohexene)

3.47–3.48

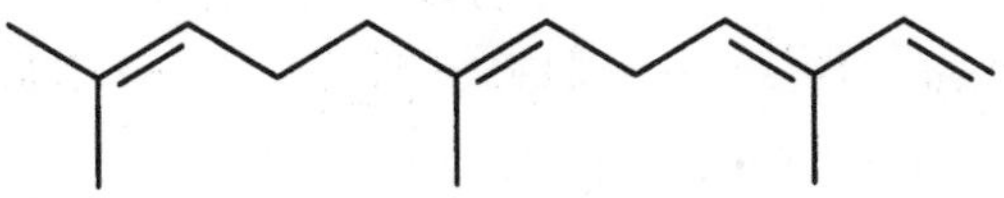

α-Farnesene

(3*E*,6*E*)-3,7,11-Trimethyldodeca-1,3,6,10-tetraene

Reactions and Reaction Coordinate Diagrams

3.49 A reaction with E_{act} = 15 kJ/mol at room temperature is likely to be fast since 15 kJ/mol is a small value for E_{act}, and a small E_{act} indicates a fast reaction.

3.50

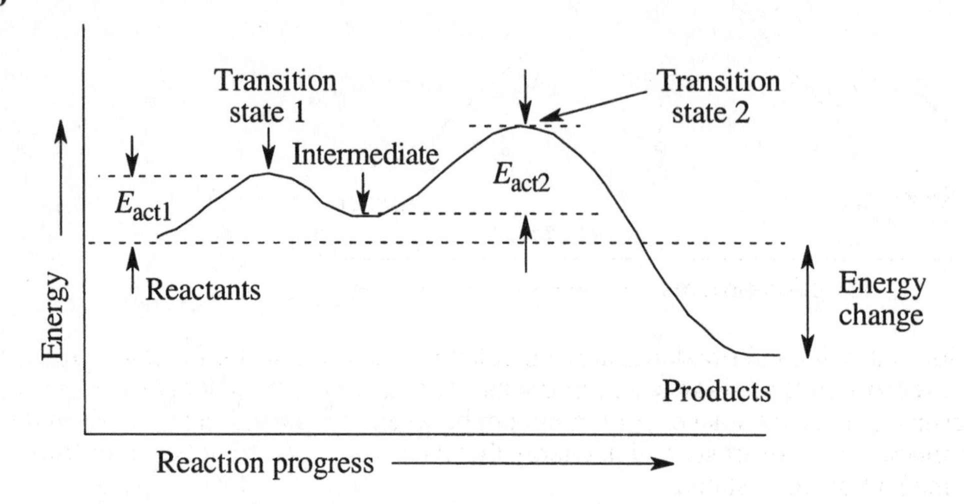

The first step is faster because E_{act1} is smaller than E_{act2}.

3.51

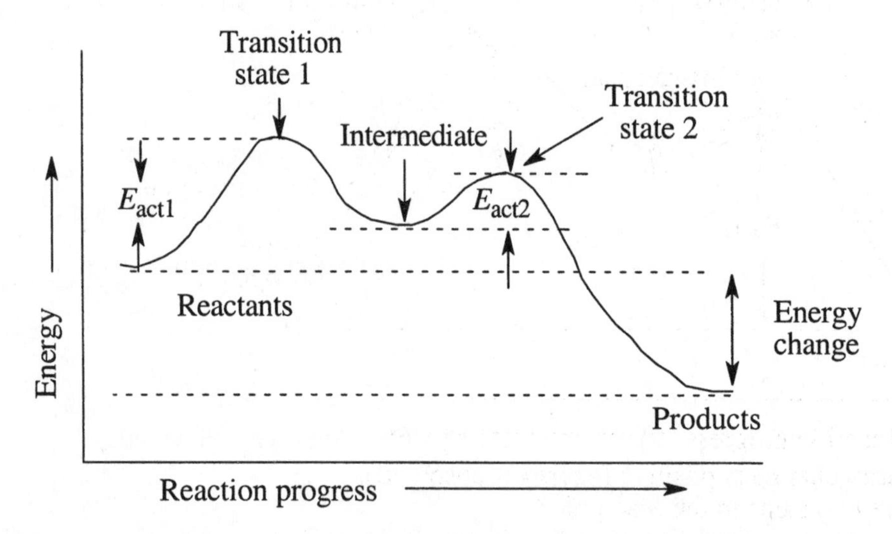

In this reaction, the second step is faster because E_{act2} is smaller than E_{act1}.

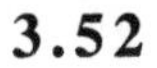

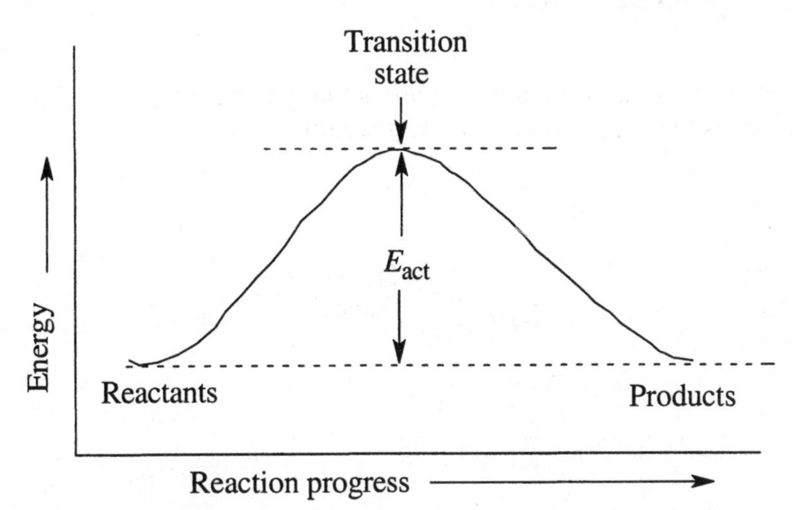

3.53 Transition states and intermediates are both relatively unstable species that are produced during a reaction. A transition state represents a structure occurring at an energy maximum. An intermediate occurs at an energy minimum between two transition states. Even though an intermediate may be of such high energy that it can't be isolated, it is still of lower energy than a transition state.

3.54

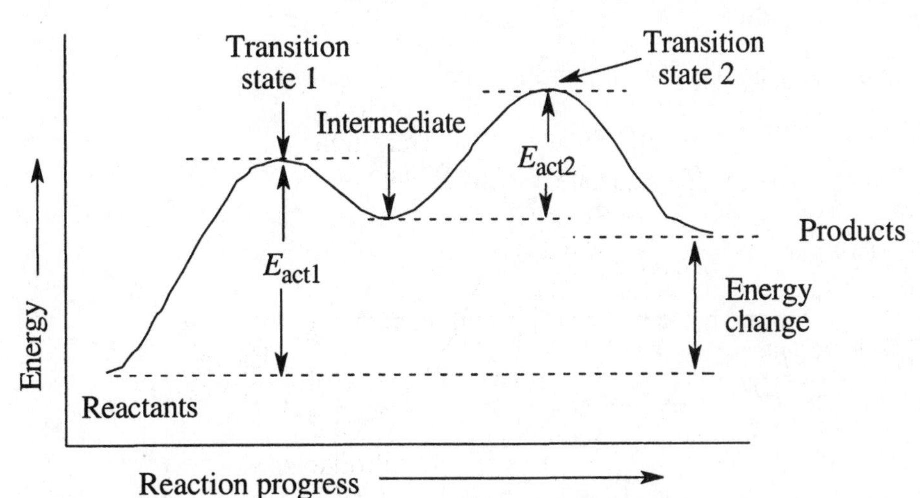

(a) The energy change is positive (energy is absorbed).
(b) There are two steps in the reaction.
(c) The second step is faster because E_{act2} is smaller than E_{act1}.
(d) There are two transition states.

Integrated Problems

3.55

(a) 4-Isopropylcycloheptene (b) 1,6-Dimethyl-cyclohexa-1,3-diene (c) *cis*-3,5-Dichlorocyclopentene

3.56

$CH_3C(CH_3){=}CH_2 + HCl \xrightarrow{?} CH_3CH(CH_3)CH_2Cl + CH_3C(CH_3)(Cl)CH_3$

2-Methylpropene 1-Chloro-2-methylpropane 2-Chloro-2-methylpropane

(2-Chloro-2-methylpropane is the observed product.)

3.57

(a) H^+ is an electrophile. (b) $:\ddot{Br}:^-$ is a nucleophile.

(c) An aldehyde is an oxygen-containing functional group.

aldehyde

3.58

bond broken bond formed

$HO:^- + H_3C{-}Cl: \longrightarrow HO{-}CH_3 + :Cl:^-$

A bond has formed between the hydroxide oxygen and the carbon of chloromethane. The bond between chlorine and carbon has been broken.

3.59

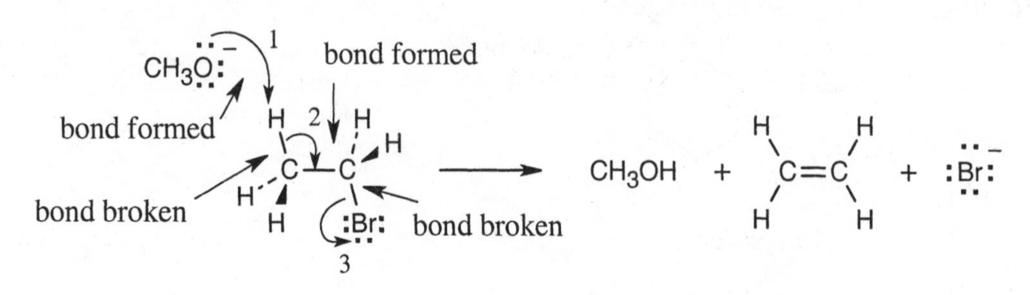

3.60

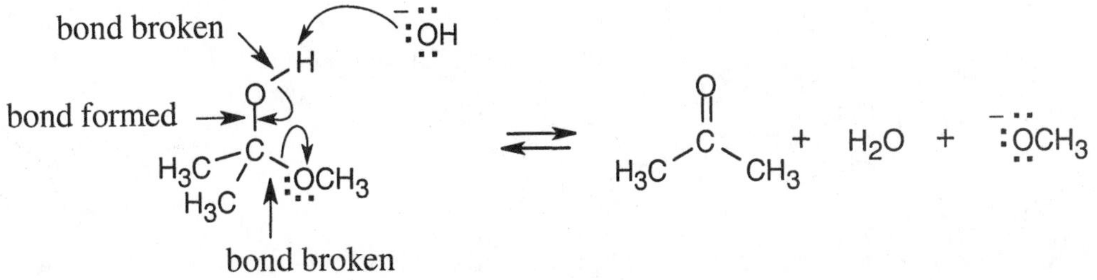

bond broken
bond formed
bond broken

$$(CH_3)_2C(O^-)OCH_3 + ^-OH \rightleftharpoons H_3C-C(=O)-CH_3 + H_2O + ^-OCH_3$$

3.61 (a) The reaction is a polar rearrangement.
(b)

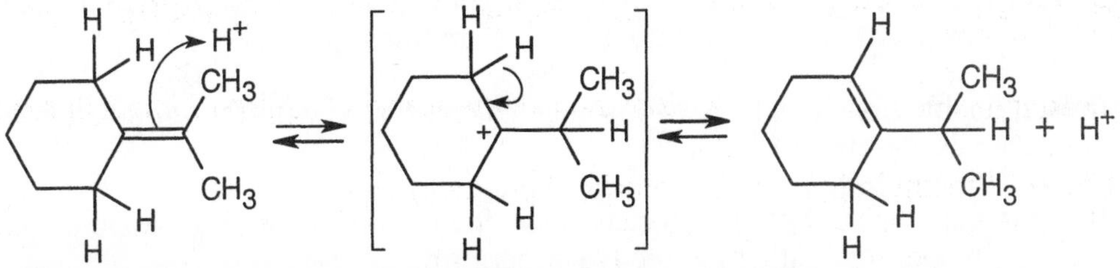

3.62

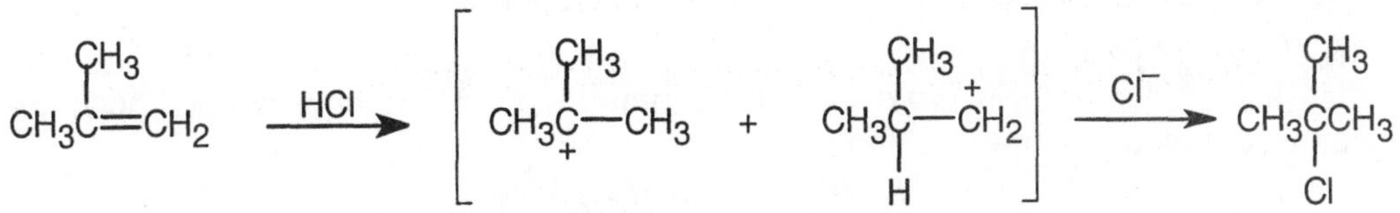

$$CH_3C(CH_3){=}CH_2 \xrightarrow{HCl} \left[CH_3\overset{+}{C}(CH_3)-CH_3 \;+\; CH_3CH(CH_3)-\overset{+}{C}H_2 \right] \xrightarrow{Cl^-} CH_3C(CH_3)(Cl)CH_3$$

The first carbocation is more stable because more alkyl groups are bonded to the positively charged carbon. The expected product is 2-chloro-2-methylpropane.

In the Medicine Cabinet

3.63

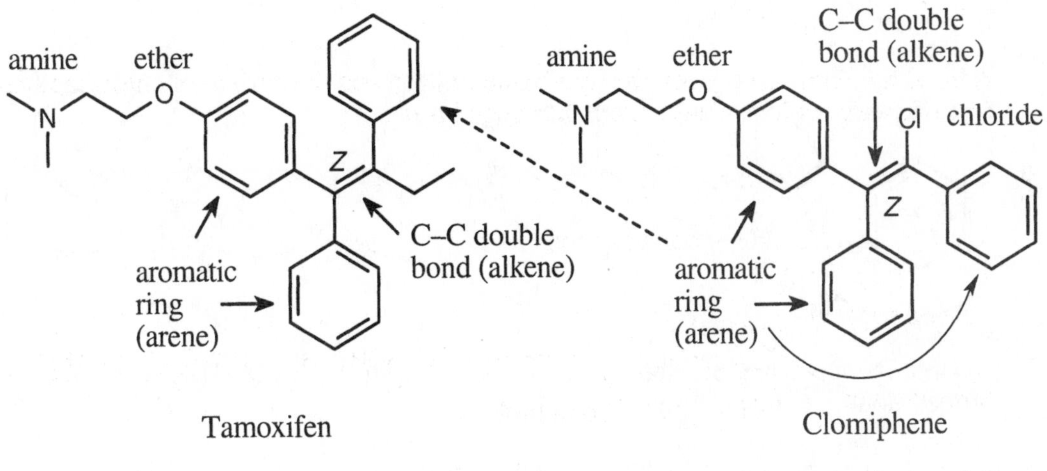

amine ether Z C–C double bond (alkene) aromatic ring (arene)

Tamoxifen

amine ether C–C double bond (alkene) Cl chloride Z aromatic ring (arene)

Clomiphene

3.64

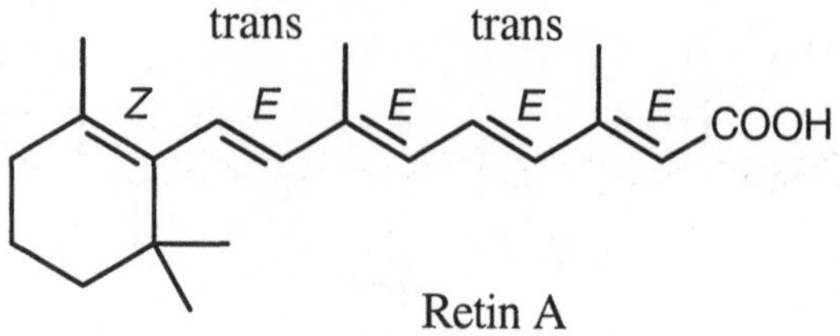

trans trans

Z E E E E COOH

Retin A

(a) Retin A contains a carboxylic acid group and five carbon–carbon double bonds.

(b) The molecular formula of Retin A is $C_{20}H_{28}O_2$, corresponding to six double bonds (including the –COOH double bond) and one ring.

(c) Four of the double bonds are *E*. and the ring double bond is *Z*. The two double bonds that can be described by the cis/trans system are trans.

(d) If each double bond can have two isomers, a compound with four double bonds can have $2^4 = 16$ isomers.

(e) It is possible that sunlight might cause isomerization of some of the double bonds of Retin A, rendering it either ineffective or harmful.

In the Field with Agrochemicals

3.65 (a) Both lycopene and carotene are formed from eight isoprene units, which are circled in the illustrated structures. Two subunits, containing four isoprenes bonded head–to–tail, are connected by a tail–to–tail bond.

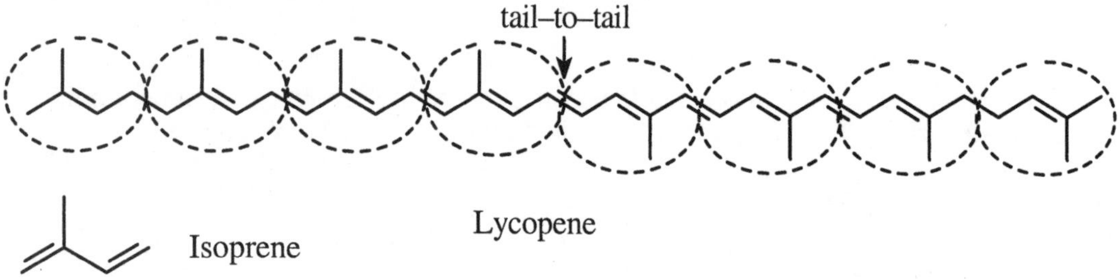

tail–to–tail

Isoprene

Lycopene

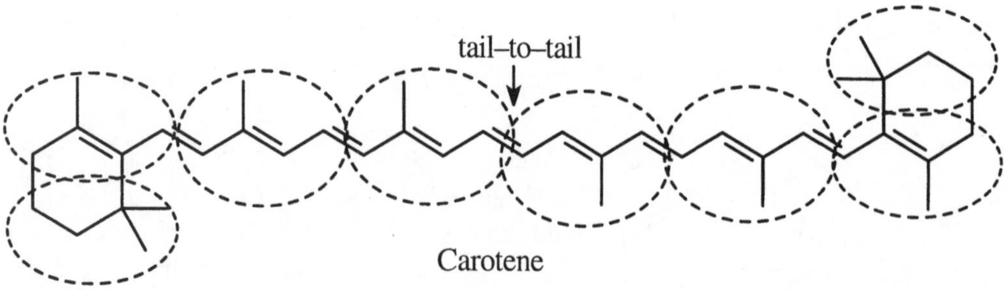

tail–to–tail

Carotene

(b) Both lycopene and carotene have the molecular formulas $C_{40}H_{56}$, corresponding to thirteen double bonds or rings.

3.66 Lycopene and carotene are constitutional isomers.

3.67

new bond

new bond

Chapter 4 – Reactions of Alkenes and Alkynes

Chapter Outline

I. Alkenes (Sections 4.1–4.8).
  A. Reactions of alkenes (Sections 4.1–4.7).
    1. Addition of HX (Sections 4.1–4.3).
      a. Orientation of addition–Markovnikov's rule (Section 4.2).
        i. In the addition of HX to a double bond, H attaches to the carbon with fewer substituents, and X attaches to the carbon with more substituents.
        ii. If the carbons have the same number of substituents, a mixture of products results.
      b. Carbocation structure and stability (Section 4.3).
        i. Carbocations are planar; the unoccupied *p* orbital extends above and below the plane.
        ii. The stability of carbocations increases with increasing substitution.
    2. Hydration of alkenes (Section 4.4).
      a. Water adds to alkenes to yield alcohols in the presence of a strong acid catalyst.
      b. Although this reaction is important industrially, reaction conditions are too severe for most molecules.
    3. Addition of halogens to alkenes (Section 4.5).
      a. Br_2 and Cl_2 react with alkenes to yield 1,2-dihaloalkanes.
      b. Reaction occurs with anti stereochemistry – both bromines come from opposite sides of the molecule.
      c. The reaction intermediate is a cyclic bromonium intermediate that is formed in a single step by interaction of an alkene with Br^+.
    4. Hydrogenation of alkenes (Section 4.6).
      a. Catalytic hydrogenation reduces alkenes to saturated hydrocarbons.
      b. Catalytic hydrogenation is a heterogeneous process that takes place on the surface of the catalyst.
      c. Hydrogenation occurs with syn stereochemistry.
    5. Oxidation of alkenes (Section 4.7).
      a. Hydroxylation: two –OH groups are added to the double bond when an alkene reacts with $KMnO_4$ under basic conditions.
        The reaction occurs with syn stereochemistry and yields a diol.
      b. Cleavage occurs when an alkene is treated with acidic $KMnO_4$.
        The products may be ketones, carboxylic acids or CO_2.
  B. Alkene polymers (Section 4.8).
    1. Many types of polymers can be formed by radical polymerization of alkene monomers.
    2. There are 3 steps in a polymerization reaction.
      a. Initiation involves homolytic cleavage of a weak bond to form a radical
        The radical adds to an alkene to generate an alkyl radical.
      b. The alkyl radical adds to another alkene molecule to yield a second radical.
        This step is repeated many, many times (up to 100,000 times for some polymers).
      c. Termination occurs when two radical fragments combine.

II. Conjugated dienes and resonance (Sections 4.9–4.11).
   A. Conjugated dienes (Section 4.9–4.10).
      1. In conjugated dienes, there is overlap of the π orbitals of the double bonds across the central single bond.
      2. This overlap leads to formation of both 1,2 and 1,4 addition products.
      3. Mechanism of 1,4-addition: allylic cations (Section 4.10).
         a. The reaction intermediate of addition to a diene is an allylic cation.
         b. An allylic cation is a resonance hybrid of two different forms.
         c. In general, the more resonance forms that can be drawn for a structure, the greater the stability.
   B. Drawing and interpreting resonance structures (Section 4.11).
      1. Resonance forms are imaginary.
      2. Resonance forms differ only in the placement of π or nonbonding electrons.
      3. Different resonance forms of a structure don't have to be equivalent.
      4. Resonance forms must be valid Lewis forms and must obey normal rules of valency.
      5. Resonance leads to stability.

III. Alkynes (Section 4.12).
   A. Naming of alkynes.
      The rules for naming alkynes are like the rules for alkenes (Sec. 6.3), with a few exceptions.
         a. The suffix *-yne* is used for an alkyne.
         b. Compounds with both double bonds and triple bonds are *enynes*.
         c. When there is a choice in numbering, double bonds receive lower numbers than triple bonds.
   B. Reactions of alkynes.
      1. Addition of H_2.
      2. Addition of HX and X_2.
      3. Hydration of alkynes.
         a. The –OH group adds to the more substituted carbon to give Markovnikov product.
         b. The intermediate enol product tautomerizes to a ketone.
         c. A mixture of products is formed from an internal alkyne, but a terminal alkyne yields a methyl ketone.
      4. Formation of acetylide anions.
         a. Terminal alkynes are weakly acidic ($pK_a = 25$).
         b. Very strong bases ($^{-}NH_2$) can deprotonate terminal alkynes.
         c. Acetylide anions can react with haloalkanes to form substitution products.
            i. The nucleophilic acetylide anion attacks the electrophilic carbon of a haloalkane to produce a new alkyne.
            ii. This reaction is called an alkylation reaction.
         d. Acetylide alkylations are limited to primary alkyl bromides and iodides.

Solutions to Problems

4.1

(a) one alkyl group ↓ (at CH); no alkyl groups ↙ (at CH_2)

$$CH_3CH_2CH{=}CH_2 + HCl \longrightarrow CH_3CH_2CH(Cl)CH_3$$

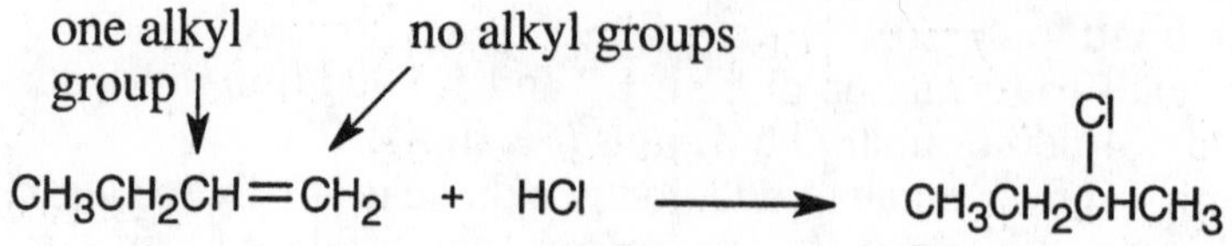

H^+ adds to the carbon with fewer alkyl groups.

(b)

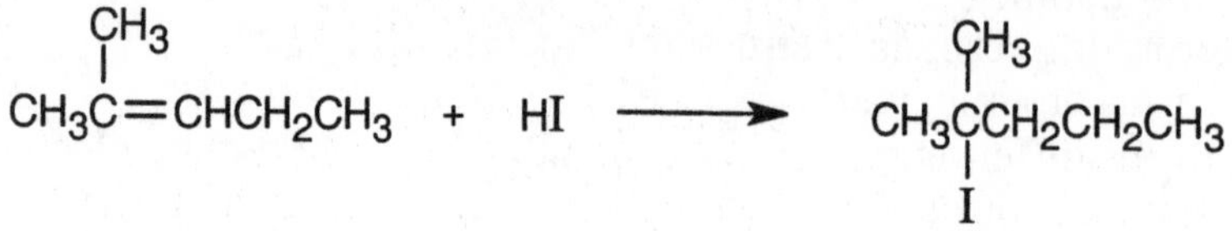

$$CH_3C(CH_3){=}CHCH_2CH_3 + HI \longrightarrow CH_3C(CH_3)(I)CH_2CH_2CH_3$$

(c) Cyclohexene + HCl ⟶ Chlorocyclohexane (Cl)

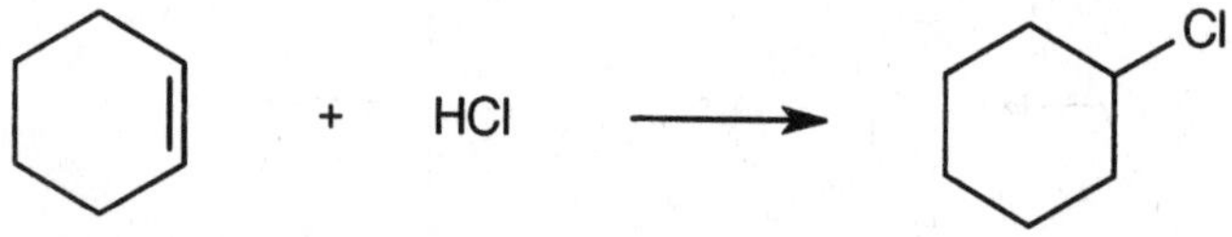

4.2

(a) Cyclopentene + HBr ⟶ Bromocyclopentane (Br)

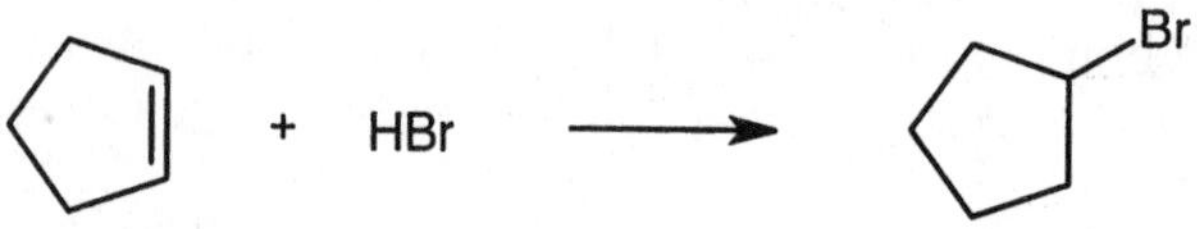

Cyclopentene   Bromocyclopentane

(b)

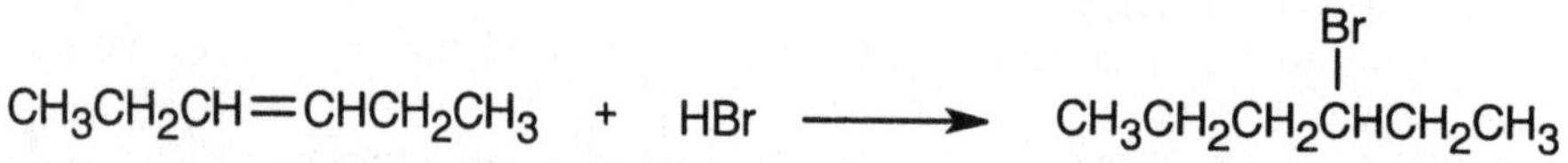

$$CH_3CH_2CH{=}CHCH_2CH_3 + HBr \longrightarrow CH_3CH_2CH_2CH(Br)CH_2CH_3$$

Hex-3-ene   3-Bromohexane

(c) 1-Isopropylcyclohexene + HI ⟶ 1-Iodo-1-isopropylcyclohexane (I)

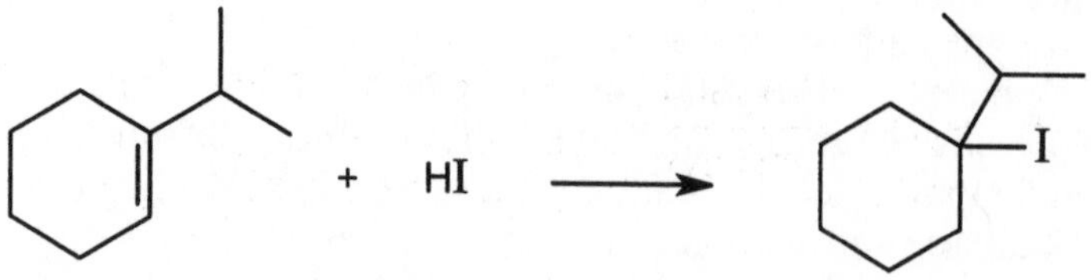

1-Isopropylcyclohexene   1-Iodo-1-isopropylcyclohexane

(d) Vinylcyclohexane + HBr ⟶ (1-bromoethyl)cyclohexane (Br)

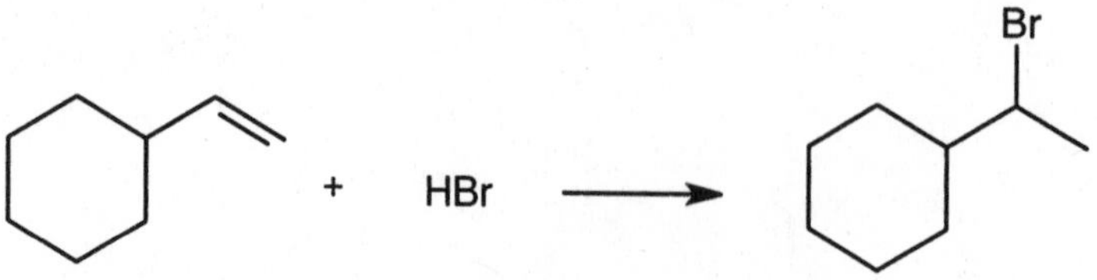

This is the only alkene that yields the desired product.

4.3

(a)

$$CH_3CH_2\overset{CH_3}{\overset{|}{C}}{=}CH\overset{CH_3}{\overset{|}{C}H}CH_3 + HBr$$

→ $\left[CH_3CH_2\overset{CH_3}{\overset{|}{\underset{+}{C}}}CH_2\overset{CH_3}{\overset{|}{C}H}CH_3\right]$ Br^- → $CH_3CH_2\underset{Br}{\overset{CH_3}{\overset{|}{C}}}CH_2\overset{CH_3}{\overset{|}{C}H}CH_3$

tertiary carbocation

→ $\left[CH_3CH_2\overset{CH_3}{\overset{|}{C}H}\underset{+}{C}H\overset{CH_3}{\overset{|}{C}H}CH_3\right]$

secondary carbocation
(not formed)

(b)

cyclopentylidene=$CHCH_3$ + HI

→ [cyclopentyl cation (+ on ring carbon)–CH_2CH_3] I^- → 1-iodo-1-ethylcyclopentane (I, CH_2CH_3)

tertiary carbocation

→ [cyclopentyl(H)–$\overset{+}{C}HCH_3$]

secondary carbocation
(not formed)

4.4

(a)

$$CH_3CH_2\overset{CH_3}{\overset{|}{C}}{=}CHCH_2CH_3 + H_2O \xrightarrow[\text{catalyst}]{H_2SO_4} CH_3CH_2\underset{OH}{\overset{CH_3}{\overset{|}{C}}}CH_2CH_2CH_3$$

(b)

1-methylcyclopentene ($-CH_3$) + H_2O $\xrightarrow[\text{catalyst}]{H_2SO_4}$ 1-methylcyclopentanol (CH_3, OH)

(c)

$$CH_3CH_2\underset{CH_3}{CH}CH_2CH{=}\overset{CH_3}{\overset{|}{C}}CH_3 + H_2O \xrightarrow[\text{catalyst}]{H_2SO_4} CH_3CH_2\underset{CH_3}{CH}CH_2CH_2\underset{OH}{\overset{CH_3}{\overset{|}{C}}}CH_3$$

4.5

(a) $CH_3CH{=}CHCH_3$ *or* $CH_3CH_2CH{=}CH_2$ + H_2O $\xrightarrow[\text{catalyst}]{H_2SO_4}$ $CH_3CH_2CH(OH)CH_3$

(b) $CH_3CH_2C(CH_3){=}CHCH_3$ *or* $CH_3CH_2C({=}CH_2)CH_2CH_3$ + H_2O $\xrightarrow[\text{catalyst}]{H_2SO_4}$ $CH_3CH_2C(OH)(CH_3)CH_2CH_3$

(c) 1,2-dimethylcyclohexene *or* 2,3-dimethylcyclohexene *or* 1-methyl-2-methylenecyclohexane (CH_2, CH_3) + H_2O $\xrightarrow[\text{catalyst}]{H_2SO_4}$ 1,2-dimethylcyclohexanol (CH_3, OH, CH_3)

4.6–4.7

H_3C, CH_3

1,2-Dimethylcyclohexene $\xrightarrow{Br_2}$ [$^+$:Br, H_3C, CH_3, :Br:$^-$]

Bromonium ion intermediate

$\longrightarrow$ Br, CH_3, H_3C, Br

trans-1,2-Dibromo-1,2-dimethylcyclohexane

4.8

(a)

$$(CH_3)_2C{=}CHCH_2CH_3 \xrightarrow[\text{catalyst}]{H_2} (CH_3)_2CHCH_2CH_2CH_3$$

2-Methylpent-2-ene → 2-Methylpentane

(b)

3,3-Dimethylcyclopentene $\xrightarrow[\text{catalyst}]{H_2}$ 1,1-Dimethylcyclopentane

4.9 In both of these reactions, the double bond is cleaved. The product in (a) contains an oxygen atom double-bonded to the carbon at each end of the original double bond.

(a) 1,2-Dimethylcyclohexene $\xrightarrow[H_3O^+]{KMnO_4}$ octane-2,7-dione

(b) 1,2-Dimethylcyclohexene $\xrightarrow[H_2O,\ ^-OH]{KMnO_4}$ cis-1,2-dimethylcyclohexane-1,2-diol (CH_3, OH; OH, CH_3)

4.10

(a)

$$(CH_3)_2C{=}CH_2 \xrightarrow[H_3O^+]{KMnO_4} (CH_3)_2C{=}O \;+\; CO_2$$

(b)

$$CH_3CH_2CH{=}CHCH_2CH_3 \xrightarrow[H_3O^+]{KMnO_4} CH_3CH_2\overset{O}{\overset{\|}{C}}OH \;+\; HO\overset{O}{\overset{\|}{C}}CH_2CH_3$$

4.11

many $F_2C{=}CF_2$ → $-(CF_2-CF_2)_n-$

Tetrafluoroethylene

4.12

$$H_2C{=}CH{-}CH{=}CH_2$$

$\downarrow Br_2$

$$\left[H_2\overset{+}{C}{-}CH{=}CH{-}CH_2Br \longleftrightarrow H_2C{=}CH{-}\overset{+}{C}H{-}CH_2Br \right]$$

$\downarrow Br^-$

$$BrH_2C{-}CH{=}CH{-}CH_2Br \qquad H_2C{=}CH{-}CHBr{-}CH_2Br$$

1,4 addition　　　　1,2 addition

4.13 Only the products that result from allylic carbocation intermediates are shown.

$CH_3CH{=}CHCH{=}CH_2$　　Penta-1,3-diene

Product	*Name*	*Results from:*
$CH_3CH{=}CHCHClCH_3$	4-Chloropent-2-ene	1,2 addition 1,4 addition
$CH_3CH_2CHClCH{=}CH_2$	3-Chloropent-1-ene	1,2 addition
$CH_3CH_2CH{=}CHCH_2Cl$	1-Chloropent-2-ene	1,4 addition

4.14

$$CH_3CH{=}CHCH{=}CH_2$$

$\xrightarrow{H^+}$ $CH_3\overset{\delta+}{C}H{=}CH{=}\overset{\delta+}{C}HCH_3$ **A** — protonation on carbon 1

$\xrightarrow{H^+}$ $CH_3CH{=}CHCH_2\overset{+}{C}H_2$ **B** — protonation on carbon 2 (not formed)

$\xrightarrow{H^+}$ $CH_3\overset{+}{C}HCH_2CH{=}CH_2$ **C** — protonation on carbon 3 (not formed)

$\xrightarrow{H^+}$ $CH_3CH_2\overset{\delta+}{C}H{=}CH{=}\overset{\delta+}{C}H_2$ **D** — protonation on carbon 4

A and **D**, which are resonance-stabilized, are formed over **B** and **C**, which are not. The positive charge of allylic carbocation **A** is delocalized over two secondary carbons, while the positive charge of carbocation **D** is delocalized over one secondary and one primary carbon. We therefore predict that carbocation **A** is the major intermediate formed, and that 4-chloropent-2-ene predominates. Note that this product results from both 1,2– and 1,4–addition.

4.15

(a) Cyclohexenylmethyl cation resonance forms: $\overset{+}{C}H_2$ ⟷ $=CH_2$ with + on ring

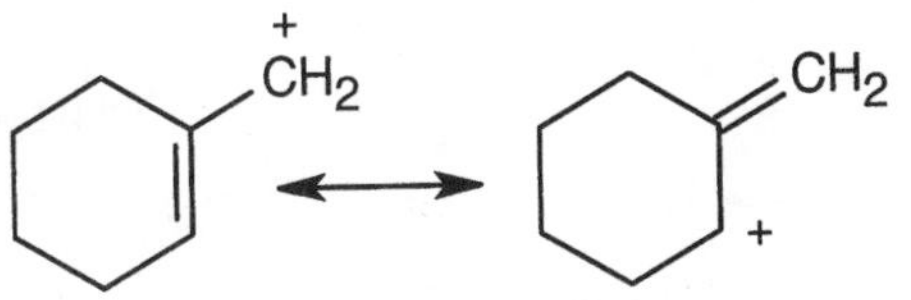

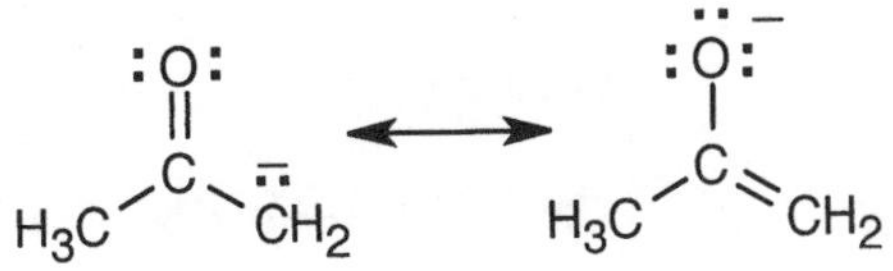

(b) $H_3C\text{–}C(=O)\text{–}\ddot{C}H_2^-$ ⟷ $H_3C\text{–}C(\text{–}O^-)=CH_2$

(c) H, + ⟷ H, + ⟷ +, H

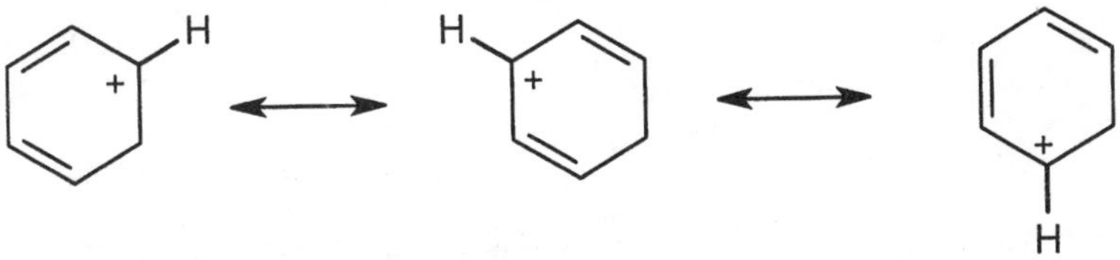

4.16

(a) $CH_3CH_2C{\equiv}CCH_2CH(CH_3)CH_3$

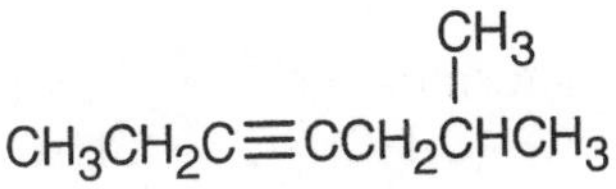

6-Methylhept-3-yne

(b) $HC{\equiv}CC(CH_3)_2CH_3$

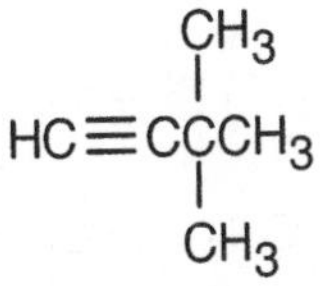

3,3-Dimethylbut-1-yne

(c) $CH_3CH(CH_3)CH_2C{\equiv}CCH_3$

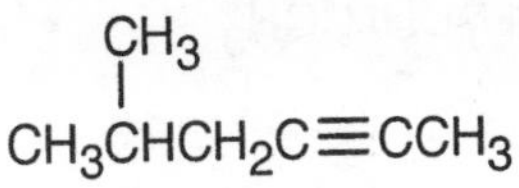

5-Methylhex-2-yne

(d) $CH_3CH{=}CHCH_2C{\equiv}CCH_3$

Hept-2-en-5-yne

4.17

(a) $CH_3CH_2CH_2C{\equiv}CH$ —1 equiv. Cl_2→ $(CH_3CH_2CH_2)(Cl)C{=}C(Cl)(H)$

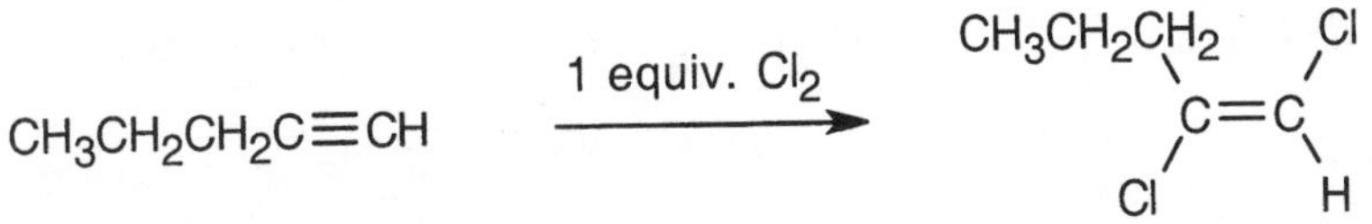

Pent-1-yne

(*E*)-1,2-Dichloropent-1-ene

(b) $CH_3CH_2CH_2C{\equiv}CCH_2CH_3$ —1 equiv. HBr→ $(CH_3CH_2CH_2)(H)C{=}C(Br)(CH_2CH_3)$ + $(CH_3CH_2CH_2)(Br)C{=}C(H)(CH_2CH_3)$

Hept-3-yne

(*Z*)-3-Bromohept-3-ene

+

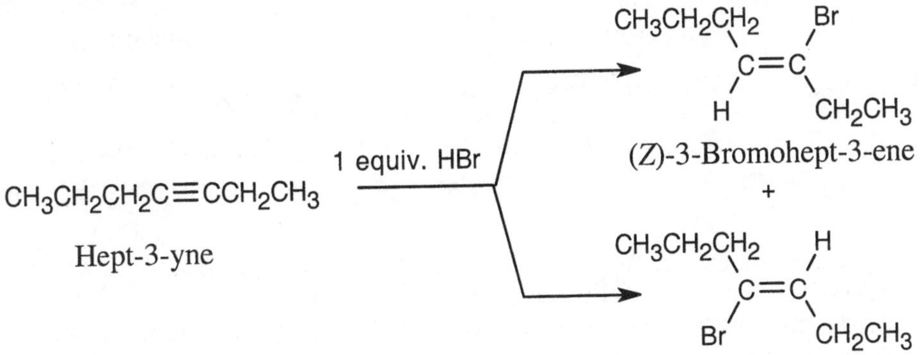

(*Z*)-4-Bromohept-3-ene

(c)

$$CH_3CH(CH_3)CH_2C\equiv CCH_2CH_3 \xrightarrow[\text{Lindlar catalyst}]{H_2} (CH_3)_2CHCH_2CH{=}CHCH_2CH_3$$

6-Methylhept-3-yne → *cis*-6-Methylhept-3-ene

4.18

$$CH_3CH_2CH_2C\equiv CCH_2CH_2CH_3 + H_2O \xrightarrow[HgSO_4]{H_2SO_4} \left[CH_3CH_2CH_2C{=}C(OH)CH_2CH_2CH_3\right] \longrightarrow CH_3CH_2CH_2CH_2C({=}O)CH_2CH_2CH_3$$

Oct-4-yne

4.19

(a)

$$CH_3CH_2CH_2C\equiv CH + H_2O \xrightarrow[HgSO_4]{H_2SO_4} CH_3CH_2CH_2C({=}O)CH_3$$

(b)

$$CH_3CH_2C\equiv CCH_2CH_3 + H_2O \xrightarrow[HgSO_4]{H_2SO_4} CH_3CH_2CH_2C({=}O)CH_2CH_3$$

4.20

(a)

$$HC\equiv CH + Na^+ \, ^-NH_2 \longrightarrow HC\equiv C^- Na^+ + NH_3$$

$$CH_3CH(CH_3)CH_2CH_2Br + HC\equiv C^- Na^+ \longrightarrow CH_3CH(CH_3)CH_2CH_2C\equiv CH$$

5-Methylhex-1-yne

(b)

$$CH_3C\equiv CH + Na^+ \, ^-NH_2 \longrightarrow CH_3C\equiv C^- Na^+ + NH_3$$

$$CH_3CH_2CH_2Br + CH_3C\equiv C^- Na^+ \longrightarrow CH_3CH_2CH_2C\equiv CCH_3$$

Hex-2-yne

or

$$CH_3CH_2CH_2C\equiv CH + Na^+ \, ^-NH_2 \longrightarrow CH_3CH_2CH_2C\equiv C^- Na^+ + NH_3$$

$$CH_3CH_2CH_2C\equiv C^- Na^+ + CH_3Br \longrightarrow CH_3CH_2CH_2C\equiv CCH_3$$

Hex-2-yne

(c)

$$CH_3\overset{\displaystyle CH_3}{\overset{|}{C}}HC{\equiv}CH \;+\; Na^+\;^-NH_2 \longrightarrow CH_3\overset{\displaystyle CH_3}{\overset{|}{C}}HC{\equiv}C^-\,Na^+ \;+\; NH_3$$

$$CH_3\overset{\displaystyle CH_3}{\overset{|}{C}}HC{\equiv}C^-\,Na^+ \;+\; CH_3Br \longrightarrow CH_3\overset{\displaystyle CH_3}{\overset{|}{C}}HC{\equiv}CCH_3$$

4-Methylpent-2-yne

Note: The reaction of $CH_3C{\equiv}C^-\,Na^+$ with $(CH_3)_2CHBr$ doesn't yield the desired product because $(CH_3)_2CHBr$ is not a primary alkyl halide.

Visualizing Chemistry

4.21

(a)

$$CH_3\overset{\displaystyle CH_3}{\overset{|}{C}}{=}CHCH_2\overset{\displaystyle CH_3}{\overset{|}{C}}HCH_2CH_3$$

2,5-Dimethylhept-2-ene

(i) $\xrightarrow[H_3O^+]{KMnO_4}$ $(H_3C)_2C{=}O \;+\; O{=}\overset{\displaystyle OH}{\overset{|}{C}}CH_2\overset{\displaystyle CH_3}{\overset{|}{C}}HCH_2CH_3$

(ii) $\xrightarrow[H_2O,\ ^-OH]{KMnO_4}$ $CH_3\underset{\displaystyle OH}{\underset{|}{\overset{\displaystyle CH_3}{\overset{|}{C}}}}-\underset{\displaystyle OH}{\underset{|}{C}}HCH_2\overset{\displaystyle CH_3}{\overset{|}{C}}HCH_2CH_3$

(b)

3,3-Dimethylcyclopentene

(i) $\xrightarrow[H_3O^+]{KMnO_4}$ $HO_2C{-}C(CH_3)_2{-}CH_2CH_2{-}CO_2H$

(ii) $\xrightarrow[H_2O,\ ^-OH]{KMnO_4}$ 3,3-dimethylcyclopentane-1,2-diol (cis; HO, H and H, OH as drawn)

4.22

(a)

4,4-Dimethylhex-1-yne: $HC{\equiv}CCH_2C(CH_3)_2CH_2CH_3$

(i) $\xrightarrow[\text{Lindlar catalyst}]{H_2}$ $H_2C{=}CHCH_2C(CH_3)_2CH_2CH_3$

(ii) $\xrightarrow[HgSO_4]{H_3O^+}$ $CH_3C(=O)CH_2C(CH_3)_2CH_2CH_3$

(b)

2,7-Dimethyloct-4-yne: $CH_3CH(CH_3)CH_2C{\equiv}CCH_2CH(CH_3)CH_3$

(i) $\xrightarrow[\text{Lindlar catalyst}]{H_2}$ cis-$(CH_3)_2CHCH_2CH{=}CHCH_2CH(CH_3)_2$ (both H on the same side of C=C)

(ii) $\xrightarrow[HgSO_4]{H_3O^+}$ $CH_3CH(CH_3)CH_2CH_2C(=O)CH_2CH(CH_3)CH_3$

4.23

(a)

2-Ethyl-3-methylbut-1-ene: $CH_3CH(CH_3)C(=CH_2)CH_2CH_3 + H_3O^+ \xrightarrow[\text{catalyst}]{H_2SO_4} CH_3CH(CH_3){-}C(OH)(CH_3)CH_2CH_3$

or

3,4-Dimethylpent-2-ene: $CH_3CH(CH_3)C(CH_3){=}CHCH_3 + H_3O^+ \xrightarrow[\text{catalyst}]{H_2SO_4} CH_3CH(CH_3){-}C(OH)(CH_3)CH_2CH_3$

(b)

4,4-Dimethylcyclopentene + $H_3O^+ \xrightarrow[\text{catalyst}]{H_2SO_4}$ 3,3-dimethylcyclopentanol (HO on ring; CH_3, CH_3 on C3)

4.24

(a)

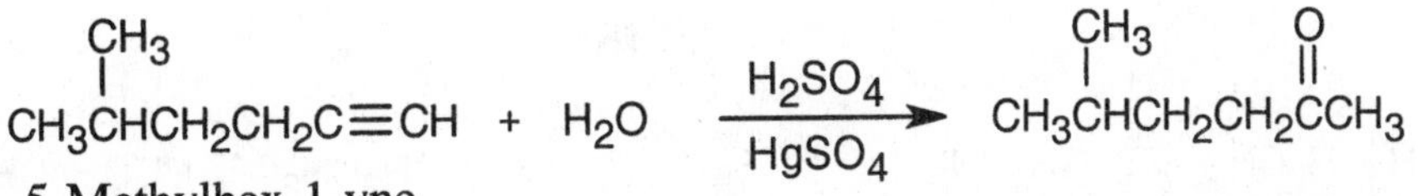

$$CH_3CH(CH_3)CH_2CH_2C{\equiv}CH + H_2O \xrightarrow[HgSO_4]{H_2SO_4} CH_3CH(CH_3)CH_2CH_2C(=O)CH_3$$

5-Methylhex-1-yne

(b)

$$\text{(CH}_3\text{-cyclohexyl)}C{\equiv}CH \xrightarrow[\text{ether}]{2\ HCl} \text{(CH}_3\text{-cyclohexyl)}CCl_2CH_3$$

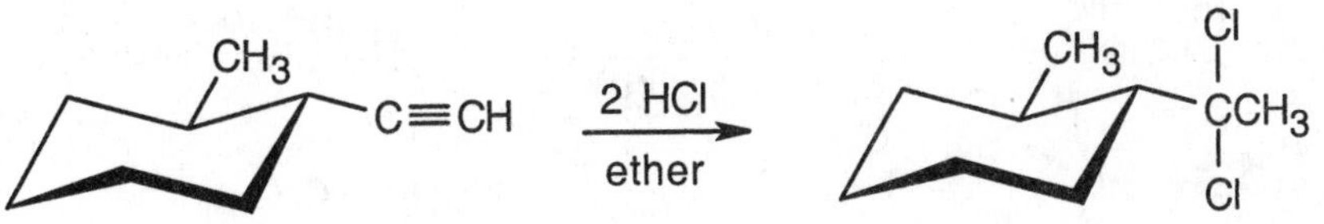

Additional Problems

Nomenclature

4.25

(a) $CH_3CH{=}CHC(CH_3){=}CHCH_3$

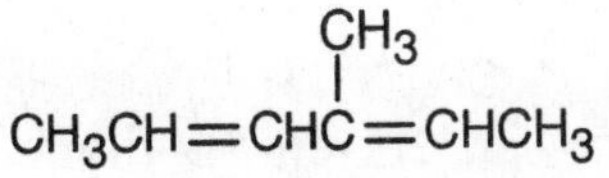

3-Methylhexa-2,4-diene

(b) $CH_3CH{=}CHCH(CH_2CH_2CH_3)CH_2C{\equiv}CH$

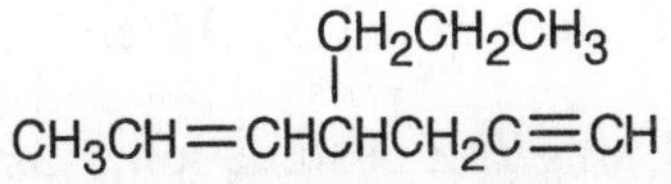

4-Propylhept-5-en-1-yne

(c) $H_2C{=}C{=}C(CH_3)CH_3$

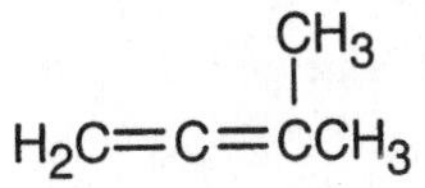

3-Methylbuta-1,2-diene

(d) $HC{\equiv}CCH_2C{\equiv}CCH(CH_3)CH_3$

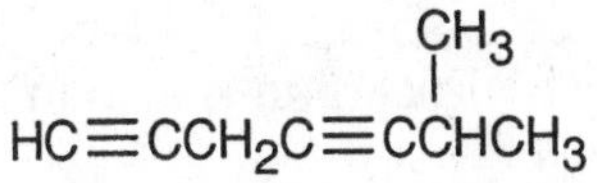

6-Methylhepta-1,4-diyne

4.26

(a) $CH_3CH_2CH_2CH_2CH(CH_2CH_3)C{\equiv}CH$

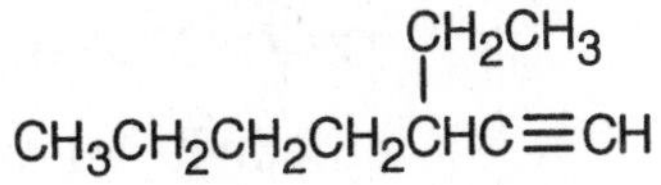

3-Ethylhept-1-yne

(b) $CH_3C(CH_3){=}CHCH(CH_3)C{\equiv}CH$

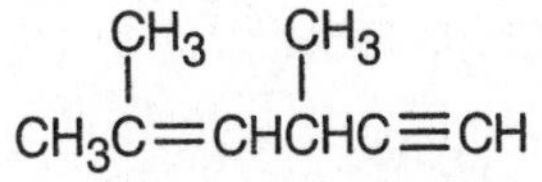

3,5-Dimethylhex-4-en-1-yne

(c) $CH_3C{\equiv}CCH_2CH_2C{\equiv}CH$

Hepta-1,5-diyne

(d) (cyclopentadiene ring)$-CH_3$

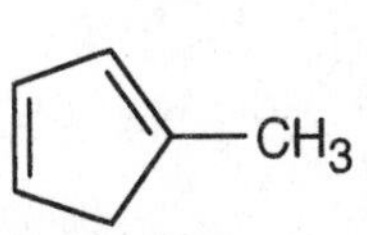

1-Methylcyclopenta-1,3-diene

4.27

(a) $CH_3CH_2CH_2C{\equiv}CCH_2CH_3$

Hept-3-yne

(b) $CH_3CH_2CH_2C{\equiv}CC(CH_3)_2CH_2CH_3$

3,3-Dimethyloct-4-yne

(c) [ring structure: $H_2C{-}C{\equiv}C{-}CH(CH_3){-}CH(CH_3){-}CH_2{-}CH_2{-}CH_2{-}CH_2{-}CH_2$, closed back to the first H_2C]

3,4-Dimethylcyclodecyne

(d) $CH_3C(CH_3)_2C{\equiv}CC(CH_3)_2CH_3$

2,2,5,5-Tetramethylhex-3-yne

4.28 (a) $CH_3CH{=}CHC{\equiv}CC{\equiv}CCH{=}CHCH{=}CHCH{=}CH_2$
Tridecatetra-1,3,5,11-en-7,9-diyne

Using *E-Z* designation: (3*E*,5*E*,11*E*)-Tridecatetra-1,3,5,11-en-7,9-diyne
The parent alkane of this hydrocarbon is tridecane.

(b) $CH_3C{\equiv}CC{\equiv}CC{\equiv}CC{\equiv}CC{\equiv}CCH{=}CH_2$
Tridec-1-en-3,5,7,9,11-pentayne
This hydrocarbon is also in the tridecane family.

Isomerism

4.29

$CH_3CH_2CH_2C{\equiv}CH$	$CH_3CH_2C{\equiv}CCH_3$	$CH_3CH(CH_3)C{\equiv}CH$
Pent-1-yne	Pent-2-yne	3-Methylbut-1-yne

4.30

Conjugated dienes:

$CH_3CH{=}CHCH{=}CH_2$	$H_2C{=}CHC(CH_3){=}CH_2$
Penta-1,3-diene	2-Methylbuta-1,3-diene

Nonconjugated dienes:

$H_2C{=}CHCH_2CH{=}CH_2$	$CH_3CH_2CH{=}C{=}CH_2$
Penta-1,4-diene	Penta-1,2-diene
$CH_3CH{=}C{=}CHCH_3$	$H_2C{=}C{=}C(CH_3)_2$
Penta-2,3-diene	3-Methylbuta-1,2-diene

4.31

(a) C_6H_8:

$CH_3C(CH_3){=}CHC{\equiv}CH$ (cyclohexadiene ring) $H_2C{=}CHCH{=}CHCH{=}CH_2$

(b) C_6H_8O:

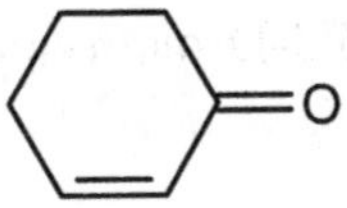

(cyclohexenone ring)=O $CH_3CH_2C{\equiv}CCH_2CH(=O)$ $H_2C{=}CHCH(CH_2OH)C{\equiv}CH$

Predict the Product

4.32

(a) $C_6H_5CH{=}CH_2 \xrightarrow{H_2/Pd} C_6H_5CH_2CH_3$

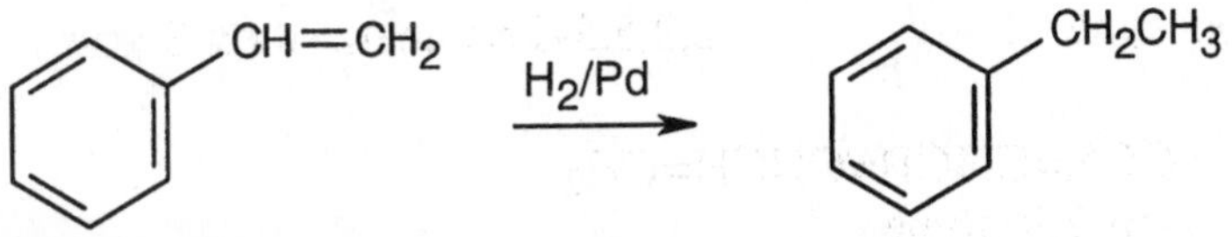

(b) $C_6H_5CH{=}CH_2 \xrightarrow{Br_2} C_6H_5CH(Br)CH_2Br$

(c) $C_6H_5CH{=}CH_2 \xrightarrow{HBr} C_6H_5CH(Br)CH_3$

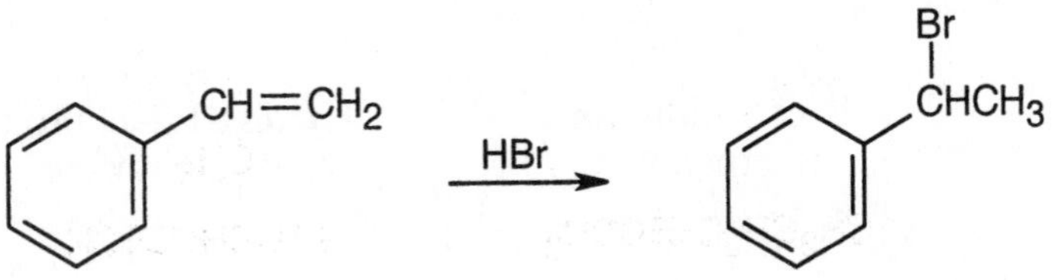

(d) $C_6H_5CH{=}CH_2 \xrightarrow[H_2O,\ ^{-}OH]{KMnO_4} C_6H_5CH(OH)CH_2OH$

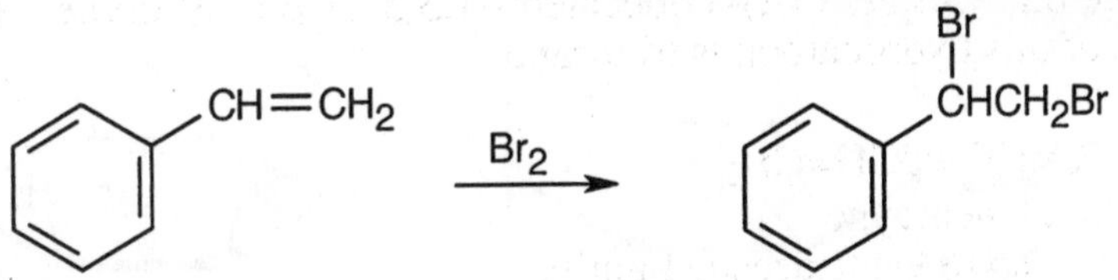

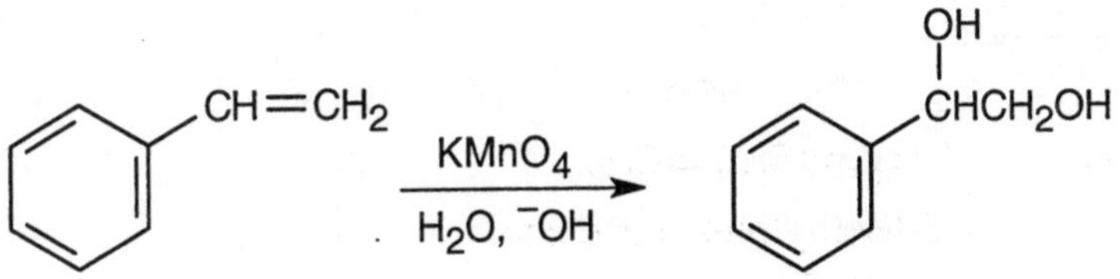

4.33

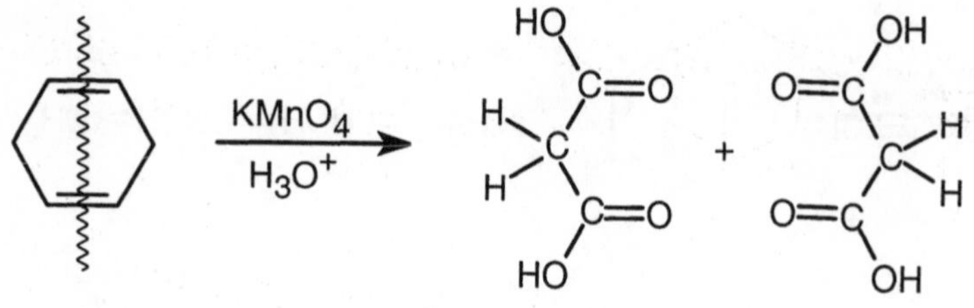

$\xrightarrow[H_3O^+]{KMnO_4}$ HO(O=)C–CH_2–C(=O)OH + HO(O=)C–CH_2–C(=O)OH

This cyclohexadiene yields only one product.

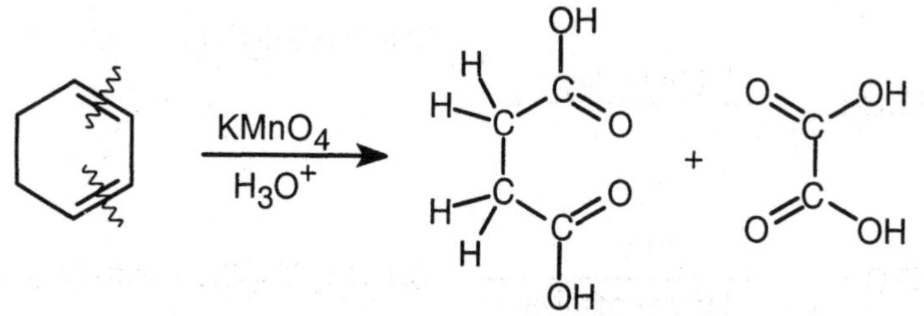

$\xrightarrow[H_3O^+]{KMnO_4}$ HO(O=)C–CH_2–CH_2–C(=O)OH + HO(O=)C–C(=O)OH

Two different products result from oxidative cleavage of this diene.

4.34

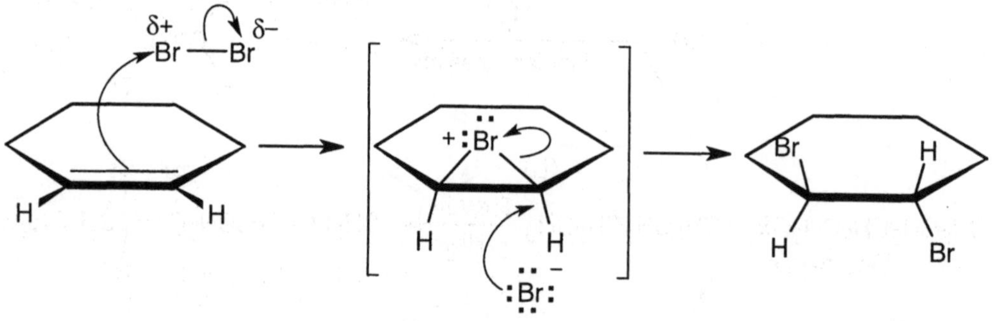

δ+ Br—Br δ−

H H

+:Br:

H H

:Br: −

Br H H Br

Cyclohexene

Bromonium ion intermediate

trans-1,2-Dibromocyclohexane

4.35

(a)

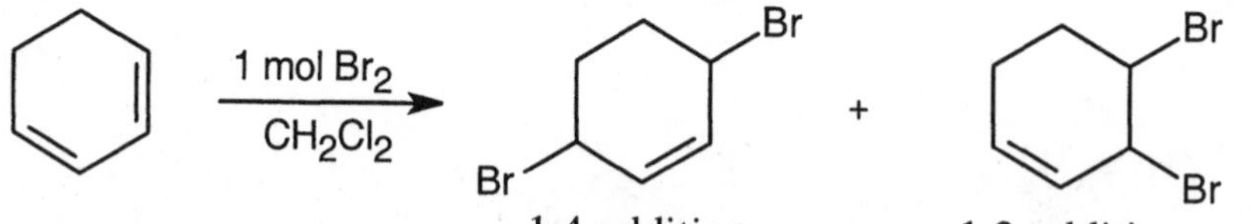

$\xrightarrow[CH_2Cl_2]{1\ mol\ Br_2}$ Br Br + Br Br

1,4-addition

1,2-addition

(b)

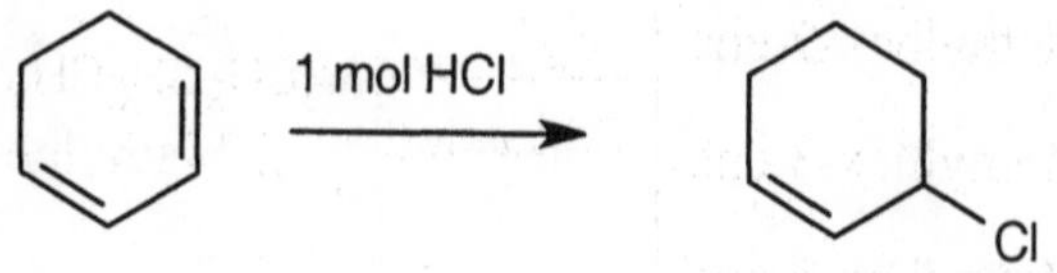

$\xrightarrow{1\ mol\ HCl}$ Cl

(c)

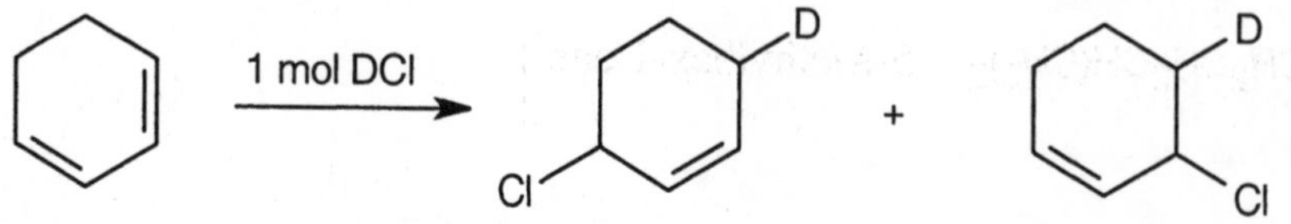

$\xrightarrow{1\ mol\ DCl}$ D Cl + D Cl

(d)

$$\text{Cyclohexene} \xrightarrow[\text{Pd catalyst}]{H_2} \text{Cyclohexane}$$

4.36

(a)

$$CH_3CH_2CH_2CH_2C{\equiv}CH \xrightarrow{\text{1 equiv. HBr}} CH_3CH_2CH_2CH_2C(Br){=}CH_2$$

(b)

$$CH_3CH_2CH_2CH_2C{\equiv}CH \xrightarrow{\text{1 equiv. } Cl_2} (CH_3CH_2CH_2CH_2)(Cl)C{=}C(Cl)(H)$$

(c)

$$CH_3CH_2CH_2CH_2C{\equiv}CH \xrightarrow[\text{Lindlar catalyst}]{H_2} CH_3CH_2CH_2CH_2CH{=}CH_2$$

4.37

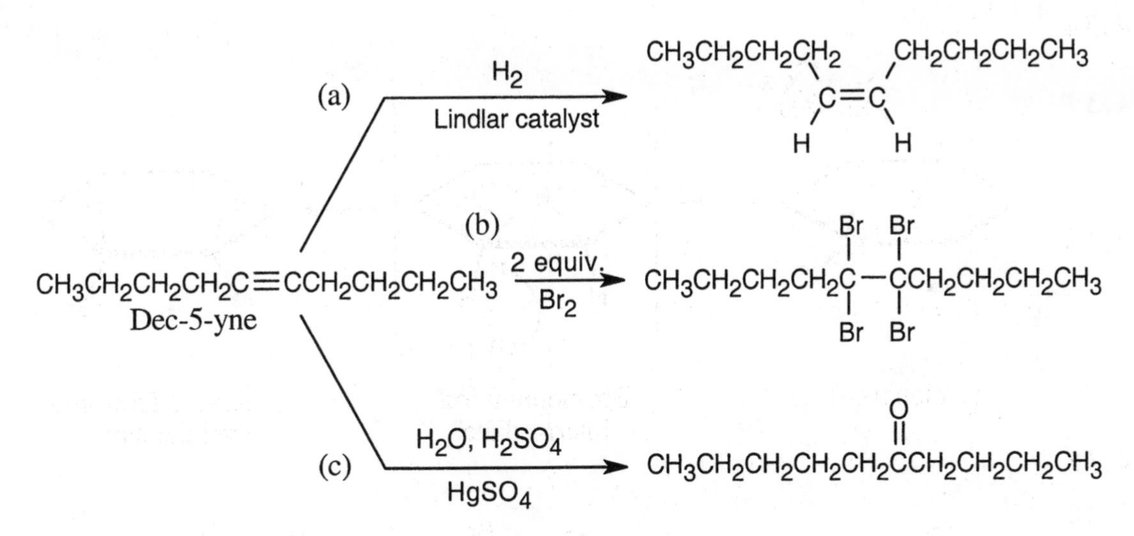

$$CH_3CH_2CH_2CH_2C{\equiv}CCH_2CH_2CH_2CH_3 \text{ (Dec-5-yne)}$$

(a)

$$\xrightarrow[\text{Lindlar catalyst}]{H_2} (CH_3CH_2CH_2CH_2)(H)C{=}C(H)(CH_2CH_2CH_2CH_3)$$

(b)

$$\xrightarrow[Br_2]{\text{2 equiv.}} CH_3CH_2CH_2CH_2CBr_2{-}CBr_2CH_2CH_2CH_2CH_3$$

(c)

$$\xrightarrow[HgSO_4]{H_2O,\ H_2SO_4} CH_3CH_2CH_2CH_2CH_2C({=}O)CH_2CH_2CH_2CH_3$$

4.38

(a)

$CH_3CH_2CH_2CH_2C(CH_3){=}CH_2$	2-Methylhex-1-ene
$CH_3CH_2CH_2CH{=}C(CH_3)_2$	2-Methylhex-2-ene
$CH_3CH_2CH{=}CHCH(CH_3)_2$	2-Methylhex-3-ene
$CH_3CH{=}CHCH_2CH(CH_3)_2$	5-Methylhex-2-ene
$H_2C{=}CHCH_2CH_2CH(CH_3)_2$	5-Methylhex-1-ene

$$\xrightarrow[\text{Pd catalyst}]{H_2} CH_3CH_2CH_2CH_2CH(CH_3)_2$$

2-Methylhexane

(b)

$$CH_3CH{=}CHCH_2CH(CH_3)_2 \xrightarrow[CH_2Cl_2]{Br_2} CH_3\underset{}{\overset{Br}{C}}H{-}\overset{Br}{C}HCH_2CH(CH_3)_2$$

5-Methylhex-2-ene → 2,3-Dibromo-5-methylhexane

(c)

$$CH_3CH_2CH_2CH_2\underset{CH_3}{C}HCH{=}CH_2 \xrightarrow{HBr} CH_3CH_2CH_2CH_2\underset{CH_3}{C}H\overset{Br}{C}HCH_3$$

3-Methylhept-1-ene → 2-Bromo-3-methylheptane

(d)

$$CH_3\overset{CH_3}{C}HCH_2CH{=}CHCH_2CH_3 \xrightarrow[H_2O,\ ^-OH]{KMnO_4} CH_3\overset{CH_3}{C}HCH_2\overset{OH}{C}H{-}\overset{OH}{C}HCH_2CH_3$$

6-Methylhept-3-ene

Predict the Reagent

4.39

$(CH_3CH_2CH_2CH_2)HC{=}CH(CH_2CH_2CH_2CH_3)$ (cis: $CH_3CH_2CH_2CH_2$ and $CH_2CH_2CH_2CH_3$ on the same side, H and H)

or

$(CH_3CH_2CH_2CH_2)HC{=}CH(CH_2CH_2CH_2CH_3)$ (trans: $CH_3CH_2CH_2CH_2$ and H on top, H and $CH_2CH_2CH_2CH_3$ below)

$$\xrightarrow[\text{Pd catalyst}]{1\ \text{mol}\ H_2} CH_3(CH_2)_8CH_3$$

$$\xrightarrow[H_3O^+]{KMnO_4} 2\ CH_3CH_2CH_2CH_2C({=}O)OH$$

4.40

$$\text{1-Methylcyclohexene} \xrightarrow[H_3O^+]{KMnO_4} HO\overset{O}{\overset{\|}{C}}CH_2CH_2CH_2CH_2\overset{O}{\overset{\|}{C}}CH_3$$

4.41

(a)

$CH_3CH{=}CHCH_3$ *or* $CH_3CH_2CH{=}CH_2$ + H_2O $\xrightarrow[\text{catalyst}]{H^+}$ $CH_3CH_2CH(OH)CH_3$

(b)

Cyclohexene + H_2O $\xrightarrow[\text{catalyst}]{H^+}$ Cyclohexanol (OH)

(c)

Cyclohexyl–$CH{=}CH_2$ + H_2O $\xrightarrow[\text{catalyst}]{H^+}$ Cyclohexyl–$CH(OH)CH_3$

4.42

(a)

$CH_3CH(CH_3)CH_2C{\equiv}CH$ + H_2O $\xrightarrow[HgSO_4]{H_2SO_4}$ $CH_3CH(CH_3)CH_2C(=O)CH_3$

(b)

Phenyl–$C{\equiv}CH$ + H_2O $\xrightarrow[HgSO_4]{H_2SO_4}$ Phenyl–$C(=O)CH_3$

4.43

Cyclooctane $\xleftarrow[\text{Pd catalyst}]{2\ H_2}$ Cycloocta-1,5-diene $\xrightarrow[H_3O^+]{KMnO_4}$ $HOOC{-}CH_2{-}CH_2{-}COOH$ + $HOOC{-}CH_2{-}CH_2{-}COOH$

Cycloocta-1,5-diene

Synthesis

4.44 (a) Reaction of 2-methylbut-2-ene with HBr gives a product in which bromine is bonded to the *more* substituted carbon.
(b) Hydroxylation of double bonds produces cis, not trans, diols.
(c) Hydration of a terminal alkyne produces a ketone, not an aldehyde.

4.45

$$CH_3C{\equiv}CH \xrightarrow[2.\ CH_3Br]{1.\ NaNH_2} CH_3C{\equiv}CCH_3 \xrightarrow[\text{Lindlar catalyst}]{H_2} \text{(H)(H}_3\text{C)C=C(H)(CH}_3\text{)}$$

cis-But-2-ene

4.46

(a)

$$CH_3CH_2C{\equiv}CH \xrightarrow[\text{Pd catalyst}]{\text{2 equiv. } H_2} CH_3CH_2CH_2CH_3$$

But-1-yne → Butane

(b)

$$CH_3CH_2C{\equiv}CH \xrightarrow[CH_2Cl_2]{\text{2 equiv. } Cl_2} CH_3CH_2CCl_2CHCl_2$$

1,1,2,2-Tetrachlorobutane

(c)

$$CH_3CH_2C{\equiv}CH \xrightarrow[\text{Lindlar catalyst}]{H_2} CH_3CH_2CH{=}CH_2 \xrightarrow{HBr} CH_3CH_2CHBrCH_3$$

$$CH_3CH_2C{\equiv}CH \xrightarrow{HBr} CH_3CH_2CBr{=}CH_2 \xrightarrow[\text{Pd catalyst}]{H_2} CH_3CH_2CHBrCH_3$$

2-Bromobutane

(d)

$$CH_3CH_2C{\equiv}CH \xrightarrow[HgSO_4]{H_2O,\ H_2SO_4} CH_3CH_2C(=O)CH_3$$

Butan-2-one

Resonance

4.47

(a) (b)

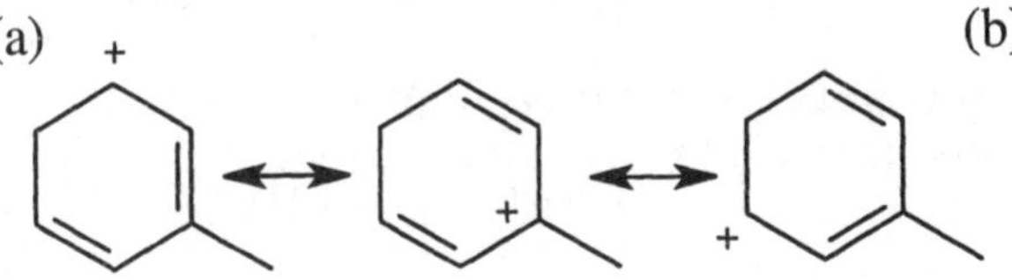

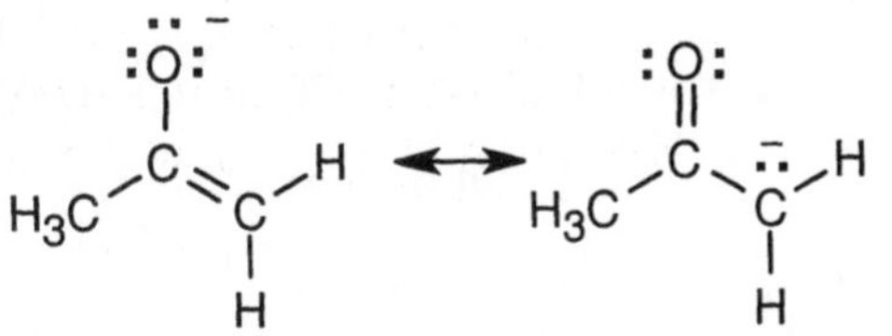

H_3C H H H_3C H H

(c)

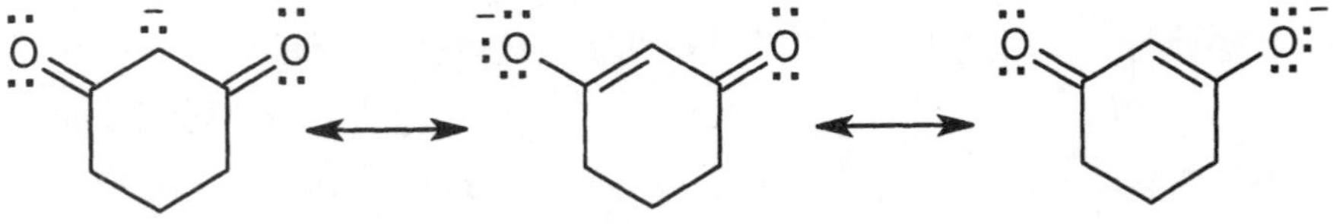

4.48 Resonance forms can't differ in the position of nuclei. The two structures in (a) are not resonance forms because the carbon and hydrogen atoms outside the ring occupy different positions in the two forms.

not resonance structures

4.49

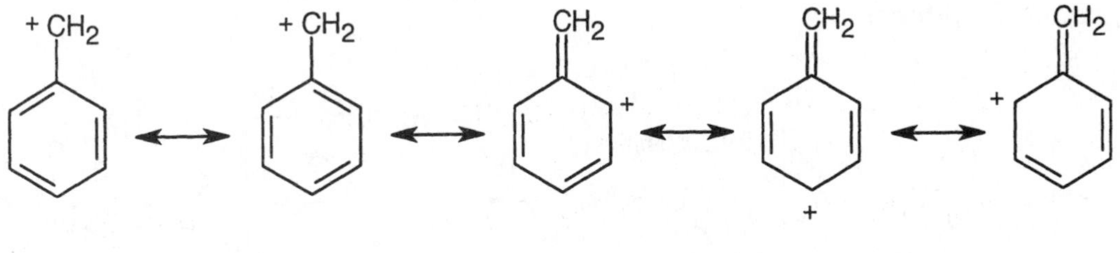

$+CH_2$ $+CH_2$ CH_2 CH_2 CH_2

4.50

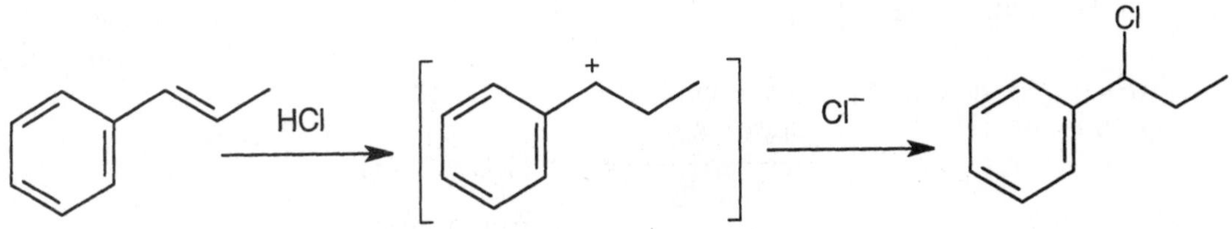

HCl Cl⁻ Cl

benzylic carbocation intermediate

The benzylic carbocation intermediate is stabilized by resonance involving the aromatic ring (as shown in the previous problem), and addition to the double bond occurs at the position indicated. Addition to the aromatic ring doesn't occur.

Polymers

4.51

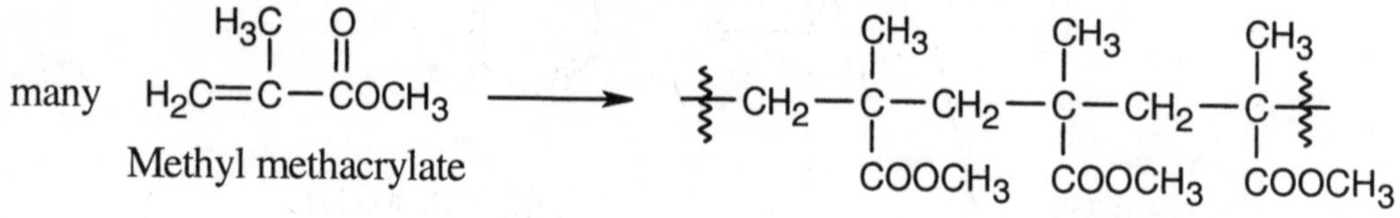

$$\text{many } H_2C{=}C(CH_3){-}COOCH_3 \longrightarrow {-}CH_2{-}C(CH_3)(COOCH_3){-}CH_2{-}C(CH_3)(COOCH_3){-}CH_2{-}C(CH_3)(COOCH_3){-}$$

Methyl methacrylate

4.52

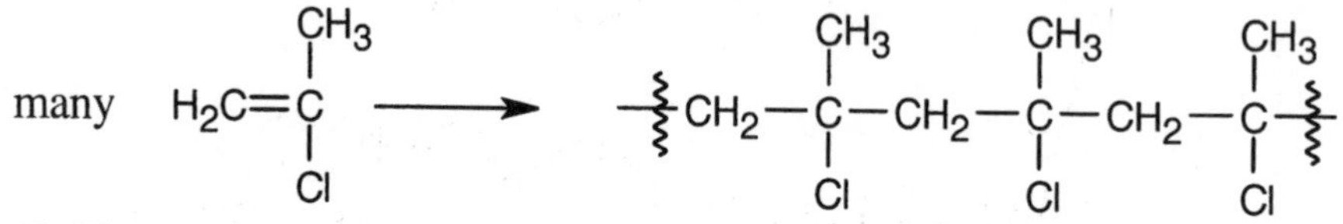

$$\text{many } H_2C{=}C(CH_3)Cl \longrightarrow {-}CH_2{-}C(CH_3)(Cl){-}CH_2{-}C(CH_3)(Cl){-}CH_2{-}C(CH_3)(Cl){-}$$

2-Chloropropene

4.53

$$\text{many } H_2C{=}CH{-}N(\text{pyrrolidone}) \longrightarrow {-}CH_2{-}CH(N)\text{-}CH_2{-}CH(N){-}CH_2{-}CH(N){-}$$

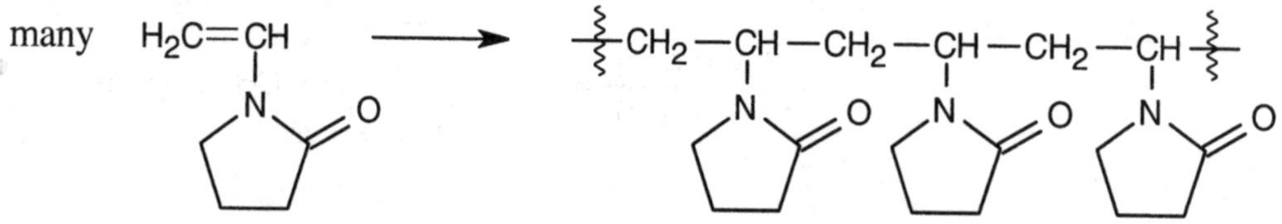

N-Vinylpyrrolidone

Poly(vinyl pyrrolidone)

Integrated Problems

4.54

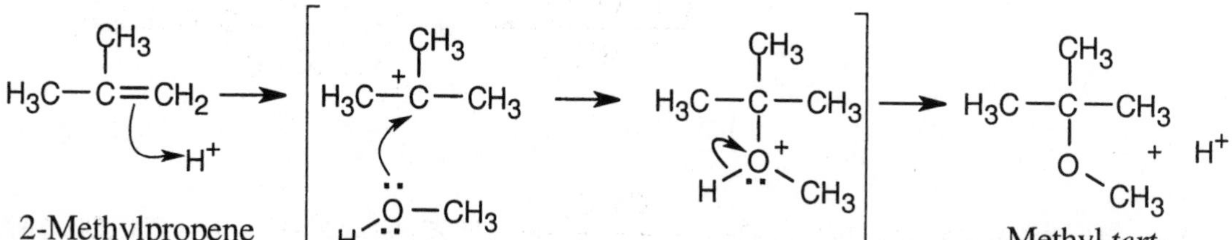

$$H_3C{-}C(CH_3){=}CH_2 \xrightarrow{H^+} \left[H_3C{-}\overset{+}{C}(CH_3){-}CH_3 \xrightarrow{H{-}\ddot{O}{-}CH_3} H_3C{-}C(CH_3)(CH_3){-}\overset{+}{O}(H){-}CH_3 \right] \longrightarrow H_3C{-}C(CH_3)(CH_3){-}O{-}CH_3 + H^+$$

2-Methylpropene

Methyl *tert*-butyl ether

reaction of double bond with H^+

nucleophilic addition of methanol

loss of H^+

4.55

Styrene (A): $C_6H_5CH{=}CH_2$

$\xrightarrow[\text{Pd catalyst}]{H_2}$ $C_6H_5CH_2CH_3$

$\xrightarrow{\text{4 equiv. } H_2}$ cyclohexyl–CH_2CH_3 (B)

$\xrightarrow[H_3O^+]{KMnO_4}$ $C_6H_5CO_2H$ (C) + CO_2

4.56

A: cyclohexylidene$=CHC{\equiv}CH$

$\xrightarrow[\text{Pd catalyst}]{\text{3 equiv. } H_2}$ cyclohexyl–$CH_2CH_2CH_3$ (B)

$\xrightarrow[H_3O^+]{KMnO_4}$ cyclohexane$=O$ + other products (Cyclohexanone)

$\xrightarrow[\text{2. } CH_3I]{\text{1. } NaNH_2}$ cyclohexylidene$=CHC{\equiv}CCH_3$ (C)

4.57

$$CH_3(CH_2)_{12}CH{=}CH(CH_2)_7CH_3 \xrightarrow[H_3O^+]{KMnO_4} CH_3(CH_2)_{12}CO_2H + CH_3(CH_2)_7CO_2H$$

Muscalure

4.58

$$HC{\equiv}CH \xrightarrow[\text{2. } CH_3(CH_2)_{11}CH_2Br]{\text{1. } NaNH_2} CH_3(CH_2)_{12}C{\equiv}CH$$

$$CH_3(CH_2)_{12}C{\equiv}CH \xrightarrow[\text{2. } CH_3(CH_2)_6CH_2Br]{\text{1. } NaNH_2} CH_3(CH_2)_{12}C{\equiv}C(CH_2)_7CH_3$$

$$CH_3(CH_2)_{12}C{\equiv}C(CH_2)_7CH_3 \xrightarrow[\text{Lindlar catalyst}]{H_2} \text{cis-}CH_3(CH_2)_{11}CH_2(H)C{=}C(H)CH_2(CH_2)_6CH_3$$

4.59

$$(CH_3)_2C{=}C(CH_3)_2 \xrightarrow[H_3O^+]{KMnO_4} (CH_3)_2C{=}O + O{=}C(CH_3)_2$$

2,3-Dimethylbut-2-ene

4.60

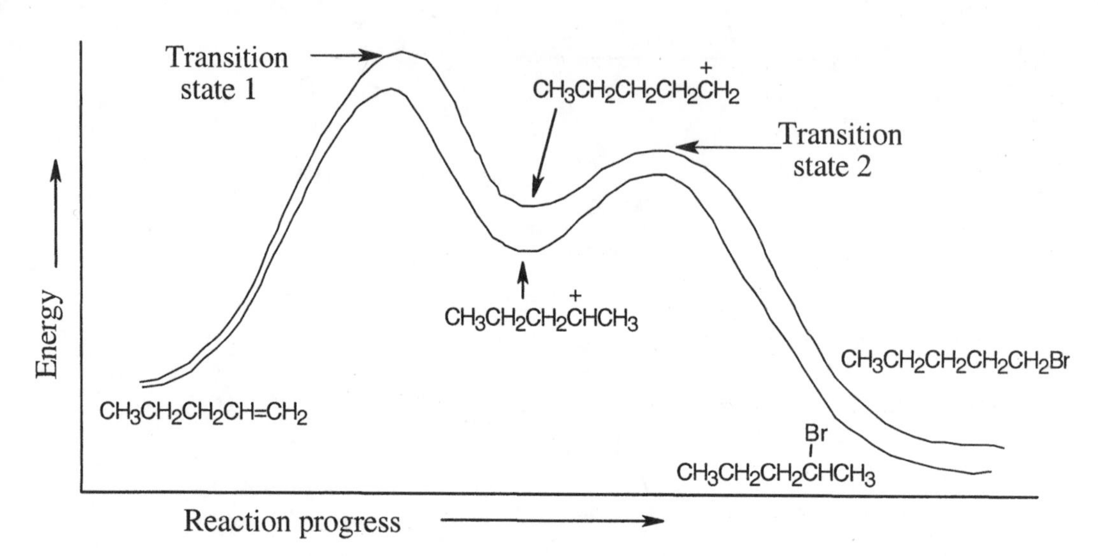

4.61

Transition State #1

H–C($\delta+$)(C$_3$H$_7$)–C(H)(H) ··· H $\delta+$

Transition State #2

H_3C, H, C^+, C_3H_7 ------- :Br:$^-$

4.62

H^+, CH_2 ⇌ [$^+$, CH_3, H, H] ⇌ CH_3, H + H^+

Methylene-cyclohexane | most stable carbocation | 1-Methylcyclohexene

Protonation occurs to produce the most stable carbocation, which can then lose H^+ to form either of two alkenes. Because 1-methylcyclohexene is the major product of this equilibrium, it must be the more stable product.

4.63

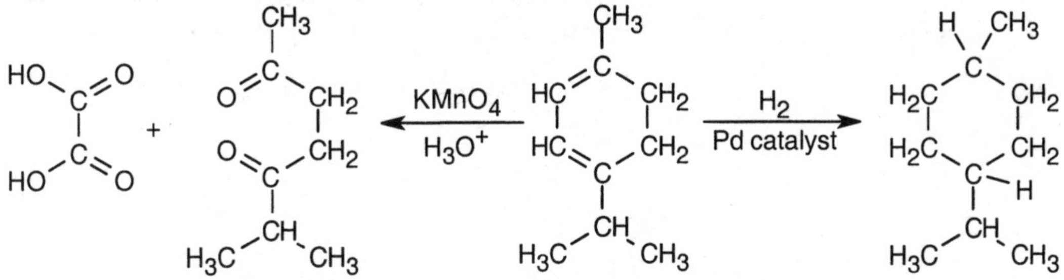

HO C O C O HO + CH$_3$ C O CH$_2$ CH$_2$ O C CH H$_3$C CH$_3$ ← KMnO$_4$ H$_3$O$^+$ ← CH$_3$ C HC CH$_2$ HC CH$_2$ C CH H$_3$C CH$_3$ → H$_2$ Pd catalyst → H CH$_3$ C H$_2$C CH$_2$ H$_2$C CH$_2$ C H CH H$_3$C CH$_3$

Oxalic acid 6-Methylhept-ane-2,5-dione α-Terpinene

4.64

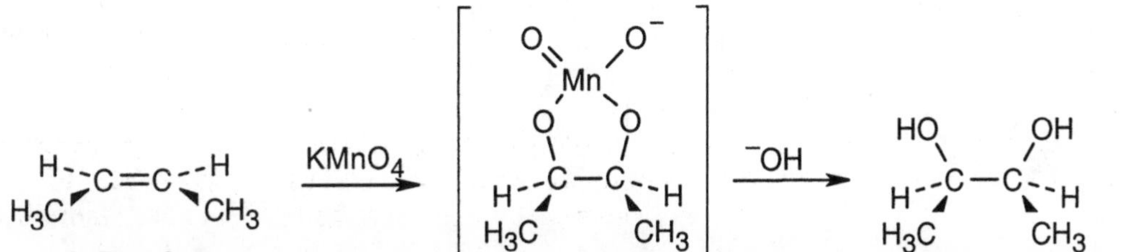

H C=C H H$_3$C CH$_3$ → KMnO$_4$ → [O Mn O$^-$ O O H C C H H$_3$C CH$_3$] → $^-$OH → HO OH H C C H H$_3$C CH$_3$

cis-But-2-ene

H C=C CH$_3$ H$_3$C H → KMnO$_4$ → [O Mn O$^-$ O O H C C CH$_3$ H$_3$C H] → $^-$OH → HO OH H C C CH$_3$ H$_3$C H

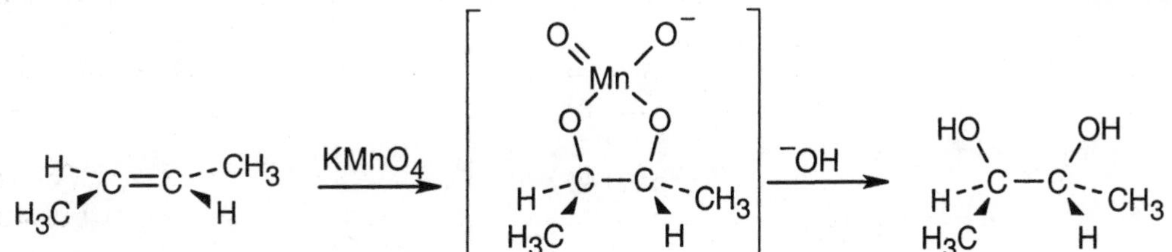

trans-But-2-ene

The intermediate retains the cis or trans stereochemistry of the alkene double bond. Treatment of the intermediate with $^-$OH does not change the bond stereochemistry. Although the products closely resemble one another, they are not superimposable and are stereoisomers.

In the Medicine Cabinet

4.65 (a)

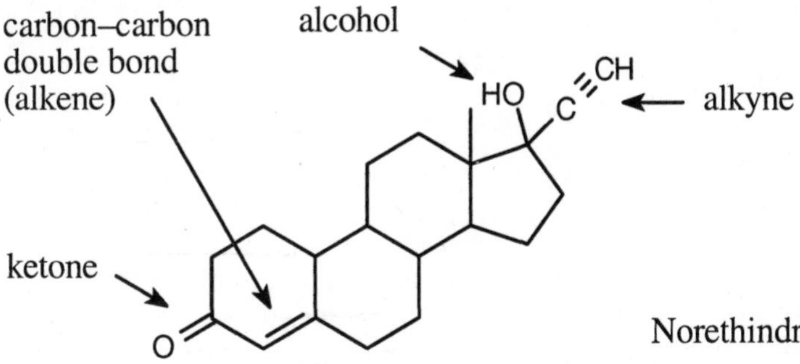

carbon–carbon double bond (alkene)

alcohol

HO C ≡CH ← alkyne

ketone

O

Norethindrone

(b) The molecular formula of Norethindrone is $C_{20}H_{26}O_2$.

(c)

$$HC{\equiv}CH + {}^{-}{:}Base \longrightarrow {}^{-}{:}C{\equiv}CH + H{:}Base$$

A

(d)

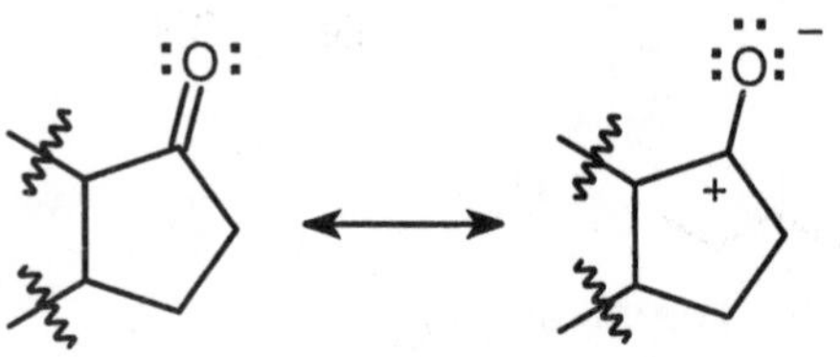

:O: ⟷ :Ö:⁻ +

(e)

:Ö:⁻ + ⁻:C≡CH ⟶ [:Ö:⁻ C≡CH H–A] ⟶ OH C≡CH

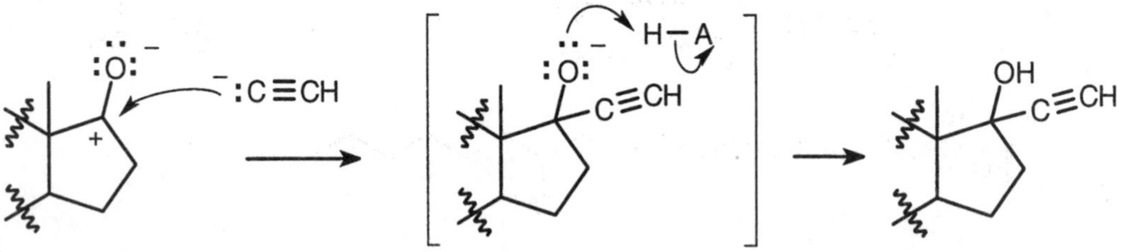

In the Field with Agrochemicals

4.66

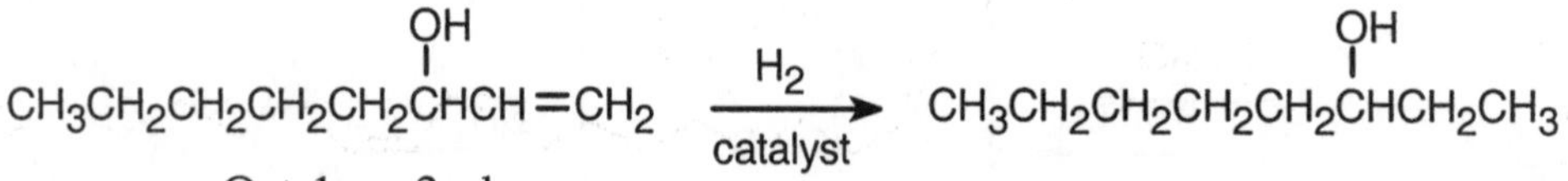

$$CH_3CH_2CH_2CH_2CH_2CH(OH)CH{=}CH_2 \xrightarrow[\text{catalyst}]{H_2} CH_3CH_2CH_2CH_2CH_2CH(OH)CH_2CH_3$$

Oct-1-en-3-ol

4.67

(a)

OH * OH * OH * O O O O O O

Castor oil

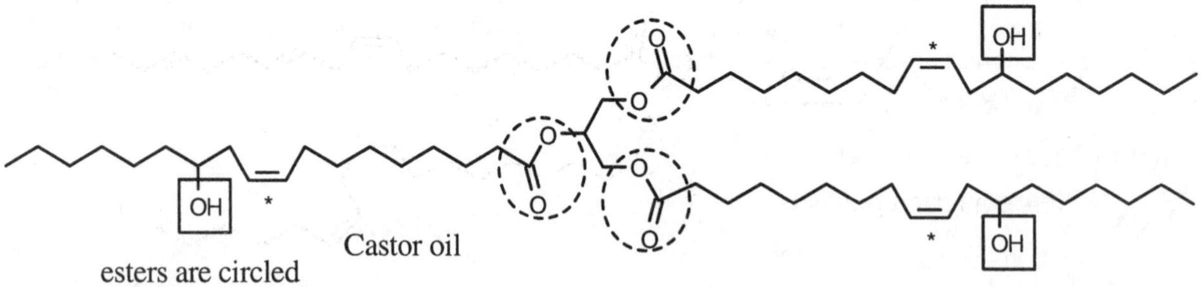

esters are circled
alcohols are boxed
alkenes are starred

(b) The molecular formula of castor oil is $C_{57}H_{104}O_9$.

(c)

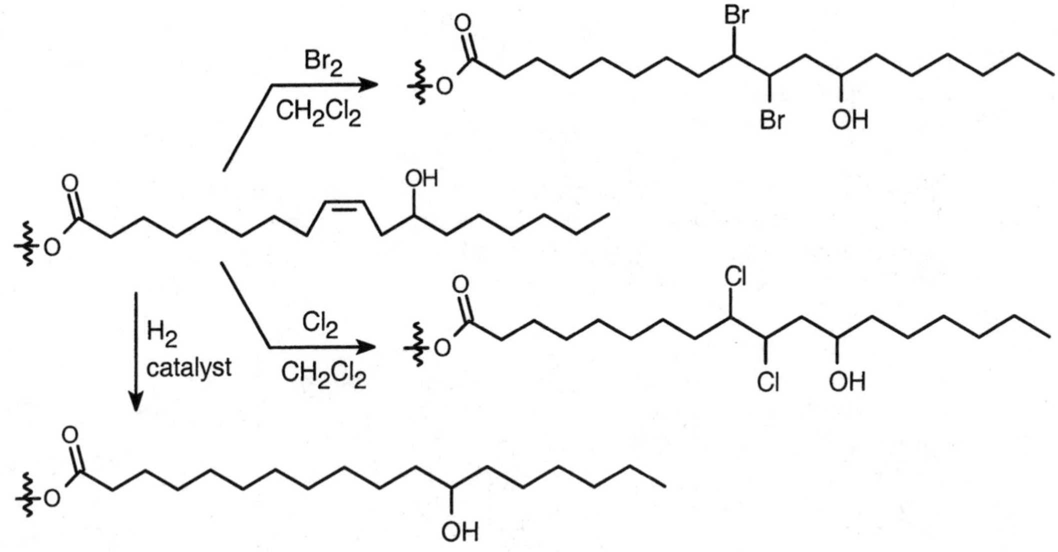

Br_2, CH_2Cl_2

Br, Br, OH

OH

H_2 catalyst

Cl_2, CH_2Cl_2

Cl, Cl, OH

OH

(d)

Conjugated products:

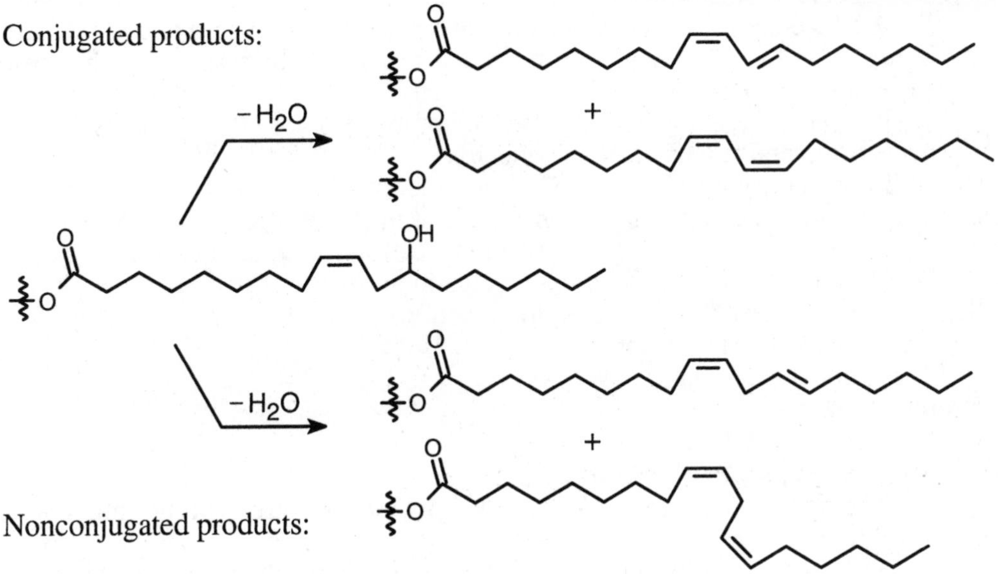

$-H_2O$

+

OH

$-H_2O$

+

Nonconjugated products:

4.68

One repeat unit of synthetic rubber: H_3C

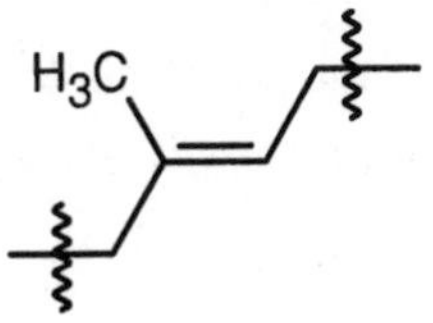

Chapter 5 – Aromatic Compounds

Chapter Outline

I. Aromatic compounds (Sections 5.1–5.3).
   A. Structure of benzene (Sections 5.1–5.2).
      1. Kekulé proposal (Section 5.1).
         a. Benzene is 1,3,5 cyclohexatriene, and all carbons are equivalent.
         b. This proposal doesn't account for the stability of benzene, or for the fact that benzene undergoes substitution, rather than addition, reactions.
         c. The proposal also doesn't explain the fact that all carbon-carbon bonds are the same length.
      2. Resonance proposal (Section 5.2).
         a. Benzene is a resonance hybrid of two contributing forms.
         b. All carbon-carbon bonds are the same length, and all bond angles are 120°.
   B. Naming aromatic compounds (Section 5.3).
      1. Many aromatic compounds have nonsystematic names.
      2. Monosubstituted benzenes are named in the same way as other hydrocarbons, with -–benzene as the parent name.
      3. Disubstituted benzenes are named by the ortho(*o*), meta(*m*), para(*p*) system.
         a. A benzene ring with two substituents in a 1,2 relationship is *o*-disubstituted.
         b. A benzene ring with two substituents in a 1,3 relationship is *m*-disubstituted.
         c. A benzene ring with two substituents in a 1,4 relationship is *p*-disubstituted.
      4. Benzenes with more than two substituents are named by numbering the position of each substituent.
         a. Number so that the lowest possible combination of numbers is used.
         b. Substituents are listed alphabetically.

II. Electrophilic aromatic substitution of benzene (Sections 5.4–5.6).
   A. General features of electrophilic aromatic substitution reactions (Section 5.4).
      1. An electron-poor reagent reacts with the electron-rich aromatic ring.
      2. Substitution, rather than addition, occurs in aromatic rings.
      3. A catalyst is needed for substitution reactions.
      4. The reaction has two steps.
      5. The product is aromatic.
   B. Bromination.
      Mechanism of bromination.
         a. Br_2 complexes with $FeBr_3$.
         b. The polarized electrophile is attacked by the π electrons of the ring in a slow, rate-limiting step.
         c. The cation intermediate is doubly allylic but is much less stable than the starting aromatic compound.
         d. The carbocation intermediate loses H^+ from the bromine-bearing carbon in a fast step to regenerate an aromatic ring.
   C. Other electrophilic aromatic substitutions (Section 5.5).
      1. Chlorination.
      2. Nitration.
      3. Sulfonation.

D. Friedel-Crafts alkylation and acylation (Section 5.6).
  1. The Friedel-Crafts reaction introduces an alkyl group onto an aromatic ring.
    a. An alkyl chloride, plus an $AlCl_3$ catalyst, produces an electrophilic carbocation.
    b. Only alkyl halides – not aryl or vinylic halides – can be used.
    c. Friedel-Crafts reactions don't succeed on rings that have amino substituents or deactivating groups.
  2. Friedel-Crafts acylation occurs when an aromatic ring reacts with a carboxylic acid chloride, plus an $AlCl_3$ catalyst.

III. Substituent effects in electrophilic aromatic substitution (Sections 5.7–5.8).
  A. General effects (Section 5.7).
    1. Substituents affect the reactivity of an aromatic ring.
    2. Substituents affect the orientation of further substitution.
    3. Substituents can be arranged in three groups:
      a. Ortho/para directing activators.
      b. Ortho/para-directing deactivators.
      c. Meta-directing deactivators.
  B. Explanation of effects (Section 5.8).
    1. Activation and deactivation.
      a. All activating groups donate electrons to an aromatic ring.
      b. All deactivating groups withdraw electrons from a ring.
      c. Electron donation or withdrawal may be due to inductive or resonance effects.
    2. Orienting effects.
      a. Ortho, para directors.
        The intermediates from *o,p* attack are more stable because they can be stabilized by resonance donation by the substituent.
      b. Any substituent with a positively polarized atom bonded to the ring is a meta director.

IV. Oxidation and reduction of aromatic compounds (Section 5.9).
  A. Side-chain oxidation occurs when an aromatic ring has been treated with $KMnO_4$. Substituted benzoic acids are the products.
  B. Aromatic rings can be reduced by catalytic hydrogenation using a powerful catalyst.

V. Polycyclic aromatic hydrocarbons consist of two or more fused benzene rings (Section 5.10) Compounds such as pyridine and pyrrole are also aromatic.

VI. Organic synthesis (Section 5.11).
  A. To synthesize substituted benzenes, it is important to introduce groups so that they have the proper orienting effects.
  B. It is best to use retrosynthetic analysis to plan a synthesis.

Solutions to Problems

5.1 According to Kekulé, four dibromobenzenes are possible.

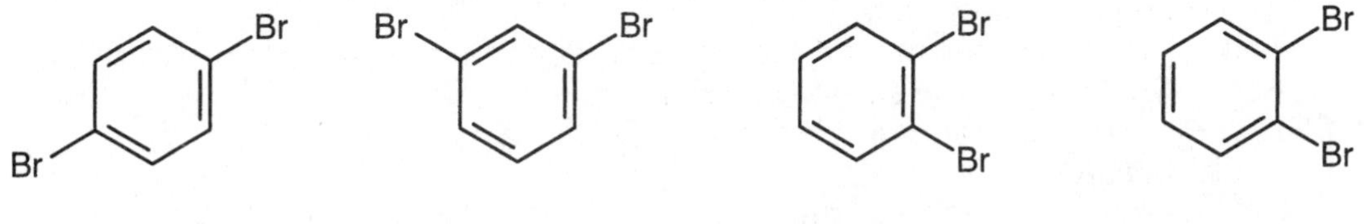

Br Br Br Br Br Br Br Br

Kekulé would say that the two isomers on the right interconvert rapidly, and only one isomer can be isolated.

5.2 According to resonance theory, *o*-dibromobenzene is not described properly by either of the two structures shown on the right in Problem 5.1, but is a resonance hybrid of the two.

5.3 Number the positions on the ring. Give number 1 to one substituent and the lowest possible number to the other substituent. Substituents at positions 1 and 2 have an ortho relationship; substituents at positions 1 and 3 have a meta relationship; substituents at positions 1 and 4 have a para relationship.

(a) Cl, CH_3 — meta disubstituted

(b) Br, NO_2 — para disubstituted

(c) SO_3H, OH — ortho disubstituted

5.4 For compounds with two substituents, determine the ortho, meta, para relationship of the substituents, and cite them in alphabetical order.

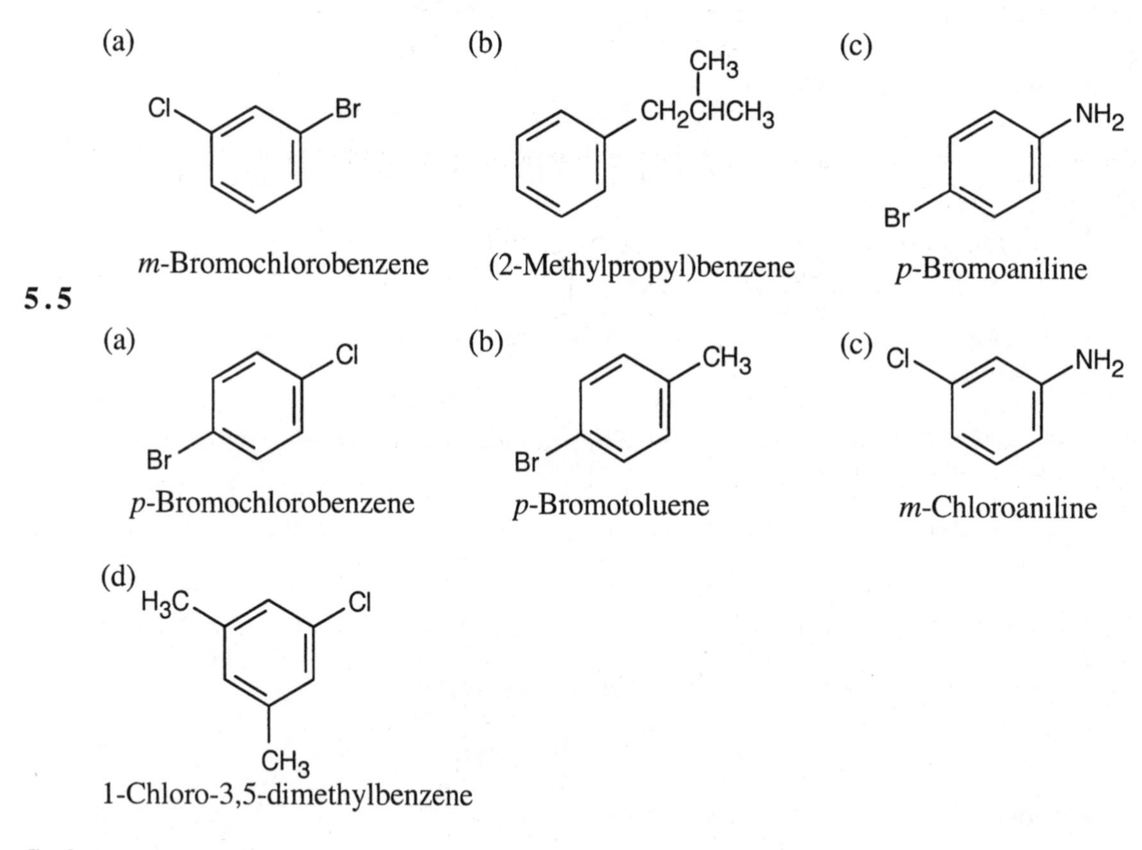

5.4 (a) *m*-Bromochlorobenzene (b) (2-Methylpropyl)benzene (c) *p*-Bromoaniline

5.5 (a) *p*-Bromochlorobenzene (b) *p*-Bromotoluene (c) *m*-Chloroaniline (d) 1-Chloro-3,5-dimethylbenzene

5.6

Toluene $\xrightarrow[FeBr_3]{Br_2}$ *o*-Bromotoluene + *m*-Bromotoluene + *p*-Bromotoluene

5.7 The mechanism of nitration is the same as the mechanism of other electrophilic aromatic substitution mechanisms we have studied.

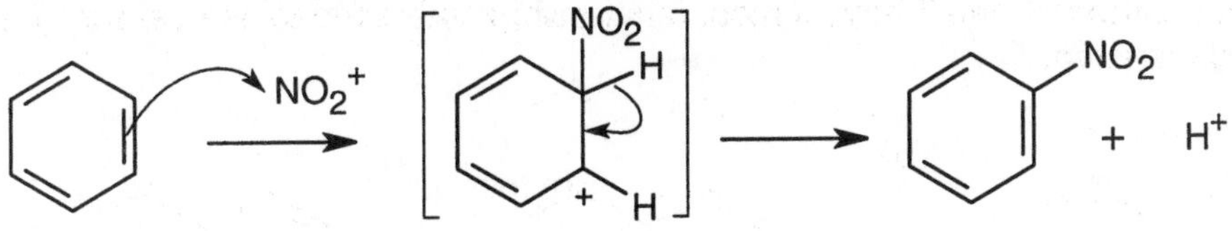

NO_2^+ ; NO_2 ; H ; + ; H ; NO_2 ; + H^+

carbocation intermediate

5.8

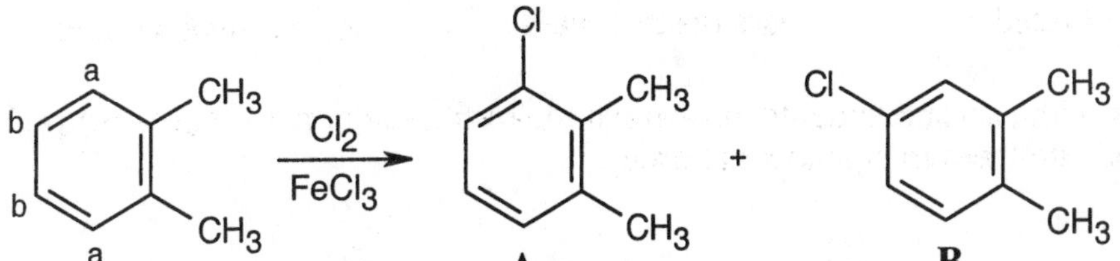

a ; b ; b ; a ; CH_3 ; CH_3 ; Cl_2 ; $FeCl_3$; Cl ; CH_3 ; CH_3 ; **A** ; + ; Cl ; CH_3 ; CH_3 ; **B**

o-Xylene

Chlorination at position "a" of *o*–xylene yields product **A**; chlorination at position "b" yields product **B**.

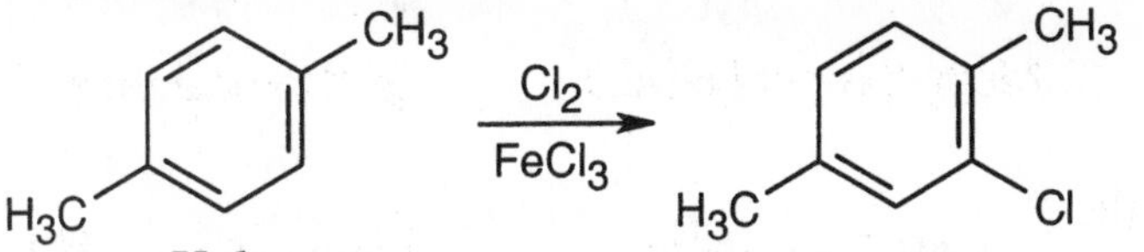

CH_3 ; H_3C ; Cl_2 ; $FeCl_3$; CH_3 ; H_3C ; Cl

p-Xylene

Only one product results from chlorination of *p*–xylene because all possible sites for chlorination are equivalent.

5.9

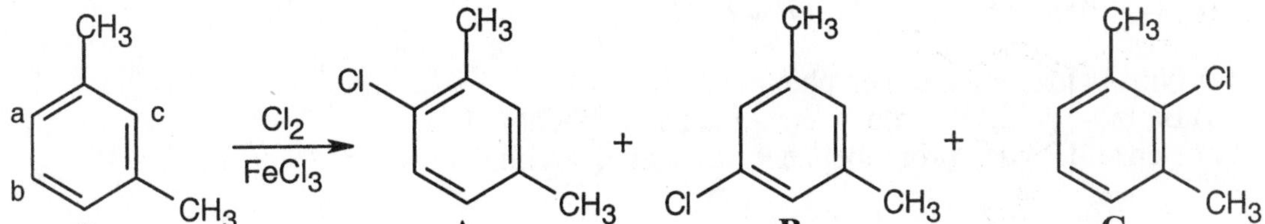

CH_3 ; a ; c ; b ; a ; CH_3 ; Cl_2 ; $FeCl_3$; CH_3 ; Cl ; CH_3 ; **A** ; + ; CH_3 ; Cl ; CH_3 ; **B** ; + ; CH_3 ; Cl ; CH_3 ; **C**

m-Xylene

Three products might form on chlorination of *m*-xylene. Product **C** is unlikely to form because substitution rarely occurs between two meta substituents.

5.10

(a)

$$C_6H_6 + CH_3CH_2Cl \xrightarrow{AlCl_3} C_6H_5CH_2CH_3$$

Ethylbenzene

(b)

$$p\text{-}CH_3C_6H_4CH_3 + CH_3CH_2Cl \xrightarrow{AlCl_3} \text{2-Ethyl-1,4-dimethylbenzene}$$

p-Xylene 2-Ethyl-1,4-dimethylbenzene

5.11

(a)

$$C_6H_6 + (CH_3)_3CCl \xrightarrow{AlCl_3} C_6H_5C(CH_3)_3$$

(b)

$$C_6H_6 + CH_3CH_2CCl(=O) \xrightarrow{AlCl_3} C_6H_5C(=O)CH_2CH_3$$

5.12 Figure 5.8 lists groups in order of the strength of their activating or deactivating effects.

Least reactive ——> *Most reactive*

(a) Nitrobenzene, toluene, phenol
(b) Benzoic acid, chlorobenzene, benzene, phenol
(c) Benzaldehyde, bromobenzene, benzene, aniline

5.13

(a)

$$C_6H_5CN + Cl_2 \xrightarrow{FeCl_3} m\text{-}ClC_6H_4CN$$

m-Chlorobenzonitrile

The cyano group is a meta-director.

(b)

$$C_6H_5Br + Cl_2 \xrightarrow{FeCl_3} o\text{-}BrC_6H_4Cl + p\text{-}BrC_6H_4Cl$$

o-Bromochlorobenzene *p*-Bromochlorobenzene

The bromo group is an ortho-para director.

5.14

(a) $C_6H_5NO_2 + SO_3 \xrightarrow{H_2SO_4}$ m-$O_2NC_6H_4SO_3H$

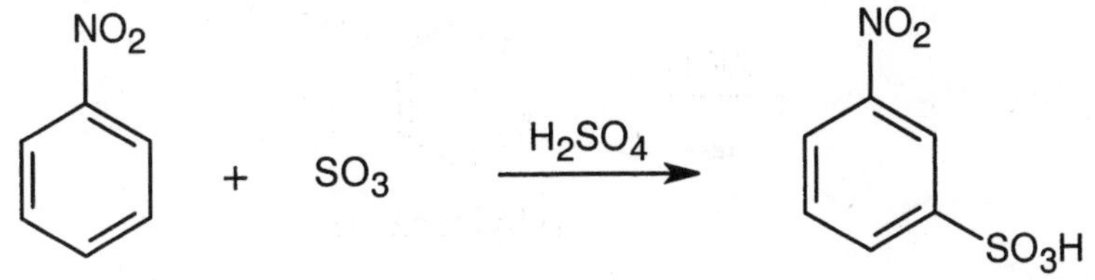

(b) $C_6H_5Br + SO_3 \xrightarrow{H_2SO_4}$ o-$BrC_6H_4SO_3H$ + p-$BrC_6H_4SO_3H$

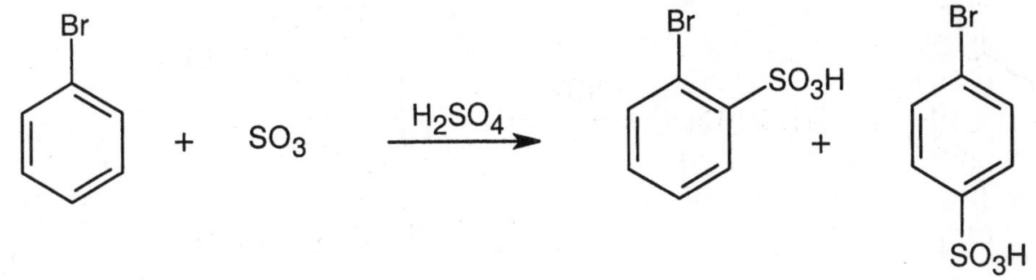

(c) $C_6H_5CH_3 + SO_3 \xrightarrow{H_2SO_4}$ o-$CH_3C_6H_4SO_3H$ + p-$CH_3C_6H_4SO_3H$

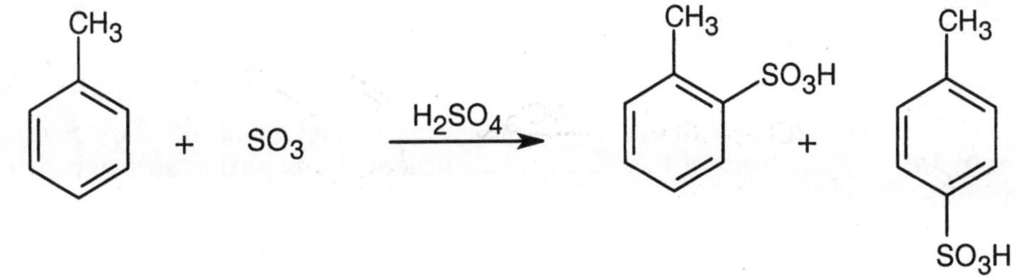

(d) $C_6H_5COOH + SO_3 \xrightarrow{H_2SO_4}$ m-$HOOCC_6H_4SO_3H$

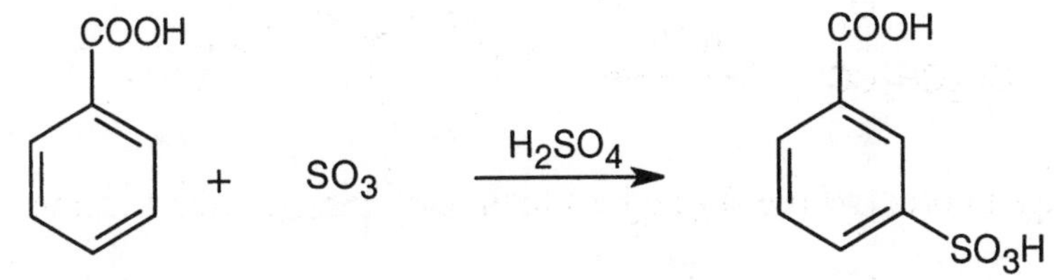

(e) $C_6H_5CN + SO_3 \xrightarrow{H_2SO_4}$ m-$NCC_6H_4SO_3H$

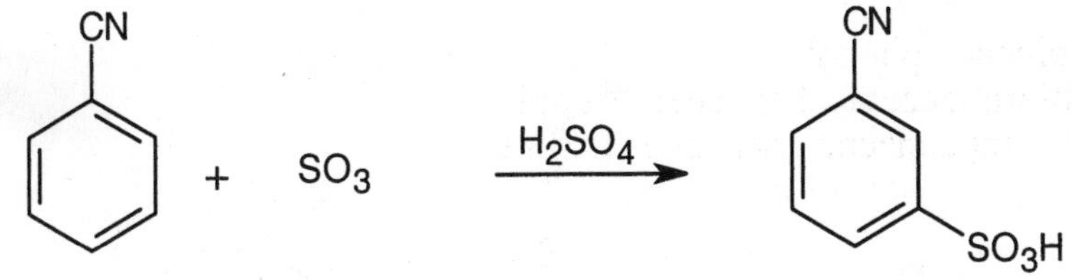

5.15

(a) Ortho attack:

$:\ddot{O}CH_3$, H, Br ⟷ $:\ddot{O}CH_3$, H, Br ⟷ $:\ddot{O}CH_3$, H, Br ⟷ $+\ddot{O}CH_3$, H, Br

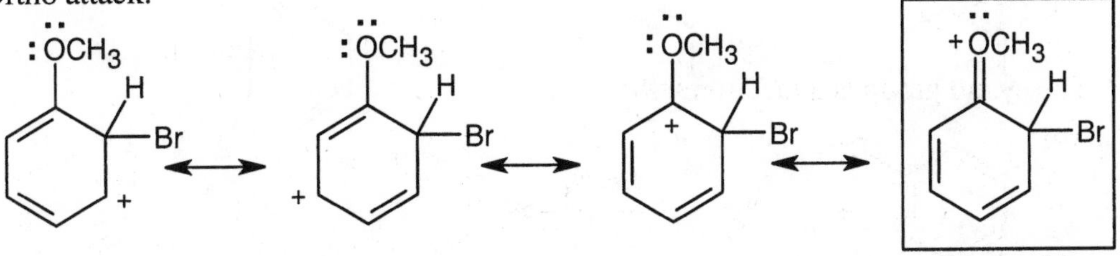

most stable

(b) Meta attack:

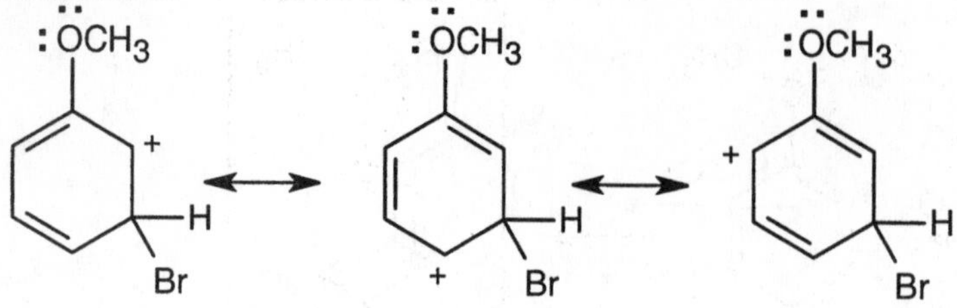

:ÖCH3 ↔ :ÖCH3 ↔ :ÖCH3

H Br H Br H Br

(c) Para attack:

:ÖCH3 ↔ :ÖCH3 ↔ :ÖCH3 ↔ +ÖCH3

Br H Br H Br H Br H

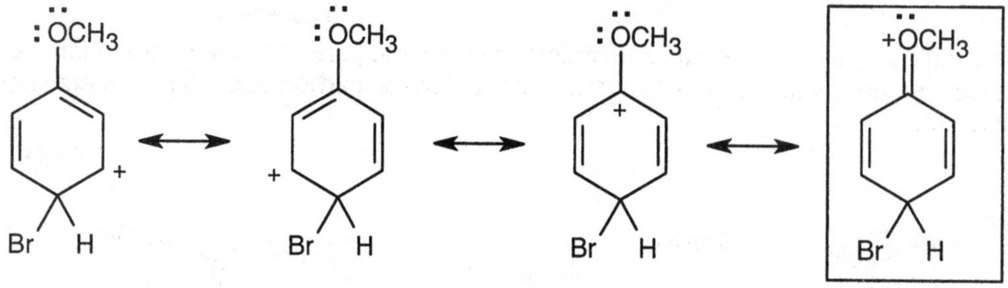

most stable

The carbocation intermediates in ortho-para substitution can be stabilized by electron donation by the oxygen atom of the $-OCH_3$ substituent. Thus, ortho-para substitution is favored.

5.16

(a) Ortho attack:

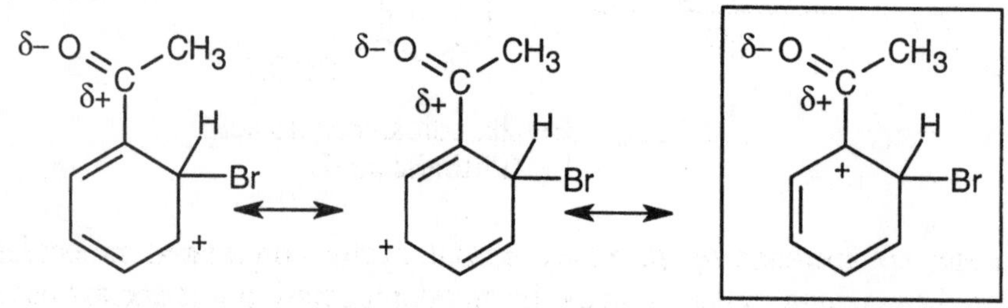

δ– O=C(δ+)–CH3 H Br ↔ δ– O=C(δ+)–CH3 H Br ↔ δ– O=C(δ+)–CH3 H Br

least stable

(b) Meta attack:

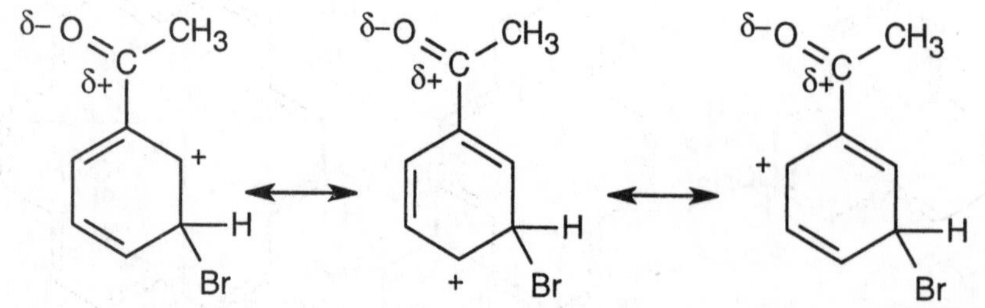

δ– O=C(δ+)–CH3 H Br ↔ δ–O=C(δ+)–CH3 H Br ↔ δ–O=C(δ+)–CH3 H Br

(c) Para attack:

least stable

The indicated carbocation intermediates of ortho and para attack are least stable because they place a positive charge next to a positively polarized carbon atom. Thus, meta substitution is favored.

5.17

(a) $\xrightarrow[H_2O]{KMnO_4}$

m-Chloroethylbenzene → *m*-Chlorobenzoic acid

(b) $\xrightarrow[H_2O]{KMnO_4}$

Tetralin → *o*-Benzenedicarboxylic acid (Phthalic acid)

5.18 Three of the molecules are aromatic because they are flat, cyclic, conjugated molecules with $(4n + 2)$ π electrons. The second molecule is cyclic and conjugated, but it doesn't have $(4n + 2)$ π electrons. Notice that the oxygen atom in the first molecule contributes two lone pair electrons to the ring π system.

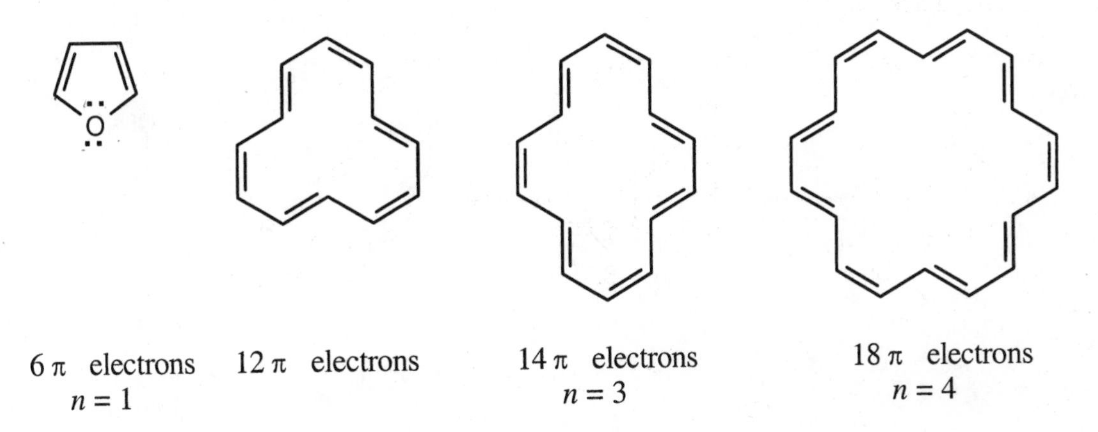

5.19

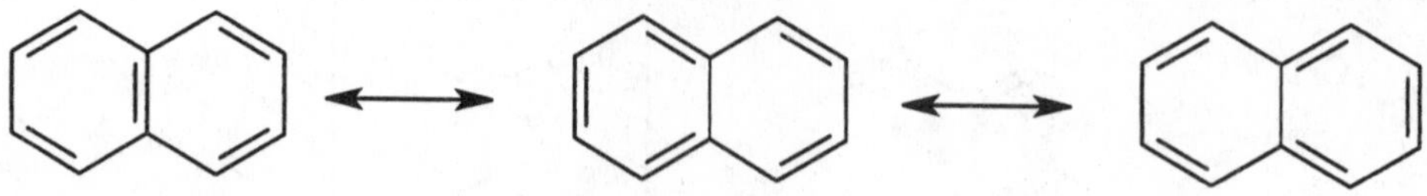

5.20 In order to obtain the desired product, you must perform the reactions in the correct order. Assume that you can separate ortho and para isomers.

(a)

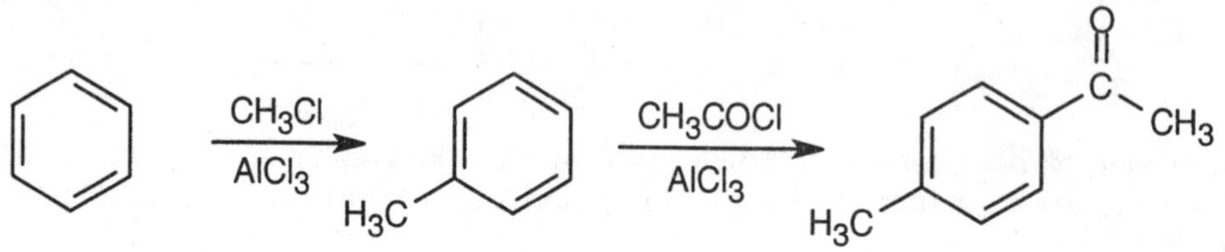

CH_3Cl / $AlCl_3$ → H_3C– (toluene) → CH_3COCl / $AlCl_3$ → H_3C–C₆H₄–C(=O)–CH_3

p-Methylacetophenone

(b)

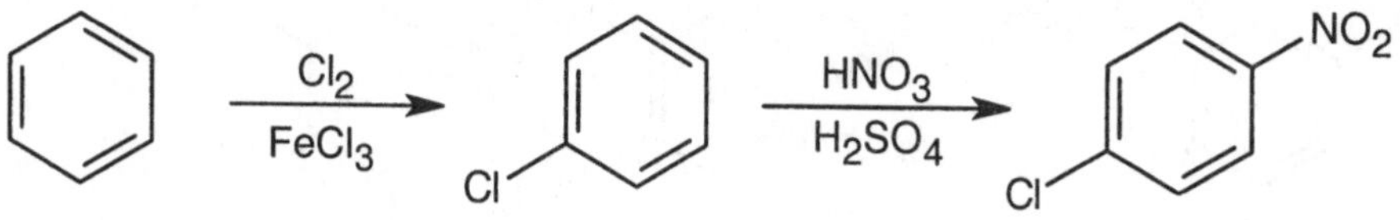

Cl_2 / $FeCl_3$ → Cl– (chlorobenzene) → HNO_3 / H_2SO_4 → Cl–C₆H₄–NO_2

p-Chloronitrobenzene

5.21 (a) Two routes can be used to synthesize *o*–bromotoluene.

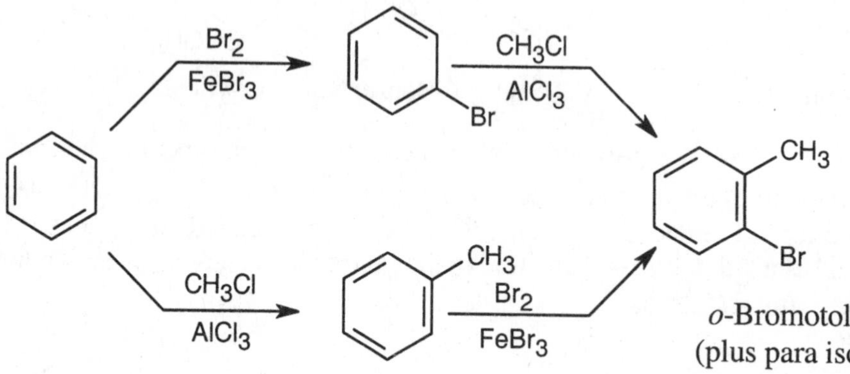

Br_2 / $FeBr_3$ → Br (bromobenzene) → CH_3Cl / $AlCl_3$ → CH_3, Br

CH_3Cl / $AlCl_3$ → CH_3 (toluene) → Br_2 / $FeBr_3$ →

o-Bromotoluene
(plus para isomer)

(b)

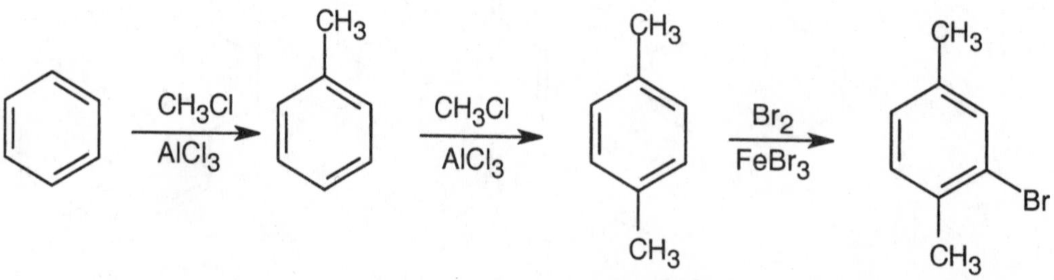

CH_3Cl / $AlCl_3$ → CH_3 → CH_3Cl / $AlCl_3$ → CH_3, CH_3 → Br_2 / $FeBr_3$ → CH_3, Br, CH_3

(plus ortho isomer)

2-Bromo-1,4-dimethylbenzene

5.22

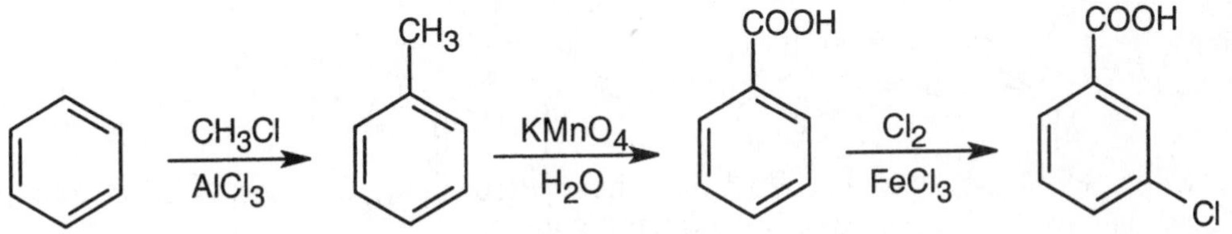

Benzene $\xrightarrow[AlCl_3]{CH_3Cl}$ toluene (CH_3) $\xrightarrow[H_2O]{KMnO_4}$ benzoic acid (COOH) $\xrightarrow[FeCl_3]{Cl_2}$ COOH, Cl

m-Chlorobenzoic acid

Visualizing Chemistry

5.23

(a) HO, CH_3, $CHCH_3$

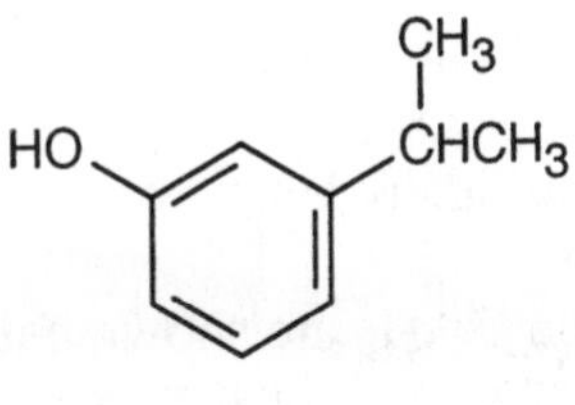

m-Isopropylphenol

(b) O, C, OH, NO_2

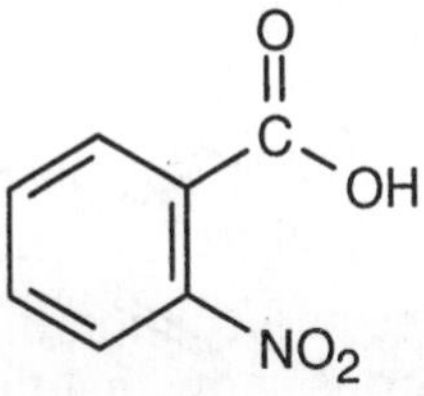

o-Nitrobenzoic acid

5.24 (a) The methoxyl group is an *o,p*-director.

(i) OCH_3 $\xrightarrow[FeBr_3]{Br_2}$ OCH_3, Br + OCH_3, Br

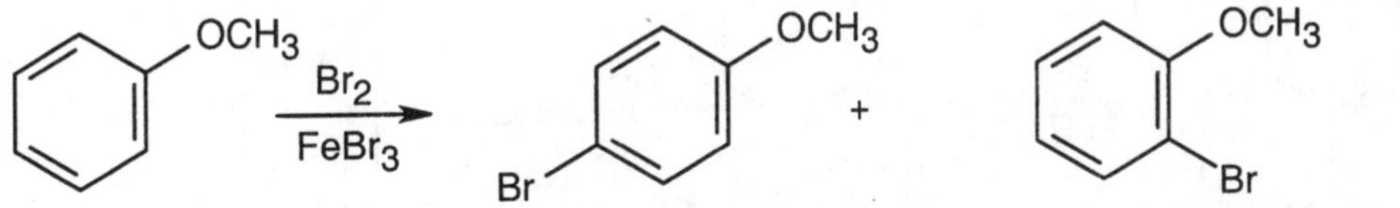

p-Bromomethoxybenzene *o*-Bromomethoxybenzene

(ii) OCH_3 $\xrightarrow[AlCl_3]{CH_3CCl\ (C{=}O)}$ H_3C, C, O, OCH_3 + OCH_3, C, CH_3, O

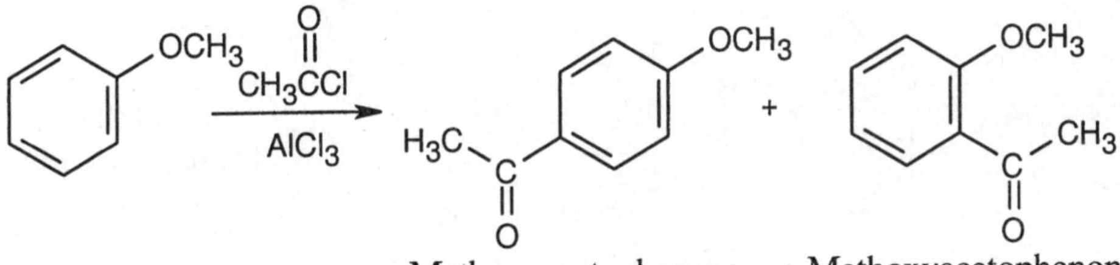

p-Methoxyacetophenone *o*-Methoxyacetophenone

(b) Both functional groups direct substituents to the same position.

(i)

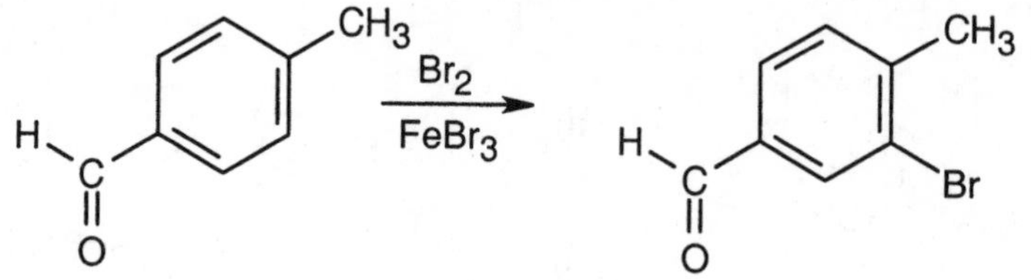

3-Bromo-4-methylbenzaldehyde

(ii)

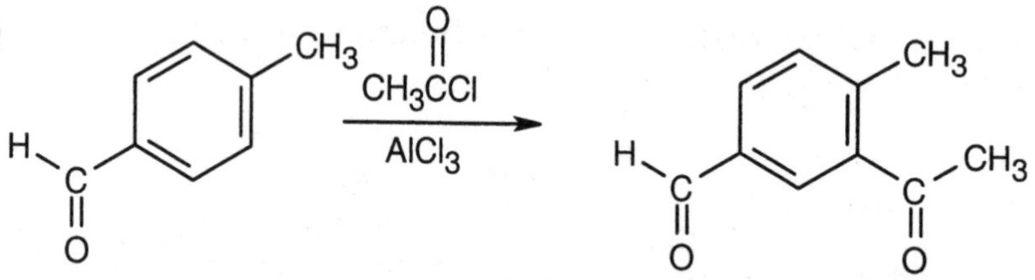

3-Acetyl-4-methylbenzaldehyde

5.25 Three resonance forms for the carbocation of the formula $C_{13}H_9$ are shown below, and more can be drawn. These forms show that the positive charge of the carbocation can be stabilized in the same way as an allylic or benzylic carbocation is stabilized – by overlap with the neighboring π electrons of the ring system.

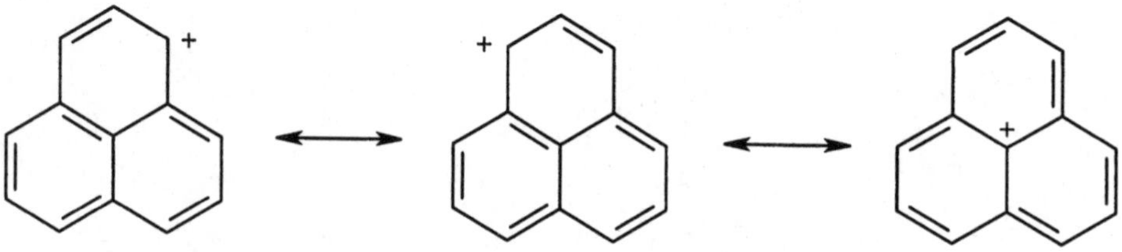

5.26

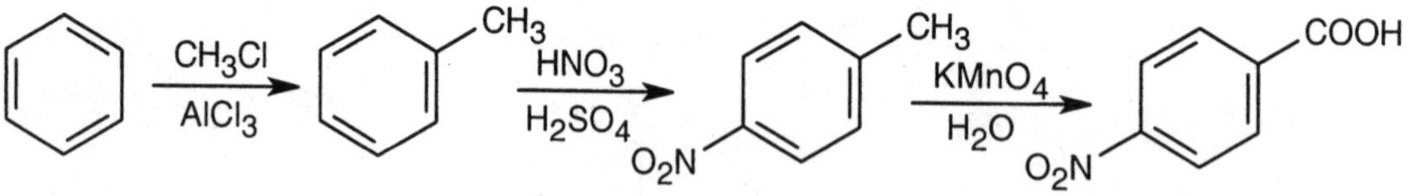

p-Nitrobenzoic acid

Additional Problems

Nomenclature

5.27

(a) (4-Methylpentyl)benzene *or* 4-Methyl-1-phenylpentane

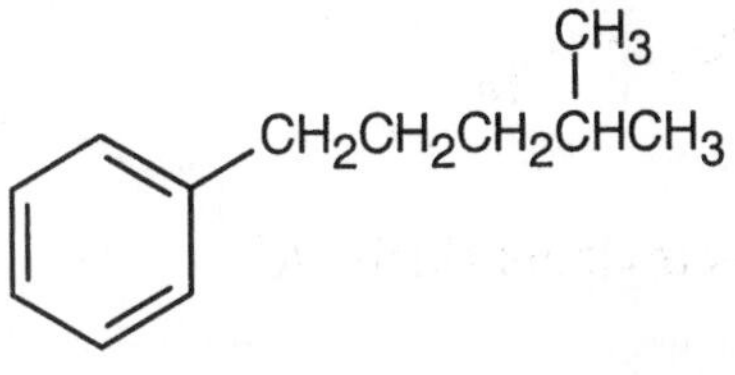

(b) *m*-Bromobenzoic acid

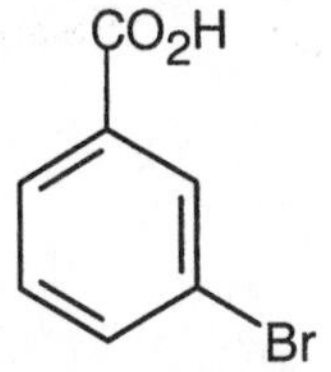

(c) 1-Bromo-3,5-dimethylbenzene

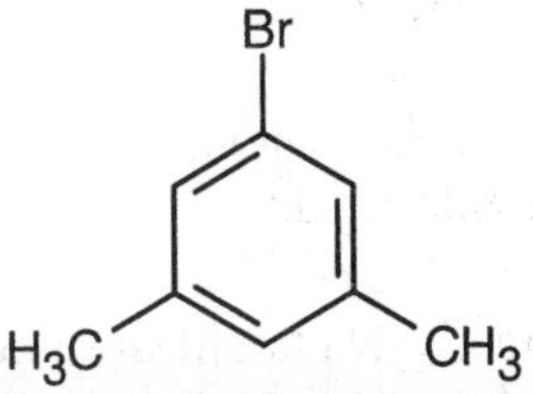

(d) *o*-Bromopropylbenzene

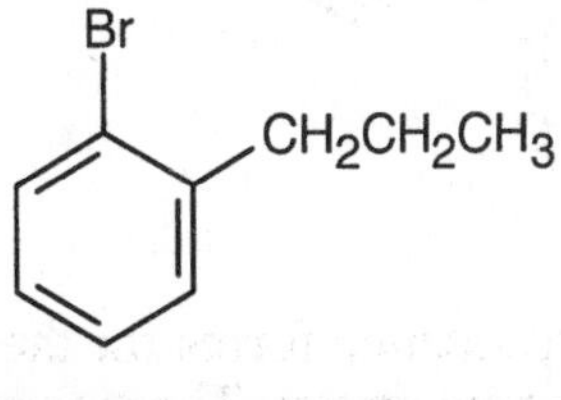

5.28

(a) *m*-Bromophenol

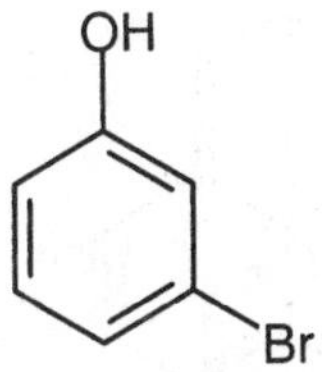

(b) 1,3,5-Benzenetriol

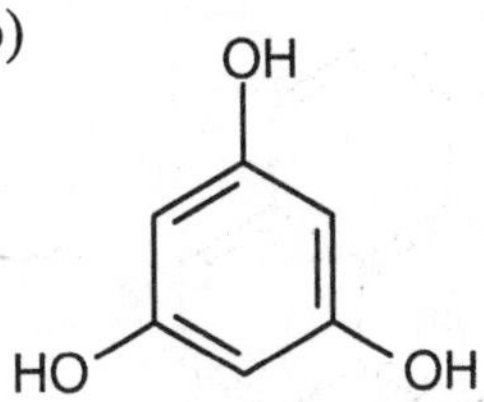

(c) *p*-Iodonitrobenzene

(d) 2,4,6-Trinitrotoluene

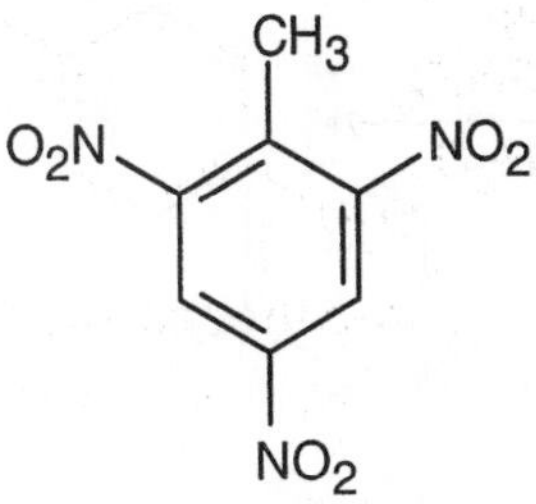

(e) *o*-Aminobenzoic acid

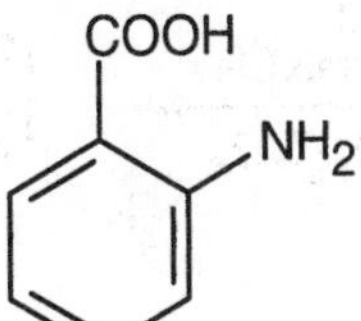

(f) 3-Methyl-2-phenylhexane

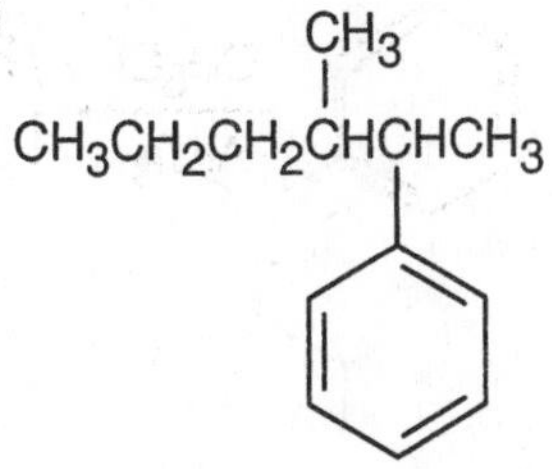

Isomers

5.29

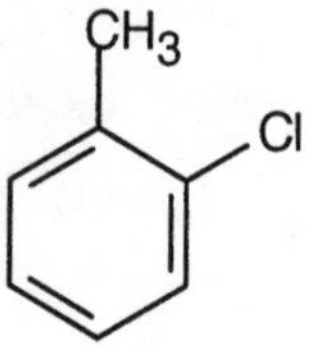

o-Chlorotoluene

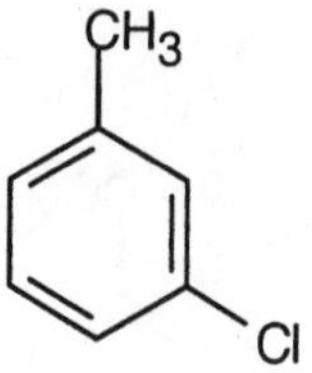

m-Chlorotoluene

p-Chlorotoluene

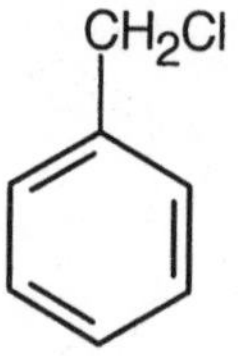

Benzyl chloride
or
(Chloromethyl)-benzene

5.30

(a)

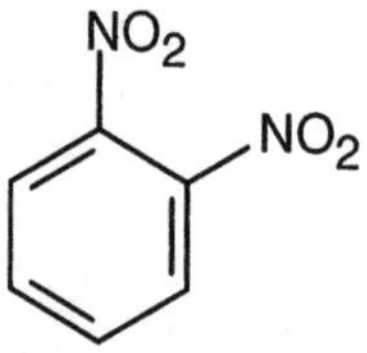

o-Dinitrobenzene

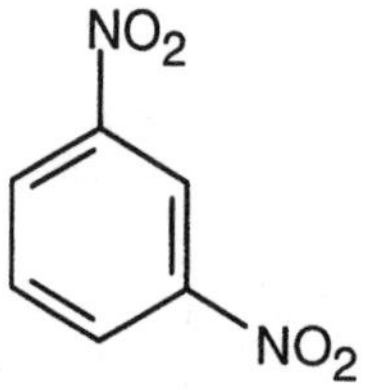

m-Dinitrobenzene

p-Dinitrobenzene

(b)

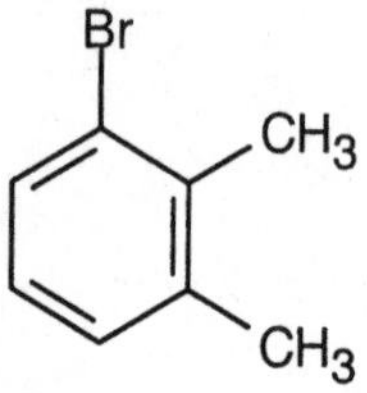

1-Bromo-2,3-dimethyl-benzene

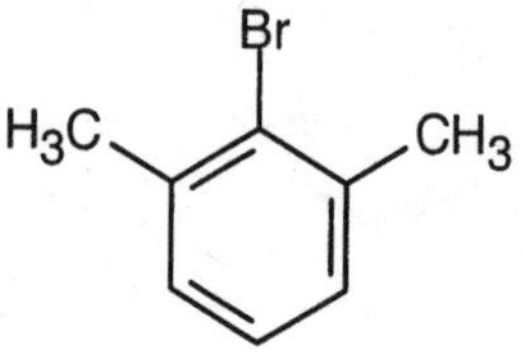

2-Bromo-1,3-dimethyl-benzene

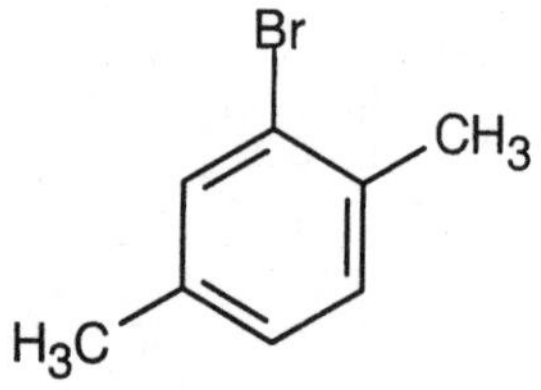

2-Bromo-1,4-dimethyl-benzene

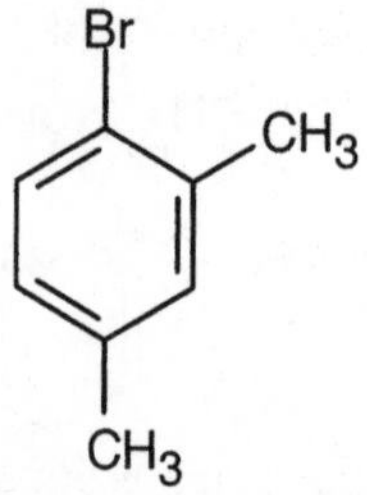

1-Bromo-2,4-dimethyl-benzene

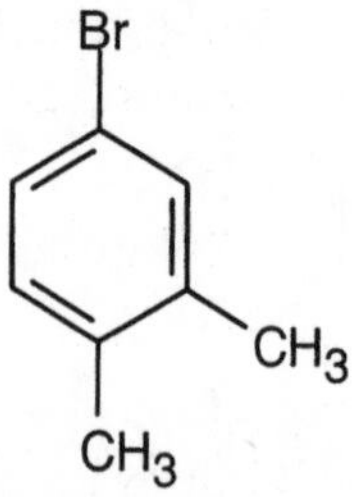

4-Bromo-1,2-dimethyl-benzene

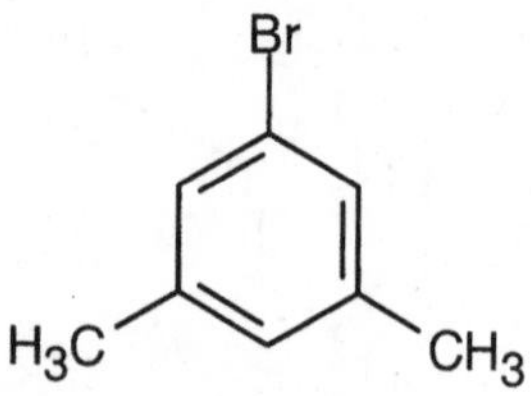

1-Bromo-3,5-dimethyl-benzene

5.31

(a)

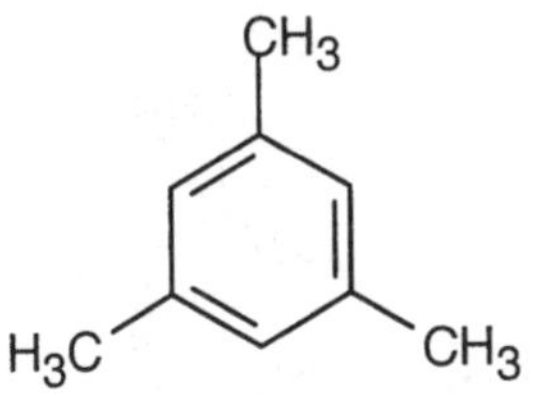

1,3,5-trimethylbenzene structure (CH_3, H_3C, CH_3)

(b)

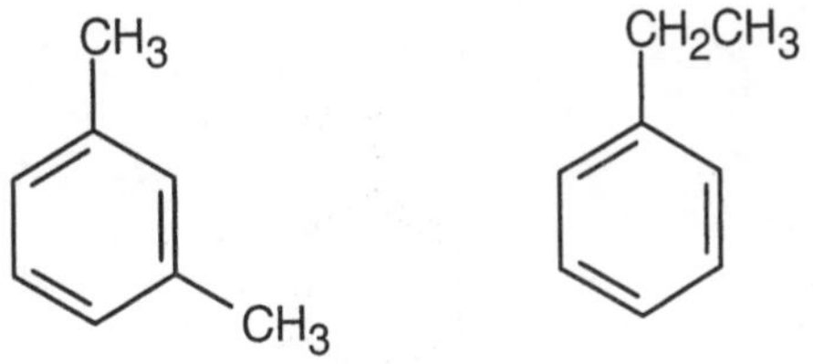

m-xylene structure (CH_3, CH_3); ethylbenzene structure (CH_2CH_3)

(c)

Structures: CH_2CH_3 / CH_2CH_3 (o-diethylbenzene); CH_2CH_3 / H_3C / CH_3; $CH(CH_3)_2$ / CH_3; CH_3 / CH_2CH_3 / CH_3; $CH_2CH_2CH_3$ / CH_3

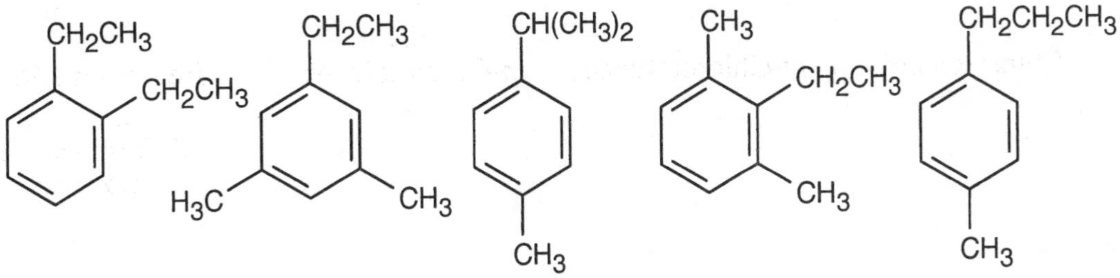

Reactions and Substituent Effects

5.32

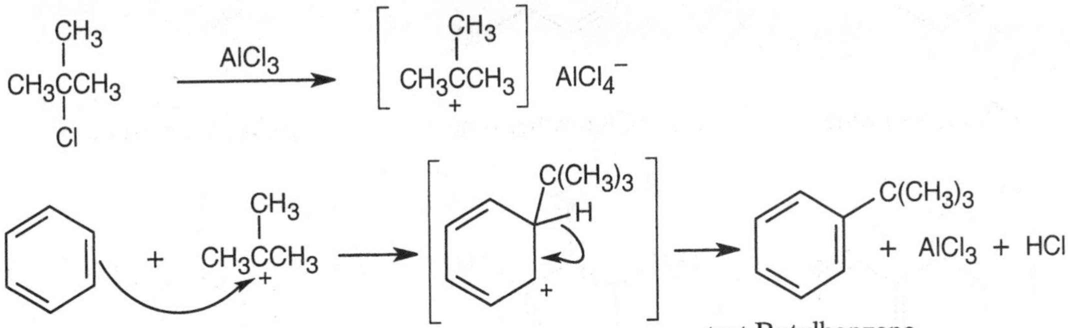

$$CH_3CCl(CH_3)CH_3 \xrightarrow{AlCl_3} [CH_3\overset{+}{C}(CH_3)CH_3]\ AlCl_4^-$$

$$C_6H_6 + CH_3\overset{+}{C}(CH_3)CH_3 \longrightarrow [\text{C}(CH_3)_3, \text{H carbocation intermediate}] \longrightarrow C_6H_5C(CH_3)_3 + AlCl_3 + HCl$$

tert-Butylbenzene

5.33

	Group	*Effect*	*Reason*
(a)	$-\ddot{N}(CH_3)_2$	*o,p*-activator	Groups that have lone-pair electrons adjacent to the aromatic ring are *o,p*-activators.
(b)	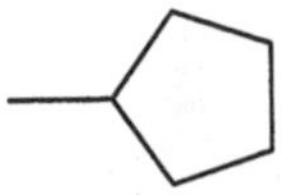 —cyclopentyl	*o,p*-activator	Alkyl groups are *o,p*-activators.
(c)	$-\ddot{O}CH_2CH_3$	*o,p*-activator	Groups that have lone-pair electrons adjacent to the aromatic ring are *o,p*-activators.
(d)	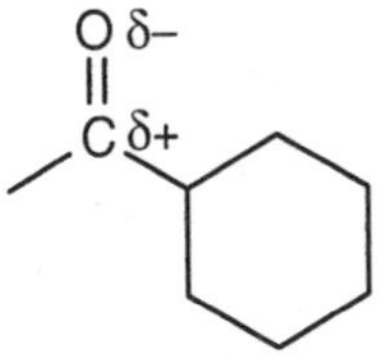 $-C(=O^{\delta-})^{\delta+}$-cyclohexyl	*m*-deactivator	Groups that have a positively polarized atom adjacent to the aromatic ring are *m*-deactivators.

5.34

(a)

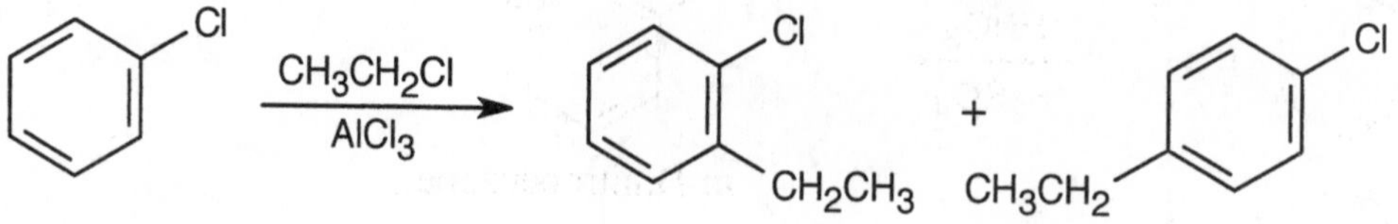

Chlorobenzene $\xrightarrow[AlCl_3]{CH_3CH_2Cl}$ o-ethylchlorobenzene (Cl, CH_2CH_3) + p-ethylchlorobenzene (CH_3CH_2, Cl)

(b)

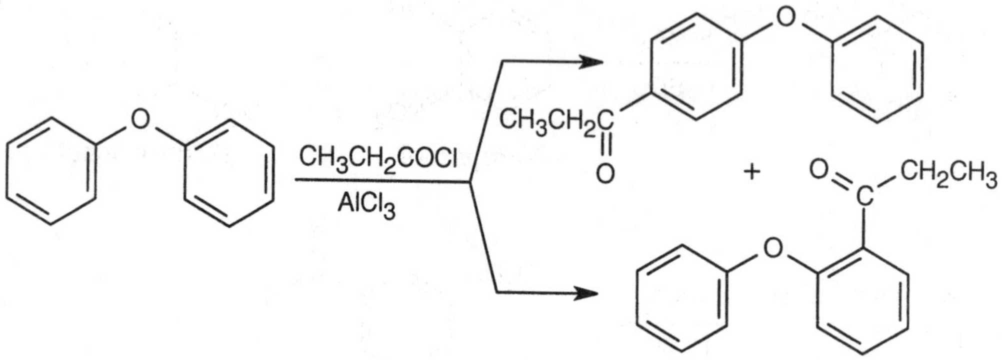

Diphenyl ether $\xrightarrow[AlCl_3]{CH_3CH_2COCl}$ p-propanoyl diphenyl ether ($CH_3CH_2C(=O)$–) + o-propanoyl diphenyl ether ($O{=}C$–CH_2CH_3)

(c)

Benzoic acid ($COOH$) $\xrightarrow[H_2SO_4]{HNO_3}$ O_2N–C_6H_4–$COOH$ (meta)

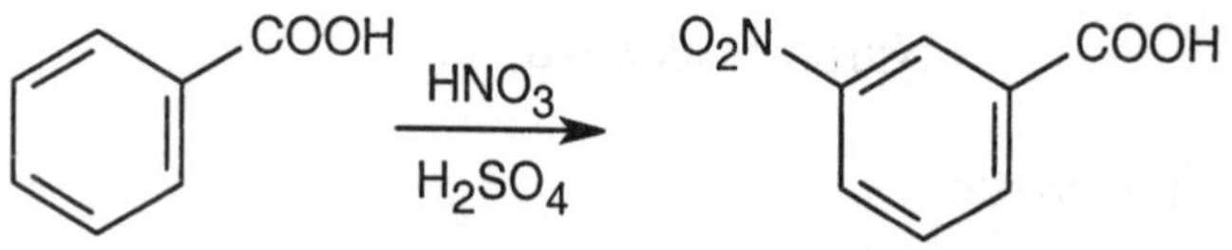

(d)

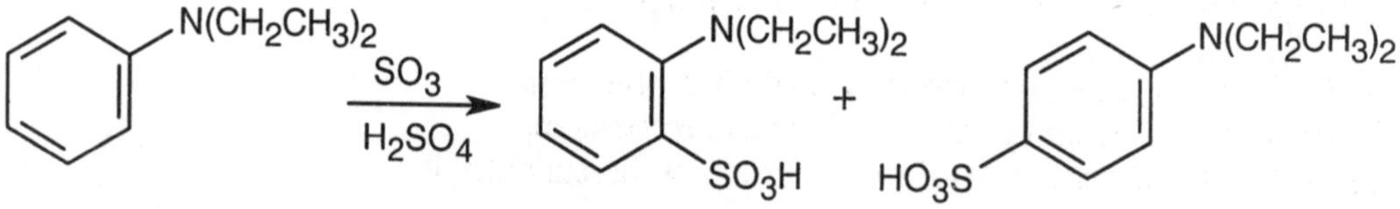

$C_6H_5N(CH_2CH_3)_2$ $\xrightarrow[H_2SO_4]{SO_3}$ o-$N(CH_2CH_3)_2$, SO_3H + p-HO_3S, $N(CH_2CH_3)_2$

5.35

(a)

Bromobenzene (Br) $\xrightarrow[H_2SO_4]{HNO_3}$ (Br, NO_2) + (O_2N, Br)

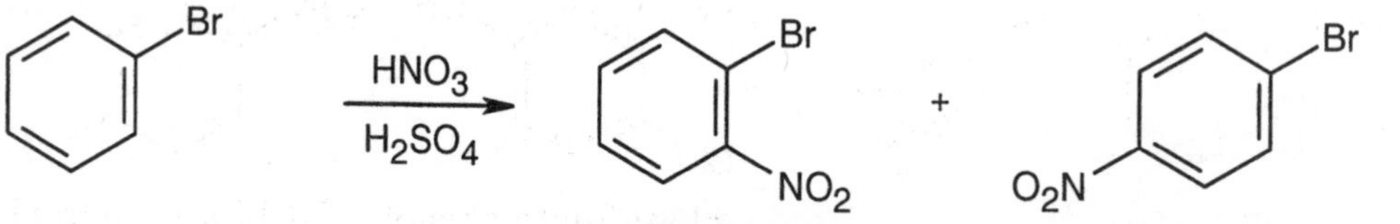

o-Bromonitrobenzene *p*-Bromonitrobenzene

(b)

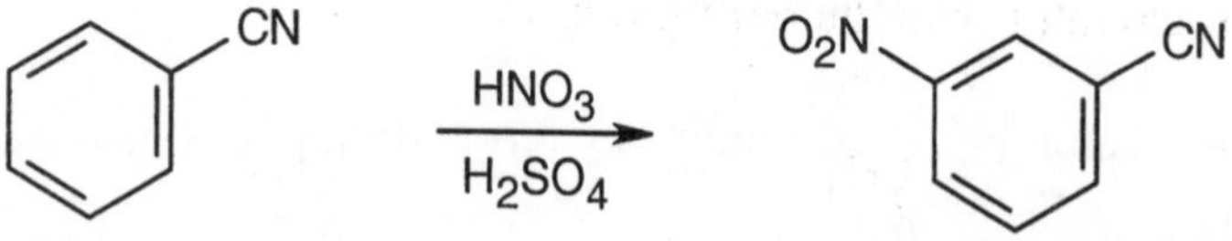

Benzonitrile (CN) $\xrightarrow[H_2SO_4]{HNO_3}$ (O_2N, CN)

m-Nitrobenzonitrile

(c)

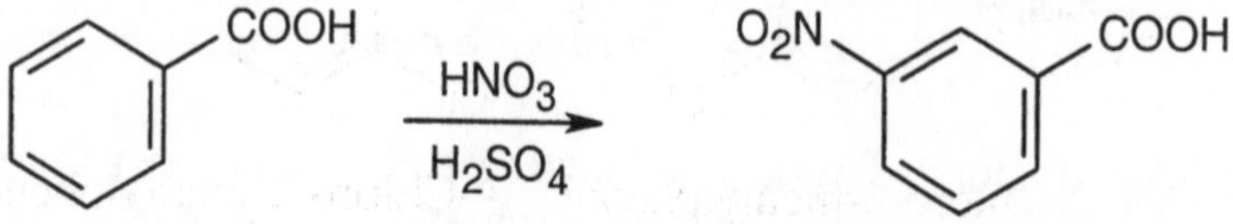

Benzoic acid ($COOH$) $\xrightarrow[H_2SO_4]{HNO_3}$ (O_2N, $COOH$)

m-Nitrobenzoic acid

(d)

$$C_6H_5NO_2 \xrightarrow[H_2SO_4]{HNO_3} C_6H_4(NO_2)_2$$

m-Dinitrobenzene

(e)

$$C_6H_5OH \xrightarrow[H_2SO_4]{HNO_3} o\text{-}O_2NC_6H_4OH + p\text{-}O_2NC_6H_4OH$$

o-Nitrophenol *p*-Nitrophenol

(f)

$$C_6H_5CHO \xrightarrow[H_2SO_4]{HNO_3} m\text{-}O_2NC_6H_4CHO$$

m-Nitrobenzaldehyde

5.36 Only phenol (e) reacts faster than benzene.

5.37 *Most reactive* ———> *Least reactive*

(a) Benzene > chlorobenzene > *o*-dichlorobenzene
(b) Phenol > nitrobenzene > *p*-bromonitrobenzene
(c) *o*-Dimethylbenzene > fluorobenzene > benzaldehyde

5.38

(a)

$$m\text{-}O_2NC_6H_4OH \xrightarrow[FeCl_3]{Cl_2} \text{4-Chloro-3-nitrophenol} + \text{2-Chloro-5-nitrophenol}$$

m-Nitrophenol 4-Chloro-3-nitrophenol 2-Chloro-5-nitrophenol

The –OH group is a stronger activator than the nitro group.

(b)

$$o\text{-}CH_3C_6H_4OH \xrightarrow[FeCl_3]{Cl_2} \text{2-Chloro-6-methylphenol} + \text{4-Chloro-2-methylphenol}$$

o-Methylphenol 2-Chloro-6-methylphenol 4-Chloro-2-methylphenol

The –OH group is a stronger director than the methyl group.

(c)

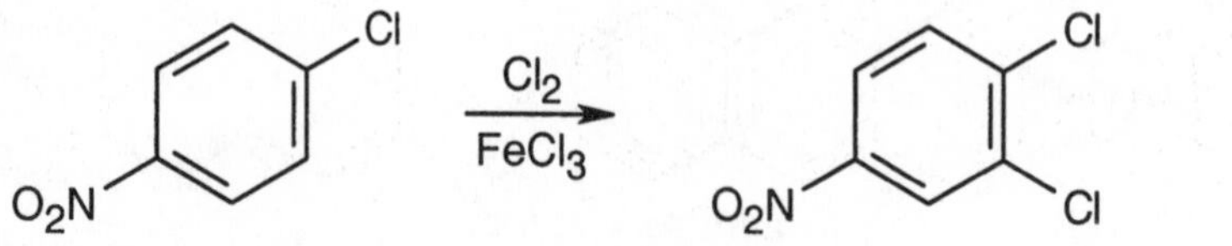

p-Chloronitrobenzene 1,2-Dichloro-4-nitrobenzene

Both groups direct chlorination to the same position.

5.39

(a)

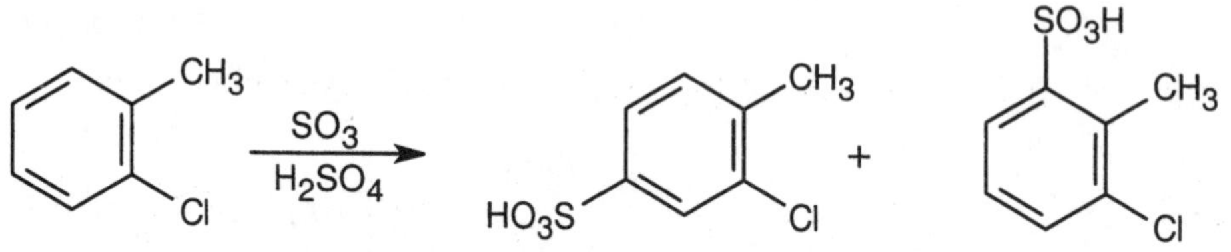

(b)

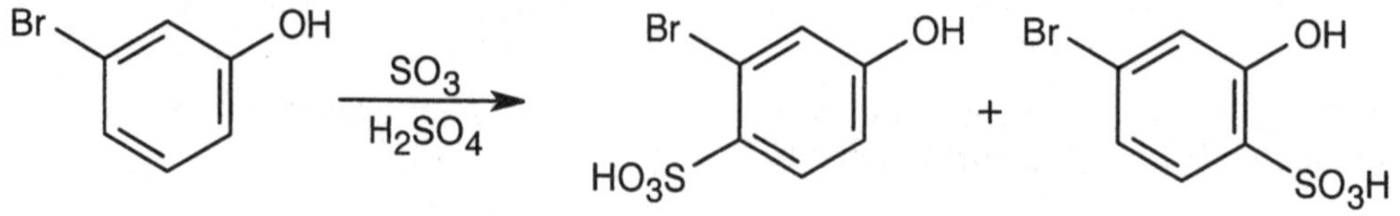

(c)

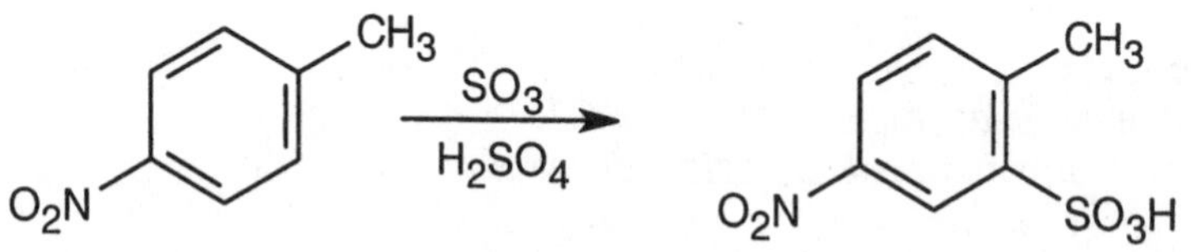

5.40 *Most reactive* ——————————> *Least reactive*

Anisole > Toluene > p-Bromotoluene > Bromobenzene

Nitrobenzene does not undergo Friedel-Crafts acylation.

5.41

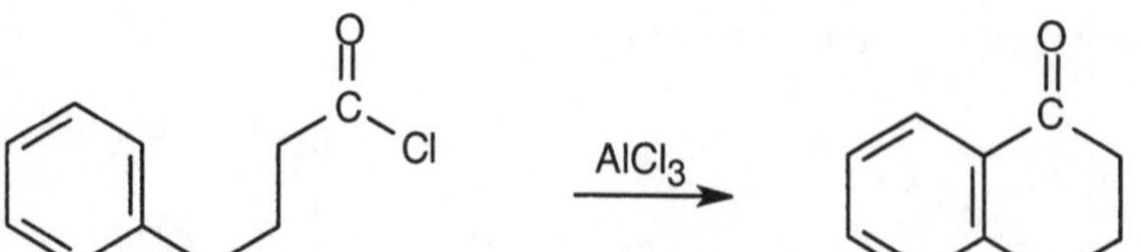

5.42

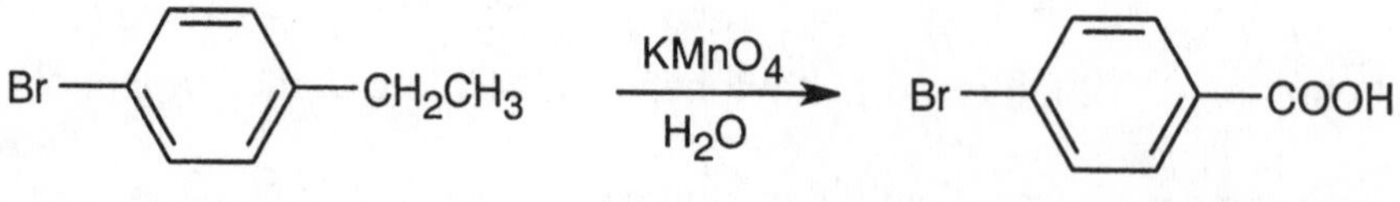

p-Bromoethylbenzene p-Bromobenzoic acid

5.43

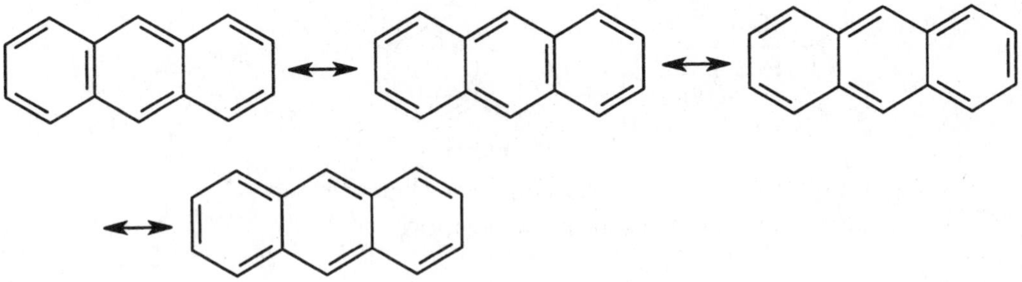

↔ ↔ ↔

5.44

$CH_3Cl + AlCl_3 \longrightarrow [CH_3^+ AlCl_4^-]$ Formation of carbocation

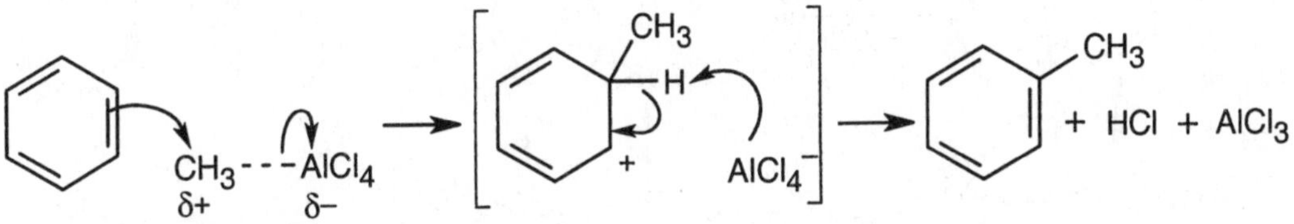

CH_3---$AlCl_4$ δ+ δ− → [CH_3, H, +, $AlCl_4^-$] → CH_3 + HCl + $AlCl_3$

Attack of ring π electrons on carbocation

Loss of proton

5.45

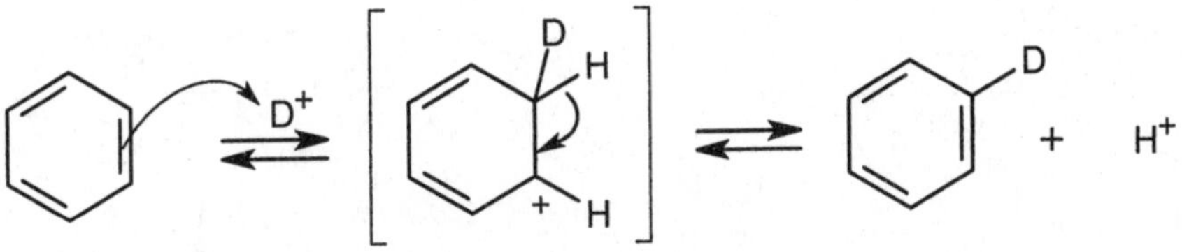

D^+ ⇌ [D, H, +, H] ⇌ D + H^+

Benzene can be protonated by strong acids. The resulting intermediate can lose either deuterium or hydrogen. If H^+ is lost, deuterated benzene is produced. Attack on D^+ can occur at all positions of the ring and leads to eventual replacement of all hydrogens by deuterium.

5.46

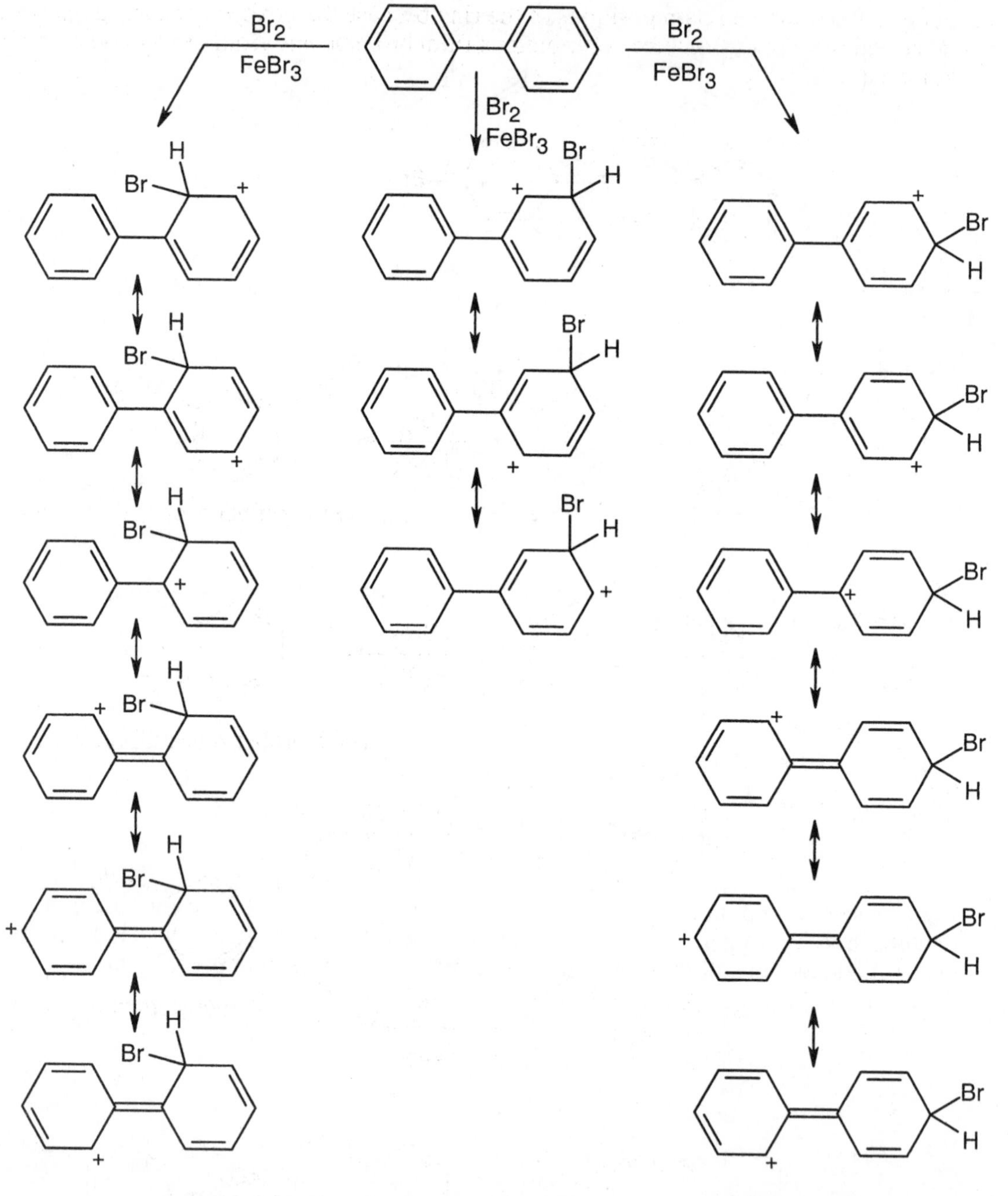

Br_2 / $FeBr_3$

Resonance structures show that bromination occurs in the ortho and para positions of the rings. The positively charged intermediate formed from ortho or para attack can be stabilized by resonance contributions from the second ring of biphenyl, but this stabilization is not possible for meta attack. Note that one of the intermediates from ortho/para attack is a benzylic carbocation.

5.47 Attack occurs on the unsubstituted ring because bromine is a deactivating substituent. Attack occurs at the ortho and para positions of the ring because the positively charged intermediate can be stabilized by resonance contributions from bromine and from the second ring (see Problem 5.46).

O_2N H $\overset{+}{Br}$

Synthesis

5.48

(a) SO_3 / H_2SO_4 → SO_3H ; Br_2 / $FeBr_3$ → Br, SO_3H

m-Bromobenzenesulfonic acid

(b) Cl_2 / $FeCl_3$ → Cl ; SO_3 / H_2SO_4 → SO_3H, Cl

o-Chlorobenzenesulfonic acid

(c) Cl_2 / $FeCl_3$ → Cl ; CH_3Cl / $AlCl_3$ → CH_3, Cl

CH_3Cl / $AlCl_3$ → CH_3 ; Cl_2 / $FeCl_3$ →

p-Chlorotoluene

5.49

(a) CH_3Cl / $AlCl_3$ → CH_3 ; HNO_3 / H_2SO_4 → CH_3, NO_2 ; $KMnO_4$ / H_2O → COOH, NO_2

o-Nitrobenzoic acid

(b) CH_3 ; $(CH_3)_3CCl$ / $AlCl_3$ → CH_3, $(CH_3)_3C$; $KMnO_4$ / H_2O → COOH, $(CH_3)_3C$

p-*tert*-Butylbenzoic acid

5.50 Resonance forms for the intermediate from attack at C1:

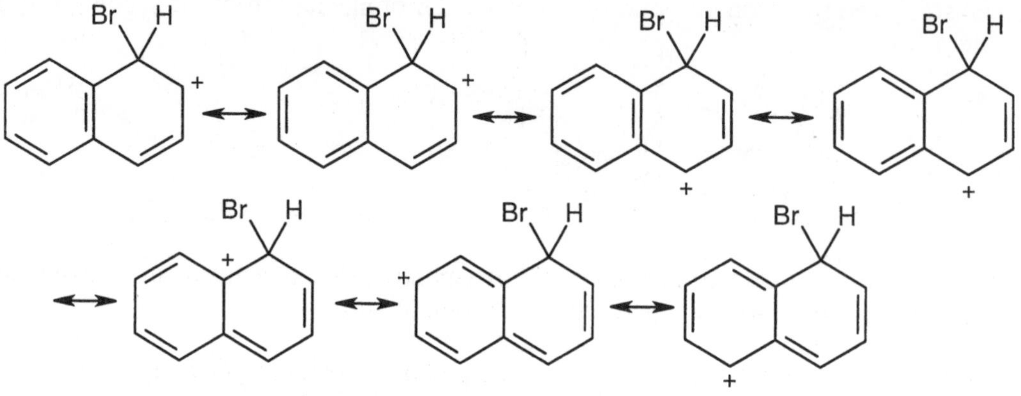

Resonance forms for the intermediate from attack at C2:

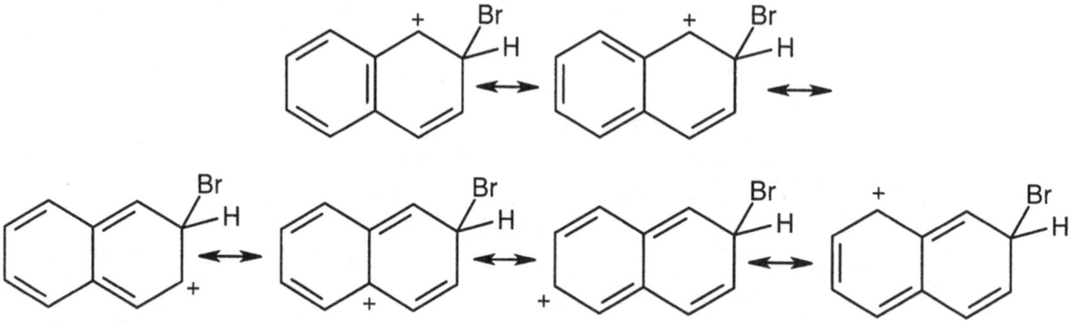

There are seven resonance forms for attack at C1 and six for attack at C2. For C1 attack, the second ring is fully aromatic in four of the resonance forms. In the other three forms, the positive charge has been delocalized into the second ring, destroying the ring's aromaticity. For C2 attack, the second ring is fully aromatic in only the first two forms. Since stabilization is lost when aromaticity is disturbed, the intermediate from C2 attack is less stable than the intermediate from C1 attack, and C1 attack is favored.

Integrated Problems

5.51

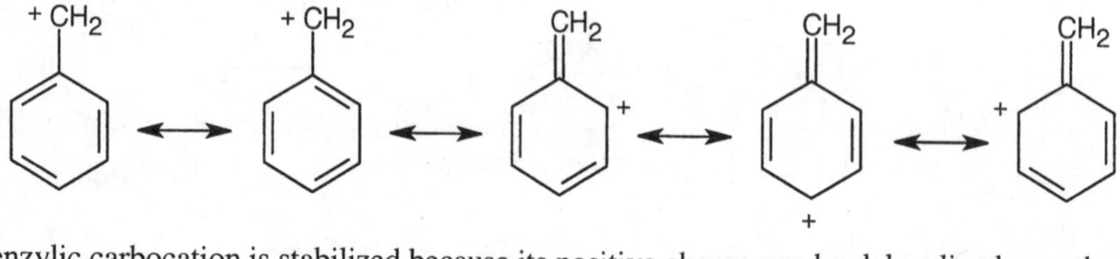

A benzylic carbocation is stabilized because its positive charge can be delocalized over the π system of the aromatic ring.

5.52

$C_6H_5CH{=}CHCH_3 \xrightarrow{H-Br} [C_6H_5\overset{+}{C}HCH_2CH_3 \;\; :\ddot{Br}:^-] \longrightarrow C_6H_5CH(Br)CH_2CH_3$

Protonation of the double bond at carbon 2 of 1-phenylpropene leads to an intermediate that can be stabilized by resonance involving the aromatic ring.

5.53 (a) Chlorination of toluene occurs at the ortho and para positions. To synthesize the given product, first oxidize toluene to benzoic acid and then chlorinate.

(b) A *tert*-butyl group can't be oxidized by $KMnO_4$ to a –COOH group because it has no benzylic hydrogens. To obtain the desired compound, alkylate chlorobenzene with CH_3Cl and $AlCl_3$ and then oxidize.

5.54

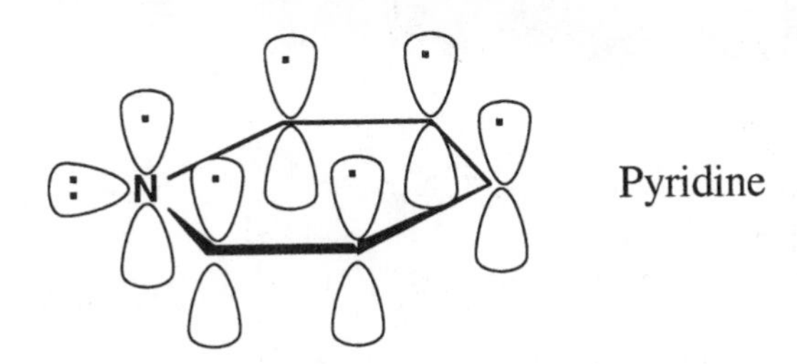

The electronic descriptions of pyridine and benzene are very similar. The pyridine ring is formed by the σ overlap of carbon and nitrogen sp^2 orbitals. In addition, six *p* orbitals, perpendicular to the plane of the ring, hold six electrons. These six *p* orbitals allow electrons to be delocalized over the π system of the pyridine ring. The lone pair of nitrogen electrons occupies an sp^2 orbital that lies in the plane of the ring.

5.55

$C_6H_5N(CH_3)_3^+\ Br^-$ Phenyltrimethylammonium bromide

The trimethylammonium group is deactivating because it is positively charged and because it has no lone-pair electrons to donate to the aromatic ring.

5.56

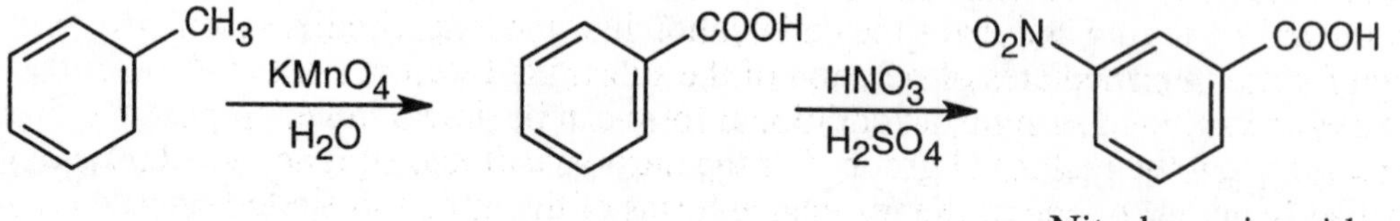

Toluene (CH_3) $\xrightarrow[H_2O]{KMnO_4}$ benzoic acid (COOH) $\xrightarrow[H_2SO_4]{HNO_3}$ O_2N, COOH

m-Nitrobenzoic acid

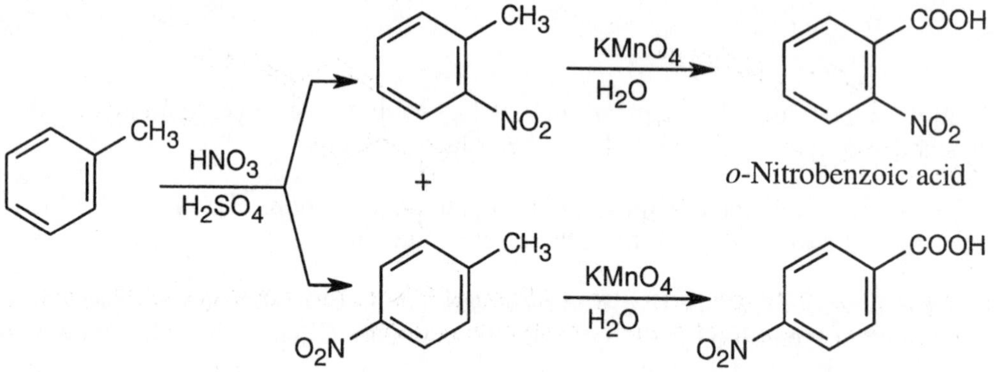

Toluene (CH_3) $\xrightarrow[H_2SO_4]{HNO_3}$ CH_3, NO_2 + CH_3, O_2N

CH_3, NO_2 $\xrightarrow[H_2O]{KMnO_4}$ COOH, NO_2

o-Nitrobenzoic acid

CH_3, O_2N $\xrightarrow[H_2O]{KMnO_4}$ COOH, O_2N

p-Nitrobenzoic acid

5.57

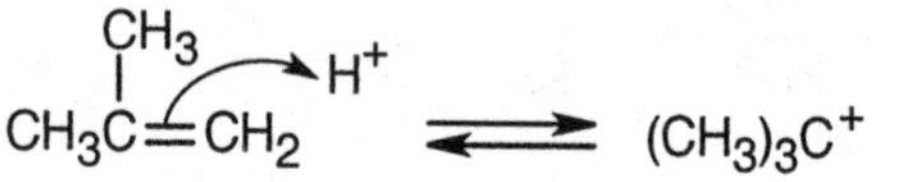

$$CH_3\overset{CH_3}{\overset{|}{C}}{=}CH_2 + H^+ \rightleftharpoons (CH_3)_3C^+$$

Acid protonates 2-methylpropene, forming a *tert*-butyl carbocation.

OH, CH_3 + $(CH_3)C+$ $\longrightarrow$ [OH, $C(CH_3)_3$, H, +, CH_3] $\longrightarrow$ OH, $C(CH_3)_3$, CH_3 + H^+

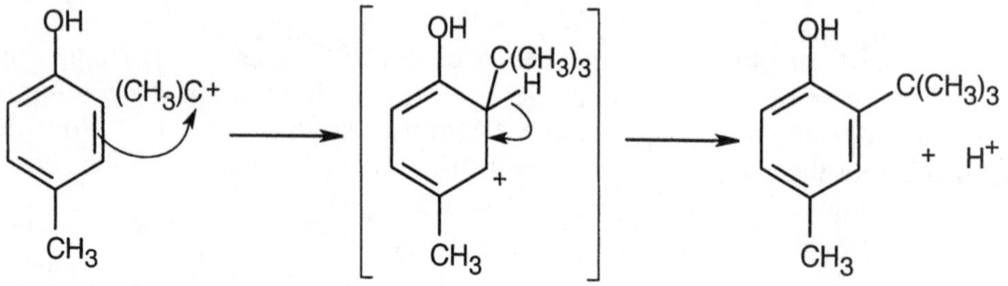

The *tert*-butyl carbocation acts as an electrophile and alkylates *p*-cresol. Alkylation occurs ortho to the –OH group for both steric and electronic reasons.

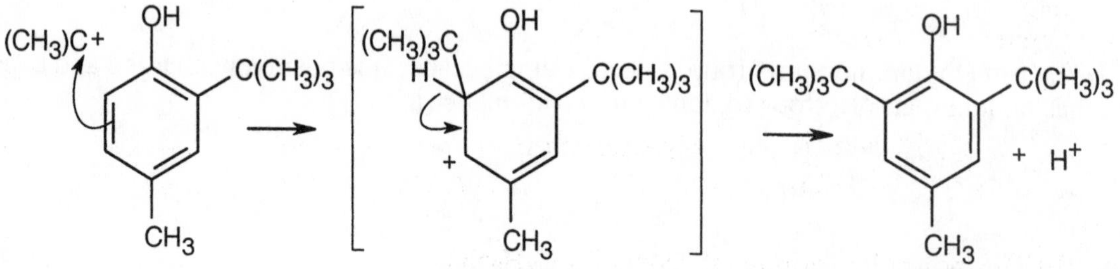

$(CH_3)C+$ + OH, $C(CH_3)_3$, CH_3 $\longrightarrow$ [$(CH_3)_3C$, H, OH, $C(CH_3)_3$, +, CH_3] $\longrightarrow$ $(CH_3)_3C$, OH, $C(CH_3)_3$, CH_3 + H^+

BHT

A second alkylation by a *tert*-butyl carbocation forms BHT.

5.58 Problem 5.52 shows the mechanism of the addition of HBr to 1-phenylpropene and shows how the aromatic ring stabilizes the carbocation intermediate. Similar resonance forms can be drawn for the intermediates of reaction of the substituted styrenes with HBr. For the methoxyl-substituted styrene, an additional form can be drawn in which positive charge can be stabilized by the methoxyl group. For the nitro-substituted styrene, no additional form is possible. In addition, one of the resonance forms of the nitro-substituted styrene intermediate is not important because it places two positive charges next to each other.

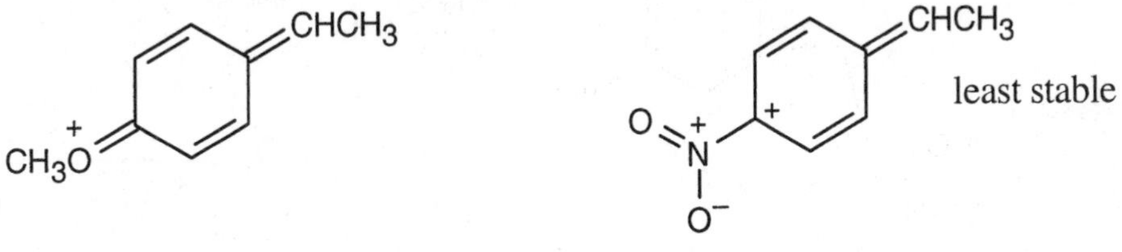

$CH_3\overset{+}{O}$ CHCH$_3$

least stable

Thus, the intermediate resulting from addition of HBr to the methoxyl-substituted styrene is more stable, and reaction of *p*-methoxystyrene is faster.

5.59

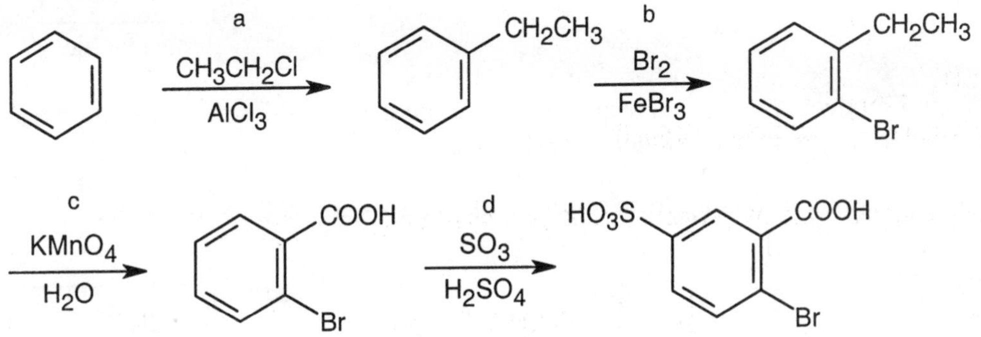

a: CH_3CH_2Cl, $AlCl_3$ → CH_2CH_3; b: Br_2, $FeBr_3$ → CH_2CH_3, Br

c: $KMnO_4$, H_2O → COOH, Br; d: SO_3, H_2SO_4 → HO_3S, COOH, Br

In the Medicine Cabinet

5.60

(a)

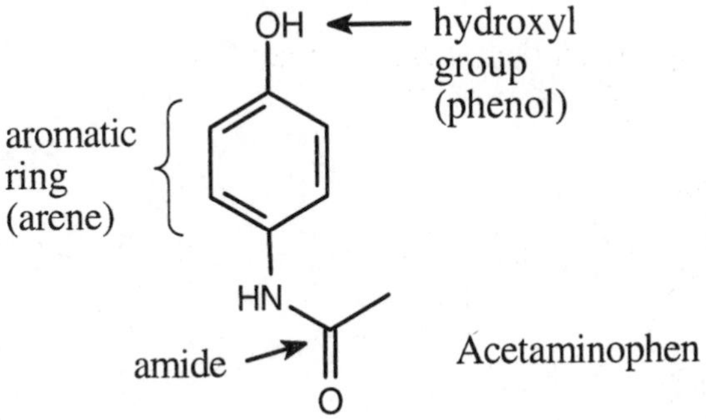

OH ← hydroxyl group (phenol)

aromatic ring (arene)

HN

amide →

O

Acetaminophen

(b) The molecular formula of Tylenol is $C_8H_9NO_2$.

(c)

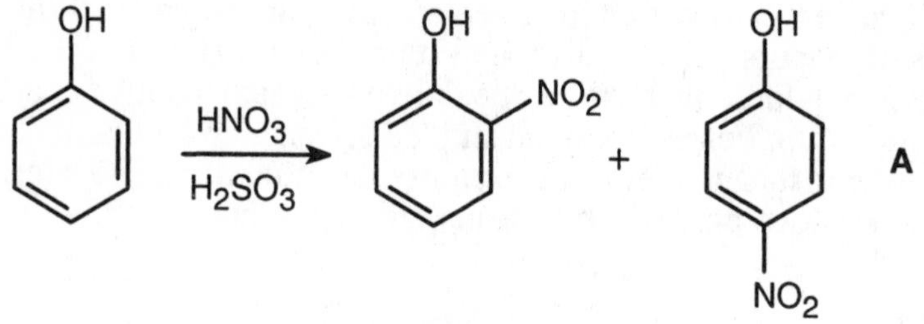

OH → (HNO_3, H_2SO_3) → OH, NO_2 + OH, NO_2 **A**

Para attack:

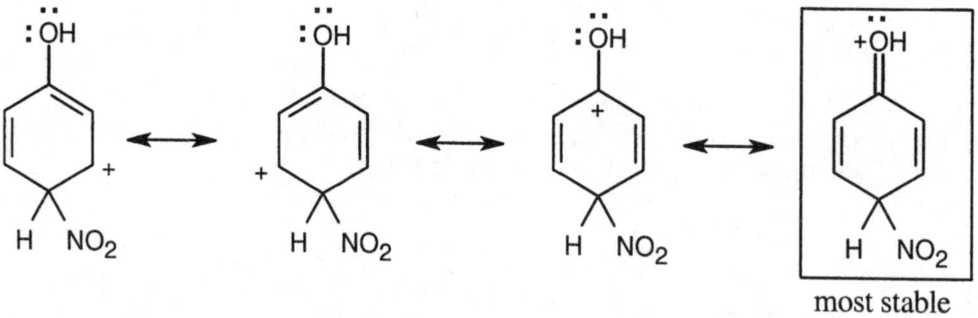

:ÖH   :ÖH   :ÖH   +ÖH

H NO_2 H NO_2 H NO_2 H NO_2

most stable

The desired product **A** is formed because the intermediate resulting from nitration can be stabilized by resonance involving the –OH group.

(d) The intermediate resulting from ortho attack is also stabilized by resonance.

Ortho attack:

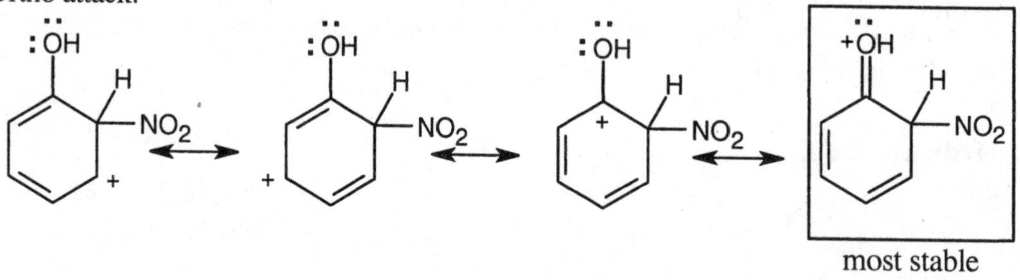

:ÖH H NO_2 :ÖH H NO_2 :ÖH H NO_2 +ÖH H NO_2

most stable

(e) The reagent that delivers the acetyl group is an anhydride.

5.61

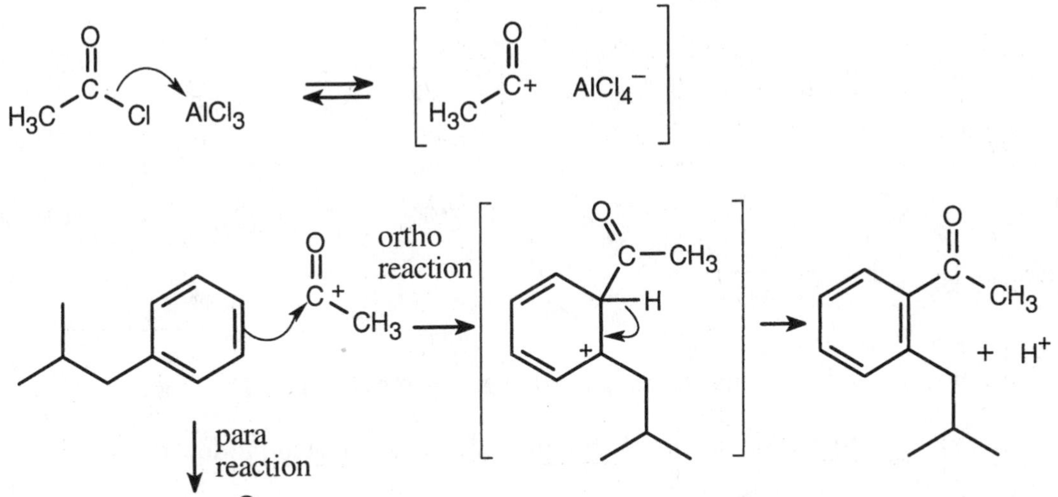

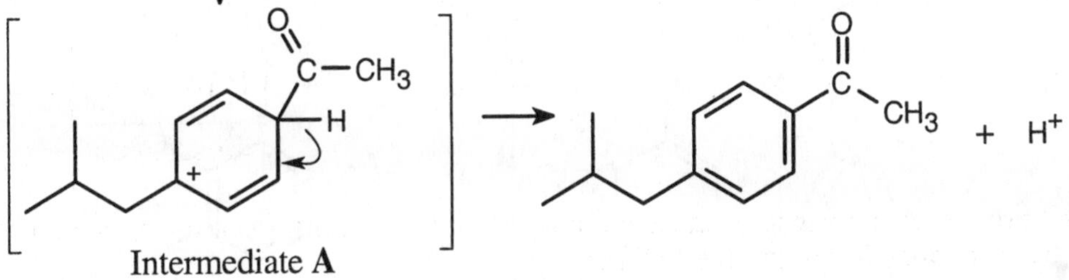

$H_3C-C(=O)-Cl$ + $AlCl_3$ ⇌ [$H_3C-C^+(=O)$ $AlCl_4^-$]

ortho reaction

para reaction

$C-CH_3$

H

CH_3 + H^+

Intermediate **A**

Friedel-Crafts acylation takes place at ring positions ortho and para to the isobutyl group because the positive charge of the intermediate can be stabilized by the inductive effect of the alkyl group. The two resonance forms that show alkyl group stabilization are pictured above. Other resonance forms can be drawn, but these are the only two that place the positive charge in the most favorable position.

5.62

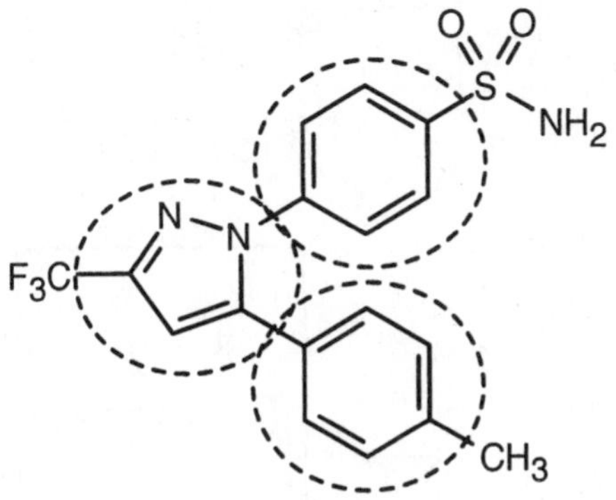

F_3C, N, N, S, NH_2, CH_3

Celecoxib

O, S, CH_3

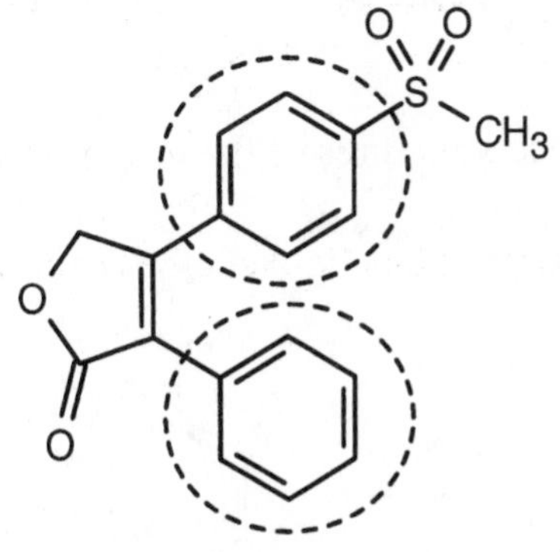

Rofecoxib

The circled five-membered ring of celecoxib is aromatic for the same reason as pyrrole is aromatic: It has six electrons in a cyclic, conjugated π system. One of the nitrogens contributes two electrons to this system.

In the Field with Agrochemicals

5.63

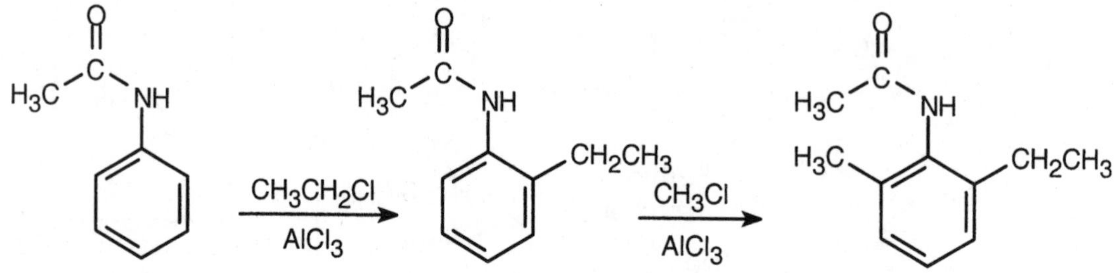

$$\text{Acetanilide} \xrightarrow[AlCl_3]{CH_3CH_2Cl} \text{2-ethylacetanilide} \xrightarrow[AlCl_3]{CH_3Cl} \text{2-ethyl-6-methylacetanilide}$$

The amine protecting group can be cleaved to give 2-ethyl-6-methylaniline.

5.64 Monochlorination gives a mixture of ortho-and para-chlorinated products.

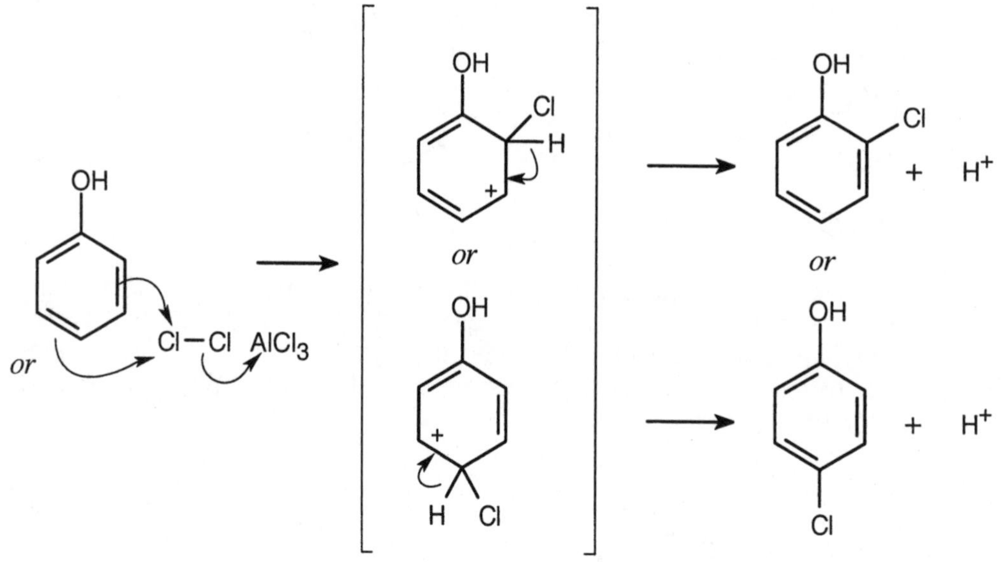

Phenol + Cl–Cl AlCl$_3$ → [ortho-chlorinated cation intermediate *or* para-chlorinated cation intermediate] → o-chlorophenol + H$^+$ *or* p-chlorophenol + H$^+$

Resonance structures of the intermediates are shown below.

Ortho attack:

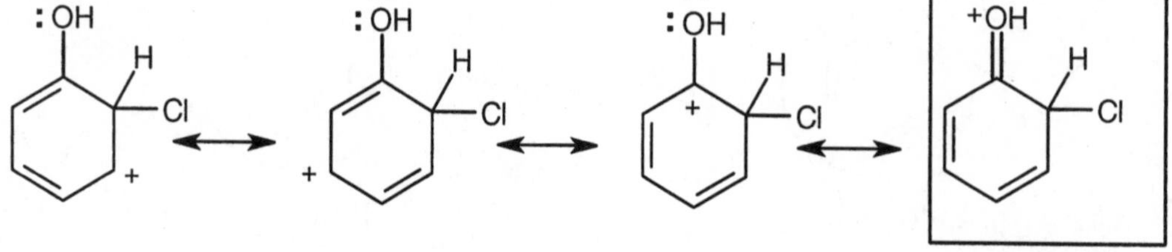

most stable

Para attack:

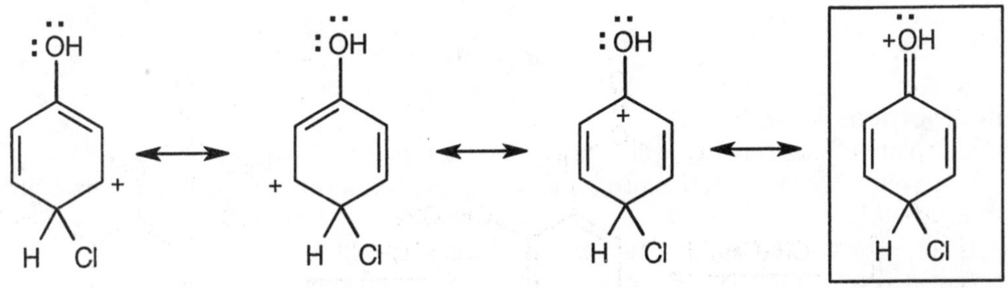

most stable

The second chlorine is directed to the correct location because the orienting effect of –OH outweighs the effect of –Cl.

Chapter 6 – Stereochemistry

Chapter Outline

I. Chirality (Sections 6.1–6.6).
   A. Enantiomers and tetrahedral carbon (Section 6.1).
      When four different groups are bonded to a carbon atom, two different arrangements are possible.
         a. These arrangements are mirror images.
         b. The two mirror-image molecules are enantiomers.
   B. Finding handedness in molecules (Section 6.2).
      1. Molecules that are not superimposable on their mirror-images are chiral.
         a. A molecule is not chiral if it contains a plane of symmetry.
         b. A molecule with no plane of symmetry is chiral.
      2. A carbon bonded to four different groups is a stereocenter (chirality center).
      3. Any $-CH_2-$ or $-CH_3$ carbon is achiral.
   C. Optical activity (Section 6.3).
      1. Solutions of certain substances rotate the plane of plane-polarized light. These substances are said to be optically active.
      2. The amount of rotation can be measured with a polarimeter.
      3. The direction of rotation can also be measured.
         a. A compound whose solution rotates plane-polarized light to the right is dextrorotatory.
         b. A compound whose solution rotates plane-polarized light to the left is levorotatory.
   D. Specific rotation (Section 6.4).
      1. The amount of rotation depends on concentration, path length and wavelength.
      2. Specific rotation is the observed rotation of a sample with concentration = 1 g/mL, sample path length of 1 dm, and light of wavelength = 589 nm.
      3. Specific rotation is a physical constant characteristic of a given optically active compound.
   E. Pasteur's discovery of enantiomerism (Section 6.5).
      1. Pasteur discovered two different types of crystals in a solution that he was evaporating.
      2. The crystals were mirror images.
      3. Solutions of each of the two types of crystals were optically active, and their specific rotations were equal in magnitude but opposite in sign.
      4. A 50:50 mixture of the crystals is called a racemate.
      5. Pasteur postulated that some molecules are handed and thus discovered the phenomenon of enantiomerism.
   F. Specification of configurations of stereoisomers (Section 6.6).
      1. Rules for assigning configurations at a stereocenter:
         a. Assign priorities to each group bonded to the carbon by using Cahn–Ingold–Prelog rules (Section 6.6).
         b. Orient the molecule so that the group of lowest priority is pointing to the rear.
         c. Draw a curved arrow from group 1 to group 2 to group 3.
         d. If the arrow is clockwise, the stereocenter has *R* configuration, and if the arrow is counterclockwise, the stereocenter has *S* configuration.
      2. The sign of optical rotation is not related to *R,S* designation.

II. Compounds with more than one stereocenter (Sections 6.7–6.11).
  A. Enantiomers and diastereomers (Section 6.7).
    1. A molecule with two chirality centers can have four possible stereoisomers.
      a. The stereoisomers group into two pairs of enantiomers.
      b. A stereoisomer from one pair is the diastereomer of a stereoisomer from the other pair.
    2. Diastereomers are stereoisomers that are not mirror images.
  B. Meso compounds (Section 6.8).
    1. A meso compound occurs when a compound with two chirality centers possesses a plane of symmetry.
    2. A meso compound is achiral despite having two chirality centers.
    3. The physical properties of meso compounds, diastereomers and racemic mixtures differ from each other and from the properties of enantiomers.
  C. Molecules with more than two chirality centers (Section 6.9).
    A molecule with n chirality centers can have a maximum of 2^n stereoisomers.
  D. The chiral environment (Section 6.10).
    1. A chiral environment makes enantiomers behave as if they were diastereomers.
    2. Some racemic mixtures can be resolved into their component enantiomers.
      a. If a racemic mixture of a carboxylic acid reacts with a chiral amine, the product ammonium salts are diastereomers.
      b. The diastereomeric salts differ in chemical and physical properties and can be separated.
      c. The original enantiomers can be recovered by acidification.

III. A review of isomerism (Section 6.11).
  A. Constitutional isomers differ in connections between atoms.
    1. Skeletal isomers have different carbon skeletons.
    2. Functional isomers contain different functional groups.
    3. Positional isomers have functional groups in different positions.
  B. Stereoisomers have the same connections between atoms, but different geometry.
    1. Enantiomers have a mirror-image relationship.
    2. Diastereomers are non-mirror-image stereoisomers.
      a. Configurational diastereomers.
      b. Cis-trans isomers differ in the arrangement of substituents in a double bond or ring.

IV. Chirality in Nature (Section 6.12).
  A. Different stereoisomers have different biological properties.
  B. Often, only one enantiomer of a drug is effective.

Solutions to Problems

6.1 Chiral: bean stalk, screw, shoe
Not chiral: screwdriver

6.2 Draw each compound, and identify all carbons that are *not* stereocenters. These carbons include:

CH_3—, —CH_2—, —CX_2—, —C=C—, —C≡C—, —C=O, all benzene ring carbons

Cross out these carbons. If all carbons are crossed out, the compound is achiral. If any carbons remain, they should be bonded to four different groups, and the compound is chiral. (Carbon stereocenters are starred.)

(a) $CH_3CH_2CH(Br)CH_2CH_3$
achiral

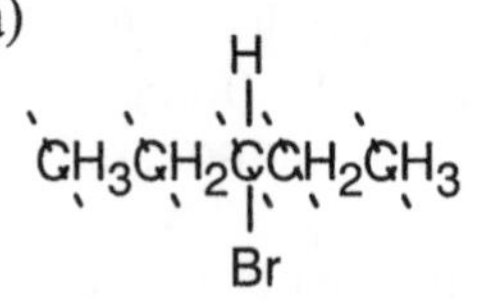

(b) $CH_3CH_2C^{*}H(Br)CH_2CH_2Br$
chiral
Carbon 3 is bonded to four different groups: —CH_2CH_3, —CH_2CH_2Br, —Br, —H

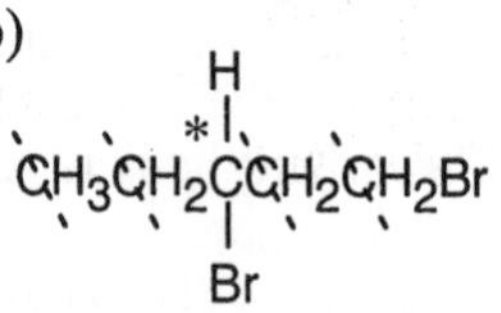

(c) $CH_3CH_2CH_2C^{*}H(CH_3)CH{=}CH_2$
chiral

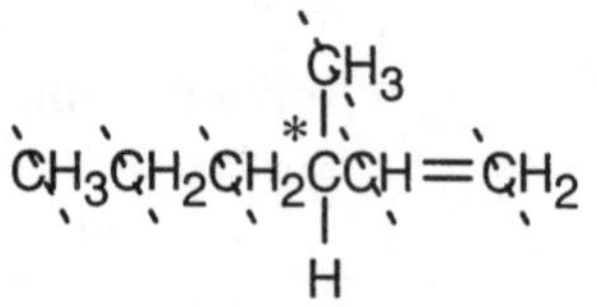

(d) H_3C, H, CH_3, H
achiral
All carbons are bonded to at least two identical groups.

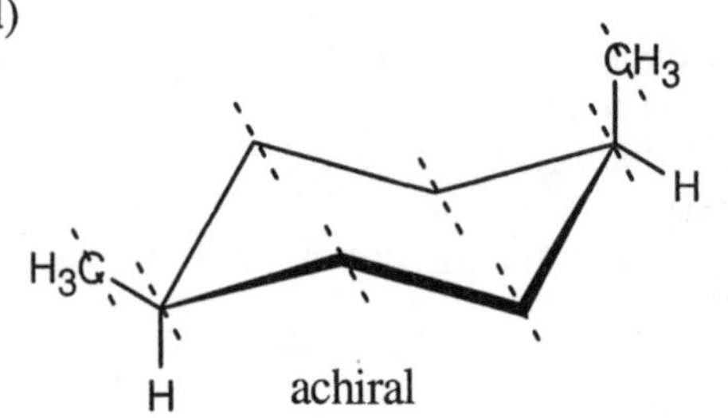

6.3 Refer to Problem 6.2 for a list of carbons that can't be stereocenters.

(a) CH_3
Toluene
achiral

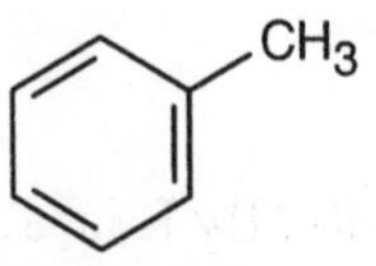

(b) * $CH_2CH_2CH_3$, N, H
Coniine
chiral

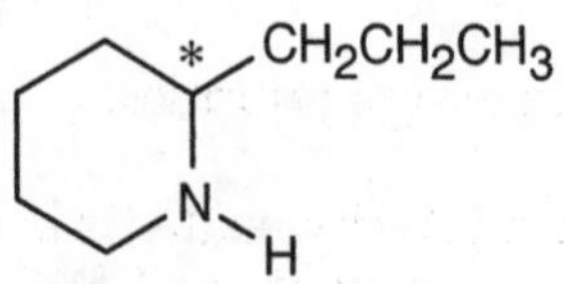

(c) CH_3CH_2, O, O, N, H, N, H, O
Phenobarbital
achiral

6.4

(a)

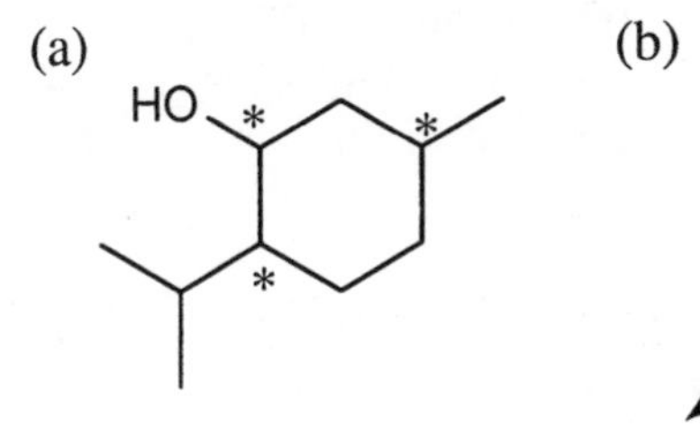

Menthol

(b)

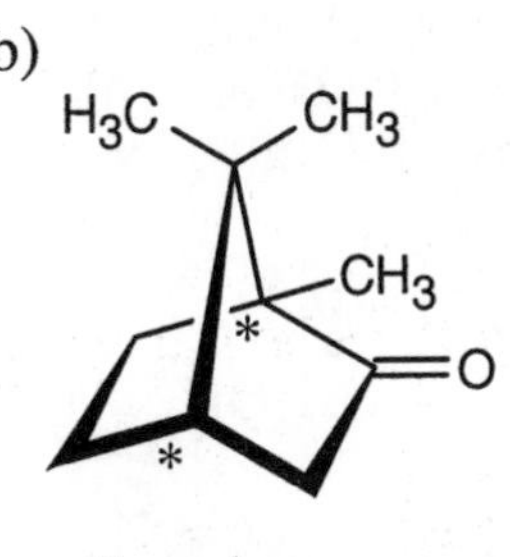

Camphor

(c)

Dextromethorphan

6.5

Alanine

6.6 Cocaine is levorotatory. (Levorotatory compounds have a minus sign in front of the degree of rotation.)

6.7

Use the formula $[\alpha]_D = \dfrac{\alpha}{l \times C}$, where

$[\alpha]_D$ = specific rotation

α = observed rotation

l = path length of cell (in dm)

C = concentration (in g/mL)

In this problem: $\alpha = 1.21°$

l = 5.00 cm = 0.500 dm

C = 1.50 g/10.0 mL = 0.150 g/mL

$$[\alpha]_D = \frac{+1.21°}{0.500 \text{ dm} \times 0.150 \text{ g/mL}} = +16.1°$$

6.8 Use the rules in Section 6.6 to assign priorities.

(a) By Rule 1, –H is of lowest priority, and –Br is of highest priority. By Rule 2, $–CH_2CH_2OH$ is of higher priority than $–CH_2CH_3$.

Highest ⟶ *Lowest*

$–Br, –CH_2CH_2OH, –CH_2CH_3, –H$

(b) By Rule 3, –COOH can be considered as having three O atoms singly bonded to the carbon. Since three oxygens are attached to a –COOH carbon and only one oxygen is attached to a $-CH_2OH$ carbon, –COOH is of higher priority than $-CH_2OH$. $-CO_2CH_3$ is of higher priority than –COOH by Rule 2, and –OH is of highest priority by Rule 1.

Highest ⟶ *Lowest*

—OH, $-CO_2CH_3$, —COOH, $-CH_2OH$

(c) —Br, —Cl, $-CH_2Br$, $-CH_2Cl$

6.9 The following scheme may be used to assign *R, S* configurations to stereocenters:

Step 1. For each stereocenter, rank substituents by the priority rules. Give the number 4 to the lowest priority substituent. For (a):

Substituent	*Priority*
–Br	1
–COOH	2
$-CH_3$	3
–H	4

Step 2. Imagine yourself looking at the molecule, with the group of lowest priority pointing to the back. Your viewpoint would be behind the plane of the paper, looking out. From that viewpoint, you would see –Br on the right, –COOH on the left, and the methyl group at the bottom. Draw the molecule as you see it, and note the direction of the arrow that travels from group 1 to group 2 to group 3. The direction of rotation is counterclockwise, and the configuration is *S*.

(a) 1 Br; 3 H_3C; H 4; COOH 2 = *S*

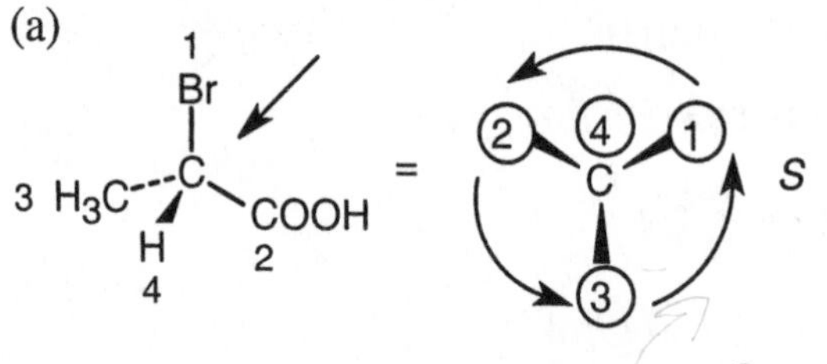

(b) 4 H; 1 HO; H_3C 3; COOH 2 = *S*

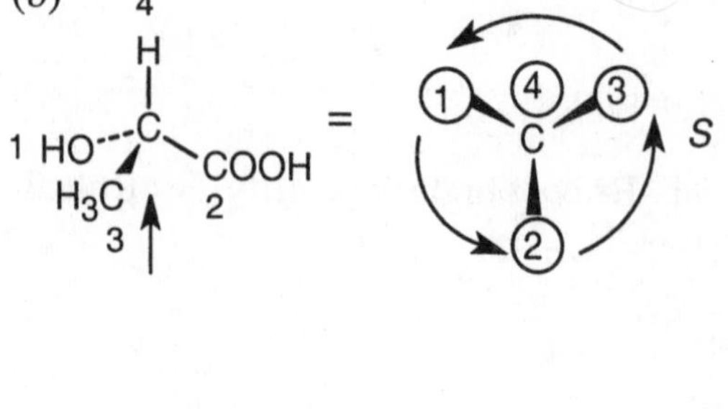

(c) 1 NH_2; 2 NC; H 4; CH_3 3 = *R*

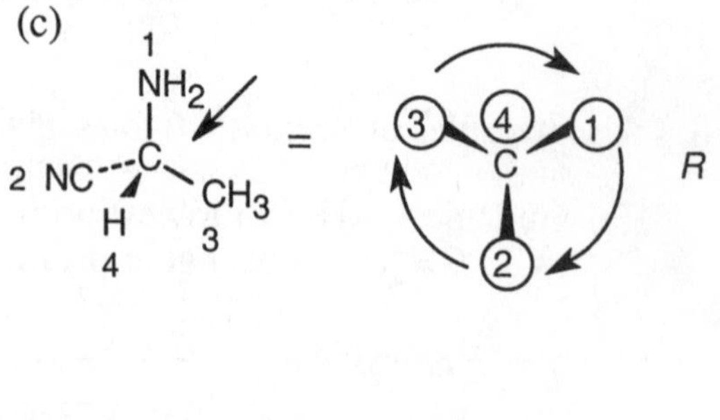

6.10 As in the previous problem, assign priorities to the substituents, giving the number 4 to the lowest priority substituent. Orient the lowest priority group toward the back, and arrange the three other groups, as we have done previously, as spokes on a steering wheel, with a counterclockwise rotation (because we need to draw an *S* enantiomer). Then, tilt the drawing until it is a tetrahedral representation.

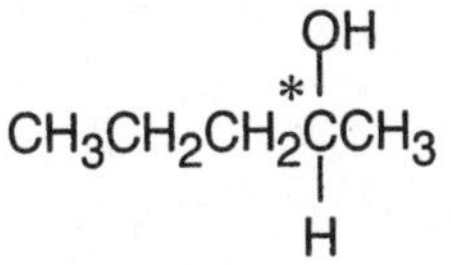

$CH_3CH_2CH_2\overset{*}{C}(OH)(H)CH_3$ (*S*)-Pentan-2-ol

Substituent	*Priority*
$-OH$	1
$-CH_2CH_2CH_3$	2
$-CH_3$	3
$-H$	4

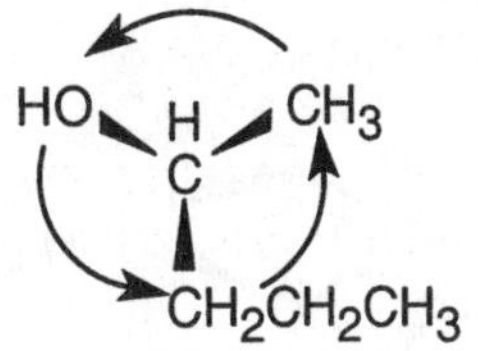

HO, H, CH_3, C, $CH_2CH_2CH_3$

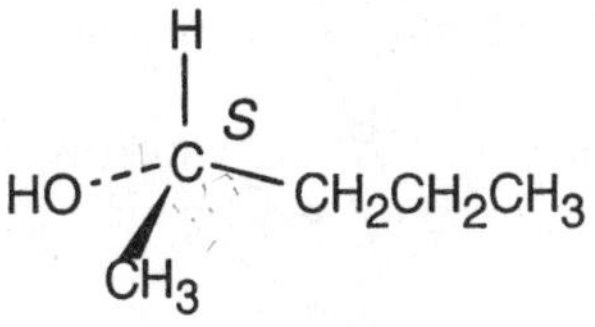

H, *S*, HO, C, $CH_2CH_2CH_3$, CH_3

6.11 Fortunately, the drawing of methionine is shown in the correct orientation.

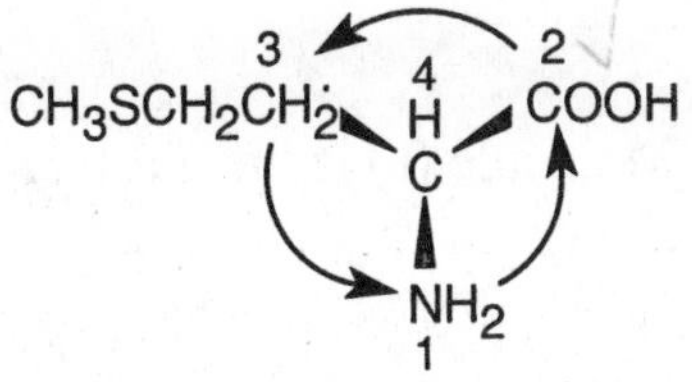

3 $CH_3SCH_2CH_2$, 4 H, 2 COOH, C, NH_2 1 (*S*)-Methionine

6.12 *R, S* assignments for more complicated molecules can be made by using the same method used in Problem 6.9. It is especially important to use molecular models when a compound has more than one stereocenter.

Step 1. Assign priorities to groups at the *top* stereocenter.

Substituent	*Priority*
–Br	1
$–CH(OH)CH_3$	2
$–CH_3$	3
–H	4

Step 2. Orient the model so that the lowest priority group of the first stereocenter points to the rear. If you aren't using a model, orient yourself so that you are 180° from the lowest priority group. You would be looking out of the page, upward to the left.

Step 3. Note the direction of rotation of the arrow that travels from to group 1 to group 2 to group 3. The rotation is clockwise, and the configuration is *R*.

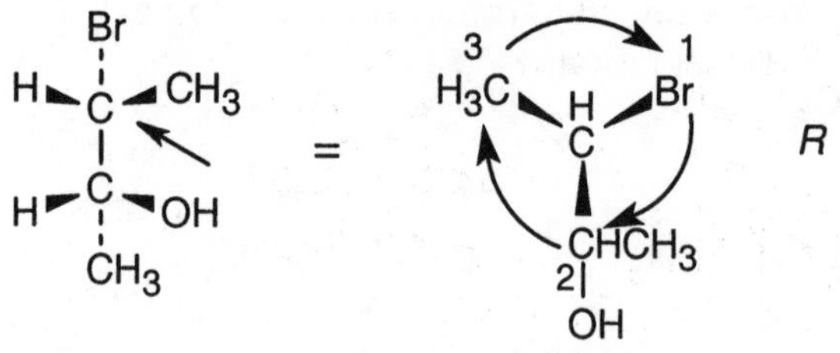

Step 4. Repeat steps 1–3 for the next stereocenter.

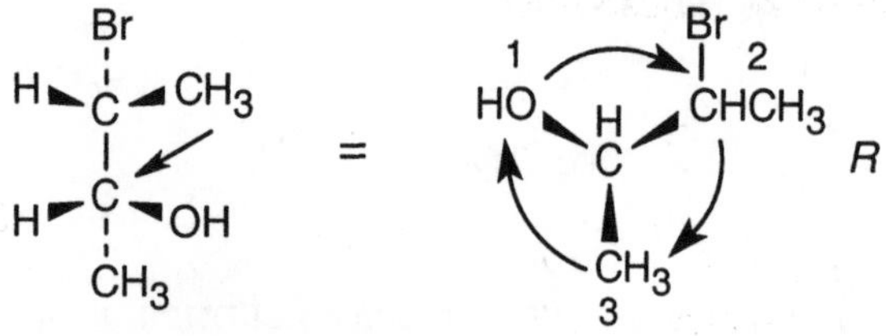

(a) *R,R* (b) *S,R* (c) *R,S*

6.13 Molecules (b) and (c) are enantiomers (mirror images). Molecule (a) is the diastereomer of (b) and (c).

6.14

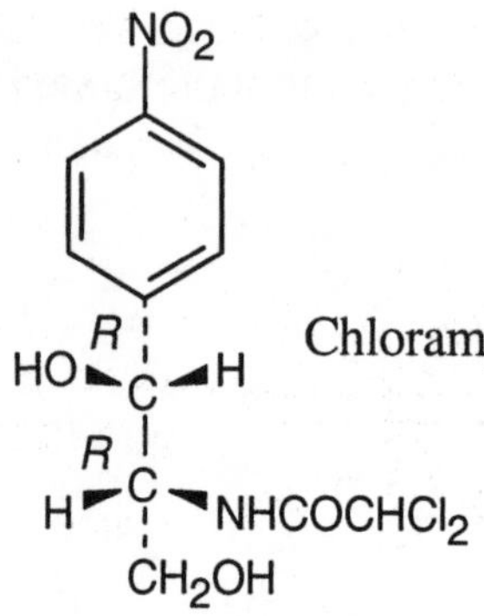

Chloramphenicol

6.15

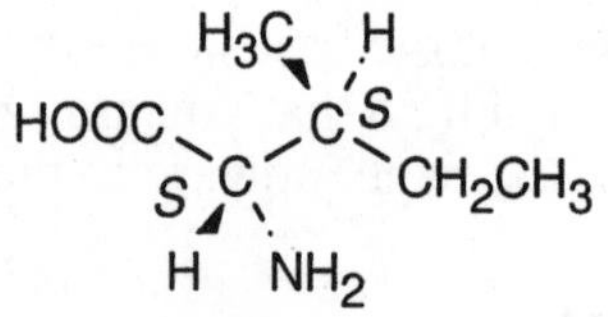

Isoleucine

6.16 To have a meso form, a molecule must have a plane of symmetry. 2,3-Dibromobutane can exist as a pair of enantiomers *or* as a meso compound, depending on the configurations at carbons 2 and 3.

(a)

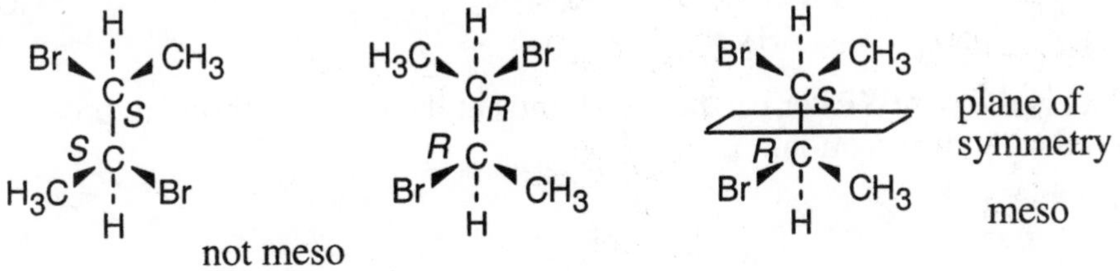

not meso

plane of symmetry

meso

(b) 2,3-Dibromopentane has no symmetry plane and thus can't exist in a meso form.
(c) 2,4-Dibromopentane can exist in a meso form.

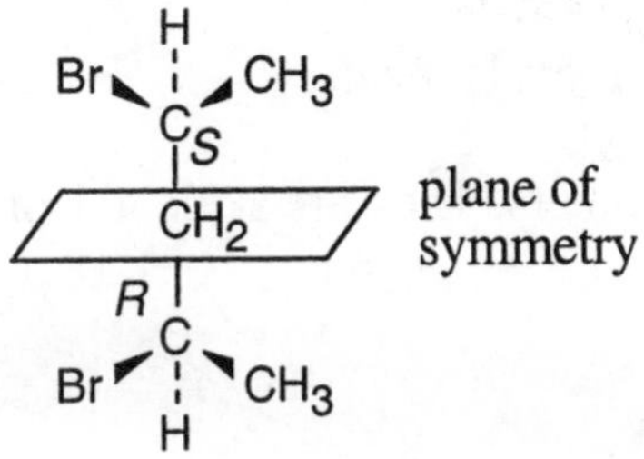

plane of symmetry

2,4-Dibromopentane can also exist as a pair of enantiomers [(2*R*,4*R* and 2*S*,4*S*)] that are not meso compounds.

6.17 To decide if a structure represents a meso compound, try to locate a plane of symmetry that divides the molecule into two halves that are mirror images. Molecular models may be helpful.

(a) plane of symmetry

meso

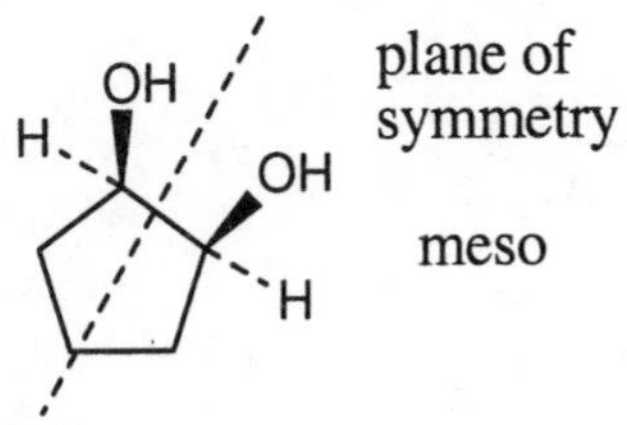

(c) plane of symmetry

meso

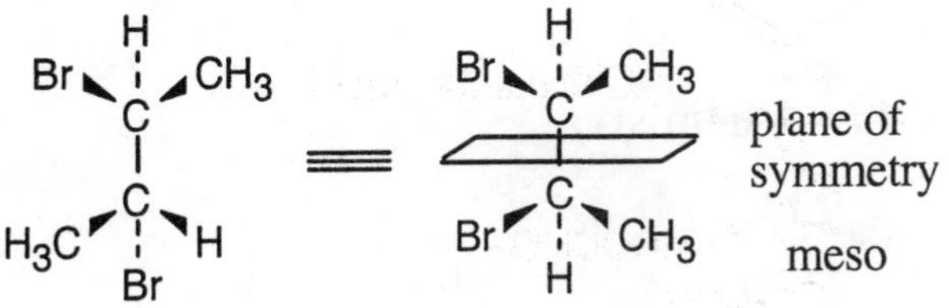

Structure (b) does not represent a meso compound.

6.18

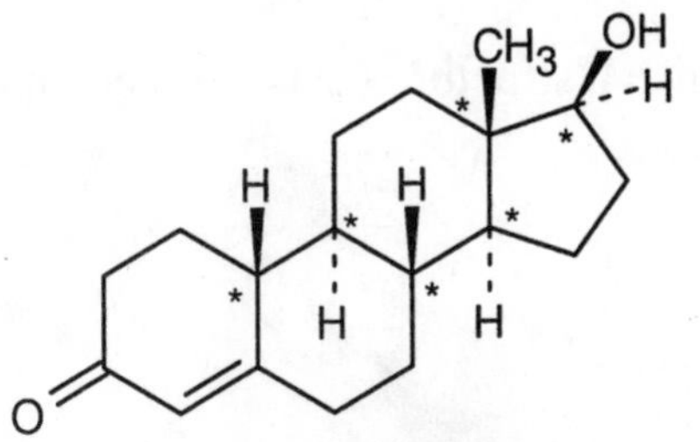

Nandrolone

Nandrolone has six stereocenters (starred) and can have, in principle, $2^6 = 64$ stereoisomers.

6.19

O
H
‖
| S
C
C
H3C OH + HO CH2CH3 acid catalyst
CH3

O H
‖ | S
C C
H3C O CH2CH3
CH3

The product is the pure *S* ester because no bonds at the chiral carbon are broken or rearranged during the reaction..

6.20

(a)

CH3
|
H C
Cl CH2CH=CHCH3

Cl

(*S*)-5-Chlorohex-2-ene Chlorocyclohexane

These two compounds are constitutional isomers. They are skeletal isomers because they have different carbon skeletons.

(b)

CH3
R
Br C H
R
H C Br
CH2CH3

CH3
S
H C Br
R
H C Br
CH2CH3

(2*R*,3*R*)-2,3-Dibromopentane (2*S*,3*R*)-2,3-Dibromopentane

The two dibromopentane stereoisomers are diastereomers.

Visualizing Chemistry

6.21

(a)
Cl
| R
H3C C H
COOH

(b)
H Cl
R
C
HOOC CH3

(c)
CH3
| S
C
HOOC Cl
H

(d)
H
R
HOOC C CH3
|
Cl

Structures (a), (b), and (d) are identical (*R* enantiomer); structure (c) is the *S* enantiomer.

6.22

(a)

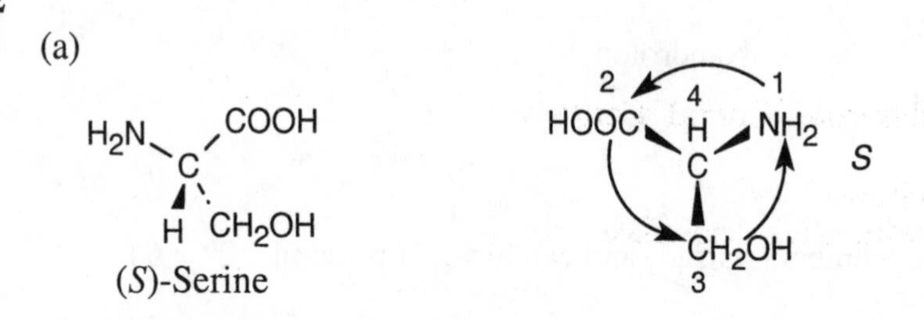

(b)

(R)-Adrenaline

R = ring

6.23

(a)

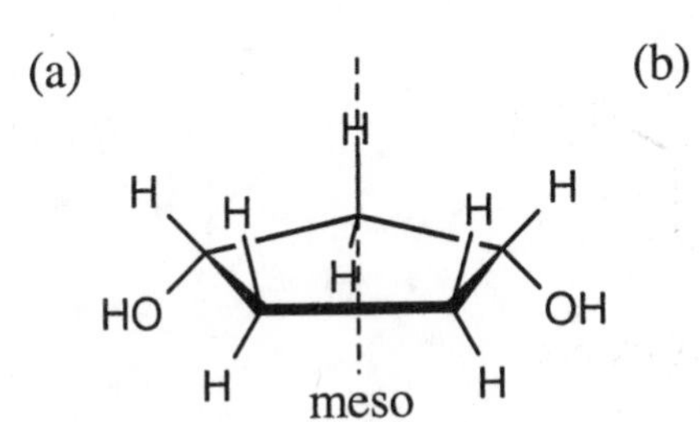

meso

(b)

meso

(c)

not meso

6.24

Pseudoephedrine

6.25

(a)

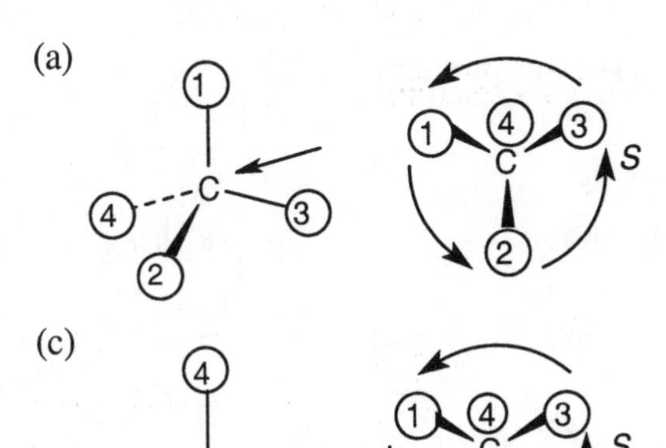

(b)

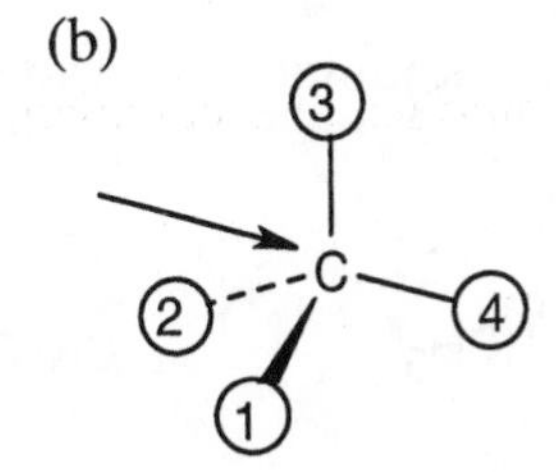

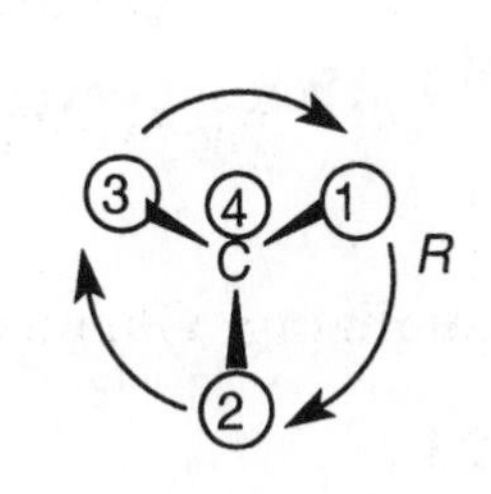

(c)

Additional Problems

Identifying Chiral Molecules and Chiral Centers

6.26 Chiral: ear, coin, scissors
Achiral: basketball, wine glass, snowflake

6.27

(a)

$$CH_3CH_2CH_2\overset{*}{C}H(CH_3)CH_2CH(CH_3)CH_3$$

2,4-Dimethylheptane
chiral

(b)

$$CH_3CH_2C(CH_3)_2-CH_2CH(CH_2CH_3)CH_2CH_3$$

5-Ethyl-3,3-dimethylheptane
achiral

(c)

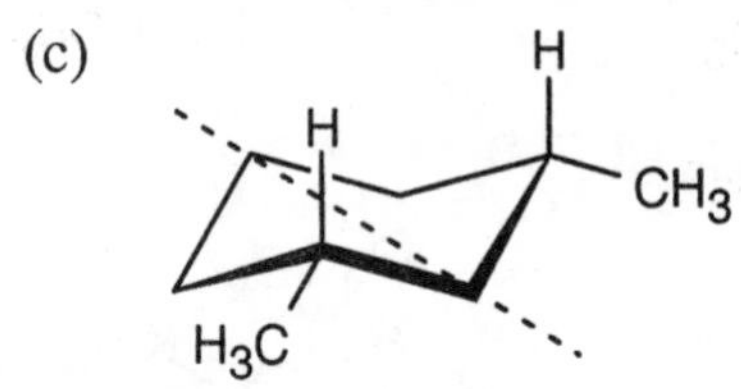

cis-1,3-Dimethylcyclohexane
achiral

6.28

Penicillin V
3 stereocenters

6.29 There are several possibilities for most parts of this problem.

(a)

$$CH_3CH_2CH_2\overset{*}{C}H(Cl)CH_3 \qquad CH_3CH_2\overset{*}{C}H(CH_3)CH_2Cl \qquad CH_3\overset{*}{C}H(Cl)CH(CH_3)CH_3$$

(b)

$$CH_3CH_2CH_2CH_2\overset{*}{C}H(OH)CH_3 \qquad CH_3CH_2CH_2\overset{*}{C}H(OH)CH_2CH_3 \qquad CH_3CH_2CH_2\overset{*}{C}H(CH_3)CH_2OH$$

$$CH_3CH_2\overset{*}{C}H(OH)CH(CH_3)CH_3 \qquad CH_3\overset{*}{C}H(OH)CH_2CH(CH_3)CH_3 \qquad CH_3CH_2\overset{*}{C}H(CH_3)CH_2CH_2OH$$

$$CH_3CH_2\overset{*}{C}H(CH_3)\overset{*}{C}H(OH)CH_3 \qquad CH_3\overset{*}{C}H(OH)C(CH_3)_3 \qquad CH_3\overset{*}{C}H(CH_3)CH(CH_3)CH_2OH$$

(c)

$$CH_3CH_2\overset{*}{C}H(CH_3)CH{=}CH_2$$

(d)

$$CH_3CH_2CH_2CH_2\overset{*}{C}H(CH_3)CH_2CH_3 \qquad CH_3CH_2CH_2\overset{*}{C}H(CH_3)CH(CH_3)CH_3 \qquad CH_3CH_2\overset{*}{C}H(CH_3)CH_2CH(CH_3)CH_3$$

$$CH_3CH_2\overset{*}{C}H(CH_3)\overset{*}{C}H(CH_3)CH_2CH_3 \qquad CH_3CH_2\overset{*}{C}H(CH_3)C(CH_3)_3$$

6.30

(a)

$$CH_3CH_2\overset{*}{C}H(CH_3){-}C(CH_3)_2CH_2CH_3$$

chiral

(b) 3-methylcyclopentanone (* at C-3)

chiral

(c) 4-methylcyclohexanone

achiral

(d)

$$BrCH_2\overset{*}{C}H(C_6H_5)\overset{*}{C}H(C_6H_5)CH_2Br$$

This compound exists as a pair of chiral enantiomers and an achiral meso compound.

(e) 1,4-dimethylcyclohexane

achiral

Optical Rotation

6.31 Refer to Problem 6.7 for the formula for calculating $[\alpha]_D$. In this problem, C = 3.00 g ÷ 5.00 mL = 0.600 g/mL, and l = 1.00 cm = 0.100 dm.

For cholic acid: $$[\alpha]_D = \frac{+2.22°}{0.100 \text{ dm} \times 0.600 \text{ g/mL}} = \frac{+2.22°}{0.0600} = +37.0°$$

6.32

For ecdysone: $$[\alpha]_D = \frac{+0.087°}{0.200 \text{ dm} \times 0.00700 \text{ g/mL}} = +62°$$

6.33 (*R*)-Serine has a specific rotation of +6.83° because enantiomers differ only in the sign of their specific rotations.

Drawing Stereoisomers

6.34

- $CH_3CH_2CH_2CH_2CH_2OH$ — achiral
- OH; $CH_3CH_2CH_2CHCH_3$ — chiral
- OH; $CH_3CH_2CHCH_2CH_3$ — achiral
- CH_3; CH_3CCH_2OH; CH_3 — achiral
- CH_3; $CH_3CH_2CHCH_2OH$ — chiral
- OH; $CH_3CH_2CCH_3$; CH_3 — achiral
- OH; $CH_3CHCHCH_3$; CH_3 — chiral
- CH_3; $HOCH_2CH_2CHCH_3$ — achiral

6.35

(a) OH; $CH_3CH_2CHCH_3$

(b) CH_3; $CH_3CH_2CHCOOH$

(c) OH; $CH_3CHCHCH_3$; Br

Priority Rules for the *R,S* System and Assignment of Stereochemistry

6.36

Highest priority ⟶ *Lowest priority*

(a) $—OCH_3$, $—OH$, $—CH_3$, $—H$

(b) $—Br$, $—Cl$, $—CH_2Br$, $—CH_3$

(c) $—C(CH_3)_3$, $—CH{=}CH_2$, $—CH(CH_3)_2$, $—CH_2CH_3$

(d) $—OCH_3$, $—COOCH_3$, $—COCH_3$, $—CH_2OCH_3$

6.37

Highest priority ⟶ *Lowest priority*

(a) —cyclopentyl, $—CH_2C(CH_3)_2CH_3$, $—CH_2CH(CH_3)CH_2CH_3$, $—CH_2CH_2CH_2CH_2CH_3$

(b) $—SO_3H$, $—SH$, $—OCH_2CH_2OH$, $—NH_2$

6.38 The stereoisomer pictured is (*S*)-lactic acid.

(*S*)-Lactic acid (*R*)-Lactic acid

6.39

(*S*)-Serine (*R*)-Serine

6.40

(a) (b) (c)

6.41

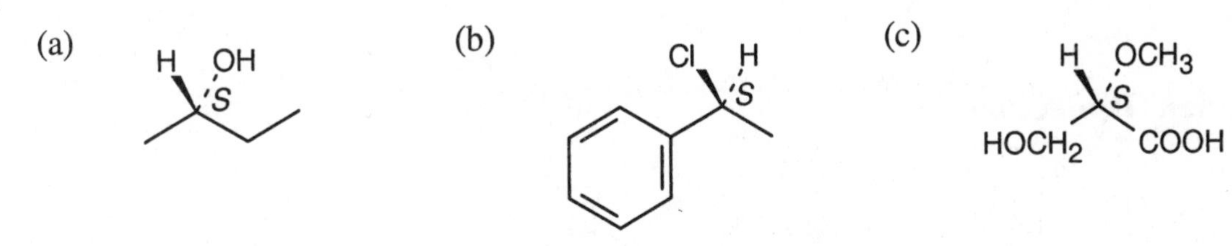

Stereochemical Relationships

6.42

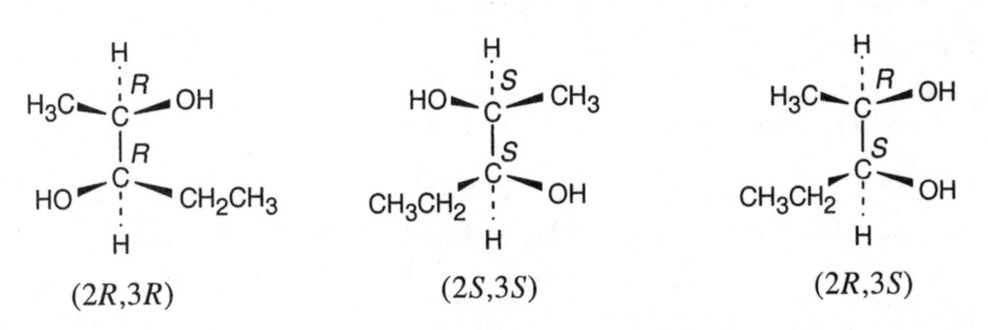

(2*R*,3*R*) (2*S*,3*S*) (2*R*,3*S*)

The specific rotations of the (2*R*,3*R*) and (2*S*,3*S*) enantiomers are equal in magnitude and opposite in sign. The specific rotations of the (2*R*,3*S*) and (2*R*,3*R*) diastereomers are not related.

6.43–6.44

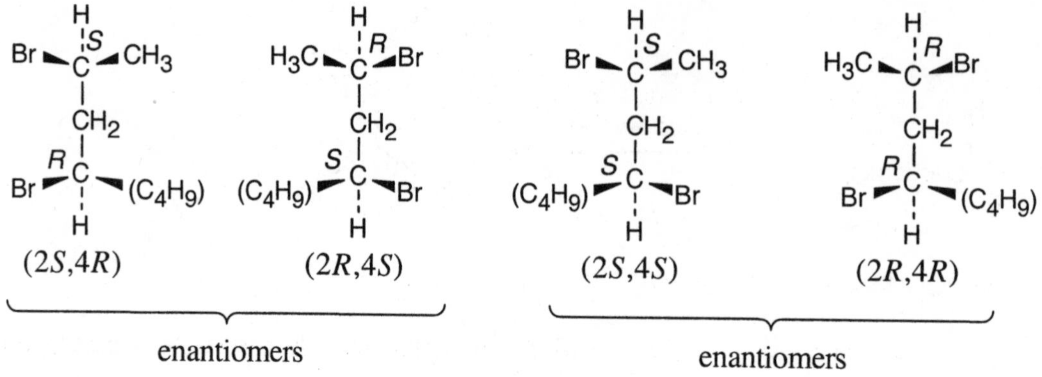

H, Br, *S*, C, CH_3, CH_2, *R*, C, Br, (C_4H_9), H
(2*S*,4*R*)

H, H_3C, *R*, C, Br, CH_2, *S*, C, (C_4H_9), Br, H
(2*R*,4*S*)

H, Br, *S*, C, CH_3, CH_2, *S*, C, (C_4H_9), Br, H
(2*S*,4*S*)

H, H_3C, *R*, C, Br, CH_2, *R*, C, Br, (C_4H_9), H
(2*R*,4*R*)

enantiomers

enantiomers

The (2*R*, 4*S*) stereoisomer is the enantiomer of the (2*S*, 4*R*) stereoisomer. The (2*S*, 4*S*) and (2*R*, 4*R*) stereoisomers are diastereomers of the (2*S*,4*R*) stereoisomer.

6.45

(a)

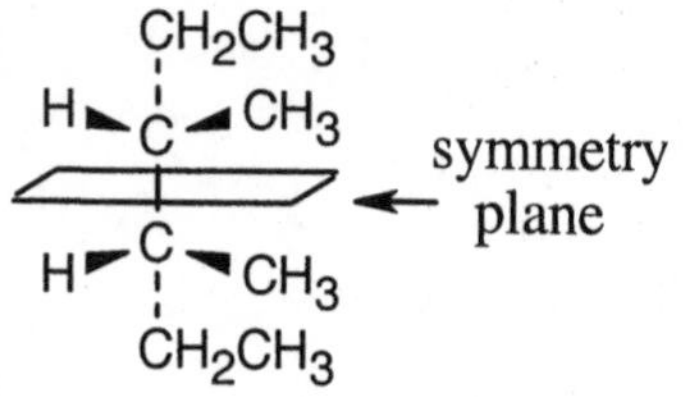

CH_2CH_3
H, C, CH_3
H, C, CH_3
CH_2CH_3

symmetry plane

(b)

CH_3
HO, C, H
R
S
Br, C, H
CH_3

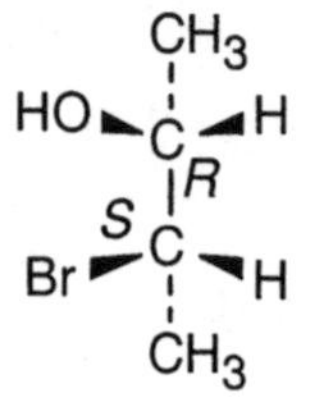

Newman Projections

6.46

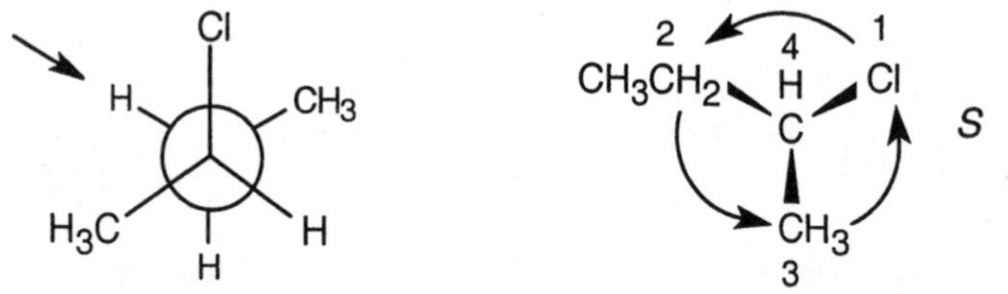

Cl
H, CH_3
H_3C, H, H

(*S*)-2-Chlorobutane

2, 4, 1
CH_3CH_2, H, Cl
C
CH_3
3
S

6.47

Cl
H_3C, H
H, H, CH_3

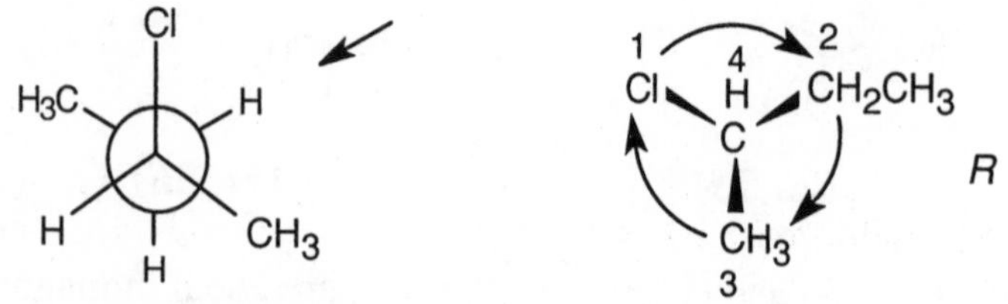

(*R*)-2-Chlorobutane

1, 4, 2
Cl, H, CH_2CH_3
C
CH_3
3
R

6.48

rotate rear carbon 180°

meso-Tartaric acid

The mirror plane of *meso*-tartaric acid is more apparent if the molecule is shown in its eclipsed conformation.

6.49

(2*R*,3*R*)-Tartaric acid

(2*S*,3*S*)-Tartaric acid

The enantiomeric tartaric acids are mirror images of each other. Unlike the meso isomer, neither of the above enantiomers contains a mirror plane.

Integrated Problems

6.50

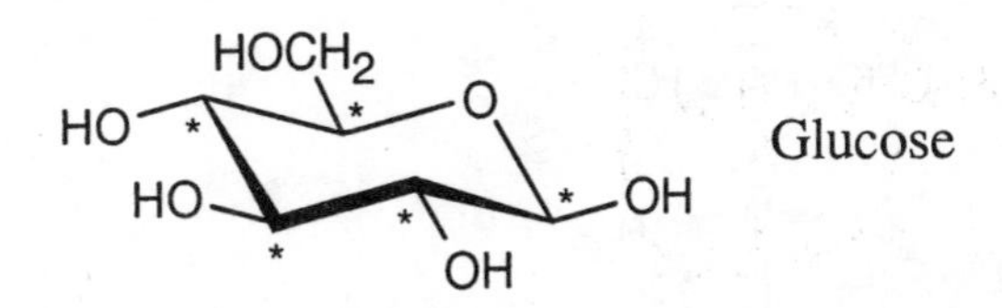

Glucose

The number of stereoisomers of a chiral compound is given by 2^n, where n equals the number of stereocenters present. Here, $n = 5$ and $2^n = 32$. Glucose thus has 32 possible stereoisomers.

6.51

(*R*)-3-Chloropent-1-ene

6.52

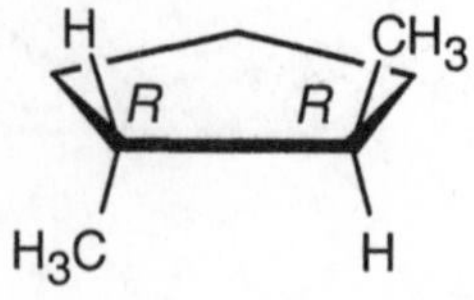

H CH3 R R H3C H **A**

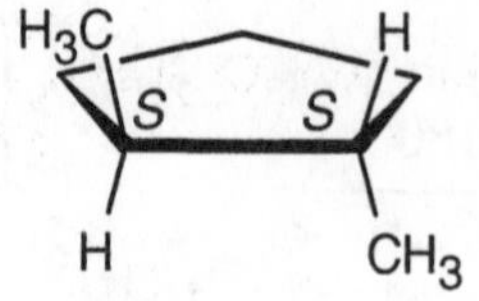

H3C H S S H CH3 **B**

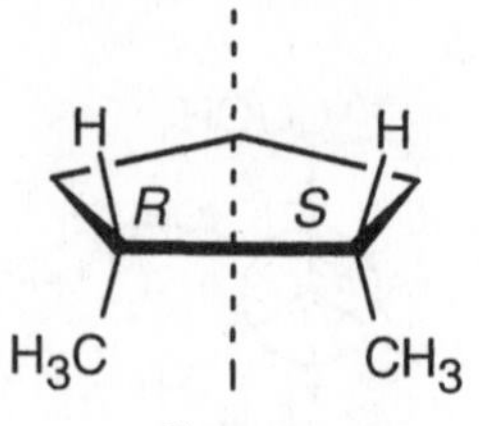

H H R S H3C CH3 **C** meso

A and **B** are enantiomers and are chiral. Compound **C** is their diastereomer and is a meso compound.

6.53

(a) CH3 H C OH S CH2 CH2 R H C OH CH3

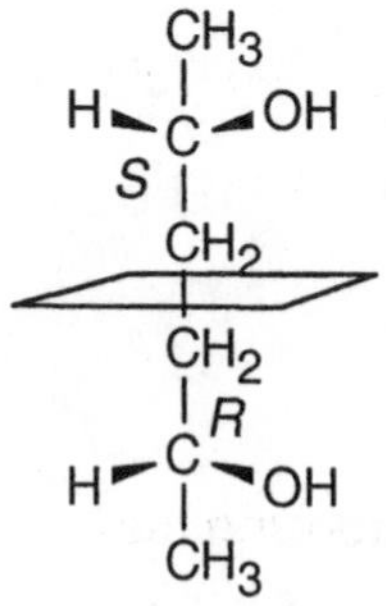

(b) H H S CH3 R CH3

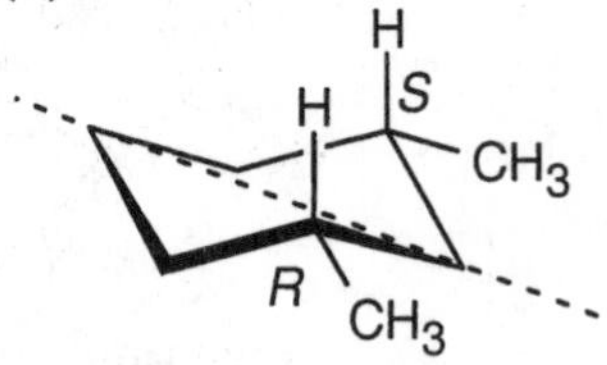

(c) OH H H R H S H3C CH3

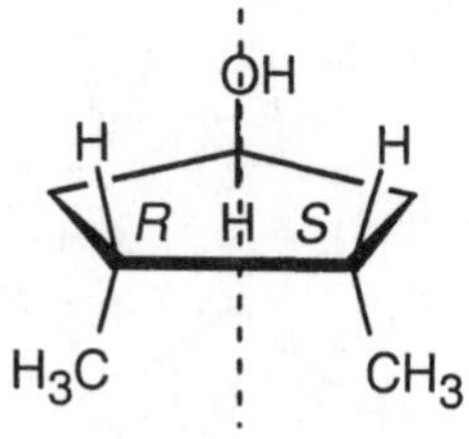

6.54

(a) COOH H C S NH2 H3C

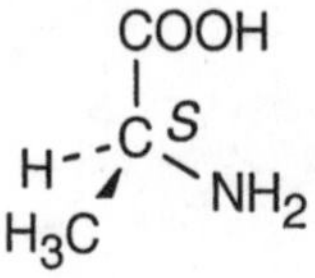

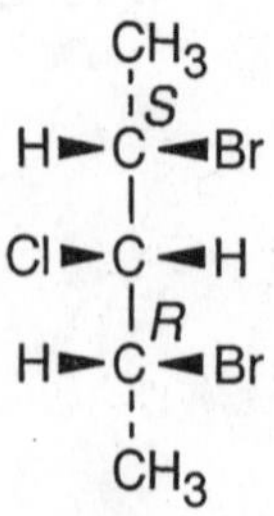

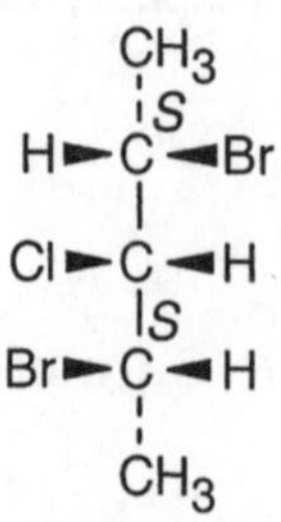

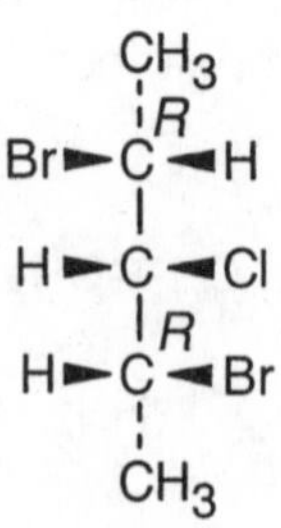

(b) R HO C H R H2N C CH3 H

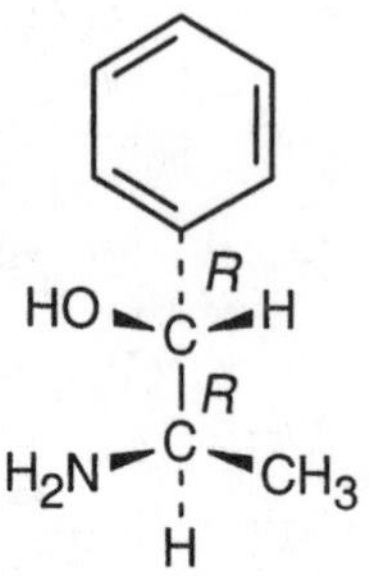

6.55

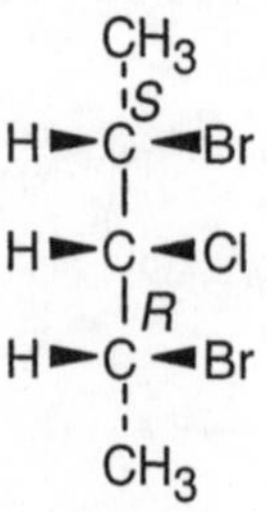

A	B	C	D
CH3	CH3	CH3	CH3
S: H–C–Br	S: H–C–Br	S: H–C–Br	R: Br–C–H
H–C–Cl	Cl–C–H	Cl–C–H	H–C–Cl
R: H–C–Br	R: H–C–Br	S: Br–C–H	R: H–C–Br
CH3	CH3	CH3	CH3

There are four stereoisomers of 2,4-dibromo-3-chloropentane. **C** and **D** are enantiomers and are optically active. **A** and **B** are optically inactive meso compounds and are diastereomers.

6.56

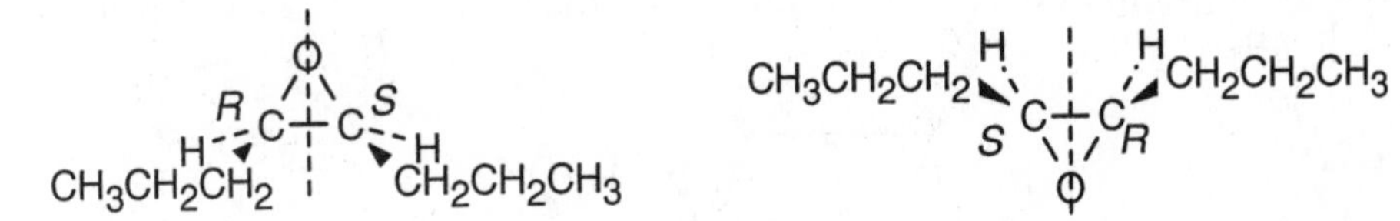

Peroxycarboxylic acids can attack either the "top" side or the "bottom" side of a double bond. The epoxide resulting from top side attack, pictured above, has two stereocenters, but because it has a plane of symmetry it is a meso compound. The epoxide resulting from bottom side attack is identical to epoxide resulting from top-side attack.

6.57–6.58

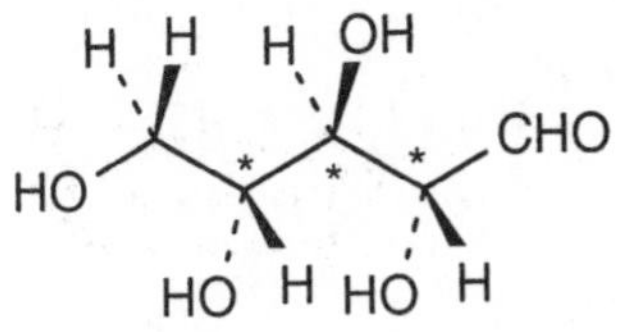

Ribose

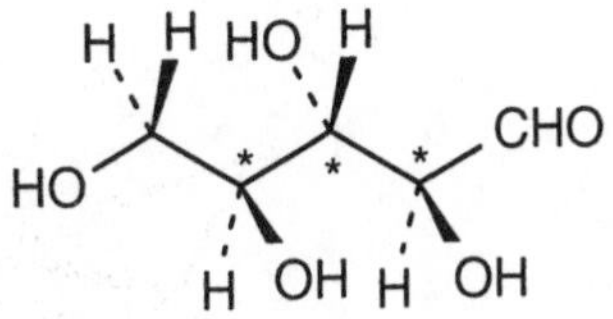

Enantiomer of ribose

Ribose has three stereocenters, which can give rise to eight stereoisomers.

6.59 Ribose has six diastereomers.

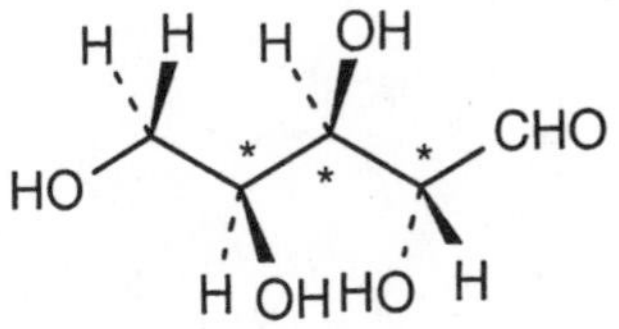

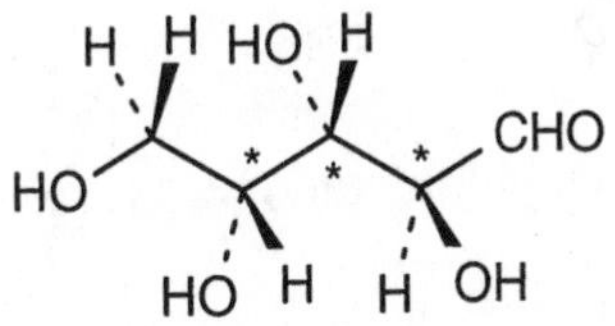

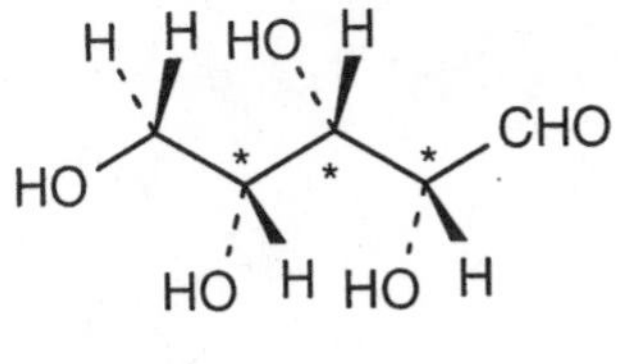

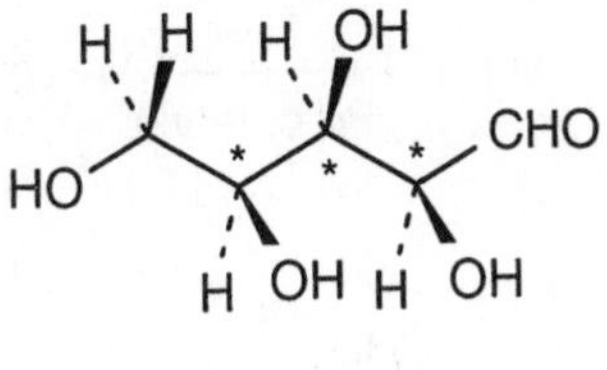

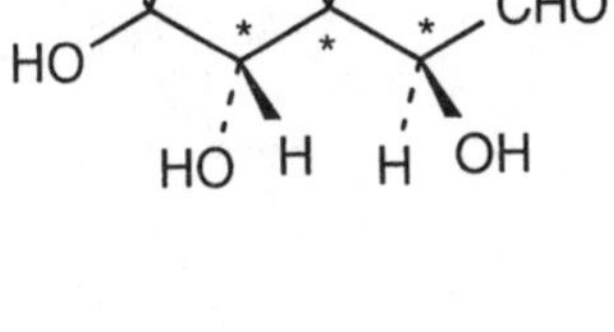

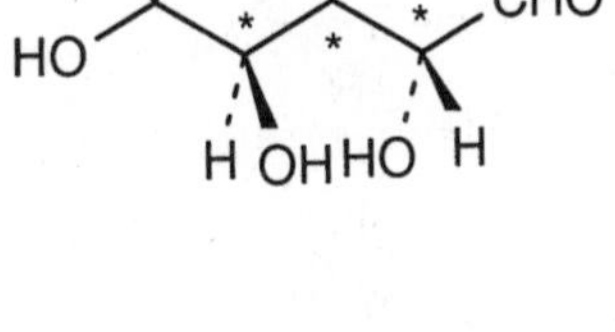

6.60 Ribitol is an optically inactive meso compound. Catalytic hydrogenation converts the aldehyde functional group into a hydroxyl group and makes the two halves of ribitol mirror images of each other.

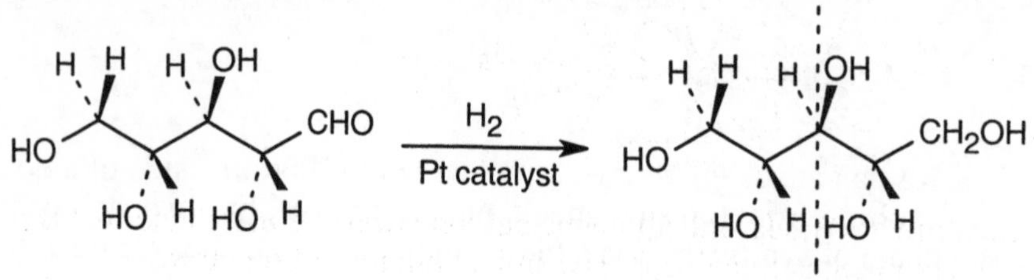

H H H OH
HO CHO
HO H HO H
H_2
Pt catalyst
H H H OH
HO CH_2OH
HO H HO H

Ribose

Ribitol

6.61

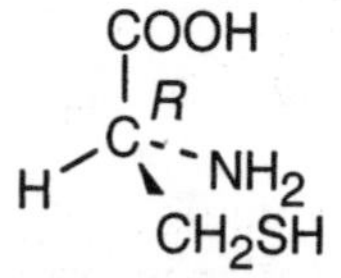

COOH
R
C
H NH_2
CH_2SH

(*R*)-Cysteine

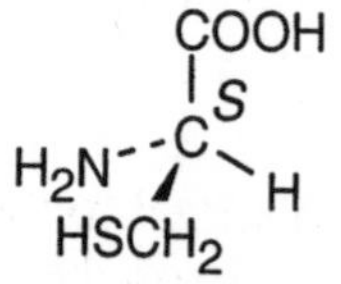

COOH
S
C
H_2N H
$HSCH_2$

(*S*)-Cysteine

6.62

O
H_3C
R
H

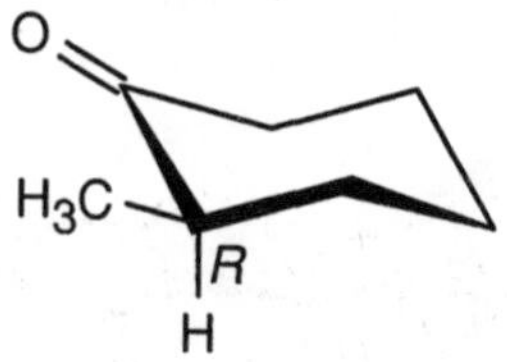

(*R*)-2-Methylcyclohexanone

6.63

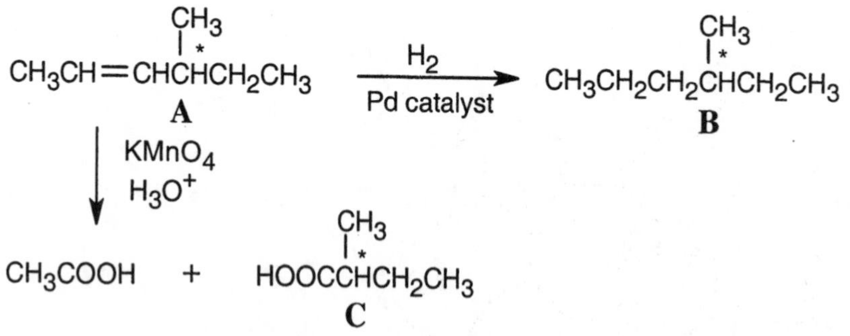

$CH_3CH{=}CHC^{*}H(CH_3)CH_2CH_3$ (**A**) $\xrightarrow[\text{Pd catalyst}]{H_2}$ $CH_3CH_2CH_2C^{*}H(CH_3)CH_2CH_3$ (**B**)

A $\xrightarrow{KMnO_4,\ H_3O^+}$ CH_3COOH + $HOOCC^{*}H(CH_3)CH_2CH_3$ (**C**)

6.64 Make a model of mycomycin. For simplicity, call –CH=CHCH=CHCH$_2$COOH "A" and –C≡CC≡CH "B". The carbon atoms in the allene group are linear and the π bonds formed are perpendicular to each other. Attach substituents to the sp^2 carbons.

Notice that the substituents A, H$_a$, and all allene carbon atoms lie in a plane that is perpendicular to the plane that contains B, H$_b$, and all allene carbon atoms.

Now, make another model identical to the first, except for an exchange of A and H$_a$. This new allene is not superimposable on the original allene; the two allenes are mirror images. The two allenes are enantiomers and are chiral because they possess no plane of symmetry.

In the Medicine Cabinet

6.65

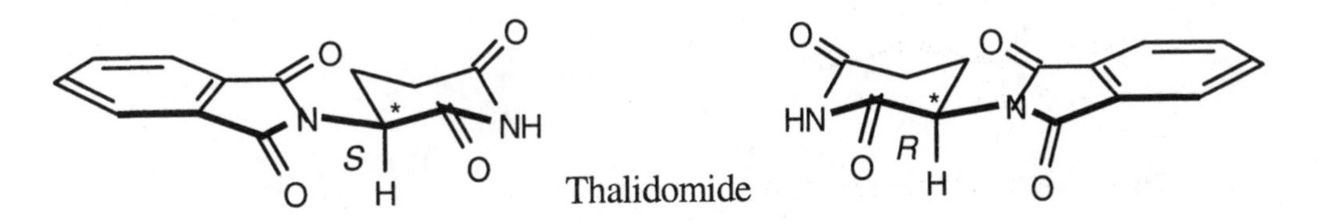

Thalidomide

6.66

(a)

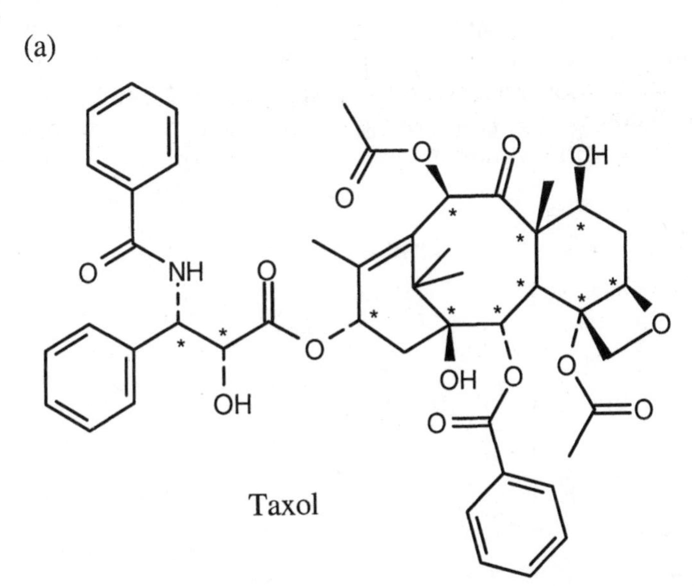

Taxol

(b) Taxol might have as many as $2^{11} = 2048$ stereoisomers!

(c) The chiral environment in which taxol is biosynthesized allows for the production of only one enantiomer.

6.67 (a)

H_3CO / COOH / H_2 metal catalyst / CH_3 / Alkene **A** → *R* COOH + *S* COOH

(b) Other than the sign of rotation, the enantiomers have identical physical properties.

(c) In a chiral environment, the enantiomers have different chemical properties and are not equally effective as drugs.

(d)

S COOH ⟹ *S* COOH, NH_2

L-Dopa

Drawing **I** shows the reaction coordinate diagram of a one-step exergonic reaction. It might also represent the reaction of alkene **A** with an achiral hydrogenation catalyst. In drawing **II**, the energy levels of reactant and product are the same as in drawing **I**, but the energy of activation is lower. Drawing **II** describes a reaction using a chiral catalyst, which can interact with one enantiomer more favorably than with the other enantiomer.

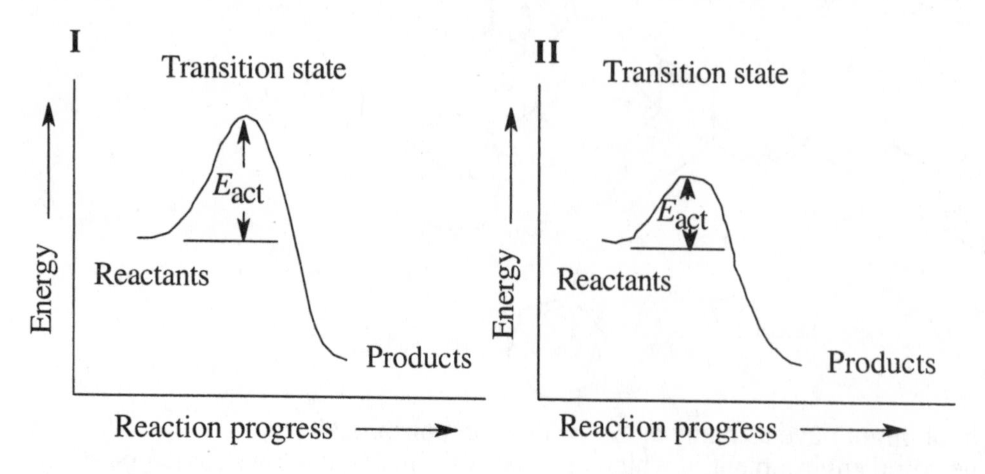

6.68

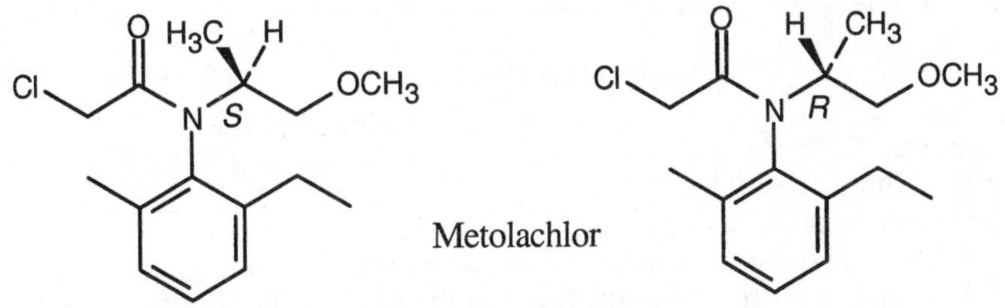

Metolachlor

6.69 (a) The graph shows that *R*-Metolachlor is ineffective at all concentrations at reducing the enzyme activity of the target enzyme. *S*-Metolachlor reduces enzyme activity to nearly zero at a concentration of 1000 pM, and it is the enantiomer of choice.
(b) The enantiomers differ in activity because they interact differently with a chiral enzyme.

6.70 If only one enantiomer is effective as an herbicide, legislation requiring that an agrochemical be sold as a pure enantiomer would reduce toxicity by 50% without reducing the effectiveness of the chemical.

Chapter 7 – Alkyl Halides

Chapter Outline

I. Alkyl halides (Sections 7.1–7.3).
   A. Naming alkyl halides (Section 7.1).
      1. Rules for naming alkyl halides:
         a. Find the longest chain and name it as the parent.
            If a double or triple bond is present, the parent chain must contain it.
         b. Number the carbon atoms of the parent chain, beginning at the end nearer the first substituent, whether alkyl or halo.
         c. Number each substituent.
            i. If more than one of the same kind of substituent is present, number each, and use the prefixes *di-*, *tri-*, *tetra-* and so on.
            ii. If different halogens are present, number all and list them in alphabetical order.
         d. If the parent chain can be numbered from either end, start at the end nearer the substituent that has alphabetical priority.
      2. Some alkyl halides are named by first citing the name of the alkyl group and then citing the halogen.
   B. Preparation of alkyl halides (Sections 7.2).
      1. Radical chlorination of alkanes.
      2. Addition of X_2 or HX to alkenes.
      3. Reaction of alcohols with HX, $SOCl_2$ or PBr_3.
   C. Grignard reagents from alkyl halides (Section 7.3).
      1. Organohalides react with Mg to produce organomagnesium halides, RMgX.
         These compounds are known as Grignard reagents.
      2. The carbon bonded to Mg is negatively polarized and is nucleophilic.
      3. Grignard reagents react with acids to form hydrocarbons.

II. Nucleophilic substitution reactions (Sections 7.4–7.6).
   A. General features (Sections 7.4).
      1. Discovery of Walden inversion.
         This discovery meant that one or more reactions must have occurred with inversion of configuration at the chirality center.
      2. Characteristics of nucleophilic substitution reactions.
         a. Nucleophilic substitutions occur by two pathways – S_N2 and S_N1.
         b. In all nucleophilic substitutions, a nucleophile reacts with a substrate and substitutes for a leaving group.
         c. Many types of products can be prepared by nucleophilic substitution reactions.
   B. S_N2 reactions (Section 7.5).
      1. A S_N2 reaction takes place in a single step without intermediates.
      2. The nucleophile attacks 180° away from the leaving group.
      3. Characteristics of the S_N2 reaction.
         a. Rates of S_N2 reactions depend on the concentration of substrate and of nucleophile.
         b. S_N2 reactions proceed with inversion of configuration.
         c. Methyl halides react fastest, followed by primary, secondary, and tertiary halides, which react slowest.
         d. Substrates with better leaving groups (stable anions) react faster.

C. S_N1 reactions (Section 7.6).
   1. An S_N1 reaction usually occurs only with tertiary substrates.
   2 An S_N1 reaction takes place in two steps.
      a. The leaving group dissociates to form a carbocation intermediate.
      b. The intermediate reacts with the nucleophile to form product.
   3. Characteristics of S_N1 reactions.
      a. The rate of an S_N1 reaction depends only on the concentration of substrate.
      b. The rate of reaction depends on the stability of the carbocation intermediate.
      c. The product of an S_N1 reaction is racemic.
      d. The best leaving groups give stable anions.

III. Elimination reactions (Sections 7.7–7.8).
   A. Characteristics of elimination reactions (Section 7.7).
      1. An alkene can form when an alkyl halide reacts with a base.
      2. The major product usually has the most highly substituted double bond (Zaitsev's Rule).
   B. E2 reactions.
      1. E2 reactions take place when a primary or secondary alkyl halide is treated with a strong base.
      2. The reaction takes place in one step.
   C. El reactions (Section 7.8).
      1. An El reaction takes place in two steps.
      2. The intermediate carbocation loses H^+ to give an alkene.
      3. The best El substrates are also the best S_N1 substrates.

IV. A summary of reactivity: S_N1, S_N2, El, E2 (Section 7.9).
   A. A primary substrate reacts by either an S_N2 route (with a good nucleophile) or an E2 route (with a strong base).
   B. A secondary substrate reacts by both S_N2 and E2 routes to give a mixture of products.
   C. A tertiary substrate reacts by an E2 pathway (with a strong base) or by a mixture of S_N1 and El pathways (with neutral or acidic nucleophiles).

V. Biological substitution reactions (Section 7.10).
   A. Biological methylations are substitution reactions.
   B. Many simple compounds are toxic because they cause undesirable methylation reactions.

Solutions to Problems

7.1

(a) $CH_3CH_2CH(Br)CH_3$ — 2-Bromobutane

(b) $CH_3CH_2CH(Cl)CH(CH_3)CH_3$ — 3-Chloro-2-methylpentane

(c) $CH_3CH(CH_3)CH_2CH_2Cl$ — 1-Chloro-3-methylbutane

(d) $CH_3C(Cl)(CH_3)CH_2CH_2Cl$ — 1,3-Dichloro-3-methylbutane

(e) $BrCH_2CH_2CH_2CH_2Cl$ — 1-Bromo-4-chlorobutane

(f) $CH_3CH(Br)CH_2CH_2CH_2Cl$ — 4-Bromo-1-chloropentane

7.2

(a) $CH_3CH_2CH_2C(CH_3)_2-CH(Cl)CH_3$

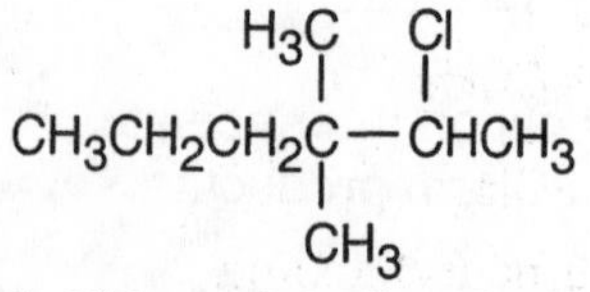

2-Chloro-3,3-dimethylhexane

(b) $CH_3CH_2CH_2CCl_2-CH(CH_3)CH_3$

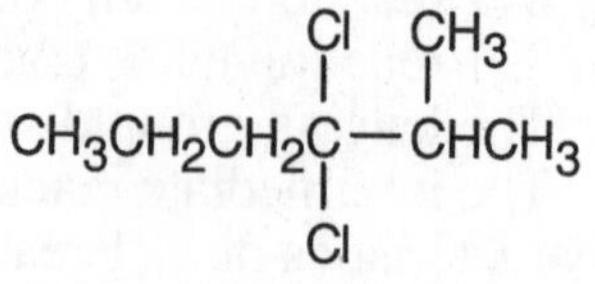

3,3-Dichloro-2-methylhexane

(c) $CH_3CH_2C(CH_2CH_3)(Br)CH_2CH_3$

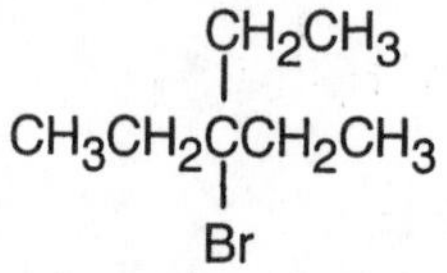

3-Bromo-3-ethylpentane

(d) $CH_3CH(Cl)CH_2CH(CH_3)CH(Br)CH_3$

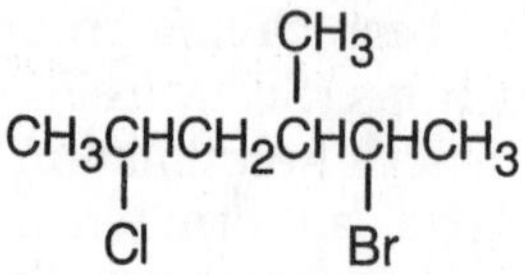

2-Bromo-5-chloro-3-methylhexane

7.3

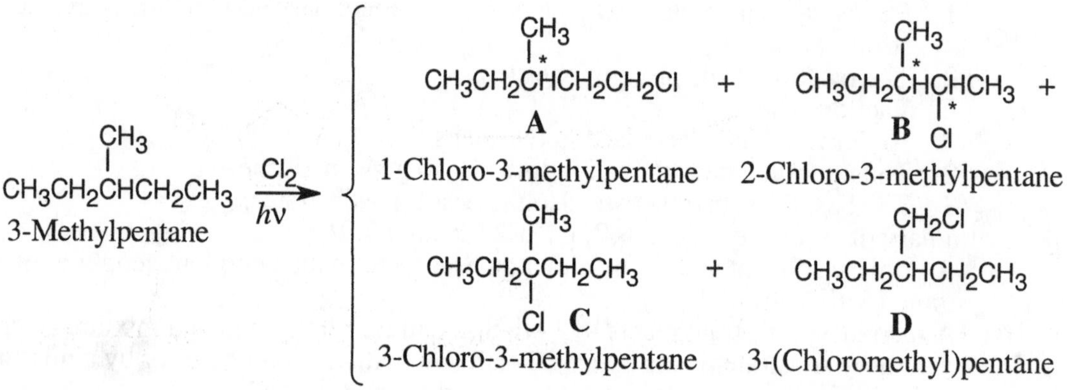

$CH_3CH_2CH(CH_3)CH_2CH_3$ (3-Methylpentane) $\xrightarrow[h\nu]{Cl_2}$

$CH_3CH_2C^*H(CH_3)CH_2CH_2Cl$ **A** 1-Chloro-3-methylpentane + $CH_3CH_2C^*H(CH_3)C^*H(Cl)CH_3$ **B** 2-Chloro-3-methylpentane +

$CH_3CH_2C(CH_3)(Cl)CH_2CH_3$ **C** 3-Chloro-3-methylpentane + $CH_3CH_2CH(CH_2Cl)CH_2CH_3$ **D** 3-(Chloromethyl)pentane

Products **A** and **B**, in which a carbon is bonded to four different groups, are chiral.

7.4 Tertiary alkyl halides can be prepared from the appropriate alcohols by reaction with HCl.

(a) $CH_3C(CH_3)_2OH \xrightarrow{HCl} CH_3C(CH_3)_2Cl$

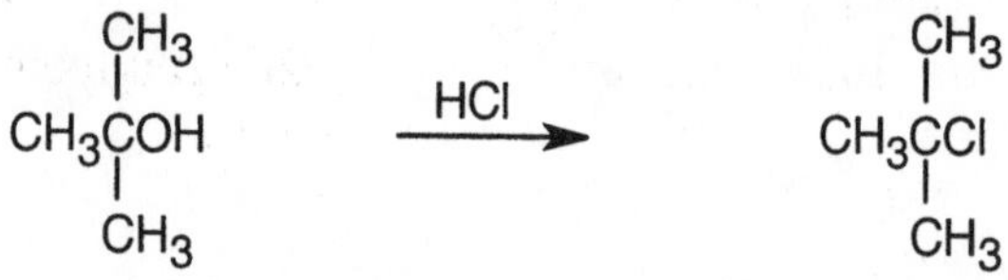

2-Chloro-2-methylpropane

(b) $CH_3CH(CH_3)CH_2CH(OH)CH_3 \xrightarrow[ether]{PBr_3} CH_3CH(CH_3)CH_2CH(Br)CH_3$

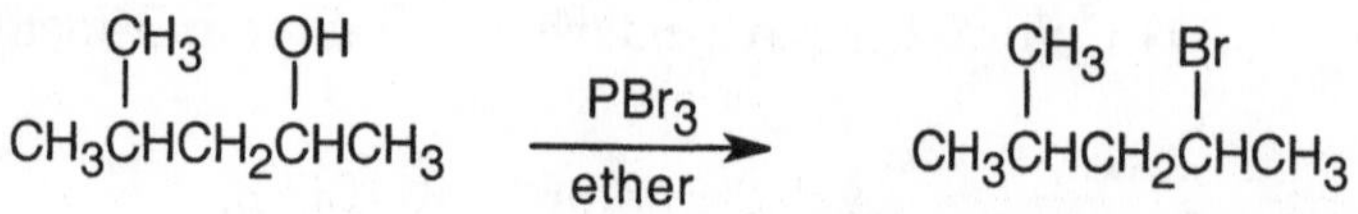

2-Bromo-4-methylpentane

(c) $HOCH_2CH_2CH_2CH_2CH(CH_3)CH_3 \xrightarrow[ether]{PBr_3} BrCH_2CH_2CH_2CH_2CH(CH_3)CH_3$

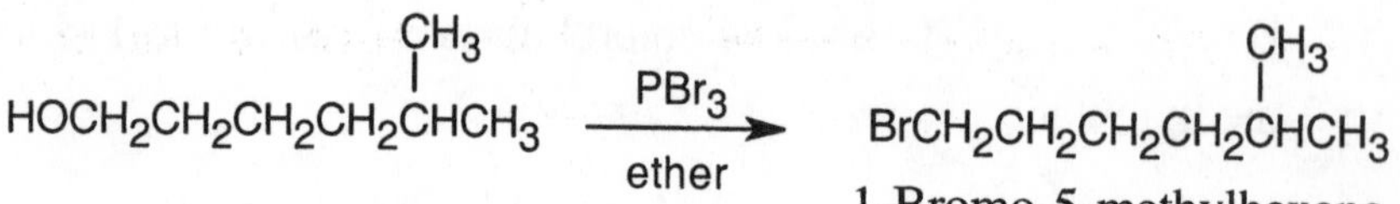

1-Bromo-5-methylhexane

(d)

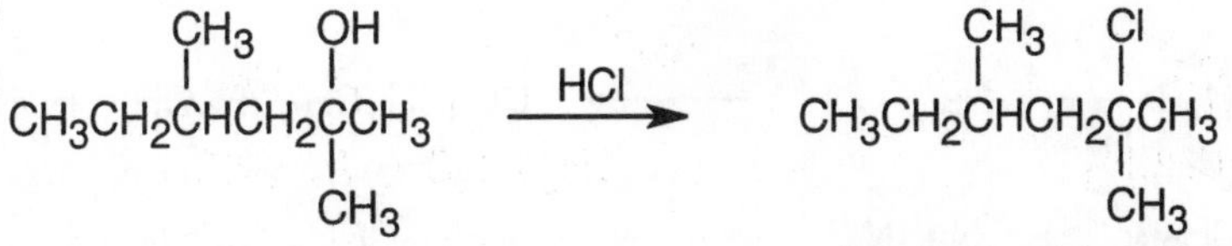

$$CH_3CH_2CH(CH_3)CH_2C(OH)(CH_3)CH_3 \xrightarrow{HCl} CH_3CH_2CH(CH_3)CH_2C(Cl)(CH_3)CH_3$$

2-Chloro-2,4-dimethylhexane

7.5

(a)

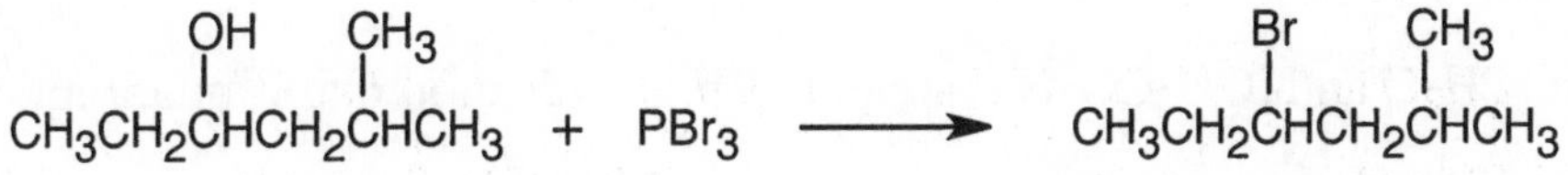

$$CH_3CH_2CH(OH)CH_2CH(CH_3)CH_3 + PBr_3 \longrightarrow CH_3CH_2CH(Br)CH_2CH(CH_3)CH_3$$

(b)

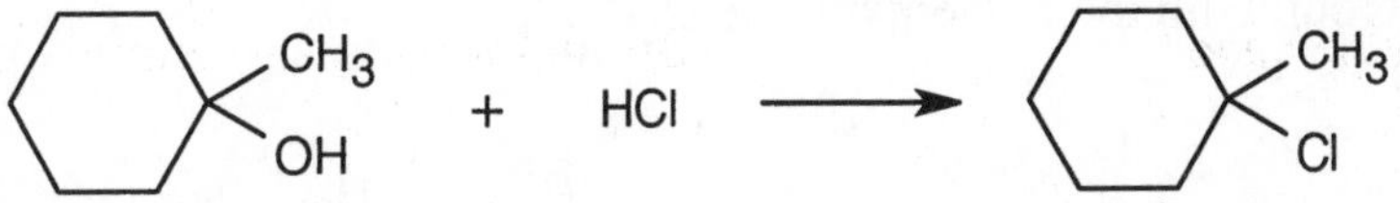

$$\text{1-methylcyclohexanol} + HCl \longrightarrow \text{1-chloro-1-methylcyclohexane}$$

(c)

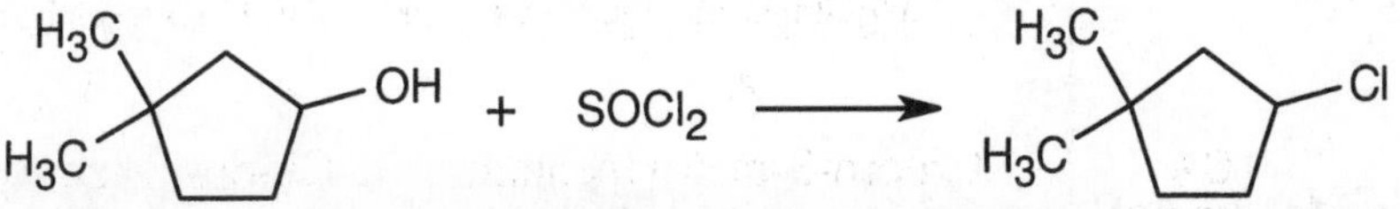

$$\text{3,3-dimethylcyclopentanol} + SOCl_2 \longrightarrow \text{1-chloro-3,3-dimethylcyclopentane}$$

7.6

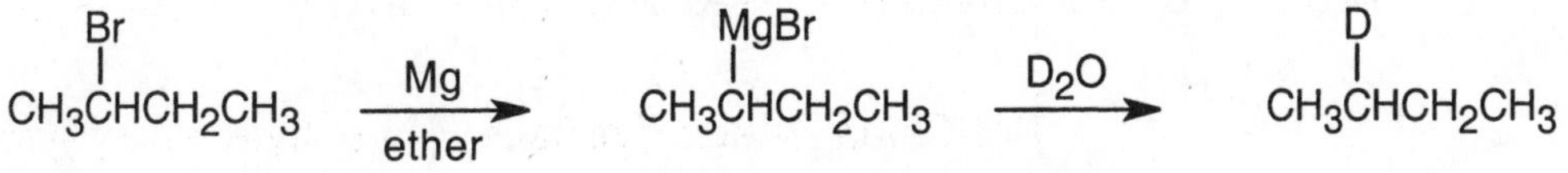

$$CH_3CH(Br)CH_2CH_3 \xrightarrow[\text{ether}]{Mg} CH_3CH(MgBr)CH_2CH_3 \xrightarrow{D_2O} CH_3CH(D)CH_2CH_3$$

7.7

$$CH_3CH(CH_3)CH_2CH_2CH_2OH \xrightarrow{PBr_3} CH_3CH(CH_3)CH_2CH_2CH_2Br \xrightarrow[\text{2. } H_2O]{\text{1. Mg, ether}} CH_3CH(CH_3)CH_2CH_2CH_3$$

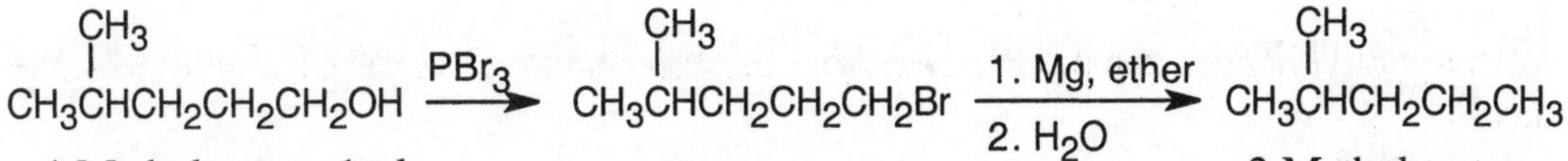

4-Methylpentan-1-ol → 2-Methylpentane

7.8

(a)

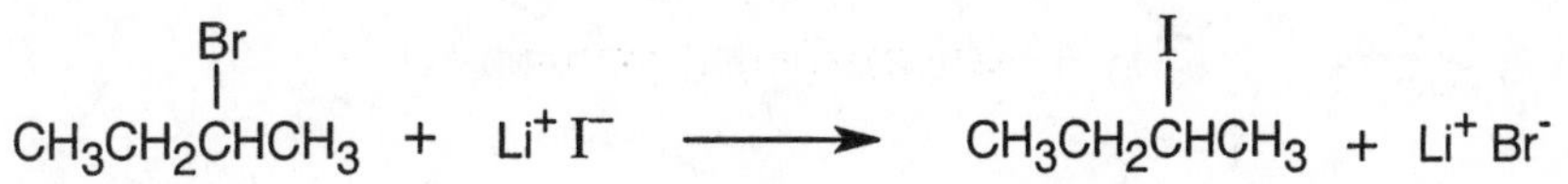

$$CH_3CH_2CH(Br)CH_3 + Li^+ I^- \longrightarrow CH_3CH_2CH(I)CH_3 + Li^+ Br^-$$

(b)

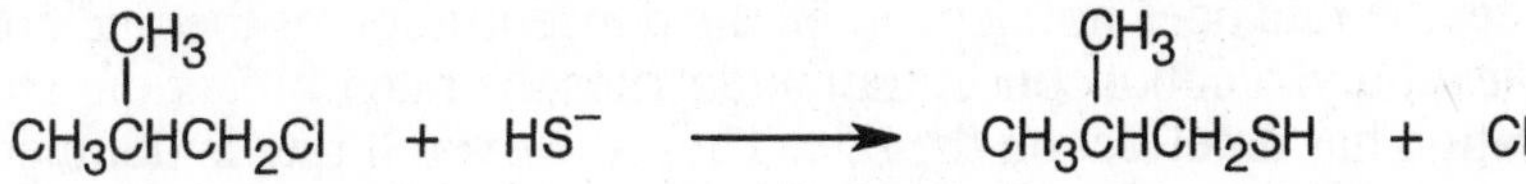

$$CH_3CH(CH_3)CH_2Cl + HS^- \longrightarrow CH_3CH(CH_3)CH_2SH + Cl^-$$

(c)

$$C_6H_5CH_2Br + NaCN \longrightarrow C_6H_5CH_2CN + Na^+ Br^-$$

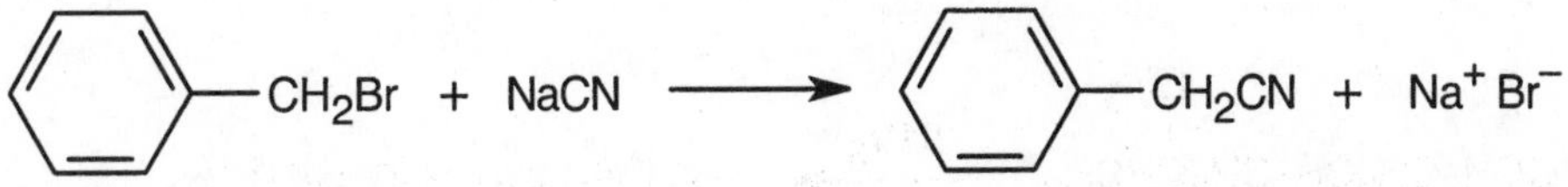

7.9

(a) $CH_3CH_2CH_2CH_2Br + Na^+ \, ^-OH \longrightarrow CH_3CH_2CH_2CH_2OH + Na^+ \, Br^-$

(b) $(CH_3)_2CHCH_2CH_2Br + Na^+ \, N_3^- \longrightarrow (CH_3)_2CHCH_2CH_2N_3 + Na^+ \, Br^-$

7.10 (a) If $[CH_3I]$ is tripled, the reaction rate is tripled.

(b) If both $[CH_3I]$ and $[CH_3CO_2Na]$ are doubled, the reaction rate is quadrupled.

7.11 As described in Practice Problem 7.5, identify the leaving group and the chirality center. Draw the product carbon skeleton, inverting the configuration at the chirality center, and replace the leaving group with the nucleophilic reactant.

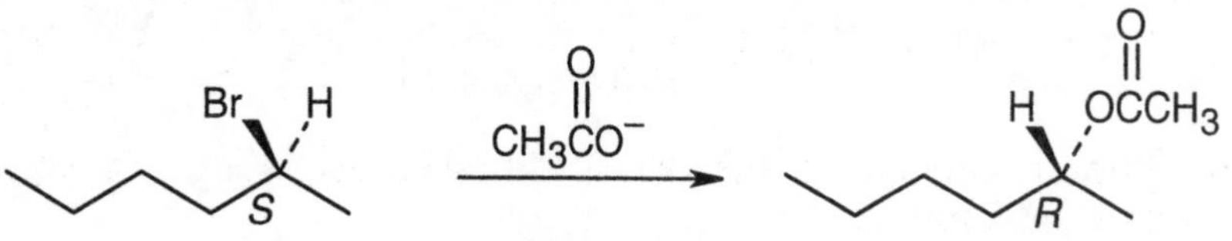

Br H S → (CH_3CO^- with C=O) → H $OCCH_3$ (C=O) R

7.12

HS^- + H_3C–CH(CH₃)(H)–CH₂–C(Br)(H)(S)–CH_3 ⟶ H_3C–CH(CH₃)(H)–CH₂–C(H)(SH)(R)–CH_3 + Br^-

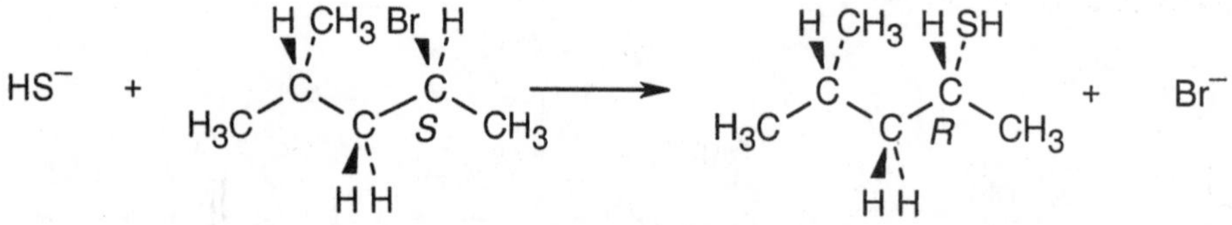

7.13 (a) Reaction of cyanide ion proceeds faster with the primary halide $CH_3CH_2CH_2Br$ than with the secondary halide $CH_3CH(Br)CH_3$.

(b) Iodide ion reacts faster with $(CH_3)_2CHCH_2Cl$. $H_2C{=}CHCl$ does not undergo S_N2 reactions.

7.14 Use the chart in Section 7.5 to identify the most reactive leaving groups.

Most reactive ⟶ *Least reactive*

$CH_3{-}I > CH_3{-}Br >> CH_3{-}F$

7.15 (a) Tripling the HBr concentration has no effect on the rate of reaction. In an S_N1 reaction such as this one, the rate does not depend on the concentration of the nucleophile.
(b) Doubling the *tert*-butyl alcohol concentration doubles the rate of reaction. Halving the HBr concentration has no effect on the rate. Thus, the overall rate is doubled.

7.16

H_3C, CH_3CH_2, $CH_2(CH_2)_3CH_3$ — *S* C—OH + H—Br ⇌ [C—OH_2^+ + :Br:⁻ ⇌ CH_3—C+ (CH_3CH_2, $CH_2(CH_2)_3CH_3$) + H_2O] ⇌ *S* C—Br (H_3C, CH_3CH_2, $CH_2(CH_2)_3CH_3$) + Br—C *R* (CH_3, CH_2CH_3, $CH_2(CH_2)_3CH_3$)

Attack by Br⁻ can occur on either side of the planar, achiral carbocation intermediate. The resulting product is a racemic mixture.

7.17 The *S* substrate reacts with water to form a mixture of *R* and *S* alcohols. The ratio of enantiomers is close to 50:50.

Br, CH_3, C (*S*), CH_2CH_3 (phenyl) $\xrightarrow{H_2O}$ HO, CH_3, C (*S*), CH_2CH_3 (phenyl) + H_3C, OH, C (*R*), CH_2CH_3 (phenyl)

7.18

(a)

$CH_3CH_2CH(Br)CH(CH_3)CH_3 \xrightarrow{\text{Base}} CH_3CH_2CH{=}C(CH_3)CH_3 + CH_3CH{=}CHCH(CH_3)CH_3$

major (trisubstituted double bond) — minor (disubstituted double bond)

(b)

$CH_3CH(CH_3)CH_2C(Cl)(CH_3){-}CH(CH_3)CH_3 \xrightarrow{\text{Base}} CH_3CH(CH_3)CH_2C(CH_3){=}C(CH_3)CH_3 + CH_3CH(CH_3)CH{=}C(CH_3)CH(CH_3)CH_3$

major (tetrasubstituted double bond) — minor (trisubstituted double bond)

$+ \; CH_3CH(CH_3)CH_2C({=}CH_2)CH(CH_3)CH_3$

minor (disubstituted double bond)

(c)

$$\text{C}_6\text{H}_{11}\text{-CH(Br)CH}_3 \xrightarrow{\text{Base}} \text{C}_6\text{H}_{10}\text{=CHCH}_3 + \text{C}_6\text{H}_{11}\text{-CH=CH}_2$$

major (trisubstituted double bond) minor (monosubstituted double bond)

7.19

(a)

$$CH_3CH(CH_3)CH_2CH_2CH(CH_3)CH_2CH_2Br \xrightarrow[\text{ethanol}]{\text{KOH}} CH_3CH(CH_3)CH_2CH_2CH(CH_3)CH{=}CH_2$$

Only this alkyl halide gives the desired alkene as the major product.

(b)

$$\text{4-bromo-1,2-dimethylcyclopentane} \xrightarrow[\text{ethanol}]{\text{KOH}} \text{3,4-dimethylcyclopentene}$$

7.20 The rate of E1 reaction would triple if the concentration of the alkyl halide were tripled.

7.21

(a)

$$CH_3CH_2CH_2CH_2Br + NaN_3 \longrightarrow CH_3CH_2CH_2CH_2N_3$$

primary substitution product

The reaction occurs by a S_N2 mechanism because the substrate is primary, the nucleophile is nonbasic, and the product is a substitution product.

(b)

$$CH_3CH_2CH(Cl)CH_2CH_3 + KOH \longrightarrow CH_3CH_2CH{=}CHCH_3$$

secondary strong base elimination product

This is an E2 reaction since a secondary halide reacts with a strong base to yield an elimination product.

(c)

$$\text{1-chloro-1-methylcyclohexane} + CH_3COOH \longrightarrow \text{1-methylcyclohexyl acetate } (\text{C}_6\text{H}_{10}(CH_3)OCOCH_3)$$

tertiary substitution product

This is an S_N1 reaction. Tertiary substrates form substitution products only by the S_N1 route.

Visualizing Chemistry

7.22

(a)

(i) $CH_3CH_2Cl + Na^+ \ {}^-SCH_3 \longrightarrow CH_3CH_2SCH_3 + NaCl$

(ii) $CH_3CH_2Cl + Na^+ \ {}^-OH \longrightarrow CH_3CH_2OH + NaCl$

Both reactions yield S_N2 substitution products because the substrate is primary and both nucleophiles are good.

(b)

(i) $CH_3CCl(CH_3)CH_2CH_3 + Na^+ \ {}^-SCH_3 \longrightarrow (H_3C)_2C{=}CHCH_3$ (major) $+ H_2C{=}C(CH_2CH_3)CH_3$ (minor) $+ HSCH_3 + NaCl$

(ii) $CH_3CCl(CH_3)CH_2CH_3 + Na^+ \ {}^-OH \longrightarrow (H_3C)_2C{=}CHCH_3$ (major) $+ H_2C{=}C(CH_2CH_3)CH_3$ (minor) $+ H_2O + NaCl$

The substrate is tertiary, and the nucleophiles are basic. Two elimination products are expected; the major product has the more substituted double bond, in accordance with Zaitsev's rule.

(c)

(i) Cl-cyclohexane (chair) $+ Na^+ \ {}^-SCH_3 \longrightarrow$ SCH_3-cyclohexane (chair) $+ NaCl$

(ii) Cl-cyclohexane (chair) $+ Na^+ \ {}^-OH \longrightarrow$ OH-cyclohexane (chair) + cyclohexene $+ H_2O + NaCl$

In (i), the secondary substrate reacts with the good, but weakly basic, nucleophile to yield substitution product. In (ii), NaOH is a poorer nucleophile but a stronger base, and both substitution and elimination products are formed.

7.23

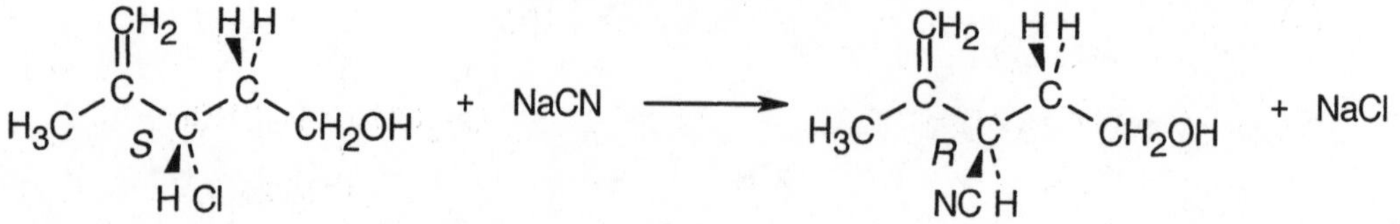

H_3C–C(=CH_2)–C(H)(Cl)[*S*]–C(H)(H)–CH_2OH + NaCN ⟶ H_3C–C(=CH_2)–C(H)(NC)[*R*]–C(H)(H)–CH_2OH + NaCl

The *S* substrate has a secondary allylic chloride group and a primary hydroxyl group. S_N2 reaction occurs at the secondary carbon to give the *R* cyano product because hydroxide is a poor leaving group.

7.24

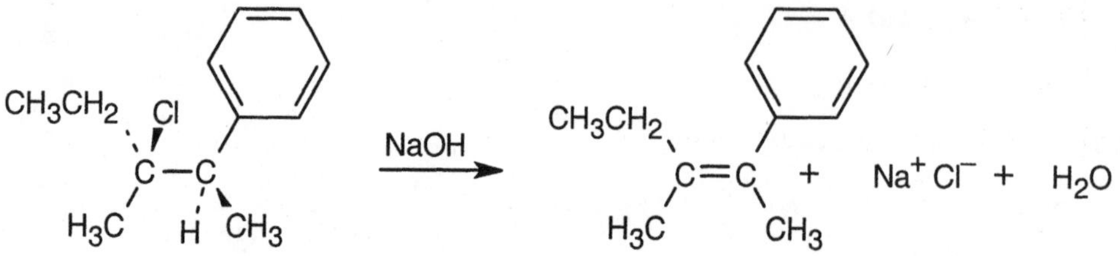

CH_3CH_2(Cl)(H_3C)C–C(H)(CH_3)(Ph) $\xrightarrow{NaOH}$ CH_3CH_2(H_3C)C=C(CH_3)(Ph) + $Na^+ Cl^-$ + H_2O

According to Zaitsev's rule, the product with the most substituted double bond is most likely to form. Notice also that the double bond that results from elimination is conjugated with the aromatic ring.

7.25

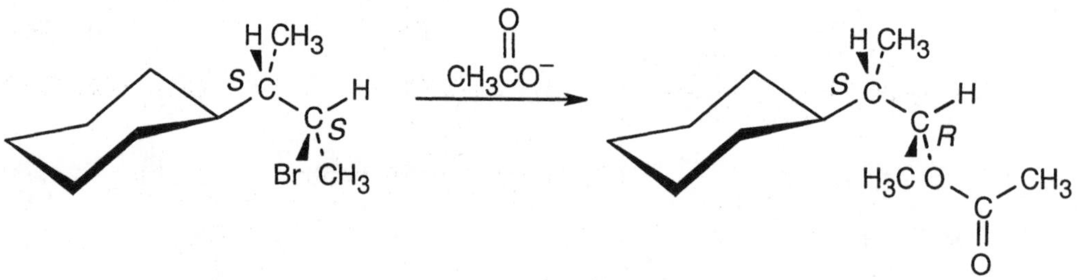

Cyclohexyl–C(H)(CH_3)[*S*]–C(H)(Br)(CH_3)[*S*] $\xrightarrow{CH_3CO_2^-}$ Cyclohexyl–C(H)(CH_3)[*S*]–C(H)(CH_3)(O–C(=O)–CH_3)[*R*]

Reaction of the secondary bromide with the weakly basic acetate nucleophile occurs by an S_N2 route, with inversion of configuration, to produce the *R* acetate.

Additional Problems

Nomenclature

7.26

(a) $CH_3CH(CH_3)CH(Br)CH(Br)CH_2CH(CH_3)CH_3$

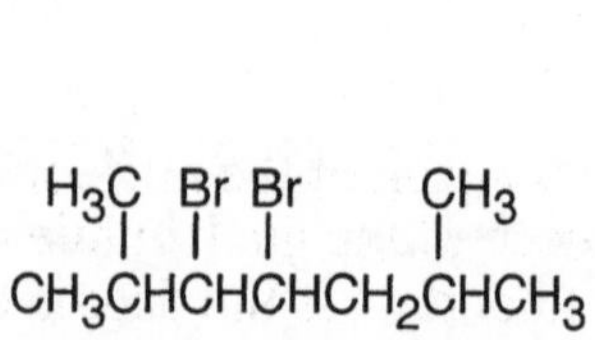

3,4-Dibromo-2,6-dimethylheptane

(b) $CH_3CH{=}CHCH_2CH(I)CH_3$

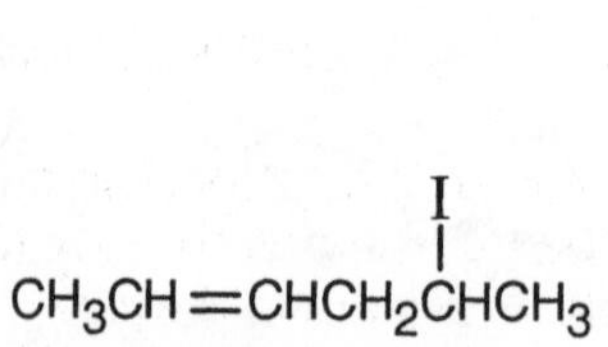

5-Iodohex-2-ene

(c) $CH_3C(Br)(CH_3)CH_2CH_2CH(Cl)CH(CH_3)CH_3$

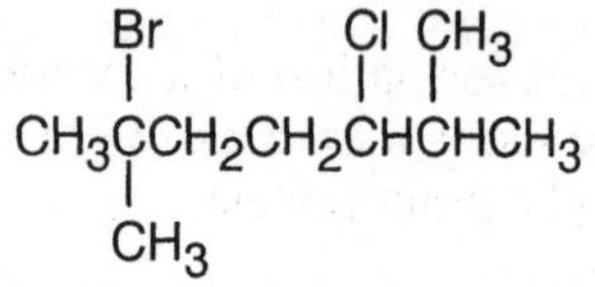

2-Bromo-5-chloro-2,6-dimethylheptane

(d) $CH_3CH_2CH(CH_2Br)CH_2CH_2CH_3$

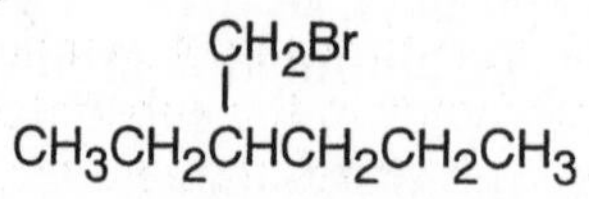

3-(Bromomethyl)hexane

7.27

(a)

$$\begin{array}{c} CH_3\ \ Cl \\ CH_3CH_2CHCHCHCH_3 \\ Cl \end{array}$$

2,3-Dichloro-4-methylhexane

(b)

$$\begin{array}{c} Br\ \ CH_3 \\ CH_3CH_2CCH_2CHCH_3 \\ CH_2CH_3 \end{array}$$

4-Bromo-4-ethyl-2-methylhexane

(c)

$$\begin{array}{c} H_3C\ I\ \ CH_3 \\ CH_3CCHCCH_3 \\ H_3C\ \ \ CH_3 \end{array}$$

3-Iodo-2,2,4,4-tetramethylpentane

7.28

$$\begin{array}{c} CH_3 \\ CH_3CH_2CH_2CHCH_3 \end{array} \xrightarrow[h\nu]{Cl_2}$$

2-Methylpentane

$$\begin{array}{c} CH_3 \\ CH_3CH_2CH_2\overset{*}{C}HCH_2Cl \end{array} \ + \ \begin{array}{c} CH_3 \\ CH_3CH_2CH_2CCH_3 \\ Cl \end{array} \ +$$

1-Chloro-2-methylpentane 2-Chloro-2-methylpentane

$$\begin{array}{c} CH_3 \\ CH_3CH_2\overset{*}{C}HCHCH_3 \\ Cl \end{array} \ + \ \begin{array}{c} CH_3 \\ CH_3\overset{*}{C}HCH_2CHCH_3 \\ Cl \end{array} \ +$$

3-Chloro-2-methylpentane 2-Chloro-4-methylpentane

$$\begin{array}{c} CH_3 \\ ClCH_2CH_2CH_2CHCH_3 \end{array}$$

1-Chloro-4-methylpentane

Three of the above products are chiral (stereocenters are starred). Each product occurs as a racemic mixture of enantiomers.

Characteristics of S_N1 and S_N2 Reactions

7.29 (a) *S_N2*: In an S_N2 reaction, the rate-limiting step involves attack of the nucleophile on the substrate. Consequently, any factor that makes approach of the nucleophile more difficult slows down the rate of reaction. Especially important is the degree of crowding at the reacting carbon atom. Tertiary carbon atoms are too crowded to allow S_N2 substitution to occur. Even steric hindrance one carbon atom away from the reacting site causes a drastic slowdown in the rate of reaction.

S_N1: The rate-limiting step in an S_N1 reaction involves formation of a carbocation. Any structural factor in the substrate that stabilizes carbocations increases the rate of reaction. Substrates that are tertiary, allylic, or benzylic react fastest.

(b) Good leaving groups (stable anions) increase the rates of both S_N1 and S_N2 reactions.

7.30 Use the table in Section 7.5 if you need help.

	Better leaving group	*Poorer leaving group*
(a)	Br^-	F^-
(b)	Cl^-	NH_2^-
(c)	I^-	OH^-

7.31

	Reacts faster	*Reacts slower*	*Reason*
(a)	$C_6H_5CH_2Br$ (benzene ring–CH_2Br)	C_6H_5Br (benzene ring–Br)	Primary, benzylic halides react rapidly in S_N2 reactions. Aryl halides are unreactive in S_N2 reactions.
(b)	CH_3Cl	$(CH_3)_3CCl$	Primary halides react faster than tertiary halides in S_N2 reactions.
(c)	$H_2C{=}CHCH_2Br$	$CH_3CH{=}CHBr$	Primary, allylic halides react rapidly in S_N2 reactions. Vinyl halides are unreactive in S_N2 reactions.

7.32 Remember that the rate of an S_N2 reaction depends on the concentrations of both the substrate and the nucleophile.

(a) If $[CH_3Br]$ is tripled and $[CN^-]$ is halved, the rate increases by a factor of 1.5.
(b) If $[CH_3Br]$ is halved and $[CN^-]$ is tripled, the rate increases by a factor of 1.5.
(c) If $[CH_3Br]$ is tripled and $[CN^-]$ is doubled, the rate increases by a factor of 6.
(d) If the reaction temperature is raised, the rate increases; the size of the increase depends on the temperature change and on other factors.
(e) Doubling the volume of the reacting solution halves the concentrations of both CH_3Br and CN^- and decreases the rate by a factor of 4.

7.33 In an S_N1 reaction, the rate depends on the concentration of the substrate and is independent of the concentration of the nucleophile.

(a) If $[(CH_3)_3CBr]$ is doubled and $[CH_3OH]$ is halved, the rate increases by a factor of 2 (remember that $[CH_3OH]$ doesn't affect the reaction rate).
(b) If $[(CH_3)_3CBr]$ is halved and $[CH_3OH]$ is doubled, the rate is halved.
(c) If both $[(CH_3)_3CBr]$ and $[CH_3OH]$ are tripled, the rate is tripled.
(d) If the reaction temperature is lowered, the rate decreases.

7.34

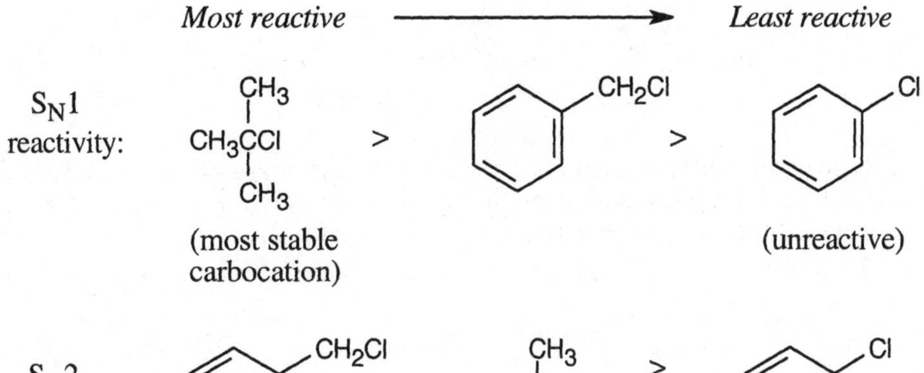

S_N2 reactivity:

CH_2Cl > CH_3CCl (CH_3, CH_3) > Cl

(primary) (tertiary) (unreactive)

7.35 S_N2 reactivity:

Most reactive → *Least reactive*

(a)

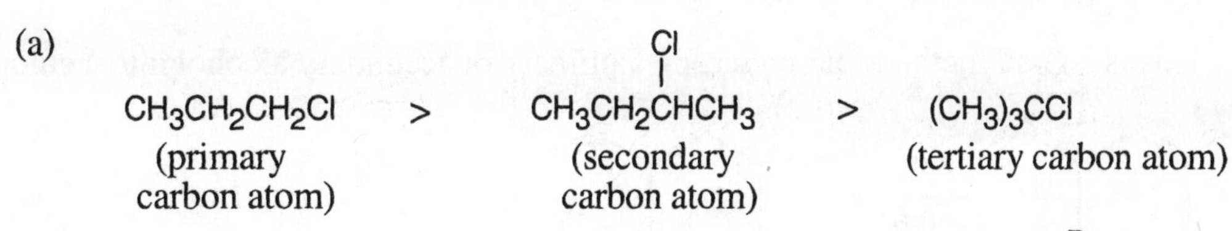

(b)

CH_3Br > $(CH_3)_2CHCH_2Br$ (primary carbon atom) > $(CH_3)_2CHCHCH_3$ with Br (secondary carbon atom)

Synthesis

7.36

(a) HCl → Cl

Chlorocyclopentane

(b) H_2O, H^+ catalyst → OH

Cyclopentanol

(c) Cl, Mg, ether → MgCl

from (a)

Cyclopentylmagnesium chloride

(d)

$$\text{(cyclopentyl)-MgCl} \xrightarrow{H_2O} \text{Cyclopentane}$$

from (c) *or* Cyclopentane

$$\text{Cyclopentene} \xrightarrow[\text{Pd catalyst}]{H_2} \text{Cyclopentane}$$

7.37

(a)

$$\text{1-methylcyclohexanol } (H_3C, OH) \xrightarrow[\text{ether}]{HBr} \text{1-bromo-1-methylcyclohexane } (H_3C, Br) + H_2O$$

This is a good method for converting a tertiary alcohol into a bromide.

(b)

$$CH_3CH_2CH_2CH_2OH \xrightarrow{SOCl_2} CH_3CH_2CH_2CH_2Cl$$

This is a good method for converting a primary or secondary alcohol into a chloride.

(c)

$$\text{Cyclohexanol (OH)} \xrightarrow[\text{ether}]{PBr_3} \text{Bromocyclohexane (Br)}$$

This is a good method for converting a primary or secondary alcohol into a bromide.

(d)

$$CH_3CH_2CH(Br)CH_3 \xrightarrow[\text{Ether}]{Mg} \underset{\textbf{A}}{CH_3CH_2CH(MgBr)CH_3} \xrightarrow{H_2O} \underset{\textbf{B}}{CH_3CH_2CH_2CH_3}$$

7.38 All of these reactions proceed by S_N2 substitution.

(a) $CH_3CH_2OH + PBr_3 \longrightarrow CH_3CH_2Br$

(b) $CH_3CH_2CH_2CH_2Br + Na^{+}\,{}^{-}CN \longrightarrow CH_3CH_2CH_2CH_2CN + Na^{+}\,Br^{-}$

(c) $CH_3Br + Na^{+}\,{}^{-}OC(CH_3)_3 \longrightarrow CH_3OC(CH_3)_3 + Na^{+}\,Br^{-}$

The reaction of CH_3O^- with $(CH_3)_3CBr$ causes elimination, not substitution.

(d) $CH_3CH_2CH_2I + Na^{+}\,{}^{-}N{=}\overset{+}{N}{=}N^{-} \longrightarrow CH_3CH_2CH_2N{=}\overset{+}{N}{=}N^{-} + Na^{+}\,I^{-}$

(e) $CH_3CH_2I + Na^{+}\,{}^{-}SH \longrightarrow CH_3CH_2SH + Na^{+}\,I^{-}$

(f)

$$CH_3Br + Na^{+}\,{}^{-}O\overset{O}{\overset{\|}{C}}CH_3 \longrightarrow CH_3O\overset{O}{\overset{\|}{C}}CH_3 + Na^{+}\,Br^{-}$$

7.39

(a) $CH_3CH_2CH_2Br$ + NaI $\longrightarrow$ $CH_3CH_2CH_2I$ + NaBr

(b) $CH_3CH_2CH_2Br$ + NaCN $\longrightarrow$ $CH_3CH_2CH_2CN$ + NaBr

(c) $CH_3CH_2CH_2Br$ + NaOH $\longrightarrow$ $CH_3CH_2CH_2OH$ + NaBr

(d) $CH_3CH_2CH_2Br$ + Mg, then H_2O $\longrightarrow$ $CH_3CH_2CH_3$ + MgBrOH

(e) $CH_3CH_2CH_2Br$ + $NaOCH_3$ $\longrightarrow$ $CH_3CH_2CH_2OCH_3$ + NaBr

7.40 (a) The reaction of cyanide anion with a tertiary halide is more likely to yield the elimination product 3-methylpent-2-ene than substitution product.

(b) Use PBr_3 to convert a primary alcohol into a primary bromide.

(c) Reaction of a tertiary alcohol with HBr gives mainly substitution product along with a lesser amount of elimination product.

Elimination Reactions and the Elimination/Substitution Continuum

7.41

$$CH_3CH_2CH_2C(CH_3)(Br)CH_2CH_3 \xrightarrow{:Base} CH_3CH_2CH{=}C(CH_3)CH_2CH_3 + CH_3CH_2CH_2C(CH_3){=}CHCH_3$$

3-Bromo-3-methylhexane 3-Methylhex-3-ene 3-Methylhex-2-ene

$$+\ CH_3CH_2CH_2C(CH_2CH_3){=}CH_2$$

2-Ethylpent-1-ene

7.42 Both *tert*-butyl chloride and *tert*-butyl bromide dissociate to form the same carbocation intermediate. In ethanol, this carbocation yields the same mixture of products in the same ratio, regardless of the starting material.

7.43

$$CH_3CH_2CH(CH_3)CH_2I + {}^{-}CN \longrightarrow CH_3CH_2CH(CH_3)CH_2CN + I^{-}$$

(primary halide)

This is a S_N2 reaction, in which reaction rate depends on the concentration of both alkyl halide and nucleophile.

(a) Halving the concentration of cyanide and doubling the concentration of alkyl halide does not change the reaction rate.

(b) Tripling the concentrations of both cyanide and alkyl halide causes a ninefold increase in reaction rate.

7.44

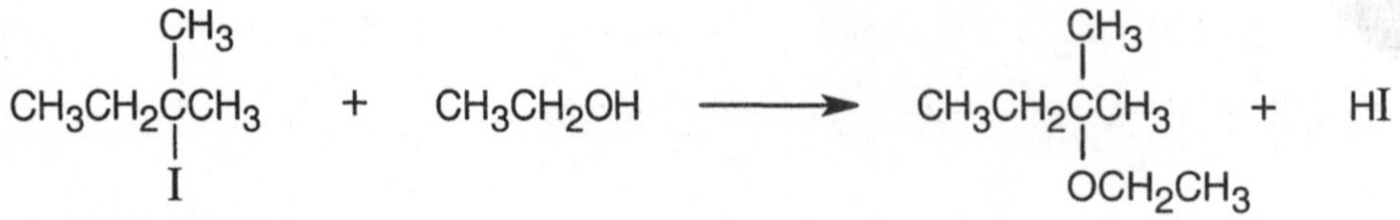

$$CH_3CH_2C(CH_3)(I)CH_3 + CH_3CH_2OH \longrightarrow CH_3CH_2C(CH_3)(OCH_2CH_3)CH_3 + HI$$

(tertiary halide)

This is an S_N1 reaction, whose reaction rate depends only on the concentration of 2-iodo-2-methylbutane. Tripling the concentration of alkyl halide triples the rate of reaction.

7.45

(a)

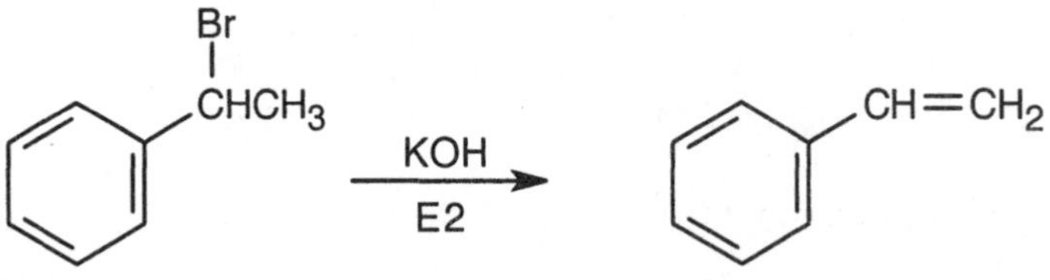

$$C_6H_5CH(Br)CH_3 \xrightarrow[E2]{KOH} C_6H_5CH{=}CH_2$$

(secondary benzylic halide) (elimination product)

This is an E2 reaction.

(b)

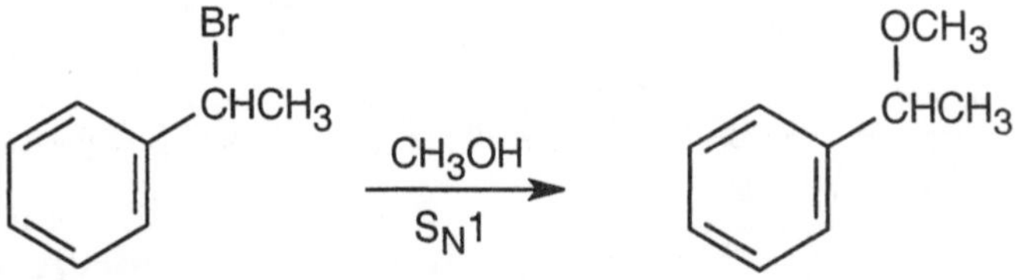

$$C_6H_5CH(Br)CH_3 \xrightarrow[S_N1]{CH_3OH} C_6H_5CH(OCH_3)CH_3$$

(secondary benzylic halide) (substitution product)

This is an S_N1 reaction.

7.46

(a)

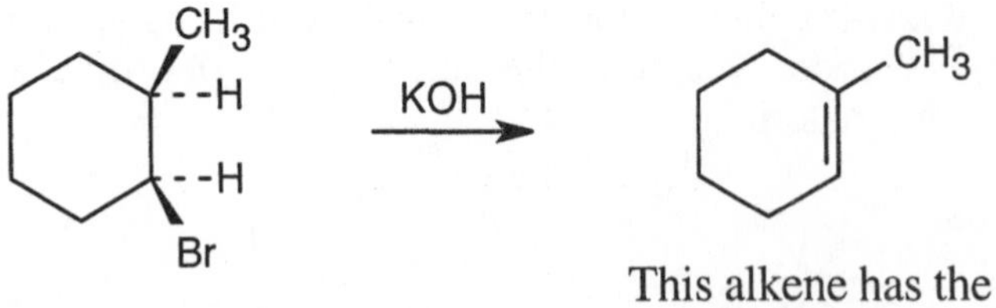

$$\text{1-bromo-2-methylcyclohexane (CH}_3\text{, H, H, Br)} \xrightarrow{KOH} \text{1-methylcyclohexene}$$

This alkene has the more substituted double bond.

(b)

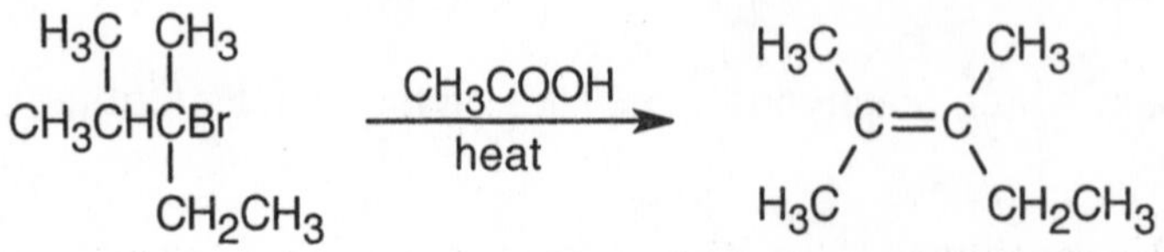

$$CH_3CH(CH_3)C(CH_3)(Br)CH_2CH_3 \xrightarrow[heat]{CH_3COOH} (H_3C)_2C{=}C(CH_3)CH_2CH_3$$

This alkene has the more substituted double bond.

7.47

$$CH_3CH_2CH(Cl)CH_3 \xrightarrow{:Base} CH_3CH_2CH{=}CH_2 + \underset{H_3C\quad H}{\overset{H\quad CH_3}{C{=}C}} + \underset{H\quad H}{\overset{H_3C\quad CH_3}{C{=}C}}$$

All other isomers of C_4H_9Cl yield only one alkene product on treatment with strong base.

Integrated Problems

7.48

H_3C, H, C_6H_{13} — C—Br

(*R*)-2-Bromooctane

(*R*)-2-Bromooctane is a secondary bromoalkane, which undergoes S_N2 substitution. Since S_N2 reactions proceed with inversion of configuration, the configuration at the carbon stereocenter is inverted. (This does not necessarily mean that all *R* isomers become *S* isomers after an S_N2 reaction. The *R,S* designation refers to the priorities of groups,which may change when the nucleophile is varied.)

$$Nu: + H_3C(H)(C_6H_{13})C{-}Br \longrightarrow Nu{-}C(H)(CH_3)(C_6H_{13}) + Br^-$$

	Nucleophile	*Product*
(a)	^-CN	$NC{-}C(H)(CH_3)(C_6H_{13})$, *S*
(b)	CH_3COO^-	$CH_3COO{-}C(H)(CH_3)(C_6H_{13})$, *S*
(c)	Br^-	$Br{-}C(H)(CH_3)(C_6H_{13})$, *S* + $H_3C(H)(C_6H_{13})C{-}Br$, *R*

2-Bromooctane is 100% racemized after 50% of the original (*R*)-2-bromooctane has reacted with Br.

7.49 S_N2 reactivity:

Most reactive ⟶ *Least reactive*

$CH_3CH_2CH_2CH_2Br$ > $CH_3CH(CH_3)CH_2Br$ > $CH_3CH_2CH(Br)CH_3$ > $CH_3C(Br)(CH_3)CH_3$

1-Bromobutane > 1-Bromo-2-methylpropane > 2-Bromobutane > 2-Bromo-2-methylpropane

7.50

$$H_2C{=}CH\dot{C}H_2 \longleftrightarrow H_2\dot{C}CH{=}CH_2$$

Two resonance forms contribute to the relative stability of the allyl radical. Because it is more stable, this radical is formed in preference to other radicals.

7.51

Five resonance forms contribute to the stability of the benzyl radical.

7.52

$$C_6H_{11}O^- + CH_3Br \longrightarrow C_6H_{11}OCH_3$$

This is an excellent method of ether preparation since bromomethane is very reactive in S_N2 substitutions.

$$CH_3O^- + C_6H_{11}Br \longrightarrow C_6H_{11}OCH_3 + C_6H_{10}$$

Reaction of a secondary haloalkane with a basic nucleophile yields both substitution and elimination products. This is obviously a less satisfactory method of ether preparation.

7.53

(1) $CH_3CH_2OH + PBr_3 \longrightarrow CH_3CH_2Br$

(2) $CH_3CH_2O^- + CH_3CH_2Br \longrightarrow CH_3CH_2OCH_2CH_3 + Br^-$

7.54

3-Bromocyclohexene $\xrightarrow[\text{Pd catalyst}]{H_2}$ Bromocyclohexane $\xrightarrow[\text{ether}]{Mg}$ $C_6H_{11}MgBr$ $\xrightarrow{H_2O}$ Cyclohexane

7.55

$$\ce{BrCH2CH2CH2CH2OH ->[:Base] [BrCH2CH2CH2CH2O^-] ->[S_N2] \text{Tetrahydrofuran} + :Br:^-}$$

Tetrahydrofuran

The product is the cyclic ether tetrahydrofuran.

7.56

cis

trans diaxial

ring flip

KOH

The expected Zaitsev product would be formed if the atoms to be eliminated could have a cis relationship. Because the non-Zaitsev cycloalkene is formed instead, E2 elimination must require the two atoms that are eliminated to have a trans diaxial relationship.

7.57

$:Br:^-$ + (R)-$CH_3CH(Br)CH_2CH_2CH_2CH_3$ ⟶ (S)-$BrCH(CH_3)CH_2CH_2CH_2CH_3$ + $:Br:^-$

Reaction of one molecule of (*R*)-2-bromohexane with one bromide ion produces one molecule of (*S*)-2-bromohexane. Reaction of 50% of the *R* starting material gives a mixture of 50% *S* enantiomer plus 50% unreacted *R* starting material – a racemic mixture. Thus, after 50% of the *R* starting material has reacted, the product is 100% racemized.

7.58 The chiral tertiary alcohol (*R*)-3-methylhexan-3-ol reacts with HBr by an S_N1 pathway. HBr protonates the hydroxyl group, which dissociates to yield a planar, achiral carbocation. Attack by the nucleophilic bromide anion can occur with equal probability from either side of the carbocation to produce (±)3-bromo-3-methylhexane.

7.59 Since butan-2-ol is a secondary alcohol, substitution can occur by either a S_N1 or a S_N2 route, depending on reaction conditions. Two factors favor an S_N1 mechanism in this case. (1) The reaction is run under solvolysis (solvent as nucleophile) conditions. (2) Dilute acid converts a poor leaving group (^-OH) into a good leaving group (H_2O), which departs easily.

Protonation of the hydroxyl oxygen..

is followed by loss of water to form a planar carbocation.

Reaction with water at either side of the planar carbocation yields racemic product.

H_3O^+

$+ H_2O$

$-H_2O$

$-H^+$

$-H^+$

7.60 Two optically inactive structures are possible for compound **A**. Any other structure of the formula $C_{16}H_{16}Br_2$ that undergoes the series of reactions is optically active.

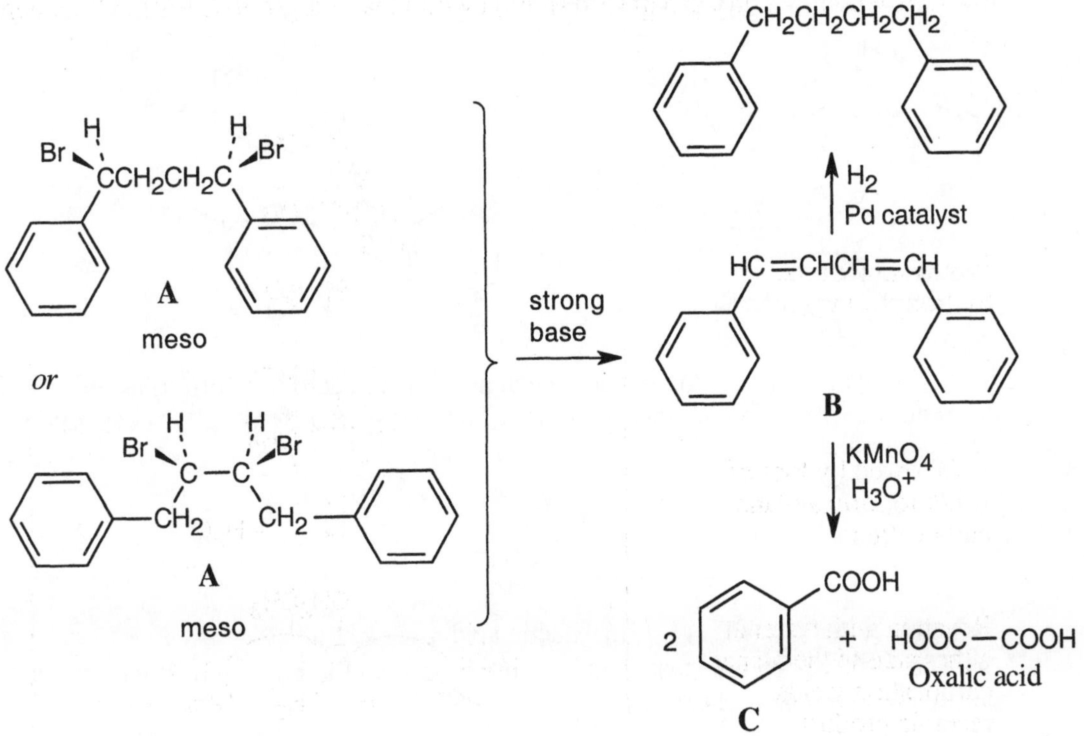

$CH_2CH_2CH_2CH_2$

H_2
Pd catalyst

HC=CHCH=CH

A
meso

or

strong base

B

$KMnO_4$
H_3O^+

A

COOH

meso

2 + HOOC—COOH
Oxalic acid

C

7.61 A Grignard reagent can't be prepared from a compound containing an acidic functional group because the Grignard reagent is immediately quenched by the proton source. For example, the –COOH, –OH, $–NH_2$, and RC≡CH functional groups are too acidic to allow for preparation of a Grignard reagent.

In the Medicine Cabinet

7.62

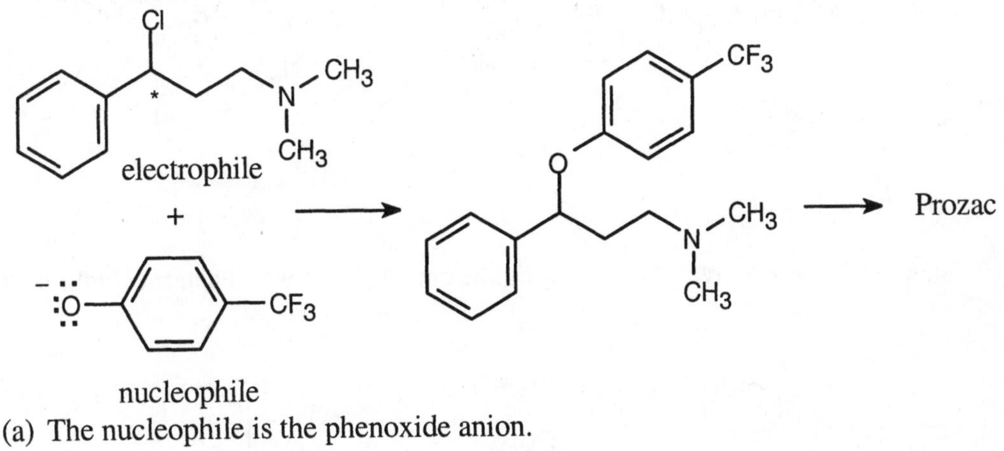

Cl
CH_3
N
CH_3
electrophile
+
CF_3
O
CH_3
N
CH_3
Prozac
:O:⁻
CF_3
nucleophile

(a) The nucleophile is the phenoxide anion.

(b) The electrophile is the alkyl chloride.

(c) This is an S_N2 reaction since the rate depends on the concentrations of both the phenoxide and alkyl chloride.
(d) The chirality center is starred.
(e) The product is the (*S*) enantiomer.

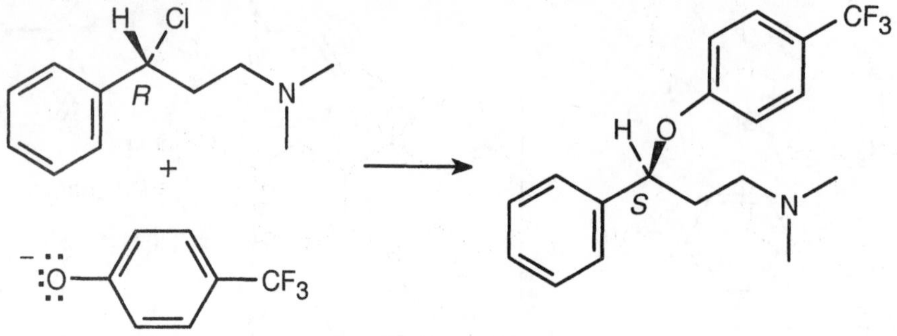

H Cl R N + :O– CF3 → CF3 H O S N

(f) An optically pure carboxylic acid such as (*S*)-lactic acid can form diastereomeric ammonium salts with racemic fluoxetine. The two salts differ in physical properties and can be separated.

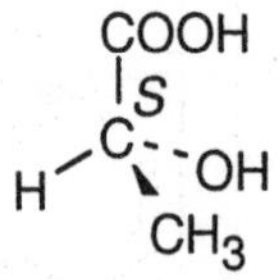

COOH S C H OH CH_3

(*S*)-Lactic acid

(g) The positive bromine test is an indication of a double bond and suggests that E2 elimination has taken place.

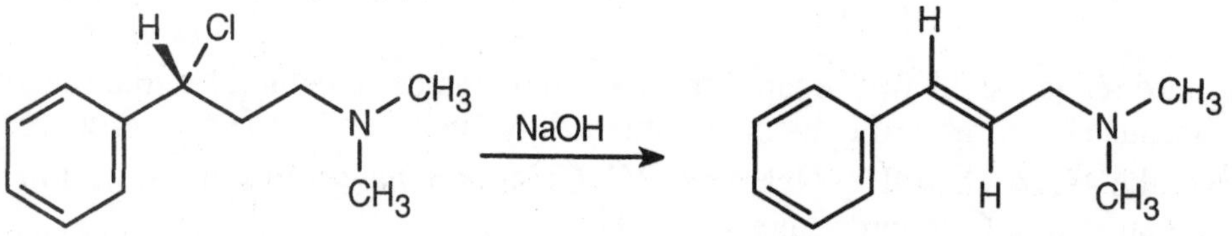

H Cl N CH_3 CH_3 NaOH → H H N CH_3 CH_3

7.63 (a)

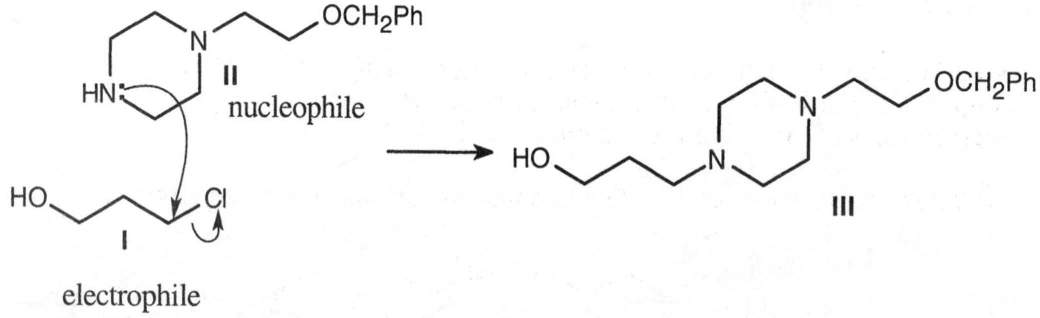

N OCH_2Ph HN: II nucleophile HO Cl I electrophile → HO N N OCH_2Ph III

(b) Since this is an S_N2 reaction, doubling the concentrations of each reactant quadruples the rate.
(c)

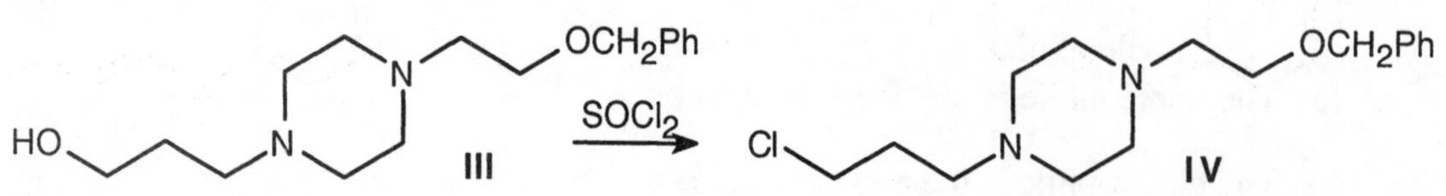

HO N N OCH_2Ph III $SOCl_2$ → Cl N N OCH_2Ph IV

(d) A Grignard reaction is used to convert **IV** to **V**.

(e)

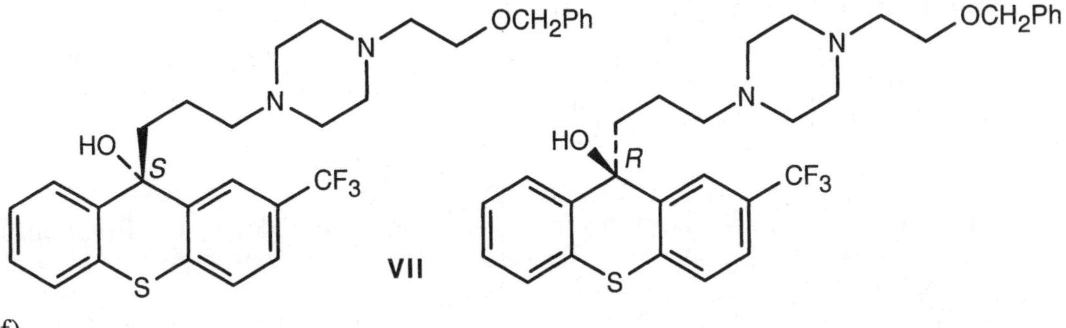

OCH2Ph OCH2Ph
N N
N N
HO S HO R
CF3 CF3
S VII S

(f)

OH OH
H—Cl N
HO: N N
1. $H_2\overset{+}{O}$:
CF3 CF3
S VIII S

2.

OH OH
:Base
N N
N H N
CF3 3. CF3
+
IX + H_2O
S S

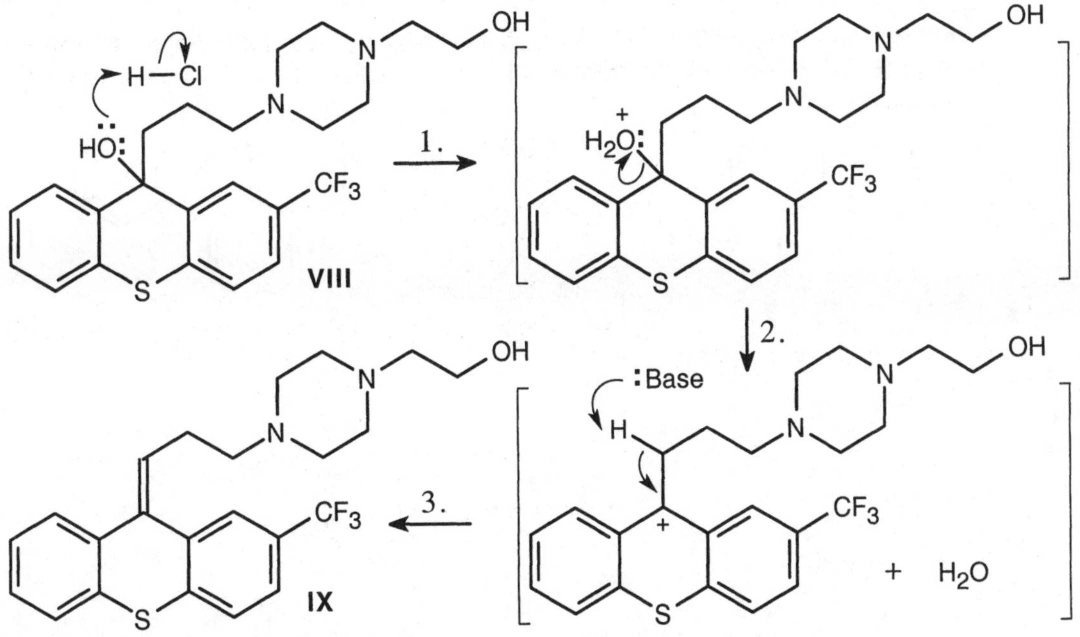

Step 1: Protonation of –OH to make it a better leaving group
Step 2: Loss of H_2O to form a stable carbocation.
Step 3: Loss of H^+ to form the double bond.

(g) The stereoisomers are *E,Z* double bond isomers and are diastereomers.

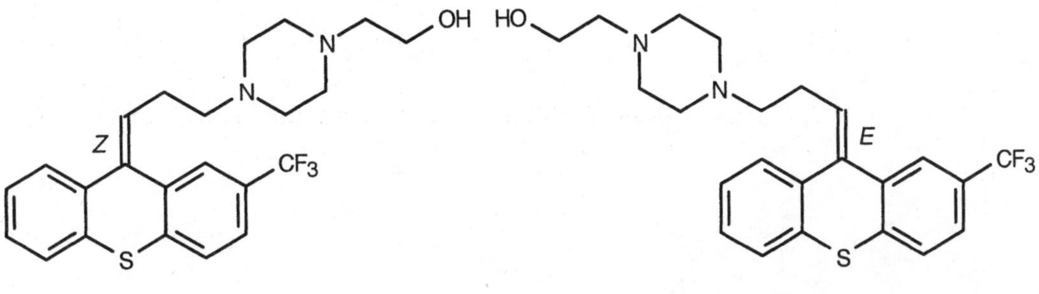

OH HO
N N
N N
Z E
CF3 CF3
S S

In the Field with Agrochemicals

7.64 (a) The CF_3 carbon is electron-deficient due to the electronegativity of the three fluorine atoms.

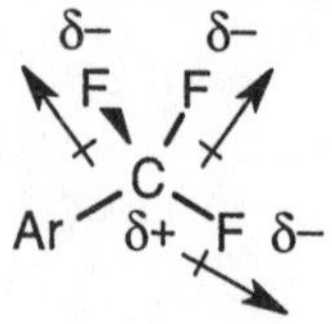

(b) It is harder to remove electrons from a benzene ring bonded to a $-CF_3$ group. The electron-withdrawing group makes the benzene ring more electron-poor and less able to be oxidized.

(c) The first two compounds have a (trifluoromethyl)phenyl group bonded to a heterocyclic ring containing nitrogen and oxygen. One would expect them to function similarly - perhaps to inhibit pigment synthesis. The other three groups have the structural similarities that are circled and inhibit lipid synthesis.

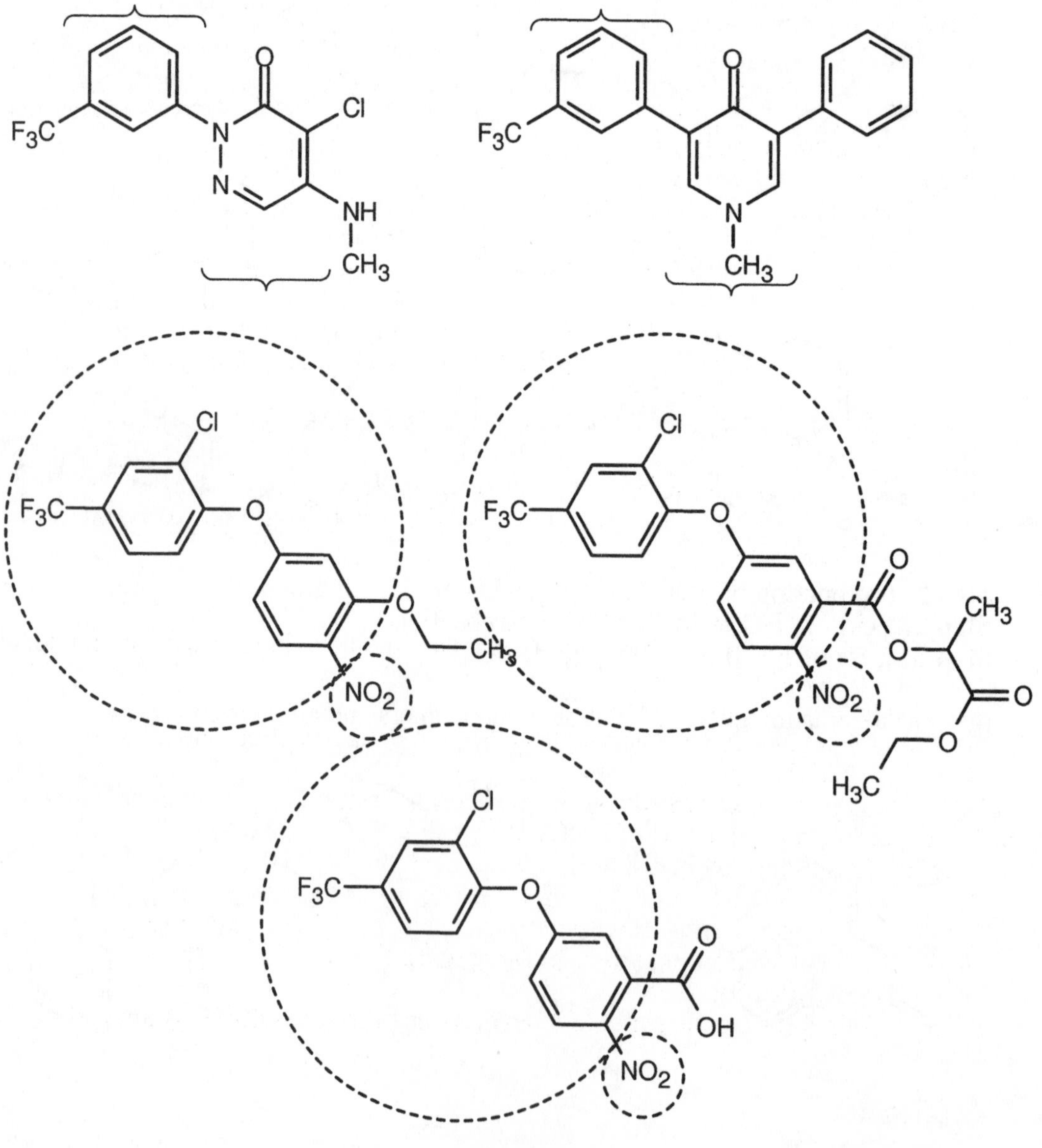

Chapter 8 – Alcohols, Phenols, and Ethers

Chapter Outline

I. Naming alcohols, phenols, and ethers (Section 8.1).
   A. Alcohols are classified as primary, secondary or tertiary, depending on the number of organic groups bonded to the –OH carbon.
   B. Rules for naming alcohols.
      1. The longest chain containing the –OH group is the parent chain, and the parent name replaces *-e* with *-ol*.
      2. Numbering begins at the end of the chain nearer the –OH group.
         The number of the carbon bonded to the –OH group is placed immediately before the suffix *-ol*.
      3. The substituents are numbered according to their position on the chain and cited in alphabetical order.
   C. Phenols are named according to rules discussed in Section 5.3.
   D. Naming ethers.
      1. Ethers with no other functional groups are named by citing the two organic substituents and adding the word "ether".
      2. When other functional groups are present, the ether is an alkoxy substituent.

II. Properties of alcohols, phenols and ethers (Sections 8.2–8.3).
   A. Hydrogen bonding (Section 8.2).
      1. The oxygen atoms of alcohols and ethers have sp^3 hybridization and a nearly tetrahedral bond angle.
      2. Alcohols and phenols have elevated boiling points, relative to hydrocarbons, due to hydrogen bonding.
         a. In hydrogen bonding, an –OH hydrogen is attracted to a lone pair of electrons on another molecule, resulting in a weak electrostatic force that holds the molecules together.
         b. These weak forces must be overcome in boiling.
   B. Acidity (Section 8.3).
      1. Alcohols and phenols are both weakly acidic and weakly basic.
      2. Alcohols and phenols dissociate to a slight extent to form alkoxide ions and phenoxide ions.
      3. Acidity of alcohols.
         a. Alcohols are similar in acidity to water.
         b. Alcohols don't react with weak bases, but they do react with alkali metals and strong bases.
      4. Acidity of phenols.
         a. Phenols are a million times more acidic than alcohols and are soluble in dilute NaOH.
         b. Phenol acidity is due to resonance stabilization of the phenoxide anion.

III. Alcohols and phenols (Sections 8.4–8.6).
   A. Preparation (Sections 8.4, 8.6).
      1. Alcohols (Section 8.4).
         a. By hydration of alkenes.
         b. By reduction of carbonyl compounds.
      2. Phenols (Section 8.6).
         Phenols can be synthesized by reacting benzenesulfonic acids with NaOH.

- B. Reactions (Sections 8.5–8.6).
  - 1. Alcohols (Section 8.5).
    - a. Dehydration to yield alkenes.
      - i. Dehydration follows Zaitsev's Rule.
      - ii. Dehydration proceeds by an E1 mechanism.
    - b. Oxidation to yield carbonyl compounds.
      - i. Primary alcohols can be oxidized to carboxylic acids with CrO_3.
      - ii. Secondary alcohols can be oxidized to ketones with CrO_3.
      - iii. Primary alcohols can be oxidized to aldehydes with PCC.
    - c. Conversion into ethers – Williamson ether synthesis.
      - i. Reaction of an alkoxide with a primary alkyl halide produces an ether.
      - ii. The alkoxide ion is formed by treating an alcohol with sodium metal.
      - iii. The reaction proceeds by an S_N2 mechanism.
  - 2. Phenols (Section 8.6).
    - a. Williamson ether synthesis.
    - b. Electrophilic aromatic substitution.
    - c. Oxidation to yield quinones.
      - i. Oxidation of a phenol with $Na_2Cr_2O_7$ yields a quinone.
      - ii. Quinones can be reduced to hydroquinones.
      - iii. Oxidation-reduction reactions of quinones are important in living systems.

IV. Ethers (Section 8.7–8.8).

- A. Preparation by Williamson ether synthesis.
- B. Reactions – acidic cleavage.
  - 1. Cleavage reactions proceed either by a S_N1 or a S_N2 mechanism.
  - 2. With primary and secondary ethers, the ether oxygen stays with the more substituted alkyl group.
  - 3. With tertiary ethers, oxygen stays with the less substituted alkyl group.
- C. Epoxides (Section 8.8).
  - 1. Epoxides are prepared by reaction of an alkene with a peroxyacid.
  - 2. Ring-opening reactions of epoxides.
    - a. Dilute aqueous acid converts an epoxide to a glycol.
    - b. Reaction takes place by S_N2 attack of H_2O on a protonated epoxide.

V. Thiols and sulfides (Section 8.9).

- A. Naming thiols and sulfides.
  - 1. Thiols (sulfur analogs of alcohols) are named by the same system as alcohols, with the suffix *-thiol* replacing *-ol.*
    The –SH group is a mercapto- group.
  - 2. Sulfides (sulfur analogs of ethers) are named by the same system as ethers, with *sulfide* replacing *ether.*
    The –SR group is an alkylthio- group.
- B. Preparation.
  - 1. Thiols may be prepared by S_N2 displacement with a sulfur nucleophile.
  - 2. Treatment of a thiol with base yields a thiolate anion, which can react with an alkyl halide to form a sulfide.
- C. Reactions.
  Thiols can be oxidized by Br_2 or I_2 to yield disulfides, RSSR.
  The reaction can be reversed by treatment with zinc and acid.

Solutions to Problems

8.1–8.2

(a) $CH_3CH(OH)CH_2CH(OH)CH(CH_3)CH_3$

5-Methylhexane-2,4-diol
secondary alcohol

(b) $C_6H_5CH_2CH_2C(OH)(CH_3)CH_3$

2-Methyl-4-phenylbutan-2-ol
tertiary alcohol

(c) HO–cyclohexane with two CH_3 groups at C-4

4,4-Dimethylcyclohexanol
secondary alcohol

(d) cyclopentane with H, Br and H, OH

cis-2-Bromocyclopentanol
secondary alcohol

8.3

(a) $CH_3CH_2CH_2CH_2C(OH)(CH_3)_2$

2-Methylhexan-2-ol

(b) $CH_3CH(OH)CH_2CH_2CH_2CH_2OH$

Hexane-1,5-diol

(c) $CH_3CH{=}C(CH_2CH_3)CH_2OH$

2-Ethylbut-2-en-1-ol

(d) Cyclohex-3-en-1-ol

(e) *o*-Bromophenol

(f) OH, O_2N, NO_2, NO_2

2,4,6-Trinitrophenol

8.4

(a) $CH_3CH(CH_3)OCH(CH_3)CH_3$

2-Isopropoxypropane
or
Diisopropyl ether

(b) cyclopentane–$OCH_2CH_2CH_3$

Propoxycyclopentane
or
Cyclopentyl propyl ether

(c) benzene ring with OCH_3 and Br

p-Bromoanisole
or
p-Bromomethoxybenzene
or
Methyl *p*-bromophenyl ether

(d)

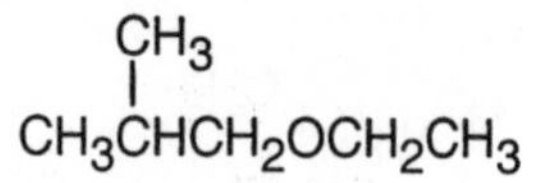

$$\begin{array}{c} CH_3 \\ | \\ CH_3CHCH_2OCH_2CH_3 \end{array}$$

1-Ethoxy-2-methylpropane
or
Ethyl isobutyl ether

8.5

(a)

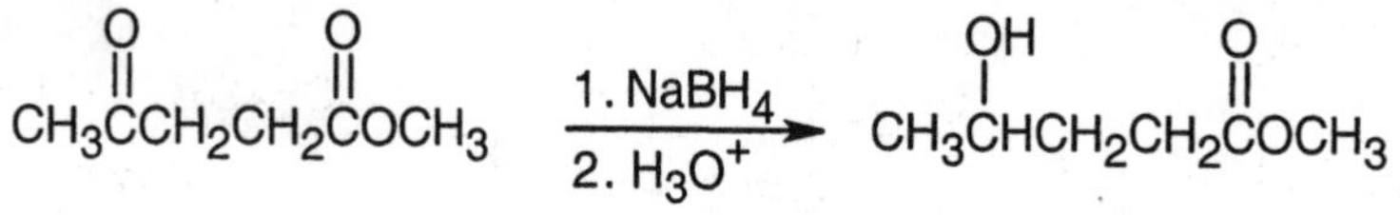

$$CH_3\overset{O}{\overset{||}{C}}CH_2CH_2\overset{O}{\overset{||}{C}}OCH_3 \xrightarrow[2.\ H_3O^+]{1.\ NaBH_4} CH_3\overset{OH}{\overset{|}{C}}HCH_2CH_2\overset{O}{\overset{||}{C}}OCH_3$$

$NaBH_4$ reduces aldehydes and ketones without interfering with other functional groups.

(b)

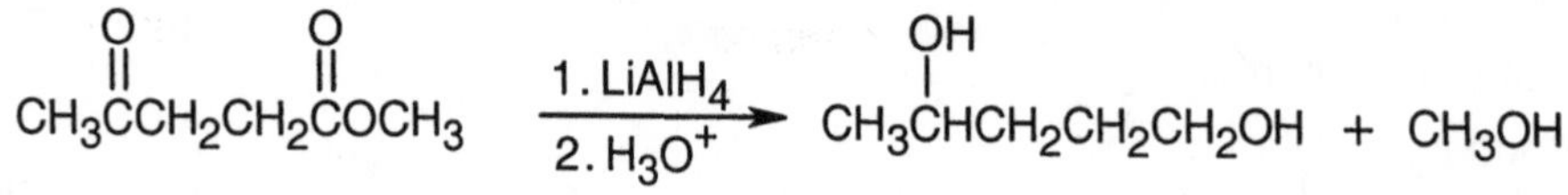

$$CH_3\overset{O}{\overset{||}{C}}CH_2CH_2\overset{O}{\overset{||}{C}}OCH_3 \xrightarrow[2.\ H_3O^+]{1.\ LiAlH_4} CH_3\overset{OH}{\overset{|}{C}}HCH_2CH_2CH_2OH + CH_3OH$$

$LiAlH_4$ reduces both ketones and esters.

8.6

(a)

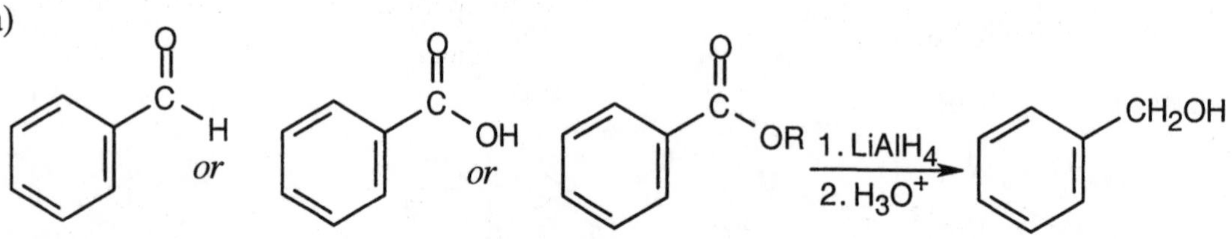

$$C_6H_5CHO \text{ or } C_6H_5COOH \text{ or } C_6H_5COOR \xrightarrow[2.\ H_3O^+]{1.\ LiAlH_4} C_6H_5CH_2OH$$

Benzyl alcohol is a product of reduction of an aldehyde, a carboxylic acid, or an ester.

(b)

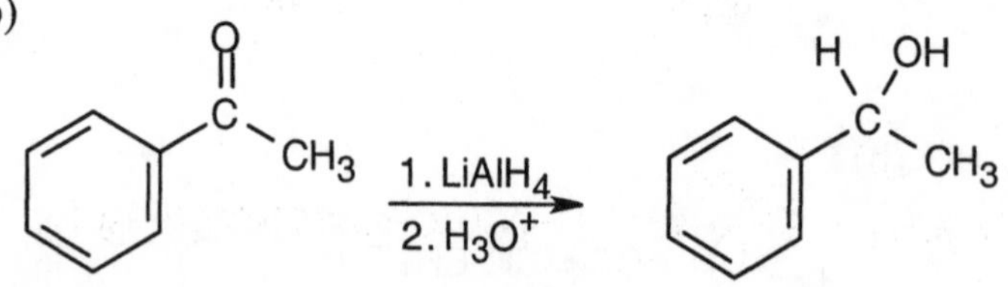

$$C_6H_5COCH_3 \xrightarrow[2.\ H_3O^+]{1.\ LiAlH_4} C_6H_5CH(OH)CH_3$$

Reduction of a ketone yields this secondary alcohol.

(c)

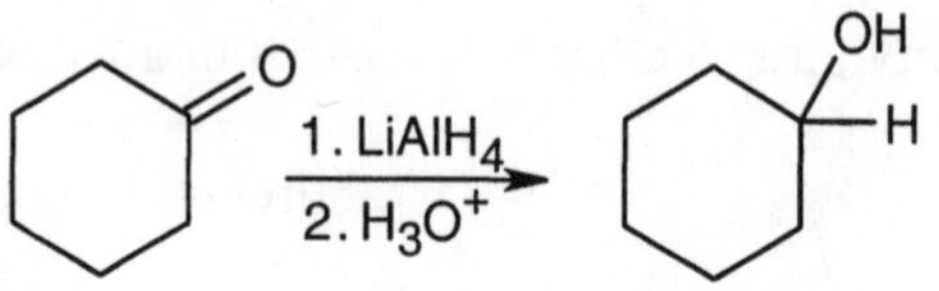

$$\text{Cyclohexanone} \xrightarrow[2.\ H_3O^+]{1.\ LiAlH_4} \text{Cyclohexanol (C(OH)H)}$$

8.7

(a)

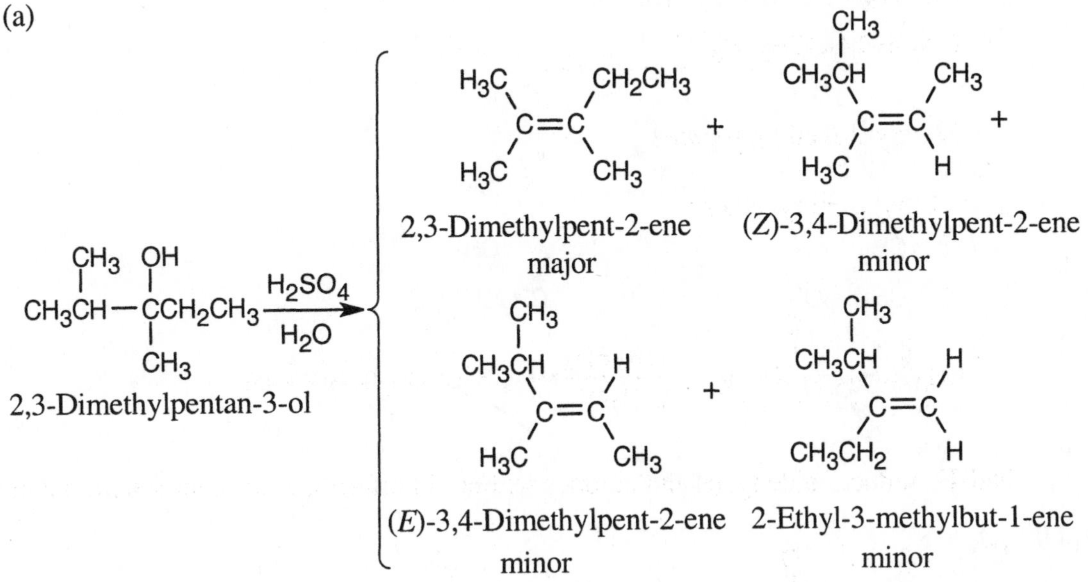

$$CH_3CH(CH_3)-C(OH)(CH_3)CH_2CH_3 \xrightarrow[H_2O]{H_2SO_4}$$

2,3-Dimethylpentan-3-ol

$(CH_3)_2C=C(CH_3)CH_2CH_3$ + $(CH_3)_2CH(CH_3)C=CH(CH_3)$ +

2,3-Dimethylpent-2-ene, major; (*Z*)-3,4-Dimethylpent-2-ene, minor

$(CH_3)_2CH(CH_3)C=CH(CH_3)$ + $(CH_3)_2CH(CH_3CH_2)C=CH_2$

(*E*)-3,4-Dimethylpent-2-ene, minor; 2-Ethyl-3-methylbut-1-ene, minor

(b)

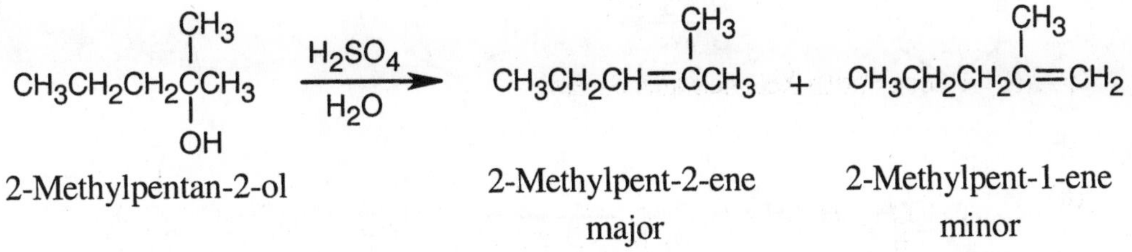

$$CH_3CH_2CH_2C(CH_3)(OH)CH_3 \xrightarrow[H_2O]{H_2SO_4} CH_3CH_2CH=C(CH_3)CH_3 + CH_3CH_2CH_2C(CH_3)=CH_2$$

2-Methylpentan-2-ol → 2-Methylpent-2-ene (major) + 2-Methylpent-1-ene (minor)

8.8

(a)

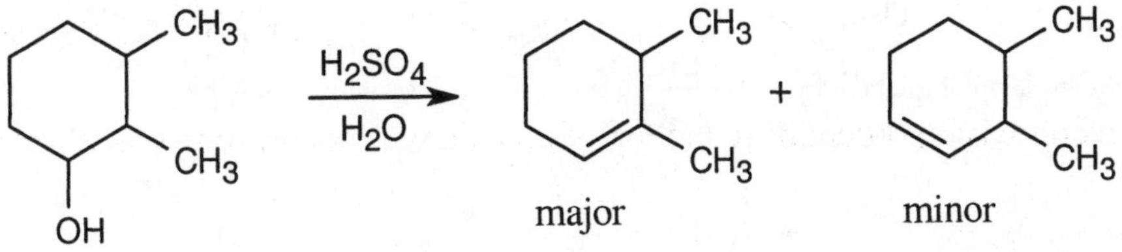

2,3-dimethylcyclohexanol (OH, CH_3, CH_3) $\xrightarrow[H_2O]{H_2SO_4}$ 1,6-dimethylcyclohexene (CH_3, CH_3) + 3,4-dimethylcyclohexene (CH_3, CH_3)

major; minor

(b)

$$CH_3CH_2CH_2CH(OH)CH_2CH_2CH_3 \xrightarrow[H_2O]{H_2SO_4} CH_3CH_2CH_2{=}CHCH_2CH_2CH_3$$

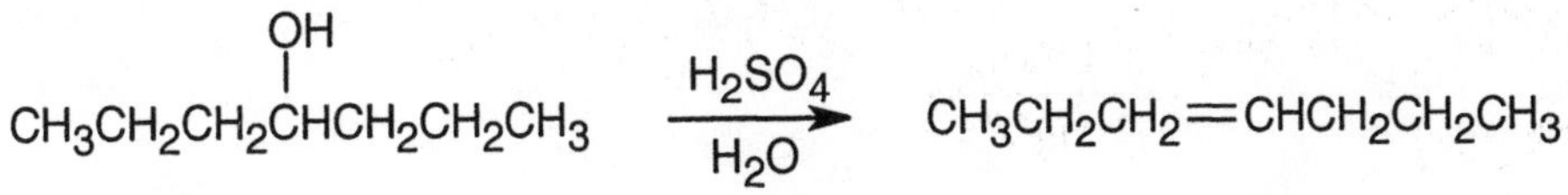

Whenever possible, start with an alcohol that gives only one product (the desired product).

8.9 Aldehydes are synthesized by oxidation of primary alcohols, and ketones are synthesized by oxidation of secondary alcohols.

(a)

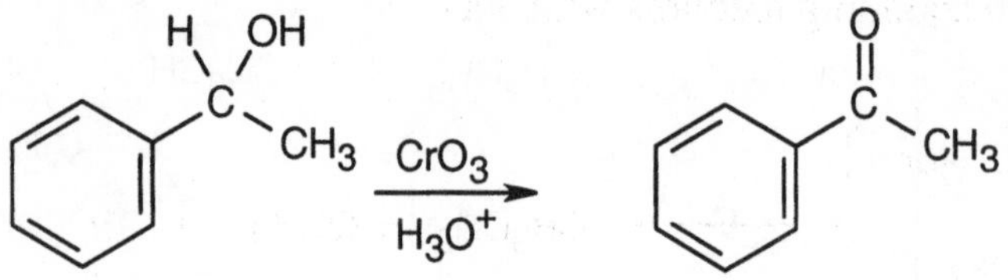

$$C_6H_5CH(OH)CH_3 \xrightarrow[H_3O^+]{CrO_3} C_6H_5COCH_3$$

(b)

$$CH_3CH(CH_3)CH_2OH \xrightarrow[CH_2Cl_2]{PCC} CH_3CH(CH_3)CHO$$

(c)

$$\text{cyclopentanol} \xrightarrow[H_3O^+]{CrO_3} \text{cyclopentanone}$$

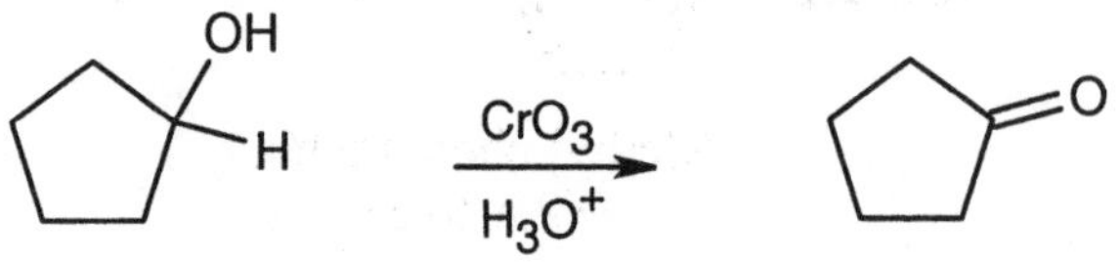

8.10

(a)

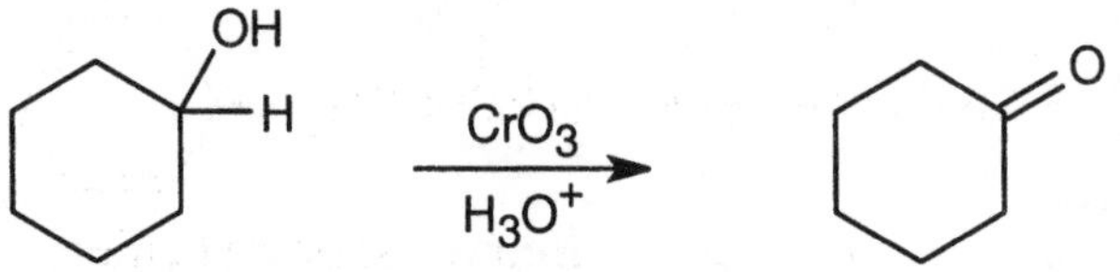

$$\text{cyclohexanol} \xrightarrow[H_3O^+]{CrO_3} \text{cyclohexanone}$$

(b)

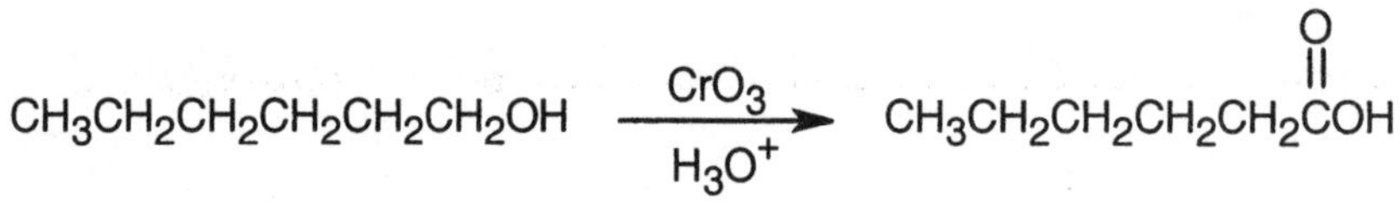

$$CH_3CH_2CH_2CH_2CH_2CH_2OH \xrightarrow[H_3O^+]{CrO_3} CH_3CH_2CH_2CH_2CH_2COOH$$

(c)

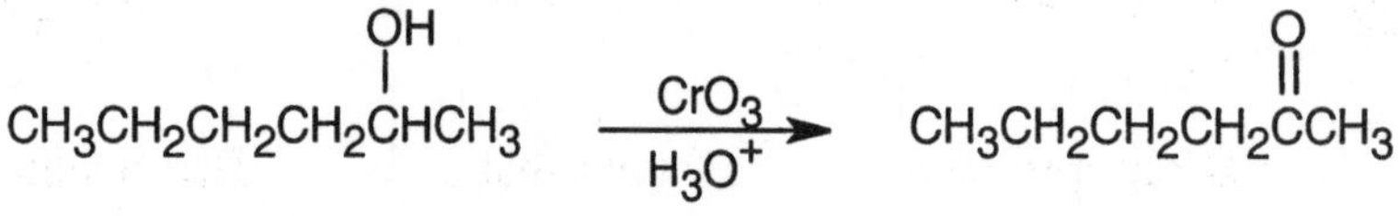

$$CH_3CH_2CH_2CH_2CH(OH)CH_3 \xrightarrow[H_3O^+]{CrO_3} CH_3CH_2CH_2CH_2COCH_3$$

8.11

(a) cyclohexanone (b) $CH_3CH_2CH_2CH_2CH_2CHO$ (c) $CH_3CH_2CH_2CH_2COCH_3$

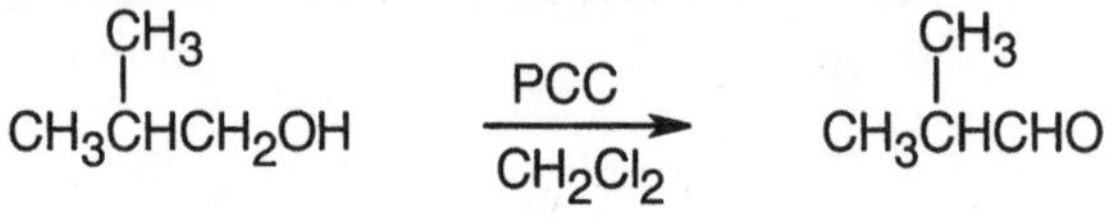

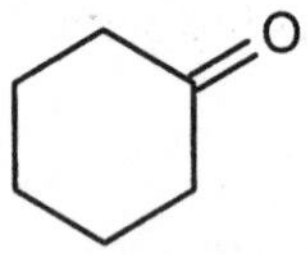

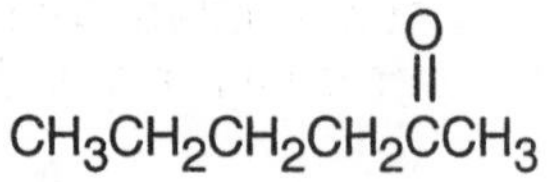

8.12

$$\text{cyclohexanol} + Na \xrightarrow[\text{alkoxide formation}]{} \text{cyclohexyl-}O^- Na^+ + 1/2\ H_2$$

$$\text{cyclohexyl-}O^- Na^+ + CH_3CH_2{-}I \xrightarrow[S_N2 \text{ substitution}]{} \text{cyclohexyl-}OCH_2CH_3 + NaI$$

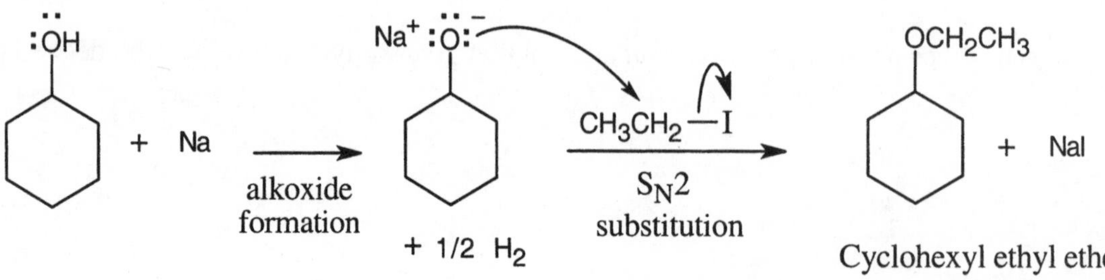

Cyclohexyl ethyl ether

8.13 Remember that the alkyl halide in the Williamson ether synthesis should be primary or methyl, in order to avoid competing elimination reactions. The alkoxide anions shown are formed by treating the corresponding alcohols with Na.

(a)

$$\left.\begin{array}{l} CH_3CH_2CH_2O^- \ + \ CH_3Br \\ \quad\quad\quad\quad\textit{or} \\ CH_3CH_2CH_2Br \ + \ CH_3O^- \end{array}\right\} \longrightarrow CH_3CH_2CH_2OCH_3 + Br^-$$

Methyl propyl ether

(b)

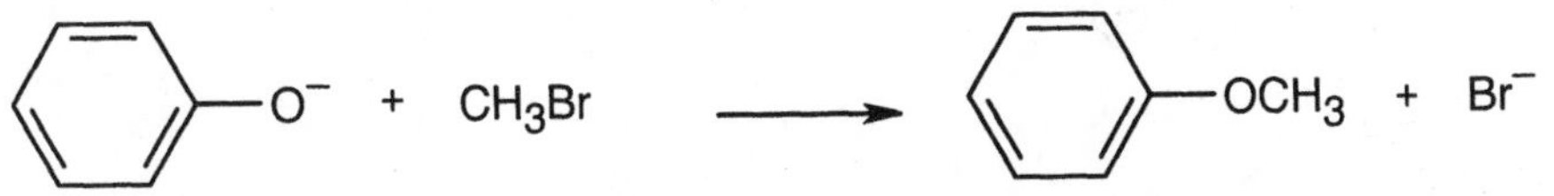

$$C_6H_5O^- + CH_3Br \longrightarrow C_6H_5OCH_3 + Br^-$$

Methyl phenyl ether (Anisole)

(c)

$$(CH_3)_2CHO^- + C_6H_5CH_2Br \longrightarrow (CH_3)_2CHOCH_2C_6H_5 + Br^-$$

Benzyl isopropyl ether

8.14

Least reactive ⟶ *Most reactive*

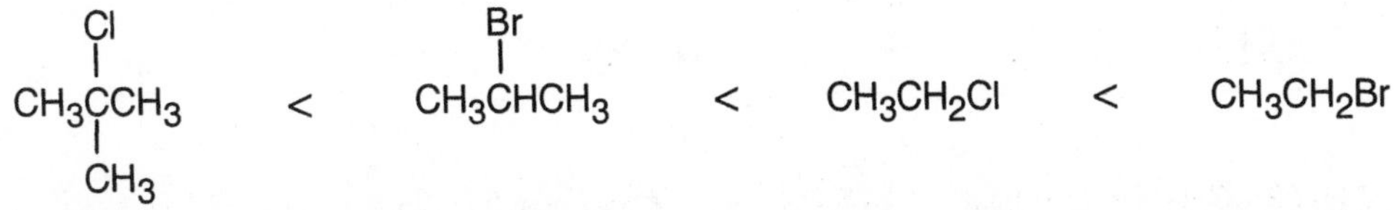

$$(CH_3)_3CCl \ < \ CH_3CHBrCH_3 \ < \ CH_3CH_2Cl \ < \ CH_3CH_2Br$$

2-Chloro-2-methylpropane < 2-Bromopropane < Chloroethane < Bromoethane

The reactivity of alkyl halides in the Williamson ether synthesis is the same as their reactivity in any S_N2 reaction.

8.15

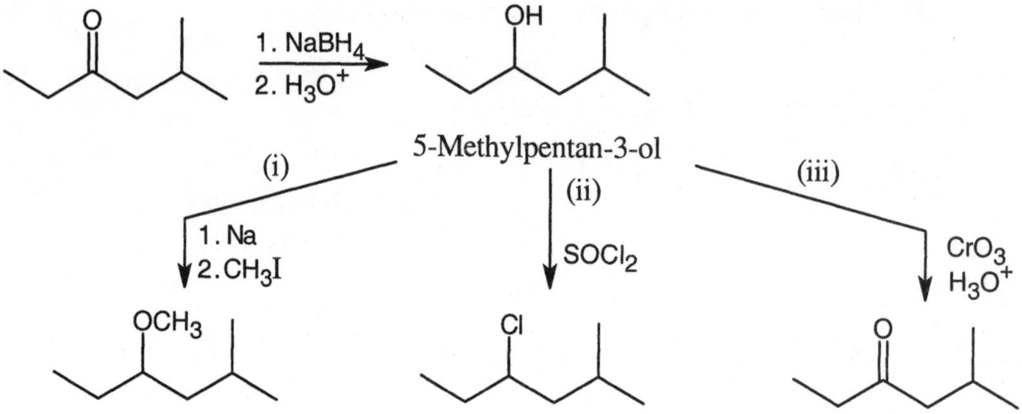

$$\text{CH}_3\text{CH}_2\text{COCH}_2\text{CH}(\text{CH}_3)_2 \xrightarrow[\text{2. } H_3O^+]{\text{1. NaBH}_4} \text{CH}_3\text{CH}_2\text{CH(OH)CH}_2\text{CH}(\text{CH}_3)_2$$

5-Methylpentan-3-ol

(i) 1. Na; 2. CH_3I → $\text{CH}_3\text{CH}_2\text{CH(OCH}_3)\text{CH}_2\text{CH}(\text{CH}_3)_2$

(ii) $SOCl_2$ → $\text{CH}_3\text{CH}_2\text{CHClCH}_2\text{CH}(\text{CH}_3)_2$

(iii) CrO_3, H_3O^+ → $\text{CH}_3\text{CH}_2\text{COCH}_2\text{CH}(\text{CH}_3)_2$

8.16

$$C_6H_6 \xrightarrow[AlCl_3]{CH_3Cl} C_6H_5CH_3 \xrightarrow[H_2SO_4]{SO_3} p\text{-}CH_3C_6H_4SO_3H \xrightarrow[\text{2. } H_3O^+]{\text{1. NaOH, 300°C}} p\text{-}CH_3C_6H_4OH$$

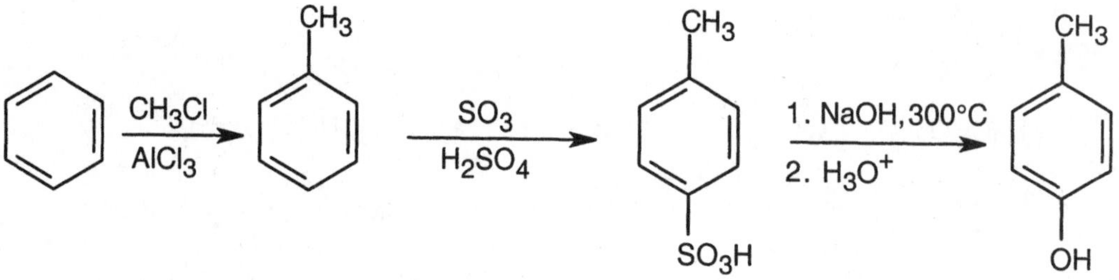

p-Cresol

8.17

(a) $$CH_3CH_2OCH_2CH_3 \xrightarrow{HI} CH_3CH_2OH + ICH_2CH_3$$

(b) $$C_6H_{11}\text{–O–}CH_2CH_3 \xrightarrow{HI} C_6H_{11}OH + ICH_2CH_3$$

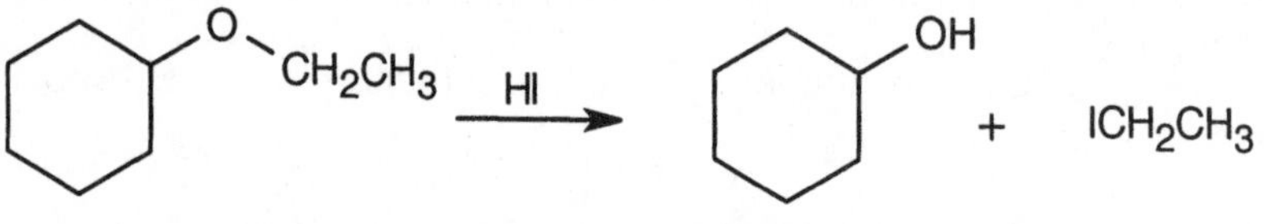

Remember that oxygen stays with the more hindered alkyl group in an S_N2 cleavage.

(c) $$(CH_3)_3C\text{–O–}CH_2CH_3 \xrightarrow{HI} (CH_3)_3C\text{–I} + HOCH_2CH_3$$

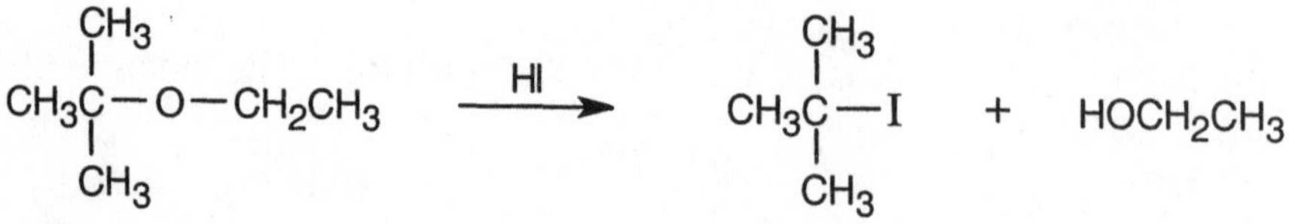

Oxygen stays with the less hindered alkyl group in an S_N1 cleavage.

8.18

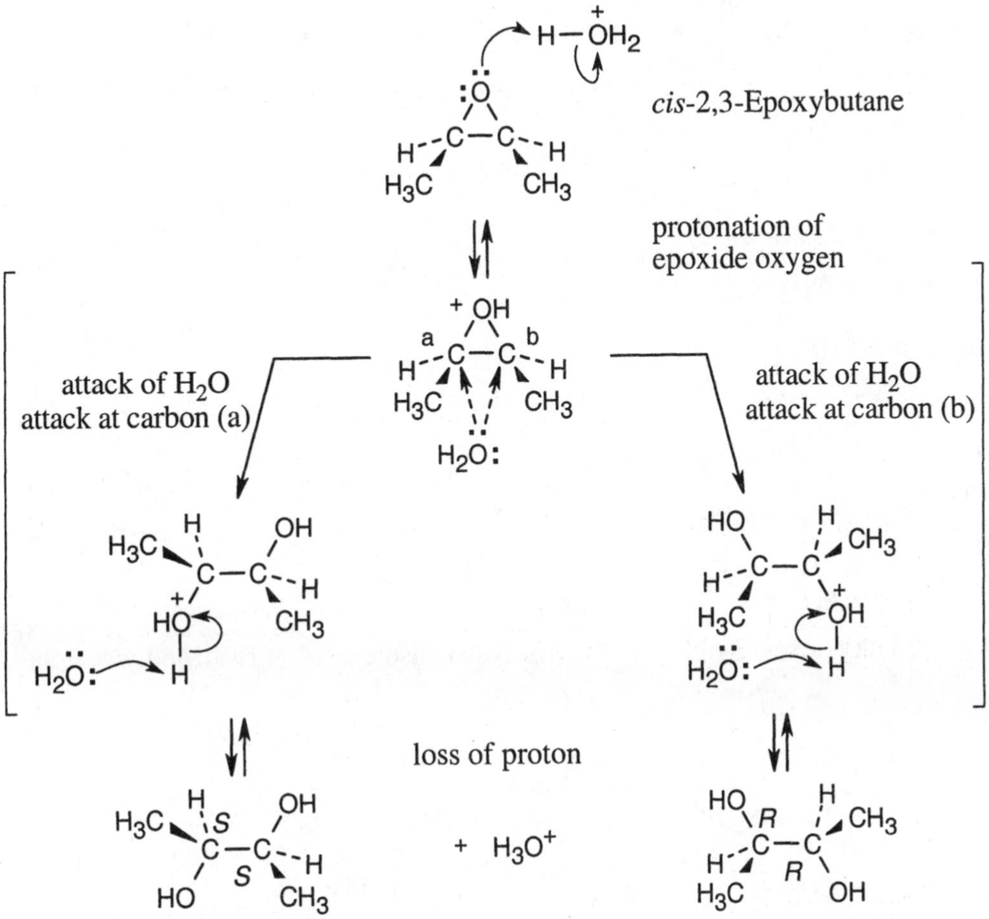

cis-2,3-Epoxybutane

protonation of epoxide oxygen

attack of H_2O attack at carbon (a)

attack of H_2O attack at carbon (b)

loss of proton

$+ \ H_3O^+$

The product of acid hydrolysis of *cis*-2,3-epoxybutane is a racemic mixture of the *R,R* and *S,S* diol enantiomers.

8.19

(a) $CH_3CH_2CH(CH_3)SH$

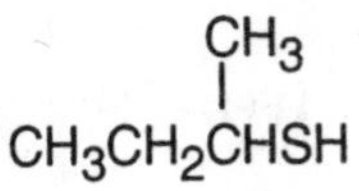

Butane-2-thiol

(b) $CH_3C(CH_3)_2CH_2CH(SH)CH_2CH(CH_3)CH_3$

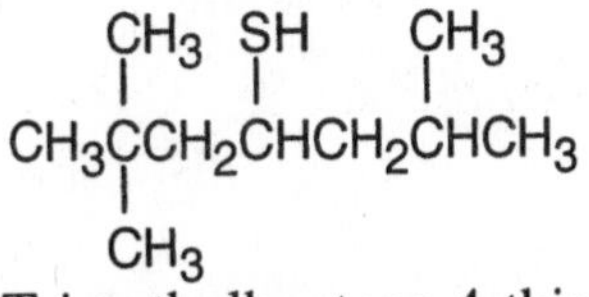

2,2,6-Trimethylheptane-4-thiol

(c) Cyclopent-3-ene-1-thiol

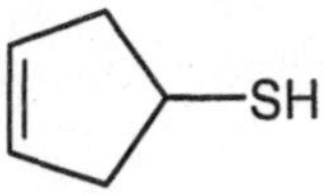

8.20

(a) $CH_3CH_2SCH_3$

Ethyl methyl sulfide

(b) $CH_3C(CH_3)_2SCH_2CH_3$

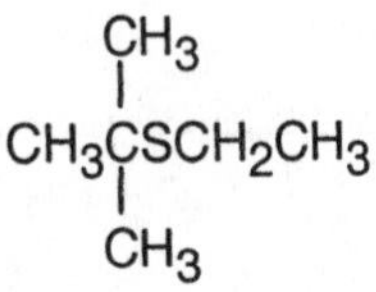

tert-Butyl ethyl sulfide

(c) *o*-Di(methylthio)benzene

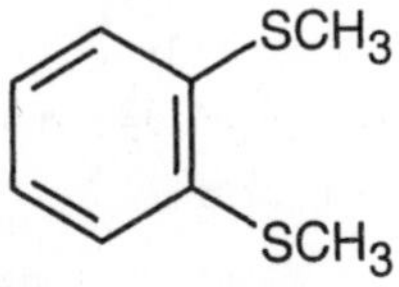

8.21

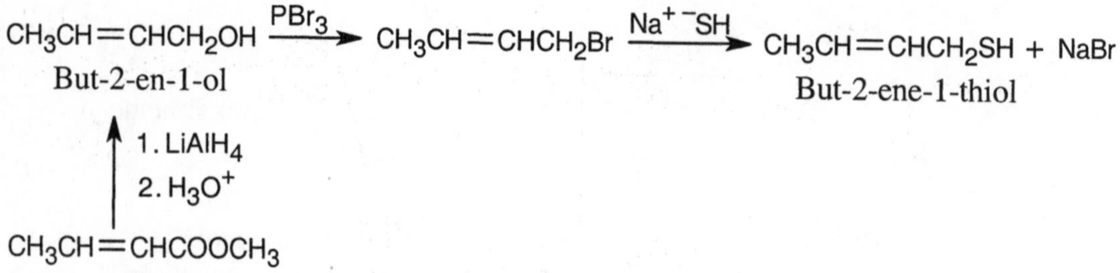

$$CH_3CH{=}CHCH_2OH \xrightarrow{PBr_3} CH_3CH{=}CHCH_2Br \xrightarrow{Na^+\ ^-SH} CH_3CH{=}CHCH_2SH + NaBr$$

But-2-en-1-ol → But-2-ene-1-thiol

$$CH_3CH{=}CHCOOCH_3 \xrightarrow[2.\ H_3O^+]{1.\ LiAlH_4} CH_3CH{=}CHCH_2OH$$

Methyl 2-butenoate → But-2-en-1-ol

Visualizing Chemistry

8.22

(a) CH_3; CH_3CHCH_2–C–CH_2CH_3; H OH

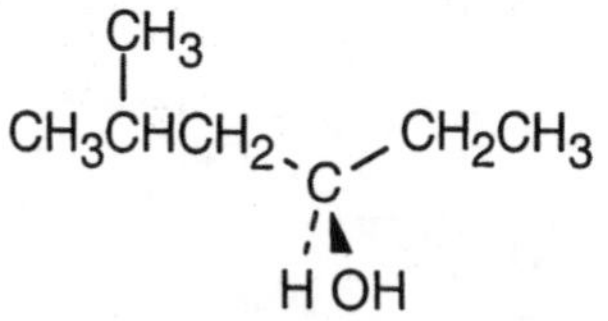

(*R*)-5-Methylhexan-3-ol

(b) O_2N, OCH_3

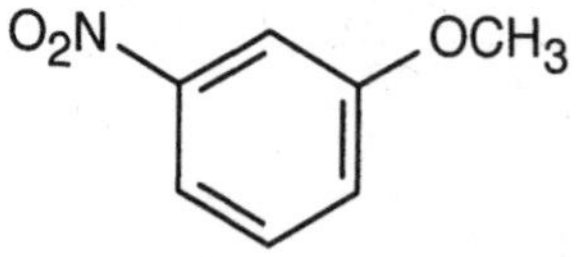

m-Nitroanisole

(c) H_3C, HO

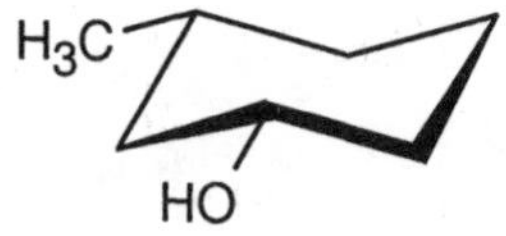

cis-3-Methylcyclohexanol

8.23

(a) H, *R*, O, H_3C, C, C, *R*, CH_3 $\xrightarrow{HBr}$ OH, *R*, C, H, H_3C, C, *S*, Br CH_3

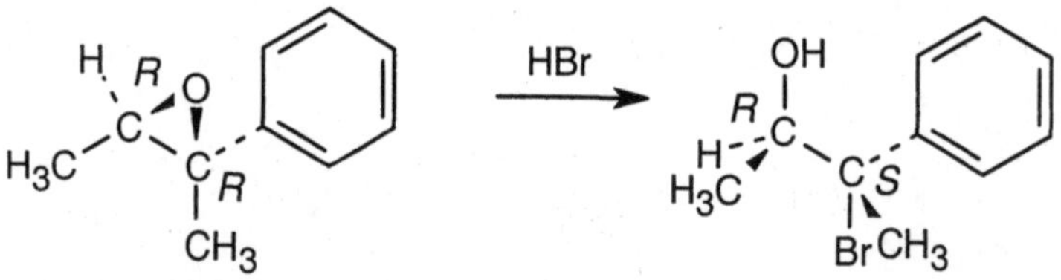

(b) H H, O, H_3C, C, C, C, C, CH_3, H_3C H, H H

1. $NaBH_4$
2. H_3O^+
3. Na
4. CH_3CH_2Br

→ H H, OCH_2CH_3, H, H_3C, C, C, C, C, CH_3, H_3C H, H H

\+

H H, H OCH_2CH_3, H_3C, C, C, C, C, CH_3, H_3C H, H H

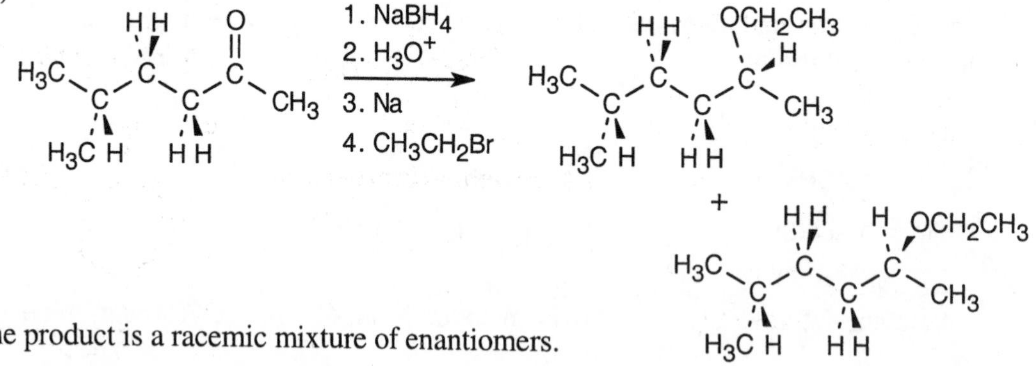

The product is a racemic mixture of enantiomers.

8.24 The products are diastereomers.

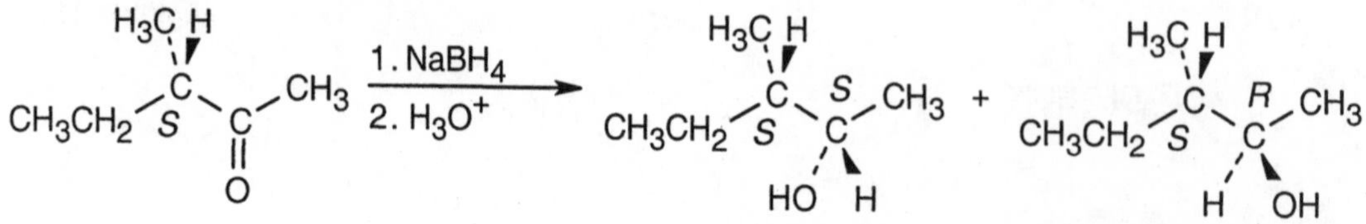

H_3C H, C, CH_3CH_2 *S*, C, CH_3, O

1. $NaBH_4$
2. H_3O^+

→ H_3C H, C, *S*, CH_3, CH_3CH_2 *S*, C, HO H + H_3C H, C, *R*, CH_3, CH_3CH_2 *S*, C, H OH

(2*S*,3*S*)-3-Methyl-
pentan-2-ol

(2*R*,3*S*)-3-Methyl-
pentan-2-ol

8.25

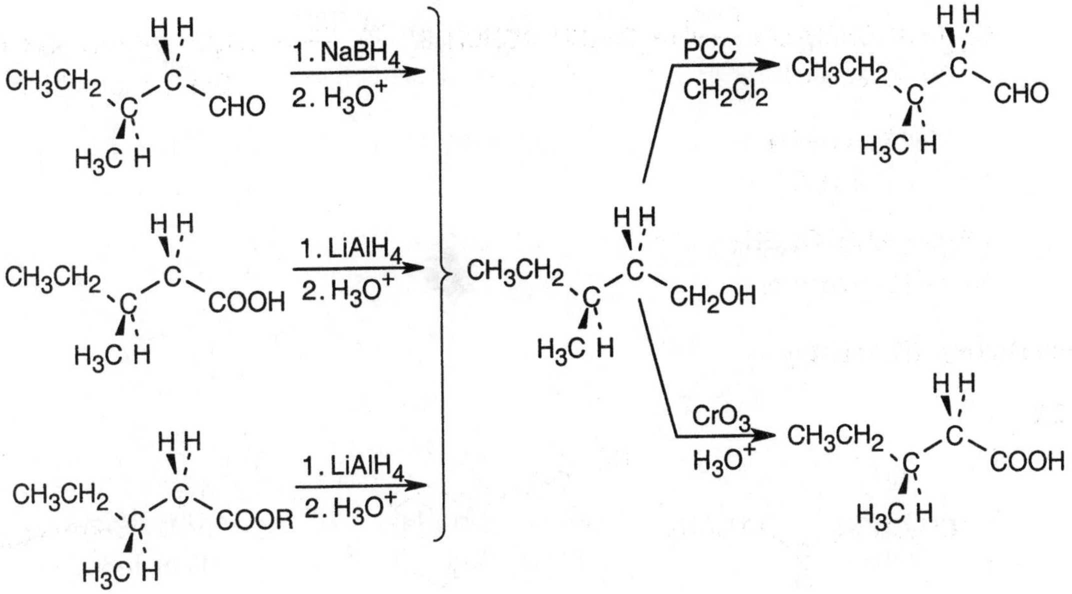

$CH_3CH_2CH(CH_3)CH_2CHO$ —(1. $NaBH_4$; 2. H_3O^+)→

$CH_3CH_2CH(CH_3)CH_2COOH$ —(1. $LiAlH_4$; 2. H_3O^+)→

$CH_3CH_2CH(CH_3)CH_2COOR$ —(1. $LiAlH_4$; 2. H_3O^+)→

→ $CH_3CH_2CH(CH_3)CH_2CH_2OH$

$CH_3CH_2CH(CH_3)CH_2CH_2OH$ —(PCC, CH_2Cl_2)→ $CH_3CH_2CH(CH_3)CH_2CHO$

$CH_3CH_2CH(CH_3)CH_2CH_2OH$ —(CrO_3, H_3O^+)→ $CH_3CH_2CH(CH_3)CH_2COOH$

Additional Problems

Nomenclature and Isomerism

8.26

(a) $CH_3CH(CH_3)OCH_2CH_3$
Ethyl isopropyl ether

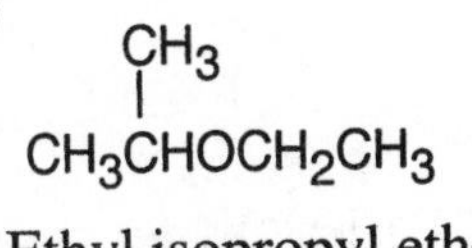

(b) COOH, OCH_3, OCH_3
3,4-Dimethoxybenzoic acid

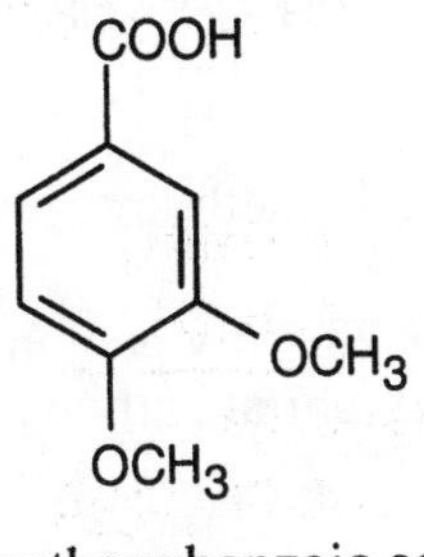

(c) $CH_3CH_2CH(OH)CH_2CH_2C(OH)(CH_3)CH_3$
2-Methylheptane-2,5-diol

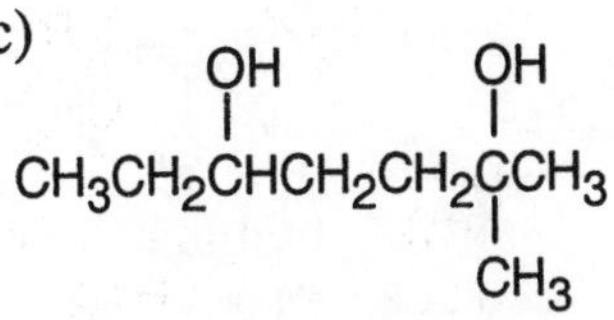

(d) H, OH, H, CH_2CH_3
trans-3-Ethylcyclohexanol

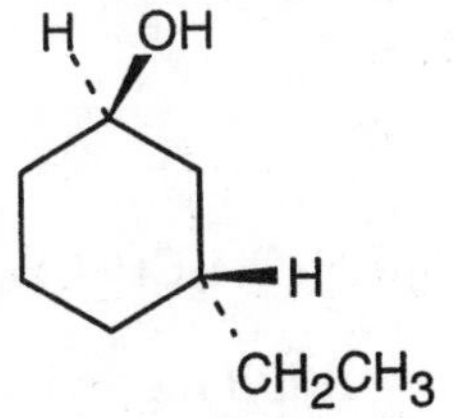

(e) OH, OCH_3, $CH_2CH{=}CH_2$
4-Allyl-2-methoxyphenol
(Eugenol)

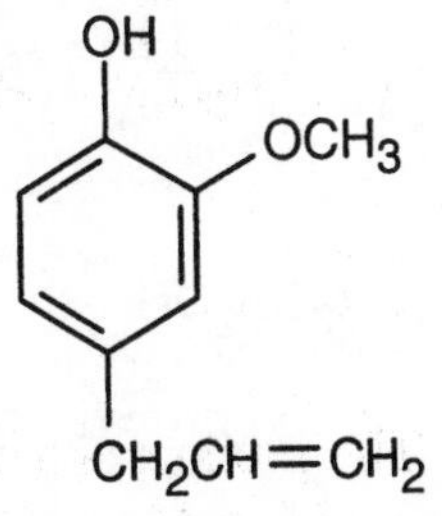

8.27

(a)

$$\mathrm{HOCH_2CH_2\overset{\overset{\displaystyle CH_3}{|}}{C}HCH_2OH}$$

2-Methylbutane-1,4-diol

(b)

$$\mathrm{CH_3\overset{\overset{\displaystyle OH}{|}}{C}H-\overset{\overset{\displaystyle CH_2CH_2CH_3}{|}}{C}HCH_2CH_3}$$

3-Ethylhexan-2-ol

(c) Ph, H, OH, H

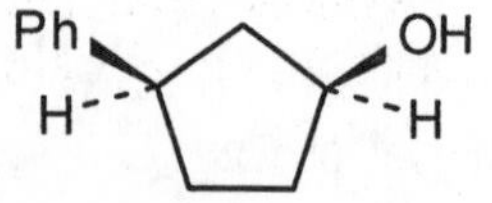

cis-3-Phenylcyclopentanol

(d)

$$\mathrm{(CH_3)_2CH\underset{\underset{\displaystyle CH_3}{|}}{\overset{\overset{\displaystyle SH}{|}}{C}}CH_2CH_2CH_3}$$

2,3-Dimethylhexane-3-thiol

8.28

$CH_3CH_2CH_2CH_2CH_2OH$
Pentan-1-ol

$$\mathrm{CH_3CH_2CH_2\overset{\overset{\displaystyle OH}{*|}}{C}HCH_3}$$
Pentan-2-ol

$$\mathrm{CH_3CH_2\overset{\overset{\displaystyle OH}{|}}{C}HCH_2CH_3}$$
Pentan-3-ol

$$\mathrm{CH_3CH_2\underset{\underset{\displaystyle CH_3}{|}}{\overset{*}{C}}HCH_2OH}$$
2-Methylbutan-1-ol

$$\mathrm{CH_3CH_2\underset{\underset{\displaystyle CH_3}{|}}{\overset{\overset{\displaystyle OH}{|}}{C}}CH_3}$$
2-Methylbutan-2-ol

$$\mathrm{CH_3\underset{*}{\overset{\overset{\displaystyle OH}{|}}{C}}H\underset{\underset{\displaystyle CH_3}{|}}{C}HCH_3}$$
3-Methylbutan-2-ol

$$\mathrm{HOCH_2CH_2\underset{\underset{\displaystyle CH_3}{|}}{C}HCH_3}$$
3-Methylbutan-1-ol

$$\mathrm{CH_3\underset{\underset{\displaystyle CH_3}{|}}{\overset{\overset{\displaystyle CH_3}{|}}{C}}CH_2OH}$$
2,2-Dimethylpropan-1-ol

8.29 Of the alcohols in Problem 8.28, only pentan-2-ol, 2-methylbutan-1-ol, and 3-methylbutan-2-ol are chiral (chiral carbons are starred).

8.30

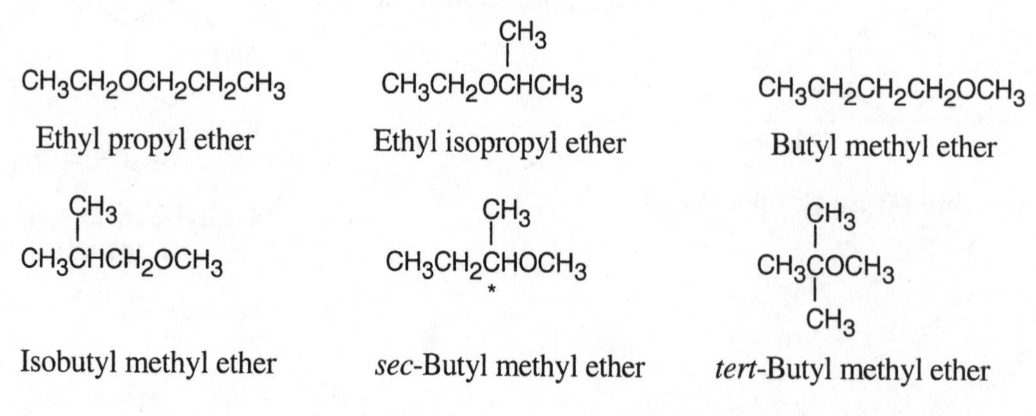

Only *sec*-butyl methyl ether is chiral.

8.31 Primary alcohols react with aqueous acidic CrO_3 to form carboxylic acids, secondary alcohols yield ketones, and tertiary alcohols are unreactive to oxidation. Of the eight alcohols in Problem 8.28, only 2-methylbutan-2-ol is unreactive to aqueous acidic CrO_3.

$$CH_3CH_2CH_2CH_2CH_2OH \xrightarrow[H_3O^+]{CrO_3} CH_3CH_2CH_2CH_2COOH$$

$$CH_3CH_2CH_2CH(OH)CH_3 \xrightarrow[H_3O^+]{CrO_3} CH_3CH_2CH_2C(=O)CH_3$$

$$CH_3CH_2CH(OH)CH_2CH_3 \xrightarrow[H_3O^+]{CrO_3} CH_3CH_2C(=O)CH_2CH_3$$

$$CH_3CH_2CH(CH_3)CH_2OH \xrightarrow[H_3O^+]{CrO_3} CH_3CH_2CH(CH_3)COOH$$

$$CH_3CH(OH)CH(CH_3)CH_3 \xrightarrow[H_3O^+]{CrO_3} CH_3C(=O)CH(CH_3)CH_3$$

$$HOCH_2CH_2CH(CH_3)CH_3 \xrightarrow[H_3O^+]{CrO_3} HOOCCH_2CH(CH_3)CH_3$$

$$CH_3C(CH_3)_2CH_2OH \xrightarrow[H_3O^+]{CrO_3} CH_3C(CH_3)_2COOH$$

Reactions and Synthesis

8.32

$$CH_3CH_2CH_2OCH_2CH_3 \xrightarrow{HI} CH_3CH_2CH_2OH + ICH_2CH_3$$

$$CH_3CH_2CH_2I + HOCH_2CH_3$$

$$(CH_3)_2CHOCH_2CH_3 \xrightarrow{HI} (CH_3)_2CHOH + ICH_2CH_3$$

$$CH_3CH_2CH_2CH_2OCH_3 \xrightarrow{HI} CH_3CH_2CH_2CH_2OH + ICH_3$$

$$(CH_3)_2CHCH_2OCH_3 \xrightarrow{HI} (CH_3)_2CHCH_2OH + ICH_3$$

$$CH_3CH_2\overset{CH_3}{\overset{|}{C}}HOCH_3 \xrightarrow{HI} CH_3CH_2\overset{CH_3}{\overset{|}{C}}HCH_2OH + ICH_3$$

$$CH_3\underset{CH_3}{\underset{|}{\overset{CH_3}{\overset{|}{C}}}}OCH_3 \xrightarrow{HI} CH_3\underset{CH_3}{\underset{|}{\overset{CH_3}{\overset{|}{C}}}}-I + HOCH_3$$

8.33

(a)

$$\text{cyclohexyl-}CH_2OH \xrightarrow{PCC} \text{cyclohexyl-}CHO$$

(b)

$$\text{phenyl-}OCH_2\overset{CH_3}{\overset{|}{C}}HCH_3 \xrightarrow{HBr} \text{phenyl-}OH + BrCH_2\overset{CH_3}{\overset{|}{C}}HCH_3$$

(c)

$$H_2C{=}CH\overset{CH_3}{\overset{|}{C}}HCH_2\overset{O}{\overset{||}{C}}OCH_3 \xrightarrow[2.\ H_3O^+]{1.\ LiAlH_4} H_2C{=}CH\overset{CH_3}{\overset{|}{C}}HCH_2CH_2OH + CH_3OH$$

(d)

$$H_2C\overset{O}{-}CHCH_2CH_3 \text{ (epoxide)} \xrightarrow{HBr} Br-CH_2\overset{OH}{\overset{|}{C}}HCH_2CH_3$$

Attack of Br^- occurs at the less substituted carbon to give the product shown.

8.34

(a) cyclohexanol (cyclohexyl–OH) $\xrightarrow{PCC}$ cyclohexanone (cyclohexyl=O)

(b) cyclohexanol (cyclohexyl–OH) $\xrightarrow{PBr_3}$ bromocyclohexane (cyclohexyl–Br)

(c) cyclohexanol (cyclohexyl–OH) $\xrightarrow[H_2O]{H_2SO_4}$ cyclohexene

(d) cyclohexene, from (c) $\xrightarrow[\text{Pd catalyst}]{H_2}$ cyclohexane

8.35

(a) $CH_3CH_2CH_2OH \xrightarrow{PCC} CH_3CH_2CH{=}O$

(b) $CH_3CH_2CH_2OH \xrightarrow[H_3O^+]{CrO_3} CH_3CH_2C({=}O)OH$

(c) $CH_3CH_2CH_2OH \xrightarrow{Na} CH_3CH_2CH_2O^- Na^+ + 1/2\ H_2$

(d) $CH_3CH_2CH_2OH \xrightarrow{SOCl_2} CH_3CH_2CH_2Cl$

8.36 Remember: for ethers other than tertiary ethers, the ether oxygen stays with the more hindered alkyl group.

(a) $CH_3CH_2OCH(CH_3)CH_3 \xrightarrow[H_2O]{HI} CH_3CH_2I + HOCH(CH_3)CH_3$

(b) $CH_3C(CH_3)_2CH_2OCH_3 \xrightarrow[H_2O]{HI} CH_3C(CH_3)_2CH_2OH + ICH_3$

8.37

(a)

$$C_6H_5CH_2CH_2OH \xrightarrow[H_2O]{KMnO_4} C_6H_5COOH$$

2-Phenylethanol

(b)

$$C_6H_5CH_2CH_2OH \xrightarrow{PBr_3} C_6H_5CH_2CH_2Br$$

$$C_6H_5CH_2CH_2Br \xrightarrow{Mg} C_6H_5CH_2CH_2MgBr$$

$$C_6H_5CH_2CH_2MgBr \xrightarrow{H_2O} C_6H_5CH_2CH_3$$

or

$$C_6H_5CH_2CH_2OH \xrightarrow{H_2SO_4} C_6H_5CH{=}CH_2$$

$$C_6H_5CH{=}CH_2 \xrightarrow[Pd]{H_2} C_6H_5CH_2CH_3$$

(c)

$$C_6H_5CH_2CH_2OH \xrightarrow{PBr_3} C_6H_5CH_2CH_2Br$$

(d)

$$C_6H_5CH_2CH_2OH \xrightarrow[H_3O^+]{CrO_3} C_6H_5CH_2C(=O)OH$$

(e)

$$C_6H_5CH_2CH_2OH \xrightarrow{PCC} C_6H_5CH_2C(=O)H$$

8.38

(a) Phenol–OH $\xrightarrow{Br_2}$ (o-bromophenol) Br, OH + Br–(p-bromophenol)–OH

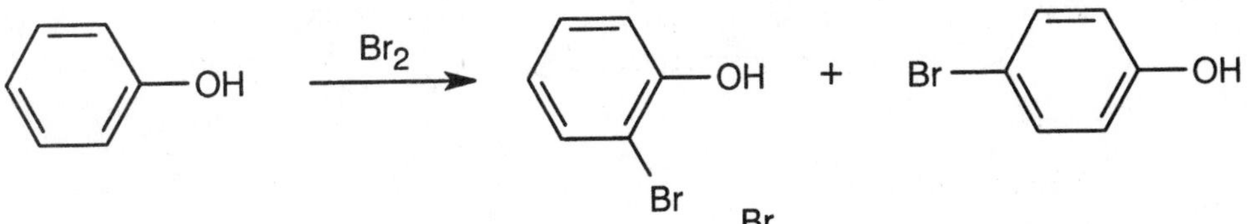

(b) Phenol–OH $\xrightarrow{3\ Br_2}$ Br, Br, Br–(2,4,6-tribromophenol)–OH

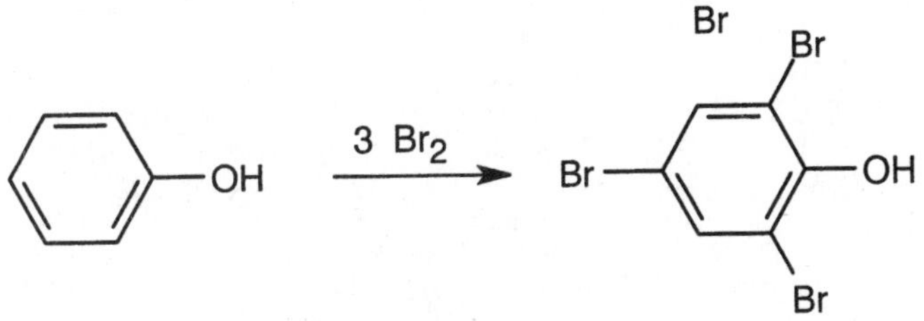

(c) Phenol–OH $\xrightarrow[2.\ CH_3I]{1.\ NaOH}$ Ph–OCH_3

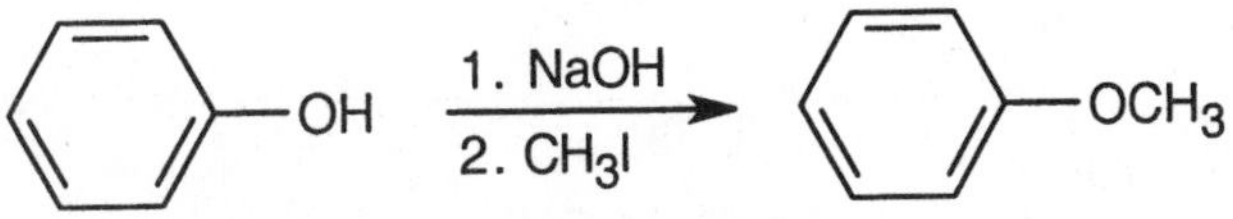

(d) Phenol–OH $\xrightarrow[H_3O^+]{Na_2Cr_2O_7}$ O=(benzoquinone)=O

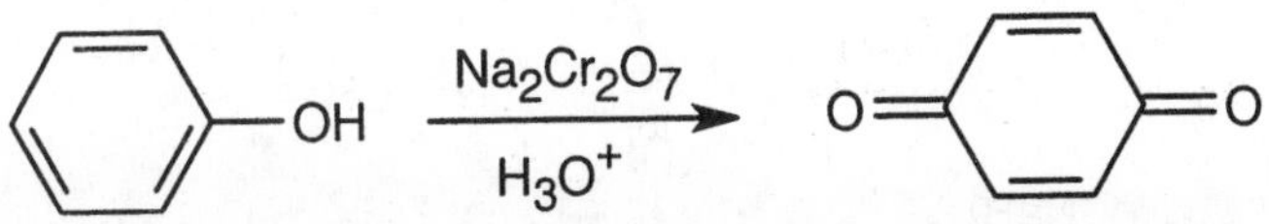

8.39

(a) $CH_3CH_2CH_2CH_2OH \xrightarrow{PBr_3} CH_3CH_2CH_2CH_2Br$

(b) $CH_3CH_2CH_2CH_2OH \xrightarrow{CrO_3,\ H_3O^+} CH_3CH_2CH_2COOH$

(c) $CH_3CH_2CH_2CH_2OH \xrightarrow{Na} CH_3CH_2CH_2CH_2O^-\ Na^+ + 1/2\ H_2$

(d) $CH_3CH_2CH_2CH_2OH \xrightarrow{PCC} CH_3CH_2CH_2CHO$

8.40

(a) (1-methylcyclohexanol) OH, CH_3 $\xrightarrow{HBr}$ Br, CH_3

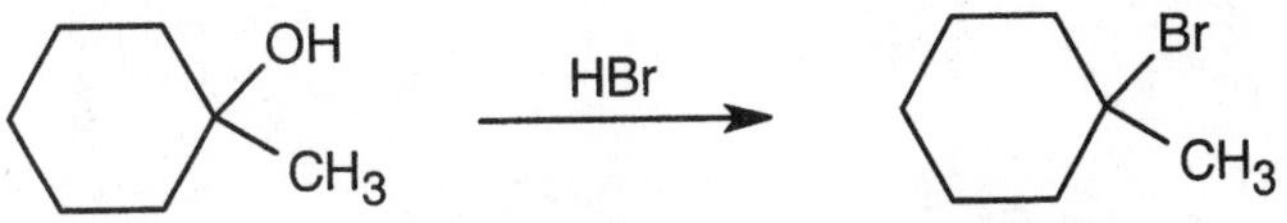

(b) OH, CH_3 $\xrightarrow{H_2SO_4}$ (1-methylcyclohexene) –CH_3

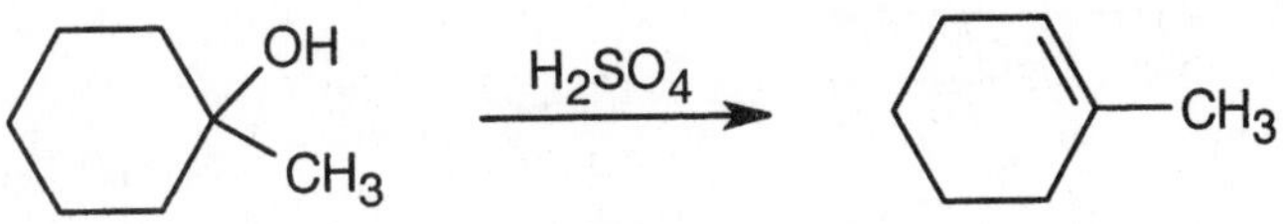

(c) OH, CH_3 $\xrightarrow[H_3O^+]{Na_2Cr_2O_7}$ no reaction

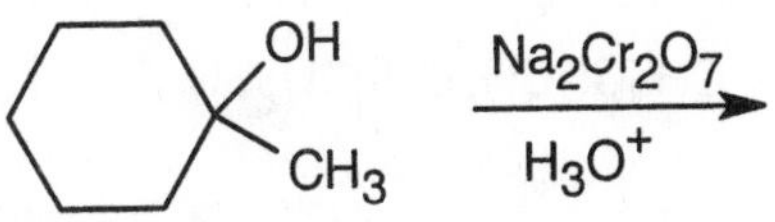

(d) OH, CH_3 $\xrightarrow{Na}$ $O^-\ Na^+$, CH_3 + 1/2 H_2

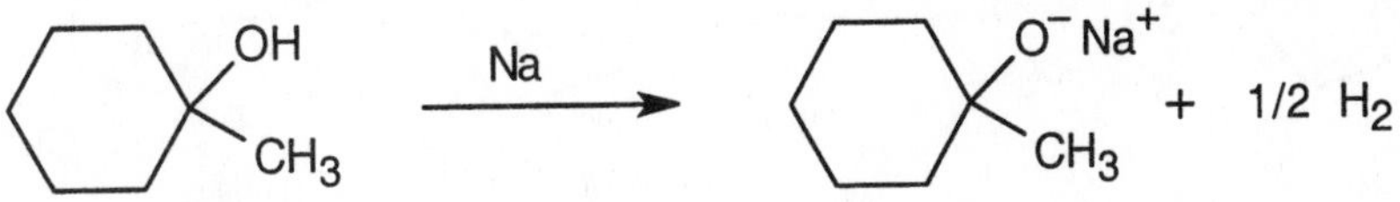

(e) 1-Methylcyclohexoxide ($O^- Na^+$, CH_3) $\xrightarrow{CH_3I}$ 1-methoxy-1-methylcyclohexane (OCH_3, CH_3)

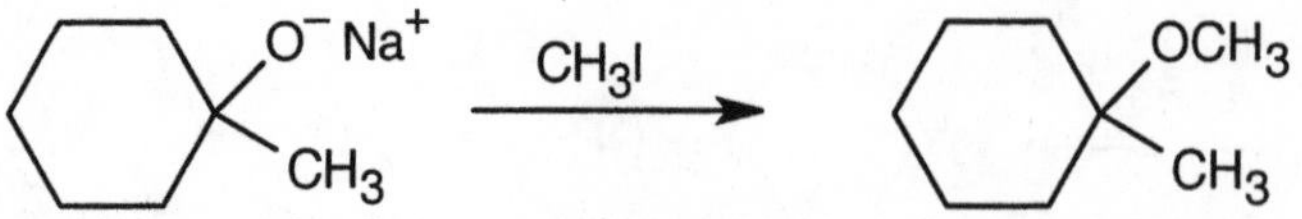

8.41

(a) Cyclopentanol (–OH) $\xrightarrow[H_3O^+]{CrO_3}$ cyclopentanone (=O)

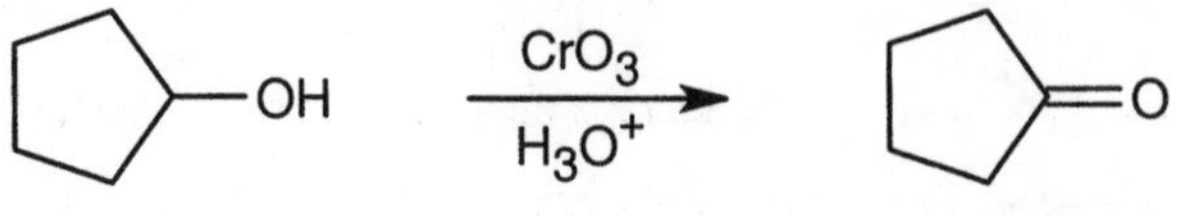

(b) Benzyl alcohol (C_6H_5–CH_2OH) $\xrightarrow{PCC}$ benzaldehyde (C_6H_5–CHO)

(c) $CH_3CH(CH_3)CH_2OH \xrightarrow[H_3O^+]{CrO_3} CH_3CH(CH_3)COOH$

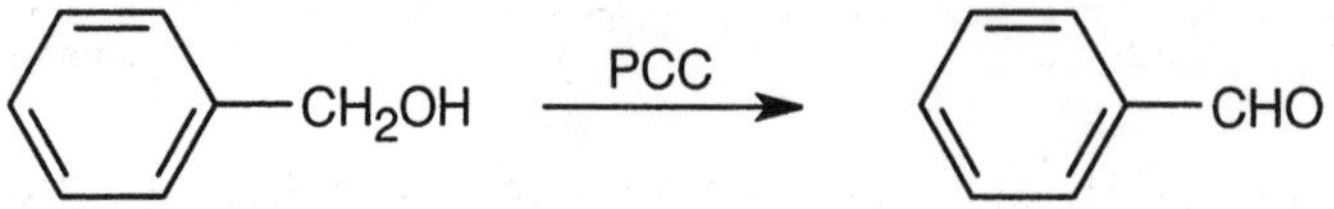

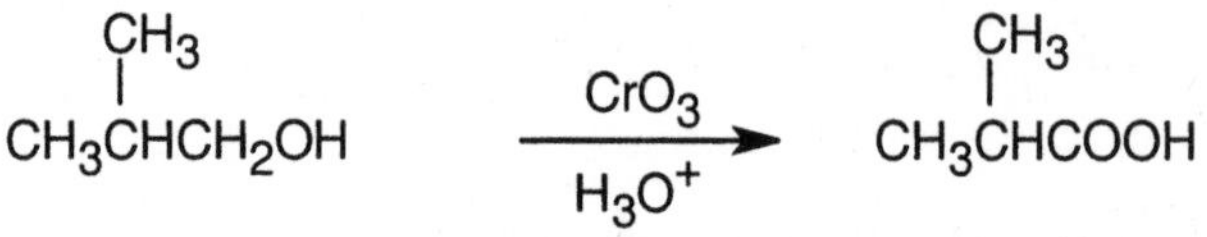

8.42

(a) $CH_3CH(CH_3)CH_2CHO \xrightarrow[2.\ H_3O^+]{1.\ NaBH_4} CH_3CH(CH_3)CH_2CH_2OH$

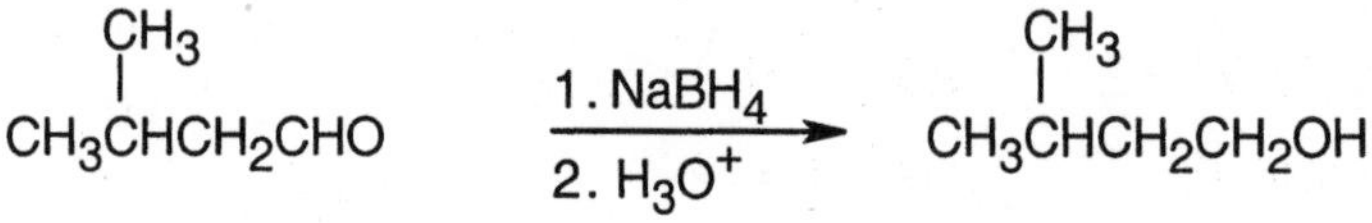

(b) Benzene-1,2-dicarboxylic acid (COOH, COOH) $\xrightarrow[2.\ H_3O^+]{1.\ LiAlH_4}$ 1,2-benzenedimethanol (CH_2OH, CH_2OH)

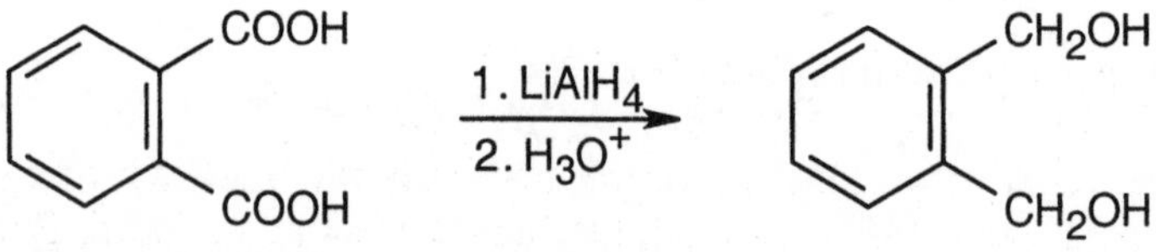

(c) $CH_3CH_2C(=O)CH_2CH(CH_3)CH_3 \xrightarrow[2.\ H_3O^+]{1.\ NaBH_4} CH_3CH_2CH(OH)CH_2CH(CH_3)CH_3$

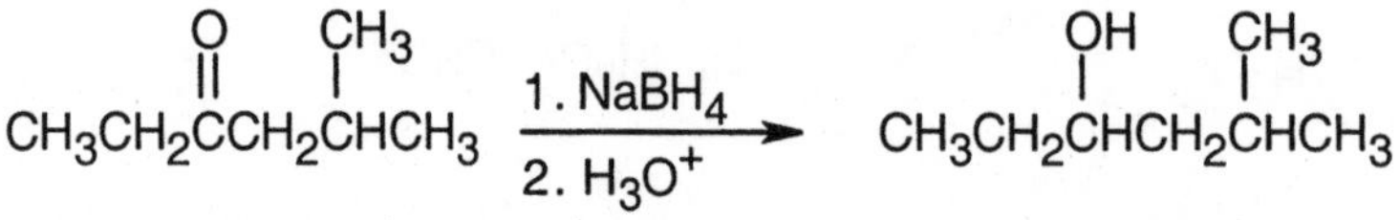

8.43

(a) Hydroquinone (HO–C_6H_4–OH) $\xrightarrow[H_3O^+]{Na_2Cr_2O_7}$ p-benzoquinone (O=C_6H_4=O)

(b) $CH_3CH(CH_3)CH_2CH_2CH_2Br \xrightarrow{Na^+\ {}^-SH} CH_3CH(CH_3)CH_2CH_2CH_2SH$

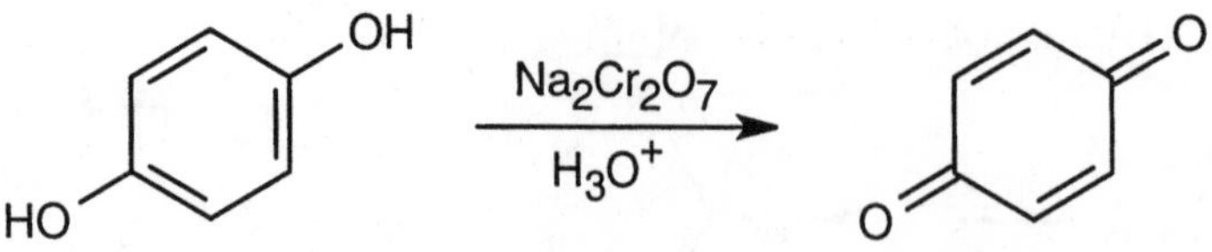

(c) 2 Cyclopentanethiol (–SH) $\xrightarrow{Br_2}$ cyclopentyl–S–S–cyclopentyl

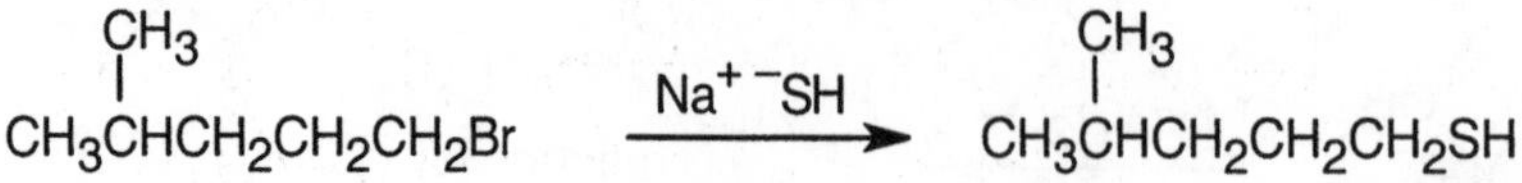

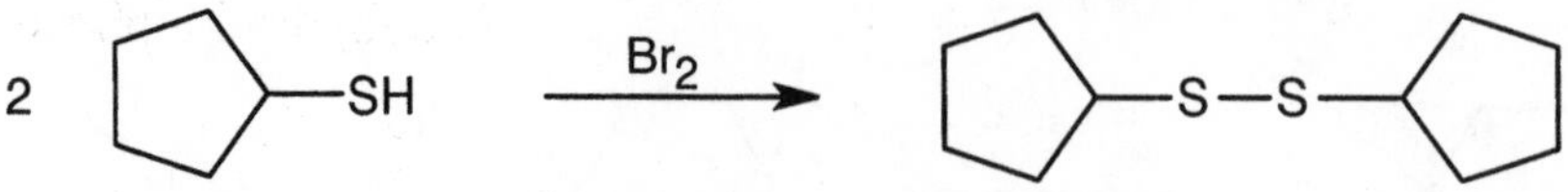

Acidity

8.44

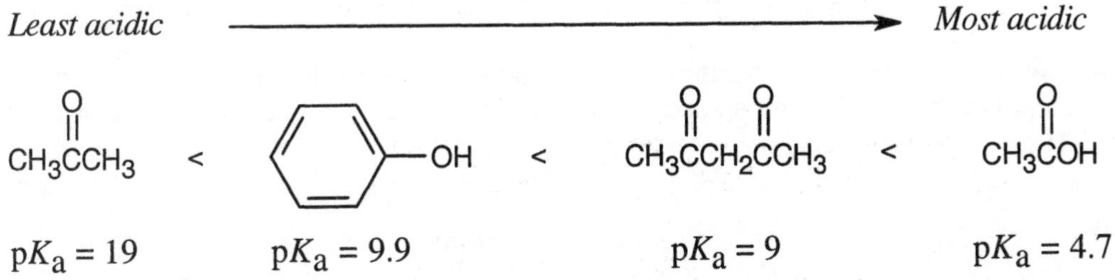

Least acidic ⟶ *Most acidic*

CH_3COCH_3 < phenol (C$_6$H$_5$OH) < $CH_3COCH_2COCH_3$ < CH_3COOH

pK_a = 19   pK_a = 9.9   pK_a = 9   pK_a = 4.7

8.45 To react completely with NaOH, an acid must have a pK_a lower than the pK_a of H_2O. Thus, all compounds in the previous problem except acetone react completely with NaOH.

8.46

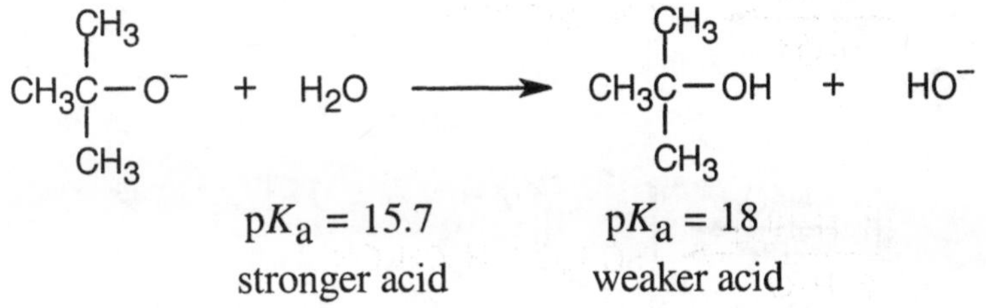

$$(CH_3)_3C{-}O^- + H_2O \longrightarrow (CH_3)_3C{-}OH + HO^-$$

pK_a = 15.7 (stronger acid)   pK_a = 18 (weaker acid)

The reaction takes place as written because water is a stronger acid than *tert*–butyl alcohol.

8.47 Only acetic acid will react with sodium bicarbonate.

8.48 Sodium bicarbonate reacts with acetic acid to produce carbonic acid, which breaks down to form CO_2. The resulting CO_2 bubbles indicate the presence of acetic acid. Phenol does not react with sodium bicarbonate.

Integrated Problems

8.49

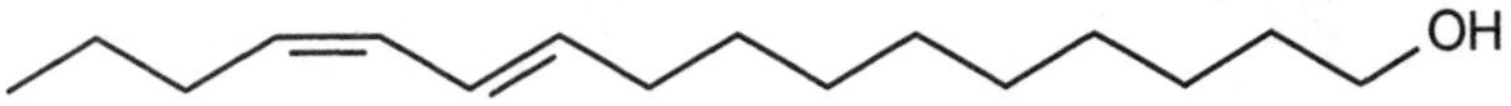

OH

(10*E*,12*Z*)-Hexadeca-10,12-dien-1-ol

8.50

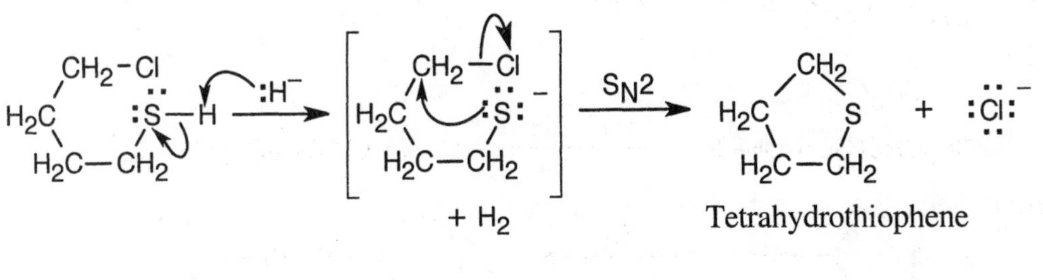

$ClCH_2CH_2CH_2CH_2SH$ + :H$^-$ ⟶ [$ClCH_2CH_2CH_2CH_2S^-$] + H_2 ⟶ (S$_N$2) Tetrahydrothiophene + :Cl:$^-$

Tetrahydrothiophene

8.51

Phenoxide ($C_6H_5O^-$) + Br–C_6H_5 ⟶ (no reaction) C_6H_5–O–C_6H_5

The Williamson ether synthesis is an S_N2 reaction between an alkoxide or phenoxide anion and an alkyl halide. It can't be used to synthesize diphenyl ether because an aryl halide (C_6H_5Br) doesn't undergo S_N2 reactions.

8.52

(a)

Benzene $\xrightarrow[AlCl_3]{CH_3Cl}$ Toluene (CH_3) $\xrightarrow[H_2O]{KMnO_4}$ Benzoic acid (COOH) $\xrightarrow[2.\ H_3O^+]{1.\ LiAlH_4}$ Benzyl alcohol (CH_2OH) $\xrightarrow{PBr_3}$ Benzyl bromide (CH_2Br)

(b)

Benzene $\xrightarrow[H_2SO_4]{SO_3}$ Benzenesulfonic acid (SO_3H) $\xrightarrow[200°]{NaOH}$ Sodium phenoxide ($O^-\ Na^+$)

(c)

C_6H_5–CH_2Br + $Na^{+}\ {}^{-}O$–C_6H_5 ⟶ C_6H_5–CH_2–O–C_6H_5

Benzyl phenyl ether

8.53 The hydroxyl group is axial in the cis isomer, which is expected to oxidize faster than the trans isomer. (Remember that the bulky *tert*-butyl group is always equatorial in the more stable isomer.)

$(CH_3)_3C$– (axial OH)

cis-4-*tert*-Butylcyclohexanol $\xrightarrow[H_3O^+]{CrO_3}$ faster

$(CH_3)_3C$– (equatorial OH)

trans-4-*tert*-Butylcyclohexanol $\xrightarrow[H_3O^+]{CrO_3}$ slower

Both give 4-*tert*-butylcyclohexanone: $(CH_3)_3C$–(cyclohexane)=O

8.54

$$2\ CH_3Br + 2\ Na^{+}\ {}^{-}SH \longrightarrow 2\ CH_3SH + 2\ NaBr$$

$$2\ CH_3SH + Br_2 \longrightarrow CH_3SSCH_3 + 2\ HBr$$

Dimethyl disulfide

8.55

$CH_3C(CH_3){=}CH_2$ + H^+ (protonation) ⇌ [$CH_3\overset{+}{C}(CH_3)CH_3$ + R–O–H (attack of alcohol oxygen on carbocation) ⇌ $CH_3C(CH_3)_2$–$\overset{+}{O}$(H)–R + :B] ⇌ (loss of proton) $CH_3C(CH_3)_2$–O–R

8.56

$(H_3C)_3C$–O–cyclohexyl + $H–OCCF_3$ (C=O) ⇌ [$(H_3C)_3C$–$\overset{+}{O}$H–cyclohexyl ⇌ $(H_3C)_2\overset{+}{C}$–CH_2–H + cyclohexanol (OH) + $^{-}$:OCCF$_3$] ⇌ $(H_3C)_2C{=}CH_2$ + $HOCCF_3$ (C=O)

The first step of acid-catalyzed cleavage of ethers is protonation of the ether oxygen. The protonated intermediate forms cyclohexanol and a tertiary carbocation, which loses a proton to form 2-methylpropene. This is an example of an E1 elimination and is the reverse of the mechanism shown in the previous problem.

8.57

(a)

Phenol (OH) $\xrightarrow{NaOH}$ $C_6H_5O^-\ Na^+$ + H_2O $\xrightarrow{CH_3CH_2Br}$ $C_6H_5OCH_2CH_3$ + NaBr

(b)

$H_3C(H)C{=}C(H)CH_3$ $\xrightarrow{RCO_3H}$ epoxide: H_3C–C(H)–(O)–C(H)–CH_3

RCO_3H = *m*-Chloroperbenzoic acid

8.58

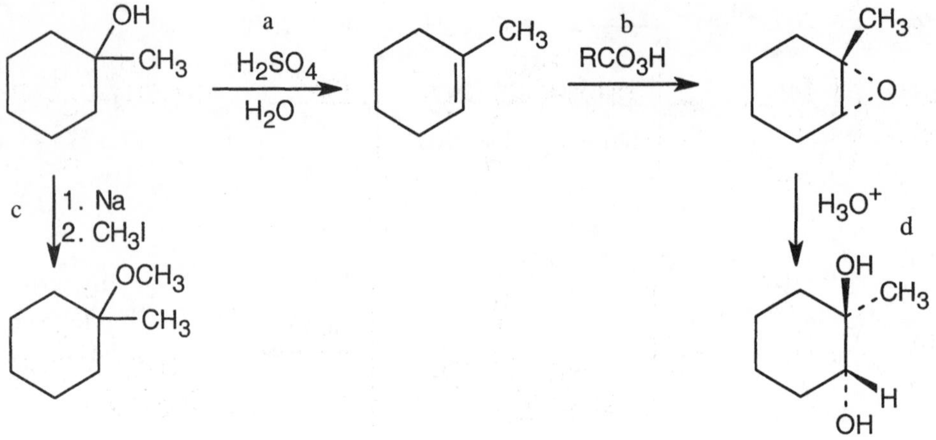

a: H_2SO_4, H_2O; b: RCO_3H; c: 1. Na, 2. CH_3I; d: H_3O^+

OH, CH_3, CH_3, CH_3, O, OCH_3, CH_3, OH, CH_3, H, OH

8.59

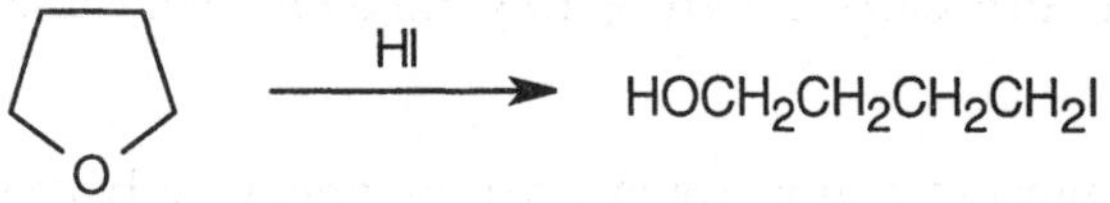

HI → $HOCH_2CH_2CH_2CH_2I$

O

8.60

O, CH_3, I, O, Li^+, + CH_3I

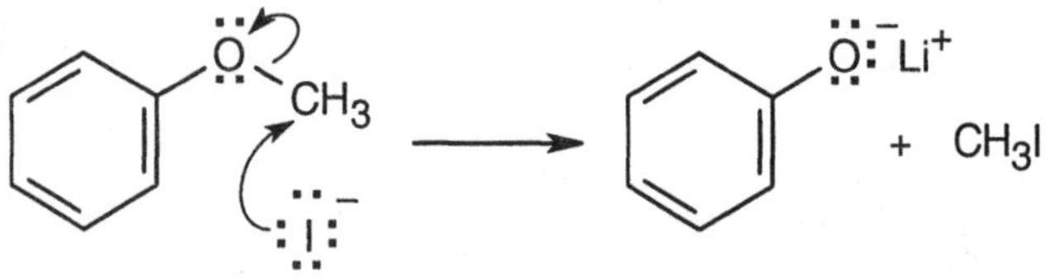

This reaction is an S_N2 displacement and can't occur at an aryl carbon.

8.61

$$\frac{1.06 \text{ g vanillin}}{152 \text{ g/mol}} = 6.97 \times 10^{-3} \text{ mol vanillin}$$

$$\frac{1.60 \text{ g AgI}}{234.8 \text{ g/mol}} = 6.81 \times 10^{-3} \text{ mol AgI}$$

$$\underset{\text{AgI}}{6.81 \times 10^{-3} \text{ mol}} \rightarrow \underset{\text{I}^-}{6.81 \times 10^{-3} \text{ mol}} \rightarrow \underset{\text{CH}_3\text{I}}{6.81 \times 10^{-3} \text{ mol}} \rightarrow \underset{-\text{OCH}_3}{6.81 \times 10^{-3} \text{ mol}}$$

Thus, 6.97×10^{-3} moles of vanillin contains 6.81×10^{-3} moles of methoxyl groups. Since the ratio of moles of vanillin to moles of methoxyl is approximately 1:1, each vanillin contains one methoxyl group.

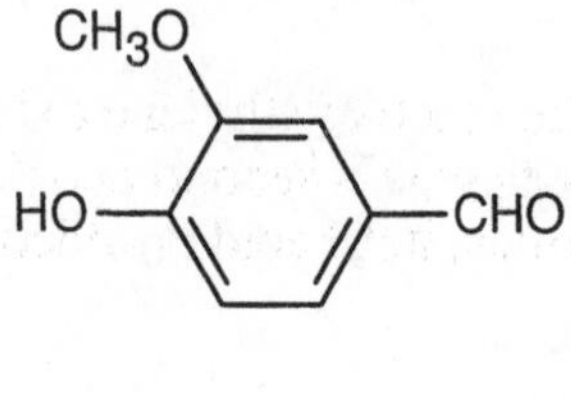

CH_3O, HO, CHO

Vanillin

8.62

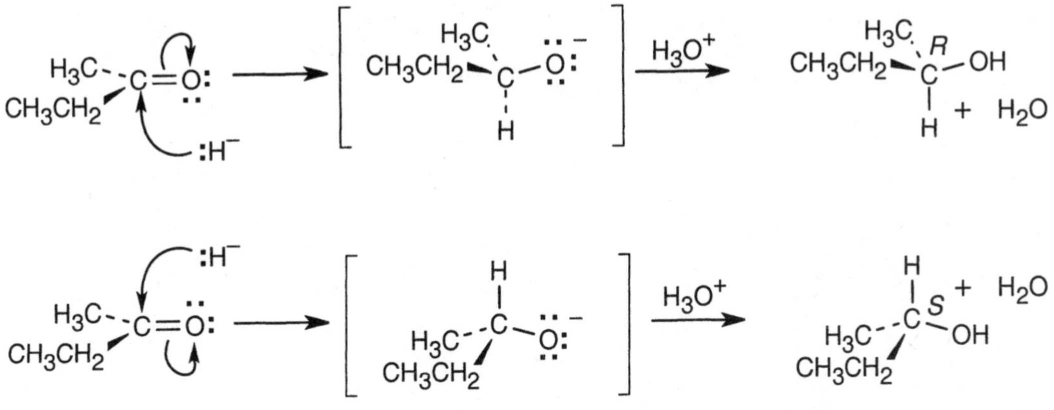

Attack occurs from both sides of the planar carbonyl group to yield a racemic mixture of enantiomers.

8.63 The negatively polarized oxygens in the center of a crown ether are able to solvate cations, which are sequestered in the middle of the cavity. A crown ether that binds Cs^+ would need to have a larger cavity, which might be formed by one or two additional ether units.

In the Medicine Cabinet

8.64

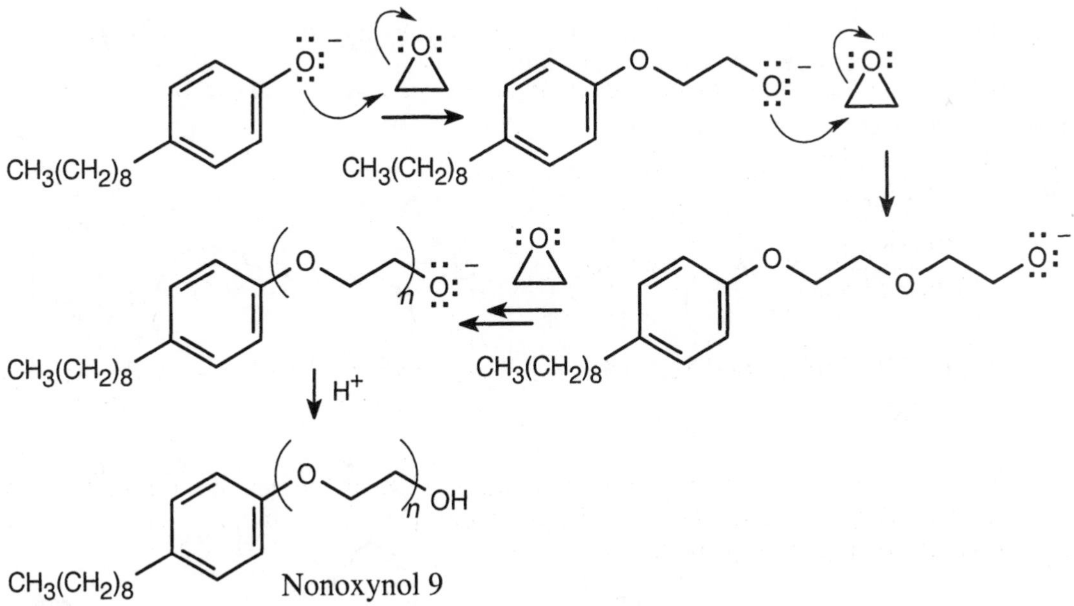

Nonoxynol 9

S_N2 attack of the phenoxide ion on the reactive ethylene oxide ring causes ring opening that yields a new alkoxide ion, which reacts with a second epoxide to form a second alkoxide ion. This process is repeated many times, until acid is added to quench the reaction.

8.65

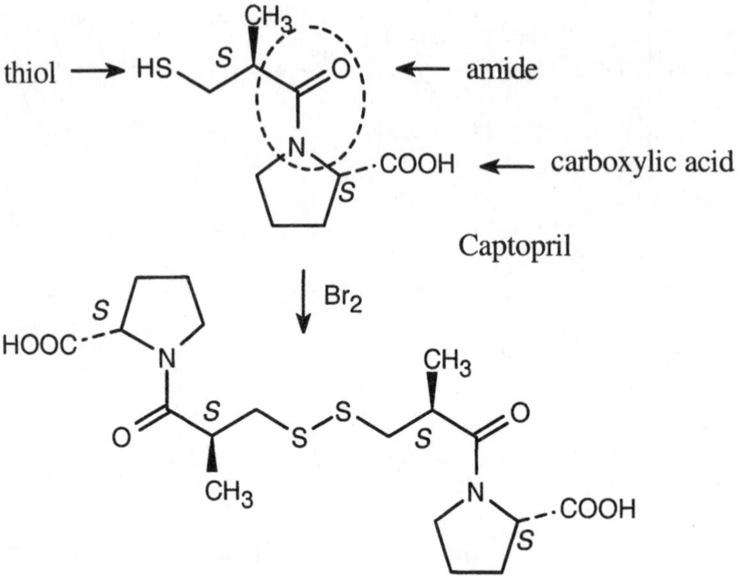

thiol → HS (S) CH$_3$ ← amide

COOH ← carboxylic acid

Captopril

$\downarrow Br_2$

8.66

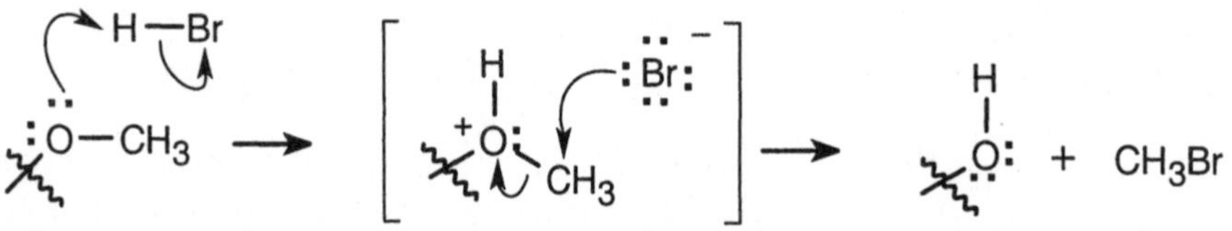

Protonation of the methoxyl oxygen is followed by S_N2 displacement by Br^-, giving methyl bromide and dopamine as products.

In the Field with Agrochemicals

8.67

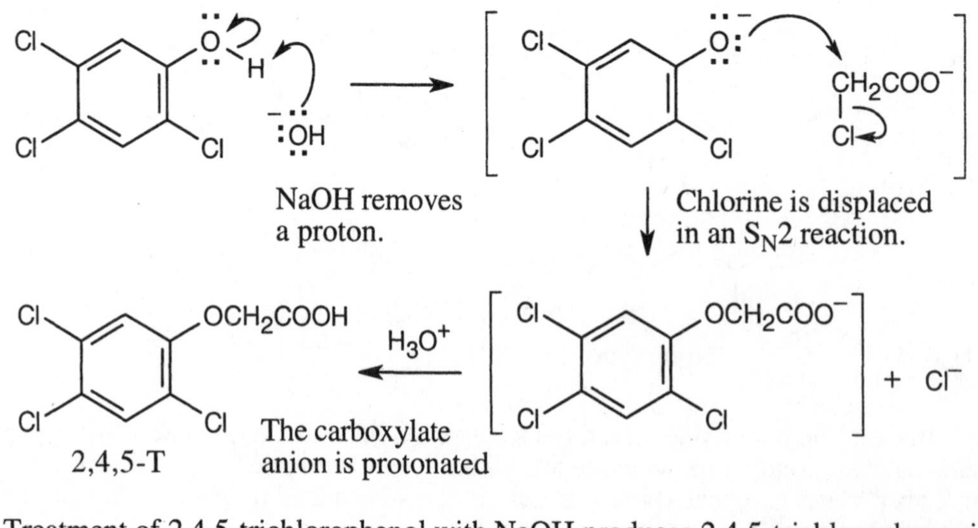

NaOH removes a proton.

Chlorine is displaced in an S_N2 reaction.

H_3O^+

The carboxylate anion is protonated

2,4,5-T

Treatment of 2,4,5-trichlorophenol with NaOH produces 2,4,5-trichlorophenoxide anion, which displaces chlorine from $ClCH_2COO^-\ {}^+Na$ in an S_N2 reaction to yield 2,4,5-T.

Chapter 9 – Aldehydes and Ketones

Chapter Outline

I. Introduction to carbonyl chemistry (Sections 9.1–9.2).
  A. The nature of carbonyl compounds (Section 9.1).
    1. All carbonyl compounds contain an acyl group, R–C=O, bonded to another group.
    2. There are two kinds of carbonyl compounds.
      a. The group bonded to the acyl group can't act as a leaving group (aldehydes and ketones).
      b. The group bonded to the acyl group can act as a leaving group.
  B. Structure and properties of the carbonyl group.
    1. The carbonyl carbon is sp^2 hybridized.
    2. The carbon-oxygen bond is polarized.
      a. The carbonyl carbon can react with nucleophiles.
      b. The carbonyl oxygen can react with electrophiles.
    3. Most carbonyl-group reactions are a result of this polarization.
  C. Naming aldehydes and ketones (Section 9.2).
    1. Naming aldehydes.
      a. Aldehydes are named by replacing the *-e* of the corresponding alkane with *-al.*
      b. The parent chain must contain the –CHO group.
      c. The aldehyde carbon is always carbon 1.
      d. When the –CHO group is attached to a ring, the suffix *-carbaldehyde* is used.
    2. Naming ketones.
      a. Ketones are named by replacing the *-e* of the corresponding alkane with *-one.*
      b. Numbering starts at the end of the carbon chain nearer to the carbonyl carbon.
      c. The word *acyl* is used when a RCO– group is a substituent.

II. Preparation of aldehydes and ketones (Section 9.3).
  A. Aldehydes.
    Oxidation of primary alcohols.
  B. Ketones.
    1. Oxidation of secondary alcohols.
    2. Hydration of terminal alkynes.
    3. Friedel-Crafts acylation of arenes.

III. Reactions of ketones and aldehydes (Sections 9.4–9.11).
  A. Oxidation of aldehydes (Section 9.4).
    Aldehydes can be oxidized to carboxylic acids with Tollens' Reagent or by atmospheric oxygen..
  B. Nucleophilic addition reactions (Sections 9.5–9.12).
    1. General features (Section 9.5).
      a. A nucleophile attacks the carbonyl carbon, with simultaneous bond formation and bond breaking.
      b. Under basic conditions, negatively charged intermediates are favored.
      c. Under acidic conditions, positively charged intermediates are favored.
    2. Reduction (Section 9.6).
      a. Aldehydes and ketones are reduced by $NaBH_4$.
      b. $H:^-$ adds to the carbonyl group, and the resulting negatively charged alkoxide intermediate is protonated.

3. Addition of water – hydration (Section 9.7).
   a. Water can add reversibly to aldehydes and ketones.
   b. The stability of the product depends on the structure of the carbonyl compound.
   c. The process can be catalyzed by both acid and base and takes place in several steps.
4. Addition of alcohols (Sections 9.8, 9.9).
   a. Acetal formation (Section 9.8).
      i. Aldehydes and ketones react with alcohols to form acetals.
      ii. The reaction is acid-catalyzed and is reversible.
      iii. The initial product is a hemiacetal.
      iv. Acetals can serve as protecting groups.
   b. Importance of acetals and hemiacetals (Section 9.9).
      Carbohydrates and other biomolecules contain acetals and hemiacetals.
5. Amines can add to aldehydes and ketones to form imines.(Section 9.10).
6. Addition of Grignard reagents – alcohol formation (Section 9.11).
   a. Grignard reagents act as carbanions and undergo nucleophilic addition to aldehydes and ketones.
      i. "R:–" adds to the carbonyl carbon to form a tetrahedral intermediate.
      ii. Water is added in a separate step to protonate the intermediate, yielding an alcohol.
   b. Types of Grignard products.
      i. Reaction of RMgX with formaldehyde yields primary alcohols.
      ii. Reaction of RMgX with aldehydes yields secondary alcohols.
      iii. Reaction of RMgX with ketones yields tertiary alcohols.
   c. Limitations of the Grignard reaction.
      i. Grignard reagents can't be prepared from reagents containing other reactive functional groups.
      ii. Grignard reagents can't be prepared from compounds having acidic hydrogens.
7. Conjugate nucleophilic addition (Section 9.12).
   a. A nucleophile can add to the double bond of an α,β-unsaturated carbonyl compound.
   b. Addition occurs at the β carbon, and the initial product is a resonance-stabilized enolate anion.
   c. Conjugate addition occurs because the carbonyl group makes the β carbon electron-poor and reactive towards nucleophiles.

Solutions to Problems

9.1

(a)

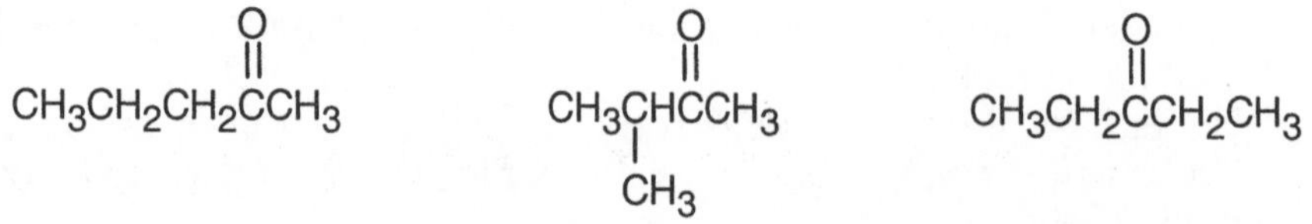

$CH_3CH_2CH_2\overset{O}{\overset{\|}{C}}CH_3$ $\quad$ $CH_3\underset{CH_3}{\underset{|}{C}}H\overset{O}{\overset{\|}{C}}CH_3$ $\quad$ $CH_3CH_2\overset{O}{\overset{\|}{C}}CH_2CH_3$

(b)

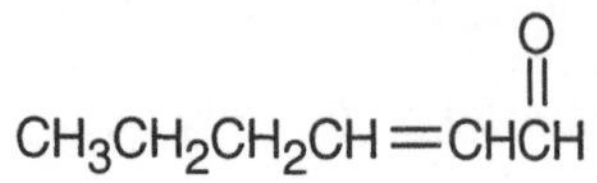

$CH_3CH_2CH_2CH{=}CH\overset{O}{\overset{\|}{C}}H$ and many other possible answers.

(c)

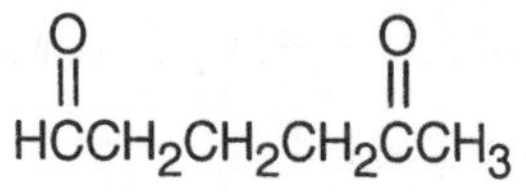

$H\overset{O}{\overset{\|}{C}}CH_2CH_2CH_2\overset{O}{\overset{\|}{C}}CH_3$ and other possible answers.

(d) 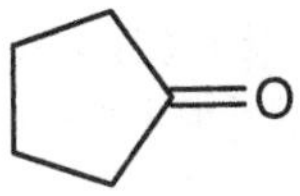 and many other possible answers.

9.2

(a) 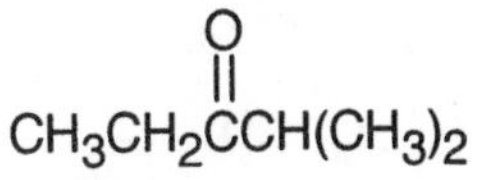

2-Methylpentan-3-one

(b) 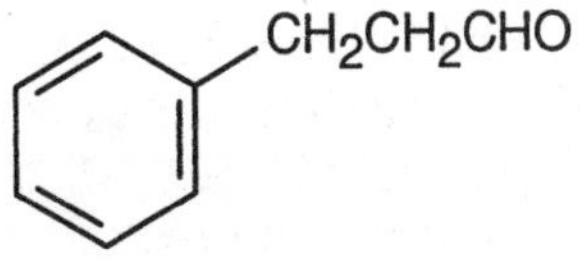

3-Phenylpropanal

(c) 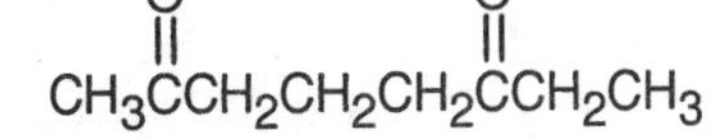

Octane-2,6-dione

(d) 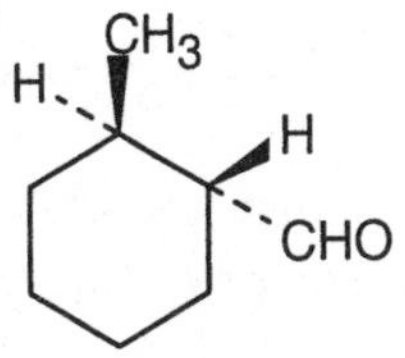

trans-2-Methylcyclohexane-carbaldehyde

(e) $OHCCH_2CH_2CH_2CHO$

Pentanedial

(f) 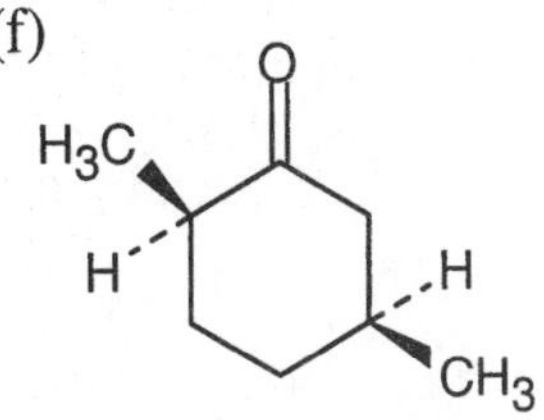

cis-2,5-Dimethyl-cyclohexanone

9.3

(a) 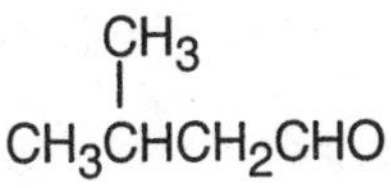

3-Methylbutanal

(b) 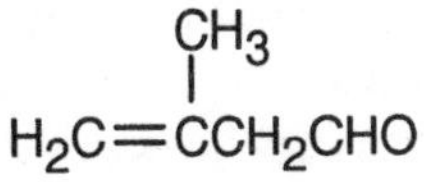

3-Methylbut-3-enal

(c) 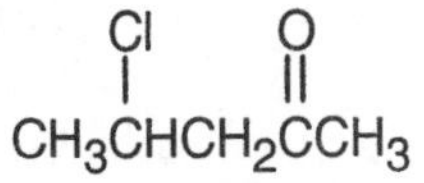

4-Chloropentan-2-one

(d) 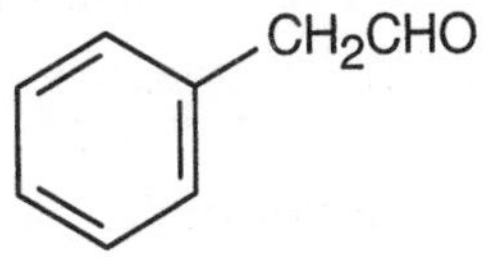

Phenylacetaldehyde

(e) 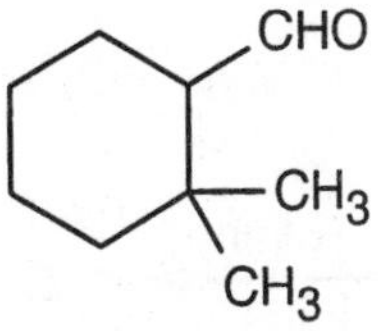

2,2-Dimethylcyclo-hexanecarbaldehyde

(f) 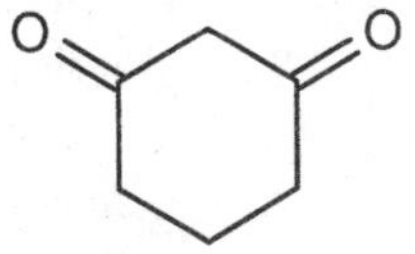

Cyclohexane-1,3-dione

9.4

(a)

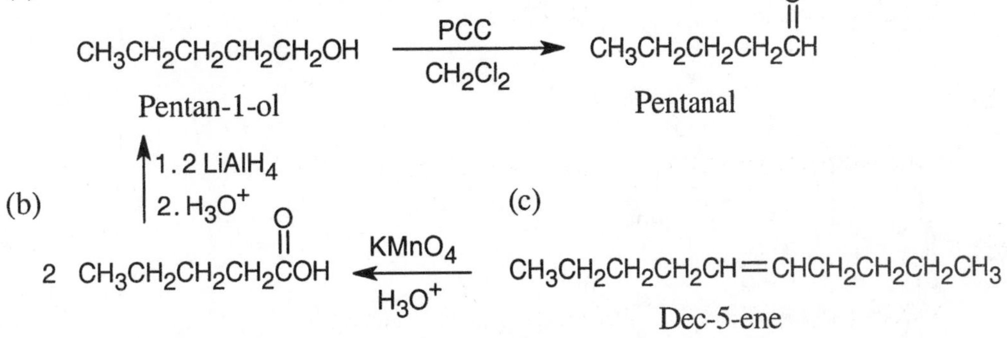
PCC
CH2Cl2
Pentan-1-ol
Pentanal
1. 2 LiAlH4
2. H3O+
(b)
(c)
KMnO4
H3O+
Dec-5-ene

9.5

(a)

$$CH_3CH_2CH_2CH_2CH(OH)CH_3 \xrightarrow[CH_2Cl_2]{PCC} CH_3CH_2CH_2CH_2C(=O)CH_3$$

Hexan-2-ol

(b)

$$CH_3CH_2CH_2CH_2C\equiv CH \xrightarrow[HgSO_4]{H_2O,\ H_2SO_4} CH_3CH_2CH_2CH_2C(=O)CH_3$$

Hex-1-yne

(c)

$$CH_3CH_2CH_2CH_2C(CH_3)=CH_2 \xrightarrow[H_3O^+]{KMnO_4} CH_3CH_2CH_2CH_2C(=O)CH_3 + CO_2$$

2-Methylhex-1-ene

9.6

(a)

$$CH_3CH_2CH=CHCH_2CH_3 \xrightarrow{H_3O^+} CH_3CH_2CH_2CH(OH)CH_2CH_3 \xrightarrow[CH_2Cl_2]{PCC} CH_3CH_2CH_2C(=O)CH_2CH_3$$

Hex-3-ene → Hexan-3-one

(b)

$$C_6H_6 \xrightarrow[AlCl_3]{CH_3C(=O)Cl} C_6H_5C(=O)CH_3 \xrightarrow[2.\ H_3O^+]{1.\ NaBH_4} C_6H_5CH(OH)CH_3$$

1-Phenylethanol

9.7

(a)

$$CH_3CH_2CH_2CH_2CHO \xrightarrow[\text{reagent}]{\text{Tollens'}} CH_3CH_2CH_2CH_2COOH$$

(b)

$$CH_3CH_2CH_2CH_2C(CH_3)_2CHO \xrightarrow[\text{reagent}]{\text{Tollens'}} CH_3CH_2CH_2CH_2C(CH_3)_2COOH$$

(c)

$$\text{Cyclohexanone} \xrightarrow[\text{reagent}]{\text{Tollens'}} \text{no reaction}$$

9.8

$:\ddot{O}:\ \delta-$ ‖ $C\ \delta+$, H_3C, CH_3, $^-:CN$ ⇌ [$:\ddot{O}:^-$, H_3C–C–CN, H_3C] $\overset{H_2O}{\rightleftharpoons}$ OH, H_3C–C–CN, H_3C + HO^-

nucleophilic addition

protonation of tetrahedral intermediate

cyanohydrin

Cyanide anion adds to the positively polarized carbonyl carbon to form a tetrahedral intermediate. This intermediate is protonated to yield acetone cyanohydrin.

9.9

$:\ddot{O}:\ \delta-$ ‖ $C\ \delta+$, H, $^-:\ddot{O}CH_3$ ⇌ [$:\ddot{O}:^-$, C, OCH_3, H] $\overset{H_2O}{\rightleftharpoons}$ OH, C, OCH_3, H + HO^-

nucleophilic addition

protonation of tetrahedral intermediate

hemiacetal

9.10

$:\ddot{O}:\ \delta-$ ‖ $C\ \delta+$, Cl_3C, H, $:\ddot{O}H_2$ ⇌ [$:\ddot{O}:^-$, Cl_3C–C–$\overset{+}{\ddot{O}}H_2$, H] ⇌ OH, Cl_3C–C–OH, H

Chloral hydrate

9.11 The labeled oxygen is starred.

$:\ddot{O}:$ ‖ C, R, R, $:\overset{*}{\ddot{O}}H_2$ ⇌ [$:\ddot{O}:^-$, R–C–$\overset{*+}{O}H_2$, R] ⇌ OH, R–C–$\overset{*}{O}H$, R

The above mechanism is similar to other nucleophilic addition mechanisms we have written. Since all of the steps are reversible, we can write the reaction in reverse to show how labeled oxygen is incorporated into an aldehyde or ketone.

OH, R–C–$\overset{*}{O}H$, R ⇌ [$\overset{+}{\ddot{O}}H_2$, R–C–$:\overset{*}{\ddot{O}}:^-$, R] ⇌ O^* ‖ C, R, R + H_2O

This exchange is very slow in water but proceeds more rapidly when either acid or base is present.

9.12

Cyclohexanone $\xrightarrow[\text{H}^+ \text{ catalyst}]{CH_3CH_2OH}$ hemiacetal (1-ethoxycyclohexanol: OH, OCH_2CH_3) $\xrightarrow[\text{H}^+ \text{ catalyst}]{CH_3CH_2OH}$ acetal (1,1-diethoxycyclohexane: OCH_2CH_3, OCH_2CH_3)

hemiacetal acetal

9.13

Benzaldehyde + $H_2C{-}CH_2$ (HO, OH) $\xrightarrow[\text{catalyst}]{H^+}$ $C_6H_5{-}CH(OH){-}OCH_2CH_2OH$

hemiacetal

Nucleophilic addition of one hydroxyl group of ethylene glycol to the carbonyl carbon of benzaldehyde, followed by a proton shift, produces the hemiacetal.

$C_6H_5{-}CH(OH){-}OCH_2CH_2OH$ $\xrightarrow[\text{catalyst}]{H^+}$ cyclic acetal (2-phenyl-1,3-dioxolane: H–C, O–CH_2, CH_2, O)

acetal

Protonation of one hydroxyl group, followed by displacement of water by the other hydroxyl group, gives the cyclic acetal.

9.14

$HCCH_2CH_2COCH_3$ (O over each C) $\xrightarrow[\text{H}^+ \text{ catalyst}]{HOCH_2CH_2OH}$ $HCCH_2CH_2COCH_3$ (aldehyde carbon as cyclic acetal O O; ester C=O)

$\xrightarrow{\text{1. LiAlH}_4 \text{ 2. H}_3\text{O}^+}$ $HCCH_2CH_2CH_2OH$ (cyclic acetal O O)

$\xrightarrow{H_3O^+}$ $HCCH_2CH_2CH_2OH$ (C=O)

9.15

Cyclohexanone:

(a) $\xrightarrow{CH_3NH_2}$ cyclohexanone N-methylimine (=N–CH_3)

(b) $\xrightarrow[\text{H}^+ \text{ catalyst}]{CH_3CH_2OH}$ 1,1-diethoxycyclohexane (OCH_2CH_3, OCH_2CH_3)

(c) $\xrightarrow{\text{1. NaBH}_4 \text{ 2. H}_3\text{O}^+}$ cyclohexanol (OH, H)

9.16 Identify the bond formed between the amine and the carbonyl compound. Add two hydrogens to the nitrogen that came from the amine, and add an oxygen to the part of the bond that came from the carbonyl compound.

H_2O + $(CH_3)_2CHC(=NCH_3)CH_2CH_3$ ⟵ $(CH_3)_2CHCOCH_2CH_3$ + CH_3NH_2

CH_3 from the amine

new bond

from the carbonyl compound

Methylamine

2-Methylpentan-3-one

9.17 All of the products have an –OH and a methyl group bonded to what was formerly a ketone carbon

(a) Cyclopentanone $\xrightarrow[2.\ H_3O^+]{1.\ CH_3MgBr}$ 1-methylcyclopentanol (HO, CH_3)

(b) Benzophenone $\xrightarrow[2.\ H_3O^+]{1.\ CH_3MgBr}$ $(C_6H_5)_2C(OH)CH_3$ (HO, CH_3)

(c) $CH_3CH_2CH_2\overset{O}{\overset{\|}{C}}CH_2CH_3 \xrightarrow[2.\ H_3O^+]{1.\ CH_3MgBr} CH_3CH_2CH_2\underset{CH_3}{\overset{OH}{C}}CH_2CH_3$

9.18 When choosing the starting materials for a Grignard reaction, two reminders may be helpful.

1. Ketone + Grignard reagent —> tertiary alcohol
   Aldehyde + Grignard reagent —> secondary alcohol
   Formaldehyde + Grignard reagent —> primary alcohol

2. More than one combination of carbonyl compound plus Grignard reagent may yield the same product.

(a) 2-Methyl-2-propanol is a tertiary alcohol. To synthesize a tertiary alcohol, start with a ketone.

$CH_3\overset{O}{\overset{\|}{C}}CH_3 \xrightarrow[2.\ H_3O^+]{1.\ CH_3MgBr} CH_3\underset{CH_3}{\overset{OH}{C}}CH_3$

(b) Since 1-methylcyclohexanol is a tertiary alcohol, start with a ketone.

Cyclohexanone $\xrightarrow[2.\ H_3O^+]{1.\ CH_3MgBr}$ 1-Methylcyclohexanol

1-Methylcyclohexanol

(c) 3-Methyl-3-pentanol is a tertiary alcohol that can be synthesized from a ketone by two different routes.

$CH_3CH_2COCH_2CH_3 \xrightarrow[2.\ H_3O^+]{1.\ CH_3MgBr}$

or

$CH_3CH_2COCH_3 \xrightarrow[2.\ H_3O^+]{1.\ CH_3CH_2MgBr}$ $CH_3CH_2C(OH)(CH_3)CH_2CH_3$

3-Methylpentan-3-ol

9.19 Two combinations of Grignard reagent and ketone can be used to form this tertiary alcohol.

Cyclopentyl–C(=O)–CH_3 $\xrightarrow[2.\ H_3O^+]{1.\ CH_3MgBr}$

or

H_3C–C(=O)–CH_3 $\xrightarrow[2.\ H_3O^+]{1.\ \text{cyclopentyl}MgBr}$ Cyclopentyl–C(OH)(CH_3)–CH_3

9.20 Identify the group that has added to the double bond. Its location is at the ß carbon of the conjugated system. Draw a double bond conjugated with the carbonyl group to complete the structure of the α,β-unsaturated component.

3-(Isopropoxy)-6-methylcyclohexanone (ring carbons numbered 1–4; $OCH(CH_3)_2$ circled) $\longleftarrow$ 6-methylcyclohex-2-enone (α, β labeled) + $HOCH(CH_3)_2$

Visualizing Chemistry

9.21

(a) $CH_3CH(CH_3)CH_2C(=O)NHCH_3$ — amide

(b) CH_3O–C(=O)–cyclohexane (with H) — ester

(c) Benzene ring bearing CH_2CHO (aldehyde) and C(=O)OH (carboxylic acid)

9.22

(a)

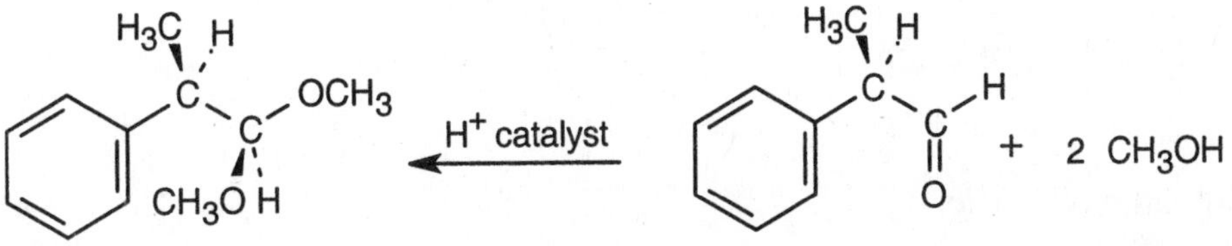

H^+ catalyst

(R)-2-Phenylpropanal + 2 CH_3OH Methanol

(b)

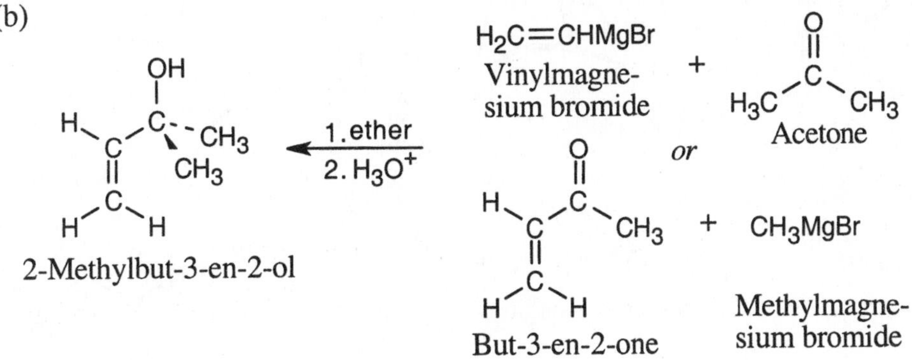

$H_2C{=}CHMgBr$ Vinylmagnesium bromide + Acetone

or

But-3-en-2-one + CH_3MgBr Methylmagnesium bromide

1. ether 2. H_3O^+

2-Methylbut-3-en-2-ol

(c)

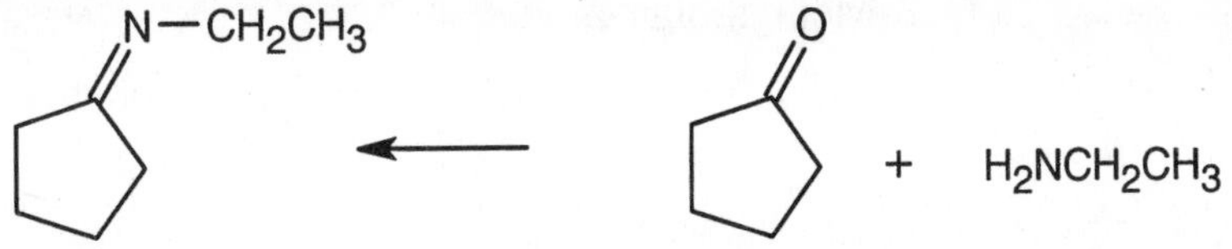

Cyclopentanone + $H_2NCH_2CH_3$ Ethylamine → imine

9.23

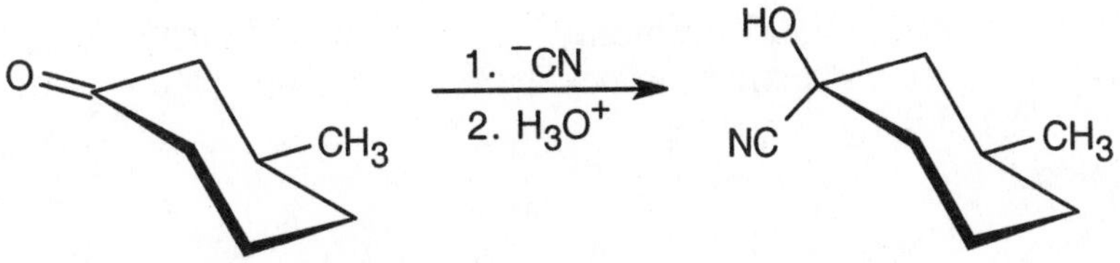

3-Methylcyclohexanone

1. ^-CN 2. H_3O^+

9.24 The intermediate results from the addition of an amine to a ketone.

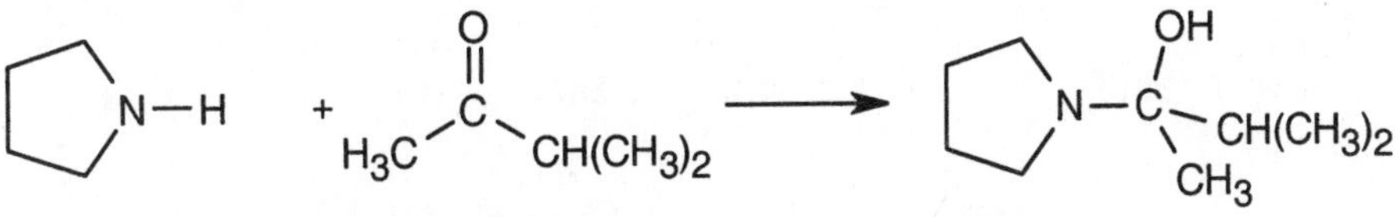

3-Methylbutan-2-one

Additional Problems

Nomenclature and Isomerism

9.25

(a)

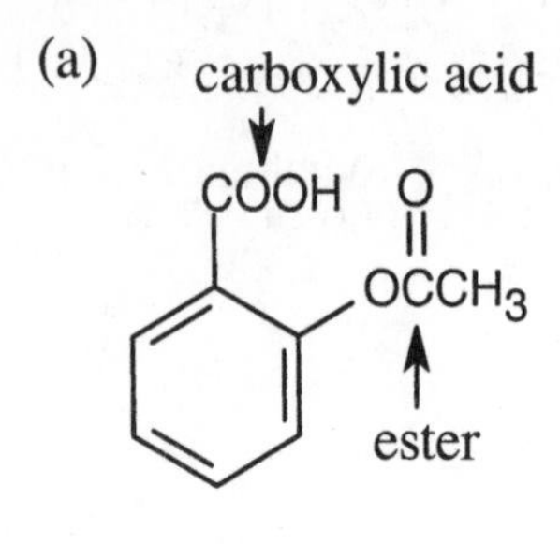

Aspirin

(b) ester; H_3C–N; $COCH_3$; H; OCC_6H_5; H; ester

Cocaine

(c) CH_2OH; HC-OH; lactone (cyclic ester); O; HO; OH

Ascorbic acid

9.26

(a) CH_3CCH_2Br (C=O)

Bromoacetone

(b) CH_3CHCCH_3 (C=O), CH_3

3-Methylbutan-2-one

(c) O_2N, CHO, NO_2

3,5-Dinitrobenzaldehyde

(d) O, H_3C, CH_3

3,5-Dimethylcyclohexanone

(e) $(CH_3)_3CCC(CH_3)_3$ (C=O)

2,2,4,4-Tetramethylpentan-3-one

(f) $HCCH_2CH_2CH$ (two C=O)

Butanedial

(g) CHO, *S*, H, C, OH, H_3C

(*S*)-2-Hydroxypropanal

(h) CH=CHCHO

3-Phenylprop-2-enal

9.27

$CH_3CH_2CH_2CH_2CHO$	$CH_3CH_2CH(CH_3)CHO$	$CH_3CH(CH_3)CH_2CHO$	$(CH_3)_3CCHO$
Pentanal	2-Methylbutanal	3-Methylbutanal	2,2-Dimethylpropanal

$CH_3CH_2CH_2C(=O)CH_3$	$CH_3CH_2C(=O)CH_2CH_3$	$CH_3CH(CH_3)C(=O)CH_3$
Pentan-2-one	Pentan-3-one	3-Methylbutan-2-one

9.28 Of the compounds shown in Problem 9.27, only 2-methylbutanal is chiral.

9.29

(a) O (b) $CH_3CH_2C(=O)CH_2C(=O)CH_3$ (c) O, C, CH_3, CH_3 (d) $CH_3CH_2CH_2CH(Br)C(=O)H$

9.30

(a) (b) (c)

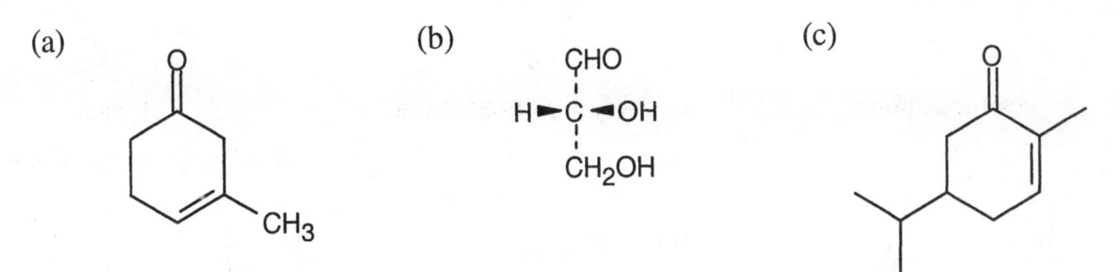

3-Methylcyclohex-3-enone (*R*)-2,3-Dihydroxypropanal (D-Glyceraldehyde) 5-Isopropyl-2-methyl-cyclohex-2-enone

(d) $CH_3CH(CH_3)C(=O)CH_2CH_3$	(e) $CH_3CH(OH)CH_2C(=O)H$	(f) OHC–C_6H_4–CHO
2-Methylpentan-3-one	3-Hydroxybutanal	*p*-Benzenedicarbaldehyde

9.31

(a) $(H_3C)_2C(OCH_3)_2$	(b) $(H_3C)_2C(OH)_2$	(c) $(H_3C)_2C{=}NCH_3$	(d) $(H_3C)_2C(OCH_3)OH$
acetal	gem diol	imine	hemiacetal

Reactions

9.32

Phenylacetaldehyde ($C_6H_5CH_2CHO$)

(a) 1. $NaBH_4$ 2. H_3O^+ → $C_6H_5CH_2CH_2OH$

(b) Tollens' reagent → $C_6H_5CH_2COOH$

(c) NH_2OH → $C_6H_5CH_2CH{=}NOH$

(d) 1. CH_3MgBr 2. H_3O^+ → $C_6H_5CH_2CH(OH)CH_3$

(e) CH_3OH, H^+ catalyst → $C_6H_5CH_2CH(OCH_3)OCH_3$

9.33

Acetophenone ($C_6H_5COCH_3$)

(a) 1. $NaBH_4$ 2. H_3O^+ → $C_6H_5CH(OH)CH_3$

(b) Tollens' reagent → no reaction

(c) NH_2OH → $C_6H_5C(=NOH)CH_3$

(d) 1. CH_3MgBr 2. H_3O^+ → $C_6H_5C(OH)(CH_3)CH_3$

(e) CH_3OH, H^+ catalyst → $C_6H_5C(OCH_3)_2CH_3$

9.34

$H_3C{-}CO{-}CH_3$

(a) 1. $NaBH_4$ 2. H_3O^+ → $CH_3CH(OH)CH_3$

(b) 1. CH_3CH_2MgBr 2. H_3O^+ → $CH_3C(OH)(CH_3)CH_2CH_3$

(c) CH_3NH_2 → $CH_3C(=NCH_3)CH_3$

(d) 1. $^-SCH_3$ 2. H_3O^+ → $CH_3C(OH)(SCH_3)CH_3$

In (a), the nucleophile is $:H^-$. In (b), the nucleophile is $^-:CH_2CH_3$.

9.35

H_3C–C(=O)–CH_2CH_3 + $^-$:CN ⇌ [H_3C–C(CN)(CH_2CH_3)–O^-] $\xrightarrow{HCN}$ H_3C–C(CN)(CH_2CH_3)–OH (*S*) + $^-$:CN

H_3C–C(=O)–CH_2CH_3 + $^-$:CN ⇌ [CH_3CH_2–C(H_3C)(CN)–O^-] $\xrightarrow{HCN}$ CH_3CH_2–C(H_3C)(CN)–OH (*R*) + $^-$:CN

Attack occurs with equal probability from either side of the planar carbonyl group to yield a racemic product mixture that is optically inactive.

9.36

H_3C–C(=O)–CH_2CH_3 + $BrMg^+$ $^-$:C_6H_5 ⇌ [H_3C–C(C_6H_5)(CH_2CH_3)–O^-] $\xrightarrow{H_3O^+}$ H_3C–C(C_6H_5)(CH_2CH_3)–OH (*S*) + H_2O

H_3C–C(=O)–CH_2CH_3 + $BrMg^+$ $^-$:C_6H_5 ⇌ [CH_3CH_2–C(H_3C)(C_6H_5)–O^-] $\xrightarrow{H_3O^+}$ CH_3CH_2–C(H_3C)(C_6H_5)–OH (*R*) + H_2O

The product is a racemic mixture of *R* and *S* enantiomers.

Synthesis

9.37

(a) 2-cyclohexenone $\xrightarrow{1.\ NaBH_4;\ 2.\ H_3O^+}$ 2-cyclohexenol (H, OH) $\xrightarrow{H_2SO_4,\ H_2O}$ 1,3-cyclohexadiene

(b)

H3C OH

1. CH_3MgBr
2. H_3O^+

H_2
Pd catalyst

H3C OH

(c)

H OH

1. $NaBH_4$
2. H_3O^+

H_2
Pd catalyst

H OH

(d)

HO C_6H_5

1. C_6H_5MgBr
2. H_3O^+

9.38

$HO:^- + C_6H_5CHBr_2 \longrightarrow HOCH(Br)C_6H_5 + Br^-$

$HO:^- + HOCH(Br)C_6H_5 \longrightarrow [HOCH(OH)C_6H_5 + Br^-] \rightleftharpoons O{=}CHC_6H_5 + H_2O$

gem diol Benzaldehyde

The initial product of two successive S_N2 reactions of $C_6H_5CHBr_2$ with hydroxide ion is a gem diol. Since the equilibrium between the gem diol and aldehyde favors the aldehyde, benzaldehyde is the observed product.

9.39

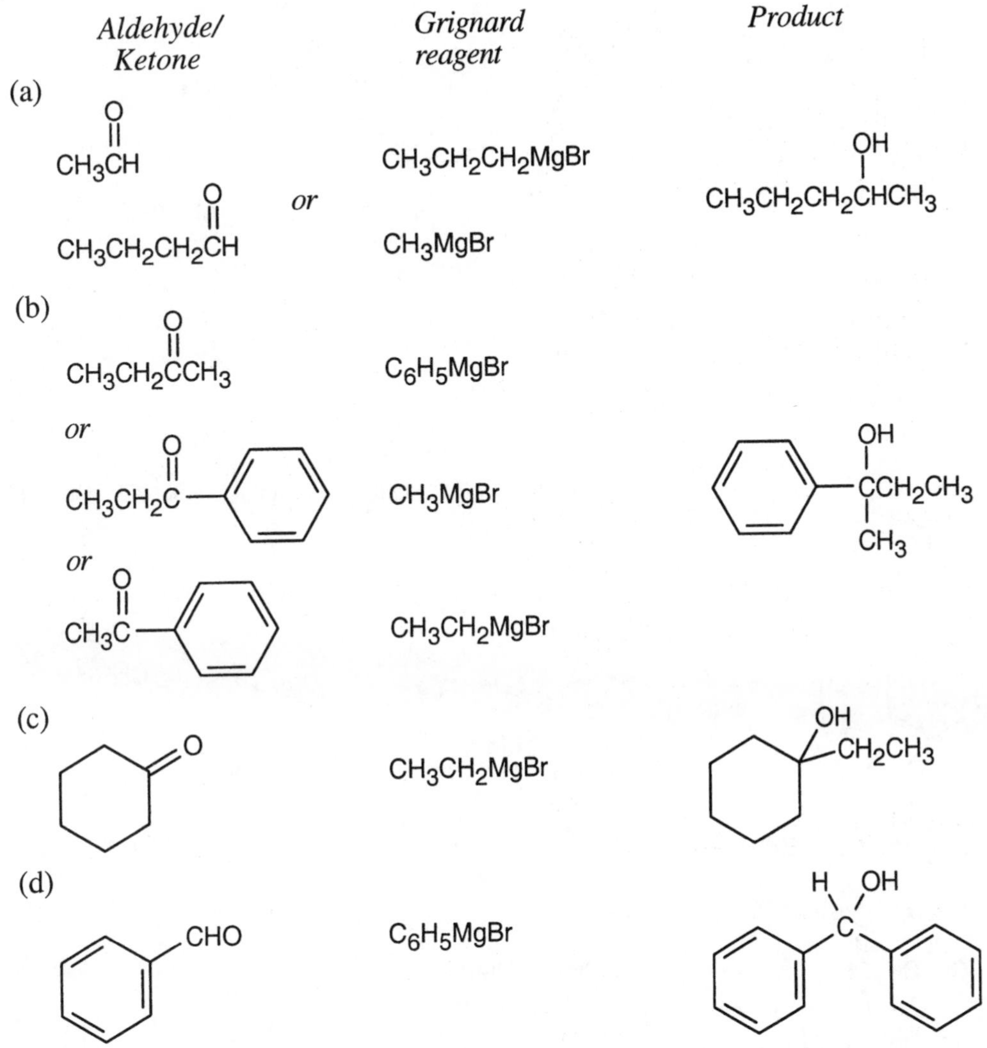

	Aldehyde/ Ketone	*Grignard reagent*	*Product*
(a)	$CH_3CH=O$	$CH_3CH_2CH_2MgBr$	$CH_3CH_2CH_2CH(OH)CH_3$
	or $CH_3CH_2CH_2CH=O$	CH_3MgBr	
(b)	$CH_3CH_2C(=O)CH_3$	C_6H_5MgBr	$C_6H_5C(OH)(CH_3)CH_2CH_3$
	or $CH_3CH_2C(=O)C_6H_5$	CH_3MgBr	
	or $CH_3C(=O)C_6H_5$	CH_3CH_2MgBr	
(c)	cyclohexanone	CH_3CH_2MgBr	1-ethylcyclohexanol (OH, CH_2CH_3)
(d)	C_6H_5CHO	C_6H_5MgBr	$(C_6H_5)_2CHOH$ (H, OH on C)

9.40

(a)

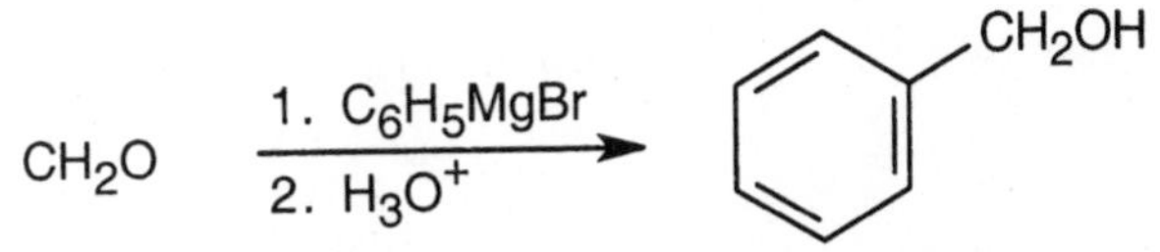

$$CH_2O \xrightarrow[2.\ H_3O^+]{1.\ C_6H_5MgBr} C_6H_5CH_2OH$$

Benzyl alcohol

(b)

$$C_6H_5C(=O)C_6H_5 \xrightarrow[2.\ H_3O^+]{1.\ C_6H_5MgBr} (C_6H_5)_3COH$$

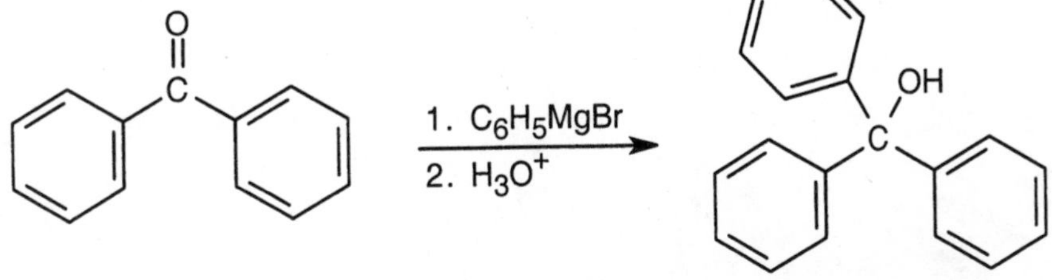

Triphenylmethanol

(c)

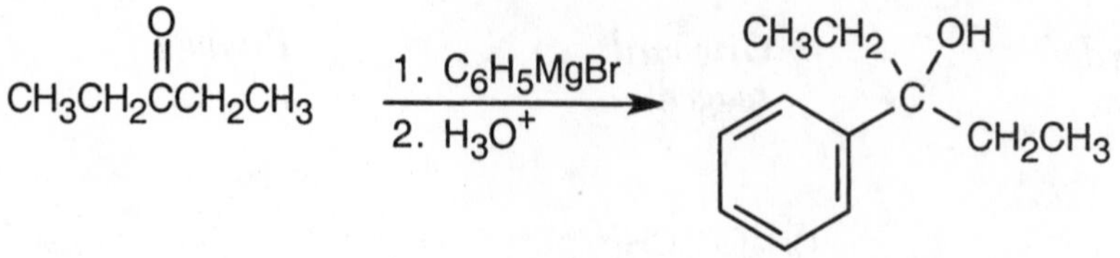

$$CH_3CH_2\overset{O}{\overset{\|}{C}}CH_2CH_3 \xrightarrow[2.\ H_3O^+]{1.\ C_6H_5MgBr} C_6H_5C(OH)(CH_2CH_3)_2$$

3-Phenylpentan-3-ol

9.41

	Aldehyde/ Ketone		*Grignard reagent*	*Product*
(a)	CH_2O		$CH_3CH(CH_3)CH_2CH_2CH_2MgBr$	$CH_3CH(CH_3)CH_2CH_2CH_2CH_2OH$
(b)	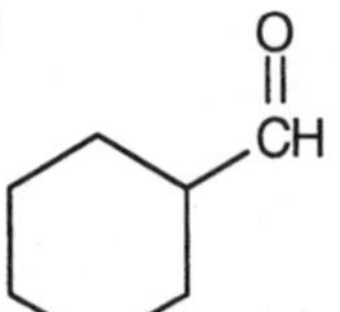 $C_6H_{11}CHO$ (cyclohexanecarbaldehyde)		CH_3MgBr	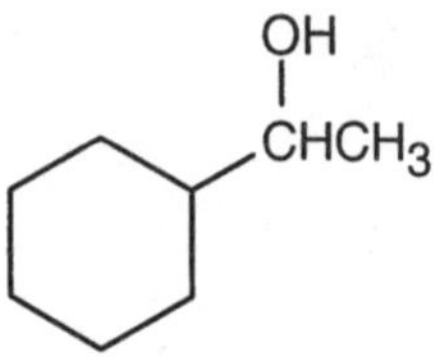 $C_6H_{11}CH(OH)CH_3$
	$CH_3\overset{O}{\overset{\|}{C}}H$	*or*	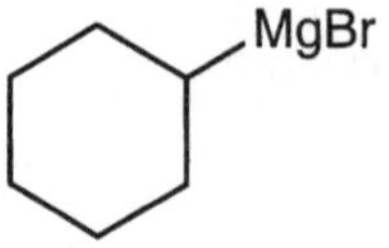 $C_6H_{11}MgBr$ (cyclohexylmagnesium bromide)	
(c)	$CH_3CH_2\overset{O}{\overset{\|}{C}}H$		$CH_3CH{=}CHMgBr$	$CH_3CH_2CH(OH)CH{=}CHCH_3$
	$CH_3CH{=}CH\overset{O}{\overset{\|}{C}}H$	*or*	CH_3CH_2MgBr	

9.42

	Aldehyde/ Ketone	*Grignard reagent*	*Product*
(a)	$H_2C{=}O$	C_6H_5MgBr	$C_6H_5CH_2OH$
(b)	$CH_3CH_2C(=O)CH_3$	C_6H_5MgBr	
	or $CH_3C(=O)C_6H_5$	CH_3CH_2MgBr	$C_6H_5C(OH)(CH_3)CH_2CH_3$
	or $CH_3CH_2C(=O)C_6H_5$	CH_3MgBr	
(c)	$C_6H_{11}C(=O)CH_3$	CH_3MgBr	$C_6H_{11}C(OH)(CH_3)CH_3$
	or $CH_3C(=O)CH_3$	$C_6H_{11}MgBr$	
(d)	$CH_3CH_2CH(CH_3)CH_2CH{=}O$	CH_3MgBr	$CH_3CH_2CH(CH_3)CH_2CH(OH)CH_3$
	or $CH_3CH{=}O$	$CH_3CH_2CH(CH_3)CH_2MgBr$	
(e)	$CH_3CH_2C(=O)CH_2CH_3$	CH_3CH_2MgBr	$CH_3CH_2C(OH)(CH_2CH_3)CH_2CH_3$
(f)	$CH_3CH_2CH{=}O$	CH_3CH_2MgBr	$CH_3CH_2CH(OH)CH_2CH_3$

9.43

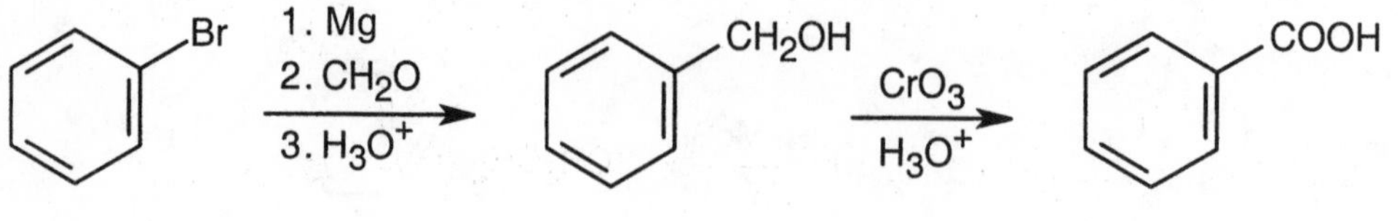

$$C_6H_5Br \xrightarrow[3.\ H_3O^+]{1.\ Mg \quad 2.\ CH_2O} C_6H_5CH_2OH \xrightarrow[H_3O^+]{CrO_3} C_6H_5COOH$$

Hemiacetals and Acetals

9.44

	Carbonyl compound	*Alcohol*	*Hemiacetal*	*Acetal*
(a)	$C_6H_5C(=O)CH_3$	$CH_3CH(OH)CH_3$	$C_6H_5C(OH)(OCH(CH_3)_2)CH_3$	$C_6H_5C(OCH(CH_3)_2)_2CH_3$
(b)	$CH_3CH_2C(=O)CH_2CH_3$	cyclopentanol	$CH_3CH_2C(OH)(O\text{-}C_5H_9)CH_2CH_3$	$CH_3CH_2C(O\text{-}C_5H_9)_2CH_2CH_3$

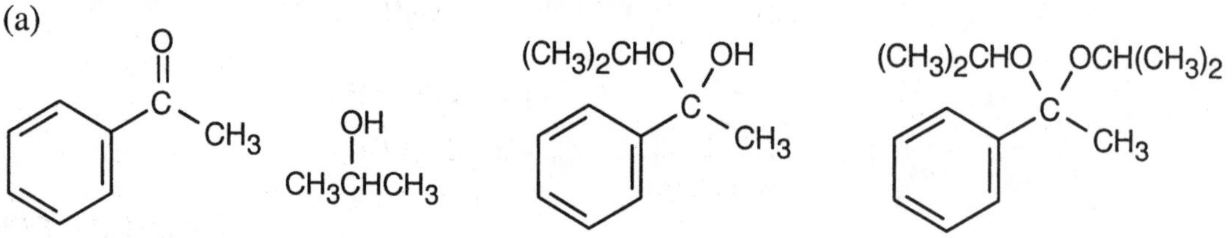

9.45

(a)

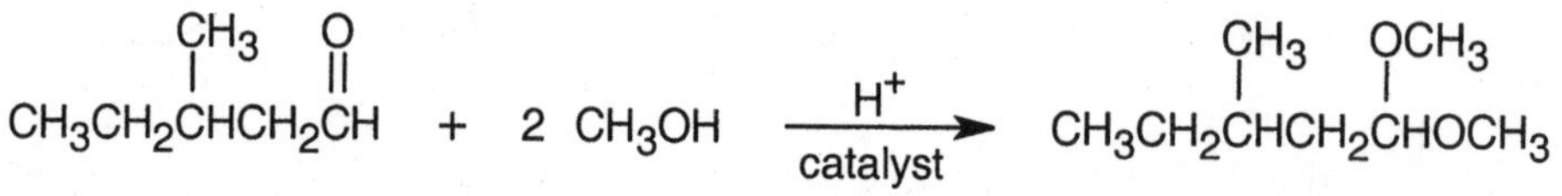

$$CH_3CH_2CH(CH_3)CH_2CHO + 2\ CH_3OH \xrightarrow[\text{catalyst}]{H^+} CH_3CH_2CH(CH_3)CH_2CH(OCH_3)OCH_3$$

(b)

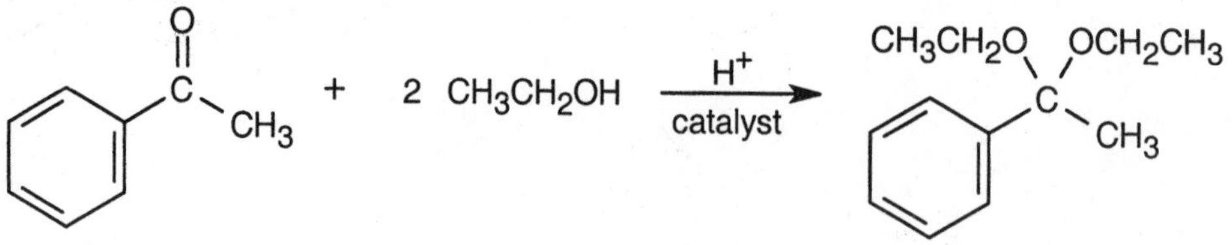

$$C_6H_5C(=O)CH_3 + 2\ CH_3CH_2OH \xrightarrow[\text{catalyst}]{H^+} C_6H_5C(OCH_2CH_3)_2CH_3$$

(c)

$$\text{cyclopentanone} + HOCH_2CH_2OH \xrightarrow[\text{catalyst}]{H^+} \text{cyclopentanone ethylene ketal (spiro 1,3-dioxolane)}$$

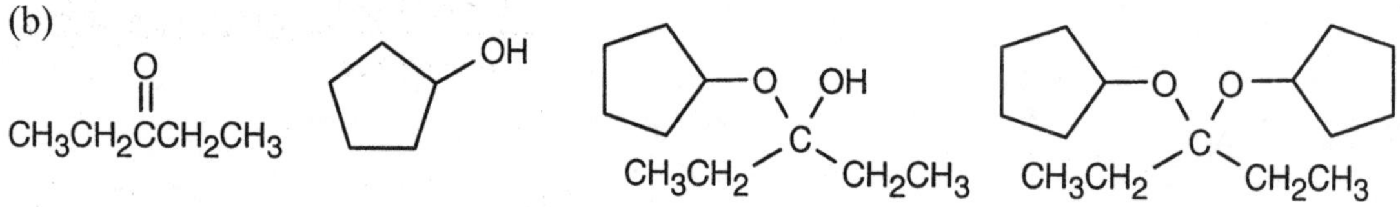

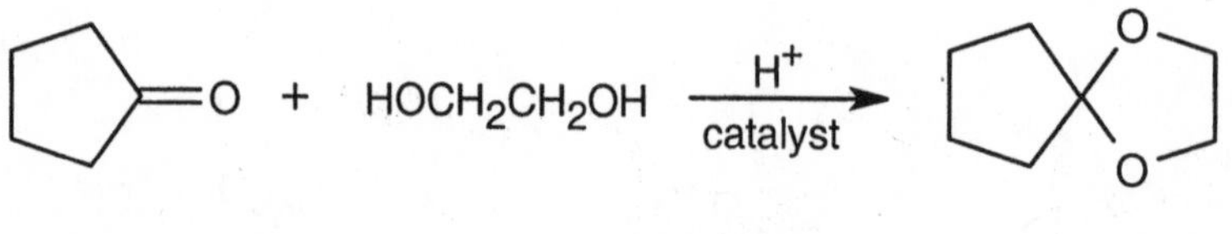

Integrated Problems

9.46

(a)

$CH_3CH_2CH_2C(=O)CH_3 + NH_2OH \longrightarrow CH_3CH_2CH_2C(=N{-}OH)CH_3$

(b)

$CH_3CH_2CH_2C(=O)CH_3$ + 2,4-(O_2N)(NO_2)$C_6H_3NHNH_2$ ⟶ 2,4-(O_2N)(NO_2)C_6H_3–NH–N=$C(CH_3)CH_2CH_2CH_3$

(c)

$CH_3CH_2CH_2C(=O)CH_3 \xrightarrow[H^+ \text{ catalyst}]{CH_3CH_2OH} CH_3CH_2CH_2C(OCH_2CH_3)(OH)CH_3 \xrightarrow[H^+ \text{ catalyst}]{CH_3CH_2OH} CH_3CH_2CH_2C(OCH_2CH_3)(OCH_2CH_3)CH_3$

9.47

(a)

cyclohexanone $\xrightarrow[2.\ H_3O^+]{1.\ CH_3MgBr}$ 1-methylcyclohexanol (OH, CH_3) $\xrightarrow[H_2O]{H_2SO_4}$ 1-methylcyclohexene (CH_3)

(b)

cyclohexanone $\xrightarrow[2.\ H_3O^+]{1.\ NaBH_4}$ cyclohexanol (OH) $\xrightarrow[H_2O]{H_2SO_4}$ cyclohexene $\xrightarrow[NaOH,\ H_2O]{KMnO_4}$ cis-1,2-cyclohexanediol (H, OH; H, OH)

(c)

1-methylcyclohexene (CH_3) $\xrightarrow{HBr}$ 1-bromo-1-methylcyclohexane (CH_3, Br)

from (a)

(d)

cyclohexanol (OH) $\xrightarrow{PBr_3}$ bromocyclohexane (Br) $\xrightarrow[ether]{Mg}$ cyclohexylmagnesium bromide (MgBr)

from (b)

cyclohexyl-MgBr + cyclohexanone (O) ⟶ 1-cyclohexylcyclohexanol (OH)

9.48

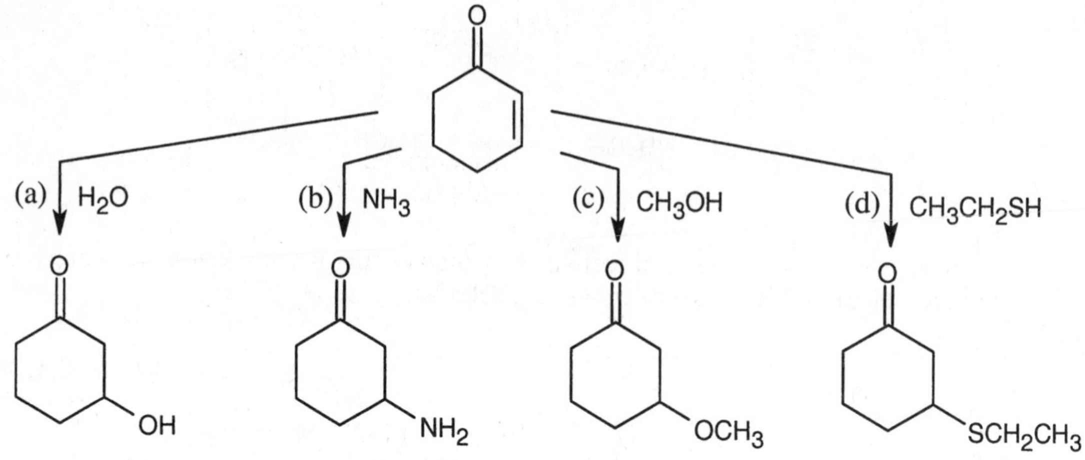

(a) H_2O (b) NH_3 (c) CH_3OH (d) CH_3CH_2SH

O O O O

OH NH_2 OCH_3 SCH_2CH_3

9.49

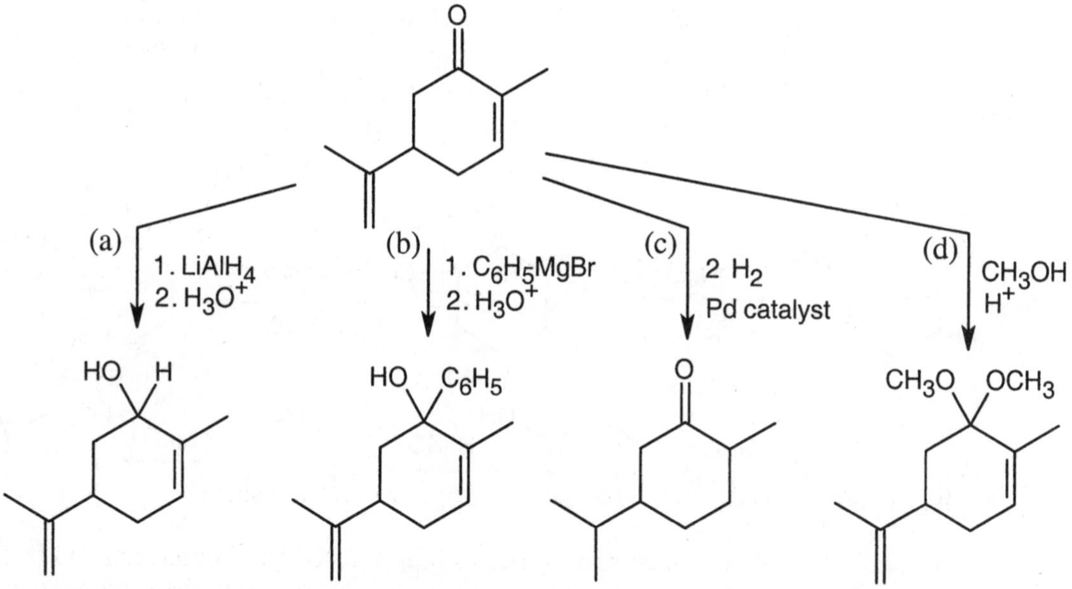

(a) 1. $LiAlH_4$ 2. H_3O^+

(b) 1. C_6H_5MgBr 2. H_3O^+

(c) 2 H_2 Pd catalyst

(d) CH_3OH H^+

HO H HO C_6H_5 O CH_3O OCH_3

9.50

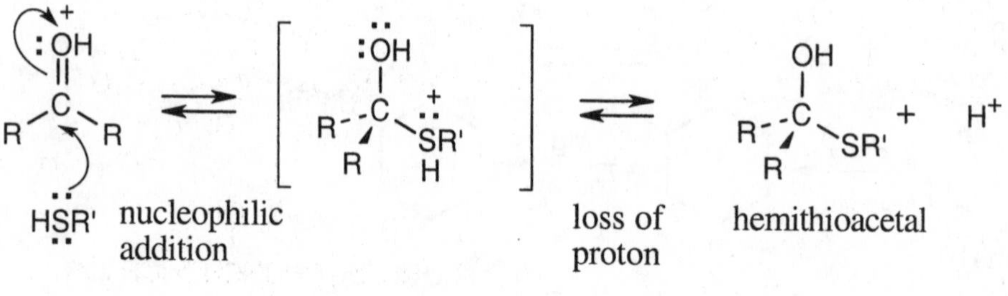

:OH R C R HSR' nucleophilic addition ⇌ [:OH R C R $\overset{+}{S}$R' H] ⇌ OH R C R SR' + H^+

loss of proton hemithioacetal

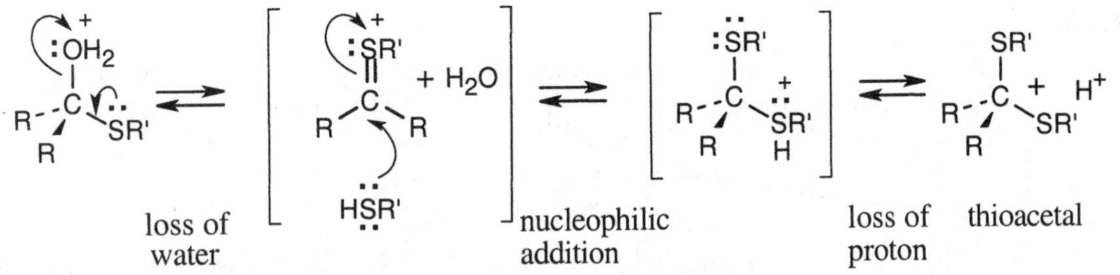

loss of water — nucleophilic addition — loss of proton — thioacetal

The reaction is similar to acetal formation, which occurs when a ketone or aldehyde reacts with an alcohol in the presence of an acid catalyst.

9.51

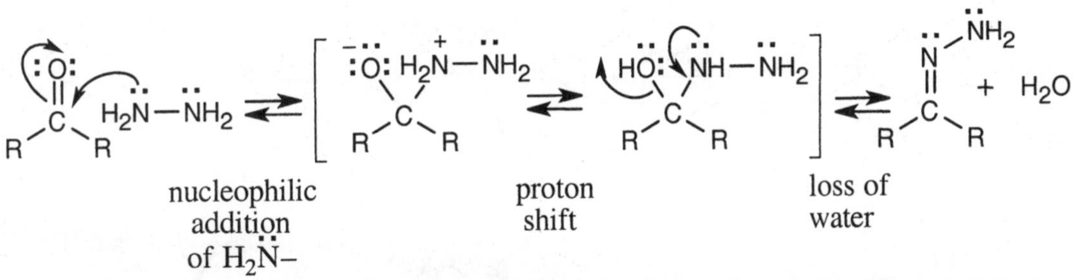

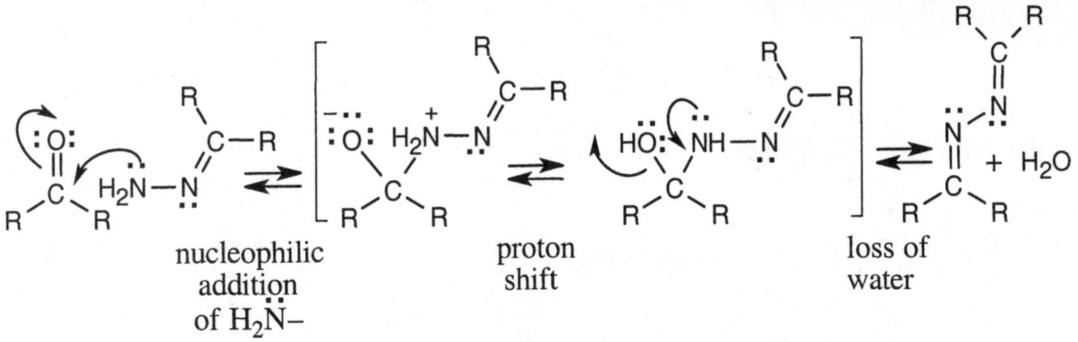

nucleophilic addition of $H_2\ddot{N}$– — proton shift — loss of water

nucleophilic addition of $H_2\ddot{N}$– — proton shift — loss of water

Each end of $H_2N–NH_2$ adds to a carbonyl group to give the azine.

9.52 Glucose is an aldehyde that exists as an equilibrium mixture of hemiacetal and free aldehyde. The open-chain aldehyde form of glucose is reduced in the same manner described in the text for other aldehydes to produce the polyalcohol sorbitol.

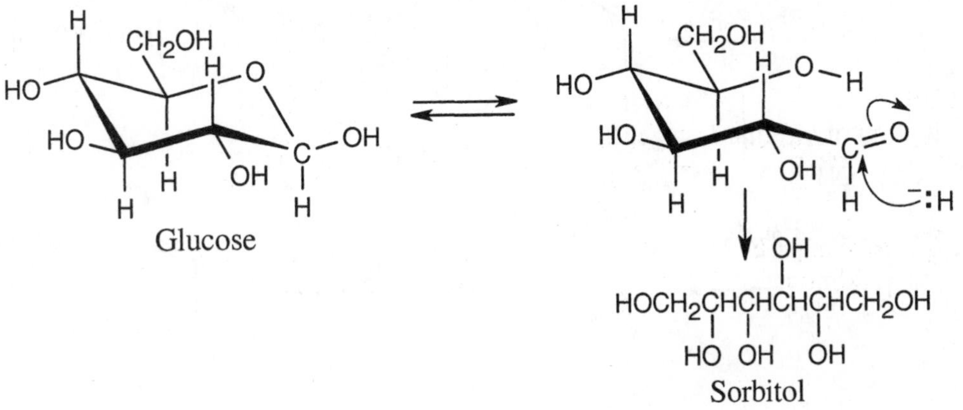

Glucose

Sorbitol

9.53

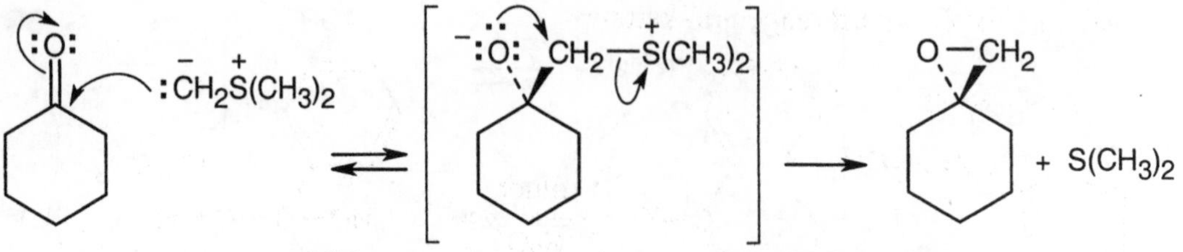

$:\ddot{O}:$ (cyclohexanone) + $:\overset{-}{C}H_2\overset{+}{S}(CH_3)_2$ ⇌ [$^{-}:\ddot{O}:\ CH_2-\overset{+}{S}(CH_3)_2$] → $O-CH_2$ (epoxide) + $S(CH_3)_2$

addition of ylide

S_N2 displacement of dimethyl sulfide by O^-

9.54

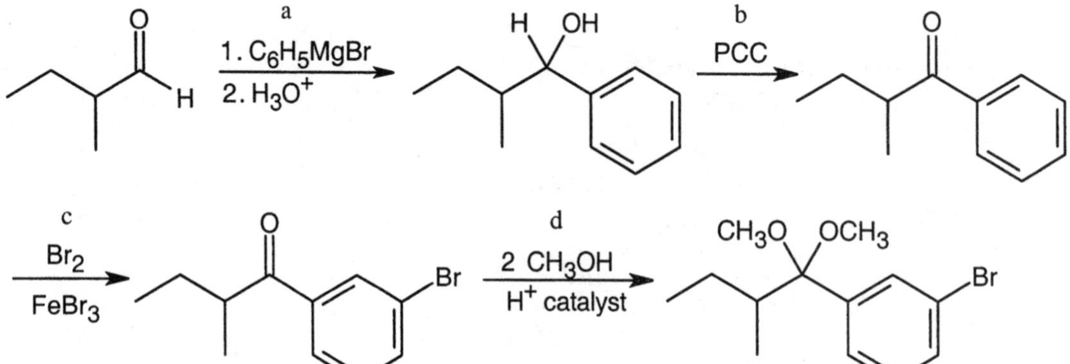

a: 1. C_6H_5MgBr 2. H_3O^+ → H OH

b: PCC

c: Br_2, $FeBr_3$ → Br

d: 2 CH_3OH, H^+ catalyst → CH_3O OCH_3, Br

In the Medicine Cabinet

9.55

Synthesis of 1-phenylpropan-1-ol:

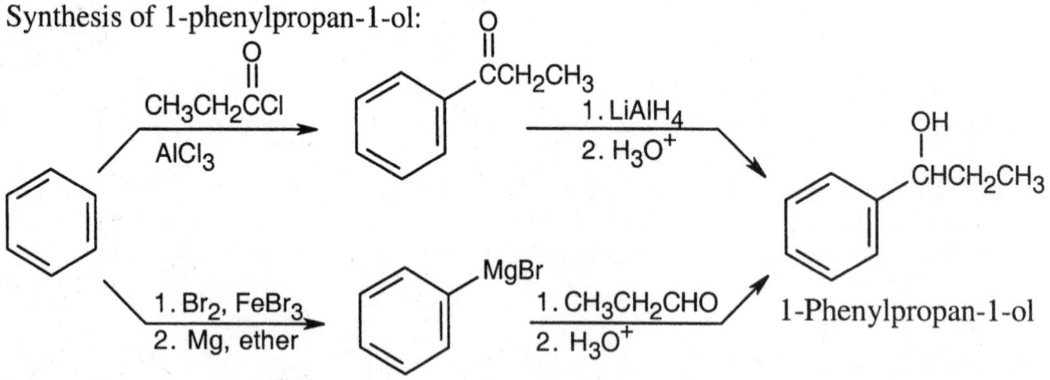

$CH_3CH_2\overset{O}{\overset{||}{C}}Cl$, $AlCl_3$ → $\overset{O}{\overset{||}{C}}CH_2CH_3$ → 1. $LiAlH_4$ 2. H_3O^+ → OH, $CHCH_2CH_3$

1. Br_2, $FeBr_3$ 2. Mg, ether → MgBr → 1. CH_3CH_2CHO 2. H_3O^+ → 1-Phenylpropan-1-ol

Formation of Grignard reagent:

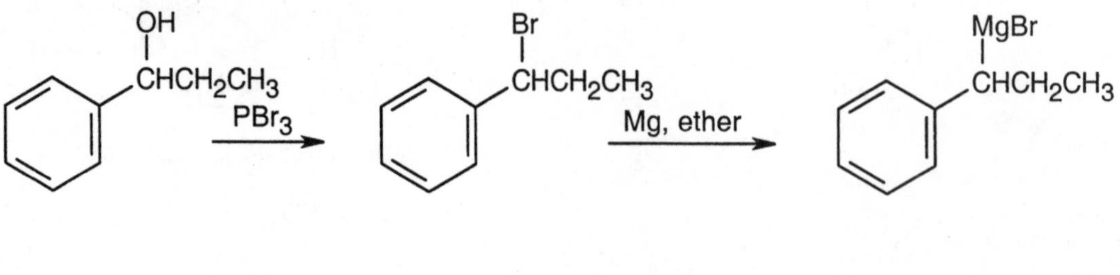

OH, $CHCH_2CH_3$ → PBr_3 → Br, $CHCH_2CH_3$ → Mg, ether → MgBr, $CHCH_2CH_3$

Addition of Grignard reagent to ketone:

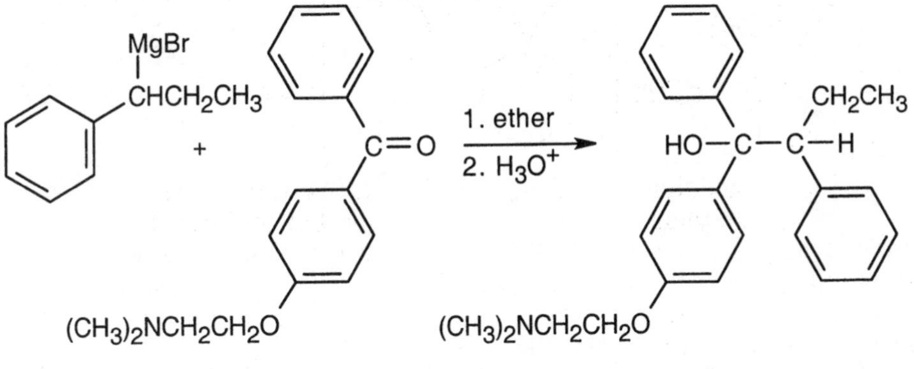

MgBr
CHCH$_2$CH$_3$
+
C=O
1. ether
2. H_3O^+
CH$_2$CH$_3$
HO—C—C—H
$(CH_3)_2NCH_2CH_2O$
$(CH_3)_2NCH_2CH_2O$

Dehydration to form Tamoxifen:

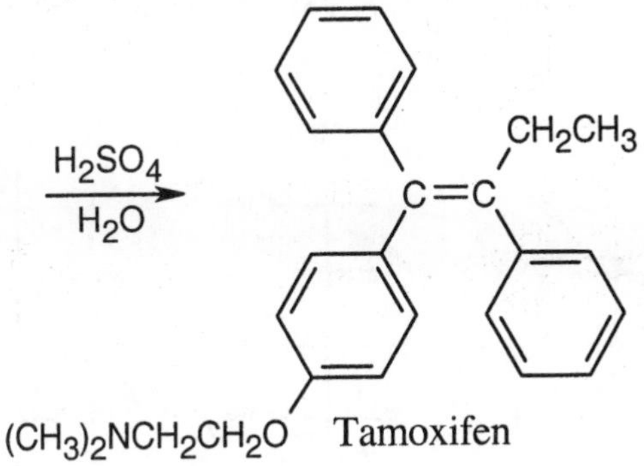

H_2SO_4
H_2O
CH$_2$CH$_3$
C=C
$(CH_3)_2NCH_2CH_2O$ Tamoxifen

9.56 (a)

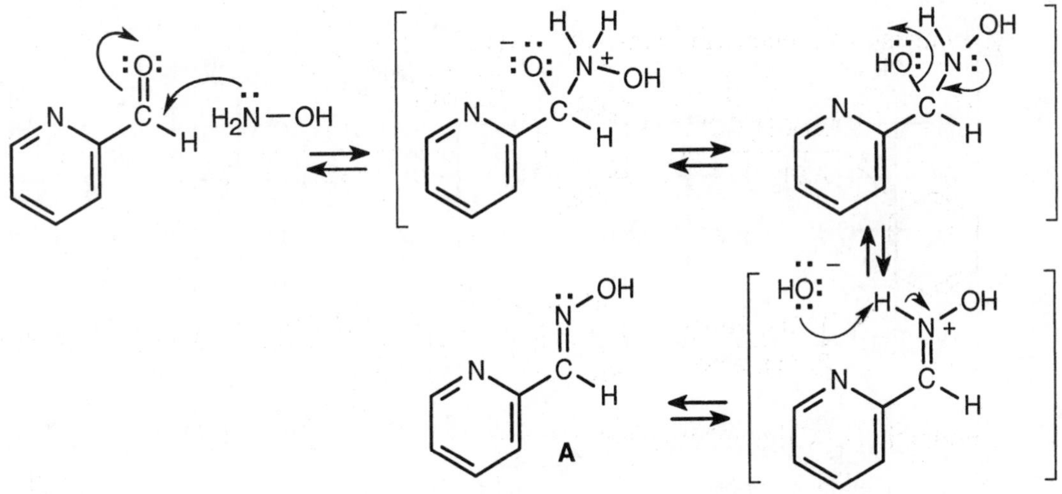

:O:
N
C
H
H_2N—OH
H H
:O: N$^+$ OH
N
C
H
H OH
HO: N:
N
C
H
HO:
H N$^+$ OH
N
C
H
N OH
N
C
H
A

(b)

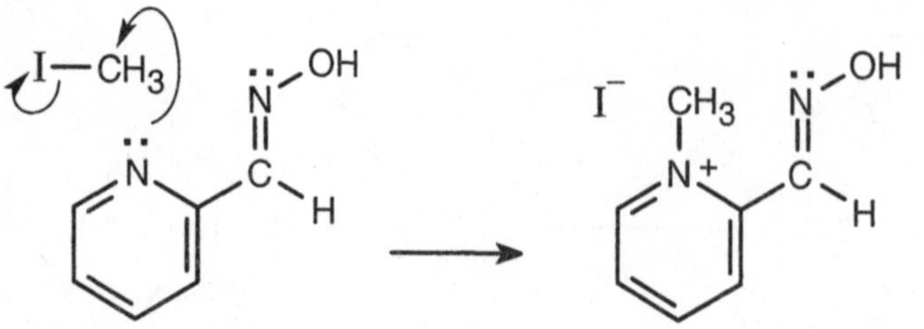

(c) Judging by the result, it seems that a pyridine nitrogen is more nucleophilic than an *N*-hydroxyl group.

9.57 (a) D-Glucose has four chirality centers.
(b) Reduction of D-glucose converts an aldehyde to a primary alcohol.
(c) Products **S** – **V** are ketones, **W** and **X** are aldehydes, and **Y** and **Z** are carboxylic acids.

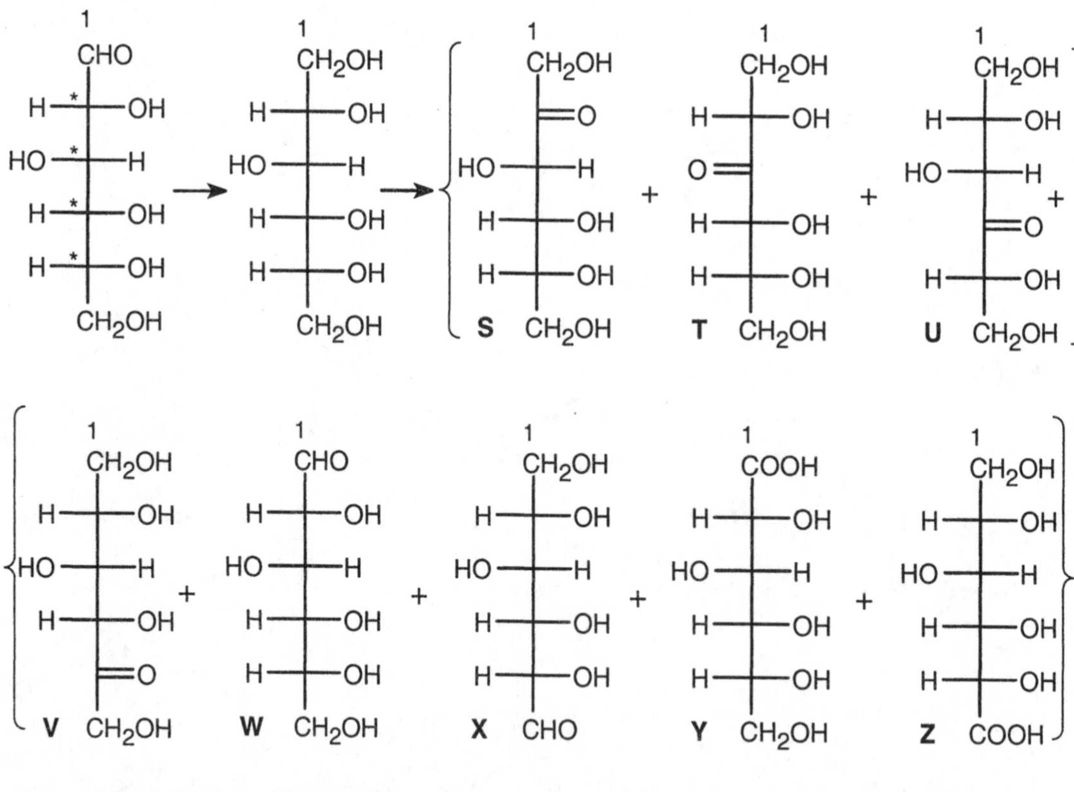

(d) An enzyme is selective and catalyzes the formation of one product, rather than eight.
(e) Acetone is used to form the two acetals.
(f) The acetals are "protecting groups" that protect several of the hydroxyl groups from being oxidized, while allowing oxidation of the desired hydroxyl group.
(g) The chirality center has the *S* configuration.

In the Field with Agrochemicals

9.58

(a) – (c)

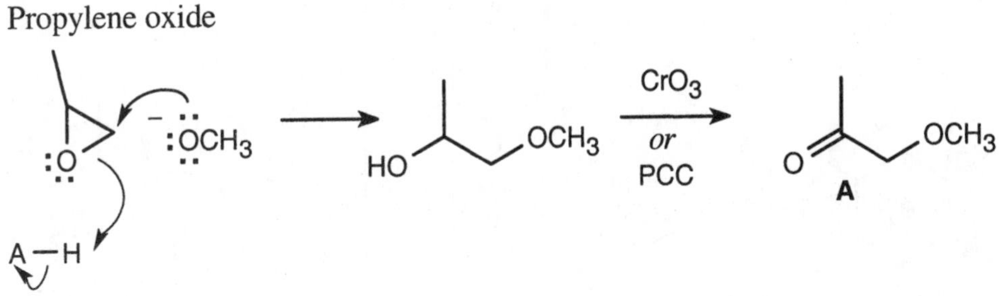

Propylene oxide

:Ö: A–H ⁻:ÖCH₃ → HO–CH(CH₃)–CH₂–OCH₃ $\xrightarrow[\text{PCC}]{CrO_3 \text{ or}}$ CH₃C(=O)CH₂OCH₃ **A**

(d) – (e)

E   Z   +   OCH₃   NH₂   HO   H–N: HO:⁻

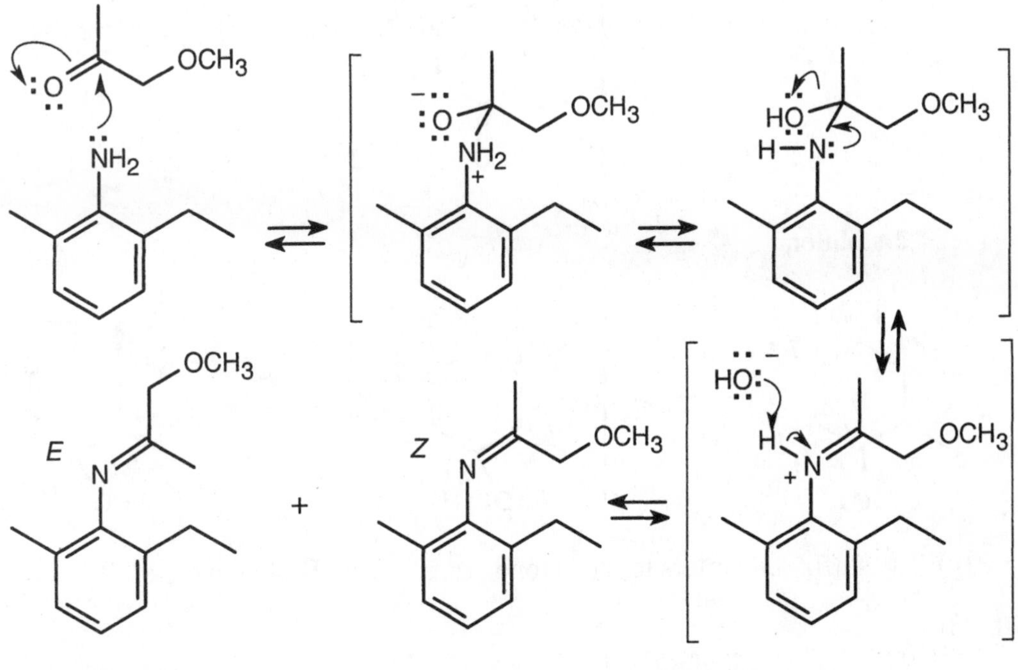

9.59

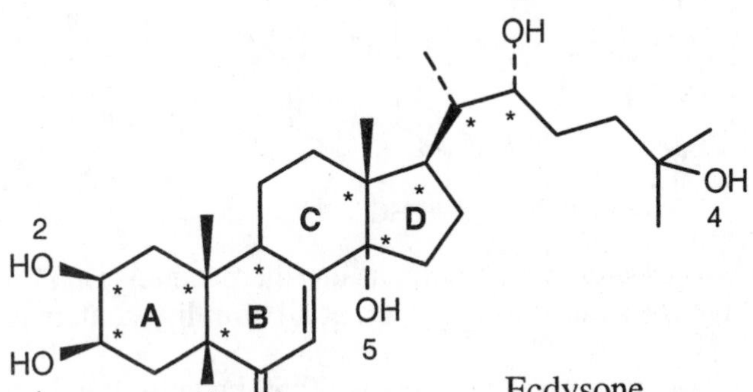

Ecdysone

(a) Hydroxyl groups 1,2,and 3 are secondary; groups 4 and 5 are tertiary.
(b) None of the hydroxyl groups will yield aldehydes on oxidation.
(c) Hydroxyl groups 1,2, and 3 yield ketones on oxidation.

(d) Hydroxyl groups 4 and 5 can't be oxidized.
(e) The ten chirality centers are starred.
(f)

1,4-addition:

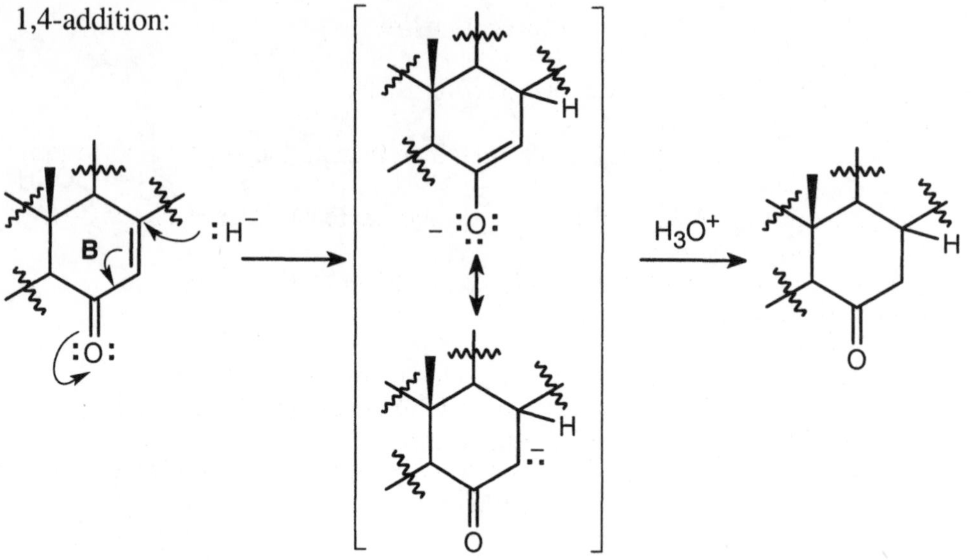

H_3O^+

1,2-addition:

H_3O^+

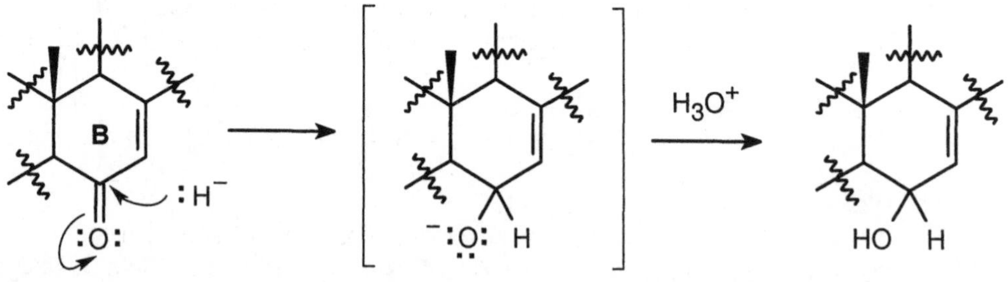

(g) The suffix *-one* refers to the ketone group in ring **B** of ecdysone.

9.60 (a)

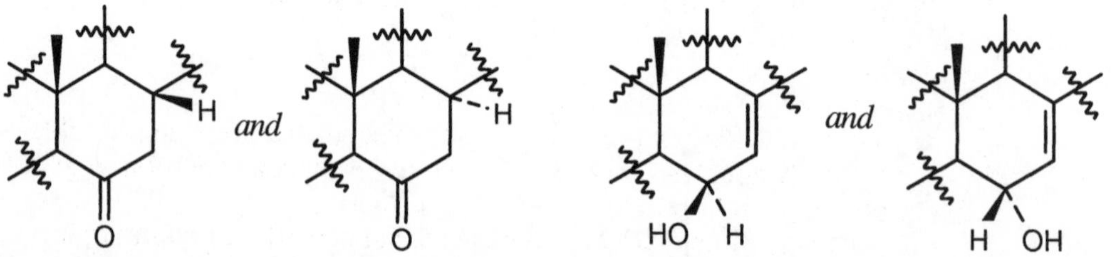

1,4-addition: and

1,2-addition: and

(b) The products of 1,4-addition are diastereomers, as are the products of 1,2-addition. Often pairs like these, that have the same configuration at all chirality centers except one, are described as epimers.
(c) The product ratio of neither pair of diastereomers is 50:50 because the diastereomers are formed in a chiral environment.

Chapter 10 – Carboxylic Acids and Derivatives

Chapter Outline

I. Naming carboxylic acids and derivatives (Section 10.1).
   A. Naming carboxylic acids.
      1. Noncyclic carboxylic acids are named by replacing the *-e* of the corresponding alkane by *-oic acid.*
      2. Compounds that have a carboxylic acid bonded to a ring are named by using the suffix *-carboxylic acid.*
      3. Many carboxylic acids have historical, nonsystematic names.
   B. Naming carboxylic acid derivatives.
      1. Acid halides.
         a. The acyl group is named first, followed by the halide.
         b. For acyclic compounds, the *-ic acid* of the carboxylic acid name is replaced by *-yl,* followed by the name of the halide.
         c. For cyclic compounds, the *-carboxylic acid* ending is replaced by *-carbonyl,* followed by the name of the halide.
      2. Acid anhydrides.
         Symmetrical anhydrides are named by replacing *acid* by *anhydride.*
      3. Amides.
         a. Amides with an unsubstituted $–NH_2$ group are named by replacing *-oic acid* by *-amide* or by replacing *-carboxylic acid* with *-carboxamide.*
         b. If nitrogen is substituted, the nitrogen substituents are named, and an *N-* is put before each.
      4. Esters.
         Esters are named by first identifying the alkyl group and then the carboxylic acid group, replacing *-oic acid* by *-oate.*
      5. Nitriles.
         a. Simple nitriles are named by adding *-nitrile* to the alkane name.
         b. More complex nitriles are named as derivatives of carboxylic acids by replacing *-oic acid* by *-onitrile* or by replacing *-carboxylic acid* by *-carbonitrile.*

II. Carboxylic acids (Sections 10.2–10.4).
   A. Occurrence, structure, and properties (Section 10.2).
      1. Carboxylic acids are abundant in nature.
      2. Carboxylic acids are strongly associated because of hydrogen-bonding, and their boiling points are elevated.
   B. Acidity (Section 10.3).
      1. Dissociation of carboxylic acids.
         a. Carboxylic acids react with bases to form salts that are water-soluble.
         b. Carboxylic acids dissociate slightly in dilute aqueous solution to give H_3O^+ and carboxylate anions.
            The K_a values for carboxylic acids are near 10^{-5}, making them weaker than mineral acids but stronger than alcohols.
         c. The relative strength of carboxylic acids is due to resonance stabilization of the carboxylate anion.
            i. Both carbon-oxygen bonds of carboxylate ions are the same length.
            ii. The bond length is intermediate between single and double bonds.
      2. Substituent effects on acidity.
         a. Carboxylic acids differ in acid strength.
            i. Electron-withdrawing groups stabilize carboxylate anions and increase acidity.

- ii. Electron-donating groups decrease acidity.
- b. These inductive effects decrease with increasing distance from the carboxyl group.

C. Synthesis of carboxylic acids (Section 10.4).
1. Oxidation of alkylbenzenes.
2. Oxidation of primary alcohols and aldehydes.
3. Hydrolysis of nitriles.

III. Nucleophilic acyl substitution reactions (Sections 10.5–10.11).

A. Mechanism of nucleophilic acyl substitution reactions (Sections 10.5–10.7).
1. A nucleophile adds to the polar carbonyl group.
2. The tetrahedral intermediate eliminates one of the two substituents originally bonded to it, resulting in a net substitution reaction.
3. The more electron-poor the carbonyl carbon, the more reactive the derivative.
4. It is possible to convert more reactive derivatives into less reactive derivatives.
   In order of decreasing reactivity: acid chlorides > acid anhydrides > esters > amides.
5. All of the reactions in this chapter can be grouped into three categories (Section 10.7).
   - a. Hydrolysis reactions.
   - b. Nucleophilic substitution reactions.
   - c. Oxidation-reduction reactions.

B. Reactions of carboxylic acids (Section 10.8).
1. Conversion to acid chlorides.
2. Reduction to alcohols using $LiAlH_4$.
3. Conversion to esters—Fischer esterification.
4. Conversion to amides.

C. Acid halides (Section 10.9).
1. Preparation from carboxylic acids, using $SOCl_2$.
2. Reactions of acid halides.
   - a. Hydrolysis to give carboxylic acids.
   - b. Reaction with alcohols to give esters.
   - c. Reaction with amines to give amides.

D. Acid anhydrides (Section 10.10).
1. Preparation from acid halides by reaction of a carboxylate with an acid chloride.
2. Reactions (same as acid halides).
   Only "half" of the anhydride molecule is used.

E. Esters (Section 10.11).
1. Preparation from acids and acid halides.
2. Reactions.
   - a. Hydrolysis to give acids.
   - b. Reaction with amines to give amides.
   - c. Reduction with $LiAlH_4$ to give primary alcohols.
   - d. Reaction with Grignard reagents to give tertiary alcohols.

F. Amides (Section 10.12).
1. Preparation from acid chlorides.
2. Reactions.
   - a. Hydrolysis to give carboxylic acids.
   - b. Reduction with $LiAlH_4$ to give amines.

G. Nitriles (Section 10.13).
1. Preparation from alkyl halides.
2. Reactions.
   - a. Hydrolysis to give carboxylic acids.
   - b. Reduction with $LiAlH_4$ to give amines.
   - c. Reaction with Grignard reagents to give ketones.

IV. Polyamides and polyesters: step-growth polymers (Section 10.14).
  A. When a diamine and a diacid chloride react, a polyamide is formed.
  B. When a diacid and a diol react, a polyester is formed.
  C. These polymers are called step-growth polymers because each bond is formed independently of the others.

V. Enzymes in organic synthesis (Section 10.15).
  Enzymes can carry out specific reactions that can't always be achieved using inorganic reagents.

Solutions to Problems

10.1

(a) $CH_3CH(CH_3)CH_2COOH$

3-Methylbutanoic acid

(b) $CH_3CH(Br)CH_2CH_2COOH$

4-Bromopentanoic acid

(c) $CH_3CH{=}CHCH_2CH_2COOH$

Hex-4-enoic acid

(d) $CH_3CH_2CH(COOH)CH_2CH_2CH_3$

2-Ethylpentanoic acid

(e) Cyclohexane ring with COOH and H on C1, CH_3 and H on C2

trans-2-Methylcyclohexanecarboxylic acid

10.2

(a) $CH_3CH_2CH_2CH(CH_3)CH(CH_3)COOH$

2,3-Dimethylhexanoic acid

(b) $CH_3CH(CH_3)CH_2CH_2COOH$

4-Methylpentanoic acid

(c) Benzene ring with COOH and OH

o-Hydroxybenzoic acid

(d) Cyclobutane ring with H and COOH on C1, H and COOH on C2

trans-Cyclobutane-1,2-dicarboxylic acid

10.3

(a) $CH_3CH(CH_3)CH_2CH_2C(=O)Cl$

4-Methylpentanoyl chloride

(b) $CH_3CH_2CH(CH_3)CN$

2-Methylbutanenitrile

(c) $H_2C{=}CHCH_2CH_2C(=O)NH_2$

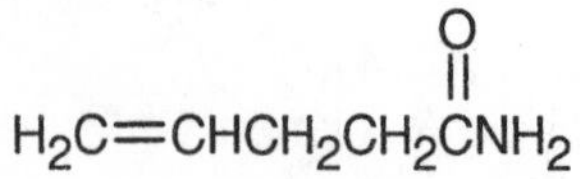

Pent-4-enamide

(d) $CH_3CH_2CH(CH_2CH_3)CN$

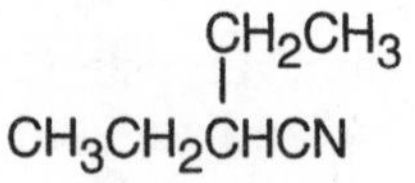

2-Ethylbutanenitrile

(e) Cyclopentyl–O–C(=O)–C(CH_3)$_3$

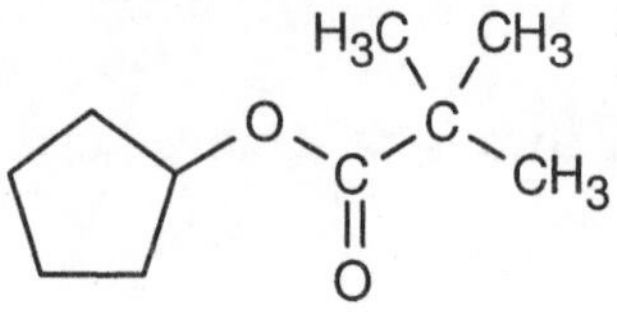

Cyclopentyl 2,2-dimethylpropanate

(f) $(H_3C)_2C{=}C(CH_3)COCl$

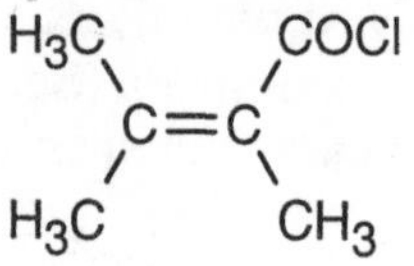

2,3-Dimethylbut-2-enoyl chloride

(g) $C_6H_5C(=O)OC(=O)C_6H_5$

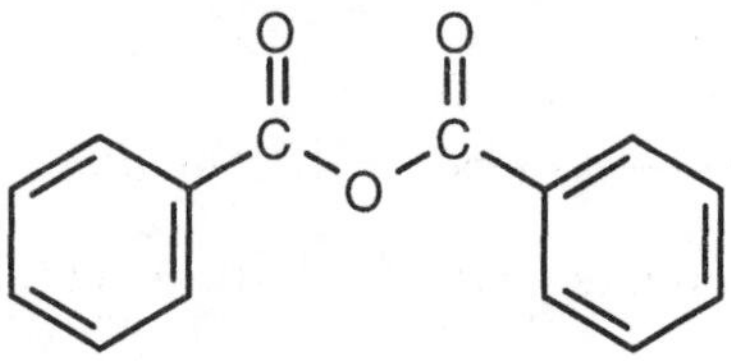

Benzoic anhydride

(h) Cyclopentyl–C(=O)$OCH(CH_3)CH_3$

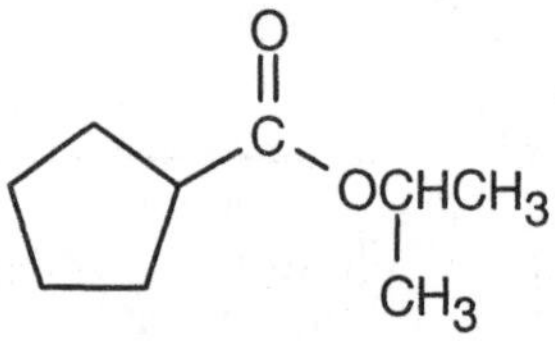

Isopropyl cyclopentanecarboxylate

10.4

(a) $CH_3C(CH_3)_2C(=O)Cl$

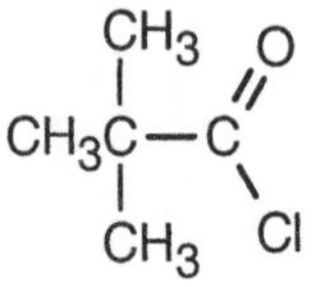

2,2-Dimethylpropanoyl chloride

(b) $C_6H_5C(=O)NHCH_3$

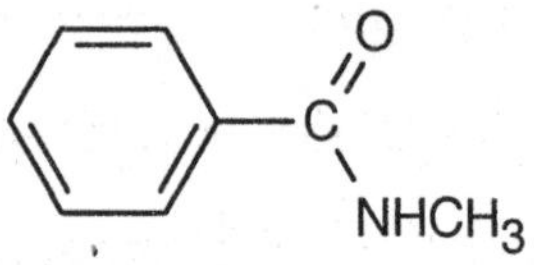

N-Methylbenzamide

(c) $CH_3C(CH_3)_2CH_2CH_2CH_2CN$

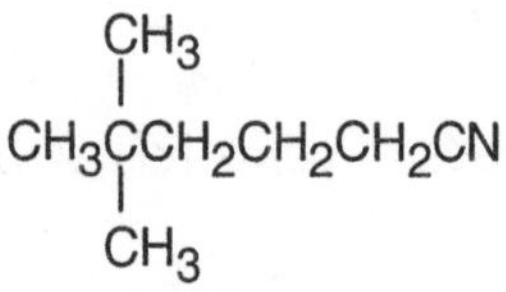

5,5-Dimethylhexanenitrile

(d) $CH_3CH_2CH_2C(=O){-}OC(CH_3)_2CH_3$

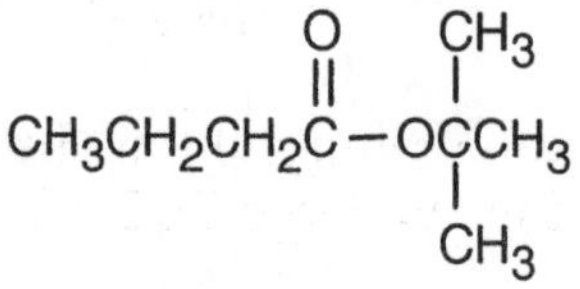

tert-Butyl butanoate

(e) Cyclohexane with C(=O)NH_2 and CH_3 on adjacent carbons (trans)

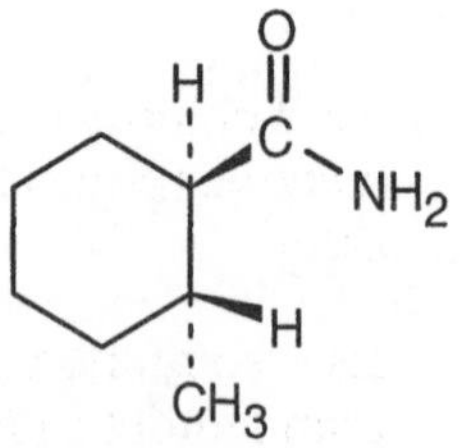

trans-2-Methylcyclohexanecarboxamide

(f) $H_3C{-}C_6H_4{-}C(=O)OC(=O){-}C_6H_4{-}CH_3$

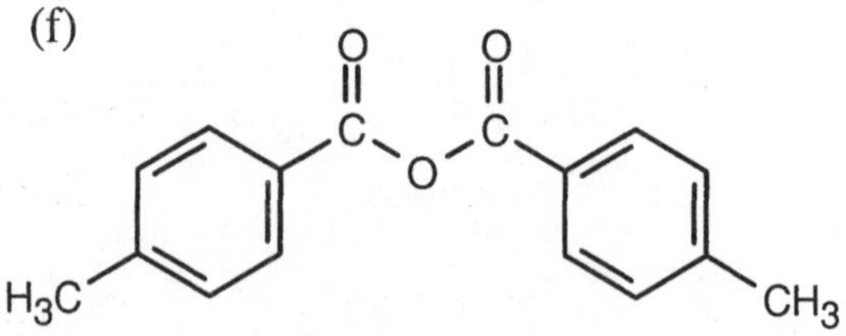

p-Methylbenzoic anhydride

(g) Cyclohexane with H_3C and COBr at 1,3 positions (cis)

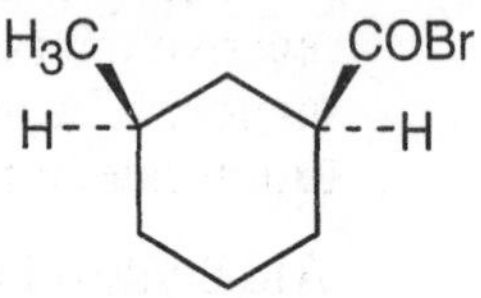

cis-3-Methylcyclohexanecarbonyl bromide

(h) Br–C_6H_4–CN

p-Bromobenzonitrile

10.5

(a)

$$C_6H_5COOH \xrightarrow{CH_3O^-Na^+} C_6H_5COO^-Na^+ + CH_3OH$$

(b)

$$(CH_3)_3CCOOH \xrightarrow{KOH} (CH_3)_3CCOO^-K^+ + H_2O$$

10.6

Least acidic ⟶ *Most acidic*

Methanol < Phenol < *p*–Nitrophenol < Acetic acid < Sulfuric acid

10.7 Remember that an electron-withdrawing group increases acidity by stabilizing the carboxylate anion. Also note that the effect of a substituent decreases with distance from the carboxyl group.

Least acidic ⟶ *Most acidic*

(a) CH_3CH_2COOH < $BrCH_2CH_2COOH$ < $BrCH_2COOH$

(b) Ethanol < Benzoic acid < *p*-Cyanobenzoic acid

10.8

$$CH_3I \xrightarrow{NaCN} CH_3CN \xrightarrow[2.\ H_3O^+]{1.\ ^-OH,\ H_2O} CH_3COOH$$

This reaction sequence can't be used to convert iodobenzene to benzoic acid because aryl halides don't undergo S_N2 substitution.

10.9

	More reactive	*Less reactive*	*Reason*
(a)	CH_3COCl Acid chloride	CH_3COOCH_3 Ester	Acid chlorides are more reactive (polar) than esters.
(b)	$CH_3CH_2COOCH_3$ Ester	$(CH_3)_2CHCONH_2$ Amide	Esters are more reactive than amides in nucleophilic acyl substitution reactions.
(c)	$CH_3COOCOCH_3$ Acid anhydride	CH_3COOCH_3 Ester	Acid anhydrides are more reactive than esters in nucleophilic acyl substitution reactions.
(d)	CH_3COOCH_3 Ester	CH_3CHO Aldehyde	Aldehydes don't undergo nucleophilic acyl substitution reactions.

10.10

$F_3C-\overset{\overset{\delta-}{O}}{\overset{\|}{C}}-OCH_3$ with partial charges $\delta-$ (on F_3C), $\delta+$ (on C), $\delta-$ (on OCH_3), $\delta-$ (on O)

$H_3C-\overset{\overset{\delta-}{O}}{\overset{\|}{C}}-OCH_3$ with partial charges $\delta+$ (on C), $\delta-$ (on OCH_3), $\delta-$ (on O)

The strongly electron-withdrawing trifluoromethyl group makes the carbonyl carbon more electron-poor and more reactive toward nucleophiles than the methyl acetate carbonyl carbon. Methyl trifluoroacetate is thus more reactive than methyl acetate in nucleophilic acyl substitution reactions.

10.11

Benzoic acid (C_6H_5COOH):

(a) $SOCl_2$ → C_6H_5COCl

(b) CH_3OH, HCl → $C_6H_5COOCH_3$ + H_2O

(c) 1. $LiAlH_4$ 2. H_3O^+ → $C_6H_5CH_2OH$

(d) $NaOH$ → $C_6H_5COO^-Na^+$ + H_2O

10.12 Arrange the acid and the ester so that the parts of the water molecule to be lost are next to each other. Remove water, and draw the new bond.

(a)

$$CH_3\overset{O}{\overset{\|}{C}}-OH + H-OCH_2CH_2CH_2CH_3 \underset{\text{catalyst}}{\overset{H^+}{\rightleftharpoons}} CH_3\overset{O}{\overset{\|}{C}}-OCH_2CH_2CH_2CH_3$$

Acetic acid, Butanol → Butyl acetate + H_2O

(b)

$$CH_3CH_2CH_2\overset{O}{\overset{\|}{C}}-OH + H-OCH_3 \underset{\text{catalyst}}{\overset{H^+}{\rightleftharpoons}} CH_3CH_2CH_2\overset{O}{\overset{\|}{C}}-OCH_3$$

Butanoic acid, Methanol → Methyl butanoate + H_2O

(c)

$$C_6H_5\overset{O}{\overset{\|}{C}}-OH + H-O\overset{CH_3}{\overset{|}{C}}HCH_3 \underset{\text{catalyst}}{\overset{H^+}{\rightleftharpoons}} C_6H_5\overset{O}{\overset{\|}{C}}-O\overset{CH_3}{\overset{|}{C}}HCH_3 + H_2O$$

Benzoic acid, 2-Propanol → Isopropyl benzoate

10.13

(a)

$$CH_3CH_2C(=O){-}Cl + H{-}OCH_3 \xrightarrow{\text{pyridine}} CH_3CH_2C(=O){-}OCH_3$$

Methyl propanoate

(b)

$$CH_3C(=O){-}Cl + H{-}OCH_2CH_3 \xrightarrow{\text{pyridine}} CH_3C(=O){-}OCH_2CH_3$$

Ethyl acetate

(c)

$$CH_3C(=O){-}Cl + H{-}O{-}C_6H_{11} \xrightarrow{\text{pyridine}} CH_3C(=O){-}O{-}C_6H_{11}$$

Cyclohexyl acetate

10.14

$$(CH_3)_2CHC(=O)Cl + :NH_3 \rightleftharpoons \left[(CH_3)_2CH{-}C(O^-)(Cl){-}NH_3^+ \underset{\text{deprotonation}}{\overset{NH_3}{\rightleftharpoons}} (CH_3)_2CH{-}C(O^-)(Cl){-}NH_2 + NH_4^+ \right]$$

2-Methylpropanoyl chloride

nucleophilic addition of ammonia

deprotonation

loss of Cl^-

$$\downarrow$$

$$(CH_3)_2CH{-}C(=O){-}NH_2 + Cl^-$$

2-Methylpropanamide

10.15

	Acid chloride	*Amine*		*Amide*
(a)	$CH_3CH_2C(=O){-}Cl$	NH_3	$\longrightarrow$	$CH_3CH_2C(=O){-}NH_2$ Propanamide
(b)	$CH_3CH(CH_3)CH_2C(=O){-}Cl$	CH_3NH_2	$\longrightarrow$	$CH_3CH(CH_3)CH_2C(=O){-}NHCH_3$ *N*-Methyl-3-methylbutanamide
(c)	$CH_3CH_2C(=O){-}Cl$	$HN(CH_3)_2$	$\longrightarrow$	$CH_3CH_2C(=O){-}N(CH_3)_2$ *N,N*-Dimethylpropanamide

Acid chloride *Amine* *Amide*

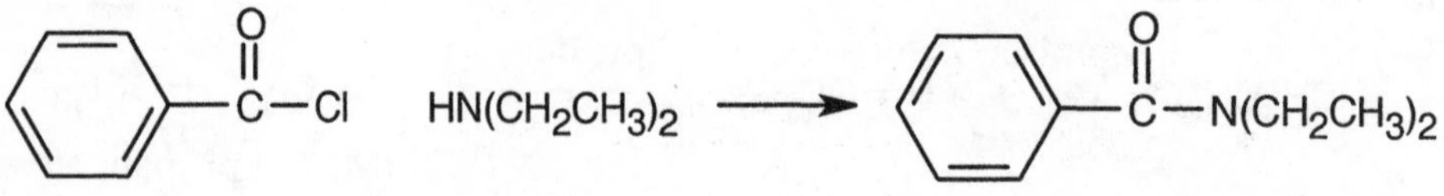

(d) C_6H_5COCl + $HN(CH_2CH_3)_2$ ⟶ $C_6H_5C(=O)N(CH_2CH_3)_2$

N,N-Diethylbenzamide

10.16

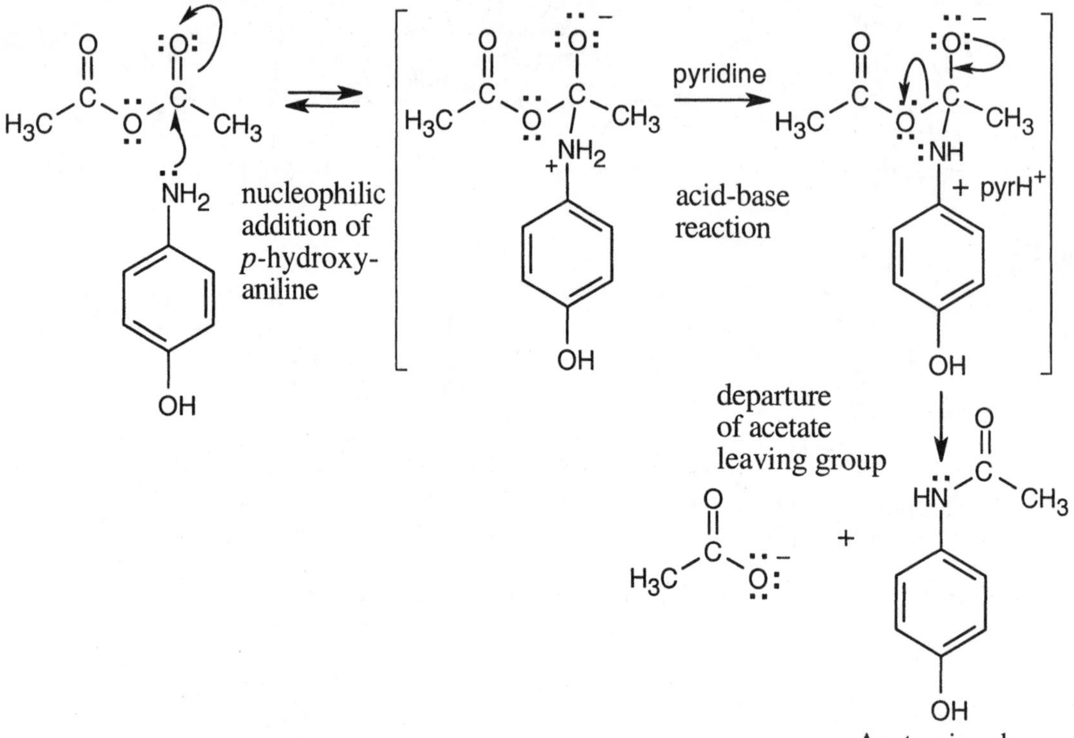

$H_3C-C(=O)-O-C(=O)-CH_3$ + NH_2–C_6H_4–OH ⇌ nucleophilic addition of *p*-hydroxy-aniline

[$H_3C-C(=O)-O-C(O^-)(CH_3)-NH_2^+$–$C_6H_4$–OH —pyridine, acid-base reaction→ $H_3C-C(=O)-O-C(O^-)(CH_3)-NH$–$C_6H_4$–OH + $pyrH^+$]

departure of acetate leaving group

↓

$H_3C-C(=O)-O^-$ + $HN(C(=O)CH_3)$–C_6H_4–OH

Acetaminophen

10.17

Phthalic anhydride + CH_3OH ⟶ $C_6H_4(COOCH_3)(COOH)$

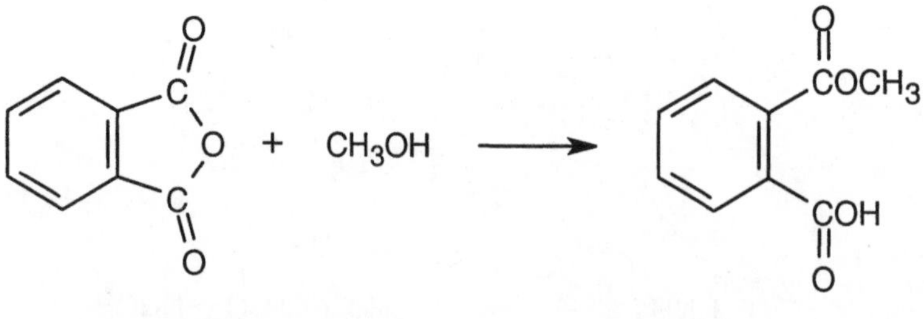

The second half of a cyclic anhydride becomes a carboxylic acid functional group.

10.18 Locate the bond to be broken, break it, add –OH to the carbonyl carbon, and add –H to the alcohol oxygen.

(a)

$$CH_3C(=O)\text{—}OCH(CH_3)_2 \xrightarrow[2.\ H_3O^+]{1.\ NaOH,\ H_2O} CH_3C(=O)OH + HOCH(CH_3)_2$$

bond to be broken

Isopropyl acetate → Acetic acid + Isopropyl alcohol

(b)

bond to be broken

Methyl cyclohexanecarboxylate $\xrightarrow[2.\ H_3O^+]{1.\ NaOH,\ H_2O}$ Cyclohexanecarboxylic acid + $HOCH_3$ (Methanol)

OCH_3 · OH

Methyl cyclohexanecarboxylate

Cyclohexane-
carboxylic acid

Methanol

10.19

$$RC(=O)OH + R'O^- \rightleftharpoons RC(=O)O^- + R'OH$$

The principal reaction of a carboxylic acid and an alkoxide is an acid-base reaction, which yields an alcohol and a carboxylate anion. The negatively charged carboxylate group is not reactive toward nucleophiles, and thus the reverse of saponification does not occur.

10.20 Reduction of an ester with $LiAlH_4$ produces two alcohols.

Ester *Alcohols*

(a)

$$CH_3CH_2CH_2CH(CH_3)C(=O)OCH_3 \xrightarrow[2.\ H_3O^+]{1.\ LiAlH_4} CH_3CH_2CH_2CH(CH_3)CH_2OH + HOCH_3$$

(b)

$$C_6H_5C(=O)OC_6H_5 \xrightarrow[2.\ H_3O^+]{1.\ LiAlH_4} C_6H_5CH_2OH + HOC_6H_5$$

10.21

$$\text{Butyrolactone} \xrightarrow[2.\ H_3O^+]{1.\ LiAlH_4} HOCH_2CH_2CH_2CH_2OH$$

Butyrolactone

Butane-1,4-diol

10.22

Ester + *Grignard Reagent* ⟶ *Tertiary alcohol*

(a) $C_6H_5C(=O)OR$ + CH_3MgBr ⟶ $C_6H_5C(OH)(CH_3)_2$

2-Phenylpropan-2-ol

(b) $H_3CC(=O)OR$ + C_6H_5MgBr ⟶ $(C_6H_5)_2C(OH)CH_3$

1,1-Diphenylethanol

(c) $CH_3CH_2CH_2CH_2C(=O)OR$ + CH_3CH_2MgBr ⟶ $CH_3CH_2CH_2CH_2C(OH)(CH_2CH_3)_2$

3-Ethylheptan-3-ol

10.23

N-Ethylbenzamide ($C_6H_5CNHCH_2CH_3$ with C=O)

(a) H^+, H_2O, heat ⟶ $C_6H_5C(=O)OH$ Benzoic acid

(b) Benzoic acid: 1. $LiAlH_4$ 2. H_3O^+ ⟶ $C_6H_5CH_2OH$ Benzyl alcohol

(c) 1. $LiAlH_4$ 2. H_2O ⟶ $C_6H_5CH_2NHCH_2CH_3$ *N*-Ethylbenzylamine

10.24

a lactam (5,5-dimethylpyrrolidin-2-one; H_3C, H_3C, C=O, N–H) —1. $LiAlH_4$ 2. H_2O⟶ a cyclic amine (2,2-dimethylpyrrolidine; H_3C, H_3C, N–H)

10.25

Nitrile	+	*Grignard Reagent*	⟶	*Ketone*
(a) CH_3CH_2CN		CH_3CH_2MgBr	⟶	$CH_3CH_2C(=O)CH_2CH_3$
(b) CH_3CH_2CN		$(CH_3)_2CHMgBr$	⟶	$CH_3CH_2C(=O)CH(CH_3)_2$
$(CH_3)_2CHCN$	*or*	CH_3CH_2MgBr		
(c) C_6H_5CN		CH_3MgBr	⟶	$C_6H_5C(=O)CH_3$
CH_3CN	*or*	C_6H_5MgBr		
(d) $C_6H_{11}CN$		$C_6H_{11}MgBr$	⟶	$C_6H_{11}C(=O)C_6H_{11}$

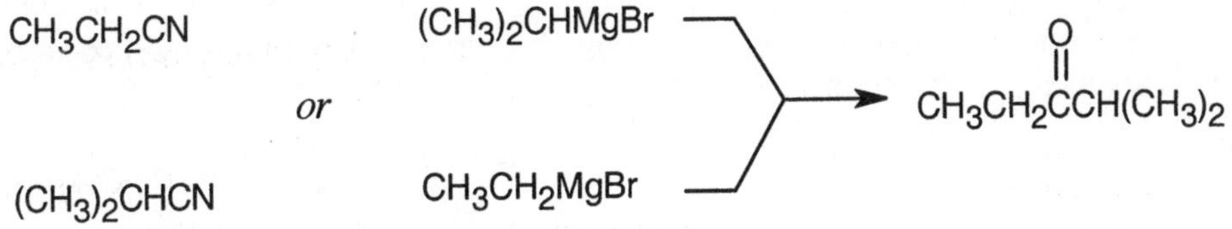

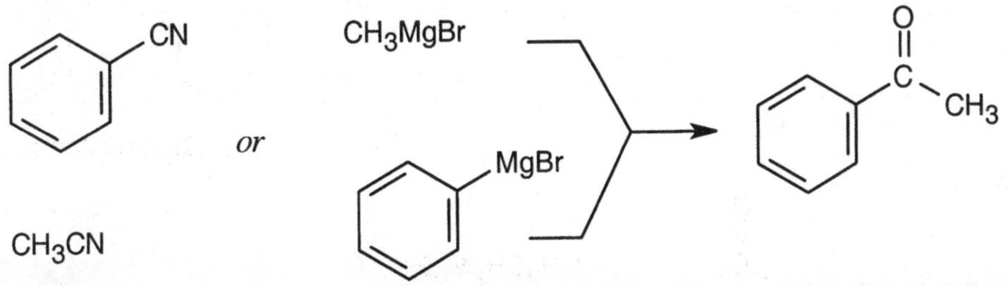

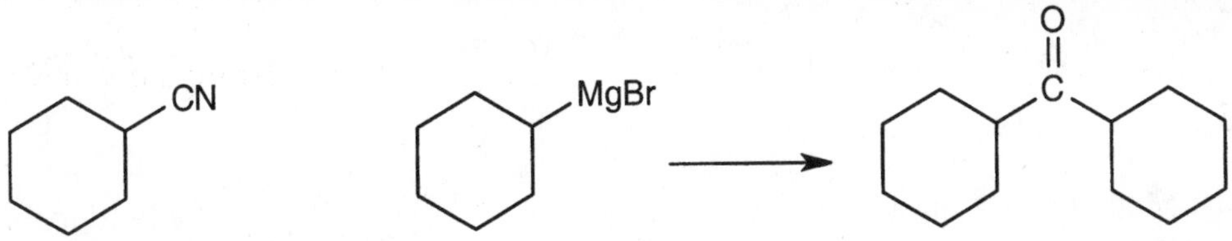

10.26

$HO_2C{-}C_6H_4{-}CO_2H$ + $H_2N{-}C_6H_4{-}NH_2$ ⟶ $\left[\,C(=O){-}C_6H_4{-}C(=O){-}NH{-}C_6H_4{-}NH\,\right]_n$

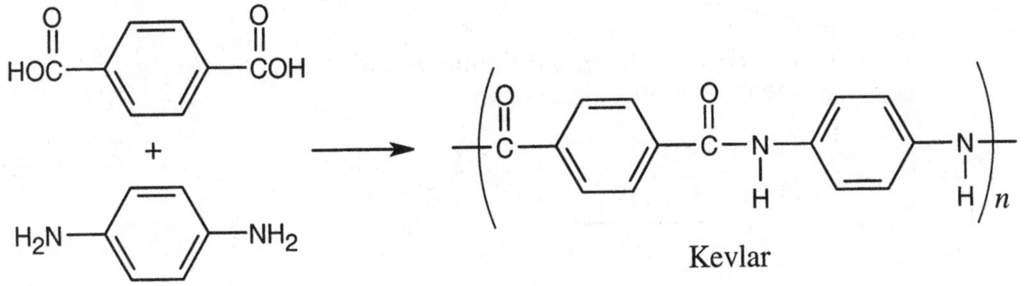

Kevlar

Visualizing Chemistry

10.27

(a) $(CH_3)_2CHCH_2C(=O)N(CH_3)_2$

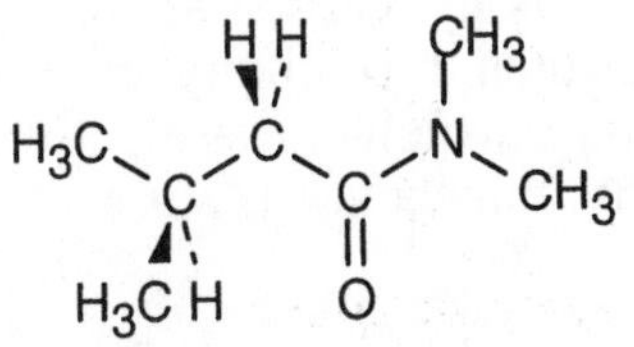

N,N-Dimethyl-3-methylbutanamide

(b) $(CH_3)_2CHCH_2CH_2OC(=O)C_6H_5$

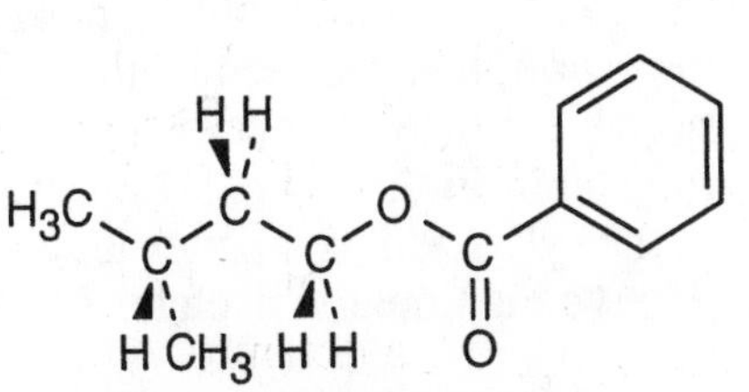

3-Methylbutyl benzoate

10.28

(a)

o-Bromobenzoic acid $\xrightarrow{SOCl_2}$ (acid chloride) $\xrightarrow[\text{pyridine}]{CH_3CH(OH)CH_3}$ Isopropyl o-bromobenzoate

o-Bromobenzoic acid Propan-2-ol Isopropyl *o*-bromobenzoate

This compound can also be synthesized by Fischer esterification of *o*-bromobenzoic acid with propan-2-ol and an acid catalyst.

(b)

Cyclopentylacetic acid $\xrightarrow{SOCl_2}$ cyclopentyl–CH_2COCl $\xrightarrow{2\ NH_3}$ cyclopentyl–CH_2CONH_2

Cyclopentylacetic acid Cyclopentylacetamide

10.29

$$H_2C{=}CHCH(CH_3)CH_2COCl \xrightarrow{2\ NH_3} \left[H_2C{=}CHCH(CH_3)CH_2C(O^-)(Cl)NH_2\right] \rightarrow H_2C{=}CHCH(CH_3)CH_2CONH_2 + NH_4^+\ Cl^-$$

3-Methylpent-4-enoyl chloride 3-Methylpent-4-enamide

The starting material is 3-methylpent-4-enoyl chloride. Ammonia adds to give the observed tetrahedral intermediate, which eliminates Cl^- to yield the above amide.

10.30 The structure represents the tetrahedral intermediate in the reaction of methyl cyclopentanecarboxylate with hydroxide, a nucleophile. The products are cyclopentanecarboxylate anion and methanol.

$$\text{cyclopentyl–}CH_2COOCH_3 \underset{\text{hydroxide}}{\overset{^-OH}{\rightleftharpoons}} \left[\text{cyclopentyl–}CH_2C(O^-)(OH)OCH_3\right] \rightarrow \text{cyclopentyl–}CH_2COO^- + HOCH_3$$

Methyl cyclopentylacetate Cyclopentylacetate Methanol

10.31 The electrostatic potential map shows that an ester oxygen is more electron-rich than the corresponding thioacetal sulfur. Consequently, the carbonyl carbon of an ester should be more electron-poor and reactive toward nucleophiles than the carbonyl carbon of a thioacetal. In this case, however, the thiomethyl group, which is less electron-rich than the methoxyl group, is a better leaving group, and methyl thioacetate is somewhat more reactive than methyl acetate.

Additional Problems

Nomenclature

10.32

(a)

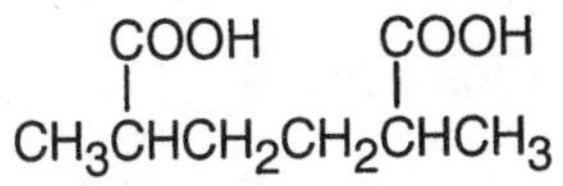

$\text{COOH}\quad\text{COOH}$

$CH_3CHCH_2CH_2CHCH_3$

2,5-Dimethylhexanedioic acid

(b)

$(CH_3)_3CCOOH$

2,2-Dimethylpropanoic acid

(c)

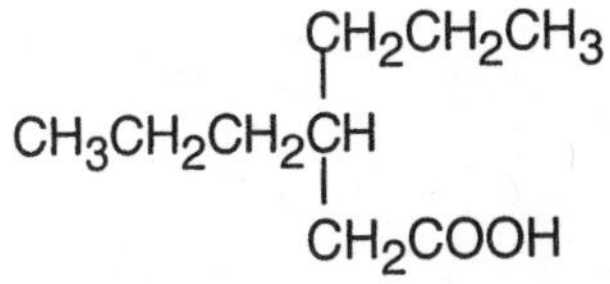

$CH_2CH_2CH_3$

$CH_3CH_2CH_2CH$

CH_2COOH

3-Propylhexanoic acid

(d)

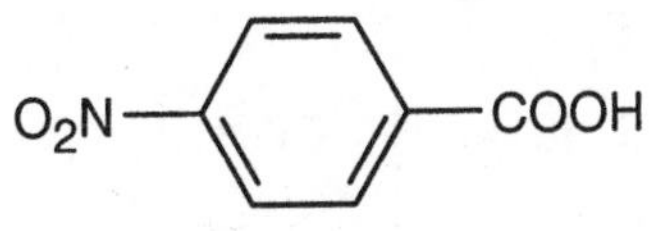

O_2N—C₆H₄—COOH

p-Nitrobenzoic acid

(e)

COOH

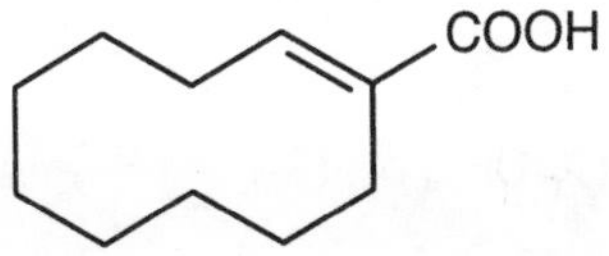

E-Cyclodec-1-enecarboxylic acid

(f)

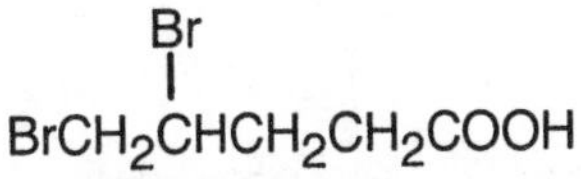

Br

$BrCH_2CHCH_2CH_2COOH$

4,5-Dibromopentanoic acid

10.33

(a)

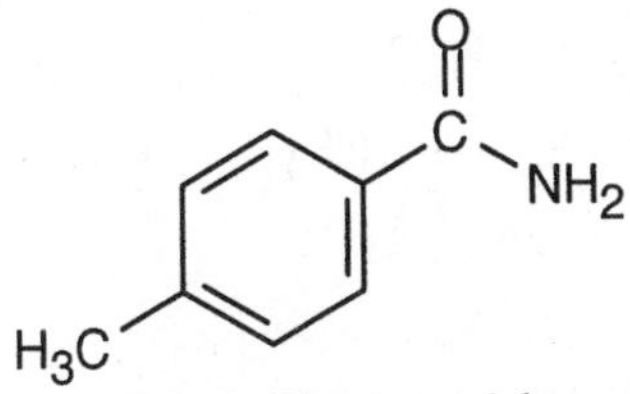

O

C

NH_2

H_3C

p-Methylbenzamide

(b)

CH_2CH_3

$CH_3CH_2CHCH{=}CHCN$

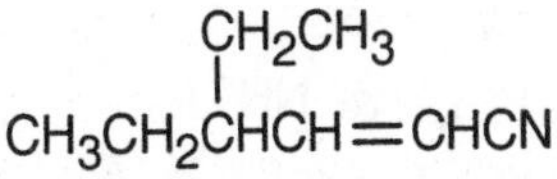

4-Ethylhex-2-enenitrile

(c)

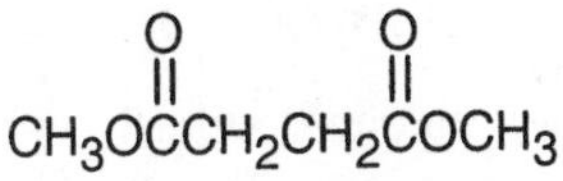

O O

$CH_3OCCH_2CH_2COCH_3$

Dimethyl butanedioate

or

Dimethyl succinate

(d)

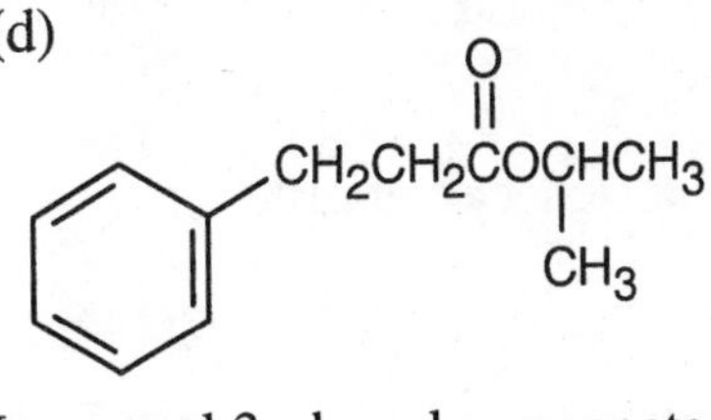

O

$CH_2CH_2COCHCH_3$

CH_3

Isopropyl 3-phenylpropanoate

(e)

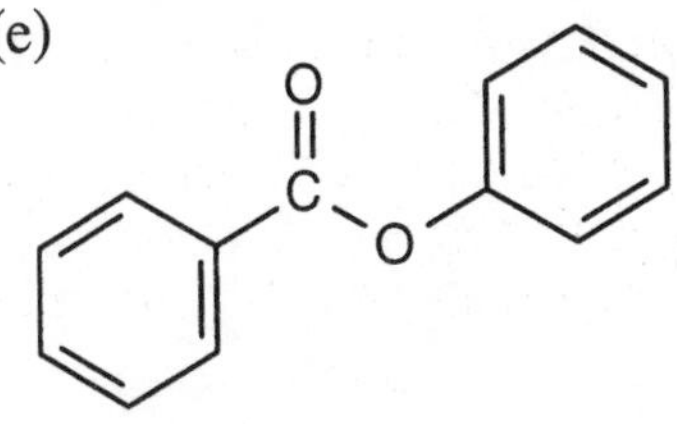

O

C

O

Phenyl benzoate

(f)

O

$CH_3CHCH_2CNHCH_3$

Br

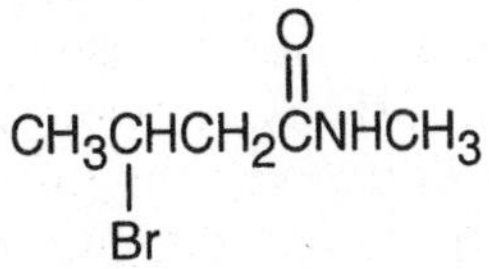

3-Bromo-*N*-methylbutanamide

(g)

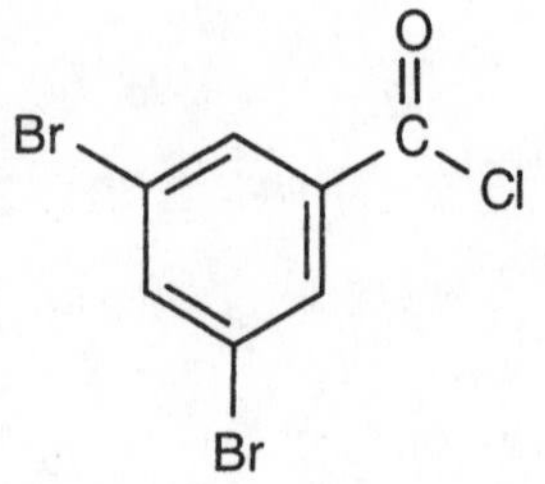

Br, Br, $C(=O)Cl$

3,5-Dibromobenzoyl chloride

(h)

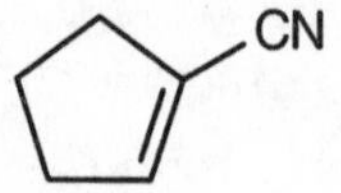

CN

Cyclopent-1-enecarbonitrile

10.34

(a)

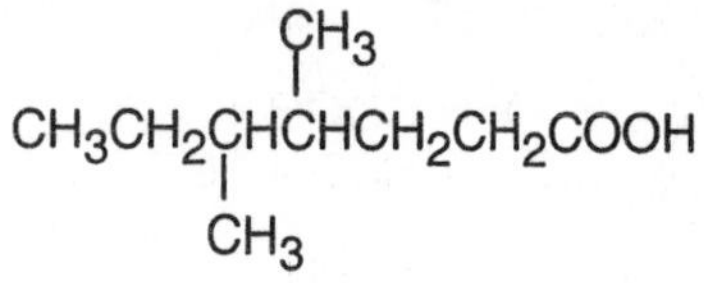

$CH_3CH_2CH(CH_3)CH(CH_3)CH_2CH_2COOH$

4,5-Dimethylheptanoic acid

(b)

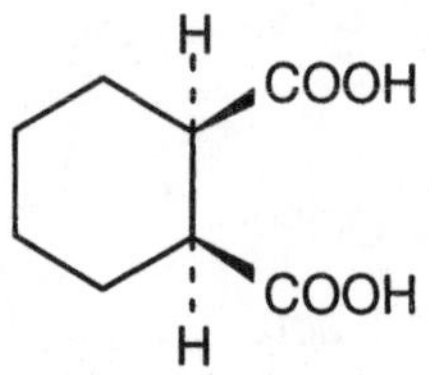

H, COOH, COOH, H

cis-Cyclohexane-1,2-dicarboxylic acid

(c)

$HOOCCH_2CH_2CH_2CH_2CH_2COOH$

Heptanedioic acid

(d)

$(C_6H_5)_3CCOOH$

Triphenylacetic acid

(e)

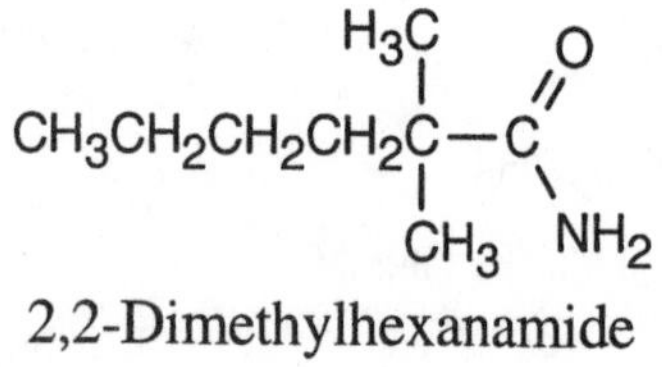

$CH_3CH_2CH_2CH_2C(CH_3)_2C(=O)NH_2$

2,2-Dimethylhexanamide

(f)

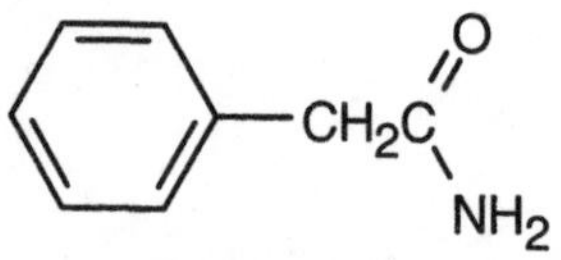

$C_6H_5CH_2C(=O)NH_2$

Phenylacetamide

(g)

CN

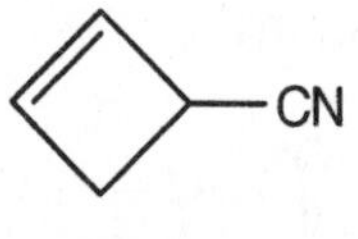

Cyclobut-2-enecarbonitrile

(h)

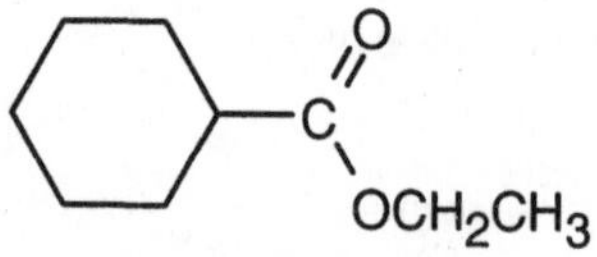

$C(=O)OCH_2CH_3$

Ethyl cyclohexanecarboxylate

10.35

$CH_3CH_2CH_2CH_2CH_2COOH$	Hexanoic acid
$CH_3CH_2CH_2CH(CH_3)COOH$	2-Methylpentanoic acid
$CH_3CH_2CH(CH_3)CH_2COOH$	3-Methylpentanoic acid
$CH_3CH(CH_3)CH_2CH_2COOH$	4-Methylpentanoic acid
$CH_3CH_2CH(CH_2CH_3)COOH$	2-Ethylbutanoic acid
$CH_3CH_2C(CH_3)_2COOH$	2,2-Dimethylbutanoic acid
$CH_3CH(CH_3)CH(CH_3)COOH$	2,3-Dimethylbutanoic acid
$CH_3C(CH_3)_2CH_2COOH$	3,3-Dimethylbutanoic acid

2-Methylpentanoic acid, 3-methylpentanoic acid and 2,3-dimethylbutanoic acid are chiral.

10.36 Many other compounds having these formulas can be drawn.

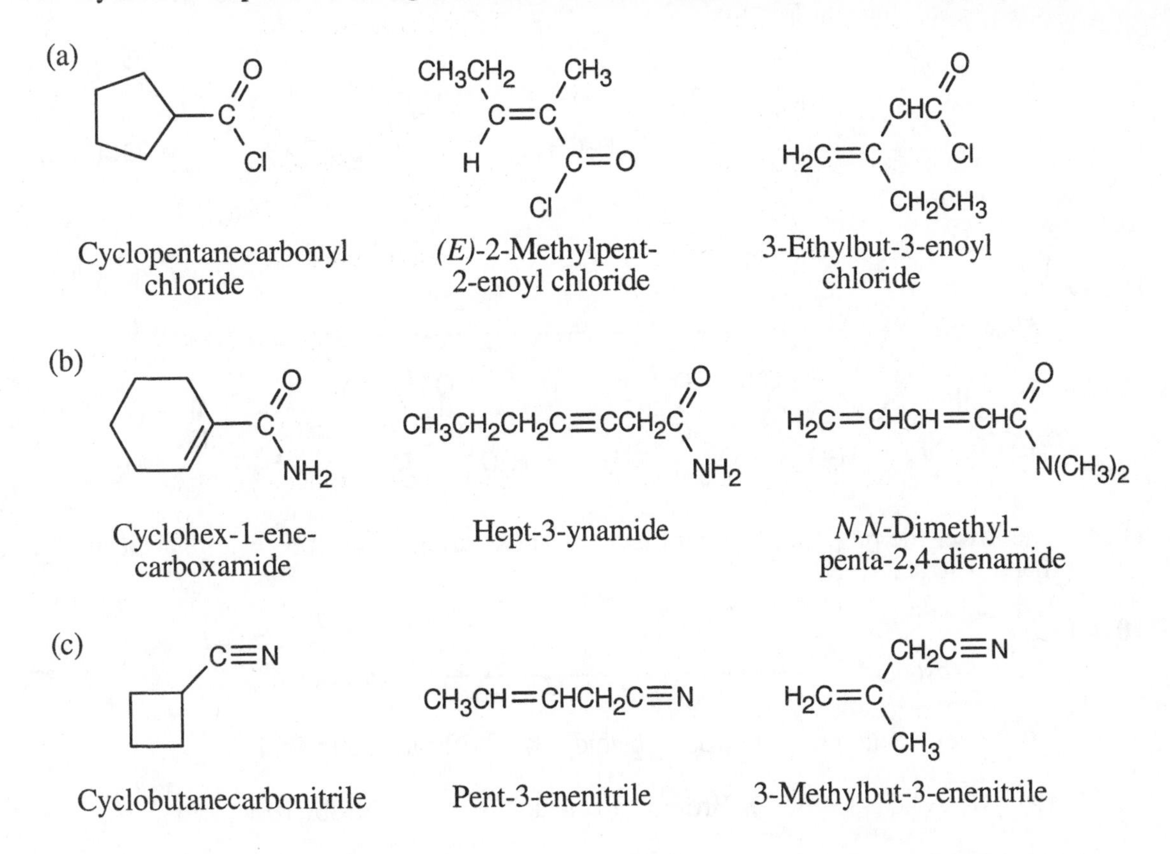

(d)

Methyl cyclopropanecarboxylate: $C_3H_5COOCH_3$ — Cyclopropyl acetate: $C_3H_5OC(O)CH_3$ — Ethyl propenoate: $H_2C{=}CHC(O)OCH_2CH_3$

Methyl cyclopropane-carboxylate | Cyclopropyl acetate | Ethyl propenoate

Physical Properties

10.37 Acetic acid molecules are strongly associated because of hydrogen bonding. Molecules of the ethyl ester are much more weakly associated, and less heat is required to overcome the attractive forces between molecules. Thus, the ethyl ester boils at a lower temperature, even though it has a higher molecular weight than the acid.

10.38 The reactivity of esters in saponification reactions is influenced by steric factors. Branching in both the acyl and alkyl portions of an ester makes it harder for the hydroxide nucleophile to approach the carbonyl carbon. This effect is less important in the alkyl portion of the ester than in the acyl portion (because alkyl branching is one atom farther away from the site of attack), but it is still measurable. The reactivity order for saponification of alkyl acetates:

Most reactive ⟶ *Least reactive*

$CH_3C(O)OCH_3 > CH_3C(O)OCH_2CH_3 > CH_3C(O)OCH(CH_3)_2 > CH_3C(O)OC(CH_3)_3$

Acidity

10.39

Most reactive ⟶ *Least reactive*

$H_3C{-}C(O){-}Cl > H_3C{-}C(O){-}O{-}C(O){-}CH_3 > H_3C{-}C(O){-}OCH_3 > H_3C{-}C(O){-}NH_2$

10.40 The lower the pK_a, the stronger the acid. Thus, tartaric acid (pK_a = 2.98) is a stronger acid than citric acid (pK_a = 3.14).

10.41

Least acidic ⟶ *Most acidic*

(a) Acetic acid < Chloroacetic acid < Trifluoroacetic acid

(b) Benzoic acid < *p*-Bromobenzoic acid < *p*-Nitrobenzoic acid

(c) Cyclohexanol < Phenol < Acetic acid

10.42 The pK_a of 2-chlorobutanoic acid is lowered by the electronegative chlorine atom, which stabilizes the adjacent carboxylate group. Since this effect decreases with increasing distance from the reaction site, the chlorine atom in 3-chlorobutanoic acid lowers the pK_a much less. The chlorine atom in 4-chlorobutanoic acid is so far from the carboxylic acid group that it has a very small effect on pK_a.

Reactions

10.43

(a) cyclohexyl–C(=O)OCH$_2$CH$_3$ $\xrightarrow[\text{2. } H_3O^+]{\text{1. 2 } CH_3CH_2MgBr}$ cyclohexyl–C(OH)(CH$_2$CH$_3$)CH$_2$CH$_3$ + CH$_3$CH$_2$OH

(b) $CH_3CH(CH_3)CH_2CH_2C{\equiv}N$ $\xrightarrow[\text{2. } H_2O]{\text{1. } LiAlH_4}$ $CH_3CH(CH_3)CH_2CH_2CH_2NH_2$

(c) cyclopentyl–C(=O)Cl $\xrightarrow{\text{2 } CH_3NH_2}$ cyclopentyl–C(=O)NHCH$_3$ + $CH_3NH_4^+\ Cl^-$

(d) cyclohexane with COOH / H and CH$_3$ / H $\xrightarrow[H_2SO_4]{CH_3OH}$ cyclohexane with COOCH$_3$ / H and CH$_3$ / H + H$_2$O

(e) $H_2C{=}CHCH(CH_3)CH_2C(=O)OCH_3$ $\xrightarrow[\text{2. } H_3O^+]{\text{1. } LiAlH_4}$ $H_2C{=}CHCH(CH_3)CH_2CH_2OH$ + CH_3OH

(f) cyclohexyl–OH $\xrightarrow[\text{pyridine}]{CH_3CO_2COCH_3}$ cyclohexyl–O–C(=O)–CH$_3$ + CH_3COO^-

(g) 2-methylbenzamide $\xrightarrow[\text{2. } H_2O]{\text{1. } LiAlH_4}$ 2-methylbenzylamine (CH_2NH_2, CH_3)

(h) (2-bromophenyl)acetic acid $\xrightarrow{SOCl_2}$ (2-bromophenyl)acetyl chloride

10.44

(a) p-toluic acid $\xrightarrow[\text{2. } H_3O^+]{\text{1. } LiAlH_4}$ 4-methylbenzyl alcohol (CH_2OH, CH_3)

(b) p-toluic acid $\xrightarrow[HCl]{CH_3OH}$ methyl p-toluate ($O{=}C{-}OCH_3$, CH_3)

(c) p-toluic acid $\xrightarrow{SOCl_2}$ p-toluoyl chloride ($O{=}C{-}Cl$, CH_3)

(d) p-toluic acid $\xrightarrow[\text{2. } CH_3I]{\text{1. } NaOH}$ methyl p-toluate ($O{=}C{-}OCH_3$, CH_3)

10.45 Substitution of chloride by cyanide proceeds by an S_N2 mechanism. In this case, however, 2-chloro-2-methylpentane, a tertiary chloride, is more likely to undergo elimination to yield 2-methyl-2-pentene.

$$CH_3CH_2CH_2CCl(CH_3)CH_3 \xrightarrow{NaCN} CH_3CH_2CH{=}C(CH_3)_2 \quad \left(\textit{not} \quad CH_3CH_2CH_2C(CN)(CH_3)CH_3\right)$$

10.46

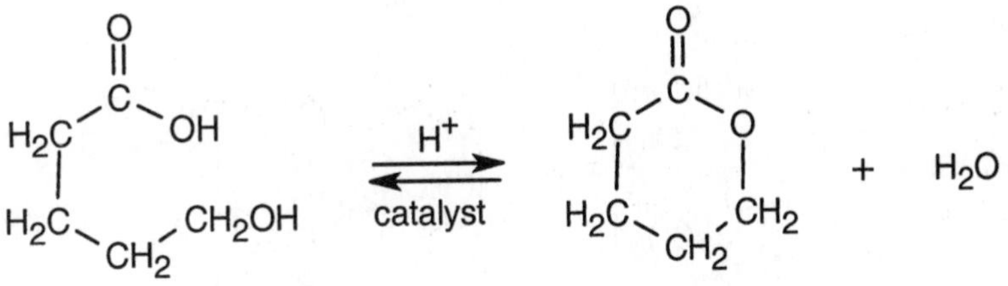

O
‖
H_2C–C–OH
|
H_2C–CH_2–CH_2OH

5-Hydroxypentanoic acid

$\xrightleftharpoons[\text{catalyst}]{H^+}$

O
‖
H_2C–C–O
| |
H_2C–CH_2–CH_2

A *lactone*

+ H_2O

10.47

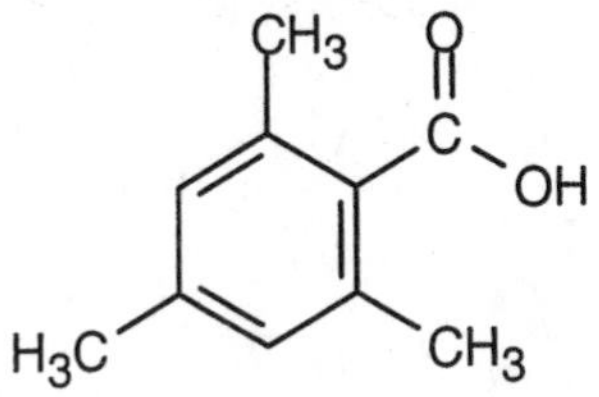

CH_3, C(=O)OH, H_3C, CH_3 (2,4,6-trimethylbenzoic acid)

2,4,6-Trimethylbenzoic acid has two methyl groups ortho to the carboxylic acid functional group. These bulky methyl groups block the approach of the alcohol and prevent esterification from occurring under Fischer esterification conditions.

10.48

(a)

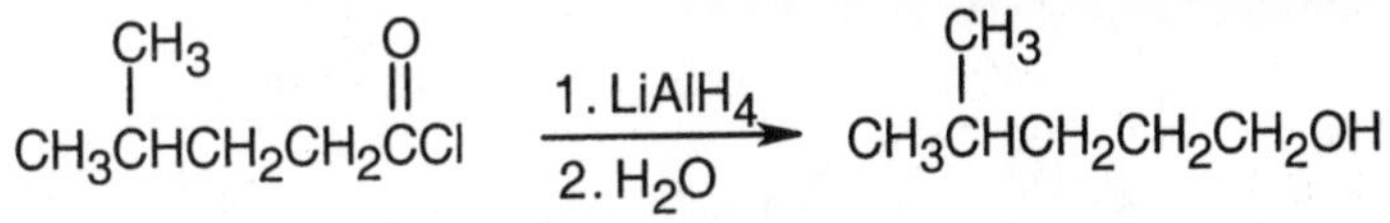

$$\underset{CH_3CHCH_2CH_2CCl}{\overset{CH_3\qquad\quad O}{|\qquad\qquad ‖}} \xrightarrow[2.\ H_2O]{1.\ LiAlH_4} \underset{CH_3CHCH_2CH_2CH_2OH}{\overset{CH_3}{|}}$$

(b)

Cyclopentane–C(CCl=O)(CH_3) $\xrightarrow[2.\ H_2O]{1.\ LiAlH_4}$ Cyclopentane–C(CH_2OH)(CH_3)

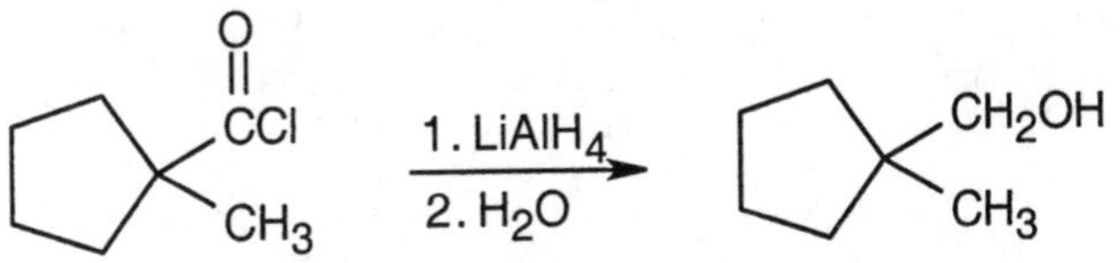

Mechanisms

10.49

:Ö: ‖ H_3C–C–Cl + :H^- $\longrightarrow$ [:Ö:$^-$ | H_3C–C(H)–Cl] $\longrightarrow$ O ‖ H_3C–C–H + Cl^-

:Ö: ‖ H_3C–C–H + :H^- $\longrightarrow$ [:Ö:$^-$ | H_3C–C(H)–H] $\xrightarrow{H_2O}$ OH | H_3C–C(H)–H + ^-OH

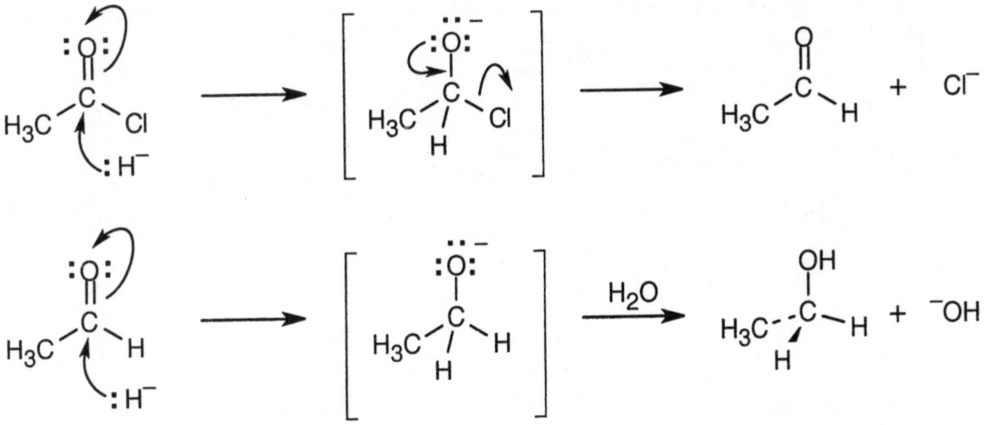

10.50

$$CH_3CH_2CH(CH_3)CH_2C(=O)Cl + {}^{-}OC(=O)CH_3 \xrightleftharpoons{\text{nucleo-philic addition}} \left[CH_3CH_2CH(CH_3)CH_2C(O^-)(Cl)OC(=O)CH_3 \right] \xrightarrow{\text{loss of } Cl^-} CH_3CH_2CH(CH_3)CH_2C(=O)OC(=O)CH_3 + Cl^-$$

10.51

$$RC(=O)Cl + :R'^- \, MgBr \longrightarrow \left[RC(O^-)(R')Cl \right] \longrightarrow RC(=O)R' + Cl^-$$

10.52 (a) The first step of the reaction of a Grignard reagent with an acid chloride is pictured in Problem 10.51. A second molecule of Grignard reagent adds to the ketone to yield a tertiary alcohol.

$$RC(=O)R' + :R'^- \, MgBr \longrightarrow \left[RC(O^- \; {}^{+}MgBr)(R')R' \right] \xrightarrow{H_3O^+} RC(OH)(R')R'$$

(b)

$$CH_3CH(CH_3)CH_2CH_2C(=O)Cl \xrightarrow[\text{2. } H_3O^+]{\text{1. 2 } CH_3MgBr} CH_3CH(CH_3)CH_2CH_2C(OH)(CH_3)CH_3$$

$$\text{1-methylcyclopentane-}C(=O)Cl \xrightarrow[\text{2. } H_3O^+]{\text{1. 2 } CH_3MgBr} \text{1-methylcyclopentane-}C(CH_3)_2OH$$

Synthesis

10.53

(a)

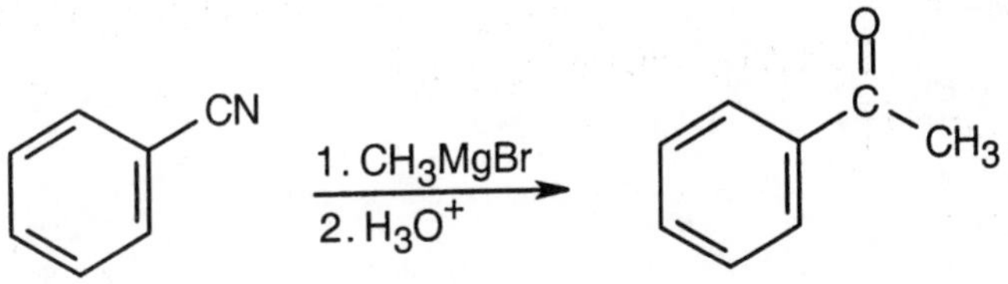

Benzonitrile (C_6H_5CN) → 1. CH_3MgBr, 2. H_3O^+ → $C_6H_5COCH_3$

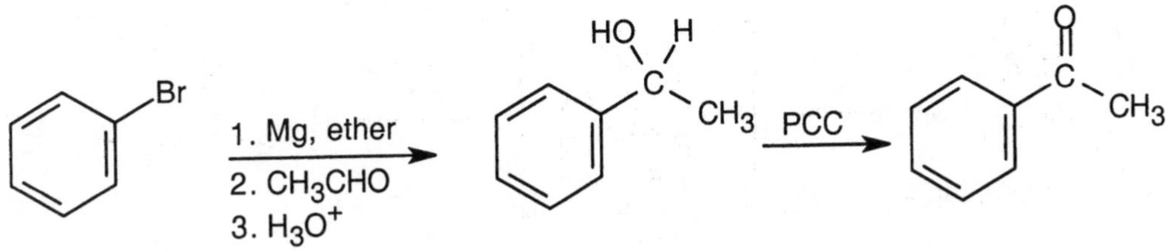

(b)

Bromobenzene (C_6H_5Br) → 1. Mg, ether, 2. CH_3CHO, 3. H_3O^+ → $C_6H_5CH(OH)CH_3$ → PCC → $C_6H_5COCH_3$

(c)

$C_6H_5COOCH_3$ → 1. $LiAlH_4$, 2. H_3O^+ → $C_6H_5CH_2OH$ → PCC → C_6H_5CHO

→ 1. CH_3MgBr, 2. H_3O^+ → $C_6H_5CH(OH)CH_3$ → PCC → $C_6H_5COCH_3$

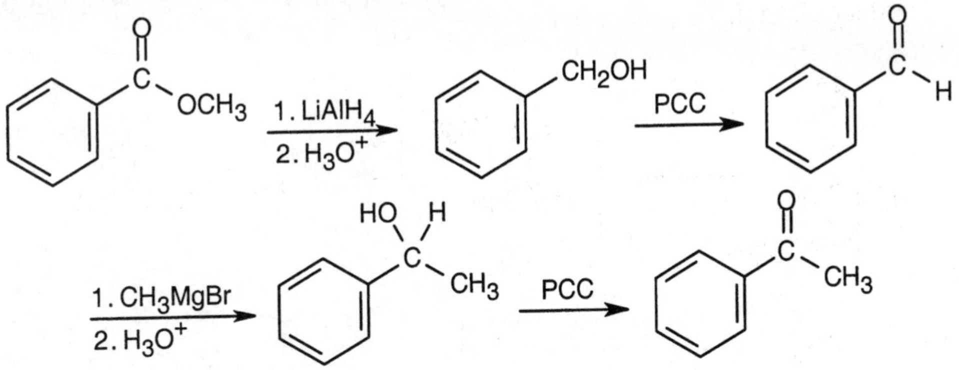

(d)

Benzene → CH_3COCl, $AlCl_3$ → $C_6H_5COCH_3$

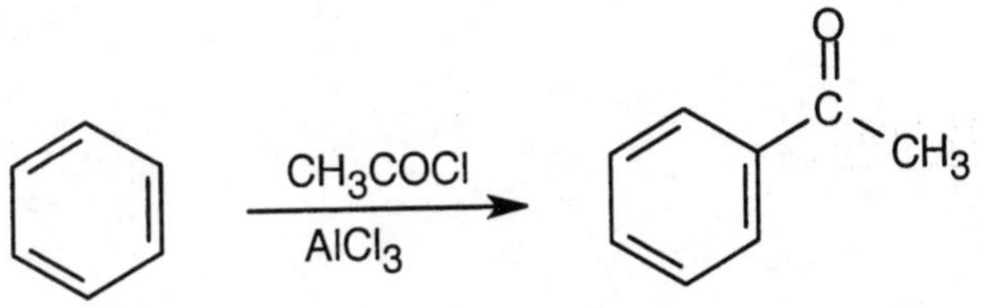

10.54

(a)

$$CH_3CH_2CH_2COOH \xrightarrow[2.\ H_3O^+]{1.\ LiAlH_4} CH_3CH_2CH_2CH_2OH$$

(b)

$$\underset{\text{from (a)}}{CH_3CH_2CH_2CH_2OH} \xrightarrow{PCC} CH_3CH_2CH_2CHO$$

(c)

$$\underset{\text{from (a)}}{CH_3CH_2CH_2CH_2OH} \xrightarrow{PBr_3} CH_3CH_2CH_2CH_2Br$$

(d)

$$\underset{\text{from (c)}}{CH_3CH_2CH_2CH_2Br} \xrightarrow{NaCN} CH_3CH_2CH_2CH_2CN$$

(e)

$$\underset{\text{from (c)}}{CH_3CH_2CH_2CH_2Br} \xrightarrow[\text{ethanol}]{KOH} CH_3CH_2CH{=}CH_2$$

(f)

$$CH_3CH_2CH_2COOH \xrightarrow{SOCl_2} CH_3CH_2CH_2COCl \xrightarrow{2\ NH_3} CH_3CH_2CH_2CONH_2 \xrightarrow[2.\ H_2O]{1.\ LiAlH_4} CH_3CH_2CH_2CH_2NH_2$$

10.55

Isobutylbenzene $\xrightarrow[AlCl_3]{ClCCH_3\ (\text{C=O})}$ 4-isobutylacetophenone $\xrightarrow[2.\ H_3O^+]{1.\ NaBH_4}$

1-(4-isobutylphenyl)ethanol (OH) $\xrightarrow{PBr_3}$ bromide (Br) $\xrightarrow{NaCN}$

nitrile (CN) $\xrightarrow[\text{acid or base}]{H_2O}$ CO_2H

Ibuprofen

Integrated Problems

10.56 Dimethyl carbonate is a diester. Use your knowledge of the Grignard reaction to work your way through this problem.

CH_3O–C(=O)–OCH_3 + $^-$C$_6$H$_5$ $^+$MgBr → (addition of Grignard reagent) → [$:\ddot{O}:^-$ $^+$MgBr ; CH_3O–C(C$_6$H$_5$)–OCH_3] → (elimination of methoxide) → C$_6$H$_5$–C(=O)–OCH_3 + $^-OCH_3$

C$_6$H$_5$–C(=O)–OCH_3 + $^-$C$_6$H$_5$ $^+$MgBr → (addition of Grignard reagent) → [$:\ddot{O}:^-$ $^+$MgBr ; (C$_6$H$_5$)$_2$C–OCH_3] → (elimination of methoxide) → $^-OCH_3$ + (C$_6$H$_5$)$_2$C=O

(C$_6$H$_5$)$_2$C=O + $^-$C$_6$H$_5$ $^+$MgBr → (addition of Grignard reagent) → [$:\ddot{O}:^-$ $^+$MgBr ; (C$_6$H$_5$)$_3$C–O$^-$] → (H_3O^+, protonation) → (C$_6$H$_5$)$_3$C–OH

Triphenylmethanol

The overall reaction consists of three additions of phenylmagnesium bromide, two eliminations of methoxide and one protonation.

10.57

(a)

$$CH_3CH_2C(=O)Cl \xrightarrow[2.\ H_3O^+]{1.\ 2\ CH_3MgBr} CH_3CH_2C(OH)(CH_3)CH_3$$

(b)

$$CH_3CH_2C(=O)Cl \xrightarrow[H_2O]{NaOH} CH_3CH_2C(=O)O^-\ Na^+$$

(c)

$$CH_3CH_2C(=O)Cl \xrightarrow{2\ CH_3NH_2} CH_3CH_2C(=O)NHCH_3 + CH_3NH_3^+\ Cl^-$$

(d)

$$CH_3CH_2C(=O)Cl \xrightarrow[2.\ H_3O^+]{1.\ LiAlH_4} CH_3CH_2CH_2OH$$

(e)

$$CH_3CH_2C(=O)Cl + HO\text{-cyclohexyl} \xrightarrow{\text{pyridine}} CH_3CH_2C(=O)O\text{-cyclohexyl}$$

(f)

$$CH_3CH_2C(=O)Cl \xrightarrow{CH_3COO^-Na^+} CH_3CH_2\overset{O}{\overset{\|}{C}}O\overset{O}{\overset{\|}{C}}CH_3$$

10.58 In general, esters are less reactive than acid chlorides (Problem 10.57).

(a)

$$CH_3CH_2C(=O)OCH_3 \xrightarrow[2.\ H_3O^+]{1.\ 2\ CH_3MgBr} CH_3CH_2C(OH)(CH_3)CH_3 + CH_3OH$$

(b)

$$CH_3CH_2C(=O)OCH_3 \xrightarrow[H_2O]{NaOH} CH_3CH_2C(=O)O^-\ Na^+ + CH_3OH$$

(c)

$$CH_3CH_2C(=O)OCH_3 \xrightarrow{CH_3NH_2} CH_3CH_2C(=O)NHCH_3 + CH_3OH$$

(d)

$$CH_3CH_2C(=O)OCH_3 \xrightarrow[2.\ H_3O^+]{1.\ LiAlH_4} CH_3CH_2CH_2OH + CH_3OH$$

(e)

$$CH_3CH_2C(=O)OCH_3 + HO\text{-cyclohexyl} \xrightarrow{\text{pyridine}} \text{no reaction}$$

(f)

$$CH_3CH_2C(=O)OCH_3 \xrightarrow{CH_3COO^-Na^+} \text{no reaction}$$

10.59

Ester	+	*Grignard Reagent*	→	*Alcohol*
(a) $CH_3CH_2CH_2C(=O)OR$		CH_3MgBr		$CH_3CH_2CH_2C(OH)(CH_3)CH_3$
(b) $CH_3C(=O)OR$		C_6H_5MgBr		$(C_6H_5)_2C(OH)CH_3$

Remember that the Grignard reagent contributes the two identical groups to the tertiary alcohol.

10.60

(a)

$$CH_3CH(CH_3)CH_2CH_2C(=O)OH + HOCH_2CH_3 \xrightarrow{HCl} CH_3CH(CH_3)CH_2CH_2C(=O)OCH_2CH_3$$

$$CH_3CH(CH_3)CH_2CH_2C(=O)Cl + HOCH_2CH_3 \xrightarrow{pyridine} CH_3CH(CH_3)CH_2CH_2C(=O)OCH_2CH_3$$

(b)

$$CH_3C(=O)OH + HOCH_2\text{–}C_5H_9 \xrightarrow{HCl} CH_3C(=O)OCH_2\text{–}C_5H_9$$

$$CH_3C(=O)Cl + HOCH_2\text{–}C_5H_9 \xrightarrow{pyridine} CH_3C(=O)OCH_2\text{–}C_5H_9$$

10.61

(a)

$$(CH_3)_2CHCH_2I \xrightarrow{NaCN} (CH_3)_2CHCH_2CN \xrightarrow[2.\ H_2O]{1.\ LiAlH_4} (CH_3)_2CHCH_2CH_2NH_2$$

(b)

$$C_6H_5CH_2Br \xrightarrow{NaCN} C_6H_5CH_2CN \xrightarrow[2.\ H_3O^+]{1.\ CH_3CH_2MgBr} C_6H_5CH_2C(=O)CH_2CH_3$$

10.62

(a)

$$\text{Br}-C_6H_4-C(=O)OCH(CH_3)_2 \xrightarrow[H_2O]{NaOH} \text{Br}-C_6H_4-C(=O)O^-Na^+ + HOCH(CH_3)_2$$

(b)

$$CH_3CH_2C(=O)O-C_6H_{11} \xrightarrow[H_2O]{NaOH} CH_3CH_2C(=O)O^-Na^+ + HO-C_6H_{11}$$

10.63

$H_3C-C(=O)-OCH_3$ + H^+ ⇌ (protonation) $[H_3C-C(=\overset{+}{O}H)-OCH_3$ + $HOCH_2CH_3$ ⇌ (addition of ethanol) $H_3C-C(OH)(OCH_3)(\overset{+}{O}(H)CH_2CH_3)$ ⇌ (proton transfer) $H_3C-C(OH)(\overset{+}{O}(H)CH_3)(OCH_2CH_3)$ ⇌ (loss of methanol) $H_3C-C(=\overset{+}{O}H)-OCH_2CH_3$ + $HOCH_3]$ ⇌ (loss of proton) $H_3C-C(=O)-OCH_2CH_3$ + H^+

The above mechanism is very similar to the mechanism of Fischer esterification, illustrated in Section 10.6. Conversion of the methyl ester to the ethyl ester occurs because the large excess of the solvent, ethanol, shifts the equilibrium to favor formation of the ethyl ester.

10.64

$(CH_3)_3CO-C(=O)-Cl$ + $:N_3^-$ ⇌ (addition of azide) $[(CH_3)_3CO-C(O^-)(N_3)(Cl)]$ → (elimination of chloride) $(CH_3)_3CO-C(=O)-N_3$ + Cl^-

10.65

$$\gamma\text{-butyrolactone} \xrightarrow[2.\ H_3O^+]{1.\ 2\ C_6H_5MgBr} HOCH_2CH_2CH_2C(OH)(C_6H_5)_2$$

10.66

H_3C–C$_6$H$_4$–Br (meta) $\xrightarrow[\text{ether}]{\text{Mg}}$ H_3C–C$_6$H$_4$–MgBr $\xrightarrow[2.\ H_3O^+]{1.\ CH_2O}$ H_3C–C$_6$H$_4$–CH_2OH

$\xrightarrow{CrO_3,\ H_3O^+}$ H_3C–C$_6$H$_4$–COH (C=O) $\xrightarrow{SOCl_2}$ H_3C–C$_6$H$_4$–CCl (C=O) $\xrightarrow[\text{NaOH}]{HN(CH_2CH_3)_2}$ H_3C–C$_6$H$_4$–$CN(CH_2CH_3)_2$ (C=O)

N,N-Diethyl-*m*-toluamide (DEET)

10.67

$RC(=O)CI_3$ + $^-$:OH $\rightleftharpoons$ [$R-C(O^-)(OH)-CI_3$ $\longrightarrow$ $RC(=O)OH$ + $^-$:CI_3] $\rightleftharpoons$ $RC(=O)O^-$ + HCI_3

addition   elimination   acid-base reaction

I_3C:$^-$ can act as a leaving group because the electron-withdrawing iodine atoms stabilize the carbanion.

10.68

$$BrCH_2CO{-}H + H_2O \underset{}{\overset{K_a}{\rightleftharpoons}} BrCH_2CO^- + H_3O^+$$

$$K_a = \frac{[BrCH_2COO^-][H_3O^+]}{[BrCH_2COOH]} = 1 \times 10^{-3}$$

	Initial Molarity	*Molarity after dissociation*
$BrCH_2COOH$	0.1	0.1 – x
$BrCH_2COO^-$	0	x
H_3O^+	~0	x

$$K_a = \frac{(x)(x)}{(0.1 - x)} = 1 \times 10^{-3}$$

Using the quadratic formula to solve for x, we find that

$$x = 0.01 - 0.0005 \sim 0.01$$

$$\text{Percent dissociation} = \frac{0.01}{0.1} \times 100\% = 10\%$$

10.69

Step 1: Water opens the caprolactam ring to form the amino acid intermediate.

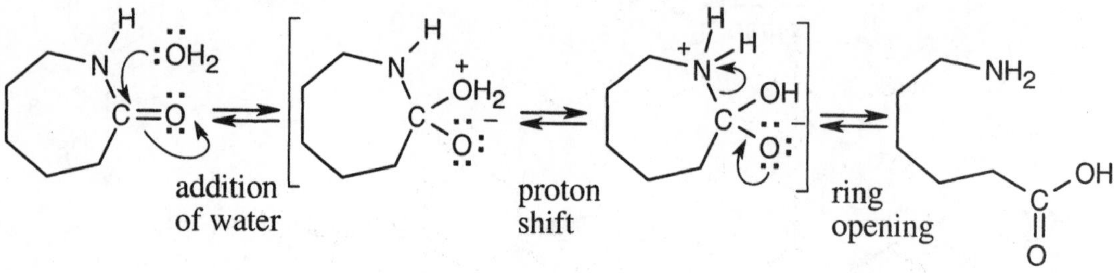

addition of water

proton shift

ring opening

Step 2: Reaction of the intermediate with a second molecule of caprolactam forms a dimer.

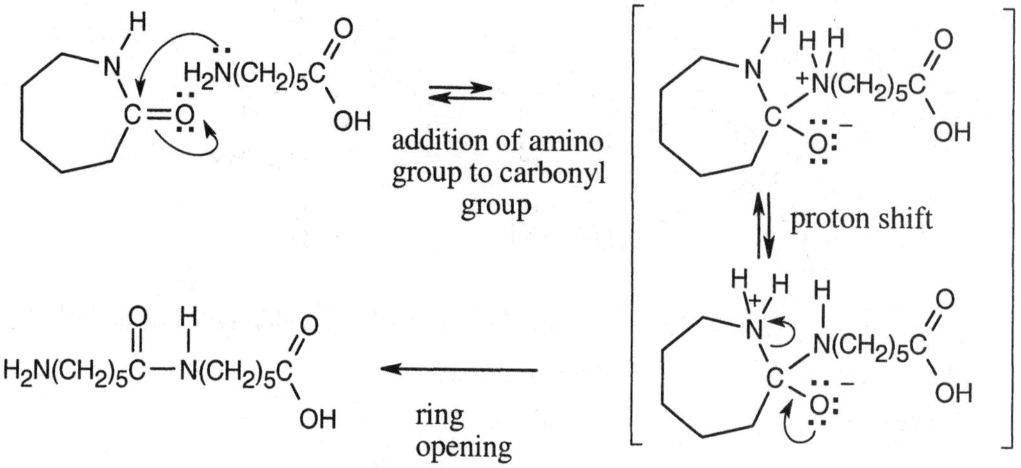

addition of amino group to carbonyl group

proton shift

ring opening

$H_2N(CH_2)_5C(=O)-N(H)(CH_2)_5C(=O)OH$

Steps 3 and beyond: Reaction of the dimer with caprolactam. This process repeats itself many, many times until the polymer stops growing. Remember that each new bond is formed in a discrete step. Heat forces the equilibrium in the direction of polymer formation.

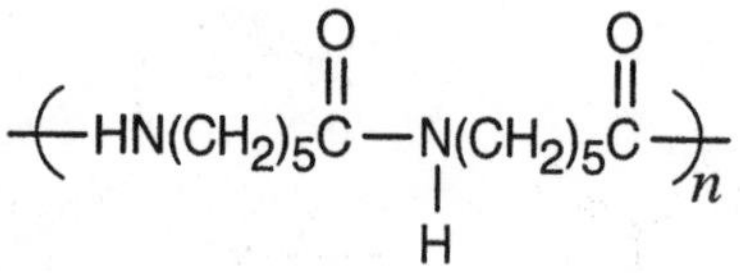

$$\left(HN(CH_2)_5\overset{O}{\overset{\|}{C}}-\underset{H}{\underset{|}{N}}(CH_2)_5\overset{O}{\overset{\|}{C}} \right)_n$$ Nylon 6

10.70

$$n\ HO(CH_2)_5OH + n\ HO\overset{O}{\overset{\|}{C}}(CH_2)_3\overset{O}{\overset{\|}{C}}OH \longrightarrow \left(O(CH_2)_5O-\overset{O}{\overset{\|}{C}}(CH_2)_3\overset{O}{\overset{\|}{C}} \right)_n + 2n\ H_2O$$

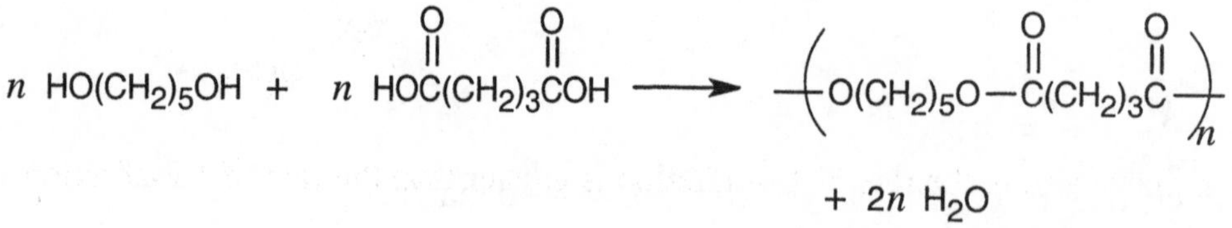

10.71

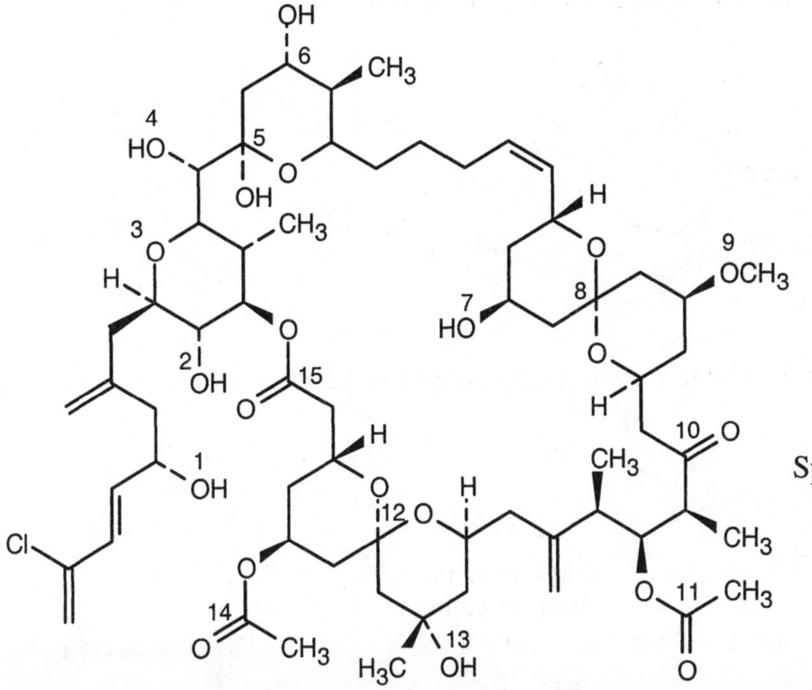

Spongistatin

alcohols: 1, 2, 4, 6, 7, 13	esters: 11, 14, 15	no aldehydes or carboxylic acids
ethers: 3, 9	hemiacetals: 5	
ketones: 10	acetals: 8, 12	

10.72

H_2N–CH_2CH_2–NH_2 + 4 $CH_2{=}CHCO_2CH_3$ → $[(CH_3OCOCH_2CH_2)_2NCH_2]_2$

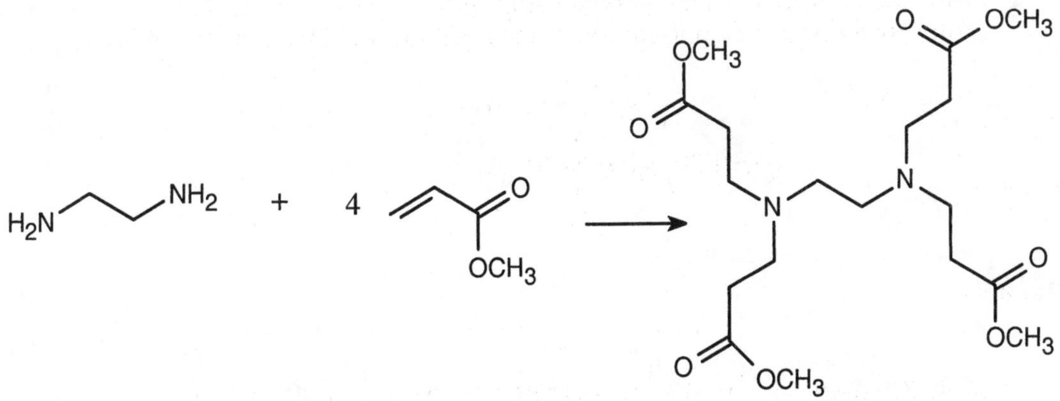

(a) In this reaction, the rate of 1,4-addition is greater than the rate of 1,2-addition.

(b)

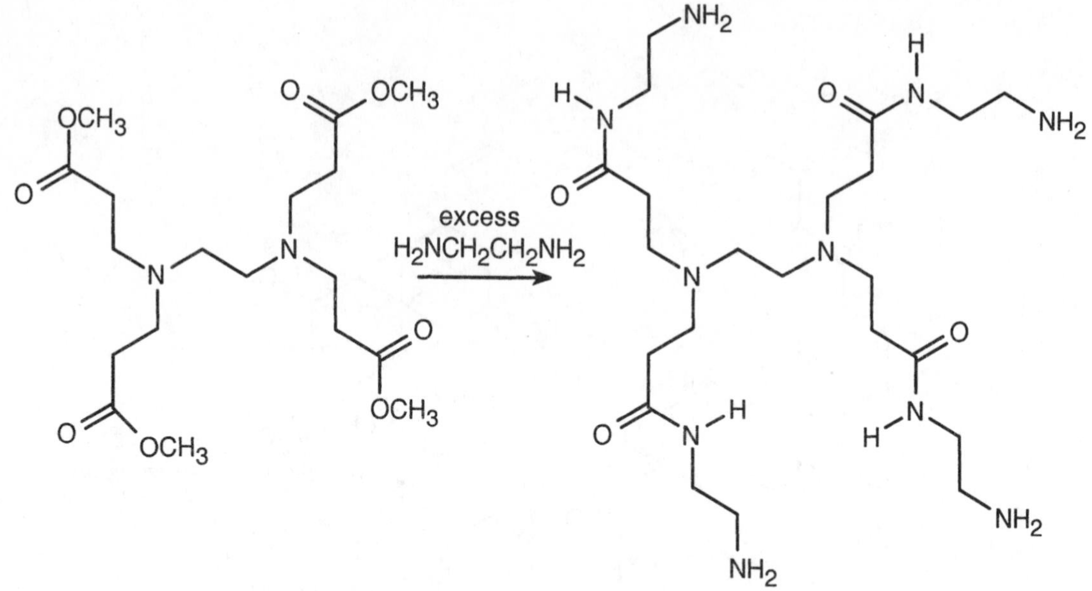

excess
$H_2NCH_2CH_2NH_2$

(c) The product of reaction with an excess of methyl acrylate has eight ester groups.

In the Medicine Cabinet

10.73

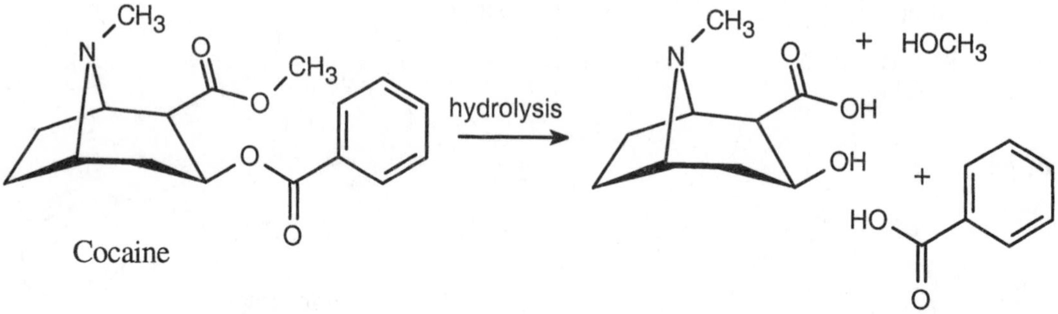

hydrolysis

$+ HOCH_3$

Cocaine

10.74

Nucleophilic acyl substitution:

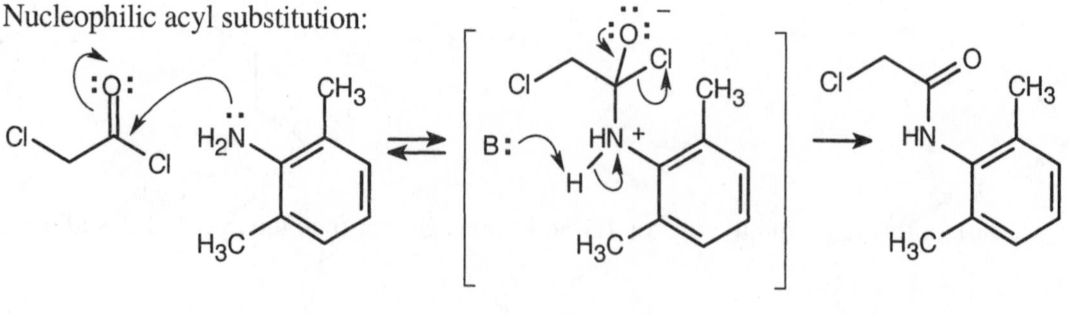

S_N2 Displacement

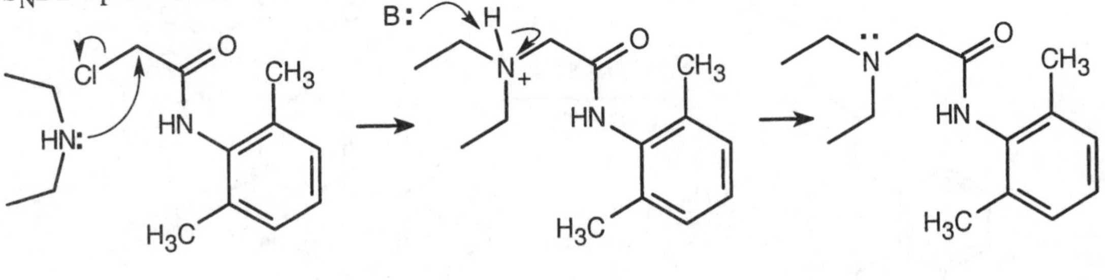

10.75 (a)

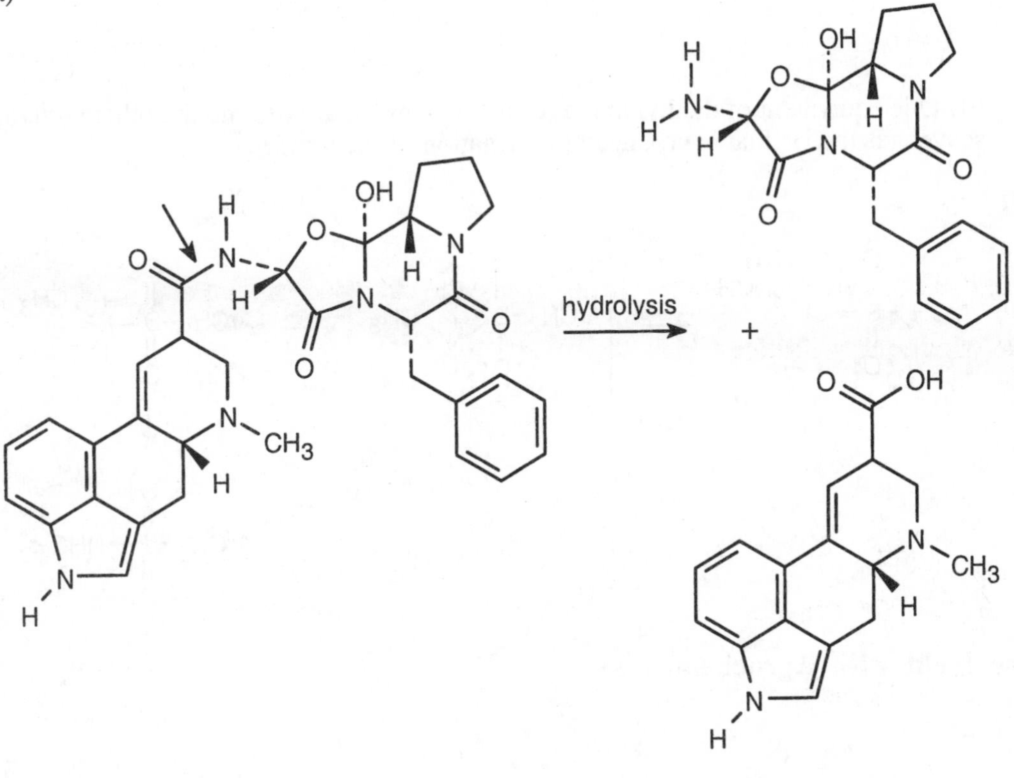

hydrolysis

(b)

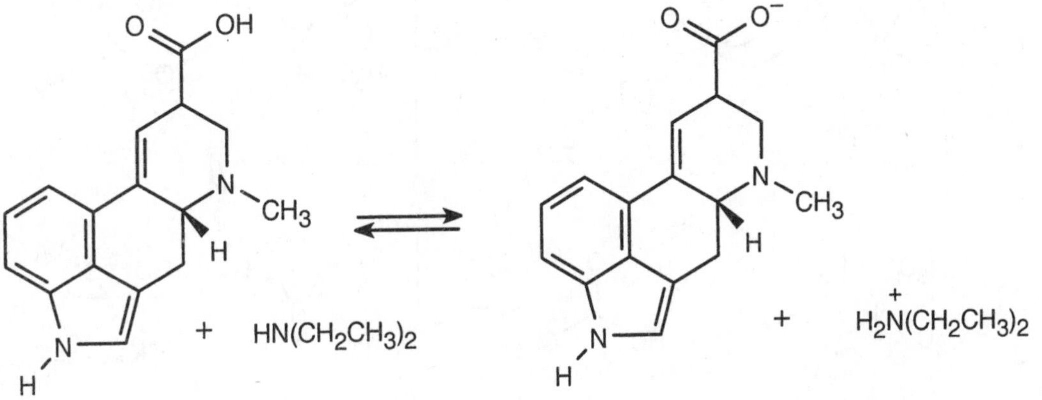

(c)

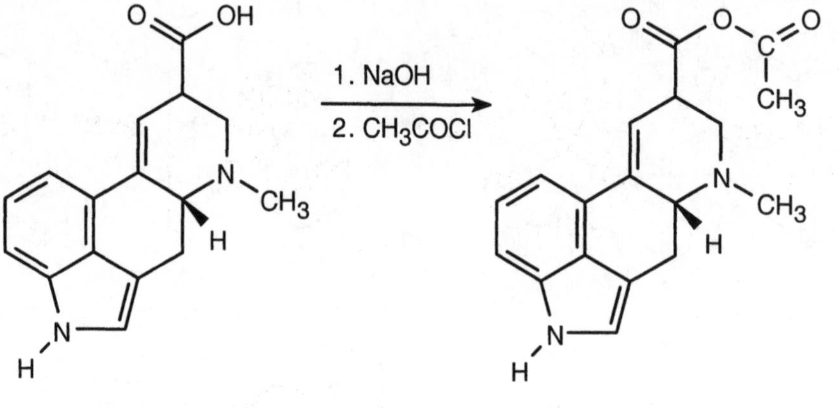

1. NaOH

2. CH_3COCl

(d) One equivalent of diethylamine reacts to form the amide, and the other equivalent scavenges the H^+ that is produced in formation of the amide.

10.76

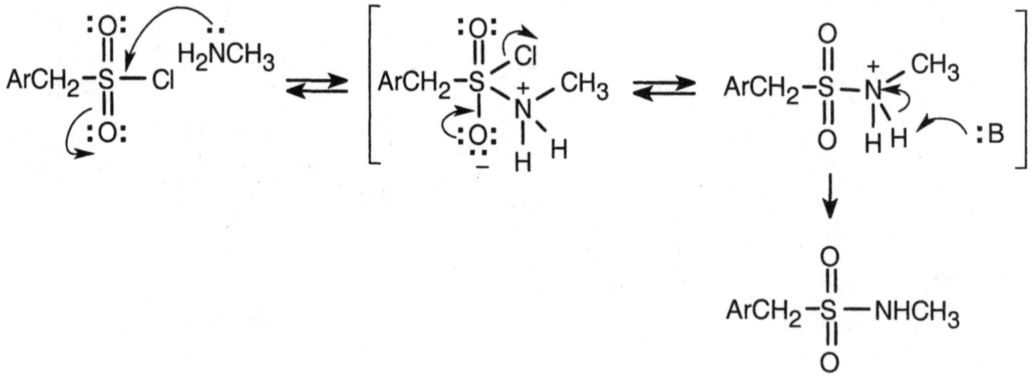

$ArCH_2-S(O)_2-Cl$ + H_2NCH_3 ⇌ [$ArCH_2-S(O)(O^-)(Cl)-N^+H_2CH_3$ ⇌ $ArCH_2-S(O)_2-N^+H_2CH_3$:B] → $ArCH_2-S(O)_2-NHCH_3$

In the Field with Agrochemicals

10.77 (a)

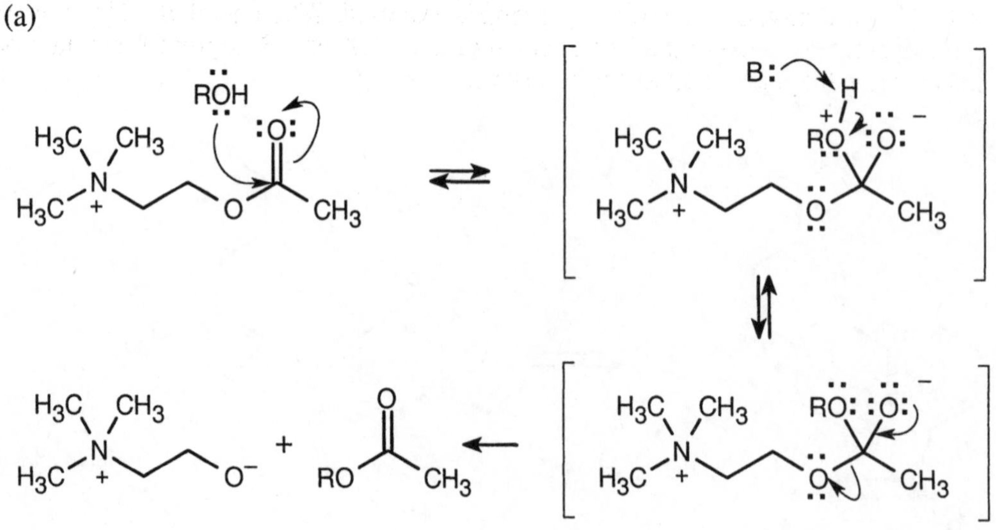

ROH

B:

RO

CH3

+

RO

(b)

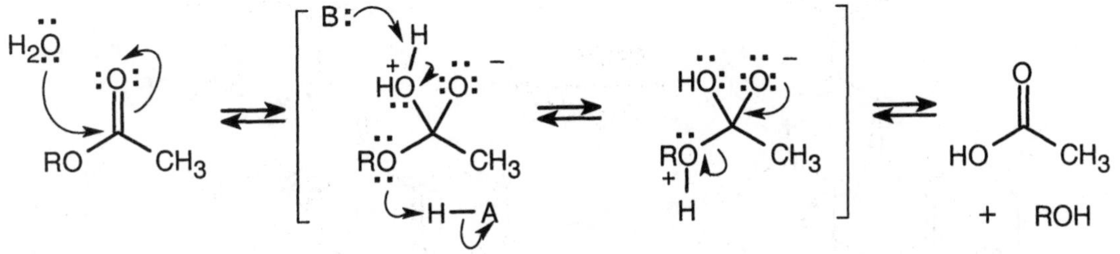

$H_2\ddot{O}$:O: RO CH₃ ⇄ [B: H +HO :O: − RO CH₃ H—A ⇄ HO: :O: − RO + H CH₃] ⇄ O HO CH₃ + ROH

10.78

ROH :S: H_3C O P O O H_3C NO_2 ⇄ [B: H + RO :S: − H_3C O P O O H_3C NO_2] ⇅ [RO :S: − H_3C O P O O H_3C NO_2] ⟶

S H_3C O P OR O H_3C + ^{-}O NO_2

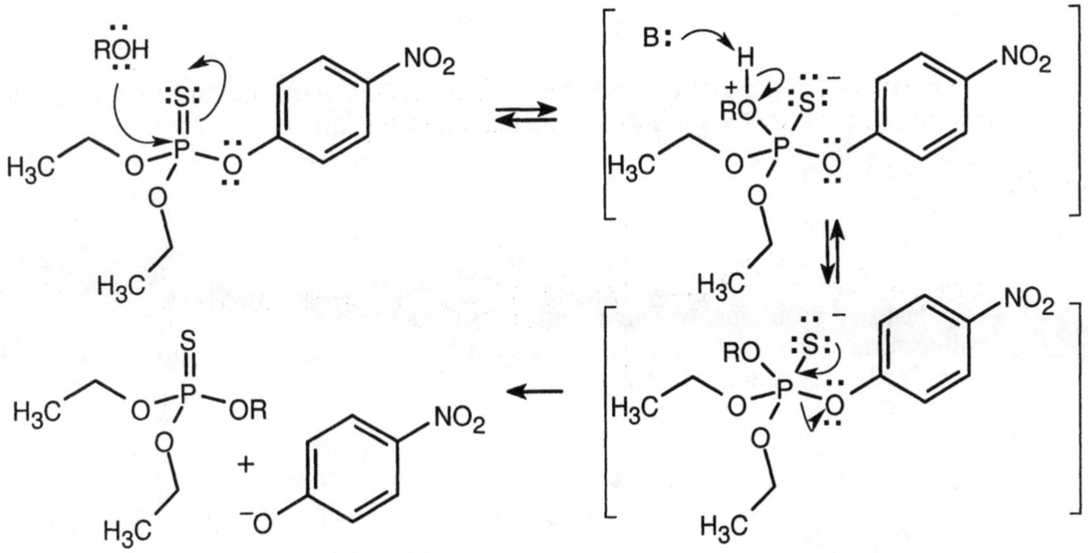

The mechanisms shown in this problem and in the previous one are very similar. A nucleophile (the enzyme hydroxyl group) adds to the positively polarized end of a double bond, $-H^+$ is removed, and a leaving group is expelled. When needed, $-H^+$ may also protonate a group to make it a better leaving group. HA and B: represent any number of acids or bases that can catalyze the reactions.

10.79

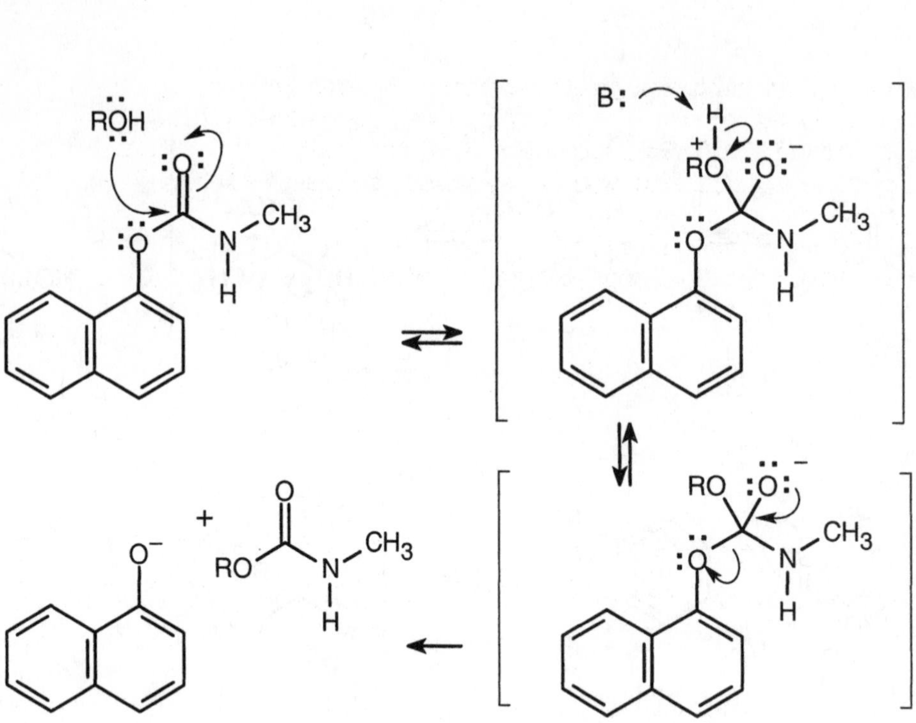

In the last two problems, we have seen the effects of pesticides on the enzyme cholinesterase. In both cases, groups bond to the enzyme, forming products that can't be hydrolyzed to regenerate the enzyme.

10.80

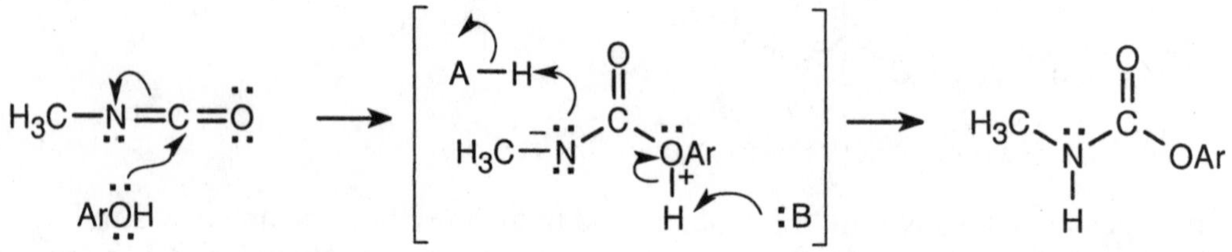

Chapter 11 – Carbonyl Alpha-Substitution Reactions and Condensation Reactions

Chapter Outline

I. Keto-enol tautomerism (Section 11.1).
   A. Nature of tautomerism.
      1. Carbonyl compounds with hydrogens bonded to their α carbons equilibrate with their corresponding enols.
      2. This rapid equilibration is called tautomerism, and the individual isomers are tautomers.
      3. Unlike resonance forms, tautomers are different isomers.
      4. Despite the fact that very little of the enol isomer is present at room temperature, enols are very important because they are reactive.
   B. Mechanism of tautomerism.
      1. In acid catalysis, the carbonyl oxygen is protonated to form an intermediate that can lose a hydrogen from its α carbon to yield a neutral enol.
      2. In base-catalyzed enol formation, an acid-base reaction occurs between a base and an α hydrogen.
         a. The resultant enolate anion is protonated to yield an enol.
         b. Protonation can occur either on carbon or on oxygen.
      3. Only hydrogens on the α positions of carbonyl compounds are acidic.

II. Enols and enolates (Sections 11.2 - 11.6).
   A. Enols (Sections 11.2 - 11.3).
      1. Reactivity of enols (Section 11.2).
         a. The electron-rich double bonds of enols cause them to behave as nucleophiles. The electron-donating enol –OH groups make enols more reactive than alkenes.
         b. When an enol reacts with an electrophile, the initial adduct loses a proton from oxygen to give a substituted carbonyl compound.
      2. Reactions of enols - alpha halogenation of aldehydes and ketones (Section 11.3).
         a. Aldehydes and ketones can be halogenated at their α positions by reaction of X_2 in acidic solution.
         b. α-Bromoketones are useful in syntheses because they can be dehydrobrominated by base treatment to form α,β-unsaturated ketones.
   B. Enolates (Sections 11.4 - 11.6).
      1. Enolate anion formation (Section 11.4).
         a. Hydrogens α to a carbonyl group are weakly acidic.
            i. This stability is due to overlap of a vacant *p* orbital with the carbonyl group *p* orbitals, allowing the carbonyl group to stabilize the negative charge by resonance.
            ii. The two resonance forms aren't equivalent; the form with the negative charge on oxygen is the major contributor to the resonance hybrid.
         b. Strong bases are needed for enolate anion formation.
         c. When a hydrogen is flanked by two carbonyl groups, it is much more acidic.
      2. Reactivity of enolate anions (Section 11.5).
         a. Enolates are more useful than enols for two reasons:
            i. Unlike enols, stable solutions of enolates are easily prepared.
            ii. Enolates are more reactive than enols because they are more nucleophilic.
         b. Enolates can react either at carbon or at oxygen.
            i. Reaction at carbon yields an α-substituted carbonyl compound.
            ii. Reaction at oxygen yields an enol derivative.

3. Alkylation reactions of enolates (Sections 11.6 - 11.7).
   a. General features.
      i. Alkylations are useful because they form a new carbon–carbon bond.
      ii. Alkylations have the same limitations as S_N2 reactions; the alkyl groups must be methyl, primary, allylic or benzylic.
   b. The malonic ester synthesis.
      i. The malonic ester synthesis is used for preparing a carboxylic acid from a halide while lengthening the carbon chain by two atoms.
      ii. Diethyl malonate is useful because its enolate is easily prepared by reaction with sodium ethoxide.
      iii. Since diethyl malonate has two acidic hydrogens, two alkylations can take place.
      iv. Heating in aqueous HCl causes hydrolysis and decarboxylation of the alkylated malonate.
         Decarboxylations are common only to β-keto acids and malonic acids.
   c. Barbiturates are synthesized by the reaction of diethyl malonate and urea (Section 11.7)

III. Carbonyl condensation reactions (Sections 11.8 - 11.12).
   A. Mechanism of carbonyl condensation reactions (Section 11.8).
      1. Carbonyl condensation reactions take place between two carbonyl components.
         a. One component (the nucleophilic donor) is converted to its enolate and undergoes an α–substitution reaction.
         b. The other component (the electrophilic acceptor) undergoes nucleophilic addition.
      2. Many kinds of carbonyl compounds undergo carbonyl condensation reactions.
   B. The aldol reaction (Sections 11.9 - 11.10).
      1. Characteristics of the aldol reaction (Section 11.9).
         a. The aldol condensation is a base-catalyzed dimerization of two aldehydes or ketones.
         b. The reaction can occur between two components that have α hydrogens.
         c. For simple aldehydes, the equilibrium favors products, but for other aldehydes and ketones, the equilibrium favors reactants.
      2. Dehydration of aldol products (Section 11.10).
         a. Aldol products are easily dehydrated to yield conjugated enones.
         b. Dehydration is catalyzed by both acid and base.
         c. Reaction conditions for dehydration are only slightly more severe than for condensation.
         d. Often, dehydration products are isolated directly from condensation reactions.
   C. The Claisen condensation (Section 11.11).
      1. Treatment of an ester with 1 equivalent of base yields a β-keto ester.
      2. The reaction is reversible and has a mechanism similar to that of the aldol reaction.
      3. A major difference from the aldol condensation is the expulsion of an alkoxide ion from the tetrahedral intermediate of the initial Claisen adduct to yield an acyl substitution product.
   D. Biological carbonyl reactions (Section 11.12).
      1. Many biomolecules are synthesized by carbonyl condensation reactions.
      2. Glycolysis (breakdown of glucose) involves several carbonyl-group reactions.
      3. Acetyl CoA is the major building block for synthesis of biomolecules.
         a. Acetyl CoA can act as an electrophilic acceptor by being attacked at its carbonyl group.
         b. Acetyl CoA can act as a nucleophilic donor by loss of its acidic α hydrogen.

Solutions to Problems

11.1–11.2 Acidic hydrogens in the keto form of each of these compounds are bold.

	Keto Form	*Enol Form*	*Number of Acidic Hydrogens*
(a)	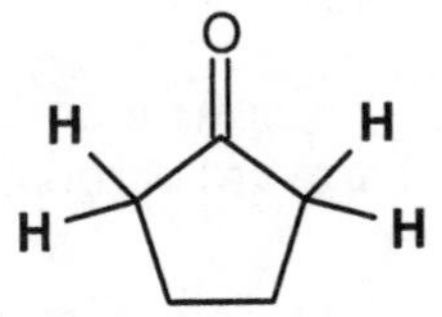	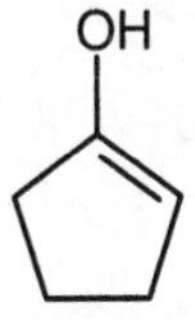	4
(b)	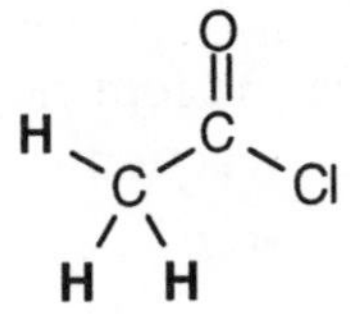	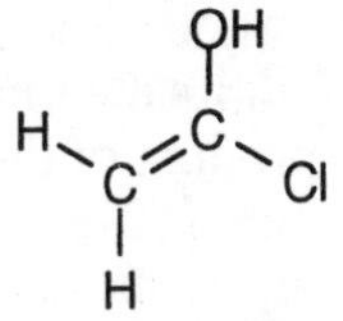	3
(c)	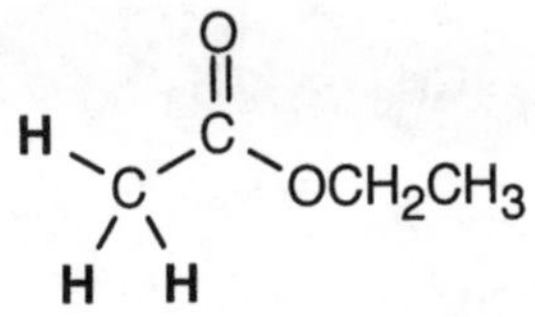	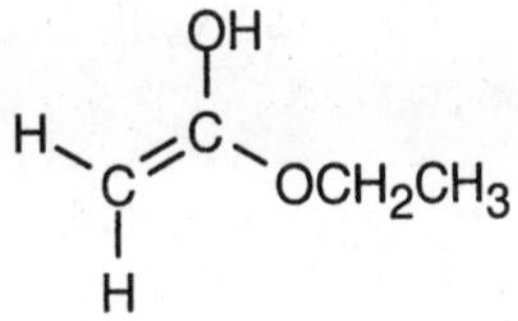	3
(d)	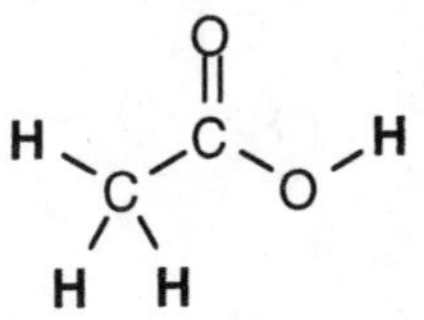	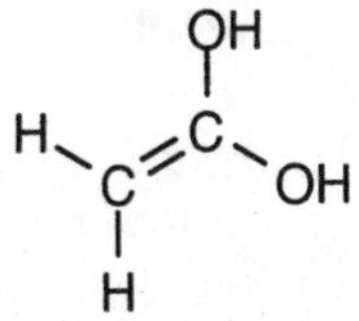	4
(e)	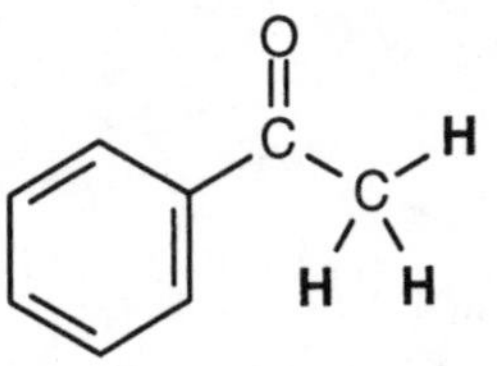	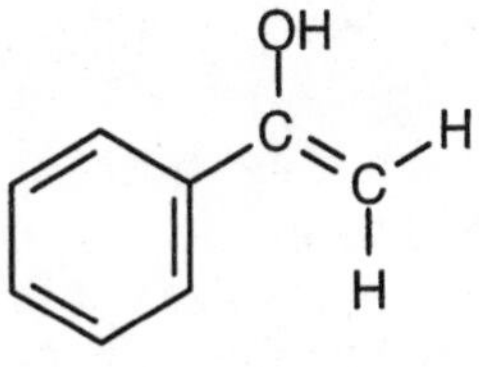	3

11.3

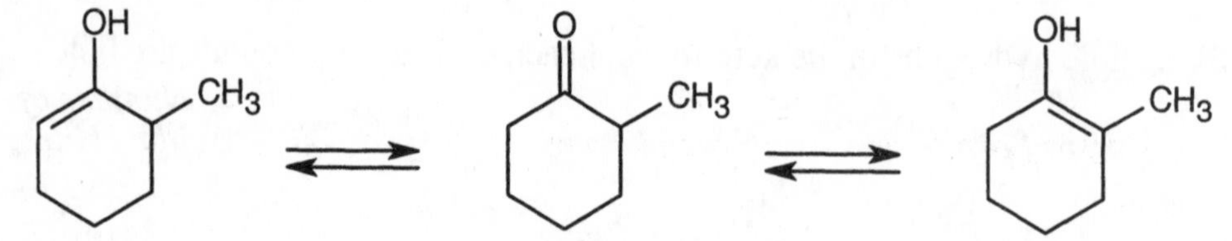

Enolization toward carbon 6	$\rightleftharpoons$	2-Methylcyclohexanone	$\rightleftharpoons$	Enolization toward carbon 2

Enolization can occur in either direction from the carbonyl group. Two different enols are formed because carbon 2 has a methyl substituent and carbon 6 does not.

11.4 As suggested in Practice Problem 11.2, locate the acidic protons and replace one of them with a halogen in order to show the product of the α-substitution reaction.

(a)

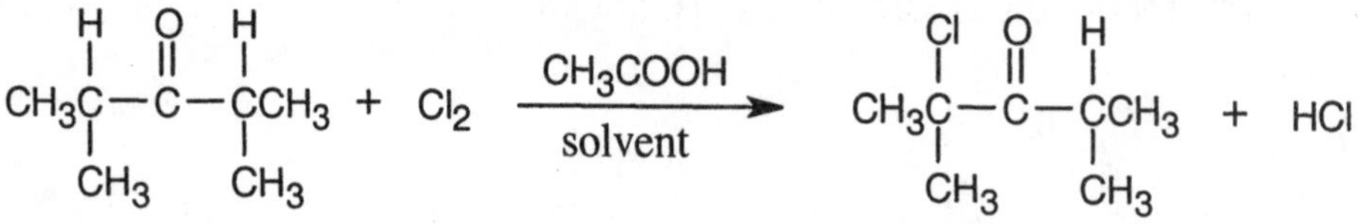

$$CH_3CH(CH_3)-C(=O)-CH(CH_3)CH_3 + Cl_2 \xrightarrow[\text{solvent}]{CH_3COOH} CH_3CCl(CH_3)-C(=O)-CH(CH_3)CH_3 + HCl$$

(b)

$$\text{2,2-dimethylcyclohexanone} + Br_2 \xrightarrow[\text{solvent}]{CH_3COOH} \text{6-bromo-2,2-dimethylcyclohexanone} + HBr$$

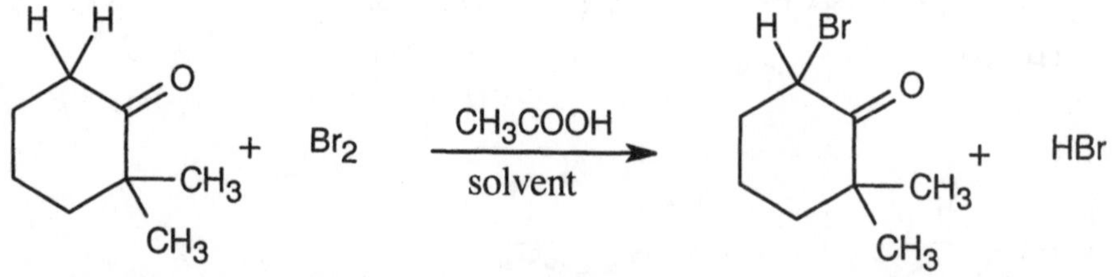

11.5 *Step 1.* Treat 3-pentanone with Br_2 in acetic acid to form the α-bromo ketone.

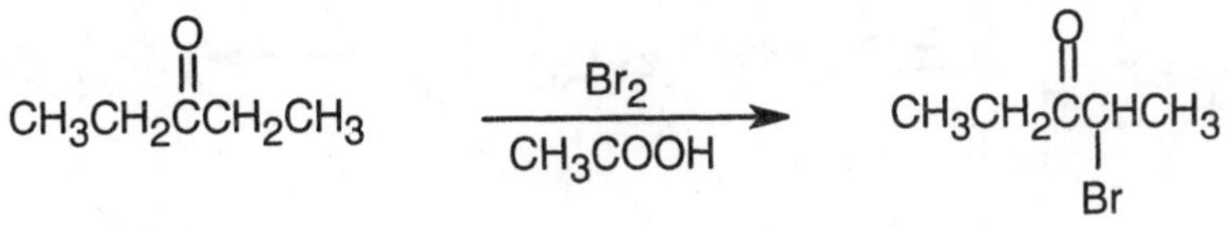

$$CH_3CH_2\overset{O}{\overset{\|}{C}}CH_2CH_3 \xrightarrow[CH_3COOH]{Br_2} CH_3CH_2\overset{O}{\overset{\|}{C}}\underset{Br}{\underset{|}{C}}HCH_3$$

Step 2. Heat the α-bromo ketone in pyridine to form 1-penten-3-one.

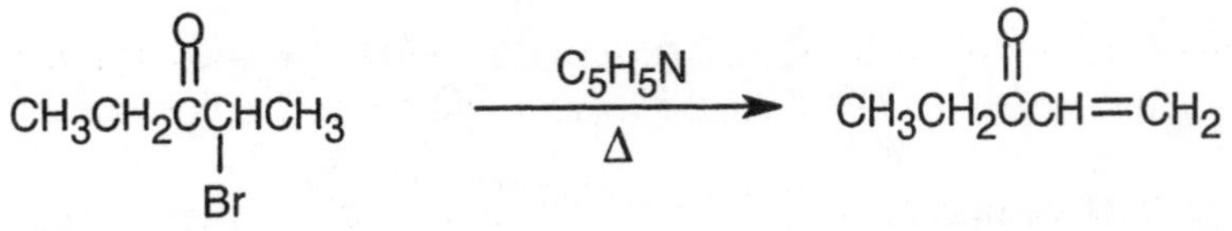

$$CH_3CH_2\overset{O}{\overset{\|}{C}}\underset{Br}{\underset{|}{C}}HCH_3 \xrightarrow[\Delta]{C_5H_5N} CH_3CH_2\overset{O}{\overset{\|}{C}}CH{=}CH_2$$

11.6 Hydrogens α to one carbonyl group are weakly acidic. Hydrogens α to two carbonyl groups are much more acidic, but they are not as acidic as carboxylic acid protons.

(a)

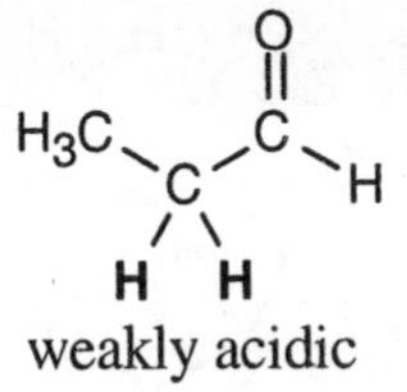

O
H_3C C H
C
H H
weakly acidic

(b)

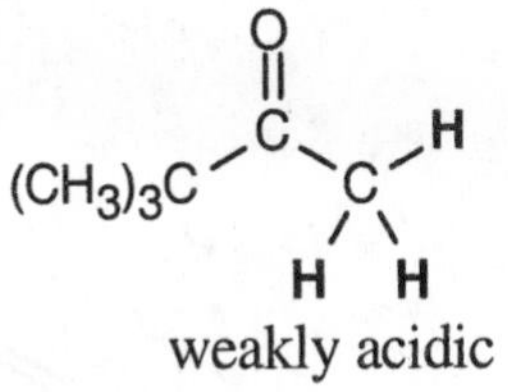

O
$(CH_3)_3C$ C H
C
H H
weakly acidic

(c)

O
H C O H
C
H H
weakly acidic
most acidic

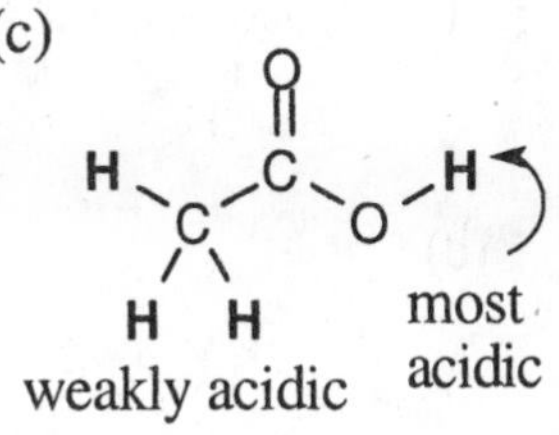

(d)

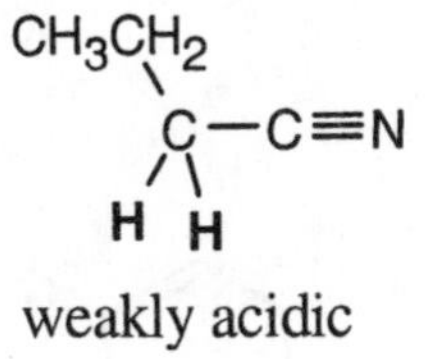

CH_3CH_2
C—C≡N
H H
weakly acidic

(e)

O
H H most acidic
H— —H
weakly acidic
O
H H
weakly acidic

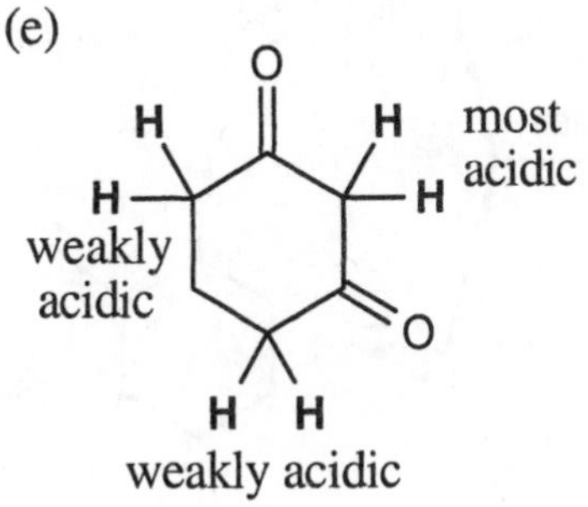

Protons between the two carbonyl groups are much more acidic than protons next to only one carbonyl group.

11.7

(a)

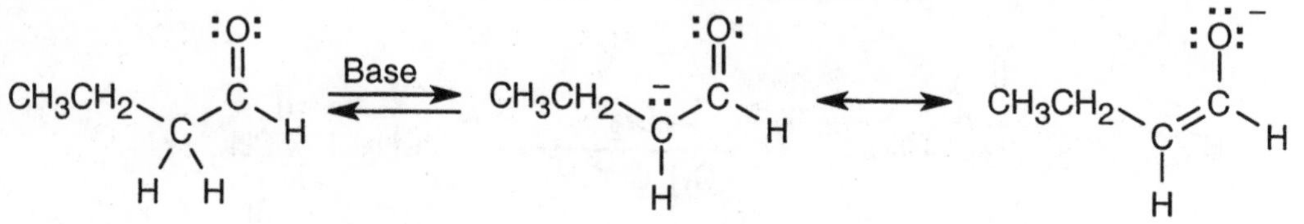

(b)

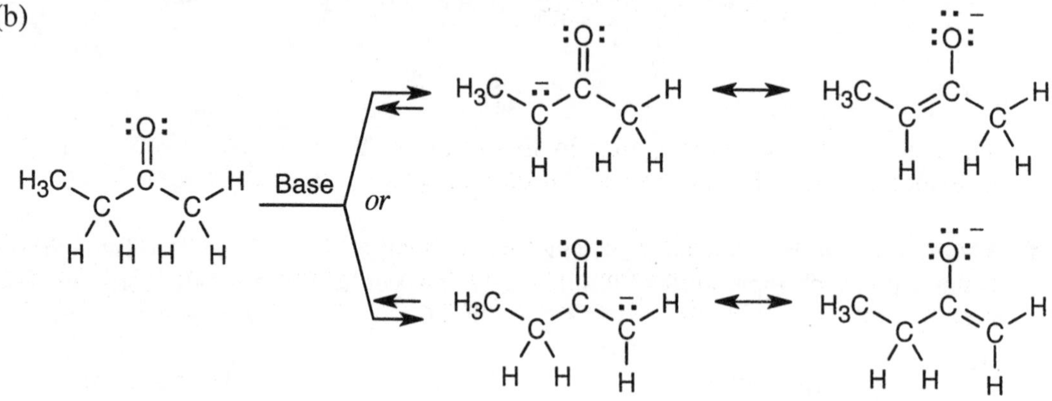

(c)

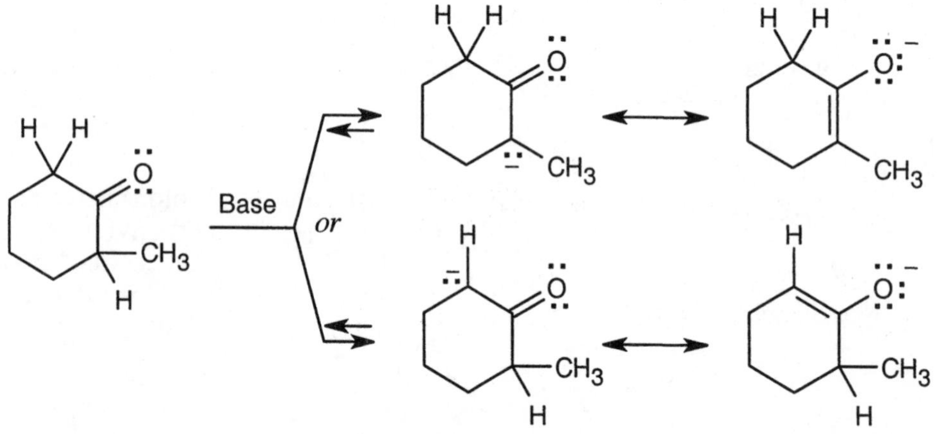

In parts (b) and (c), two different enolate ions can form. *E/Z* double bond isomers of enolate anions in (a) and (b) are also formed.

11.8

The most stable enolate is formed by loss of one of the acidic hydrogens from carbon 2. The enolate resulting from loss of a proton at carbon 4 is much less stable.

11.9 The malonic ester synthesis produces α-substituted acetic acid compounds. Look for the acetic acid component of the compound you want to synthesize. The rest of the molecule comes from the alkyl halide. Remember that the alkyl halide should be primary or methyl.

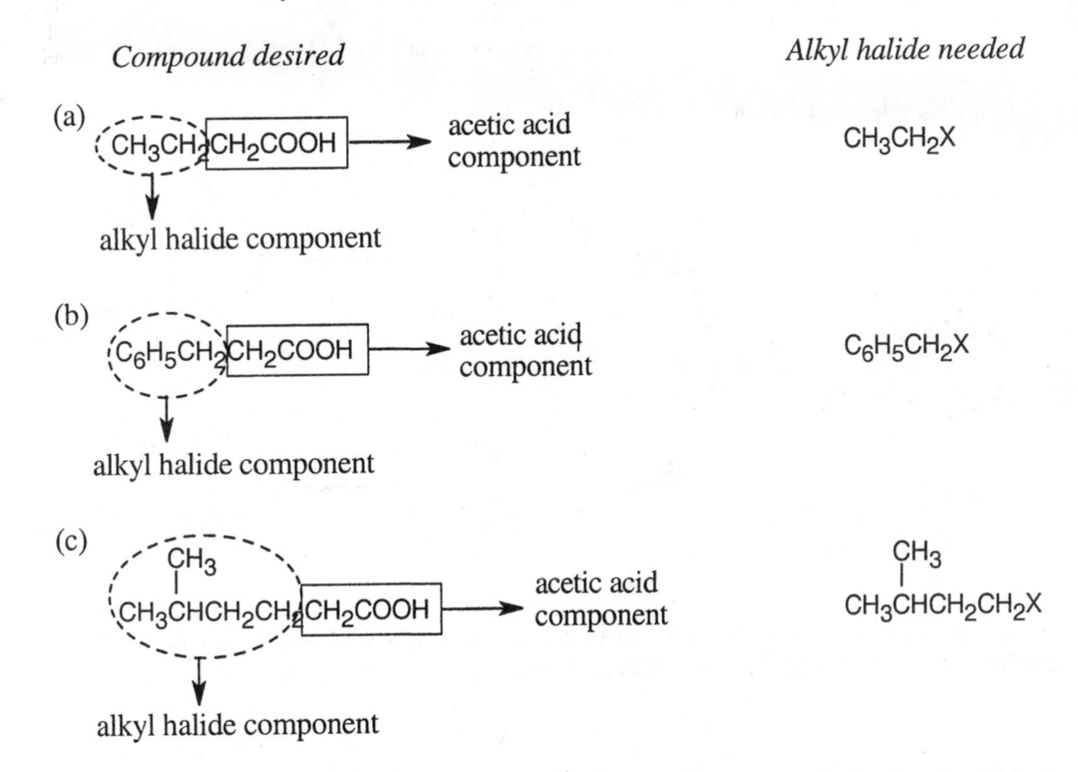

11.10 As in Problem 11.9, locate the acetic acid component of the target molecule. This fragment comes from malonic ester. The rest of the molecule comes from the alkyl halide or halides.

Notice in this problem, and in other problems that involve the malonic ester synthesis, the space-saving use of "Et" when an ethyl group is part of a molecule. Also notice that sodiomalonic ester is prepared by the following reaction:

$$CH_2(COOEt)_2 + Na^+\ ^-OEt \longrightarrow Na^+\ ^-{:}CH(COOEt)_2 + EtOH$$

(a)

from halide $(CH_3)_2CHCH_2$ | CH_2COOH from malonic ester

$$CH_2(COOEt)_2 \xrightarrow[2.\ (CH_3)_2CHCH_2Br]{1.\ Na^+\ ^-OEt} (CH_3)_2CHCH_2{-}CH(COOEt)_2 + NaBr$$

$\downarrow$ H_3O^+, heat

$$(CH_3)_2CHCH_2{-}CH_2COOH + CO_2 + 2\ EtOH$$

4-Methylpentanoic acid

(b)

$CH_3CH_2CH_2$ | $CHCOOH$ (with CH_3 substituent) from malonic ester

from halides $CH_3CH_2CH_2$, CH_3

$$CH_2(COOEt)_2 \xrightarrow[2.\ CH_3CH_2CH_2Br]{1.\ Na^+\ ^-OEt} CH_3CH_2CH_2{-}CH(COOEt)_2 + NaBr$$

$\downarrow$ 1. $Na^+\ ^-OEt$ 2. CH_3Br

$$CH_3CH_2CH_2{-}\underset{CH_3}{\underset{|}{C}}(COOEt)_2 + NaBr \xrightarrow[heat]{H_3O^+} CH_3CH_2CH_2{-}\underset{CH_3}{\underset{|}{C}}HCOOH$$

2-Methylpentanoic acid

$+\ CO_2 + 2\ EtOH$

11.11 This carboxylic acid is composed of the same alkyl groups that we used in the previous problem. The target carboxylic acid is 2,4-dimethylpentanoic acid, and the alkyl groups are methyl and 2-methylpropyl (isobutyl).

$CH_3\underset{CH_3}{\underset{|}{C}}HCH_2$ | $\underset{CH_3}{\underset{|}{C}}HCOOH$ from malonic ester

from halides

$$CH_2(COOEt)_2 \xrightarrow[2.\ CH_3CH(CH_3)CH_2Br]{1.\ Na^+\ ^-OEt} CH_3\underset{CH_3}{\underset{|}{C}}HCH_2{-}CH(COOEt)_2 + NaBr$$

$\downarrow$ 1. $Na^+\ ^-OEt$ 2. CH_3Br

$$CH_3\underset{CH_3}{\underset{|}{C}}HCH_2{-}\underset{CH_3}{\underset{|}{C}}(COOEt)_2 + NaBr \xrightarrow[heat]{H_3O^+} CH_3\underset{CH_3}{\underset{|}{C}}HCH_2{-}\underset{CH_3}{\underset{|}{C}}HCOOH$$

2,4-Dimethylpentanoic acid

$+\ CO_2 + 2\ EtOH$

11.12 Compounds that can undergo the aldol reaction must have acidic α-hydrogen atoms. Cyclohexanone (a) is the only compound in this problem capable of undergoing the aldol reaction.

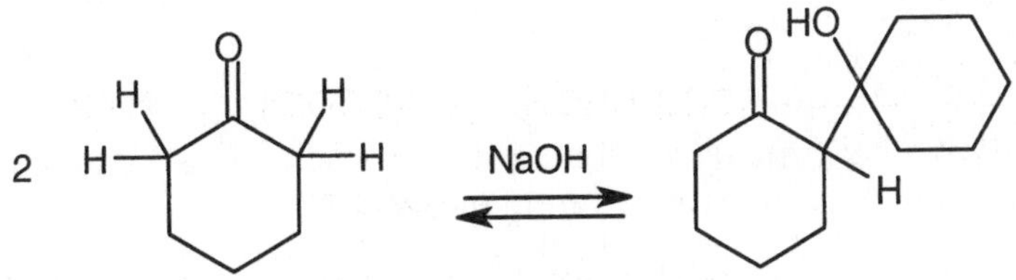

2 (cyclohexanone) $\xrightleftharpoons{NaOH}$ (aldol product)

No α-hydrogens:

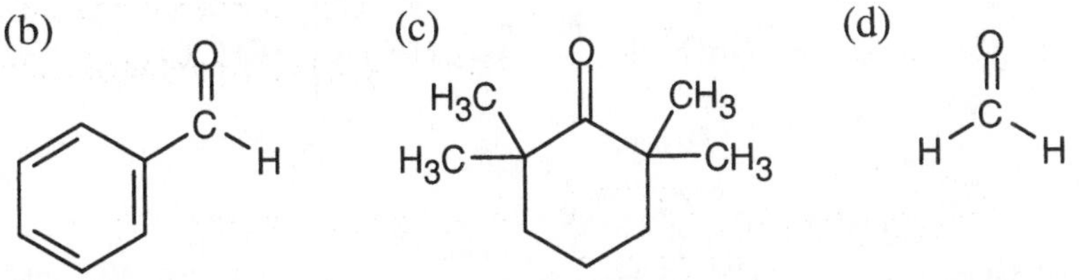

(b) Benzaldehyde

(c) 2,2,6,6-Tetramethyl-cyclohexanone

(d) Formaldehyde

11.13 It is easier to draw the product of an aldol reaction if the two components are drawn so that the new bond connects them.

(a)

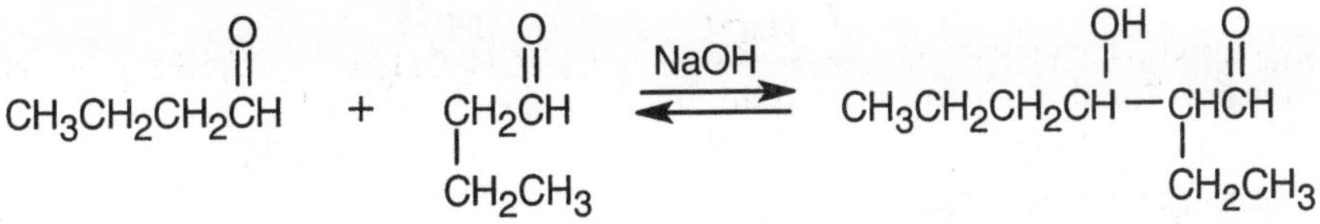

$$CH_3CH_2CH_2CHO + CH_3CH_2CH_2CHO \xrightleftharpoons{NaOH} CH_3CH_2CH_2CH(OH)-CH(CH_2CH_3)CHO$$

(b) cyclopentanone + cyclopentanone $\xrightleftharpoons{NaOH}$ aldol product

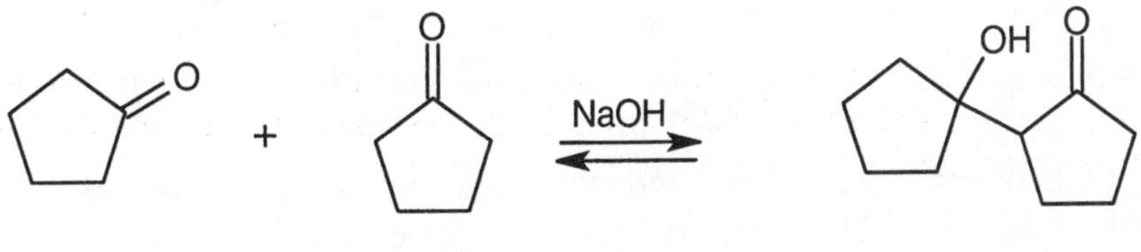

(c)

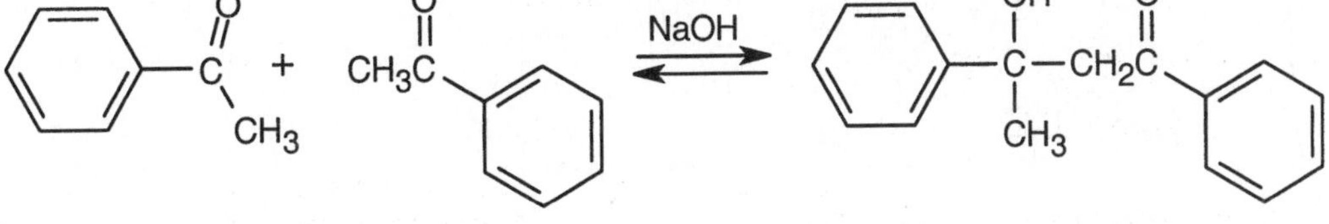

$$C_6H_5C(=O)CH_3 + CH_3C(=O)C_6H_5 \xrightleftharpoons{NaOH} C_6H_5C(OH)(CH_3)-CH_2C(=O)C_6H_5$$

11.14

(a)

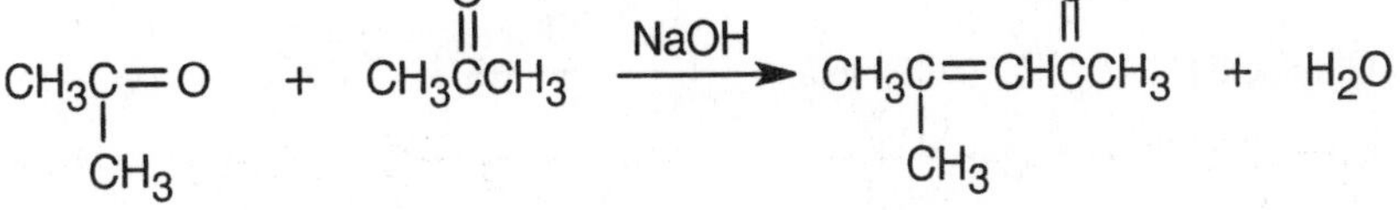

$$CH_3C(CH_3)=O + CH_3CCH_3 \text{ (with C=O)} \xrightarrow{NaOH} CH_3C(CH_3)=CHC(=O)CH_3 + H_2O$$

(b) cyclopentanone + cyclopentanone $\xrightarrow{NaOH}$ 2-cyclopentylidenecyclopentanone + H_2O

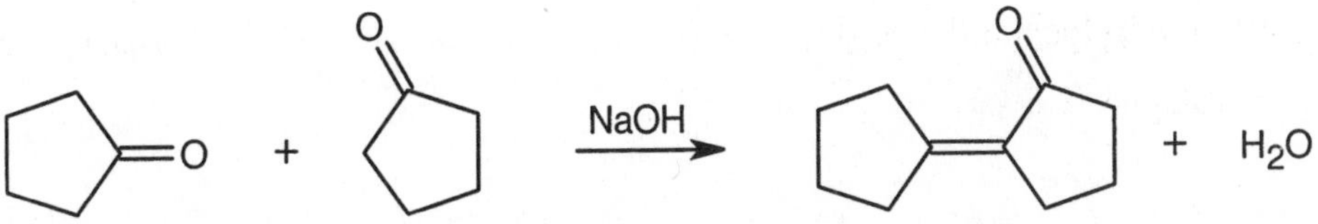

(c)

$$CH_3CH_2CH{=}O + H_2C(CH_3)CHO \xrightarrow{NaOH} CH_3CH_2CH{=}C(CH_3)CHO + H_2O$$

11.15 Two different enolate anions can be formed from butan-2-one. Each enolate can react with a second molecule of butan-2-one to yield two different enones.

$$CH_3CH_2COCH_3 + :\overset{-}{C}H(CH_3)COCH_3 \xrightarrow{NaOH} CH_3CH_2C(OH)(CH_3){-}CH(CH_3)COCH_3 \longrightarrow CH_3CH_2C(CH_3){=}C(CH_3)COCH_3 + H_2O$$

$$CH_3CH_2COCH_3 + :\overset{-}{C}H_2COCH_2CH_3 \xrightarrow{NaOH} CH_3CH_2C(OH)(CH_3){-}CH_2COCH_2CH_3 \longrightarrow CH_3CH_2C(CH_3){=}CHCOCH_2CH_3 + H_2O$$

11.16 The enone was formed from two molecules of acetophenone.

$$C_6H_5C(CH_3){=}O + H_3CC(=O)C_6H_5 \xrightarrow{NaOH} C_6H_5C(CH_3){=}CHC(=O)C_6H_5 + H_2O$$

11.17 As in the aldol condensation, compounds that undergo Claisen condensation reactions must have acidic α hydrogens. Only ester (c) yields a Claisen condensation product.

(a) $HCOOCH_3$

no α hydrogens

(b) $H_2C{=}CHCOOCH_3$

no acidic α hydrogens

(c)

$$CH_3CH(CH_3)CH_2CH_2COOCH_3 + CH_3CH(CH_3)CH_2CH_2COOCH_3 \xrightarrow[2.\ H_3O^+]{1.\ Na^+\ {}^-OCH_3} CH_3OH + CH_3CH(CH_3)CH_2CH_2C(=O){-}CH(CH_2CH(CH_3)CH_3)COOCH_3$$

11.18 Writing the two Claisen components in the correct orientation makes it easier to predict the product.

(a)

$$(CH_3)_2CHCH_2\overset{O}{\overset{\|}{C}}OCH_3 + \underset{CH(CH_3)_2}{CH_2}\overset{O}{\overset{\|}{C}}OCH_3 \xrightarrow[2.\ H_3O^+]{1.\ Na^+\ ^-OCH_3} (CH_3)_2CHCH_2\overset{O}{\overset{\|}{C}}\underset{CH(CH_3)_2}{CH}\overset{O}{\overset{\|}{C}}OCH_3 + CH_3OH$$

(b)

$$C_6H_5CH_2\overset{O}{\overset{\|}{C}}OCH_3 + \underset{C_6H_5}{CH_2}\overset{O}{\overset{\|}{C}}OCH_3 \xrightarrow[2.\ H_3O^+]{1.\ Na^+\ ^-OCH_3} C_6H_5CH_2\overset{O}{\overset{\|}{C}}\underset{C_6H_5}{CH}\overset{O}{\overset{\|}{C}}OCH_3 + CH_3OH$$

(c)

$$C_6H_{11}CH_2\overset{O}{\overset{\|}{C}}OCH_3 + \underset{C_6H_{11}}{CH_2}\overset{O}{\overset{\|}{C}}OCH_3 \xrightarrow[2.\ H_3O^+]{1.\ Na^+\ ^-OCH_3} C_6H_{11}CH_2\overset{O}{\overset{\|}{C}}\underset{C_6H_{11}}{CH}\overset{O}{\overset{\|}{C}}OCH_3 + CH_3OH$$

Visualizing Chemistry

11.19

(a)

Pentan-3-one + Pentan-3-one $\xrightarrow{NaOH}$ product

Pentan-3-one

(b)

3-Methylbutanal + 3-Methylbutanal $\xrightarrow{NaOH}$ product

3-Methylbutanal

11.20 The enolate of methyl phenylacetate adds to a second molecule of methyl phenylacetate to form the intermediate that is pictured. Elimination of methoxide (circled) and acidification give the product shown.

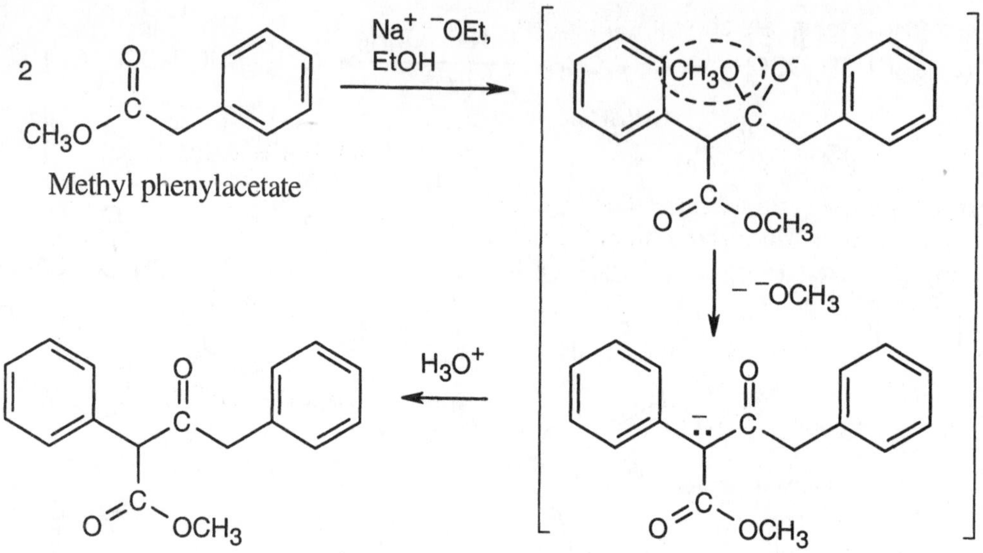
2
Na^+ ^-OEt,
EtOH
CH_3O
Methyl phenylacetate
$-\,^-OCH_3$
H_3O^+
OCH_3

11.21

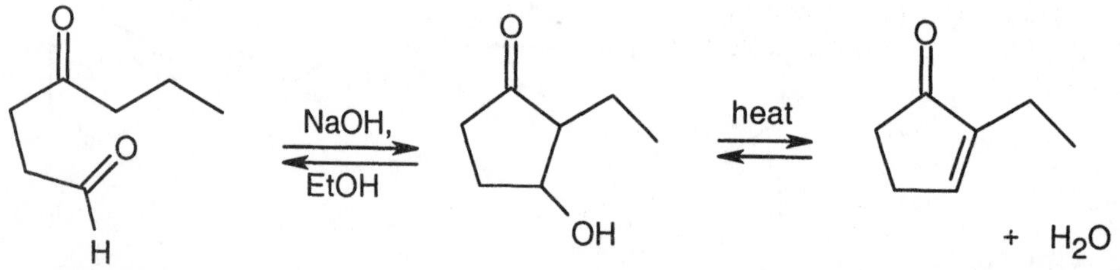
NaOH,
EtOH
heat
OH
+ H_2O

4-Oxoheptanal

11.22 This product is formed from the reaction of malonic ester with both benzyl bromide and bromomethane.

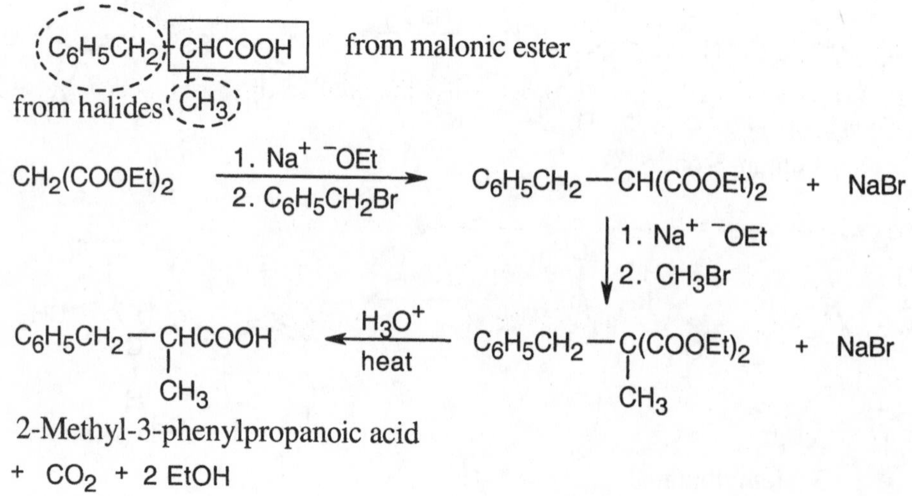
$C_6H_5CH_2$ CHCOOH from malonic ester
from halides CH_3
$CH_2(COOEt)_2$
1. Na^+ ^-OEt
2. $C_6H_5CH_2Br$
$C_6H_5CH_2—CH(COOEt)_2$ + NaBr
1. Na^+ ^-OEt
2. CH_3Br
$C_6H_5CH_2—C(COOEt)_2$ + NaBr
CH_3
H_3O^+
heat
$C_6H_5CH_2—CHCOOH$
CH_3
2-Methyl-3-phenylpropanoic acid
+ CO_2 + 2 EtOH

Additional Problems

Acidity, Tautomerism, and Reactivity

11.23 Acidic hydrogens are shown in boldface.

(a) $HOCH_2C(=O)CH_3$

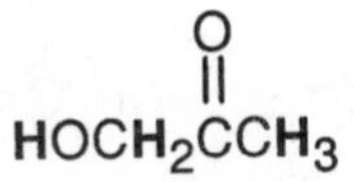

All hydrogens are acidic. The alcohol hydrogen is more acidic than the other hydrogens.

(b) $HOCH_2CH_2C(=O)C(CH_3)_3$

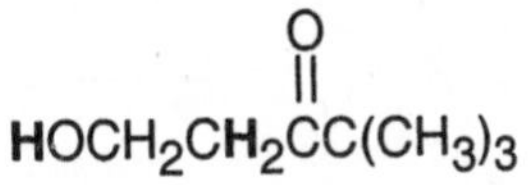

(c) (cyclopentane-1,3-dione with H H on C2, H and H on C4, H and H on C5)

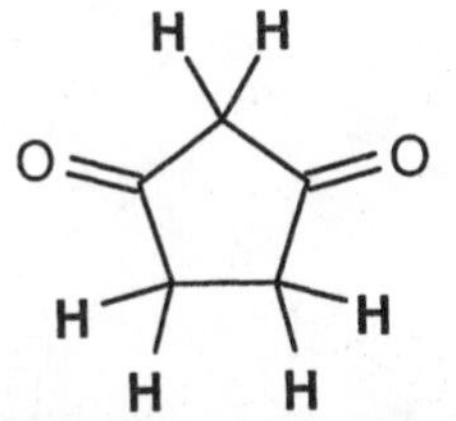

All hydrogens are acidic. The two hydrogens between the carbonyl groups are much more acidic than the others.

(d) $CH_3CH{=}CHC(=O)H$

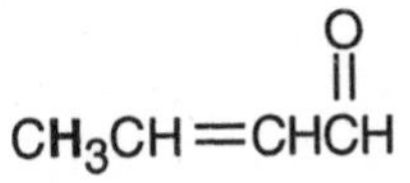

11.24

(1,3-cyclohexanedione ⇌ enol ⇌ enol ⇌ enol ⇌ enol)

OH, O, OH, OH, O

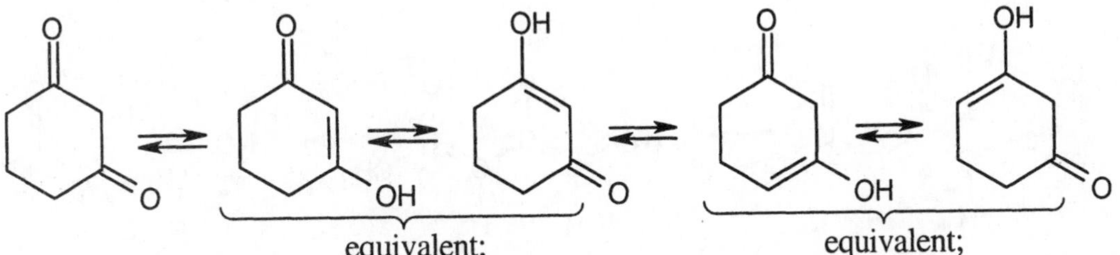

equivalent; more stable

equivalent; less stable

The first two monoenols are more stable because the enol double bond is conjugated with the carbonyl group.

11.25

Least acidic ⟶ *Most acidic*

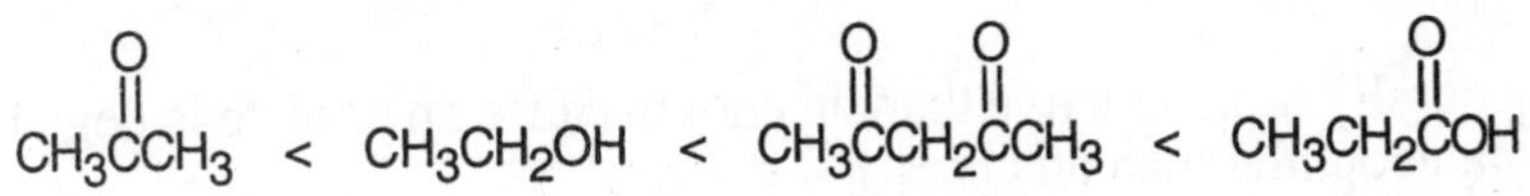

$CH_3C(=O)CH_3$ < CH_3CH_2OH < $CH_3C(=O)CH_2C(=O)CH_3$ < $CH_3CH_2C(=O)OH$

11.26 Pentane-2,4-dione is enolized to a greater extent than acetone because its enol form is more stable than the enol form of acetone. This enol stability is due to two factors: (1) conjugation of the enol double bond with the second carbonyl group, and (2) hydrogen bonding between the enol hydrogen and the second carbonyl group.

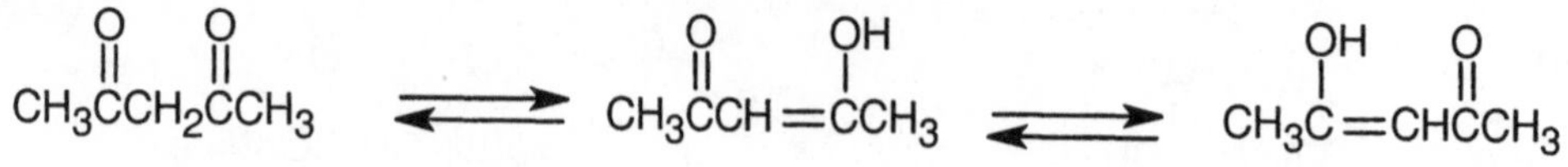

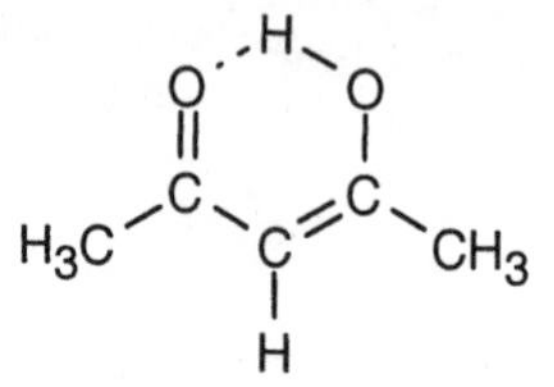

$$CH_3\overset{O}{\overset{\|}{C}}CH_2\overset{O}{\overset{\|}{C}}CH_3 \rightleftharpoons CH_3\overset{O}{\overset{\|}{C}}CH{=}\overset{OH}{\overset{|}{C}}CH_3 \rightleftharpoons CH_3\overset{OH}{\overset{|}{C}}{=}CH\overset{O}{\overset{\|}{C}}CH_3$$

11.27

(a)

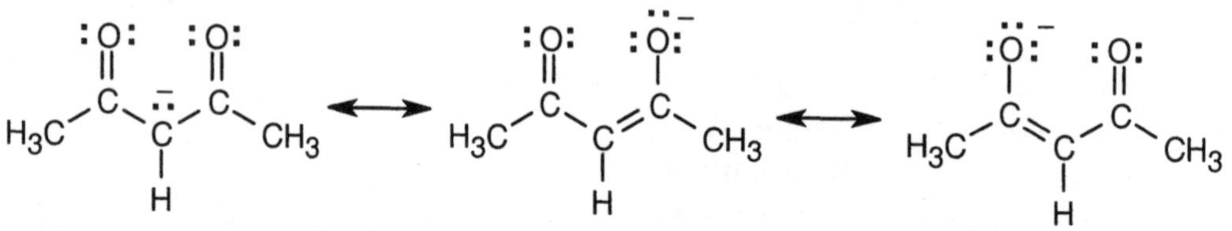

(b) $H_2\ddot{C}^{-}{-}C{\equiv}N: \longleftrightarrow H_2C{=}C{=}\ddot{N}:^{-}$

(c)

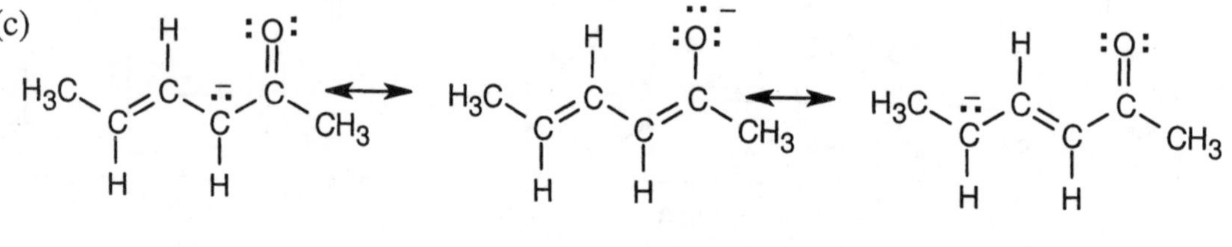

(d)

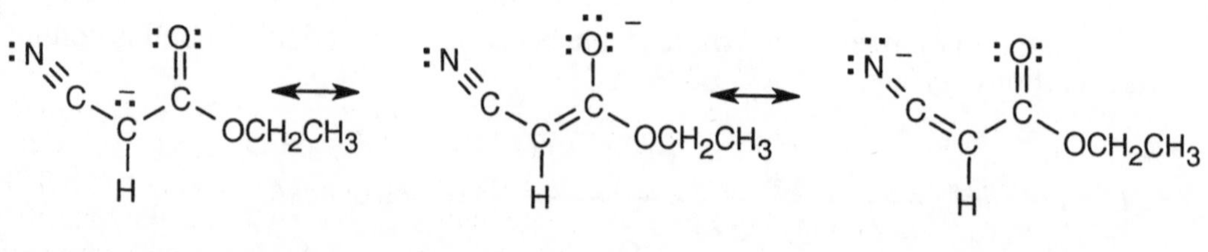

11.28 An enolate anion is generally more reactive than an enol because an enolate is negatively charged and is more nucleophilic than an enol.

Mechanisms

11.29 In *acid-catalyzed enolization*, protonation of the carbonyl oxygen is followed by the loss of a proton on the carbon α to the carbonyl group to give the enol. In *base-catalyzed enolization*, abstraction of a proton from the α carbon is followed by protonation of the negatively charged oxygen to give the enol.

11.30

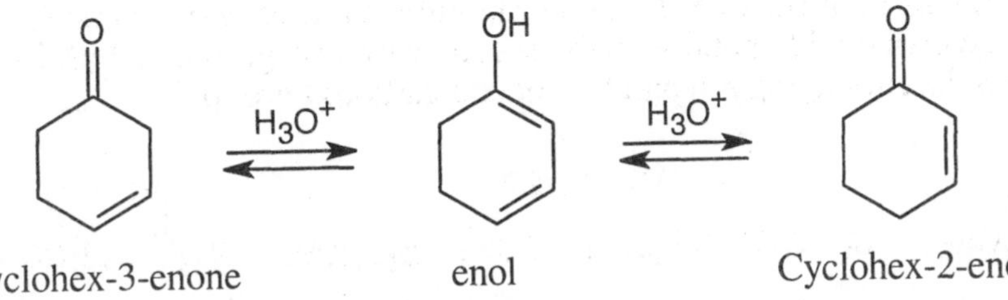

Cyclohex-3-enone ⇌ (H_3O^+) enol ⇌ (H_3O^+) Cyclohex-2-enone

3-Cyclohexenone and 2-cyclohexenone can be interconverted because they are both in equilibrium with the same enol under acidic conditions.

11.31

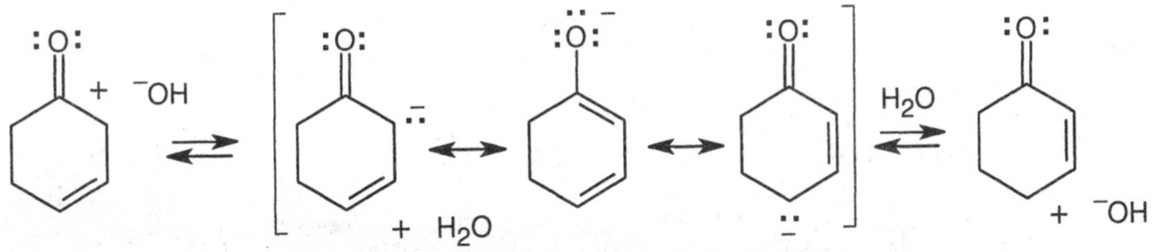

Cyclohex-3-enone + ^{-}OH ⇌ [enolate resonance forms] + H_2O ⇌ (H_2O) cyclohex-2-enone + ^{-}OH

Cyclohex-3-enone and cyclohex-2-enone form the same enolate anion on base treatment and can thus be interconverted.

11.32

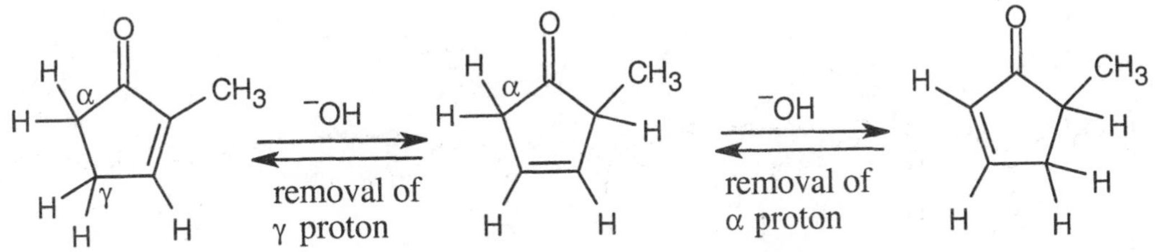

^{-}OH: removal of γ proton; ^{-}OH: removal of α proton

As in Problem 11.31, the isomers are in equilibrium through their common enolate.

11.33 Treatment of either the cis or trans isomer with base causes enolate anion formation α to the carbonyl group and results in loss of configuration at the α-position. Reprotonation at carbon 2 produces either of the diastereomeric 4-*tert*-butyl-2-methylcyclohexanones. In both diastereomers the *tert*-butyl group of carbon 4 occupies the equatorial position for steric reasons. The methyl group of the cis isomer is also equatorial; the methyl group of the trans isomer is axial. The trans isomer is less stable because of interactions of the axial methyl group with the ring hydrogens.

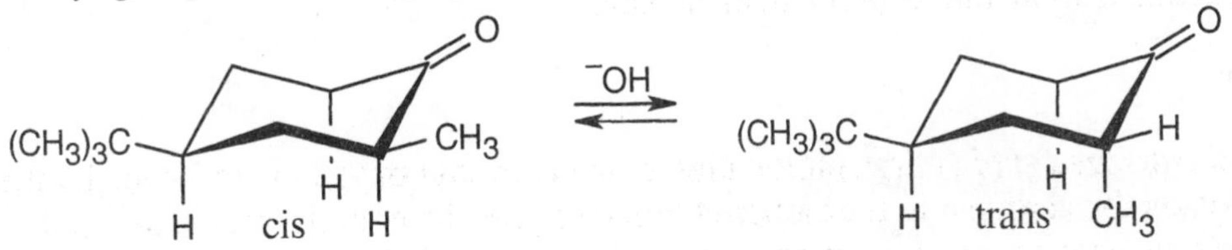

$(CH_3)_3C$ ... CH_3 (cis) ⇌ (^{-}OH) $(CH_3)_3C$ ... CH_3 (trans)

11.34

$H_3C-C(=O)-CH_3$ ⇌ (deuteration of carbonyl oxygen) [$D_2\ddot{O}$: ... $H-CH_2-C(=\overset{+}{O}D)-CH_3$] ⇌ (loss of proton at alpha position) $H_2C=C(OD)-CH_3$ (*enol*) + $H\overset{+}{O}D_2$

deuteration of carbonyl oxygen

loss of proton at alpha position

enol

$H_2C=C(OD)-CH_3$ (*enol*) + $D-\overset{+}{O}D_2$ ⇌ (deuteration of enol double bond) [$DCH_2-C(=\overset{+}{O}D)-CH_3$ + $:\ddot{O}D_2$] ⇌ (loss of deuterium on carbonyl oxygen) $DCH_2-C(=O)-CH_3$ + D_3O^+

enol

deuteration of enol double bond

loss of deuterium on carbonyl oxygen

11.35

2-methylcyclohexanone (chiral C2, H, CH_3) ⇌ (H_3O^+ or ^-OH) [*enol*: :ÖH, CH_3] ⇌ 2-methylcyclohexanone (CH_3, H) + 2-methylcyclohexanone (H, CH_3) + H_3O^+ or ^-OH

enol

Carbon 2 loses its chirality in the step in which the enol double bond is formed. Protonation occurs with equal probability from either side of sp^2-hybridized carbon 2, resulting in racemic product.

11.36

3-methylcyclohexanone (chiral C3, H, CH_3) ⇌ (H_3O^+ or ^-OH) [*enol*: :ÖH, H, CH_3, H] ⇌ 3-methylcyclohexanone (H, H, CH_3, H) + H_3O^+ or ^-OH

enol

(*R*)-3-Methylcyclohexanone is not racemized by acid or base because its chirality center is not involved in the enolization reaction.

Aldol Condensation

11.37 Compounds (a), (c) and (f) do not undergo aldol condensation because they don't have acidic protons α to the carbonyl group. Compound (e) also doesn't form aldol products; although it has an acidic α proton, the reaction site is sterically hindered, and the equilibrium favors starting material.

(b)

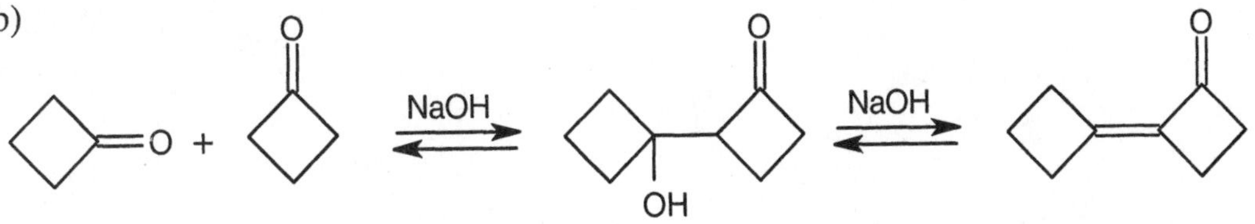

O + ⇌ (NaOH) OH ⇌ (NaOH)

(d)

$$CH_3(CH_2)_7CH_2\overset{O}{\overset{\|}{C}}H + \underset{CH_2(CH_2)_6CH_3}{CH_2\overset{O}{\overset{\|}{C}}H} \overset{NaOH}{\rightleftharpoons} CH_3(CH_2)_7CH_2\overset{OH}{CH}-\underset{CH_2(CH_2)_6CH_3}{CH\overset{O}{\overset{\|}{C}}H}$$

$$\updownarrow NaOH$$

$$CH_3(CH_2)_7CH_2CH=\underset{CH_2(CH_2)_6CH_3}{C\overset{O}{\overset{\|}{C}}H}$$

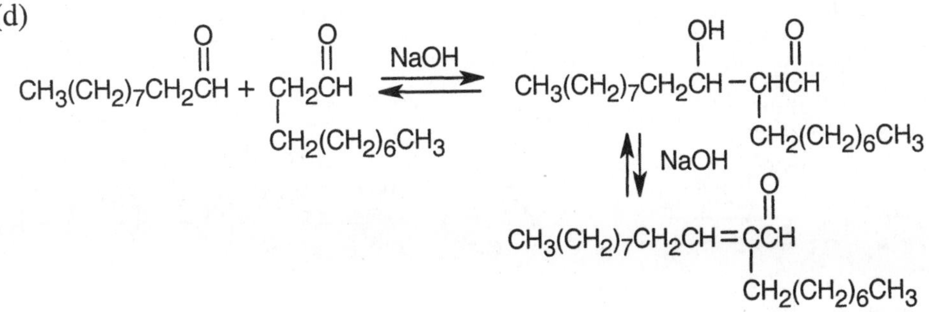

11.38 Try to think backwards from the product to recognize the aldol components. In these products, the double bond is the new bond formed.

(a)

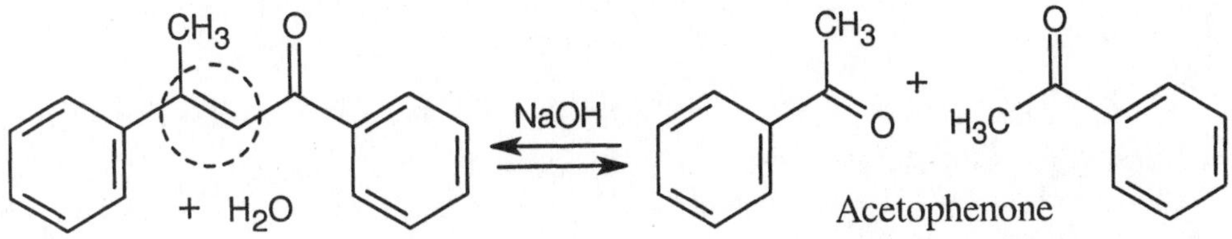

CH3 O + H2O ⇌ (NaOH) CH3 O + H3C

Acetophenone

(b)

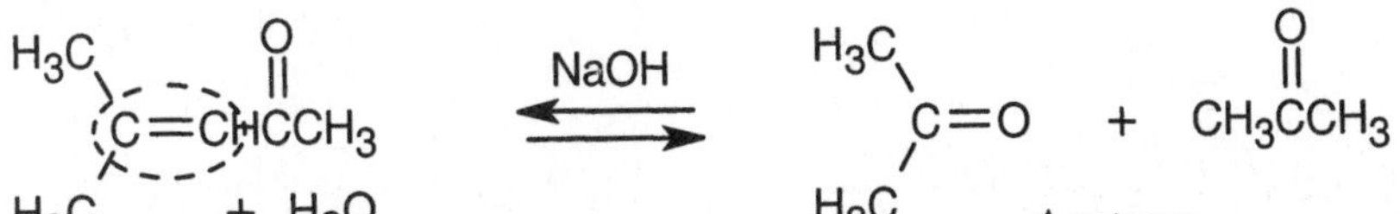

$$(H_3C)_2C=CH\overset{O}{\overset{\|}{C}}CH_3 + H_2O \overset{NaOH}{\rightleftharpoons} (H_3C)_2C=O + CH_3\overset{O}{\overset{\|}{C}}CH_3$$

Acetone

(c)

$$CH_3CH_2CH_2CH=\underset{CH_2CH_3}{C\overset{O}{\overset{\|}{C}}H} + H_2O \overset{NaOH}{\rightleftharpoons} CH_3CH_2CH_2\overset{O}{\overset{\|}{C}}H + \underset{CH_2CH_3}{CH_2\overset{O}{\overset{\|}{C}}H}$$

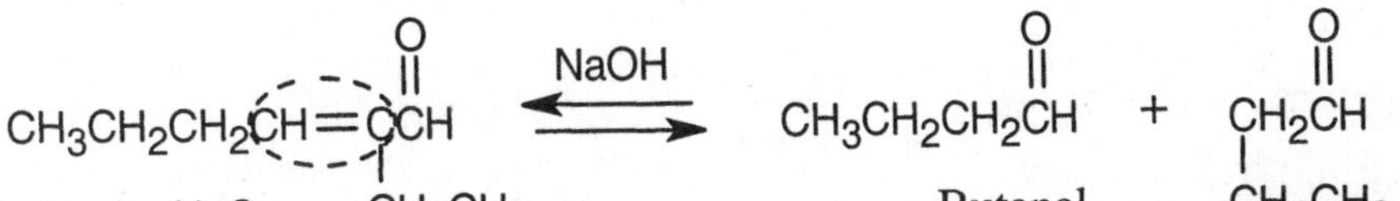

Butanal

(d)

$$\text{(2-methylcyclopentanone aldol condensation product)} + H_2O \underset{}{\overset{NaOH}{\rightleftharpoons}} \text{2-Methylcyclopentanone} + \text{2-Methylcyclopentanone}$$

2-Methylyclopentanone

11.39

$$\text{2,6-Heptanedione} \underset{ethanol}{\overset{NaOH}{\rightleftharpoons}} \text{3-hydroxy-3-methylcyclohexanone} \underset{ethanol}{\overset{NaOH}{\rightleftharpoons}} \text{3-Methylcyclohex-2-enone} + H_2O$$

2,6-Heptanedione

3-Methylcyclohex-2-enone

11.40

$$\text{Hexanedial} \underset{ethanol}{\overset{NaOH}{\rightleftharpoons}} \text{2-hydroxycyclopentanecarboxaldehyde} \underset{ethanol}{\overset{NaOH}{\rightleftharpoons}} \text{Cyclopent-1-enecarboxaldehyde} + H_2O$$

Hexanedial

Cyclopent-1-enecarboxaldehyde

Malonic Ester Synthesis

11.41 First, convert malonic ester into its sodium salt.

$$CH_2(COOEt)_2 + Na^{+}\,{}^{-}OEt \longrightarrow Na^{+}\,{}^{-}\!:CH(COOEt)_2 + EtOH$$

(a)

$$CH_3CH_2CH_2Br + Na^{+}\,{}^{-}\!:CH(COOEt)_2 \longrightarrow CH_3CH_2CH_2CH(COOEt)_2 + NaBr$$

$$\downarrow H_3O^{+}, \Delta$$

$$\underset{\text{Ethyl pentanoate}}{CH_3CH_2CH_2CH_2\overset{O}{\overset{\|}{C}}OEt} \xleftarrow[H^{+}\text{ catalyst}]{EtOH} CH_3CH_2CH_2CH_2\overset{O}{\overset{\|}{C}}OH + 2\ EtOH + CO_2$$

(b)

$$CH_2(CO_2Et)_2 \xrightarrow[2.\ (CH_3)_2CHBr]{1.\ Na^{+}\,{}^{-}OEt} (CH_3)_2CH{-}CH(CO_2Et)_2 + NaBr$$

$$\downarrow H_3O^{+}, \text{heat}$$

$$\underset{\text{Ethyl 3-methylbutanoate}}{(CH_3)_2CHCH_2CO_2Et} \xleftarrow[H^{+}\text{ catalyst}]{EtOH} (CH_3)_2CHCH_2CO_2H + CO_2 + 2\ EtOH$$

Some elimination product will also form.

(c)

$$CH_3Br + Na^+ \ ^-{:}CH(COOEt)_2 \longrightarrow CH_3CH(COOEt)_2 + NaBr$$

$$CH_3CH(COOEt)_2 \xrightarrow{Na^+ \ ^-OEt} Na^+ \ ^-{:}C(CH_3)(COOEt)_2 + EtOH$$

$$Na^+ \ ^-{:}C(CH_3)(COOEt)_2 \xrightarrow{CH_3CH_2Br} CH_3CH_2C(CH_3)(COOEt)_2 + NaBr$$

$$CH_3CH_2C(CH_3)(COOEt)_2 \xrightarrow{H_3O^+,\ \Delta} CH_3CH_2CH(CH_3)COOH + 2\ EtOH + CO_2$$

$$CH_3CH_2CH(CH_3)COOH \xrightarrow[H^+ \text{ catalyst}]{CH_3CH_2OH} CH_3CH_2CH(CH_3)COOCH_2CH_3$$

Ethyl 2-methylbutanoate

(d)

$$(CH_3)_3C{-}COOCH_2CH_3$$

Ethyl 2,2-dimethylpropanoate

This ester can't be prepared by the malonic ester route because it is trisubstituted at the α carbon.

11.42

$$CH_2(COOEt)_2 + Na^+ \ ^-OEt \longrightarrow Na^+ \ ^-{:}CH(COOEt)_2 + EtOH$$

$$Na^+ \ ^-{:}CH(COOEt)_2 + BrCH_2CH_2CH_2CH_2Br \longrightarrow BrCH_2CH_2CH_2CH_2CH(COOEt)_2 + NaBr$$

$$BrCH_2CH_2CH_2CH_2CH(COOEt)_2 \xrightarrow{Na^+ \ ^-OEt} \text{cyclo-}(CH_2)_4C(COOEt)_2 + NaBr$$

$$\text{cyclo-}(CH_2)_4C(COOEt)_2 \xrightarrow{H_3O^+,\ \Delta} \text{cyclo-}(CH_2)_4CHCOOH + 2\ EtOH + CO_2$$

Cyclopentanecarboxylic acid

11.43

$$\underset{\text{(CH}_3\text{ on C-3 and C-7)}}{CH_3\overset{CH_3}{\overset{|}{C}}=CHCH_2CH_2\overset{CH_3}{\overset{|}{C}}=CHCH_2OH} \xrightarrow{PBr_3} CH_3\overset{CH_3}{\overset{|}{C}}=CHCH_2CH_2\overset{CH_3}{\overset{|}{C}}=CHCH_2Br$$

$$CH_3\overset{CH_3}{\overset{|}{C}}=CHCH_2CH_2\overset{CH_3}{\overset{|}{C}}=CHCH_2Br \; + \; Na^{+}\,^{-}\!:CH(COOEt)_2$$

$$\downarrow$$

$$CH_3\overset{CH_3}{\overset{|}{C}}=CHCH_2CH_2\overset{CH_3}{\overset{|}{C}}=CHCH_2CH(COOEt)_2 \; + \; NaBr$$

$$\downarrow H_3O^{+},\ \Delta$$

$$CH_3\overset{CH_3}{\overset{|}{C}}=CHCH_2CH_2\overset{CH_3}{\overset{|}{C}}=CHCH_2CH_2\overset{O}{\overset{||}{C}}OH \; + \; 2\ EtOH \; + \; CO_2$$

$$\downarrow \begin{array}{l} CH_3CH_2OH \\ H^{+}\ \text{catalyst} \end{array}$$

$$CH_3\overset{CH_3}{\overset{|}{C}}=CHCH_2CH_2\overset{CH_3}{\overset{|}{C}}=CHCH_2CH_2\overset{O}{\overset{||}{C}}OCH_2CH_3$$

Ethyl geranylacetate

Claisen Condensation

11.44 Two of the products result from Claisen *self*-condensation.

$$\underset{\text{Ethyl acetate}}{CH_3\overset{O}{\overset{||}{C}}\!\!\underset{OCH_2CH_3}{\underset{|}{}}} \; + \; \underset{\text{Ethyl acetate}}{CH_3\overset{O}{\overset{||}{C}}OCH_2CH_3} \xrightarrow[2.\ H_3O^{+}]{1.\ Na^{+}\,^{-}OCH_2CH_3} CH_3\overset{O}{\overset{||}{C}}CH_2\overset{O}{\overset{||}{C}}OCH_2CH_3 \; + \; CH_3CH_2OH$$

$$\underset{\text{Ethyl propionate}}{CH_3CH_2\overset{O}{\overset{||}{C}}\!\!\underset{OCH_2CH_3}{\underset{|}{}}} \; + \; \underset{\text{Ethyl propionate}}{\underset{CH_3}{\underset{|}{CH_2}}\overset{O}{\overset{||}{C}}OCH_2CH_3} \xrightarrow[2.\ H_3O^{+}]{1.\ Na^{+}\,^{-}OCH_2CH_3} CH_3CH_2\overset{O}{\overset{||}{C}}\underset{CH_3}{\underset{|}{C}H}\overset{O}{\overset{||}{C}}OCH_2CH_3 \; + \; CH_3CH_2OH$$

The other two products are formed from *mixed* Claisen condensations.

$$\underset{\text{Ethyl acetate}}{CH_3C(=O)OCH_2CH_3} + \underset{\text{Ethyl propionate}}{CH_3CH_2COCH_2CH_3 \ (C=O)} \xrightarrow[2.\ H_3O^+]{1.\ Na^+\ ^-OCH_2CH_3} CH_3C(=O)CH(CH_3)C(=O)OCH_2CH_3 + CH_3CH_2OH$$

$$\underset{\text{Ethyl propionate}}{CH_3CH_2C(=O)OCH_2CH_3} + \underset{\text{Ethyl acetate}}{CH_3C(=O)OCH_2CH_3} \xrightarrow[2.\ H_3O^+]{1.\ Na^+\ ^-OCH_2CH_3} CH_3CH_2C(=O)CH_2C(=O)OCH_2CH_3 + CH_3CH_2OH$$

11.45 As in the previous problem, a Claisen *self* condensation between two molecules of ethyl acetate yields one product.

$$\underset{\text{Ethyl acetate}}{CH_3C(=O)OCH_2CH_3} + \underset{\text{Ethyl acetate}}{CH_3C(=O)OCH_2CH_3} \xrightarrow[2.\ H_3O^+]{1.\ Na^+\ ^-OCH_2CH_3} CH_3C(=O)CH_2C(=O)OCH_2CH_3 + CH_3CH_2OH$$

The other product results from a *mixed* Claisen condensation.

$$\underset{\text{Ethyl benzoate}}{C_6H_5C(=O)OCH_2CH_3} + \underset{\text{Ethyl acetate}}{CH_3C(=O)OCH_2CH_3} \xrightarrow[2.\ H_3O^+]{1.\ Na^+\ ^-OCH_2CH_3} C_6H_5C(=O)CH_2C(=O)OCH_2CH_3 + CH_3CH_2OH$$

Ethyl benzoate does not undergo Claisen self-condensation because it has no α protons.

11.46 As in the malonic acid synthesis, you should identify the components of the target compound. The "acetone component" comes from acetoacetic ester; the other component comes from a primary alkyl halide.

For all parts of this problem, the ethyl acetoacetate anion is prepared by the following reaction:

$$CH_3C(=O)CH_2COOEt + Na^+\ ^-OEt \longrightarrow Na^+\ ^-\!:\!CH(COOEt)C(=O)CH_3 + EtOH$$

(a)

$C_6H_5CH_2$ ┆ $CH_2\overset{O}{\overset{\|}{C}}CH_3$ from acetoacetic ester

from halide

$$C_6H_5CH_2Br + Na^+ \; ^-{:}\underset{COOEt}{\underset{|}{C}H}\overset{O}{\overset{\|}{C}}CH_3 \longrightarrow C_6H_5CH_2\underset{COOEt}{\underset{|}{C}H}\overset{O}{\overset{\|}{C}}CH_3 + NaBr \xrightarrow[\Delta]{H_3O^+} C_6H_5CH_2CH_2\overset{O}{\overset{\|}{C}}CH_3 + EtOH + CO_2$$

4-Phenylbutan-2-one

(b)

$CH_3\underset{}{\overset{CH_3}{\overset{|}{C}}}HCH_2$ ┆ $CH_2\overset{O}{\overset{\|}{C}}CH_3$ from acetoacetic ester

from halide

$$CH_3\overset{CH_3}{\overset{|}{C}}HCH_2Br + Na^+ \; ^-{:}\underset{COOEt}{\underset{|}{C}H}\overset{O}{\overset{\|}{C}}CH_3 \longrightarrow CH_3\overset{CH_3}{\overset{|}{C}}HCH_2\underset{COOEt}{\underset{|}{C}H}\overset{O}{\overset{\|}{C}}CH_3 + NaBr \xrightarrow[\Delta]{H_3O^+} CH_3\overset{CH_3}{\overset{|}{C}}HCH_2CH_2\overset{O}{\overset{\|}{C}}CH_3 + EtOH + CO_2$$

5-Methylhexan-2-one

(c)

$CH_3CH_2CH_2$ ┆ $\underset{CH_3}{\underset{|}{C}H}\overset{O}{\overset{\|}{C}}CH_3$ from acetoacetic ester

from halides

$$CH_3CH_2CH_2Br + Na^+ \; ^-{:}\underset{COOEt}{\underset{|}{C}H}\overset{O}{\overset{\|}{C}}CH_3 \longrightarrow CH_3CH_2CH_2\underset{COOEt}{\underset{|}{C}H}\overset{O}{\overset{\|}{C}}CH_3 + NaBr$$

$$\xrightarrow{Na^+ \; ^-OEt} CH_3CH_2CH_2\underset{COOEt}{\underset{|}{\overset{..}{C}}^-}\overset{O}{\overset{\|}{C}}CH_3 \; Na^+ + EtOH$$

$$\xrightarrow{CH_3Br} CH_3CH_2CH_2\underset{COOEt}{\underset{|}{\overset{H_3C}{\overset{|}{C}}}}\text{-}\overset{O}{\overset{\|}{C}}CH_3 + NaBr \xrightarrow[\Delta]{H_3O^+} CH_3CH_2CH_2\overset{H_3C}{\overset{|}{C}}H\overset{O}{\overset{\|}{C}}CH_3 + EtOH + CO_2$$

3-Methylhexan-2-one

11.47 The acetoacetic ester synthesis can be used only if the target compounds have certain characteristics.

(1) Three carbons must come from acetoacetic ester. In other words, ketones of the form $RCOCH_3$ can't be synthesized by the reaction of RX with acetoacetic ester.

(2) Alkyl halides must be primary or methyl since the acetoacetic ester synthesis is an S_N2 reaction.

(3) Trisubstitution at the α position can't be achieved by an acetoacetic ester synthesis.

(a) Butan-2-one is produced by the reaction of sodium acetoacetate with CH_3Br.

(b) Phenylacetone can't be produced by an acetoacetic acid synthesis. The necessary halide component, bromobenzene, does not undergo S_N2 reactions [see (2) above].

(c) Acetophenone can't be produced by an acetoacetic acid synthesis [see (1) above].

(d) 3,3-Dimethylbutan-2-one can't be produced by an acetoacetic acid synthesis because it is trisubstituted at the α position [see (3) above].

11.48 Aldol self-condensation:

$$CH_3CHO + CH_3CHO \xrightleftharpoons{NaOH} CH_3CH{=}CHCHO + H_2O$$

But-2-enal

Mixed aldol reaction:

$$C_6H_5CHO + CH_3CHO \xrightleftharpoons{NaOH} C_6H_5CH{=}CHCHO + H_2O$$

Cinnamaldehyde

11.49

$$CH_3CHO + CH_3CHO \xrightarrow{NaOH} CH_3CH{=}CHCHO \xrightarrow[2.\ H_3O^+]{1.\ LiAlH_4} CH_3CH{=}CHCH_2OH \xrightarrow{H_2,\ Pd\ catalyst} CH_3CH_2CH_2CH_2OH$$

Butan-1-ol

11.50 Malonic ester has two acidic protons and can be alkylated twice, to yield a compound of the structure $R_2C(COOC_2H_5)_2$. Decarboxylation then gives $R_2CHCOOH$. The α proton is only weakly acidic and is no longer activated by two adjacent ester groups. Thus, trialkylation does not occur.

Integrated Problems

11.51

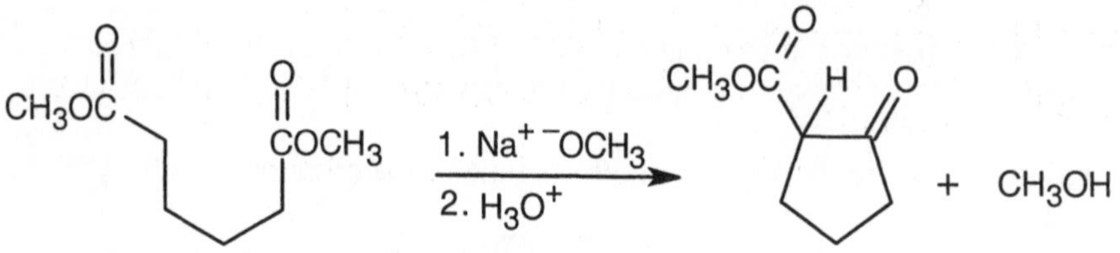

$CH_3OC(=O)CH_2CH_2CH_2CH_2COCH_3$ (1. $Na^{+}\,{}^{-}OCH_3$; 2. H_3O^+) → 2-(methoxycarbonyl)cyclopentanone + CH_3OH

11.52

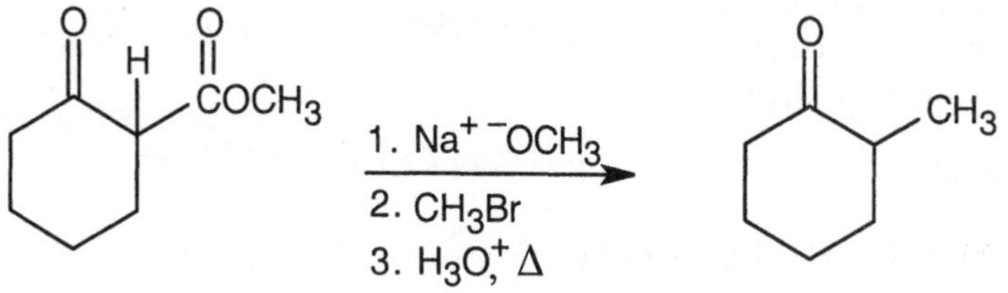

methyl 2-oxocyclohexanecarboxylate (1. $Na^{+}\,{}^{-}OCH_3$; 2. CH_3Br; 3. H_3O^+, Δ) → 2-methylcyclohexanone

11.53

addition of hydroxide

elimination of acetate

acid–base reaction

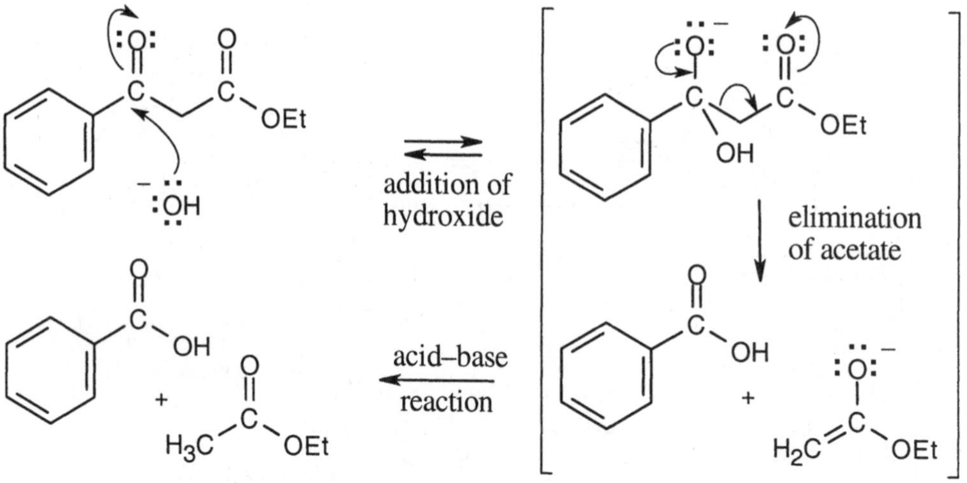

Hydroxide ion can react at two different sites of the ß-keto ester. Abstraction of the acidic α-proton is more favorable but is reversible and does not lead to product. Addition of hydroxide ion to the carbonyl group, followed by irreversible elimination of ethyl acetate, accounts for the observed products.

11.54

Enolization at the γ position produces an anion that is stabilized by delocalization of the negative charge over the π system of five atoms.

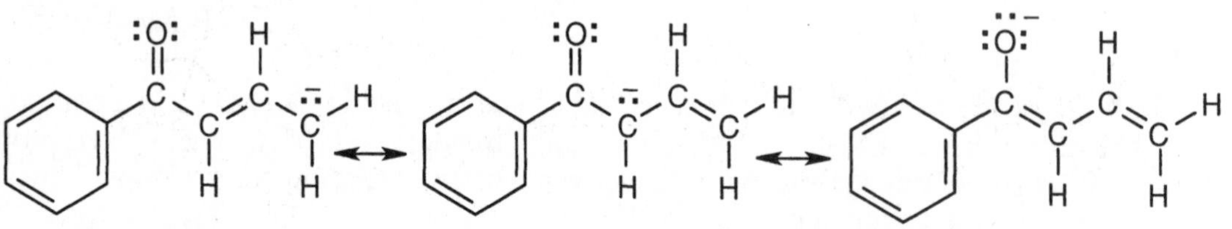

11.55

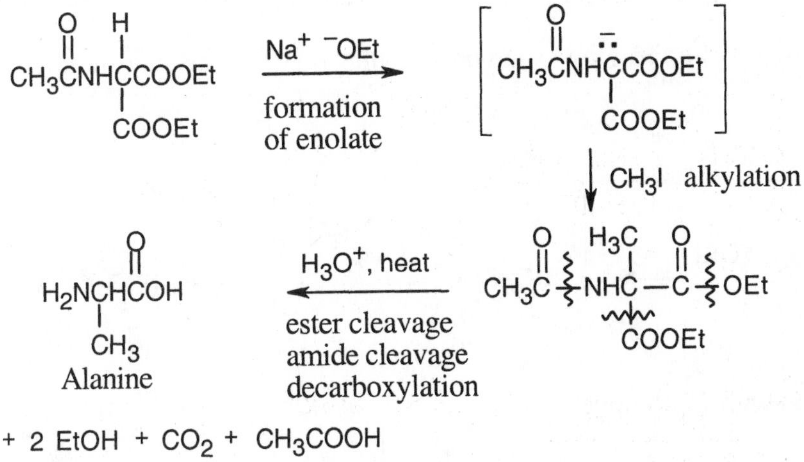

$$CH_3\overset{O}{\overset{\|}{C}}NH\overset{H}{\overset{|}{\underset{COOEt}{\underset{|}{C}}}}COOEt \xrightarrow[\text{formation of enolate}]{Na^+ \ ^-OEt} \left[CH_3\overset{O}{\overset{\|}{C}}NH\overset{\cdot\cdot}{\overset{-}{\underset{COOEt}{\underset{|}{C}}}}COOEt \right]$$

$\downarrow$ CH_3I alkylation

$$H_2N\underset{CH_3}{\underset{|}{C}}HCOH \ (\text{Alanine}) \xleftarrow[\text{ester cleavage, amide cleavage, decarboxylation}]{H_3O^+, \text{ heat}} CH_3\overset{O}{\overset{\|}{C}}\text{—}NH\overset{H_3C}{\overset{|}{\underset{COOEt}{\underset{|}{C}}}}\text{—}\overset{O}{\overset{\|}{C}}\text{—}OEt$$

$+ \ 2 \ EtOH \ + \ CO_2 \ + \ CH_3COOH$

Acid cleaves both ester bonds, as well as the amide bond.

In the Medicine Cabinet

11.56

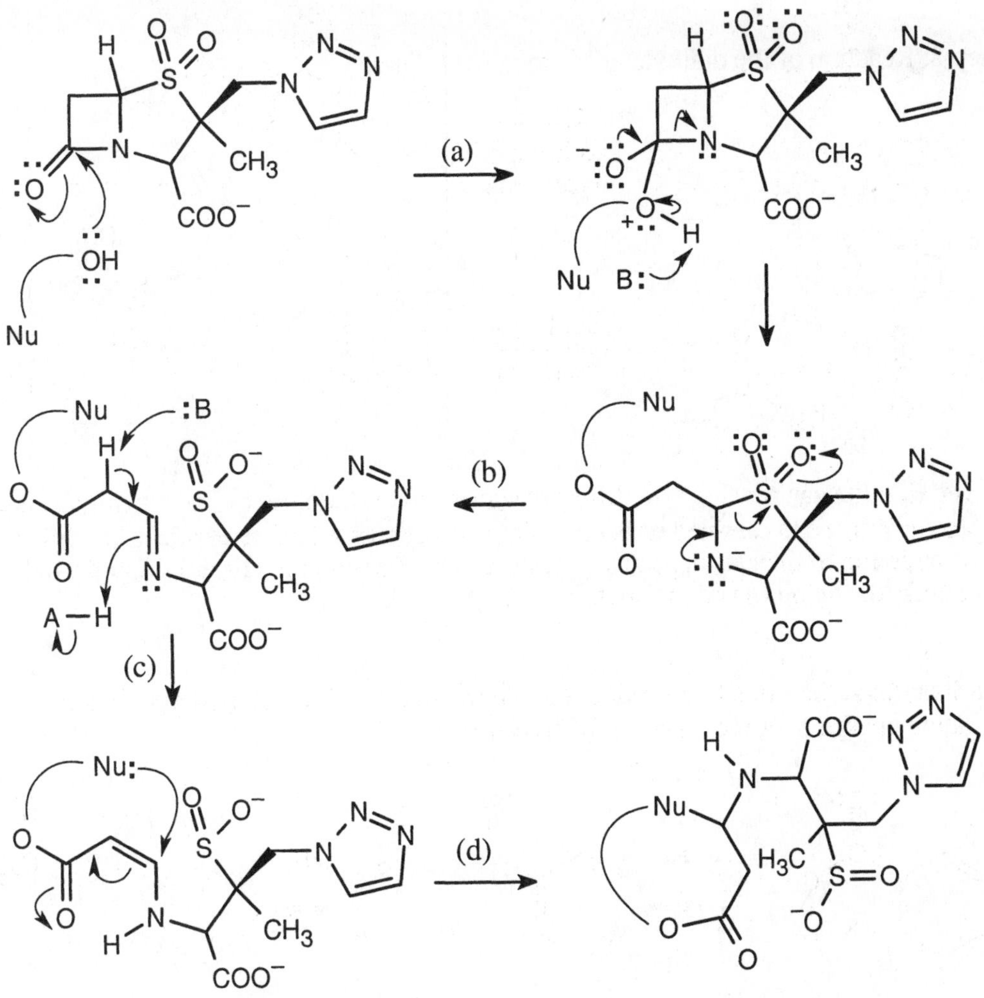

(a) (b) (c) (d)

Nu, OH, :B, A—H, Nu:, CH_3, COO^-, H_3C, S=O

11.57 Although amides are less reactive than esters, the lactam ring of Tazobactam reacts with a hydroxyl group because reaction removes the strain associated with a four membered ring.

11.58 The $-SO_2R$ group is a good leaving group because the two oxygens can stabilize a negative charge.

11.59

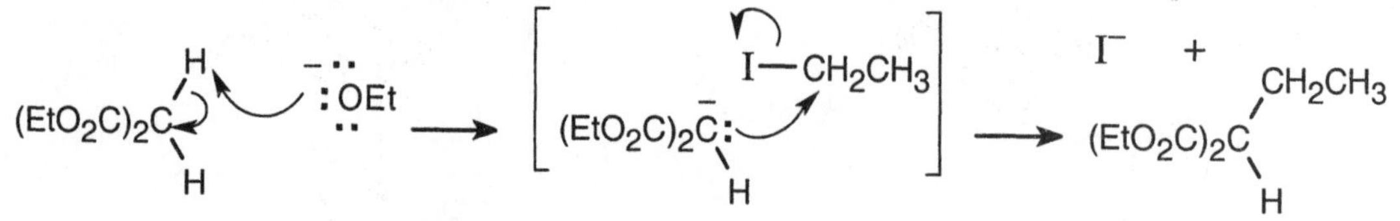

$(EtO_2C)_2CH_2$ + $:\ddot{O}Et^-$ ⟶ [$(EtO_2C)_2\bar{C}H$ + $I-CH_2CH_3$] ⟶ I^- + $(EtO_2C)_2CH(CH_2CH_3)$

11.60 Step 1: Alkylation with 2-bromopentane:

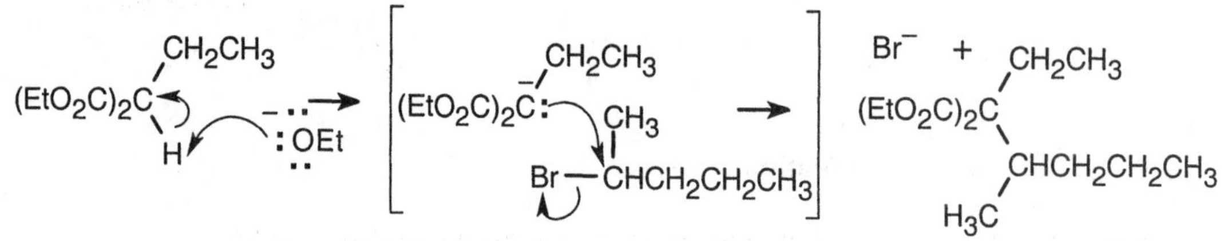

$(EtO_2C)_2CH(CH_2CH_3)$ + $:\ddot{O}Et^-$ ⟶ [$(EtO_2C)_2\bar{C}(CH_2CH_3)$ + $Br-CH(CH_3)CH_2CH_2CH_3$] ⟶ Br^- + $(EtO_2C)_2C(CH_2CH_3)(CH(CH_3)CH_2CH_2CH_3)$

Step 2: Reaction of the dialkylated malonate with urea:

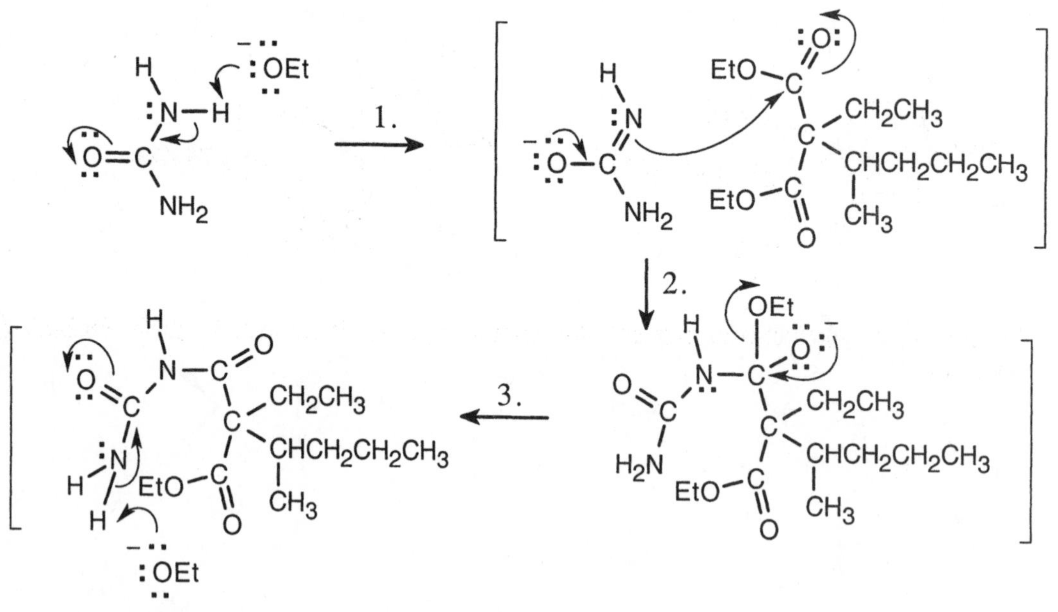

$H_2N-CO-NH_2$ + $:\ddot{O}Et^-$ —1.⟶ [$^-:\ddot{O}-C(=NH)NH_2$ + $(EtO-CO)_2C(CH_2CH_3)(CHCH_2CH_2CH_3)(CH_3)$] —2.⟶ [tetrahedral intermediate: $H_2N-C(=O)-NH-C(OEt)(O^-)-C(CH_2CH_3)(CHCH_2CH_2CH_3(CH_3))-C(=O)OEt$] —3.⟶ [$O=C(NH_2)-NH-C(=O)-C(CH_2CH_3)(CHCH_2CH_2CH_3(CH_3))-C(=O)OEt$ + $:\ddot{O}Et^-$]

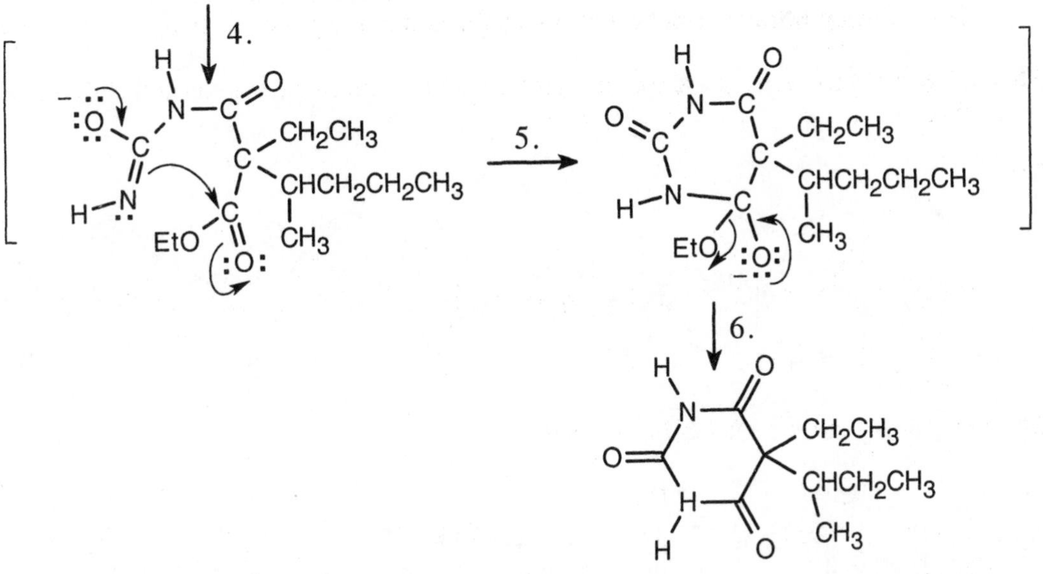

Steps 1 and 4: Deprotonation of urea.

Steps 2 and 5: Attack of urea nitrogen on the ester carbonyl group.

Steps 3 and 6: Loss of ⁻OEt.

11.61

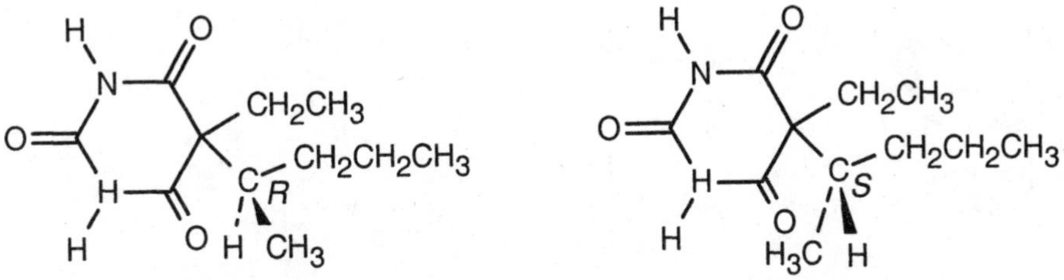

In the chiral cellular environment, the enantiomers are likely to have different biological activity.

In the Field with Agrochemicals

11.62 (a)

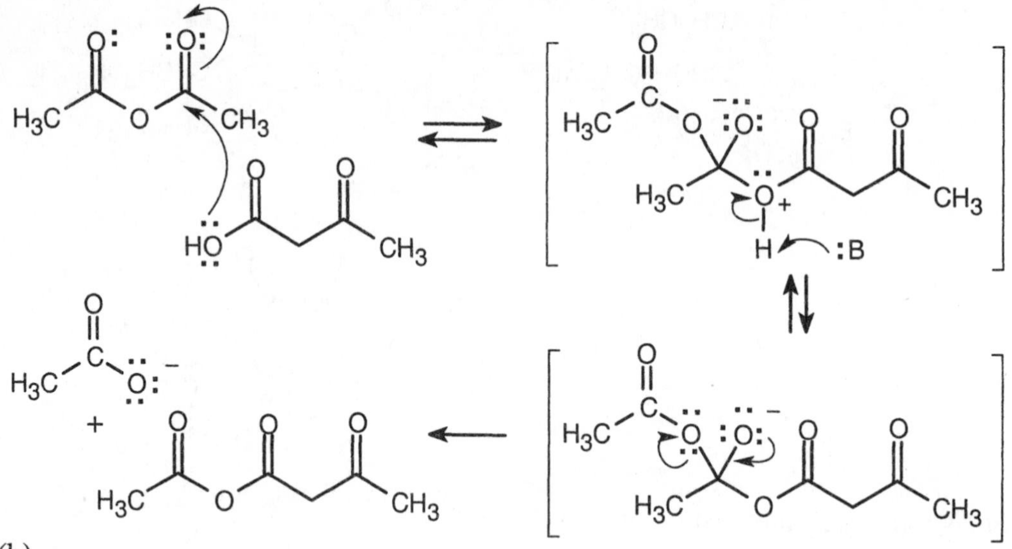

(b)

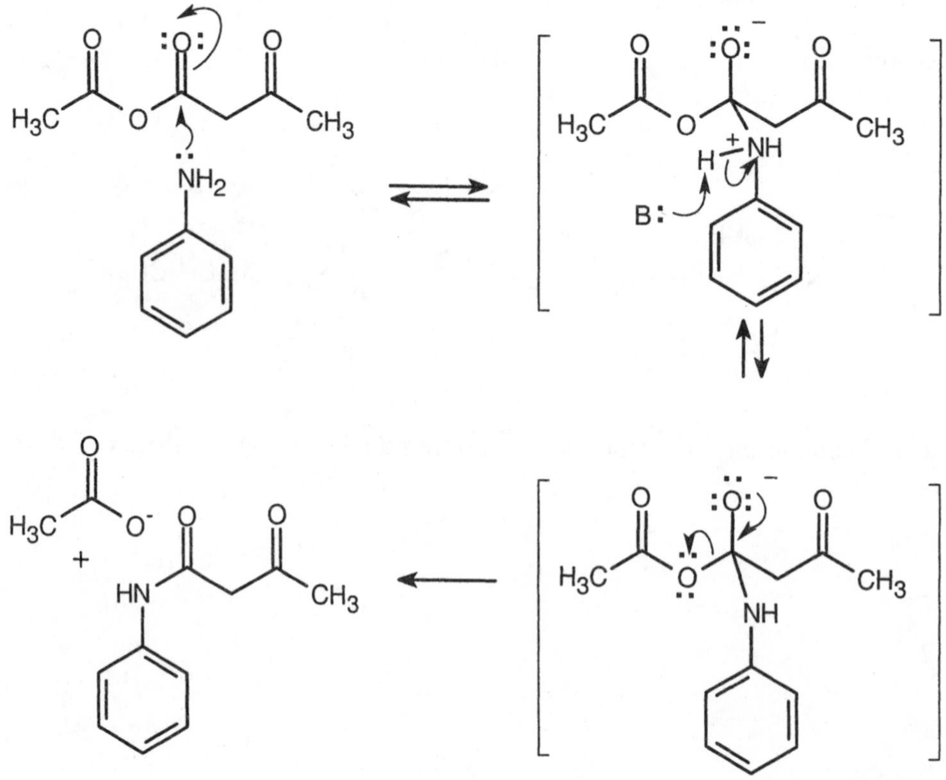

(c)

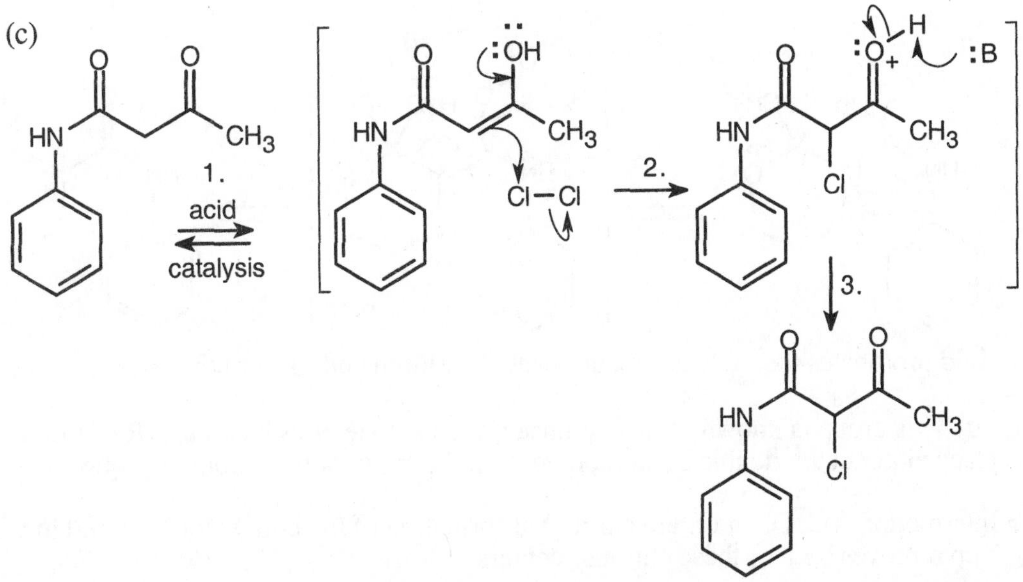

The steps: (1) acid-catalyzed enolization; (2) electrophilic addition of –Cl to the double bond; (3) deprotonation.

(d)

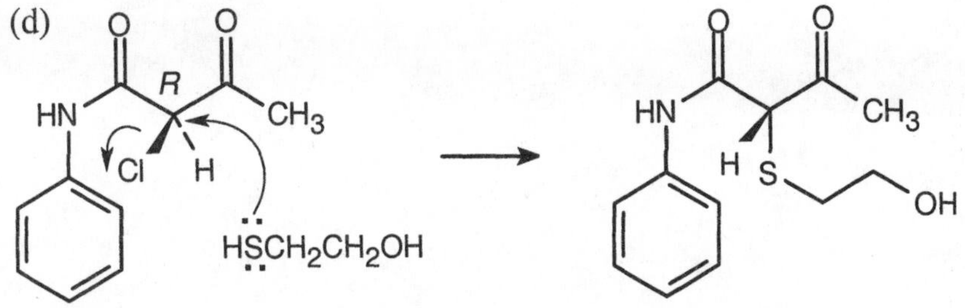

The results show that –SH is more nucleophilic than –OH.

(e)

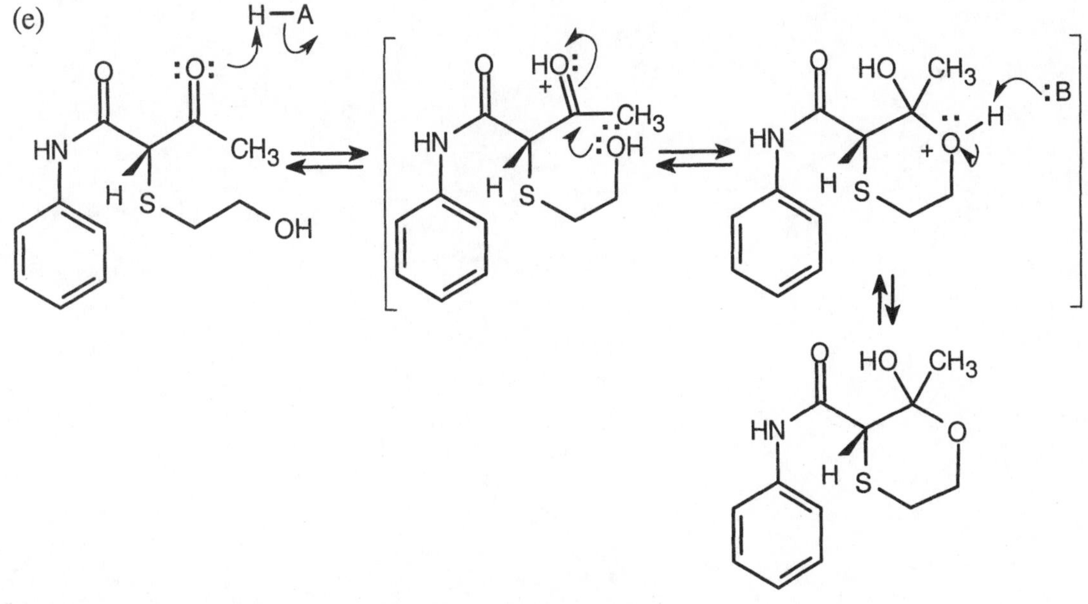

The reaction is an acid-catalyzed hemiacetal formation.

(f)

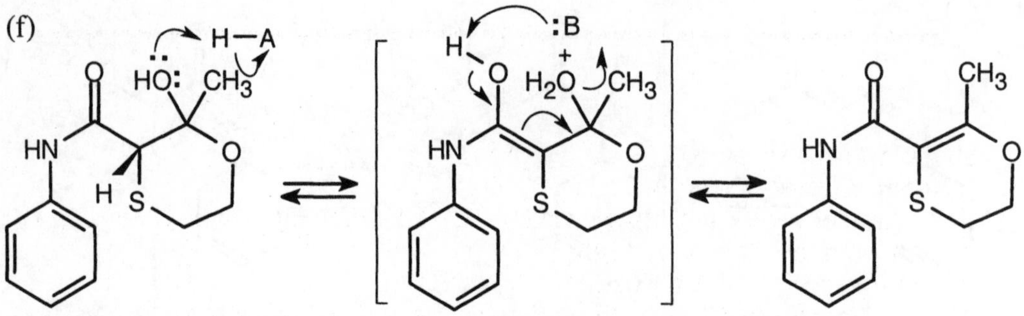

Acid protonates the –OH group, and water is eliminated to form the enone.

(g) This group is known as an α–β unsaturated ketone or as an enone. Reaction can take place either at the double bond (conjugate addition) or at the carbonyl group.

11.63 Intermediates C, D, and E are chiral, but formation of the enone double bond in the last step removes both of these chirality centers.

Chapter 12 – Amines

Chapter Outline

I. Facts about amines (Section 12.1 –12.3).
  A. Naming amines (Section 12.1).
    1. Amines are classified as primary (RNH_2), secondary (R_2NH), tertiary (R_3N) or quaternary ammonium salts (R_4N^+).
    2. Primary amines are named two ways:
      a. For simple amines, the suffix *-amine* is added to the name of the alkyl substituent.
      b. For more complicated amines, the $–NH_2$ group is an amino substituent on the parent molecule.
    3. Secondary and tertiary amines:
      a. Symmetrical amines are named by using the prefixes *di-* and *tri-* before the name of the alkyl group.
      b. Unsymmetrical amines are named as *N*-substituted primary amines. The largest group is the parent.
    4. The simplest arylamine is aniline.
    5. Heterocyclic amines (nitrogen is part of a ring) have specific parent names. Nitrogen receives the lowest possible number.
  B. Structure and properties of amines (Section 12.2).
    1. The three amine bonds and the lone pair occupy the corners of a tetrahedron.
    2. Amines with fewer than 5 carbons are water-soluble.
    3. Amines have higher boiling points than alkanes of similar molecular weight.
    4. Amines smell awful.
  C. Amine basicity (Section 12.3).
    1. The lone pair of electrons on nitrogen makes amines both nucleophilic and basic.
    2. The basicity constant K_b is the measure of the equilibrium of an amine with water. The larger the value of K_b (smaller pK_b), the stronger the base.
    3. Base strength.
      a. Alkylamines have similar basicities.
      b. Arylamines and heterocyclic aromatic amines are less basic than alkylamines.
        i. Pyridine lone pair electrons are in an sp^2 hybridized atomic orbital and are less available for bonding.
        ii. The pyrrole lone pair electrons are in a $2p$ atomic orbital and are part of the ring π system.
      c. Amides are nonbasic.
      d. Amine basicity can be used as a means of separating amines. An amine can be converted to its salt, extracted from a solution with water, neutralized, and re-extracted with an organic solvent.

II. Synthesis of amines (Section 12.4).
  A. Reduction of amides, nitriles and nitro groups.
    1. S_N2 displacement with ^-CN, followed by reduction, converts a primary alkyl halide into an amine with one more carbon atom.
    2. Amide reduction converts an amide or nitrile into an amine with the same number of carbons.
  B. S_N2 reactions of alkyl halides.
    It is possible to alkylate ammonia or an amine with RX. Unfortunately, it is difficult to avoid overalkylation.

C. Reductive amination of aldehydes and ketones.
   Treatment of an aldehyde or ketone with ammonia or an amine in the presence of a reducing agent yields an alkylated amine.

D. Reduction of nitrobenzenes.
   1. Catalytic hydrogenation can be used if no other interfering groups are present.
   2. Fe can also be used.

III. Reactions of amines (Section 12.5).
   Alkylation and acylation.
   1. Alkylation of primary and secondary amines is hard to control.
   2. Primary and secondary amines can also be acylated.

IV. Heterocyclic aromatic amines (Section 12.6).

A. Pyrrole.
   1. Although pyrrole is both an amine and a conjugated diene, its behavior is that of an aromatic ring.
   2. Pyrrole is nonbasic because all 5 of the nitrogen electrons are used in bonding.
   3. The carbon atoms in pyrrole are electron-rich and are reactive toward electrophiles.
      Substitution occurs at the position next to nitrogen.

B. Pyridine.
   1. Pyridine is the nitrogen-containing analog of benzene.
   2. The nitrogen lone pair isn't part of the π electron system .
   3. Pyridine is a stronger base than pyrrole but a weaker base than alkylamines.
   4. Pyridine is often used as a basic catalyst.

C. Fused-ring heterocycles.

V. Alkaloids (Section 24.7).

A. Alkaloids are naturally-occurring amines.

B. Many alkaloids have biological properties.

Solutions to Problems

12.1

(a) $(CH_3)_2CHNH_2$
primary amine

(b) $(CH_3CH_2)_2NH$
secondary amine

(c) cyclohexyl–$N(CH_3)_2$
tertiary amine

(d) $C_6H_5CH_2\overset{+}{N}(CH_3)_3\ I^-$
quaternary ammonium salt

12.2

(a) $CH_3CH(CH_3)NHCH_3$

(b) $C_6H_5N(CH_3)CH_2CH_3$

(c) cyclohexyl–$\overset{+}{N}(CH_3)(CH_2CH_3)(CH_2CH_2CH_3)\ Br^-$

12.3

(a) $CH_3NHCH_2CH_3$

N-Methylethylamine

(b) $(C_6H_{11})_3N$

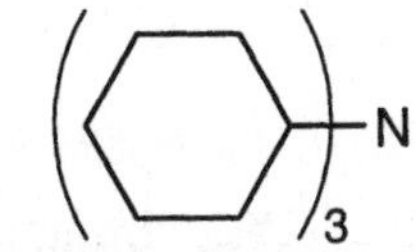

Tricyclohexylamine

(c) N—CH_3

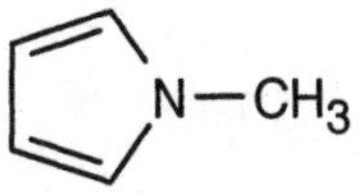

N-Methylpyrrole

(d) $C_6H_{11}N(CH_3)CH_2CH_2CH_3$

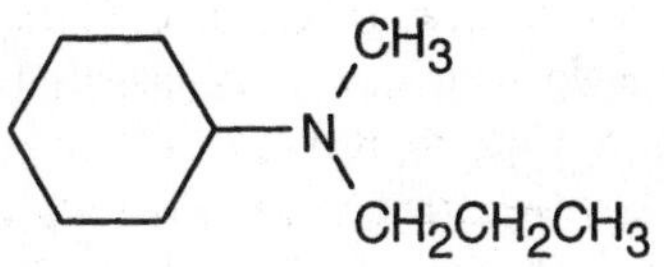

N-Methyl-*N*-propylcyclohexylamine

(e) $H_2NCH_2CH_2CH(CH_3)NH_2$

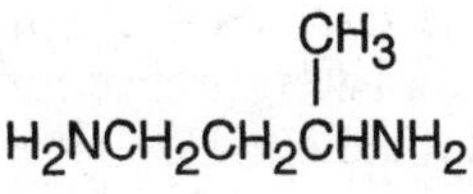

Butane-1,3-diamine

12.4

(a) $(CH_3CH_2)_3N$

Triethylamine

(b) $C_6H_5NHCH_3$

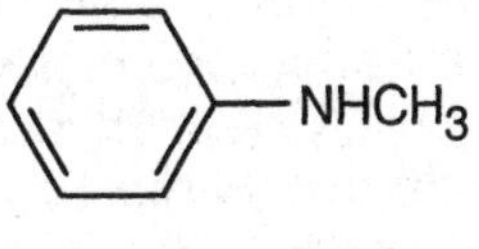

N-Methylaniline

(c) $(CH_3CH_2)_4N^+\ Br^-$

Tetraethylammonium bromide

(d) Br—C_6H_4—NH_2

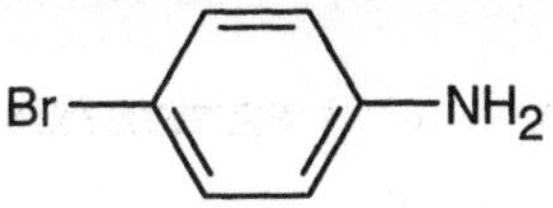

p-Bromoaniline

(e) $C_5H_9N(CH_3)CH_2CH_3$

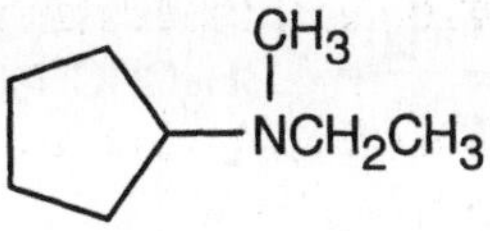

N-Ethyl-*N*-methylcyclopentylamine

12.5

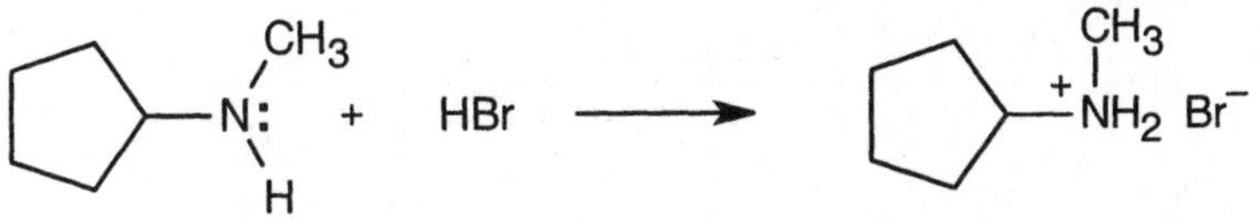

$$C_5H_9\ddot{N}(CH_3)H + HBr \longrightarrow C_5H_9\overset{+}{N}H_2(CH_3)\ Br^-$$

12.6 *More basic* | *Less basic*

(a) $CH_3CH_2NH_2$ amine | $CH_3CH_2CONH_2$ amide

(b) NaOH hydroxide | $C_6H_5NH_2$ arylamine

(c) CH_3NHCH_3 alkylamine | $CH_3NHC_6H_5$ arylamine

(d) $(CH_3)_3N$ amine | CH_3OCH_3 ether

The basicity order for the above compounds:
hydroxide > alkylamine > arylamine > amide, ether

12.7

(a)

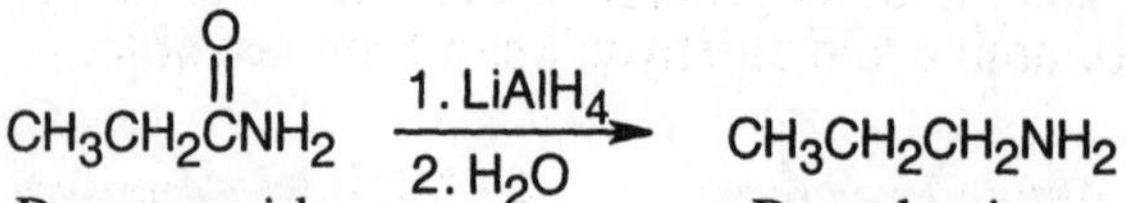

$$CH_3CH_2\overset{O}{\overset{\|}{C}}NH_2 \xrightarrow[2.\ H_2O]{1.\ LiAlH_4} CH_3CH_2CH_2NH_2$$

Propanamide → Propylamine

(b)

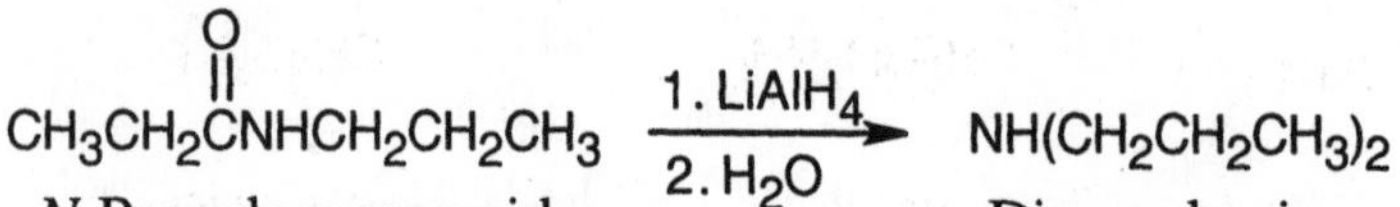

$$CH_3CH_2\overset{O}{\overset{\|}{C}}NHCH_2CH_2CH_3 \xrightarrow[2.\ H_2O]{1.\ LiAlH_4} NH(CH_2CH_2CH_3)_2$$

N-Propylpropanamide → Dipropylamine

(c)

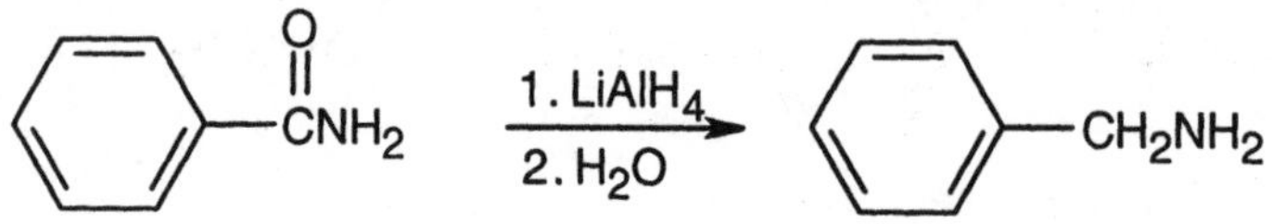

$$C_6H_5\overset{O}{\overset{\|}{C}}NH_2 \xrightarrow[2.\ H_2O]{1.\ LiAlH_4} C_6H_5CH_2NH_2$$

Benzamide → Benzylamine

12.8

(a)

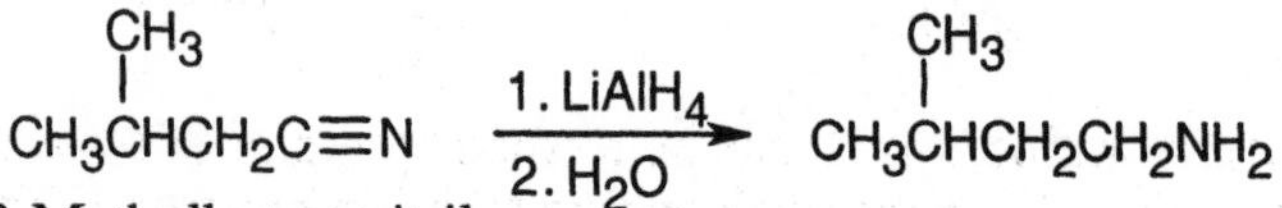

$$CH_3\overset{CH_3}{\overset{|}{C}}HCH_2C\equiv N \xrightarrow[2.\ H_2O]{1.\ LiAlH_4} CH_3\overset{CH_3}{\overset{|}{C}}HCH_2CH_2NH_2$$

3-Methylbutanenitrile

(b)

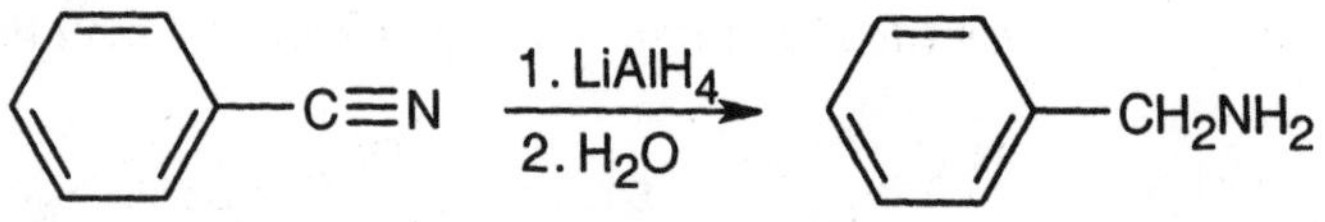

$$C_6H_5-C\equiv N \xrightarrow[2.\ H_2O]{1.\ LiAlH_4} C_6H_5-CH_2NH_2$$

Benzonitrile

12.9

(a)

$$3\ CH_3CH_2Br \xrightarrow{NH_3} (CH_3CH_2)_3N$$

Some quaternary ammonium salt is also formed.

(b)

$$4\ CH_3Br \xrightarrow{NH_3} (CH_3)_4N^+\ Br^-$$

12.10

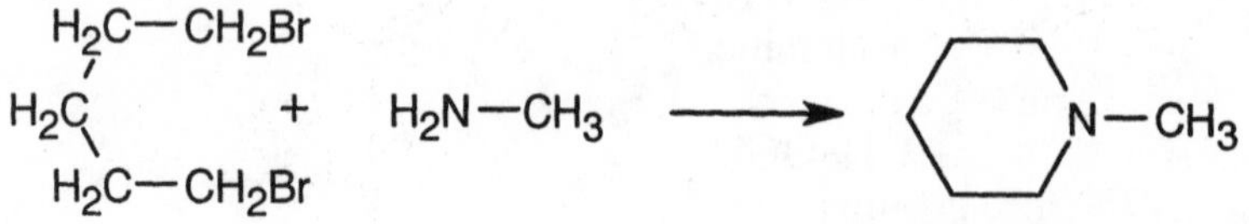

$$H_2C(CH_2CH_2Br)_2 + H_2N-CH_3 \longrightarrow \text{(piperidine ring)}N-CH_3$$

12.11 Look at the target molecule to find the groups bonded to nitrogen. One group comes from the aldehyde/ketone precursor, and the other group comes from the amine precursor. In most cases, two combinations of amine and aldehyde/ketone are possible.

	Amine	*Amine Precursor*		*Carbonyl Precursor*
(a)	$CH_3CH_2NHCH(CH_3)_2$	$CH_3CH_2NH_2$		$CH_3C(=O)CH_3$
			or	
		$H_2NCH(CH_3)_2$		CH_3CHO
(b)	C_6H_5–$NHCH_2CH_3$	C_6H_5–NH_2		$OHCCH_3$
(c)	cyclopentyl–$NHCH_3$	cyclopentyl–NH_2		$O{=}CH_2$
			or	
		H_2NCH_3		cyclopentanone (cyclopentyl=O)

12.12 The target molecule is a tertiary amine, which can be formed by either of two combinations of aldehyde plus secondary amine. After the nucleophilic addition reaction of the two components, the intermediate imine is reduced by catalytic hydrogenation, using nickel as a catalyst.

Amine	*Amine Precursor*		*Carbonyl Precursor*
$(H_3C)_2N$–CH_2–(m-C_6H_4)–CH_3	H_3C–N(H)–CH_3		OHC–(m-C_6H_4)–CH_3
		or	
	H–N(CH_3)–CH_2–(m-C_6H_4)–CH_3		$H_2C{=}O$

12.13

(a) Benzene $\xrightarrow[AlCl_3]{CH_3Cl}$ toluene ($C_6H_5CH_3$) $\xrightarrow[H_2O]{KMnO_4}$ benzoic acid (C_6H_5COOH) $\xrightarrow[H_2SO_4]{HNO_3}$ O_2N–(m-C_6H_4)–COOH $\xrightarrow[\text{Pt catalyst}]{H_2}$ H_2N–(m-C_6H_4)–COOH

m-Aminobenzoic acid

(b)

Benzene $\xrightarrow[H_2SO_4]{HNO_3}$ NO_2-benzene $\xrightarrow[\text{Pt catalyst}]{H_2}$ NH_2-benzene $\xrightarrow{3\ Br_2}$ 2,4,6-Tribromoaniline

Bromination of the reactive aniline ring requires no catalyst.

12.14

(a)

N−CH_3 (pyrrole) $\xrightarrow{Br_2}$ 2-Br, N−CH_3

(b)

N−CH_3 $\xrightarrow[AlCl_3]{CH_3Cl}$ 2-CH_3, N−CH_3

(c)

N−CH_3 $\xrightarrow[AlCl_3]{CH_3COCl}$ 2-$COCH_3$, N−CH_3

12.15

Pyrrole (N−H) $\xrightarrow{+NO_2}$ [H, NO_2 ↔ H, NO_2 ↔ H, NO_2 (N^+)] → 2-NO_2 pyrrole (N−H) + H^+

12.16

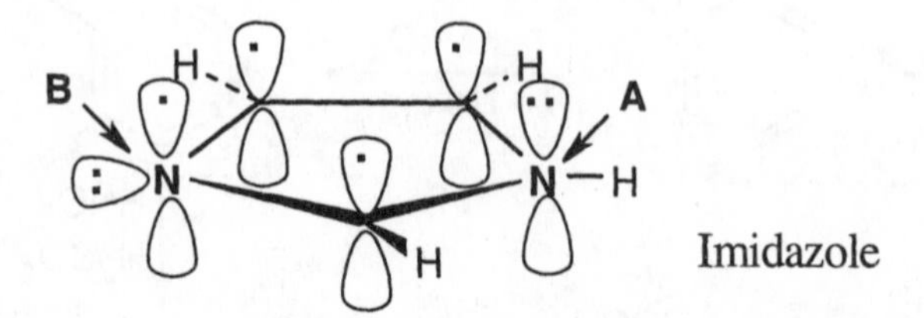

Imidazole

Nitrogen atom **B** is more "pyridine-like" because its lone pair of electrons is in an sp^2 orbital that lies in the plane of the ring. Nitrogen **A**, whose lone pair of electrons is in a *p* orbital that is part of the ring π system, is "pyrrole-like."

12.17 The electrostatic potential map shows that the nitrogen on the left (corresponding to **B** in the previous problem) is more basic. This makes sense because the lone pair of **A** is part of the ring π system and is not available for donation to a Lewis acid.

Visualizing Chemistry

12.18

(a) 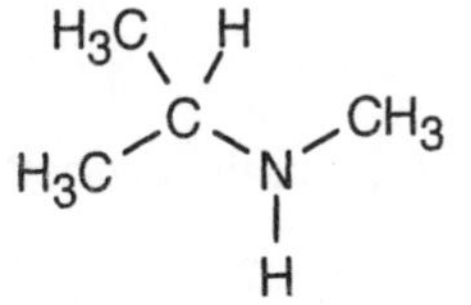 *N*-Methylisopropylamine
secondary amine

(b) 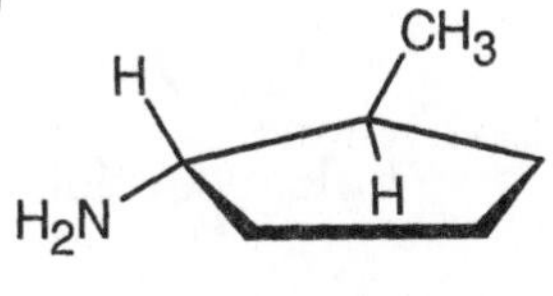 *trans*-(2-Methylcyclopentyl)amine
primary amine

(c) 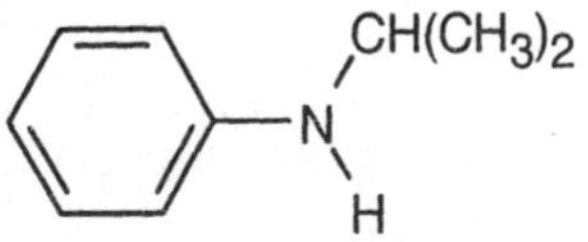 *N*-Isopropylaniline
secondary amine

12.19

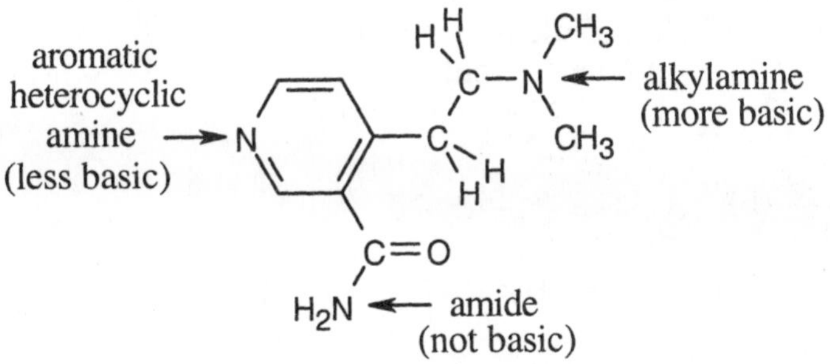

aromatic heterocyclic amine (less basic)

alkylamine (more basic)

amide (not basic)

12.20

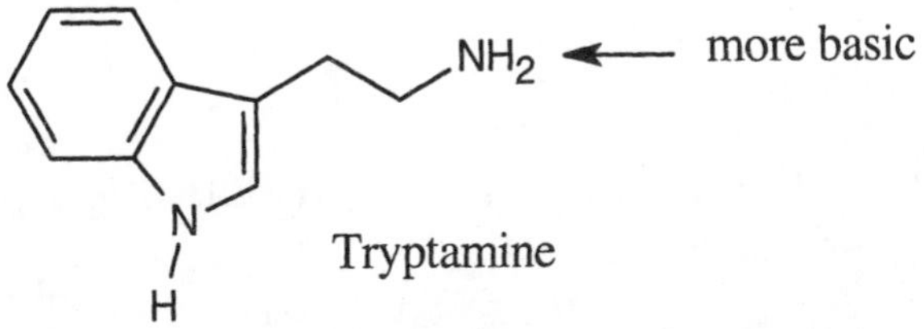

more basic

Tryptamine

The electrostatic potential map shows that the indicated nitrogen is more basic because it is more electron-rich. The electrons of the other nitrogen are part of the fused-ring π system and are less available for donation to a Lewis acid.

12.21

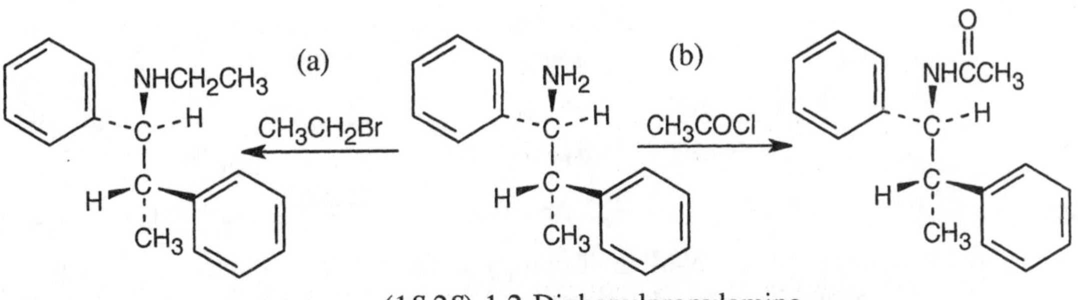

(a) CH_3CH_2Br

(b) CH_3COCl

(1*S*,2*S*)-1,2-Diphenylpropylamine

In reaction (a), polyalkylated product will also form.

Additional Problems

Nomenclature

12.22

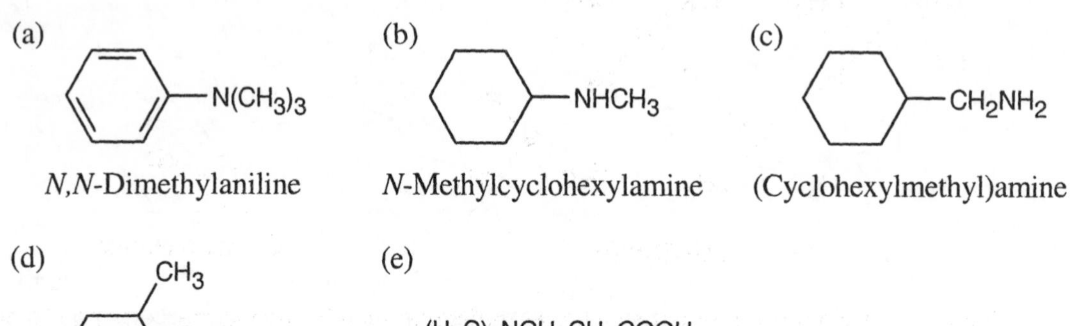

(a) *N,N*-Dimethylaniline (b) *N*-Methylcyclohexylamine (c) (Cyclohexylmethyl)amine

(d) (2-Methylcyclohexyl)amine (e) 3-(*N,N*-dimethylamino)-propanoic acid

12.23

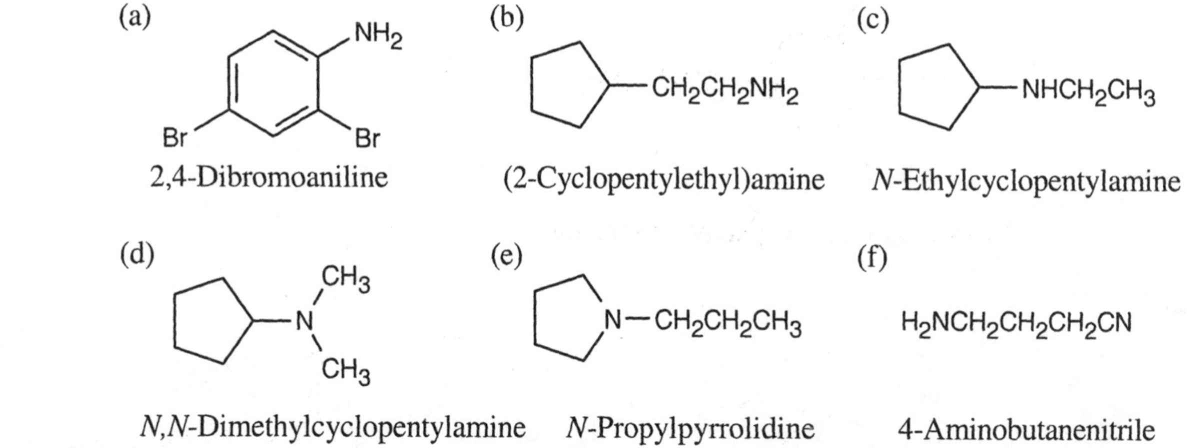

(a) 2,4-Dibromoaniline (b) (2-Cyclopentylethyl)amine (c) *N*-Ethylcyclopentylamine

(d) *N,N*-Dimethylcyclopentylamine (e) *N*-Propylpyrrolidine (f) 4-Aminobutanenitrile

12.24

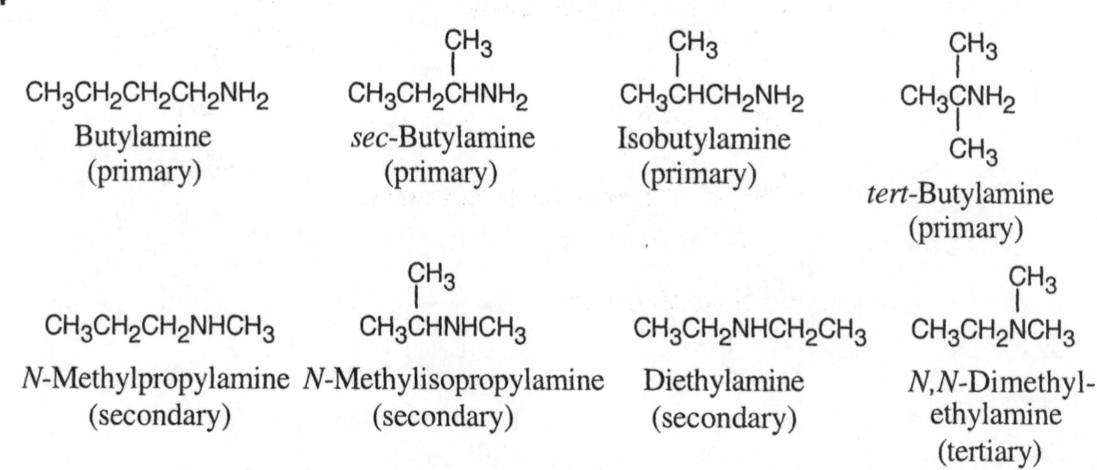

Butylamine (primary) — *sec*-Butylamine (primary) — Isobutylamine (primary) — *tert*-Butylamine (primary)

N-Methylpropylamine (secondary) — *N*-Methylisopropylamine (secondary) — Diethylamine (secondary) — *N,N*-Dimethylethylamine (tertiary)

Physical Properties

12.25

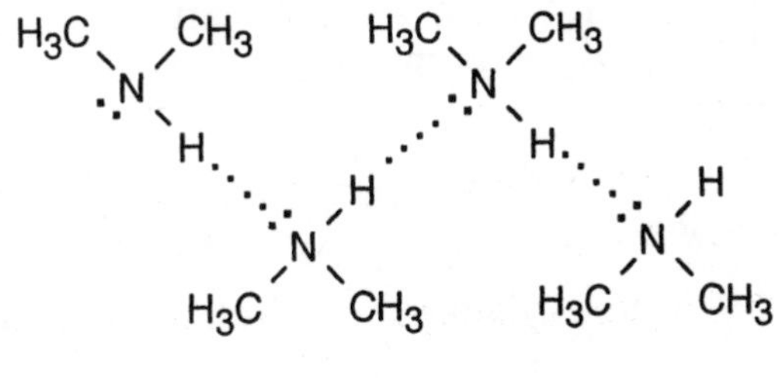

$(CH_3)_2NH$ molecules hydrogen-bonded: H_3C, CH_3, N, H ···· N

Dimethylamine

$(CH_3)_3N$:

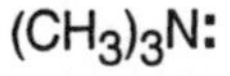

Trimethylamine

Even though dimethylamine has a lower molecular weight than trimethylamine, it boils at a higher temperature. Liquid dimethylamine forms hydrogen bonds that must be broken in the boiling process. Since extra energy must be added to break these hydrogen bonds, dimethylamine has a higher boiling point than trimethylamine, which does not form hydrogen bonds.

12.26

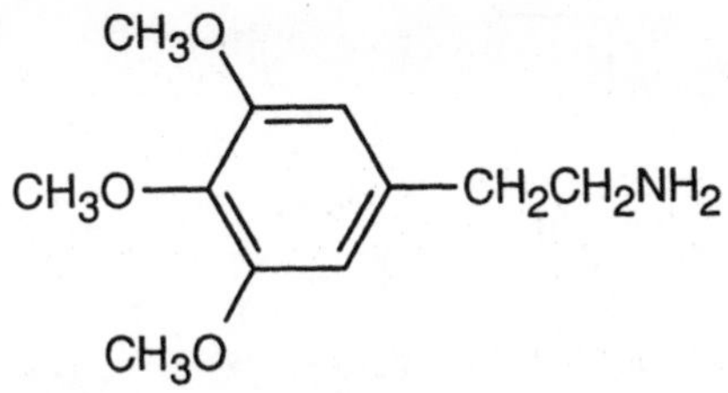

CH_3O, CH_3O, CH_3O — $CH_2CH_2NH_2$

2-(3,4,5-Trimethoxyphenyl)ethylamine

12.27

(a)

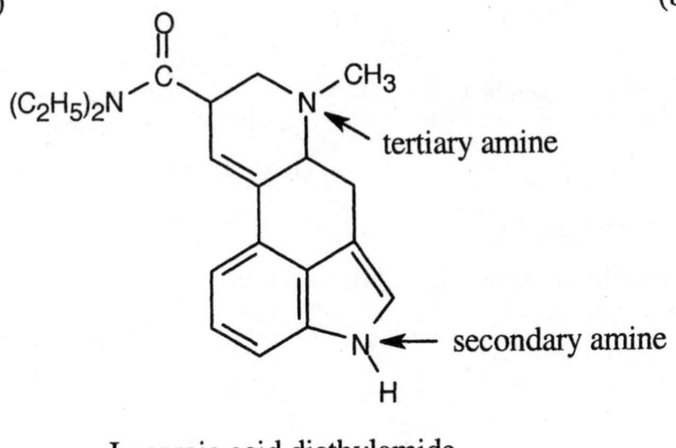

$(C_2H_5)_2N$—C(=O)— ... N—CH_3 ← tertiary amine; N—H ← secondary amine

Lysergic acid diethylamide

(b)

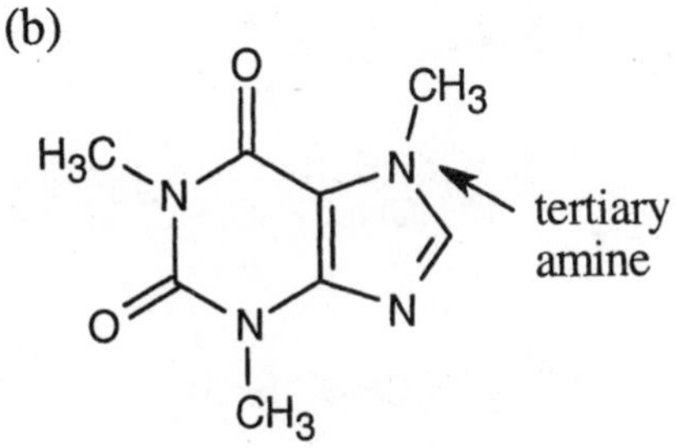

O, H_3C, N, O, N, CH_3, N (CH_3) ← tertiary amine, N

Caffeine

12.28

(a) $NHCH_3$

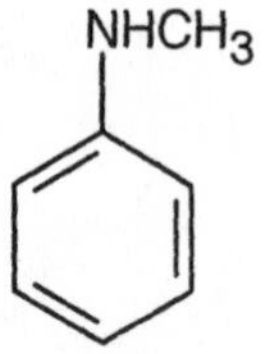

N-Methylaniline

(b) NH_2, H_3C, CH_3

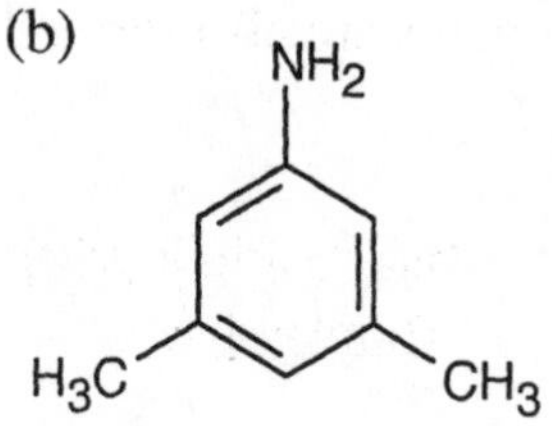

3,5-Dimethylaniline

(c) $(CH_3)_4N^+\ Br^-$

Tetramethylammonium bromide

(d) N–H

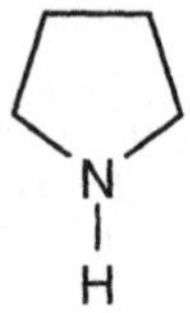

Pyrrolidine

Reactions and Synthesis

12.29

(a)

$$CH_3CH_2CH_2NH_2 + CH_3Br \longrightarrow CH_3CH_2CH_2NHCH_3$$

Polyalkylated products are also formed.

(b)

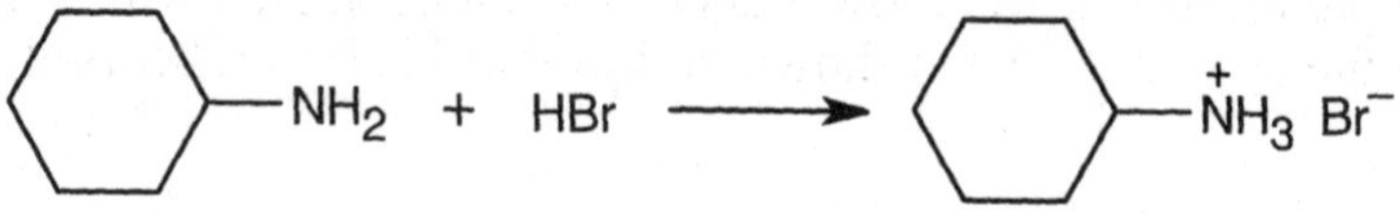

$$C_6H_{11}\text{–}NH_2 + HBr \longrightarrow C_6H_{11}\text{–}\overset{+}{N}H_3\ Br^-$$

(c)

$$CH_3CH_2\overset{O}{\overset{\|}{C}}NH_2 \xrightarrow[2.\ H_2O]{1.\ LiAlH_4} CH_3CH_2CH_2NH_2$$

(d)

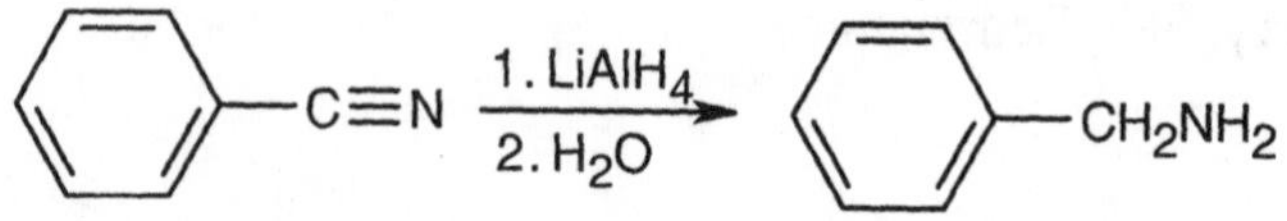

$$C_6H_5\text{–}C{\equiv}N \xrightarrow[2.\ H_2O]{1.\ LiAlH_4} C_6H_5\text{–}CH_2NH_2$$

12.30 In these reactions, polyalkylated products are also formed.

(a)

$$CH_3CH_2CH_2CH_2CH_2CH_2Br \xrightarrow{\text{excess } NH_3} CH_3CH_2CH_2CH_2CH_2CH_2NH_2$$

Hexylamine

(b)

$$4\ CH_3I + NH_3 \longrightarrow (CH_3)_4N^+\ I^-$$

Tetramethylammonium iodide

(c)

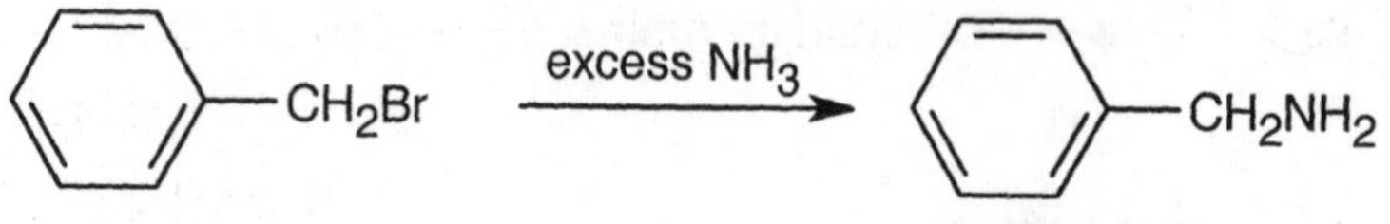

$$C_6H_5\text{–}CH_2Br \xrightarrow{\text{excess } NH_3} C_6H_5\text{–}CH_2NH_2$$

Benzylamine

(d)

$$C_6H_{11}\text{–}Br \xrightarrow{\text{excess } NH_3} C_6H_{11}\text{–}NH_2 \xrightarrow{CH_3Br} C_6H_{11}\text{–}NHCH_3$$

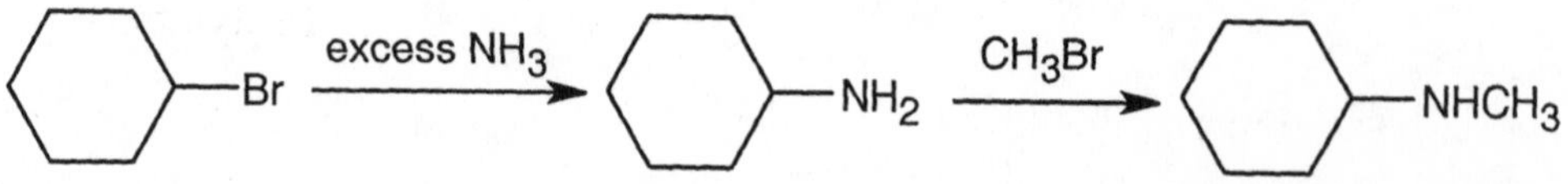

N-Methylcyclohexylamine

12.31 In parts (a) and (b), overalkylation products are also formed.

(a)

$$CH_3CH_2CH_2CH_2Br \xrightarrow{\text{excess } NH_3} \underset{\text{Butylamine}}{CH_3CH_2CH_2CH_2NH_2}$$

(b)

$$CH_3CH_2CH_2CH_2Br + \underset{\text{from (a)}}{CH_3CH_2CH_2CH_2NH_2} \longrightarrow \underset{\text{Dibutylamine}}{(CH_3CH_2CH_2CH_2)_2NH}$$

(c)

$$CH_3CH_2CH_2CH_2Br \xrightarrow{NaCN} CH_3CH_2CH_2CH_2CN \xrightarrow[2.\ H_2O]{1.\ LiAlH_4} \underset{\text{Pentylamine}}{CH_3CH_2CH_2CH_2CH_2NH_2}$$

12.32

In addition to the preparations shown below, butan-1-ol can be treated with PBr_3 to form 1-bromobutane, and the synthetic routes shown in Problem 12.31 can be used.

(a)

$$CH_3CH_2CH_2CH_2OH \xrightarrow[H_3O^+]{CrO_3} CH_3CH_2CH_2\overset{O}{\overset{\|}{C}}OH \xrightarrow{SOCl_2} CH_3CH_2CH_2\overset{O}{\overset{\|}{C}}Cl$$

$$CH_3CH_2CH_2\overset{O}{\overset{\|}{C}}Cl \xrightarrow{2\ NH_3} CH_3CH_2CH_2\overset{O}{\overset{\|}{C}}NH_2 \xrightarrow[2.\ H_2O]{1.\ LiAlH_4} \underset{\text{Butylamine}}{CH_3CH_2CH_2CH_2NH_2}$$

(b)

$$\underset{\text{from (a)}}{CH_3CH_2CH_2\overset{O}{\overset{\|}{C}}Cl} + \underset{\text{from (a)}}{CH_3CH_2CH_2CH_2NH_2} \longrightarrow CH_2CH_2CH_2\overset{O}{\overset{\|}{C}}NHCH_2CH_2CH_2CH_3$$

$$\xrightarrow[2.\ H_2O]{1.\ LiAlH_4} \underset{\text{Dibutylamine}}{(CH_3CH_2CH_2CH_2)_2NH}$$

(c)

$$CH_3CH_2CH_2CH_2OH \xrightarrow{PBr_3} CH_3CH_2CH_2CH_2Br \xrightarrow{NaCN} CH_3CH_2CH_2CH_2CN$$

$$\xrightarrow[2.\ H_2O]{1.\ LiAlH_4} \underset{\text{Pentylamine}}{CH_3CH_2CH_2CH_2CH_2NH_2}$$

12.33

(a)

$$C_6H_5CONH_2 \xrightarrow[2.\ H_2O]{1.\ LiAlH_4} C_6H_5CH_2NH_2$$

(b)

$$C_6H_5COH \xrightarrow{SOCl_2} C_6H_5CCl \xrightarrow{2\ NH_3} C_6H_5CNH_2 \xrightarrow[2.\ H_2O]{1.\ LiAlH_4} C_6H_5CH_2NH_2$$

(c)

$$C_6H_6 \xrightarrow[AlCl_3]{CH_3Cl} C_6H_5CH_3 \xrightarrow[H_2O]{KMnO_4} C_6H_5COH$$

..then proceed as in part (b)

12.34

(a)

$$CH_3CH_2CH_2CH_2CNH_2 \xrightarrow[2.\ H_2O]{1.\ LiAlH_4} CH_3CH_2CH_2CH_2CH_2NH_2$$

(b)

$$CH_3CH_2CH_2CH_2CN \xrightarrow[2.\ H_2O]{1.\ LiAlH_4} CH_3CH_2CH_2CH_2CH_2NH_2$$

(c)

$$CH_3CH_2CH_2CH_2COH \xrightarrow{SOCl_2} CH_3CH_2CH_2CH_2CCl \xrightarrow{2\ NH_3} CH_3CH_2CH_2CH_2CNH_2$$

$$\downarrow \text{1. } LiAlH_4 \text{ 2. } H_2O$$

$$CH_3CH_2CH_2CH_2CH_2NH_2$$

Basicity

12.35 $CH_3CH_2NH_2$ is more basic than $CF_3CH_2NH_2$ because the electron-withdrawing fluorine atoms make the nitrogen of $CF_3CH_2NH_2$ more electron-poor and less basic.

12.36 The aldehyde group, which is electron-withdrawing, makes *p*-aminobenzaldehyde less basic than aniline.

12.37 Triethylamine is more basic than aniline. The lone-pair electrons of the aniline nitrogen are delocalized by orbital overlap with the aromatic ring π electrons and are less available for donation to an acid. Thus, the reaction of triethylammonium chloride with aniline does not proceed as written.

Integrated problems

12.38

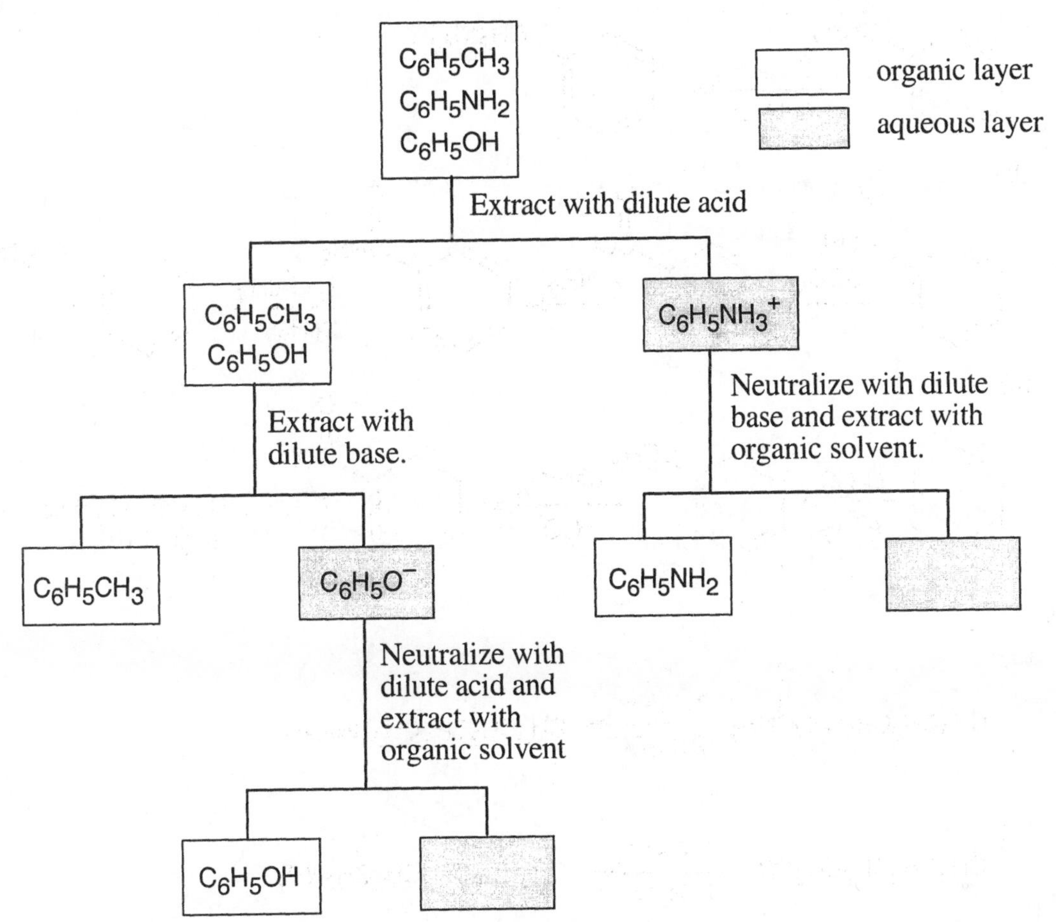

12.39 Diphenylamine is less basic than aniline. The nitrogen lone-pair electrons of diphenylamine can overlap with the π electron system of both aromatic rings, causing even greater electron delocalization than occurs for aniline.

12.40 The first step in this sequence is a 1,4 addition to a conjugated diene, as we saw in Chapter 4. An S_N2 displacement of chlorine by cyanide yields a dinitrile, which is reduced to the diamine.

$$H_2C{=}CHCH{=}CH_2 \xrightarrow{Cl_2} ClCH_2CH{=}CHCH_2Cl \xrightarrow{2\ NaCN} N{\equiv}CCH_2CH{=}CHCH_2C{\equiv}N$$

$$N{\equiv}CCH_2CH{=}CHCH_2C{\equiv}N \xrightarrow{1.\ LiAlH_4;\ 2.\ H_2O} H_2NCH_2CH_2CH{=}CHCH_2CH_2NH_2$$

$$H_2NCH_2CH_2CH_2CH_2CH_2CH_2NH_2 \xleftarrow[\text{Pt catalyst}]{H_2} H_2NCH_2CH_2CH{=}CHCH_2CH_2NH_2$$

Hexane-1,4-diamine

12.41

$$\mathrm{HO\overset{O}{\overset{\|}{C}}CH_2CH_2CH_2CH_2\overset{O}{\overset{\|}{C}}OH \xrightarrow{SOCl_2} Cl\overset{O}{\overset{\|}{C}}CH_2CH_2CH_2CH_2\overset{O}{\overset{\|}{C}}Cl}$$

$$\downarrow 4\ \mathrm{NH_3}$$

$$\mathrm{H_2NCH_2CH_2CH_2CH_2CH_2CH_2NH_2 \xleftarrow[2.\ H_2O]{1.\ LiAlH_4} H_2N\overset{O}{\overset{\|}{C}}CH_2CH_2CH_2CH_2\overset{O}{\overset{\|}{C}}NH_2}$$

1,6-Hexanediamine

12.42

(a) H_3C, NH_2 (3-methylaniline) $\xrightarrow{Br_2}$ H_3C, NH_2, Br + H_3C, NH_2, Br

(b) H_3C, NH_2 $\xrightarrow[\text{excess}]{CH_3I}$ H_3C, $\overset{+}{N}(CH_3)_3I^-$

(c) H_3C, NH_2 $\xrightarrow[\text{pyridine}]{CH_3COCl}$ H_3C, $NH\overset{O}{\overset{\|}{C}}CH_3$

12.43

(a) N, CH_2CH_3

2-Ethylpyrrole

(b) CH_3, H_3C, NH_2

2,3-Dimethylaniline

(c) CH_3, N–H

3-Methylindole

12.44 Furan, the oxygen analog of pyrrole, is aromatic because it has 6 π electrons in a cyclic, conjugated system. Oxygen contributes two lone-pair electrons from a *p* orbital perpendicular to the plane of the ring. The other oxygen lone pair is in an sp^2 orbital that lies in the plane of the furan ring.

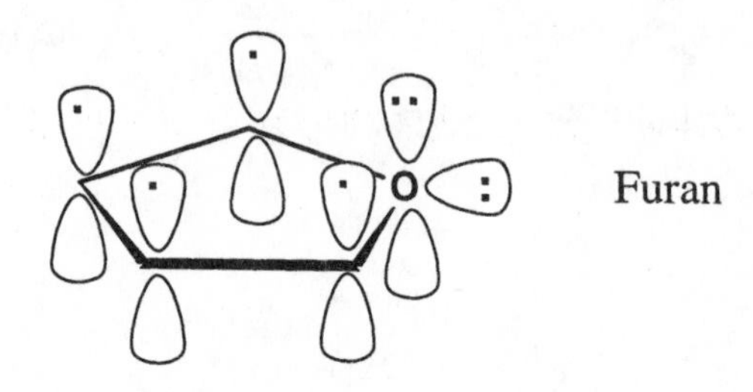

Furan

12.45

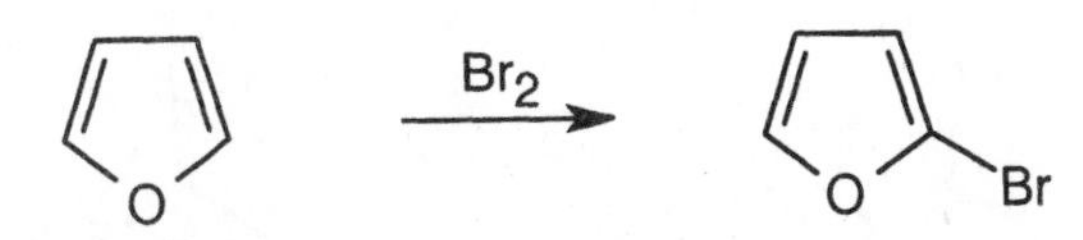

12.46 In this chapter, we learned that the lone pair electrons of an aromatic amine nitrogen are delocalized over the ring, decreasing basicity. In Chapter 5 we learned that an $-NO_2$ group withdraws electrons from an aromatic ring. Thus, a nitro group further decreases aromatic amine basicity by making the amine nitrogen less electron-rich and less reactive toward electrophiles and acids.

12.47 Just as an electron-withdrawing group decreases basicity, an electron-donating group increases basicity by making the amine nitrogen more electron-rich and more reactive toward electrophiles and Lewis acids. Thus, *p*-methoxyaniline is more basic than aniline because its methoxyl group is electron-donating.

12.48

Three resonance forms contribute to the stability of the imide anion.

12.49

Atropine $\xrightarrow[H_2O]{^-OH}$ Tropine + $HOOCCH(CH_2OH)C_6H_5$ Tropic acid

Tropine $\xrightarrow{H_2SO_4}$ Tropidene + H_2O

We know the location of the –OH group of tropine because it is stated that tropine is an optically inactive alcohol. This hydroxyl group results from basic hydrolysis of the ester that is composed of tropine and tropic acid.

12.50 The reaction of trimethylamine with ethylene oxide is an S_N2 reaction.

$(CH_3)_3N{:} + H_2C{-}CH_2 \text{ (ethylene oxide)} \longrightarrow (CH_3)_3\overset{+}{N}CH_2CH_2O^- \xrightarrow{H_2O} (CH_3)_3\overset{+}{N}CH_2CH_2OH + {}^-OH$

Choline

12.51

$:NH_2CH_3$ — conjugate addition of amine → [enolate: OCH_3, $\overset{+}{N}H_2CH_3$ ⇌ proton transfer ⇌ OCH_3, $\ddot{N}HCH_3$]

⇅ nucleophilic addition of amine

[$:\ddot{O}:^-$ OCH_3, $\overset{+}{N}H-CH_3$ ⇌ proton transfer ⇌ $:\ddot{O}:^-$ H^+ $:OCH_3$, $N-CH_3$] ⇌ elimination of methanol ⇌ $N-CH_3$ + CH_3OH

12.52

a: C_6H_5CHO —(1. $NaBH_4$; 2. H_3O^+)→ $C_6H_5CH_2OH$

b: $C_6H_5CH_2OH$ —(PBr_3)→ $C_6H_5CH_2Br$

c: —(2 NH_3)→ $C_6H_5CH_2NH_2$

d: —($CH_3CO_2COCH_3$)→ $C_6H_5CH_2NHCOCH_3$

e: —(HNO_3, H_2SO_4)→ $O_2N-C_6H_4-CH_2NHCOCH_3$

f: —(1. Fe, H_3O^+; 2. NaOH)→ $H_2N-C_6H_4-CH_2NHCOCH_3$

In the Medicine Cabinet

12.53 - 12.54

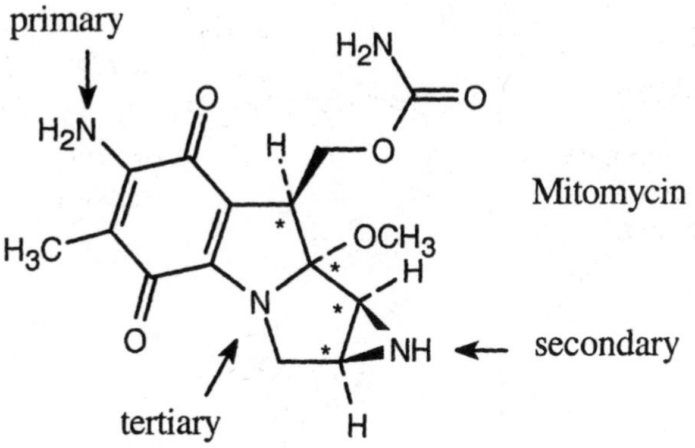

primary

H_2N O H H_2N O O

Mitomycin

H_3C OCH_3 H

O N NH ← secondary

tertiary H

12.55, 12.57

H_2N O H_2N O H O H—A H_3C $\ddot{O}CH_3$ O N NH

1. →

H_2N O H_2N O H O H_+ H_3C $\ddot{O}CH_3$ O N NH

2.

H_2N O H_2N O O H_3C O N $\ddot{N}H$ H—A

3. ←

:B H_2N O H_2N O H O H_3C O N + NH

4.

H_2N O H_2N O O H_3C O N $H_2\ddot{N}$—DNA $\overset{+}{N}H_2$

5. →

H_2N O H_2N O O :B H H H_3C N DNA + O N NH_2

6.

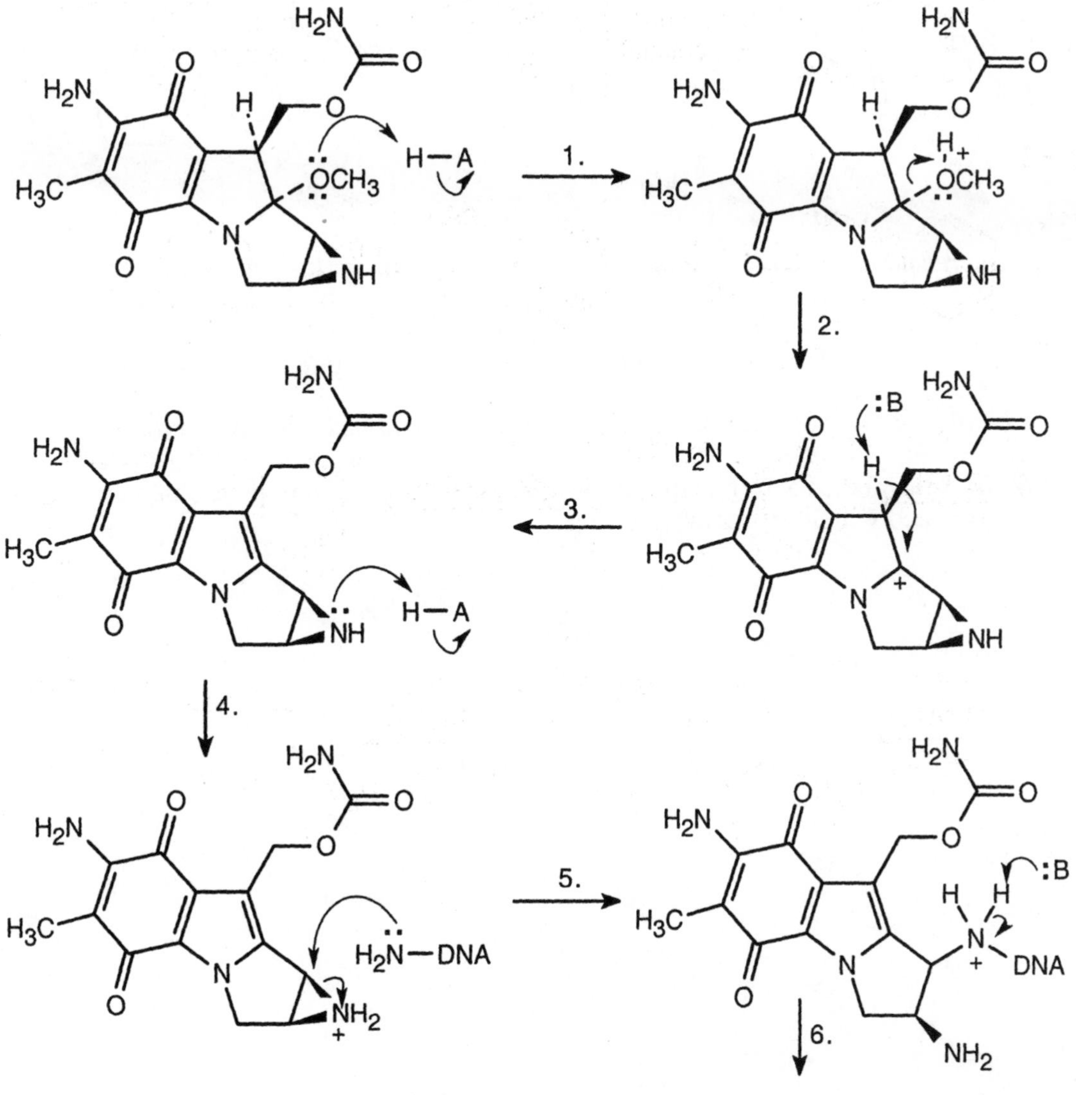

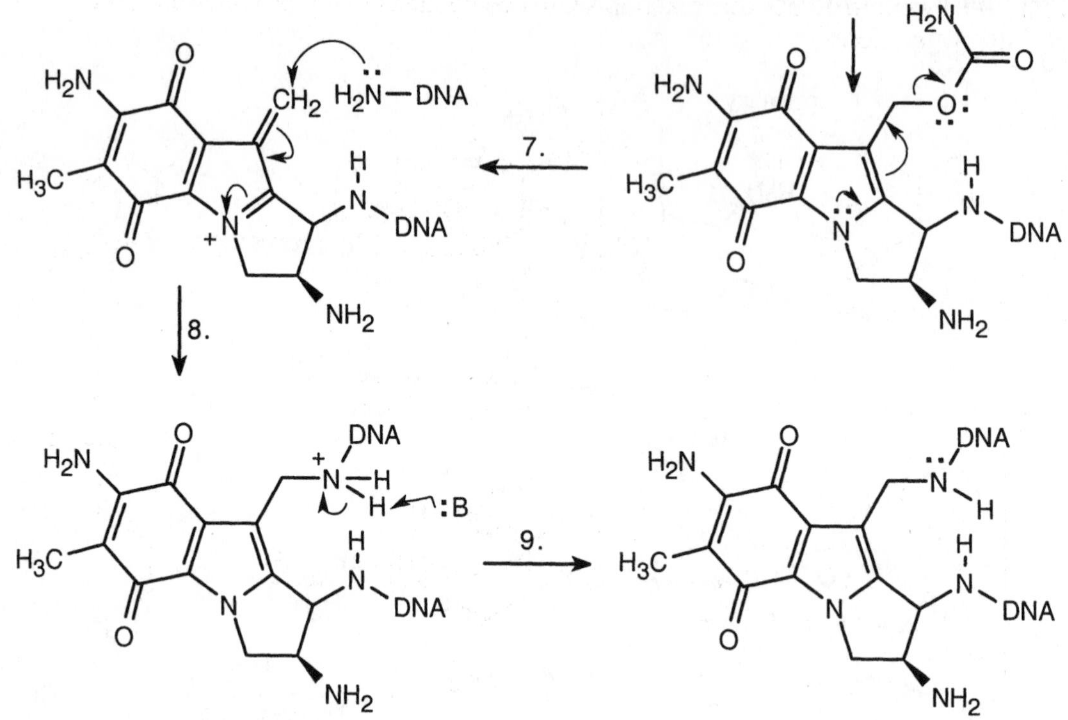

The reactions: Steps 1 – 3: E1 elimination (protonation, loss of $HOCH_3$, deprotonation)
Steps 4 – 6: S_N2 substitution (protonation, substitution, deprotonation)
Step 7: E1 elimination of carbamate
Steps 8 – 9: conjugate addition of DNA (addition, deprotonation)

Notice that five of the nine steps are either protonations or deprotonations.

12.56 The 3-membered aziridine ring is reactive because reaction removes the angle strain associated with three-membered rings.

12.58 Like Problem 12.55, this reaction sequence contains many protonations and deprotonations.

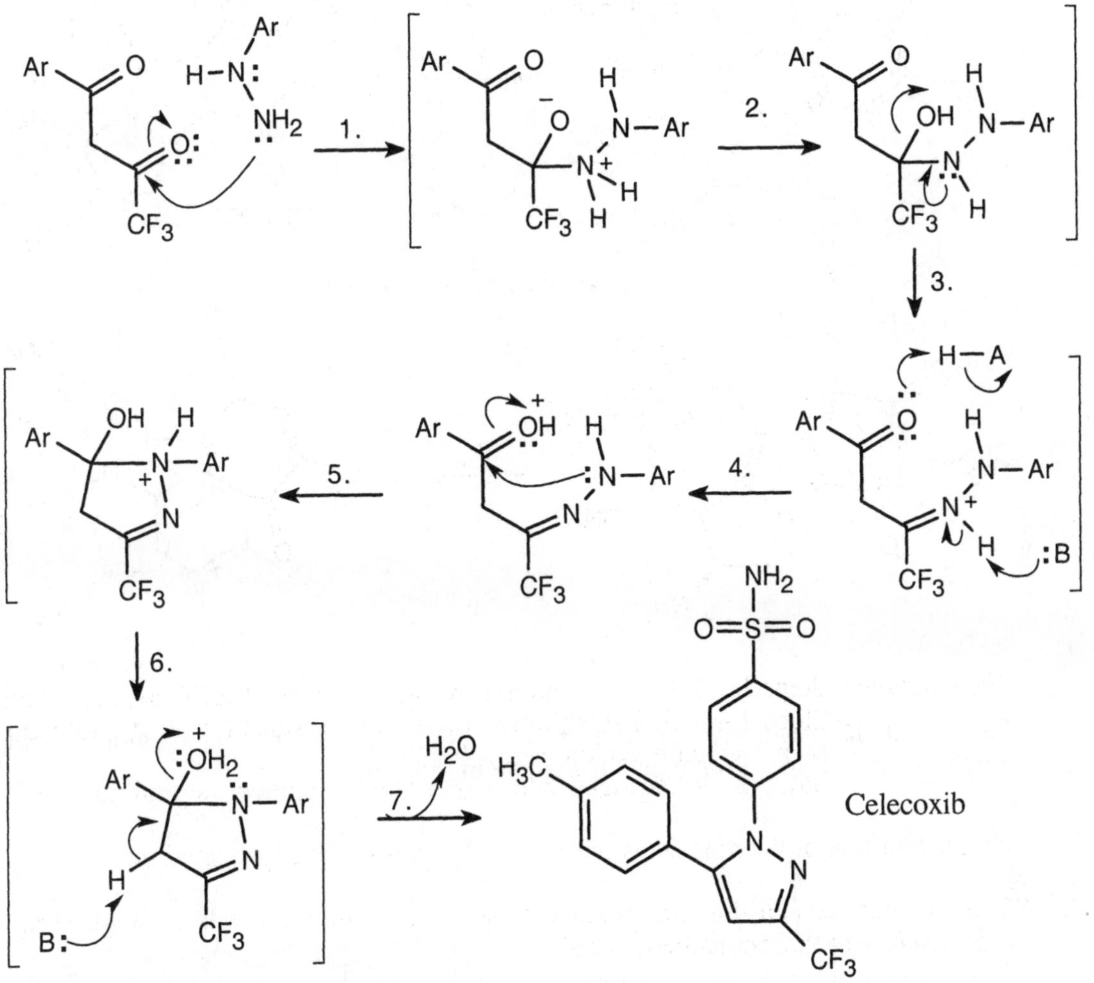

Celecoxib

Steps 1,5: Nucleophilic addition of an amine to a carbonyl group.

Steps 2,4,6: Proton transfers.

Steps 3,7: Loss of water.

In the Field with Agrochemicals

12.59

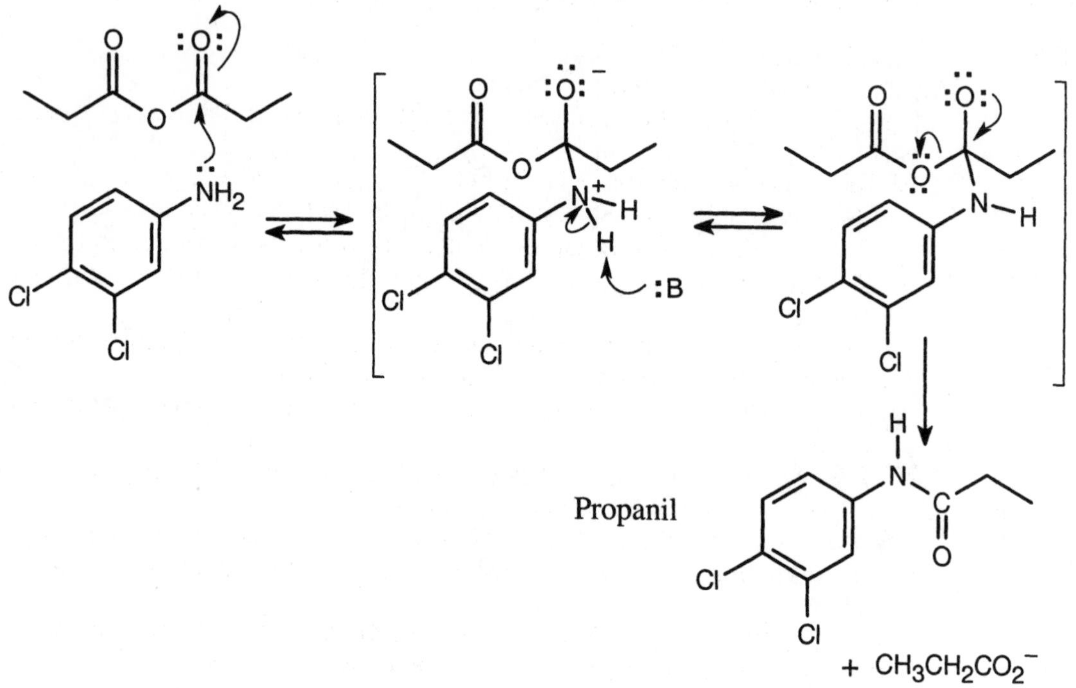

O :O: NH$_2$ Cl Cl ⇌ [O :Ö:$^-$ O N$^+$ H H :B Cl Cl ⇌ O :Ö: Ö N–H Cl Cl]

H N C O Cl Cl

Propanil

+ $CH_3CH_2CO_2^-$

This is a nucleophilic acyl substitution with 3,4 dichloroaniline, the nucleophile, adding to propanoic anhydride. Propanoate is the leaving group.

12.60

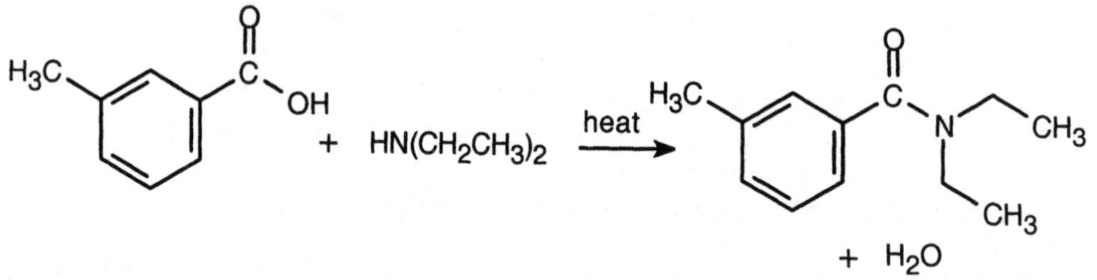

H_3C O C OH + $HN(CH_2CH_3)_2$ $\xrightarrow{\text{heat}}$ H_3C O C N CH_3 CH_3

+ H_2O

Elevated temperatures must be used to carry out this reaction.

12.61

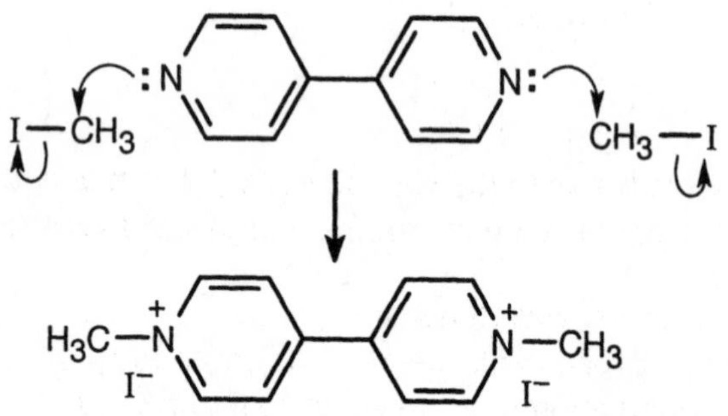

I–CH_3 :N N: CH_3–I

↓

H_3C–N^+ I^- N^+–CH_3 I^-

Chapter 13 – Structure Determination

Chapter Outline

I. Electromagnetic Radiation (Section 13.1).
   A. The nature of radiant energy.
      1. The different types of electromagnetic radiation make up the electromagnetic spectrum.
      2. Electromagnetic radiation behaves both as a particle and as a wave.
      3. Electromagnetic radiation can be characterized by three variables.
         a. The wavelength (λ) measures the distance from one wave maximum to the next.
         b. The frequency (ν) measures the number of wave maxima that pass a fixed point per unit time.
         c. The amplitude is the wave height measured from the midpoint to the maximum.
      4. Wavelength times frequency equals the speed of light.
      5. Electromagnetic energy is transmitted in discrete energy bundles called quanta.
         a. $\varepsilon = h \times \nu$
         b. Energy varies directly with frequency but inversely with wavelength.
   B. Electromagnetic radiation and organic molecules.
      1. When an organic compound is struck by a beam of electromagnetic radiation, it absorbs radiation of certain wavelengths, and transmits radiation of other wavelengths.
      2. If we determine which wavelengths are absorbed and which are transmitted, we can obtain an absorption spectrum of the compound.
   C. Types of radiation.
      1. X-ray crystallography.
      2. Mass spectrometry.
      3. Ultraviolet and visible spectroscopy.
      4. Infrared spectroscopy.
      5 Nuclear magnetic resonance spectroscopy.

II. X-ray crystallography (Section 13.2).
   A. Technique.
      1. A source of X-rays is aimed at a crystal.
      2. The periodic arrangement of atoms in the crystal diffracts the X-rays.
      3. The diffraction pattern is measured.
      4 A computer reconstructs the three dimensional structure.
   B. Results.
      The three-dimensional arrangement of very large molecules can be deduced.

III. Mass spectrometry (Section 13.3).
   Mass spectrometry provides the molecular weights of small organic molecules.
      1. Results are reported as mass to charge (*m/z*) because only ions are detected.
      2. Molecular fragments can also be identified.
      3. Isotopes are observable.

IV. Ultraviolet spectroscopy (Sections 13.4 - 13.5).
   A. Principles of ultraviolet spectroscopy (Section 13.4).
      1. The ultraviolet region of interest is between the wavelengths 200 nm and 400 nm.
      2. The energy absorbed is used to promote a π electron in a conjugated system from one molecular orbital to another.
   B. A UV spectrum is a plot of absorbance vs. wavelength.
   C. UV spectra usually consist of a single broad peak, whose maximum is λ_{max}.
   D. λ_{max} increases with the degree of conjugation of a molecule (Section 13.5).

E. Beer's law allows one to measure the concentration of a solution.
   Beer's law: $A = \varepsilon c l$

V. Infrared spectroscopy and organic molecules (Section 13.6).
   A. Infrared radiation.
      1. The infrared (IR) region of the electromagnetic spectrum extends from 7.8×10^{-7} m to 10^{-4} m.
         a. Organic chemists use the region from 2.5×10^{-6} m to 2.5×10^{-5} m.
         b. Wavelengths are usually given in μm, and frequencies are expressed in wavenumbers, which are the reciprocal of wavelength.
         c. The useful range of IR radiation is 4000 cm^{-1} – 400 cm^{-1}; this corresponds to energies of 48.0 kJ/mol – 4.80 kJ/mol.
      2. IR radiation causes bonds to stretch and bend and causes other molecular vibrations.
      3. Energy is absorbed at a specific frequency that corresponds to the frequency of the vibrational motion of a bond.
      4. If we measure the frequencies at which IR energy is absorbed, we can find out the kinds of bonds a compound contains and identify functional groups.
   B. Interpreting IR spectra.
      1. Most molecules have very complex IR spectra.
         a. This complexity means that each molecule has a unique fingerprint that allows it to be identified by IR spectroscopy.
         b. Complexity also means that not all absorptions can be identified.
      2. Most functional groups have characteristic IR absorption bands that don't change from one compound to another.
      3. The significant regions of IR absorptions :
         a. 4000 cm^{-1} - 2500 cm^{-1} corresponds to absorptions by C–H, O–H, and N–H bonds.
         b. 2500 cm^{-1} - 2000 cm^{-1} corresponds to triple-bond stretches.
         c. 2000 cm^{-1} - 1500 cm^{-1} corresponds to double bond stretches.
         d. The region below 1500 cm^{-1} is the fingerprint region, where many complex bond vibrations occur.

VI. Nuclear magnetic resonance spectroscopy (NMR) (Sections 13.7 - 13.14).
   A. General characteristics (Sections 13.7 - 13.9).
      1. Theory of NMR spectroscopy (Section 13.7).
         a. Many nuclei behave as if they were spinning about an axis.
            i. The positively charged nuclei produce a magnetic field that can interact with an externally applied magnetic field.
            ii. The ^{13}C nucleus and the ^{1}H nucleus behave in this manner.
            iii. In the absence of an external magnetic field the spins of magnetic nuclei are randomly oriented.
         b. When a sample containing these nuclei is placed between the poles of a strong magnet, the nuclei align themselves either with the applied field or against the applied field.
            The parallel orientation is slightly lower in energy and is slightly favored.

c. If the sample is irradiated with radiofrequency energy of the correct frequency, the nuclei of lower energy absorb energy and "spin-flip" to the higher energy state.
   i. The magnetic nuclei are in resonance with the applied radiation.
   ii. The frequency of the rf radiation needed for resonance depends on the magnetic field strength and on the identity of the magnetic nuclei.
   iii. In a strong magnetic field, higher frequency rf energy is needed.
   iv At a magnetic field strength of 2.62 T, rf energy of 100 MHz is needed to bring a ^{1}H nucleus into resonance, and energy of 25 MHz is needed for ^{13}C.

2. The nature of NMR absorptions (Section 13.8).
   a. Not all ^{13}C nuclei and not all ^{1}H nuclei absorb at the same frequency.
      i. Each magnetic nucleus is surrounded by electrons that set up their own magnetic fields.
      ii. These little fields oppose the applied field and shield the magnetic nuclei. $B_{effective} = B_{applied} - B_{local}$ .This expression shows that the magnetic field felt by a nucleus is less than the applied field.
      iii. These shielded nuclei absorb at slightly different values of magnetic field strength.
      iv. A sensitive NMR spectrometer can detect these small differences.
      v. Thus, NMR spectra can be used to map the carbon-hydrogen framework of a molecule.
   b. NMR spectra.
      i. The horizontal axis shows effective field strength, and the vertical axis shows intensity of absorption.
      ii. Each peak corresponds to a chemically distinct nucleus.
      iii. Zero absorption is at the bottom.
      iv. Absorptions due to both ^{13}C and ^{1}H can't both be observed at the same time.
3. Chemical Shifts (Section 13.9).
   a. Field strength increases from left (downfield) to right (upfield).
      i. Nuclei that absorb downfield require a lower field strength for resonance and are deshielded.
      ii. Nuclei that absorb upfield require a higher field strength and are shielded.
   b. TMS is used as a reference point in both ^{13}C NMR and ^{1}H NMR.
      The TMS absorption occurs upfield of other absorptions, and is set as the zero point.
   c. The chemical shift is the position on the chart where a nucleus absorbs.
   d. NMR charts are calibrated by using an arbitrary scale – the delta scale.
      i. One δ equals 1 ppm of the spectrometer operating frequency.
      ii. By using this system, all chemical shifts occur at the same value of δ, regardless of the spectrometer operating frequency.

B. ^{1}H NMR spectroscopy (Sections 13.10 - 13.13).
   1. Chemical shifts (Section 13.10).
      a. Chemical shifts are determined by the local magnetic fields surrounding magnetic nuclei.
         i. More strongly shielded nuclei absorb upfield.
         ii. Less shielded nuclei absorb downfield.

b. Most ^{1}H NMR chemical shifts are in the range 0–10 δ.
   i. Protons that are sp^3-hybridized absorb at higher field strength.
   ii. Protons that are sp^2-hybridized absorb at lower field strength.
   iii. Protons on carbons that are bonded to electronegative atoms absorb at lower field strength.
c. The ^{1}H NMR spectrum can be divided into 6 regions:
   i. Saturated (0–1.5 δ).
   ii. Allylic (1.5–2.5 δ).
   iii. H bonded to C next to an electronegative atom (2,5–4.5 δ).
   iv. Vinylic (4.5–6.5 δ).
   v. Aromatic (6.5–8.0 δ).
   vi. Aldehyde and carboxylic acid protons absorb even farther downfield.

2. Integration of ^{1}H NMR signals: proton counting (Section 13.11).
   a. The area of a peak is proportional to the number of protons causing the peak.
   b. Integrated peak areas are superimposed over a spectrum as a stair-step line.
   c. To compare two peaks, measure the relative heights of the steps.
3. Spin-spin splitting (Section 13.12).
   a. The tiny magnetic field produced by one nucleus can affect the magnetic field felt by other nuclei.
   b. Protons that have n equivalent neighboring protons show a peak in their ^{1}H NMR spectrum that is split into $n + 1$ smaller peaks (a multiplet).
   c. This splitting is caused by the coupling of spins of neighboring nuclei.
   d. The distance between peaks in a multiplet is called the coupling constant (J).
      i. The value of J is usually 0–18 Hz.
      ii. The value of J is determined by the geometry of the molecule and is independent of the spectrometer operating frequency.
      iii. The value of J is shared between both groups of hydrogens whose spins are coupled.
      iv. By comparing values of J, it is possible to know the atoms whose spins are coupled.
   e. Three rules for spin-spin splitting in ^{1}H NMR:
      i. Chemically identical protons don't show spin-spin splitting.
      ii. The signal of a proton with n equivalent neighboring protons is split into a multiplet of $n + 1$ peaks with coupling constant J.
      iii. Two groups of coupled protons have the same value of J.
4. Uses of ^{1}H NMR spectroscopy (Section 13.13).
   ^{1}H NMR can be used to identify the products of reactions.

C. ^{13}C NMR spectroscopy (Section 13.14).
1. Uses of ^{13}C NMR spectroscopy.
   ^{13}C NMR spectroscopy can show the number of nonequivalent carbons in a molecule and can identify symmetry in a molecule.
2. Characteristics of ^{13}C NMR spectroscopy.
   a. Each distinct carbon shows a single line.
   b. The chemical shift depends on the electronic environment within a molecule.
      i. Carbons bonded to electronegative atoms absorb downfield.
      ii. Carbons with sp^3 hybridization absorb in the range 0–90 δ.
      iii. Carbons with sp^2 hybridization absorb in the range 110–220 δ.
         Carbonyl carbons absorb in the range 160–220 δ.
   c. Symmetry reduces the number of absorptions.
   d. Peaks aren't uniform in size.

Solutions to Problems

13.1 At first glance, we know that: (1) energy increases as wavelength decreases, and (2) the wavelength of X-radiation is shorter than the wavelength of infrared radiation. Thus, we estimate that an X ray is of higher energy than an infrared ray.
An exact solution:
$\varepsilon = h\nu = hc/\lambda$; $h = 6.62 \times 10^{-34}$ J·s; $c = 3.00 \times 10^{8}$ m/s
for $\lambda = 1.0 \times 10^{-6}$ m (infrared radiation):

$$\varepsilon = \frac{(6.62 \times 10^{-34}\ \text{J·s})(3.00 \times 10^{8}\ \text{m/s})}{1.0 \times 10^{-6}\ \text{m}} = 2.0 \times 10^{-19}\ \text{J}$$

for $\lambda = 3.0 \times 10^{-9}$ m (X radiation):

$$\varepsilon = \frac{(6.62 \times 10^{-34}\ \text{J·s})(3.00 \times 10^{8}\ \text{m/s})}{3.0 \times 10^{-9}\ \text{m}} = 6.6 \times 10^{-17}\ \text{J}$$

Confirming our estimate, the calculation shows that an X ray is of higher energy than infrared radiation.

13.2 First, convert radiation in m to radiation in Hz by the equation:

$$\nu = \frac{c}{\lambda} = \frac{3.00 \times 10^{8}\ \text{m/s}}{9.0 \times 10^{-6}\ \text{m}} = 3.3 \times 10^{13}\ \text{Hz}$$

The equation $\varepsilon = h\nu$ says that the greater the value of ν, the greater the energy. Thus, radiation with $\nu = 3 \times 10^{13}$ Hz ($\lambda = 9 \times 10^{-6}$ m) is higher in energy than radiation with $\nu = 4 \times 10^{9}$ Hz.

13.3 Choose a typical frequency from Figure 13.1 for each type of radiation, and using the equation $\varepsilon = h\nu$, calculate the energy of one photon of radiation. To calculate energy per mole, multiply the above answer by Avogadro's Number (6.02×10^{23}).

Type of Radiation	*Frequency (Hz)*	*Energy (J/photon)*	*Energy (J/mol)*
Cell phone	9×10^{8}	6×10^{-25}	4×10^{-1}
Transistor radio	1×10^{10}	7×10^{-24}	4
Sunlight	5×10^{14}	3×10^{-19}	2×10^{5}
X-ray	1×10^{17}	7×10^{-17}	4×10^{7}

The energy of a C–C bond is about 400 KJ/mol (4×10^{5} J/mol). Thus, only X-rays have an energy capable of breaking a C–C bond.

13.4

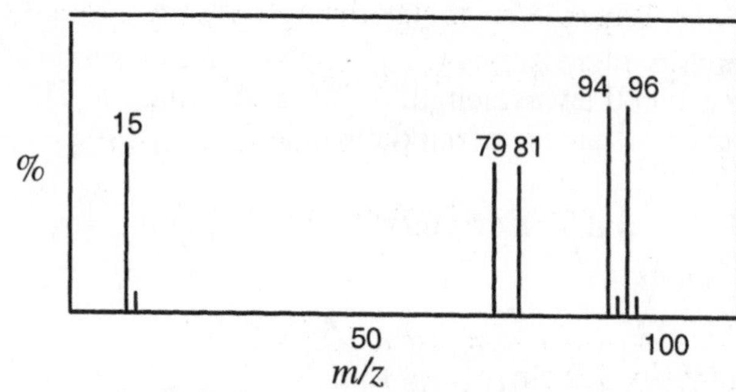

The very small peaks are due to ^{13}C. The peaks at $m/z = 96$ and 94 are due to $CH_3{}^{81}Br$ and $CH_3{}^{79}Br$. The two bromine isotopes appear at $m/z = 81$ and 79, and the methyl group appears at $m/z = 15$

13.5 The molecular weight of the alkene (56) increases by 254 when X_2 is added to the double bond, yielding a product with molecular weight of 310. Since X_2 has a molecular weight of 254, X has a molecular weight of 127, corresponding to iodine.

13.6 The trace impurity, appearing at m/z = 88, has a molecular weight that is 16 units greater than the molecular weight of butanal. Butanoic acid is the likely impurity.

13.7 Only conjugated compounds absorb in the region 200 nm–400 nm.

Compound	*Absorption at 200–400 nm?*	*Compound*	*Absorption at 200–400 nm?*
(a)	 yes	(b)	 no
(c) $H_2C{=}C(H){-}C({=}O)OCH_3$	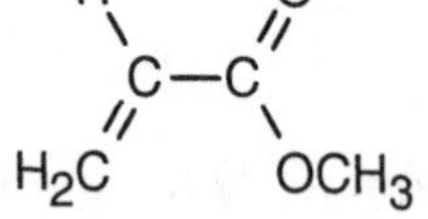 yes	(d) Br, CH_3	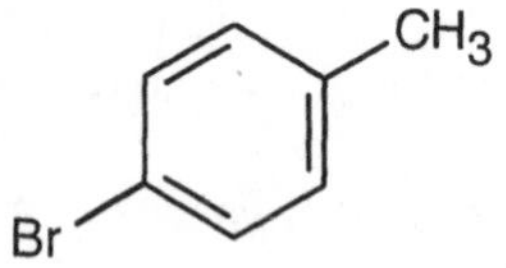 yes
(e) O, CH_3	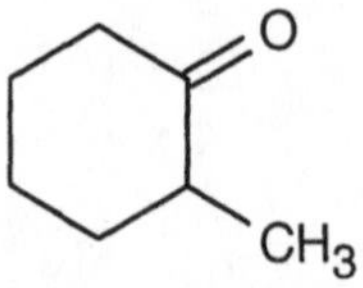 no	(f) O, CH_3	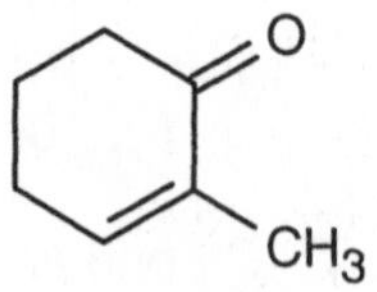 yes

13.8 Both compounds are conjugated and absorb in the ultraviolet region. Hexa-1,3,5-triene has a longer system of conjugated bonds and absorbs at a longer wavelength (lower energy) than hexa-1,3-diene.

13.9

$$\varepsilon = \frac{A}{c \times l}$$

Where ε = molar absorptivity (in $L \cdot mol^{-1} \cdot cm^{-1}$

A = absorbance

l = sample pathlength (in cm)

c = concentration (in $mol \cdot L^{-1}$)

In this problem:

$\varepsilon = 6{,}100 = 6.1 \times 10^3\ L \cdot mol^{-1}\ cm^{-1}$

$l = 1.0\ cm$

$A = 0.20$

$$c = \frac{A}{\varepsilon \times l} = \frac{0.20}{6.1 \times 10^3\ L \cdot mol^{-1} \cdot cm^{-1} \times 1.0\ cm} = 3.3 \times 10^{-5}\ M$$

13.10

$$\varepsilon = \frac{A}{c \times l} = \frac{0.63}{1.8 \times 10^{-8}\ mol \cdot L^{-1} \times 1.0\ cm} = 3.5 \times 10^7\ L \cdot mol^{-1} \cdot cm^{-1}$$

13.11 Use Table 13.3 and the descriptions in Section 13.6 to assign these absorptions.

IR absorption	*Due to:*
(a) 1715 cm^{-1}	ketone
(b) 1540 cm^{-1}	nitro group
(c) 2210 cm^{-1}	nitrile group or alkyne
(d) 1720 cm^{-1} 2500-3100 cm^{-1}	carboxylic acid
(e) 3500 cm^{-1} 1735 cm^{-1}	alcohol ester

13.12 To use IR to distinguish between isomers, find a strong IR absorption present in one isomer that is absent in the other isomer.

(a)

CH_3CH_2OH
Strong hydroxyl band
at 3400–3640 cm^{-1}

CH_3OCH_3
No band in the region
3400–3640 cm^{-1}

(b)

$CH_3CH_2CH_2CH_2CH{=}CH_2$
Alkene bands at
3020–3100 cm^{-1} and
at 1650–1670 cm^{-1}.

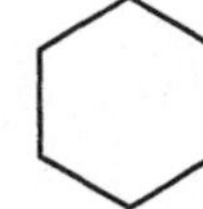

No bands in alkene region.

(c)

CH_3CH_2COOH
Strong, broad band
at 2500–3100 cm^{-1}

$HOCH_2CH_2CHO$
Strong band at
3400–3640 cm^{-1}

13.13

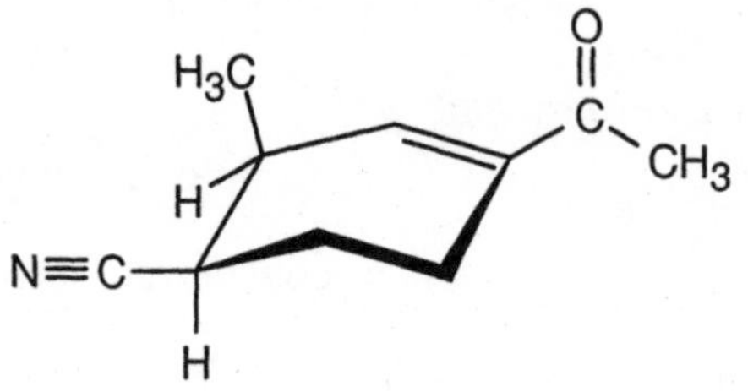

The compound contains nitrile and ketone groups, as well as a carbon-carbon double bond. The nitrile absorption occurs at 2210–2260 cm^{-1}. The ketone shows an absorption at 1690 cm^{-1}; this value is lower than the usual value because the ketone is next to the double bond. The double bond absorptions occur at 3020–3100 cm^{-1} and at 1640–1680 cm^{-1}.

13.14 From Problem 13.1, we find that $\lambda = 1 \times 10^{-6}$ m is a typical value for the wavelength of infrared radiation. Use the equation $\nu = c / \lambda$ to find the frequency of infrared radiation.

$$\nu = \frac{c}{\lambda} = \frac{3 \times 10^{8}\ \text{m/s}}{1 \times 10^{-6}\ \text{m}} = 3 \times 10^{14}\ \text{Hz}$$

Since ν for NMR radiation (1×10^{8} Hz) is less than ν for ultraviolet radiation, the amount of energy used by NMR spectroscopy is less than that used by IR spectroscopy.

13.15

	Compound	*Absorptions in ^1H NMR*	*Absorptions in ^{13}C NMR*
(a)	CH_4	1	1
(b)	CH_3CH_3	1	1
(c)	$CH_3CH_2CH_3$	2	2
(d)	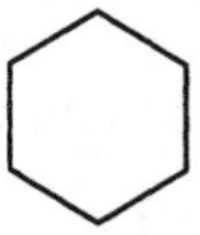 (cyclohexane)	1	1
(e)	CH_3OCH_3	1	1
(f)	 (benzene)	1	1
(g)	$(CH_3)_3COH$	2	2
(h)	CH_3CH_2Cl	2	2
(i)	$(CH_3)_2C{=}C(CH_3)_2$	1	2

13.16

```
H3C       H
   \     /
    C=C
   /    \
 Cl      H
```

2-Chloropropene

The two protons on C1 are not equivalent: one proton is on the same side of the double bond as chlorine and the other proton is on the opposite side. Since 2-chloropropene has three different types of protons, it shows three signals in its ^{1}H NMR spectrum.

13.17

```
CH3CH2      H
      \    /
       C=C
      /    \
   H3C      C—CH3
            ‖
            O
```

This molecule shows five ^{1}H NMR signals and seven ^{13}C NMR signals.

13.18 (a) 2.1 ppm x 100 MHz = 210 Hz

(b) The position of absorption in δ units is 2.1 δ for both a 100 MHz and a 220 MHz instrument. A measurement in δ units is independent of the operating frequency of the NMR spectrometer.

(c) 2.1 ppm x 220 MHz = 462 Hz

13.19

$$\delta = \frac{\text{Observed chemical shift (in Hz)}}{100 \text{ MHz}}$$

(a) $\delta = \frac{727 \text{ Hz}}{100 \text{ MHz}} = 7.27\ \delta$ for $CHCl_3$ (b) $\delta = \frac{305 \text{ Hz}}{100 \text{ MHz}} = 3.05\ \delta$ for CH_3Cl

(c) $\delta = \frac{347 \text{ Hz}}{100 \text{ MHz}} = 3.47\ \delta$ for CH_3OH (d) $\delta = \frac{530 \text{ Hz}}{100 \text{ MHz}} = 5.30\ \delta$ for CH_2Cl_2

13.20

Compound	*^{1}H Chemical Shift*
(a) CH_3CH_3	0.88 δ
(b) CH_3COCH_3	2.17 δ
(c) C_6H_6	7.17 δ
(d) $(CH_3)_3N$	2.22 δ

13.21

H_3C CH_3 *p*-Xylene

There are two peaks in the ^1H NMR spectrum of *p*-xylene. The four ring protons absorb at 7.0 δ, and the six methyl-group protons absorb at 2.3 δ. The peak ratio of methyl protons:ring protons is 3:2.

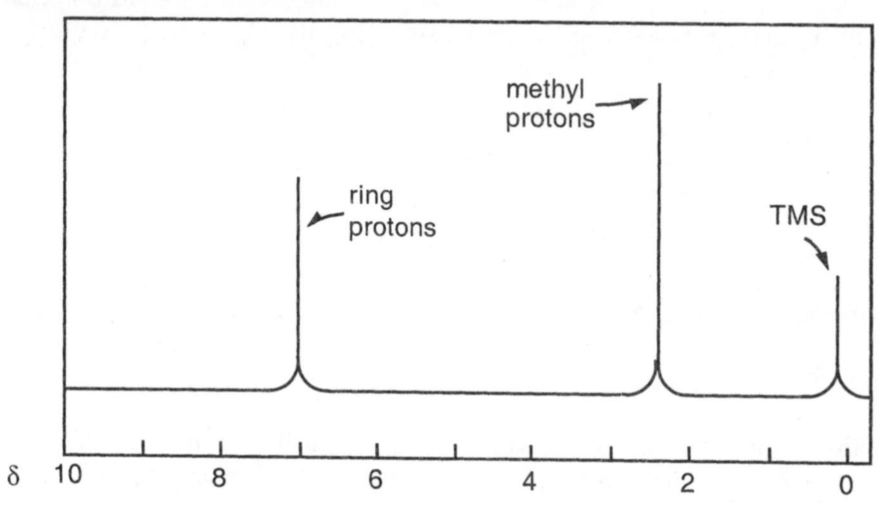

13.22

	Compound	Proton	Number of Adjacent Protons	Splitting
(a)	$(\overset{1}{CH_3})_3\overset{2}{CH}$	1	1	doublet
		2	9	multiplet
(b)	$\overset{1}{CH_3}\overset{2}{CH}Br_2$	1	1	doublet
		2	3	quartet
(c)	$\overset{1}{CH_3}O\overset{2}{CH_2}\overset{3}{CH_2}Br$	1	0	singlet
		2	2	triplet
		3	2	triplet
(d)	$\overset{1}{CH_3}\overset{2}{CH_2}CO_2\overset{3}{CH_3}$	1	2	triplet
		2	3	quartet
		3	0	singlet
(e)	$Cl\overset{1}{CH_2}\overset{2}{CH_2}\overset{1}{CH_2}Cl$	1	2	triplet
		2	4	quintet
(f)	$(\overset{1}{CH_3})_2\overset{2}{CH}CO_2\overset{3}{CH_3}$	1	1	doublet
		2	6	septet
		3	0	singlet

13.23

(a) C_2H_6O has only one type of proton, with no neighbors.

CH_3OCH_3

(b) $C_3H_6O_2$ has two types of protons; neither kind has neighbors.

CH_3COOCH_3

(c) C_3H_7Cl has two types of protons; one kind of proton has six neighbors, and the other kind has one neighbor.

$(CH_3)_2CHCl$

13.24

Proton(s)	Number of Adjacent Protons	Splitting
1	1	doublet
2	6	septet
3	0	singlet
4	3	quartet
5	1	doublet

13.25 Look for a plane of symmetry that reduces the number of peaks seen in the ^{13}C NMR spectrum.

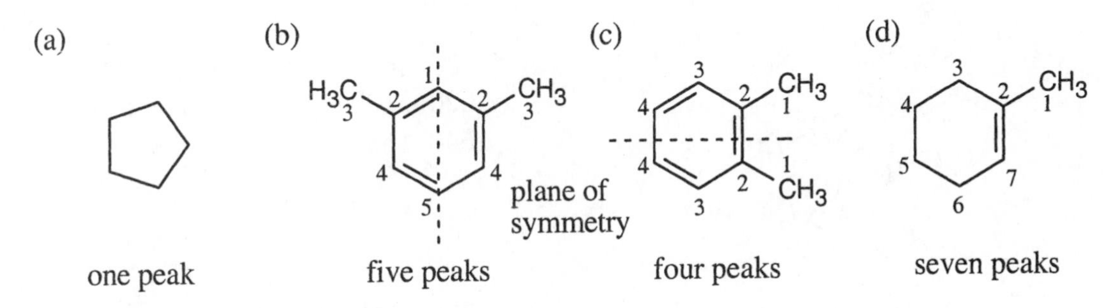

13.26 (a) $CH_3CH_2CH_2CH_2CH_2CH{=}CH_2$ seven peaks

(b) $(CH_3)_2CHCH_2CH_2CH_3$ five peaks

(c) $(CH_3)_2CHCH_2Cl$ three peaks

Other structures are possible.

Visualizing Chemistry

13.27

(a)

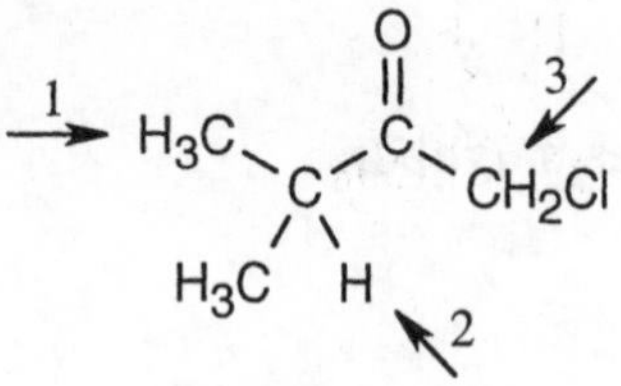

O, 3, 1, H_3C, C, C, CH_2Cl, H_3C, H, 2

1. doublet
2. septet
3. singlet

(b)

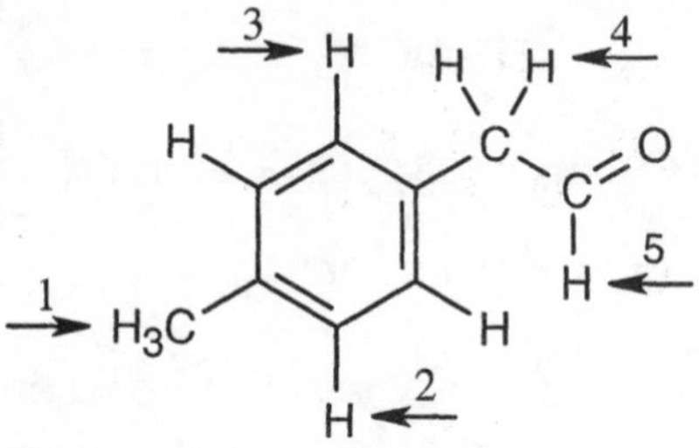

3, H, H, H, 4, H, C, C=O, H, 5, 1, H_3C, H, H, 2

1. singlet
2. doublet
3. doublet
4. doublet
5. triplet

13.28

Compound	*Significant IR Absorption*	*Due to:*
(a) H, H, 3, O, 2, C, C, H, +, N, 1, O^-, O	1540 cm^{-1}	nitro group (1)
	1725 cm^{-1}	aldehyde (2)
	3030 cm^{-1},	aromatic ring C–H(3)
	1600, 1500 cm^{-1}	aromatic ring C=C(3)
(b) O, H_3C, 1C, O, C, CH_3, H, 2, H, C, 3, CH_2, 2	1735 cm^{-1}	ester (1)
	3020–3100 cm^{-1}	alkene =C–H(2)
	1640–1680 cm^{-1}	alkene C=C (3)
(c) O, H_3C, OH, 2, 1, C, C, H_3C, C, CH_3, H H	1715 cm^{-1}	ketone (1)
	3400–3650 cm^{-1}	alcohol (2)

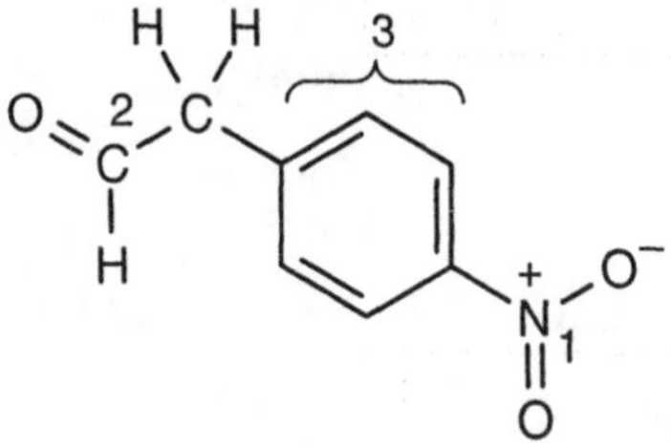

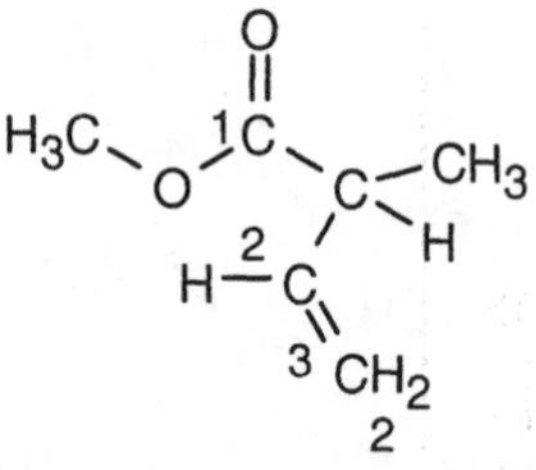

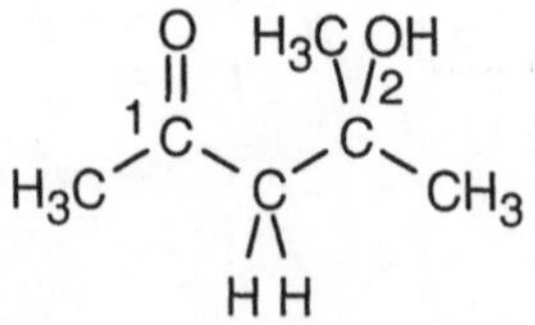

13.29 Because of symmetry, the compound shows five absorptions in its ^{13}C NMR spectrum.

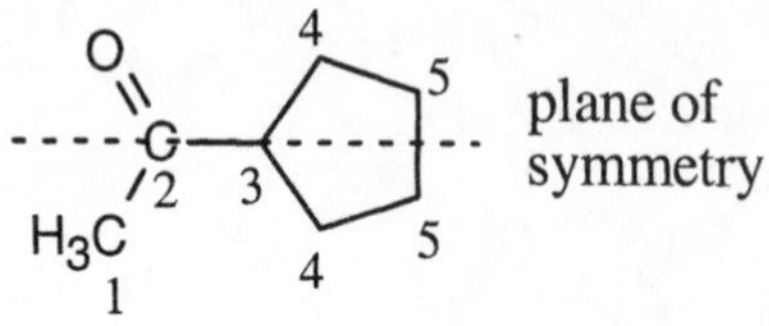

O, 4, 5, C, 2, 3, H_3C, 1, 4, 5

plane of symmetry

13.30 The compound has 5 different types of carbons and 4 different types of hydrogens.

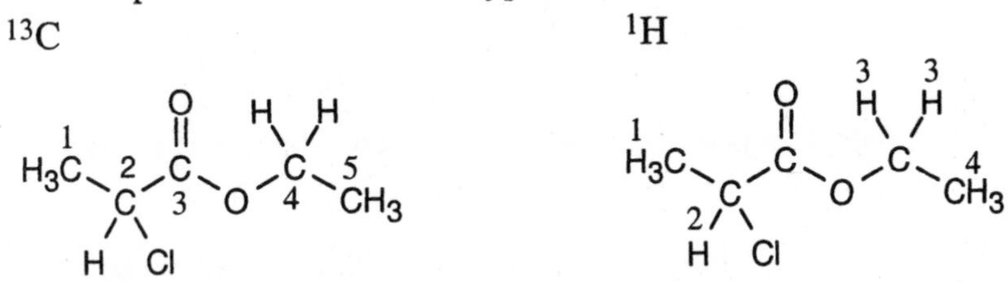

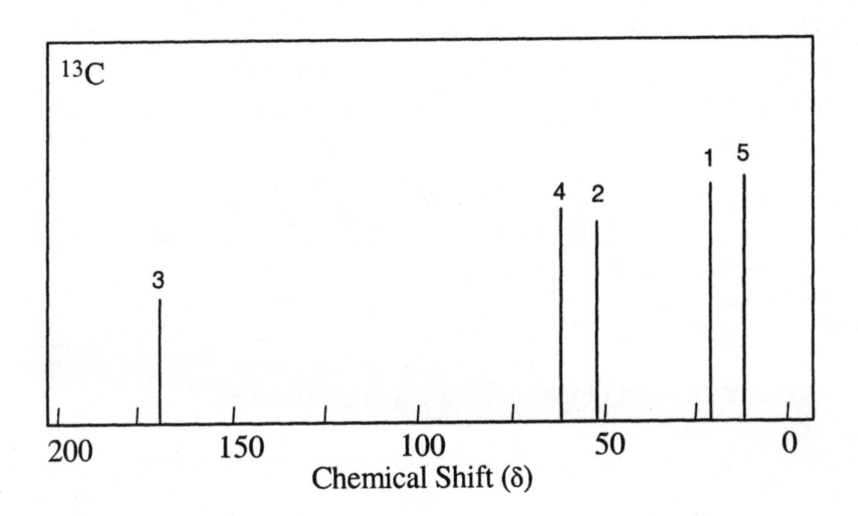

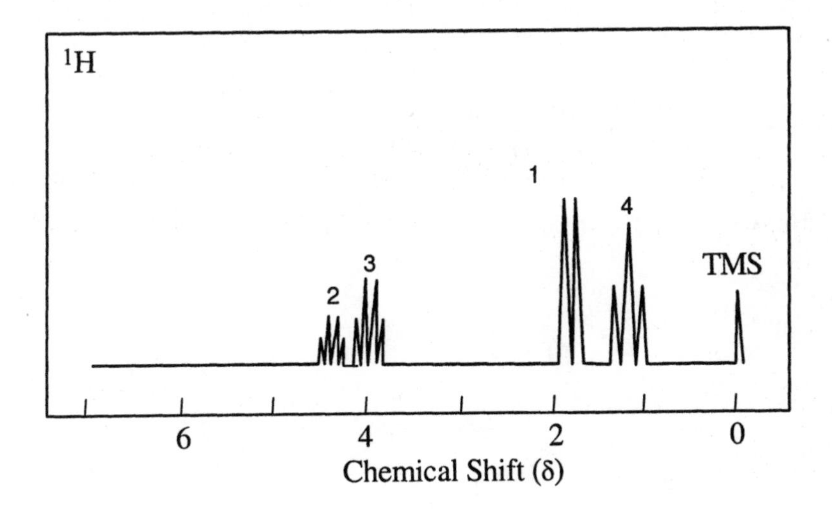

Additional Problems

General Questions

13.31 (a) The *chemical shift* is the exact position at which a nucleus absorbs rf energy in an NMR spectrum.

(b) If the NMR signal of nucleus A is split by the spin of adjacent nucleus B, there is reciprocal splitting of the signal of nucleus B by the spin of nucleus A. The spins of the two nuclei are said to be coupled. The distance between two individual peaks within the multiplet of A is the same as the distance between two individual peaks within the multiplet of B. This distance, measured in Hz, is called the *coupling constant.*

(c) λ_{max} is the wavelength in an ultraviolet spectrum at which the percent absorbance is the greatest.

(d) *Spin-spin splitting* is the splitting of a single NMR resonance into multiple lines. Spin-spin splitting occurs when the effective magnetic field felt by a nucleus is influenced by the small magnetic moments of adjacent nuclei. In ^1H NMR the signal of a proton with n neighboring protons is split into $n+1$ peaks. The magnitude of spin-spin splitting is given by the coupling constant J.

(e) A *wave number* is the reciprocal of the wavelength in centimeters.

(f) The *applied magnetic field* is the magnetic field that is externally applied to a sample by an NMR spectrometer.

(g) In mass spectrometry, the expression m/z is the ratio of mass to charge of an ion.

13.32

$$E = \frac{1.20 \times 10^{-4}\ \text{kJ/mol}}{\lambda\ (\text{in m})} = \frac{1.20 \times 10^{-4}\ \text{kJ/mol}}{6.55 \times 10^{-6}}$$

$$= 0.183 \times 10^{2}\ \text{kJ/mol} = 18.3\ \text{kJ/mol}$$

13.33

$$E = \frac{1.20 \times 10^{-4}\ \text{kJ/mol}}{\lambda\ (\text{in m})} \qquad \lambda = 217\ \text{nm} = 2.17 \times 10^{-7}\text{m}$$

$$E = \frac{1.20 \times 10^{-4}\ \text{kJ/mol}}{2.17 \times 10^{-7}} = 5.53 \times 10^{2}\ \text{kJ/mol}$$

Compare this value with the energy required for infrared excitation (Problem 13.32). More energy is required for ultraviolet excitation than for infrared excitation.

13.34

$$E = \frac{1.20 \times 10^{-4} \text{ kJ/mol}}{\lambda \text{ (in m)}}$$

To find λ, use the formula

$$\lambda = \frac{c}{\nu} \text{ where } c = 3.0 \times 10^{8} \text{ m/s}$$

In this problem, $\nu = 100$ MHz, or 10^8 Hz

and $\lambda = 3.0$ m

$$E = \frac{1.20 \times 10^{-4} \text{ kJ/mol}}{3.0} = 4 \times 10^{-5} \text{ kJ/mol}$$

for $\nu = 220$ MHz, or 2.2×10^8 Hz

$\lambda = 1.4$ m

$E = 8.6 \times 10^{-5}$ kJ/mol

Increasing the spectrophotometer frequency from 100 MHz to 220 MHz increases the amount of energy needed for resonance.

UV Spectroscopy

13.35

$\varepsilon = 13{,}260 \text{ M}^{-1} \text{ cm}^{-1} = 1.3260 \times 10^{4} \text{ M}^{-1} \text{ cm}^{-1}$

$c = 42 \text{ mM} = 4.2 \times 10^{-2} \text{ M}$

$l = 1.0$ cm

$A = \varepsilon \times c \times l = 1.3260 \times 10^{4} \text{ M}^{-1} \text{ cm}^{-1} \times 4.2 \times 10^{-2} \text{ M} \times 1.0 \text{ cm} = 5.7 \times 10^{2}$

13.36 If a concentration of 4.2×10^{-2} M (c_1) has $A = 5.7 \times 10^2$ (A_1) (Problem 13.35), a solution with $A = 0.93$ (A_2) has a concentration of:

$c_2 = A_2c_1/A_1 = 0.93 \times 4.2 \times 10^{-2} \text{ M} / 5.7 \times 10^2 = 6.9 \times 10^{-5} \text{ M}$

If this solution resulted from a 1/1000 dilution, the concentration of the original solution was 6.9×10^{-2} M.

13.37 The human metabolite has a more extensive conjugated system, and its value of λ_{max} should be greater.

13.38

O, NaBH$_4$ →, HO H

The UV absorption due to the conjugated enone disappears after reduction is complete.

IR Spectroscopy

13.39

IR Absorption	Due to:
(a) 1670 cm^{-1}	alkene or carbonyl group
(b) 1735 cm^{-1}	ester group
(c) 1540 cm^{-1}	nitro group
(d) 1715 cm^{-1} 2500-3100 cm^{-1}	carboxylic acid group

13.40

	Compound	Significant IR Absorption	Due to:
(a)	C_6H_5COOH (benzoic acid)	2500–3100 cm^{-1} 1710 cm^{-1} 1600, 1500 cm^{-1}	O–H (carboxylic acid) C=O (carboxylic acid) C=C (aromatic ring)
(b)	$C_6H_5COOCH_3$	1715 cm^{-1} 1600, 1500 cm^{-1}	C=O (conjugated ester) C=C (aromatic ring)
(c)	HO–C_6H_4–C≡N	2210–2260 cm^{-1} 3200–3650 cm^{-1} 1600, 1500 cm^{-1}	C≡N (nitrile) O–H (hydroxyl) C=C (aromatic ring)
(d)	cyclohexenone (ring with C=C and C=O)	1715 cm^{-1} 1640–1680 cm^{-1}	C=O (ketone) C=C (double bond)
(e)	$CH_3COCH_2CH_2COOCH_3$	1735 cm^{-1} 1715 cm^{-1}	C=O (ester) C=O (ketone)

13.41 Reciprocal centimeters (cm^{-1}) are units of frequency of radiation. Since frequency is directly proportional to energy, the larger the frequency, the more energy absorbed by a bond, and the stronger the bond. A carbon-oxygen double bond absorbs energy at a higher frequency (1700 cm^{-1}) than a carbon–oxygen single bond (1000 cm^{-1}) because the C=O double bond is stronger. This answer confirms what we already know about the relative strengths of single and double bonds.

13.42 (a) Absorptions at 3300 cm^{-1} and 2150 cm^{-1} are due to a terminal triple bond. Possible structures:

$CH_3CH_2CH_2C{\equiv}CH$ $(CH_3)_2CHC{\equiv}CH$

(b) IR absorption at 3400 cm^{-1} is due to a hydroxyl group. Since no double bond absorption is present, the compound must be a cyclic alcohol.

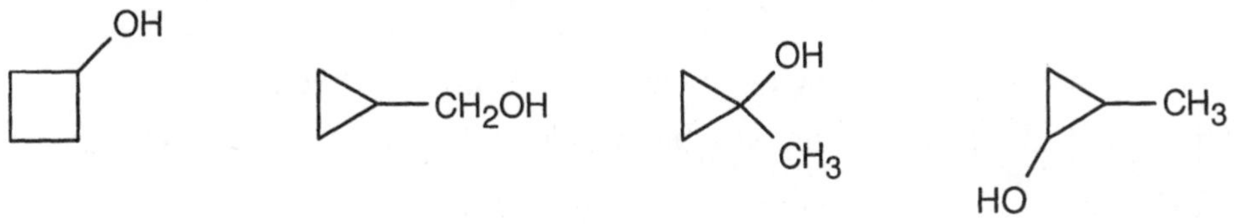

OH CH₂OH OH CH₃ CH₃ HO

(c) Absorption at 1715 cm^{-1} is due to a ketone. The only possible structure is $CH_3CH_2COCH_3$.

(d) Absorptions at 1600 cm^{-1} and 1500 cm^{-1} are due to an aromatic ring. Possible structures:

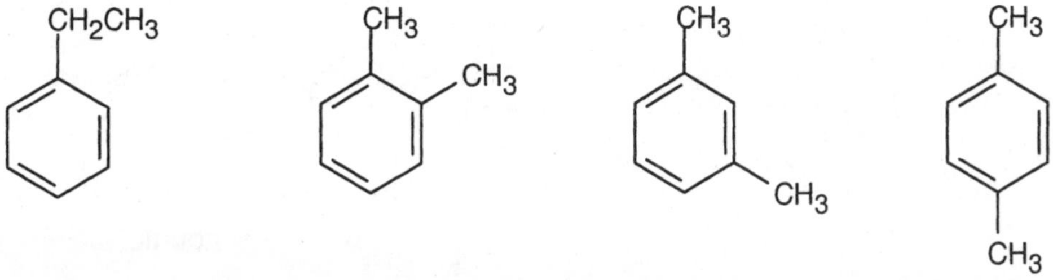

CH_2CH_3 CH_3 CH_3 CH_3 CH_3 CH_3 CH_3 CH_3

Mass Spectrometry

13.43 Each of these compounds can be detected by its molecular weight, which can be measured to four decimal places and is thus unambiguous. The second value shown for each compound is the exact mass.

Methamphetamine	Cocaine	Phencyclidine	THC	Ecstasy
$C_{10}H_{15}N$	$C_{17}H_{21}NO_4$	$C_{17}H_{25}N$	$C_{21}H_{30}O_2$	$C_{11}H_{15}NO_2$
149.23	303.35	243.39	314.46	193.24
149.120449	303.147059	243.198699	314.224580	193.110279

13.44

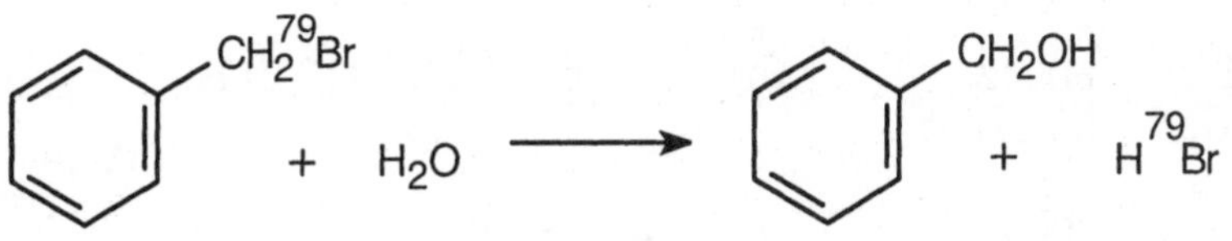

$$C_6H_5CH_2{}^{79}Br + H_2O \longrightarrow C_6H_5CH_2OH + H^{79}Br$$

m/z 170 108 80

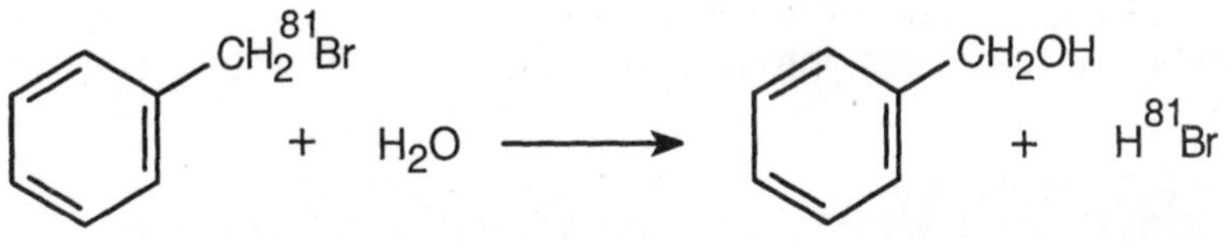

$$C_6H_5CH_2{}^{81}Br + H_2O \longrightarrow C_6H_5CH_2OH + H^{81}Br$$

m/z 172 108 82

In a humid atmosphere, benzyl bromide (*m/e* = 170, 172) undergoes a substitution reaction with water to give benzyl alcohol (*m/e* = 108) and HBr (*m/e* = 80, 82).

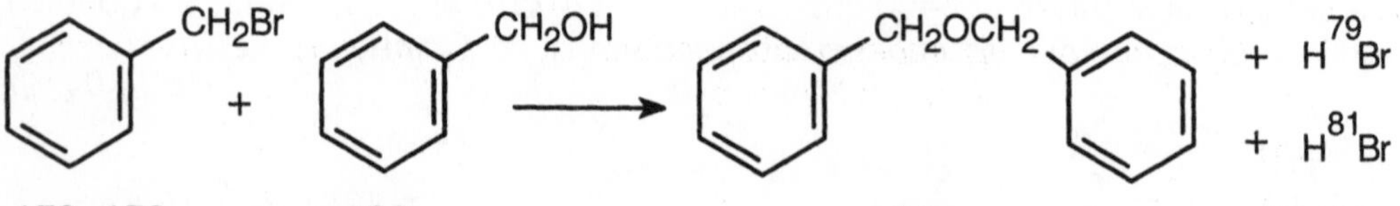

m/z 170, 172 108 198 80, 82

Benzyl bromide can also react with benzyl alcohol in an S_N1 reaction that yields dibenzyl ether (*m/z* = 198).

13.45 The molar mass of benzoylecgonine (289 g/mol) is 14 g/mol less than the molar mass of cocaine (303 g/mol), suggesting that the methyl ester of cocaine has been converted to a carboxylic acid.

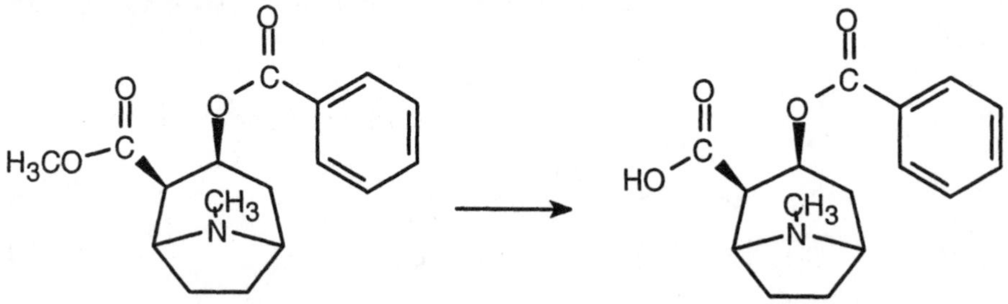

Cocaine Benzoylecgonine

13.46 The peak at *m/z* = 288 corresponds to testosterone. The peak at *m/z* = 322 shows the same molecular weight as ^{35}C clostebol, but is not accompanied by the peak at *m/z* = 324 that is due to the ^{37}Cl isotope. Thus, clostebol is not present in the sample.

NMR Spectroscopy

13.47 See Problem 13.19 for the method of solution.

(a) $\delta = \dfrac{218 \text{ Hz}}{100 \text{ MHz}} = 2.18\ \delta$ (b) $\delta = \dfrac{478 \text{ Hz}}{100 \text{ MHz}} = 4.78\ \delta$ (c) $\delta = \dfrac{751 \text{ Hz}}{100 \text{ MHz}} = 7.51\ \delta$

13.48 δ x (spectrometer frequency/10^6) = observed chemical shift (in Hz)
Here, spectrometer frequency = 220 MHz

(a) 2.18 x 220 Hz = 480 Hz (b) 4.78 x 220 Hz = 1050 Hz
(c) 7.51 x 220 Hz = 1650 Hz

13.49 (a) 2.1 δ x 80 Hz = 170 Hz (b) 3.45 δ x 80 Hz = 280 Hz
(c) 6.30 δ x 80 Hz = 500 Hz

13.50 (a) Since the symbol "δ" indicates ppm downfield from TMS, chloroform absorbs at 7.3 ppm.

(b) δ x (spectrometer frequency/10^6) = observed chemical shift (in Hz)
7.3 x 360 Hz = 2600 Hz

c) The value of δ is still 7.3 because the chemical shift measured in δ is independent of the operating frequency of the spectrometer.

13.51

	Compound	*Number of absorptions in* ^{13}C *spectrum*
(a) 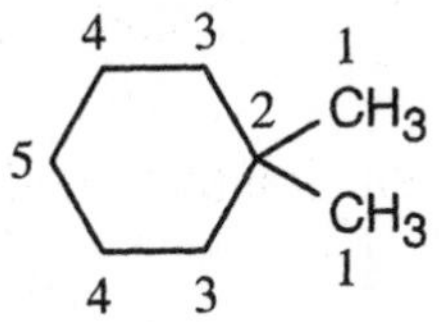	4 3 1 2 CH_3 5 CH_3 4 3 1	5
(b)	1 2 3 $CH_3CH_2OCH_3$	3
(c) 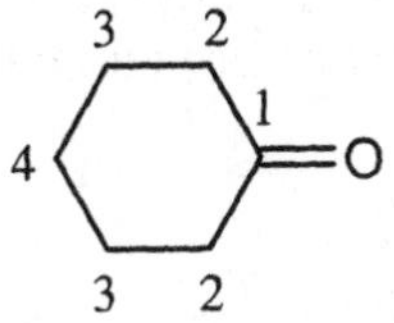	3 2 1 4 =O 3 2	4
(d) 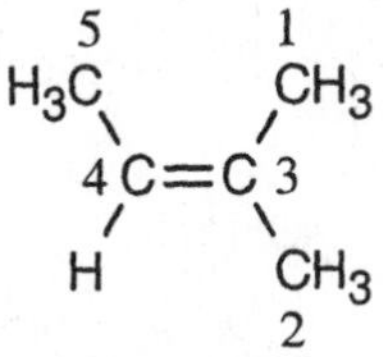	5 1 H_3C CH_3 4 C=C 3 H CH_3 2 Carbons 1 and 2 are not equivalent.	5
(e) 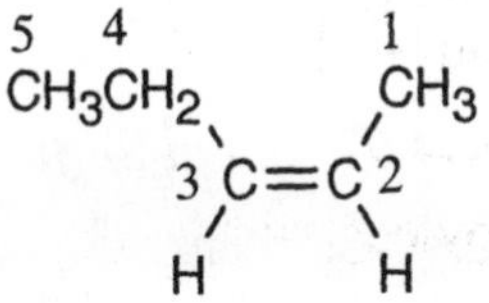	5 4 1 CH_3CH_2 CH_3 3 C=C 2 H H	5
(f) 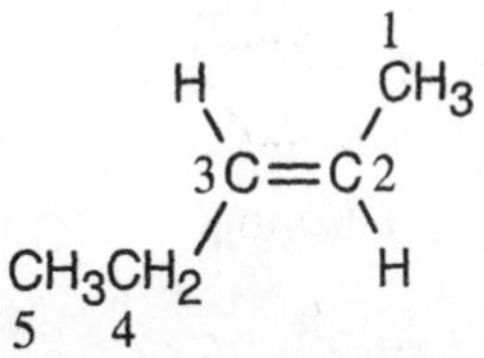	1 H CH_3 3C=C2 CH_3CH_2 H 5 4	5

13.52

	Compound	*Types of Non-equivalent Protons*
(a) 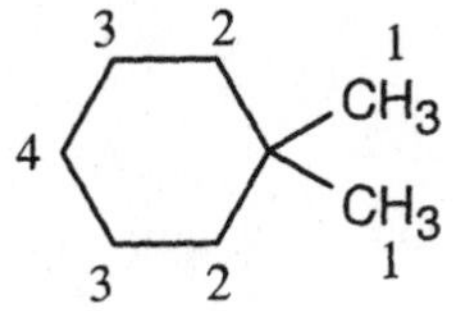	3 2 1 CH_3 4 CH_3 3 2 1	4
(b)	1 2 3 $CH_3CH_2OCH_3$	3
(c) 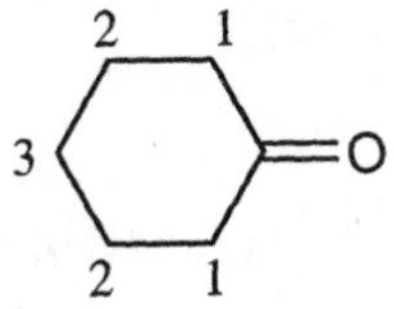	2 1 3 =O 2 1	3
(d) 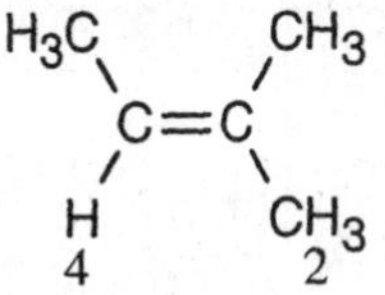	3 1 H_3C CH_3 C=C H CH_3 4 2 Protons 1 and 2 are not equivalent.	4
(e) 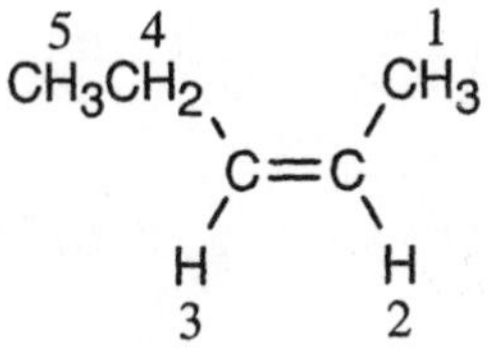	5 4 1 CH_3CH_2 CH_3 C=C H H 3 2	5
(f)	3 1 H CH_3 C=C CH_3CH_2 H 5 4 2	5

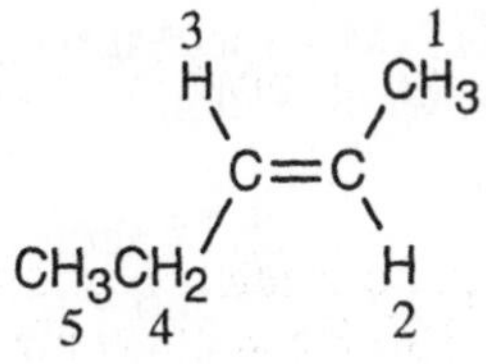

13.53

	Compound	*Protons*	*Chemical Shift*	*Rel. Peak Area*	*Splitting*
(a)	1 2 CH_3CHCl_2	1	1.0 δ	3	doublet
		2	3.9 δ	1	quartet
(b)	2 3 1 $CH_3CO_2CH_2CH_3$	1	1.2 δ	3	triplet
		2	2.0 δ	3	singlet
		3	4.1 δ	2	quartet
(c)	1 3 2 $(CH_3)_3CCH_2CH_3$	1	0.9 δ	9	singlet
		2	0.9 δ	3	triplet
		3	1.2 δ	2	quartet

The peaks due to protons 1 and 2 overlap

13.54

Lowest Chemical Shift ⟶ *Highest Chemical Shift*

CH_4 < cyclohexane < CH_3COCH_3 < CH_2Cl_2, $H_2C{=}CH_2$ < benzene

0.23 δ 1.43 δ 2.17 δ 5.30 δ 5.33 δ 7.37 δ

Integrated problems

13.55 (a) $CH_3CH_2NHCH_3$
N–H absorption at
3300–3500 cm^{-1}

$(CH_3)_3N$
No absorption
at 3300–3500 cm^{-1}

(b) CH_3COCH_3
Strong ketone absorption
at 1715 cm^{-1}

$H_2C{=}CHCH_2OH$
Strong alcohol absorption
at 3400–3640 cm^{-1}

(c) CH_3COCH_3
Strong ketone absorption
at 1715 cm^{-1}

CH_3CH_2CHO
Strong aldehyde absorption
at 1725 cm^{-1}

13.56 One isomer of each pair in Problem 13.55 shows only one peak in its ^1H NMR spectrum. In (a), $(CH_3)_3N$ absorbs at 2.12 δ; the other isomer has a more complicated ^1H NMR spectrum. In (b) and (c), the acetone absorption occurs at 2.17 δ; the other isomers, again, show more complicated spectra.

13.57 (a)The ^{13}C NMR spectrum of $(CH_3)_3N$ shows only one peak; the spectrum of the other isomer shows 3 peaks.

(b) (c)The spectrum of acetone shows two peaks, one at 30 δ and one at 208 δ. The spectra of the other isomers show three peaks.

13.58

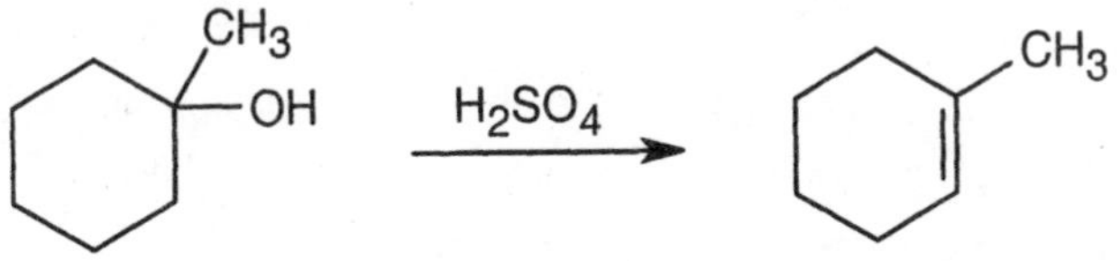

$$\text{1-Methylcyclohexanol} \xrightarrow{H_2SO_4} \text{1-Methylcyclohexene}$$

1-Methylcyclohexanol 1-Methylcyclohexene

The infrared spectrum of the starting alcohol shows a broad absorption at 3400–3640 cm^{-1}, due to an O–H stretch, and another strong absorption at 1050–1100 cm^{-1}, due to a C–O stretch. The alkene product exhibits medium intensity absorbances at 1640–1680 cm^{-1} and at 3000–3100 cm^{-1}. Monitoring the *disappearance* of one of the alcohol absorptions allows one to decide when the alcohol is totally dehydrated. It is also possible to monitor the *appearance* of one of the alkene absorbances.

13.59 The mass spectrum of 1-methylcyclohexanol shows a peak at $m/z = 114$. After reaction is complete, this peak disappears and a peak due to 1-methylcyclohexene ($m/z = 96$) appears.

13.60 The structural formula shows that the compound is most likely aromatic. The absorption near 1700 cm^{-1} is due to a carbonyl group. The absorptions between 1500–1600 cm^{-1} are due to an aromatic ring. (Peaks at 700–900 cm^{-1} are also characteristic of aromatic rings.) The compound is benzaldehyde.

13.61

1-Methylcyclohexene (**A**) Methylenecyclohexane (**B**) plane of symmetry

^{13}**C:** Symmetrical methylenecyclohexane (**B**) has only five different kinds of carbons and shows five peaks in its ^{13}C NMR spectrum. 1-Methylcyclohexene (**A**) has seven different kinds of carbons and shows seven peaks.

^{1}**H:** 1-Methylcyclohexene (**A**) has six different kinds of protons; methylenecyclohexane (**B**) has four different kinds of protons. Since several absorptions in each of the spectra overlap, it is more helpful to focus on specific absorptions in each spectrum. The ^{1}H NMR spectrum of **A** shows an unsplit methyl group and a vinylic proton signal of relative area 1. The vinylic absorption of **B** has relative area 2.

13.62

$CH_3CH_2C{\equiv}CCH_2CH_3$ — Hex-3-yne (**C**)

$CH_3CH{=}CHCH{=}CHCH_3$ — Hexa-2,4-diene (**D**)

The isomers are easily distinguished by UV spectroscopy, since only **D** is conjugated and absorbs in the UV region.

^{1}H NMR can also be used to identify the product. The spectrum of **C** consists of a quartet and a triplet. The spectrum of **D** is more complex, but shows four protons absorbing in the vinylic region of the spectrum; no **C** protons absorb in this region.

13.63

	¹H NMR	*¹³C NMR*
(a)	$ClCH_2CH_2CH_2Cl$ (b a b)	$ClCH_2CH_2CH_2Cl$ (b a b)
Number of peaks:	2	2
Chemical shift:	2.2 δ (quintet) (a) 3.7 δ (triplet) (b)	15–55 δ (a) 35–80 δ (b)
(b)	$CH_3C(=O)CH_2CH_2Cl$ (a b c)	$CH_3C(=O)CH_2CH_2Cl$ (a d b c)
Number of peaks:	3	4
Chemical shift:	2.1 δ (singlet) (a) 2.5 δ (triplet) (b) 3.7 δ (triplet) (c)	8–30 δ (a) 15–55 δ (b) 35–80 δ (c) 170–210 δ (d)

13.64

(a) CH_3
$H_3C—C—CH_3$
CH_3

(b)

(c) O O

13.65

(a) Possible structures for C_3H_6O:

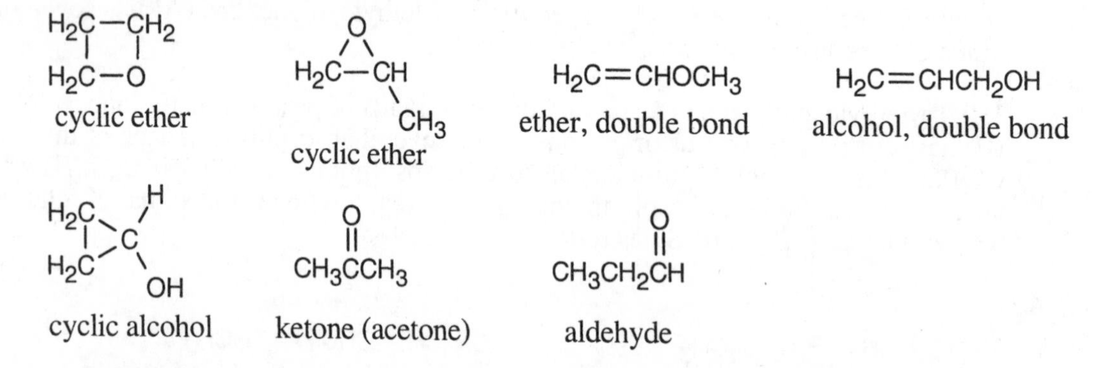

(b) An IR absorption at 1715 cm^{-1} is due to a carbonyl group. Only the last two compounds show an absorption near this region.

(c) The compound must be acetone, which has only one kind of proton and shows only one ^1H NMR absorption. (The aldehyde, with three different kinds of protons, would show three absorptions.)

13.66

	Compound	*Proton(s)*	*Chemical Shift*
(a)	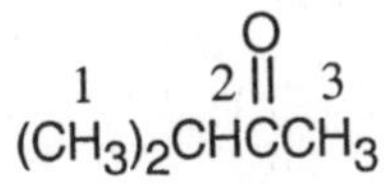 $(CH_3)_2CHCCH_3$ (with C=O; 1 = $(CH_3)_2$, 2 = CH, 3 = CH_3)	1	0.95 δ
		2	2.43 δ
		3	2.10 δ
(b)	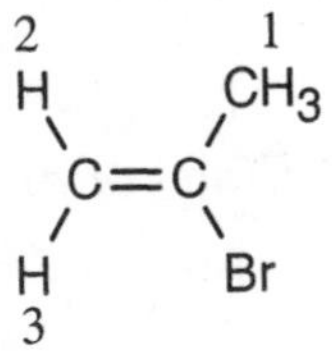 $H_2C{=}C(CH_3)Br$ (1 = CH_3, 2 = H, 3 = H)	1	2.32 δ
		2	5.25 δ
		3	5.54 δ

13.67 Either ^1H NMR or ^{13}C NMR can be used to distinguish among these isomers. In either case, it is first necessary to find the number of different kinds of protons or carbon atoms.

^{13}C NMR is the preferred method for identifying these compounds since each isomer differs in the number of absorptions in its ^{13}C NMR spectrum.

^1H NMR can also be used to distinguish among the isomers. The two isomers that show two ^1H NMR peaks differ in their splitting patterns.

Compound	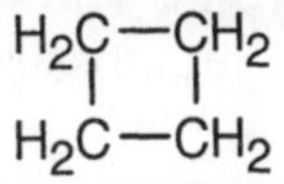 $H_2C{-}CH_2$ / $H_2C{-}CH_2$ (cyclobutane)	$H_2C{=}CHCH_2CH_3$	$CH_3CH{=}CHCH_3$	$(CH_3)_2C{=}CH_2$
Kinds of protons	1	5	2	2
Kinds of carbon atoms	1	4	2	3
Number of ^1H NMR peaks	1	5	2	2
Number of ^{13}C NMR peaks	1	4	2	3

13.68

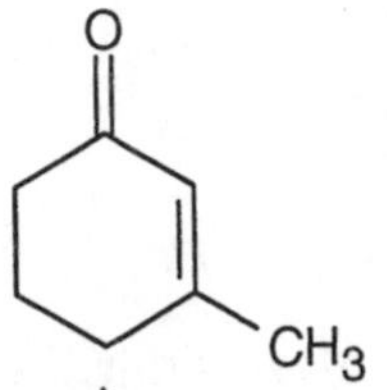

A: 3-methylcyclohex-2-enone (CH_3)

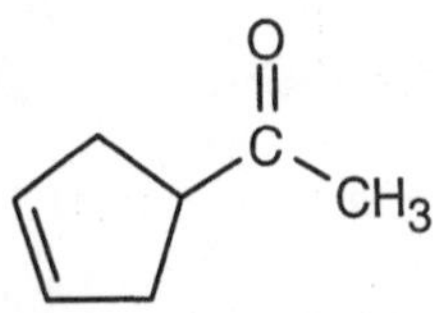

B: cyclopentenyl methyl ketone ($C(=O)CH_3$)

Distinguishing features of the ^1H NMR spectrum of **A** include one unsplit vinylic proton and a singlet methyl group. For **B**, distinguishing features of the ^1H NMR spectrum include two split vinylic protons and a singlet methyl group adjacent to a ketone.

13.69 Compound **A** has seven different kinds of carbons and shows seven lines in its ^{13}C NMR spectrum. Compound **B** has five different kinds of carbons (because of symmetry) and shows five lines in its ^{13}C NMR spectrum.

13.70 Only **A** is conjugated and shows absorption in the UV region.

13.71 The four isomers of $C_3H_6Br_2$ are shown below, along with the number of different kinds of protons for each structure.

Compound	*Kinds of protons*	*Compound*	*Kinds of protons*
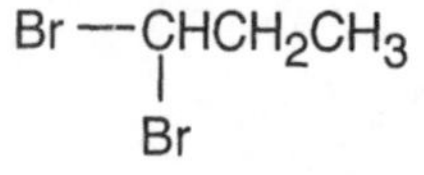 $Br{-}CH(Br)CH_2CH_3$	3	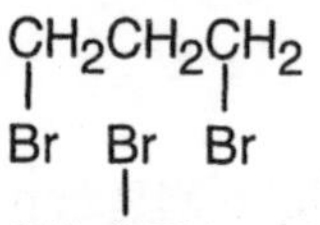 $BrCH_2CH_2CH_2Br$	2
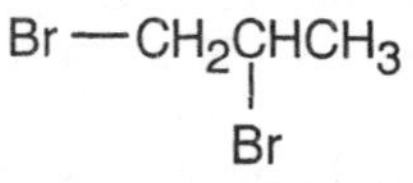 $Br{-}CH_2CH(Br)CH_3$	3	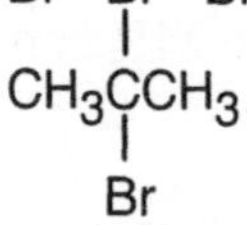 $CH_3CBr_2CH_3$	1

Because the spectrum of $C_3H_6Br_2$ shows two kinds of protons, it must represent 1,3-dibromopropane. The splitting pattern shown in the spectrum (triplet, quintet) is what is expected for 1,3-dibromopropane.

13.72 The IR absorption at 1740 cm^{-1} is due to an ester group. The splitting pattern (triplet, quartet) is caused by an ethyl group. Two structures are possible at this point.

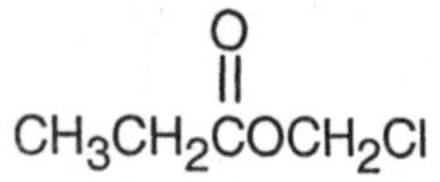

$CH_3CH_2C(=O)OCH_2Cl$ **I**

$CH_3CH_2OC(=O)CH_2Cl$ **II**

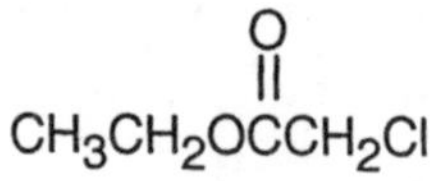

In **I**, the protons attached to the carbon bonded to both oxygen and chlorine ($-OCH_2Cl$) absorb far downfield (5.0 – 6.0 δ). Because no signal is present in this region of the ^1H NMR spectrum given, the unknown must be **II**. In addition, the quartet absorbing at 4.3 δ is typical of a CH_2 group next to an electronegative atom and coupled with a methyl group.

13.73

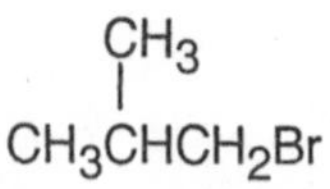

$CH_3CH(CH_3)CH_2Br$

13.74

(a) $H_3C(Cl)C{=}C(H)CH_2Cl$ Z

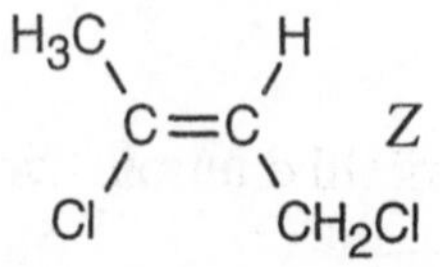

The *E* isomer is also an acceptable answer.

(b) $C_6H_5C(CH_3)_3$

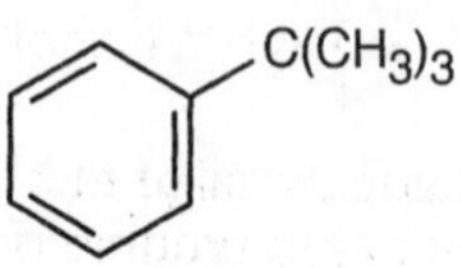

13.75

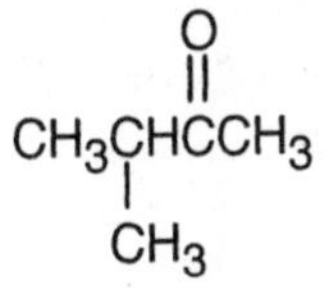

$$CH_3\underset{\displaystyle CH_3}{CH}\overset{\displaystyle O}{\overset{\|}{C}}CH_3$$

In the Medicine Cabinet

13.76

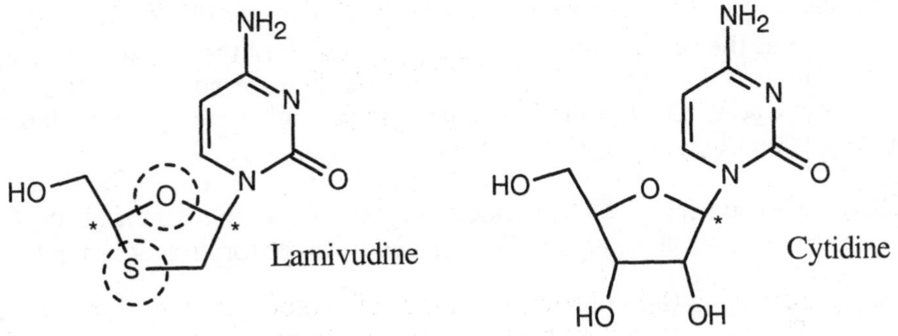

NH_2 N HO O N O S * * Lamivudine

NH_2 N HO O N O * HO OH Cytidine

(a) The two chirality centers of lamivudine are starred.
(b) The anomeric carbon of cytidine is starred.
(c) The acid-sensitive acetal-like groups are circled
(d)

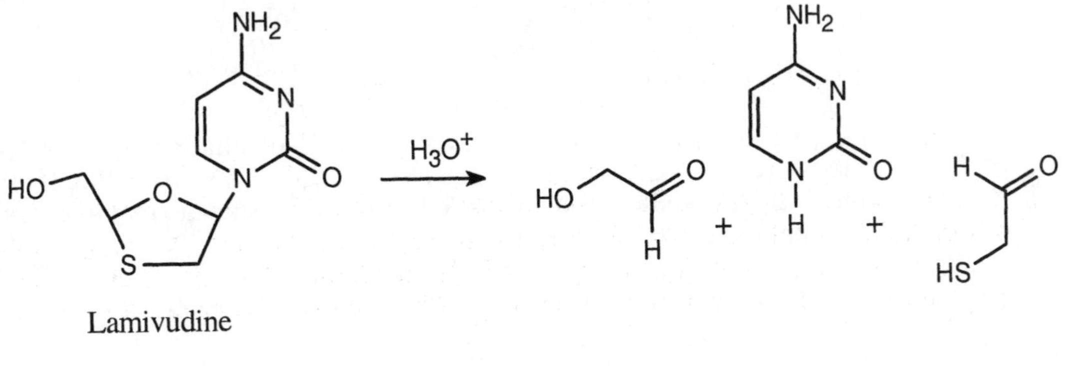

NH_2 N HO O N O S

Lamivudine $\xrightarrow{H_3O^+}$ HO O H + NH_2 N N O H + H O HS

(e)

$$\varepsilon = 8{,}600\ M^{-1}\ cm^{-1} = 8.6 \times 10^3\ L\ mol^{-1}\ cm^{-1}$$

$$l = 1.0\ cm$$

$$A = 0.195$$

$$c = \frac{A}{\varepsilon \times l} = \frac{0.195}{8.6 \times 10^3\ L{\cdot}mol^{-1}{\cdot}cm^{-1} \times 1.0\ cm} = 2.3 \times 10^{-5}\ M$$

Since the calculated value of c is the concentration after a tenfold dilution, the plasma concentration of lamivudine is 2.3×10^{-4}M, or 0.23 mM.

13.77

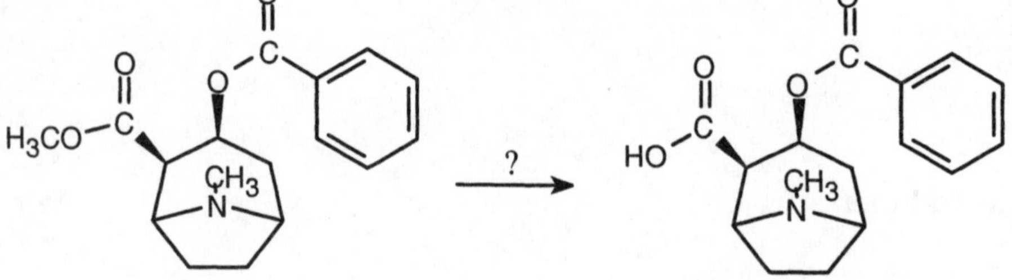

Cocaine → ? → Benzoylecgonine

IR spectroscopy is the best technique for deciding which of the two methyl groups of cocaine has been removed. If the ester methyl group has been removed, an IR absorption due to the carboxylic acid –O–H group will appear. If the amine methyl group has been removed, an absorption due to –N–H will appear.

^{1}H NMR can also be used. If the ester methyl group has been lost, the absorption at around 4 δ due to the methyl group will be replaced by an absorption at around 12 δ.

In the Field with Agrochemicals

13.78

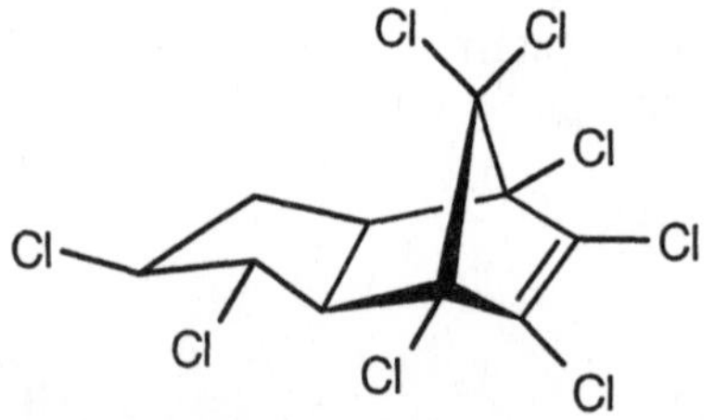

Chlordane

$C_{10}H_6Cl_8$

The mass of the carbons and hydrogens alone is $m/z = 126$ (excluding ^{2}H and ^{13}C). If all of the chlorines were ^{35}Cl, m/z for chlordane would be $126 + (8 \times 35) = 406$. If all chlorines were ^{37}Cl, m/z would equal $126 + (8 \times 37) = 422$. Thus the maximum $m/z = 422$, and the minimum is 406.

Chapter 14 – Biomolecules: Carbohydrates

Chapter Outline

I. Classification of carbohydrates (Section 14.1).
   A. Simple *vs.* complex:
      1. Simple carbohydrates can't be hydrolyzed to smaller units.
      2. Complex carbohydrates are made up of two or more simple sugars linked together.
         a. A disaccharide is composed of two monosaccharides.
         b. A polysaccharide is composed of three or more monosaccharides.
   B. Aldoses *vs.* ketoses:
      1. A monosaccharide with an aldehyde carbonyl group is an aldose.
      2. A monosaccharide with a ketone carbonyl group is a ketose.
   C. *Tri-*, *tetr-* , *pent-*, etc. indicate the number of carbons in the monosaccharide.

II. Monosaccharides (Sections 14.2 – 14.7).
   A. Configurations of monosaccharides (Section 14.2 – 14.4).
      1. Fischer projections (Section 14.2).
         a. Each stereocenter of a monosaccharide is represented by a pair of crossed lines.
         b. The carbonyl carbon is placed at or near the top of the Fischer projection.
         c. Fischer projections can be rotated by 180°, but not by 90° or 270°.
         d. Carbohydrates with more than one stereocenter are shown by stacking the centers on top of each other.
      2. D,L sugars (Section 14.3).
         a. (*R*)-Glyceraldehyde is also known as D-glyceraldehyde.
         b. In D sugars, the –OH group farthest from the carbonyl group points to the right. Most naturally-occurring sugars are D sugars.
         c. In L sugars, the –OH group farthest from the carbonyl group points to the left.
         d. D,L designations refer only to the configuration farthest from the carbonyl carbon and are unrelated to the direction of rotation of plane-polarized light.
      3. Configurations of the aldoses (Section 14.4).
         a. There are 4 aldotetroses – D and L erythrose and threose.
         b. There are 4 D,L pairs of aldopentoses: ribose, arabinose, xylose and lyxose.
         c. There are 8 D,L pairs of aldohexoses : allose, altrose, glucose, mannose, gulose, idose, galactose, and talose.
   B. Cyclic structures of monosaccharides (Sections 14.5 – 14.6).
      1. Hemiacetal formation (Section 14.5).
         a. Monosaccharides are in equilibrium with their internal hemiacetals.
            i. Glucose exists primarily as a six-membered pyranose ring, formed by the –OH group at C5 and the aldehyde.
            ii. Fructose also can exist as a five-membered furanose ring.
         b. Structure of pyranose rings.
            i. Pyranose rings have a chair-like geometry.
            ii. The hemiacetal oxygen is at the right rear for D-sugars.
            iii. An –OH group on the right in a Fischer projection is on the bottom in a pyranose ring, and an –OH group on the left is on the top.
            iv. For D sugars, the $-CH_2OH$ group is on the top.

2. Mutarotation (Section 14.6).
   a. When a monosaccharide cyclizes, a new stereocenter is generated.
      i. The two diastereomers are anomers.
      ii. The form with the anomeric –OH group trans to the $-CH_2OH$ group is the α anomer (minor anomer).
      iii. The form with the anomeric –OH group cis to the $-CH_2OH$ group is the β anomer (major anomer).
   b. When a solution of either pure anomer is dissolved in water, the optical rotation of the solution reaches a constant value.
      i. This process is called mutarotation.
      ii. Mutarotation is due to the reversible opening and recyclizing of the hemiacetal ring and is catalyzed by both acid and base.

C. Reactions of monosaccharides (Section 14.7).
   1. Ester and ether formation.
      a. Esterification occurs by treatment with an acid anhydride or acid chloride.
      b. Ethers are formed by treatment with methyl iodide and Ag_2O.
      c. Ester and ether derivatives are crystalline and easy to purify.
   2. Glycoside formation.
      a. Treatment of a hemiacetal with an alcohol and an acid catalyst yields an acetal.
         i. Acetals aren't in equilibrium with an open-chain form.
         ii. Aqueous acid reconverts the acetal to a monosaccharide.
      b. These acetals, called glycosides, occur in nature.
   3. Reduction of monosaccharides.
      Reaction of a monosaccharide with $NaBH_4$ yields an alditol.
   4. Oxidation of monosaccharides.
      a. Several mild reagents can oxidize the carbonyl group to a carboxylic acid (aldonic acid).
         i. Tollens reagent, Fehling's reagent and Benedict's reagent all serve as tests for reducing sugars.
         ii. All aldoses and some ketoses are reducing sugars, but glycosides are nonreducing.
      b. The more powerful oxidizing agent, dilute HNO_3, oxidizes aldoses to dicarboxylic acids (aldaric acids).

III. Other carbohydrates (Sections 14.8 – 14.11).
   A. Disaccharides (Section 14.8).
      1. Cellobiose and maltose.
         a. Cellobiose and maltose contain a 1,4'-glycosidic acetal bond between two glucose monosaccharide units.
            The prime (') shows that the glycosidic bond is between two different sugars.
         b. Maltose consists of two glucopyranose units joined by a 1,4'-α-glycosidic bond.
         c. Cellobiose consists of two glucopyranose units joined by a 1,4'-β-glycosidic bond.
         d. Both maltose and cellobiose are reducing sugars and exhibit mutarotation.
         e. Humans can't digest cellobiose but can digest maltose.
      2. Sucrose.
         a. Sucrose is a disaccharide that yields glucose and fructose on hydrolysis.
            i. Sucrose is called "invert sugar" because the sign of rotation changes when sucrose is hydrolyzed.
            ii. Sucrose is one of the most abundant pure organic chemicals in the world.

b. The two monosaccharides are joined by a glycosidic link between C1 of glucose and C2 of fructose.
c. Sucrose isn't a reducing sugar and doesn't exhibit mutarotation.

B. Polysaccharides (Section 14.9).
1. Polysaccharides have a reducing end and undergo mutarotation, but aren't considered to be reducing sugars because of their size.
2. Important polysaccharides.
a. Cellulose.
i. Cellulose consists of thousands of D-glucose units linked by 1,4'-β-glycosidic bonds.
ii. In nature, cellulose is used as structural material.
b. Starch.
i. Starch consists of thousands of D-glucose units linked by 1,4'-α-glycosidic bonds.
ii. Starch can be separated into amylose (water-soluble) and amylopectin (water-insoluble) fractions.
Amylopectin contains 1,6'-α-glycosidic branches.
iii. Starch is digested in the mouth by glycosidase enzymes, which only cleave α-glycosidic bonds.
c. Glycogen.
i. Glycogen is an energy-storage polysaccharide.
ii. Glycogen contains both 1,4'- and 1,6'-links.

C. Other important carbohydrates (Section 14.10).
1. Deoxy sugars have an –OH group missing and are components of nucleic acids.
2. In amino sugars, an –OH is replaced by a $-NH_2$.
Amino sugars are found in chitin and in antibiotics.

D. Cell surface carbohydrates and carbohydrate vaccines (Section 14.11).
1. Polysaccharides are involved in cell-surface recognition.
a. Polysaccharide markers on the surface of red blood cells are responsible for blood-group incompatibility.
b. Red blood cells have two types of markers (antigenic determinants) – A and B.
c. Unusual carbohydrates are components of these markers.
2. Possible anticancer vaccines have been synthesized from antibodies to cell-surface polysaccharides found on the surface of cancer cells.

E. Plant carbohydrates may be able to replace nonrenewable resources as a source of many chemicals (Section 14.12).

Solutions to Problems

14.1

(a) Threose *an aldotetrose*

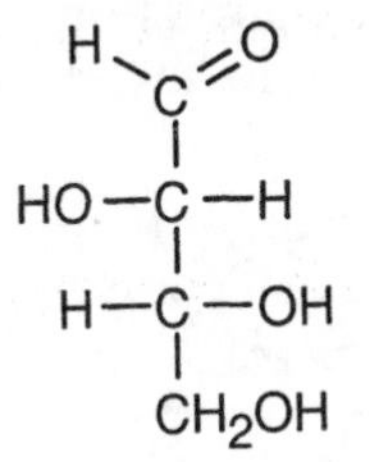

(b) Ribulose *a ketopentose*

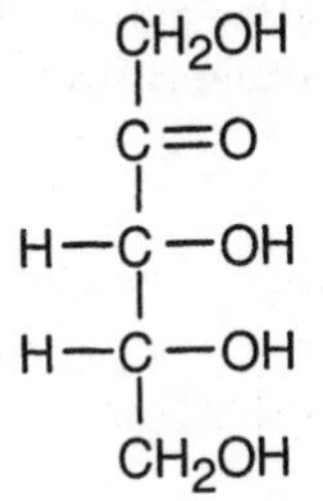

(c) Tagatose *a ketohexose*

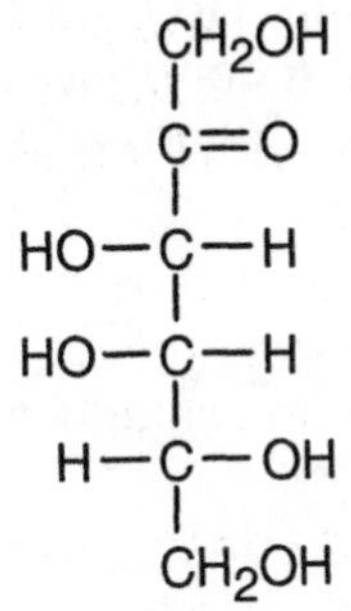

(d) 2-Deoxyribose *an aldopentose*

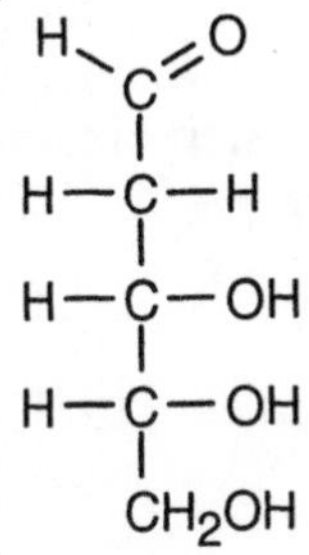

14.2 As in Practice Problem 14.2, orient the molecule so that two horizontal bonds are pointing out of the page and two vertical bonds are pointing into the page. Then draw two perpendicular lines and arrange the functional groups in the same order as in the tetrahedral projection.

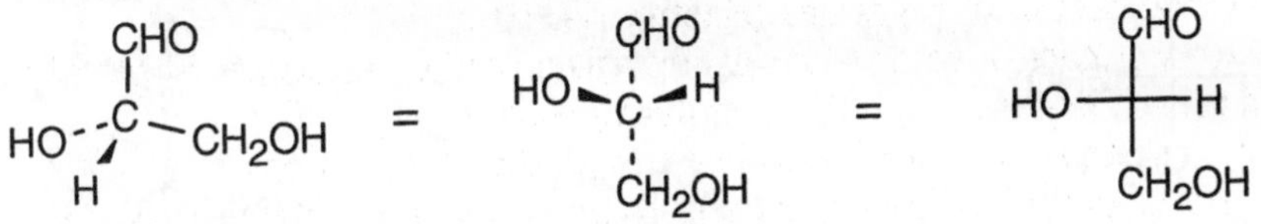

14.3 To solve this problem, first draw the correct tetrahedral representations of the 2-chlorobutane enantiomers. (If necessary, review Section 6.6 and Practice Problem 6.5.) Then, convert the tetrahedral representations to Fischer projections by the method used in Problem 14.2.

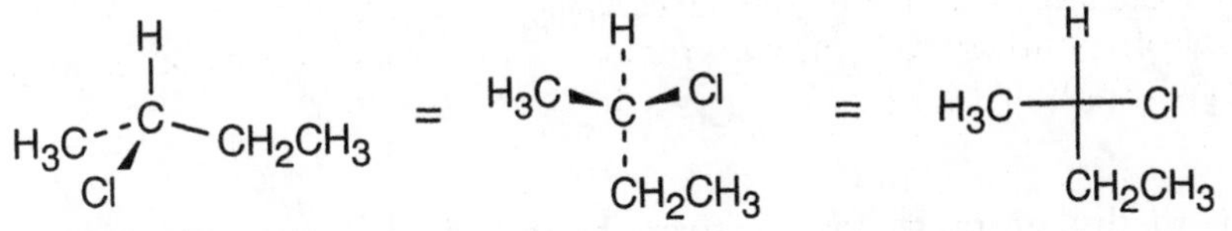

(*R*)-2-Chlorobutane

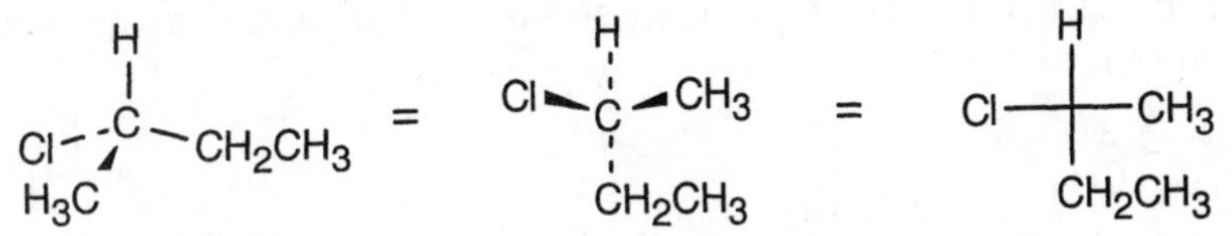

(*S*)-2-Chlorobutane

14.4 First, convert the Fischer projection to a tetrahedral representation by drawing the horizontal bonds out of the page and the vertical bonds into the page. Then use the sequence rules described in Section 6.6 to assign priorities to the four groups. Rotate the lowest priority group to the rear, and note the direction of rotation of the arrows that go from group 1 to group 2 to group 3. If the arrows indicate clockwise rotation, the stereoisomer has the *R* configuration; if the arrows indicate counterclockwise rotation, the stereoisomer has the *S* configuration.

(a)

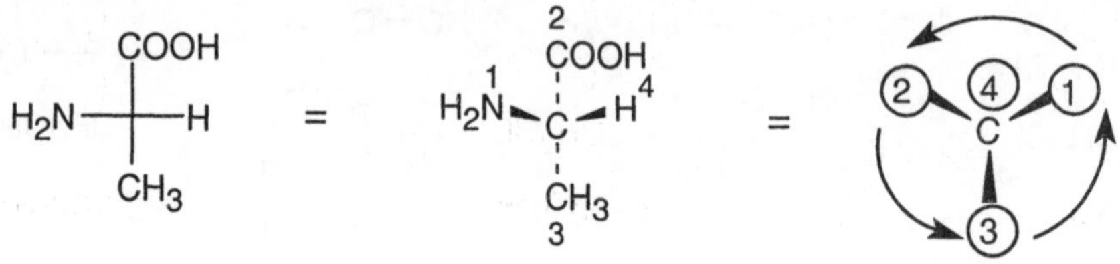

COOH
H_2N—+—H = (1) H_2N, (2) COOH, (4) H, (3) CH_3 = S
CH_3

(b)

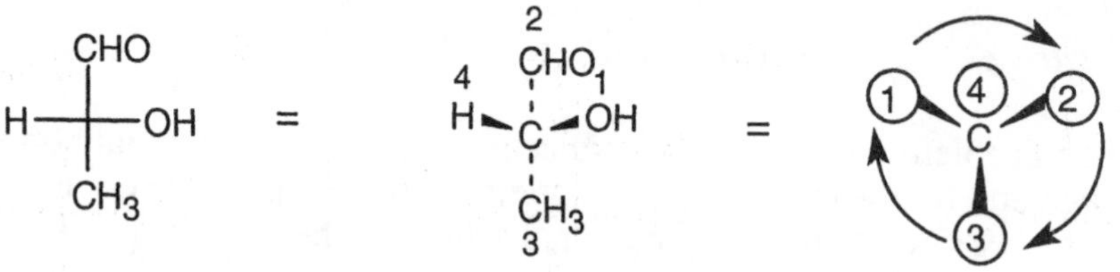

CHO
H—+—OH = (4) H, (2) CHO, (1) OH, (3) CH_3 = R
CH_3

(c)

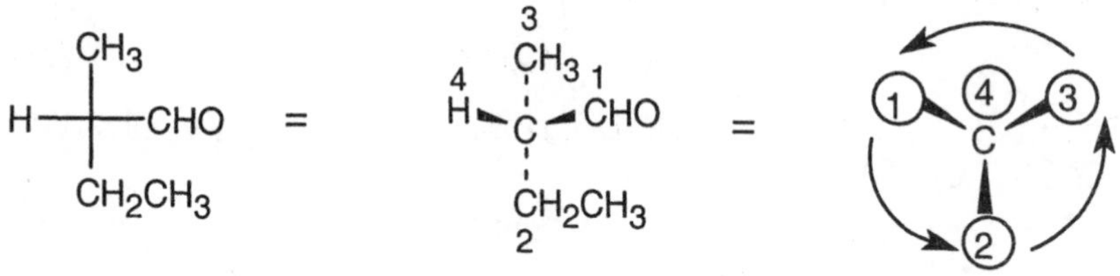

CH_3
H—+—CHO = (4) H, (3) CH_3, (1) CHO, (2) CH_2CH_3 = S
CH_2CH_3

14.5 Bring the methyl group slightly forward, so that you have the structure on the left:

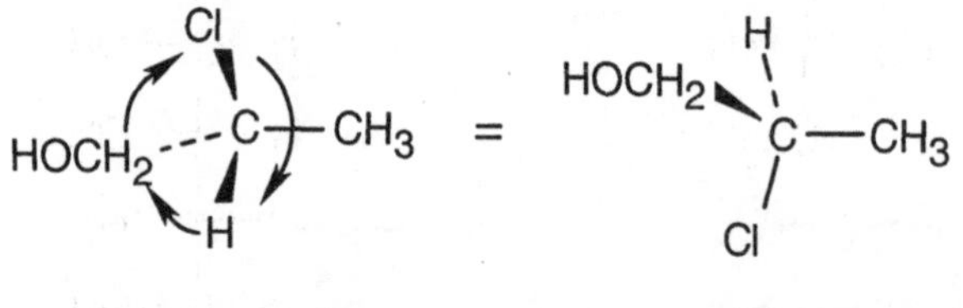

Cl, $HOCH_2$, CH_3, H = H, $HOCH_2$, C—CH_3, Cl

Holding the methyl group steady, perform the indicated rotation to arrive at the second structure shown above. This structure is close in structure to the "steering wheel" structures that we used to assign configurations in Chapter 6. Redraw the second structure above as a "steering wheel" structure, assign priorities to the four substituents, and determine the configuration (*R*).

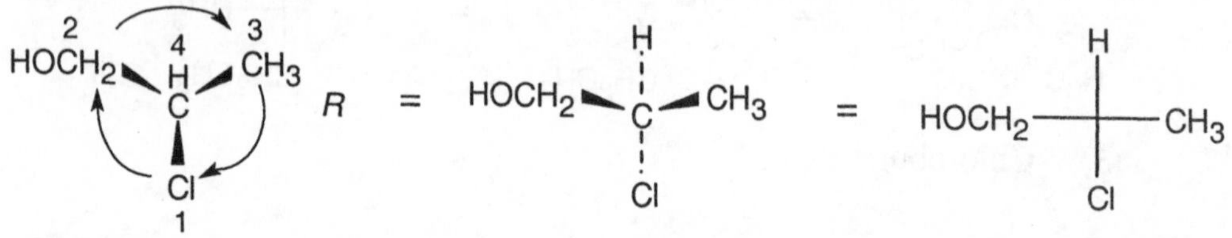

2 $HOCH_2$, 4 H, 3 CH_3, 1 Cl — *R* = H, $HOCH_2$, CH_3, Cl =

H
$HOCH_2$—+—CH_3
Cl

A small tilt of the first structure above produces the second structure, which can easily be drawn as a Fischer projection.

14.6 For a D sugar, the –OH group on the lowest carbon chirality center is on the right. For an L sugar, it is on the left.

(a)
CHO
HO—(S)—H
HO—(S)—H
CH_2OH

L-Erythrose

(b)
CHO
H—(R)—OH
HO—(S)—H
H—(R)—OH
CH_2OH

D-Xylose

(c)
CH_2OH
C=O
HO—(S)—H
H—(R)—OH
CH_2OH

D-Xylulose

14.7

(a)
CHO
H—(R)—OH
H—(R)—OH
CH_2OH

D-Erythrose

(b)
CHO
HO—(S)—H
H—(R)—OH
HO—(S)—H
CH_2OH

L-Xylose

(c)
CH_2OH
C=O
H—(R)—OH
HO—(S)—H
CH_2OH

L-Xylulose

14.8

(a)
O=C—H
H—OH
HO—H
HO—H
CH_2OH

L-Arabinose

(b)
O=C—H
H—OH
HO—H
CH_2OH

L-Threose

(c)
O=C—H
HO—H
H—OH
H—OH
HO—H
CH_2OH

L-Galactose

14.9 Thirty-two aldoheptoses are possible. Sixteen are D sugars, and sixteen are L sugars.

14.10

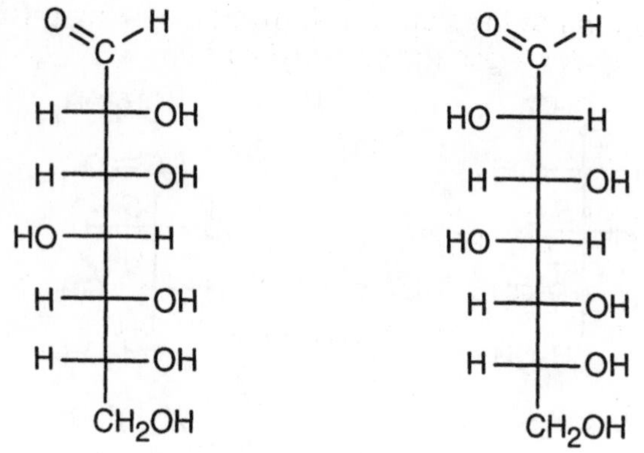

14.11

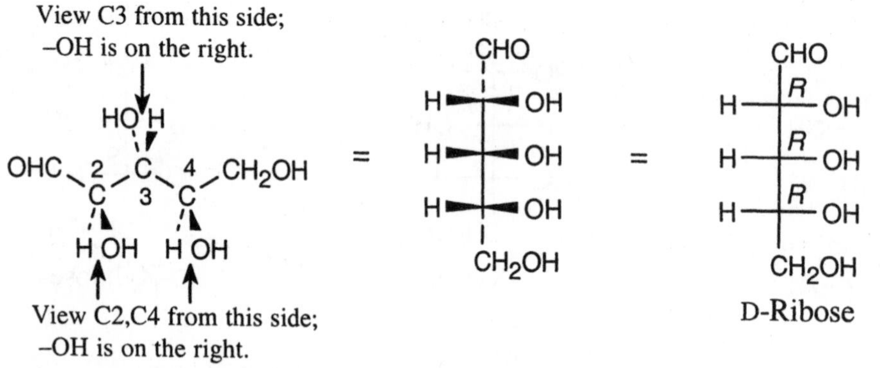

View C3 from this side;
–OH is on the right.

View C2,C4 from this side;
–OH is on the right.

D-Ribose

14.12 Follow the steps in Practice Problem 14.5. First, draw the Fischer projection of D-galactose. Lay the projection on its side, and curl it around so that the aldehyde is at the right front, and the CH_2OH is at the left rear. Connect the –OH at C5 to the C1 carbonyl group, raise the left carbon (C4), and lower the right carbon (C1) to form the chair. Note: For simplicity, hydrogens have been left off the ring structures.

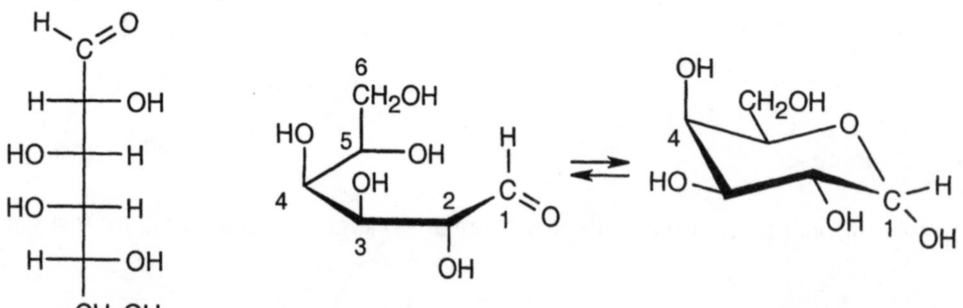

D-Galactose

Pyranose form

14.13 The steps for drawing a furanose are similar to the steps for drawing a pyranose. Ring formation occurs between the –OH group at C4 and the carbonyl carbon.

O=C–H
H–R–OH
H–R–OH
H–R–OH
CH_2OH

5
$HOCH_2$
OH
H
4 3 2 C=O
1
OH OH

D-Ribose

$HOCH_2$
O
C–H
4 1 OH
OH OH

(Furanose form)

14.14 The furanose of fructose results from ring formation between the –OH group at C5 and the ketone at C2. In the α anomer, the anomeric –OH group is trans to the C6 $-CH_2OH$ group, and in the β anomer the two groups are cis. In the pyranose form, cyclization occurs between the –OH group at C6 and the ketone.

OH
HO O
OH
OH CH_2OH

β-D-Fructopyranose

OH
HO O
CH_2OH
OH OH

α-D-Fructopyranose

cis
* *
$HOCH_2$ OH
O OH
CH_2OH
OH

β-D-Fructofuranose

trans
*
$HOCH_2$ CH_2OH
O OH
OH
OH *

α-D-Fructofuranose

14.15

a
OH
e
$HOCH_2$
O
HO OH
e e
OH
e

β-D-Galactopyranose

e a
e $HOCH_2$ OH
HO O
HO OH
e e

β-D-Mannopyranose

β-D-Galactopyranose and β-D-mannopyranose each have one hydroxyl group in the axial position. Galactose and mannose are of equal stability.

14.16

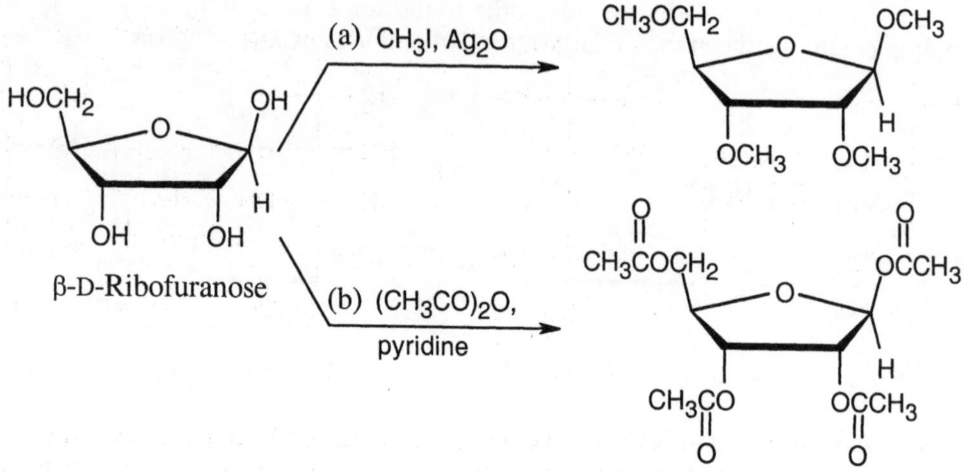

$HOCH_2$ OH O H OH OH

β-D-Ribofuranose

(a) CH_3I, Ag_2O

CH_3OCH_2 OCH_3 O H OCH_3 OCH_3

(b) $(CH_3CO)_2O$, pyridine

O ‖ CH_3COCH_2 O ‖ $OCCH_3$ O H CH_3CO ‖ O $OCCH_3$ ‖ O

14.17

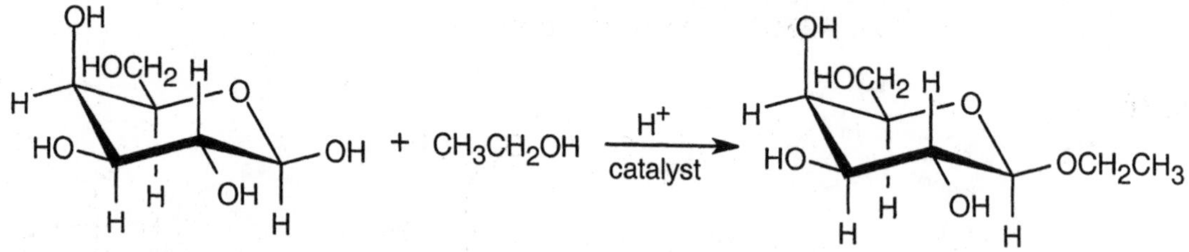

OH $HOCH_2$ H H O HO OH H OH H H

β-D-Galactopyranose

+ CH_3CH_2OH $\xrightarrow[\text{catalyst}]{H^+}$

OH $HOCH_2$ H H O HO OCH_2CH_3 H OH H H

Ethyl β-D-Galactopyranoside

14.18

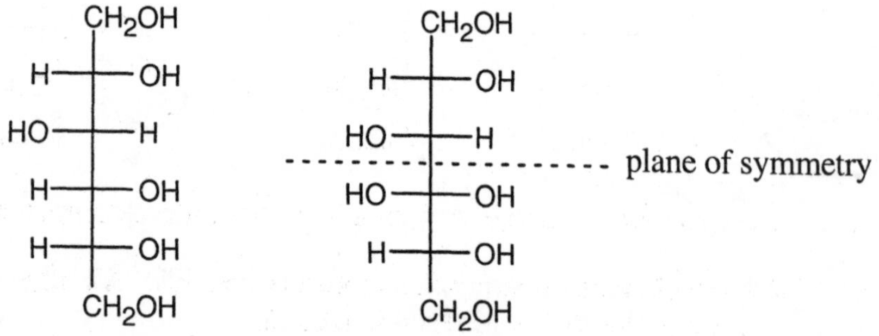

CH_2OH
H—OH
HO—H
H—OH
H—OH
CH_2OH

D-Glucitol

CH_2OH
H—OH
HO—H
- - - - - - plane of symmetry
HO—OH
H—OH
CH_2OH

Galactitol

Reduction of the aldehyde group of D-galactose yields an alditol that has a plane of symmetry and is an optically inactive meso compound.

14.19

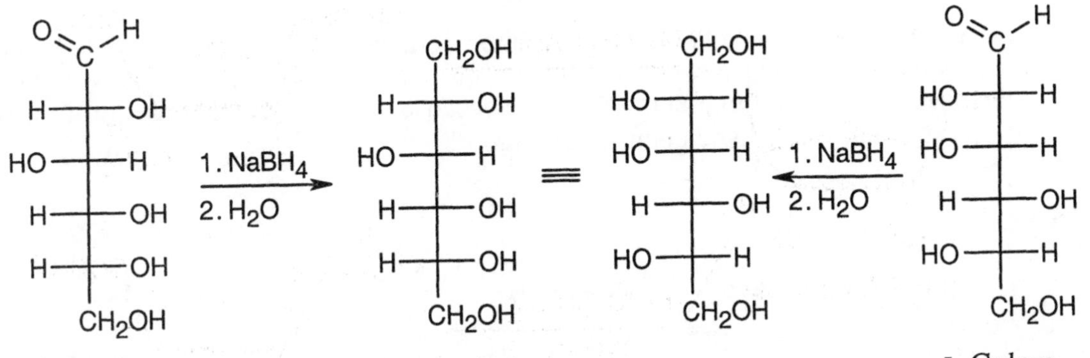

D-Glucose —(1. $NaBH_4$; 2. H_2O)→ D-Glucitol ≡ ←(1. $NaBH_4$; 2. H_2O)— L-Gulose

Reaction of an aldose with $NaBH_4$ produces a polyol (alditol). Because an alditol has the same functional group at both ends, the number of stereoisomers of an *n*-carbon alditol is one half the number of stereoisomers of the parent aldose, and two different aldoses can yield the same alditol. Here L-gulose and D-glucose form the same alditol (rotate the Fischer projection of L-gulitol 180° to see the identity).

14.20

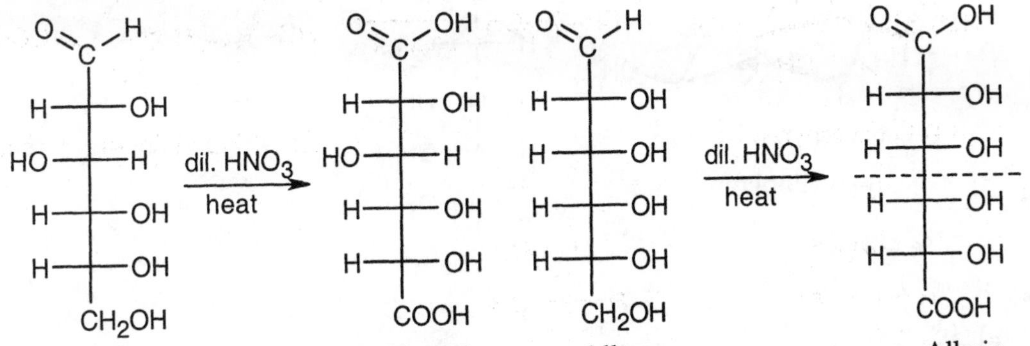

D-Glucose —(dil. HNO_3, heat)→ D-Glucaric acid

D-Allose —(dil. HNO_3, heat)→ Allaric acid

Allaric acid has a plane of symmetry and is an optically inactive meso compound.

14.21 D-Allose and D-galactose yield meso aldaric acids (optically inactive). All other D-aldohexoses form optically active aldaric acids on oxidation.

14.22

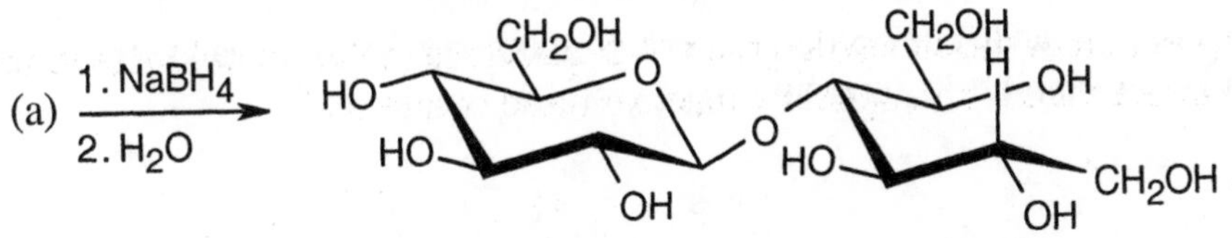

Cellobiose

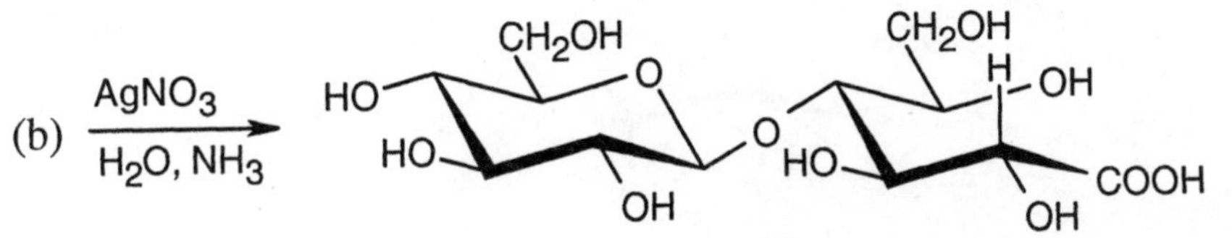

(a) 1. $NaBH_4$; 2. H_2O →

(b) $AgNO_3$; H_2O, NH_3 →

Visualizing Chemistry

14.23 (a) Convert the model to a Fischer projection, remembering that the aldehyde group is on top, pointing into the page, and that the groups bonded to the carbons below point out of the page. The model represents a D-aldose because the –OH group at the chiral carbon farthest from the aldehyde points to the right.

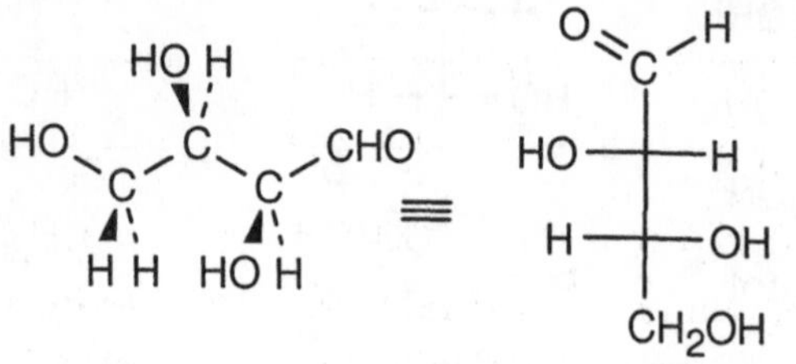

D-Threose

(b) Break the hemiacetal bond and uncoil the aldohexose. Notice that all hydroxyl groups point to the right in the Fischer projection. The model represents the β anomer of D-allopyranose.

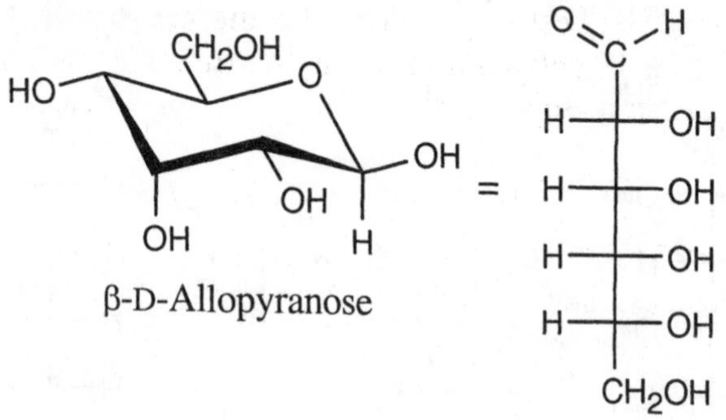

β-D-Allopyranose

14.24

(a)

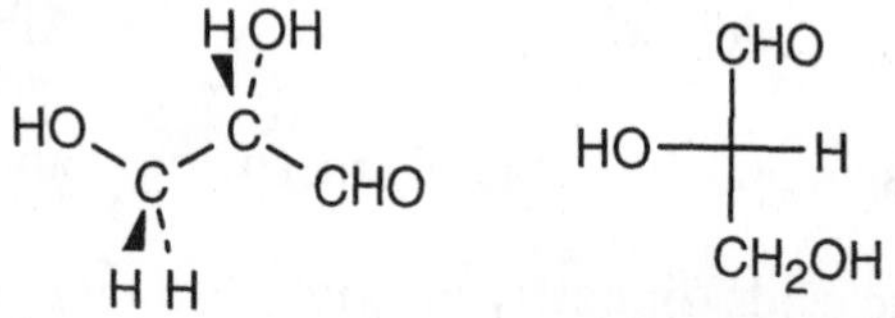

L-Glyceraldehyde

(b)

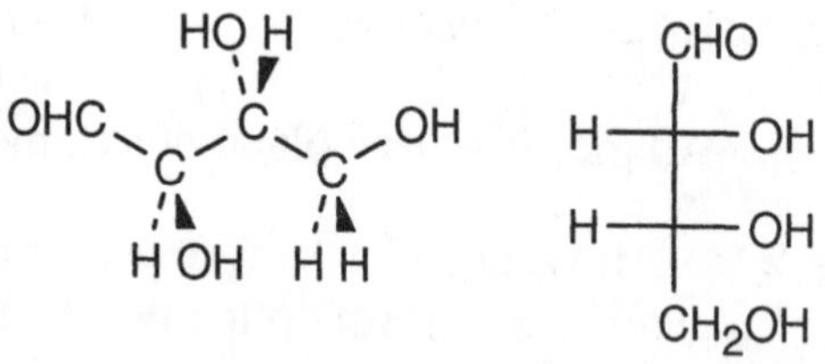

D-Erythrose

14.25 The structure represents an α anomer because the anomeric –OH group and the $-CH_2OH$ group are trans. The compound is α-L-mannopyranose because the –OH group at C2 is the only non-anomeric axial hydroxyl group.

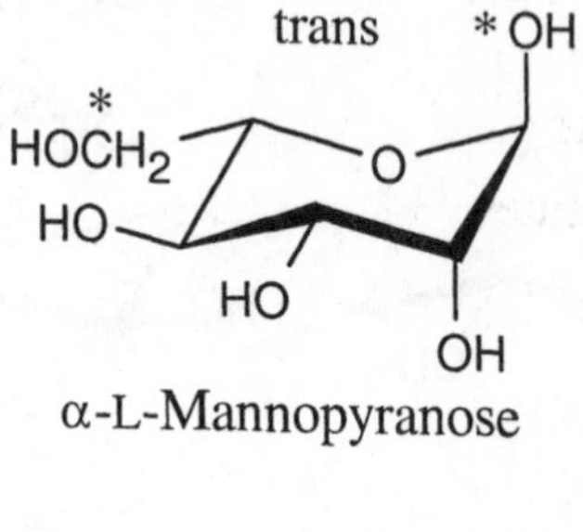

α-L-Mannopyranose

14.26

(a)

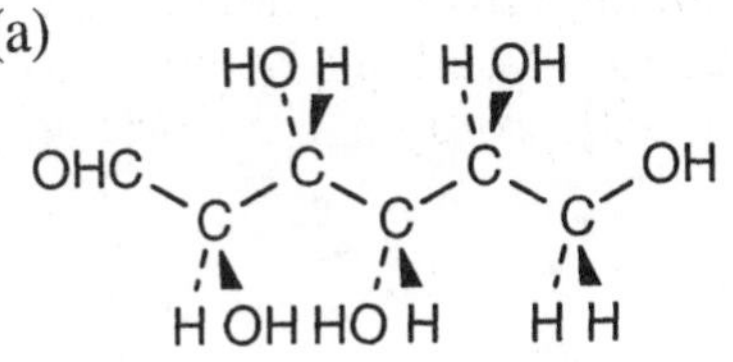

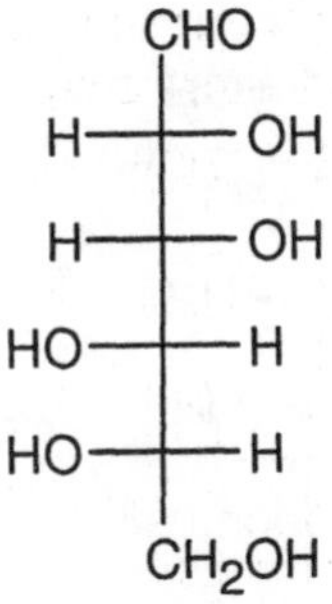

L-Mannose

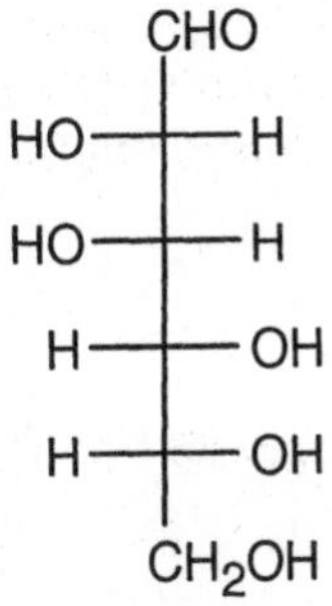

D-Mannose (enantiomer)

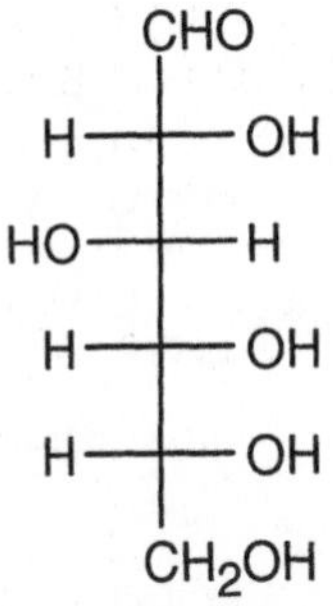

D-Glucose (diastereomer)

(b) The model represents an L-aldohexose because the hydroxyl group on the chiral carbon farthest from the aldehyde group points to the left.

(c) This is tricky! The furanose ring of an aldohexose is formed by connecting the –OH group at C4 to the aldehyde carbon. The best way to draw the anomer is to lie L-mannose on its side and form the ring. All substituents point down in the furanose, and the anomeric –OH and the $-CH(OH)CH_2OH$ group are cis.

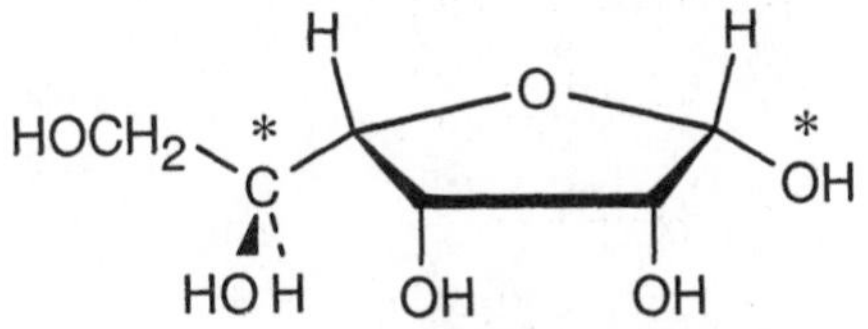

β-L-Mannofuranose

Additional Problems

Nomenclature

14.27

(a) a ketotriose

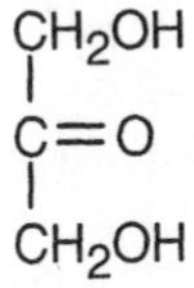

(b) a ketopentose

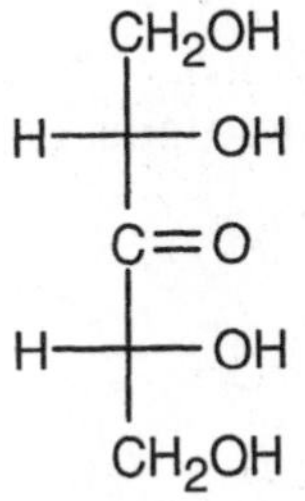

(c) an aldoheptose

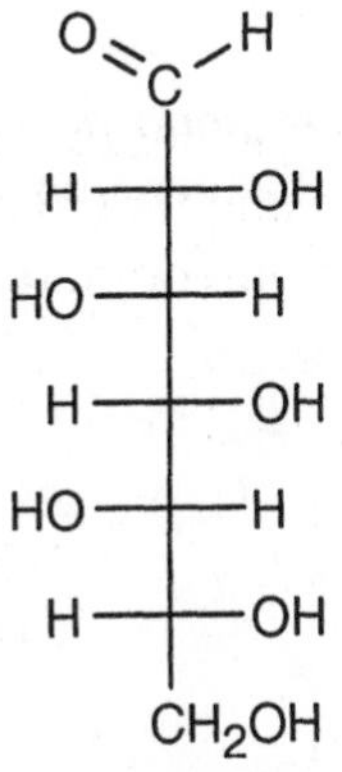

14.28

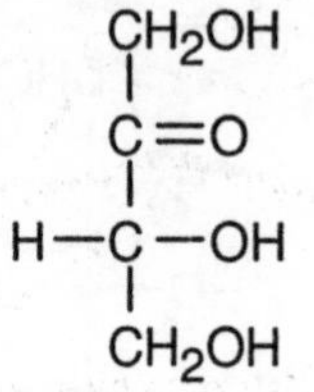

CH_2OH
|
C=O
|
H—C—OH
|
CH_2OH

a ketotetrose

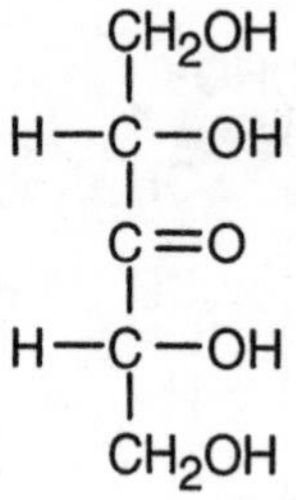

CH_2OH
|
H—C—OH
|
C=O
|
H—C—OH
|
CH_2OH

a ketopentose

14.29

O=C—H
|
HO—C—H
|
H—C—OH
|
H—C—H
|
H—C—OH
|
CH_2OH

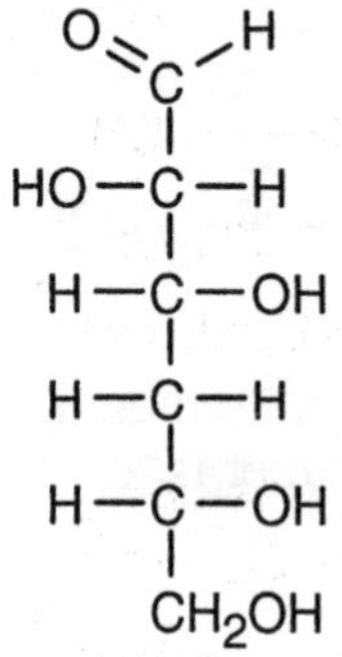

a deoxyaldohexose

14.30

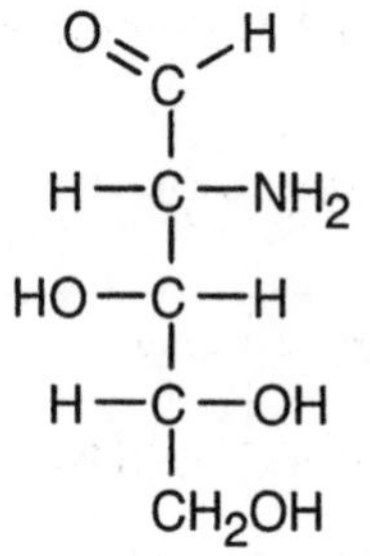

O=C—H
|
H—C—NH_2
|
HO—C—H
|
H—C—OH
|
CH_2OH

a five-carbon amino sugar

14.31

Definition	*Example*
(a) A *monosaccharide* is a carbohydrate that can't be hydrolyzed to smaller units.	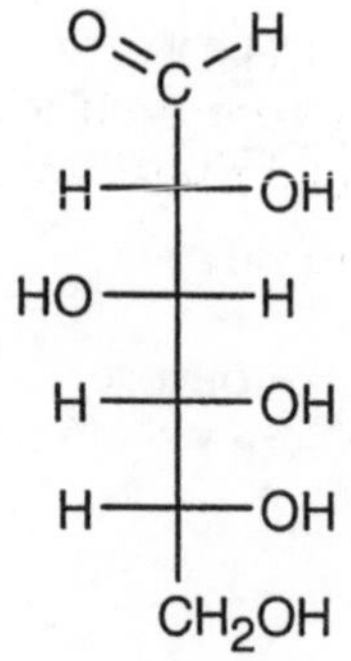 O=C—H; H—C—OH; HO—C—H; H—C—OH; H—C—OH; CH_2OH D-Glucose

(b) An *anomeric center* is a stereocenter formed when an open-chain monosaccharide cyclizes to a furanose or a pyranose ring.

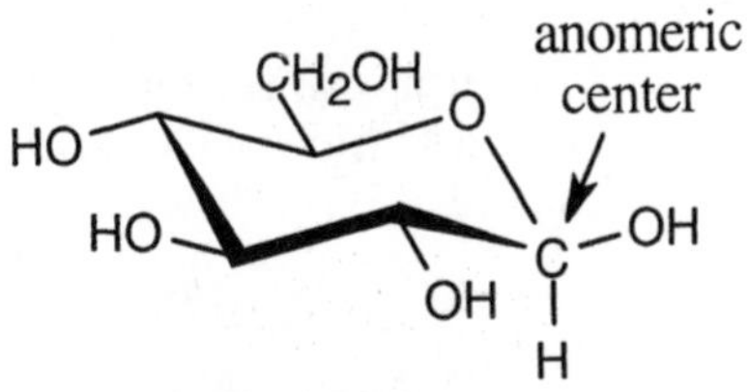

anomeric center

β-D-Glucopyranose

(c) A *Fischer projection* is a drawing of a carbohydrate in which each stereocenter is represented as a pair of perpendicular lines. Vertical lines represent bonds going into the page, and horizontal lines represent bonds coming out of the page.

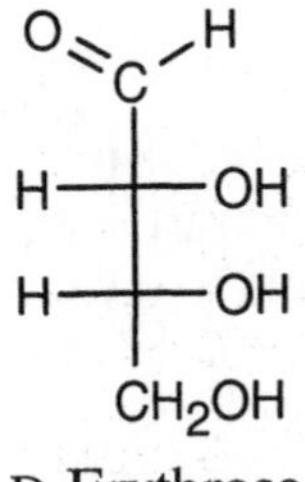

D-Erythrose

(d) A *glycoside* is an acetal of a carbohydrate, formed when an anomeric hydroxyl group reacts with another compound containing a hydroxyl group.

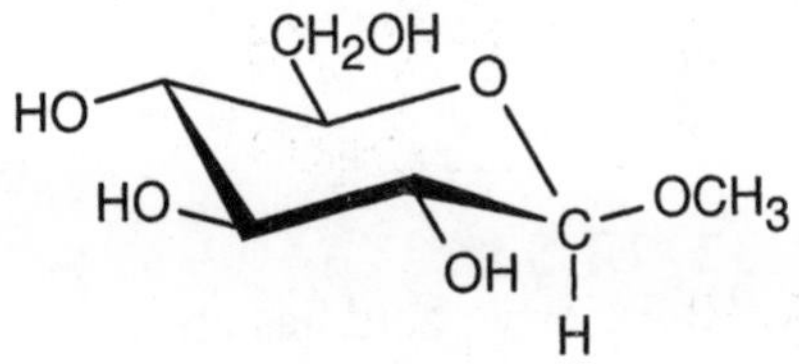

Methyl β-D-glucopyranoside

(e) A *reducing sugar* is a sugar that reacts with any of several reagents to yield an aldonic acid plus reduced reagent.

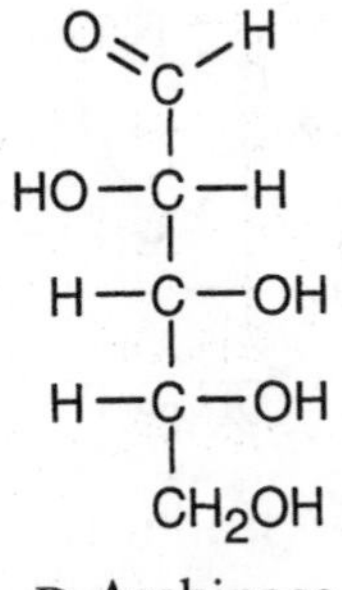

D-Arabinose

(f) A *pyranose* is a six-membered cyclic hemiacetal ring form of a monosaccharide.

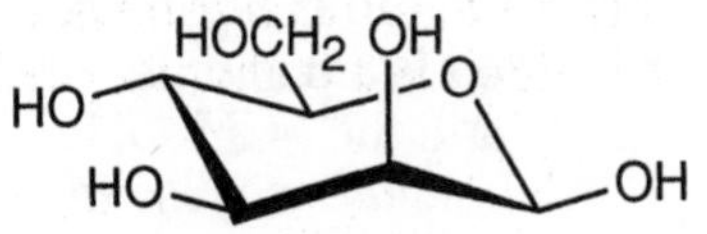

β-D-Mannopyranose

(g) A *1,4' link* occurs when the anomeric hydroxyl group (at carbon 1) of a pyranose or furanose forms a glycosidic bond with the hydroxyl group at carbon 4 of a second monosaccharide.

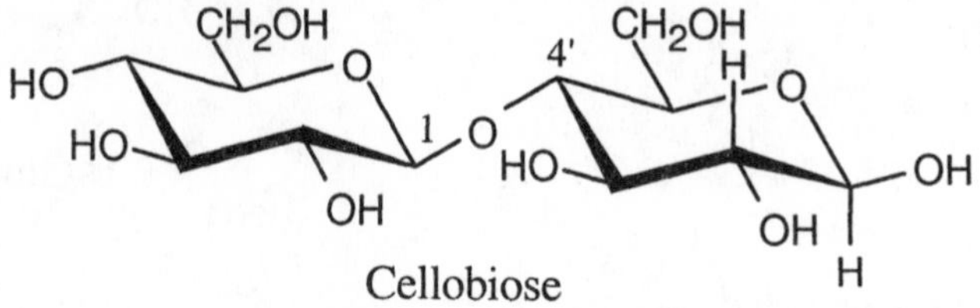

Cellobiose

(h) A D-*sugar* is a sugar in which the hydroxyl group farthest from the carbonyl group points to the right in a Fischer projection.

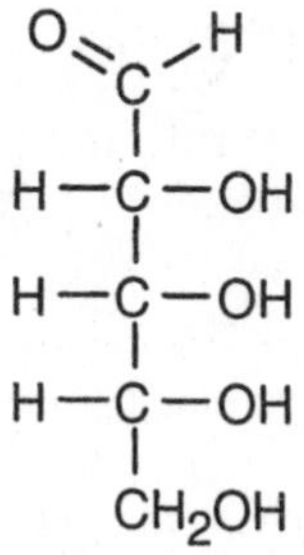

D-Ribose

14.32–14.33

Ascorbic acid has an L configuration because the hydroxyl group at the lowest stereocenter points to the left.

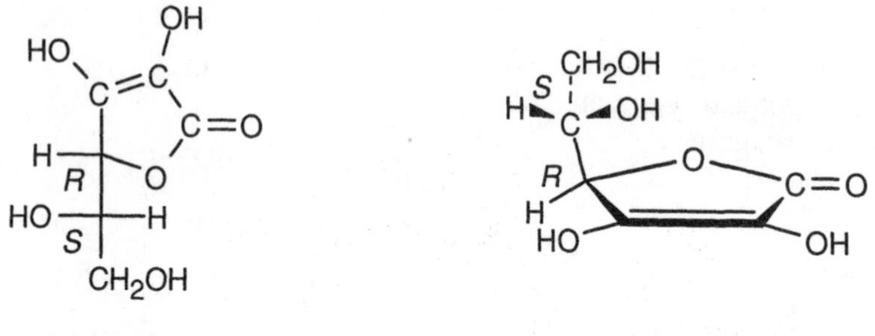

L-Ascorbic acid

14.34

The structure is a pyranose (6-membered ring) and is a β anomer (the C1 hydroxyl group and the $-CH_2OH$ groups are cis). It is a D sugar because the –O– at C5 is on the right in the uncoiled form.

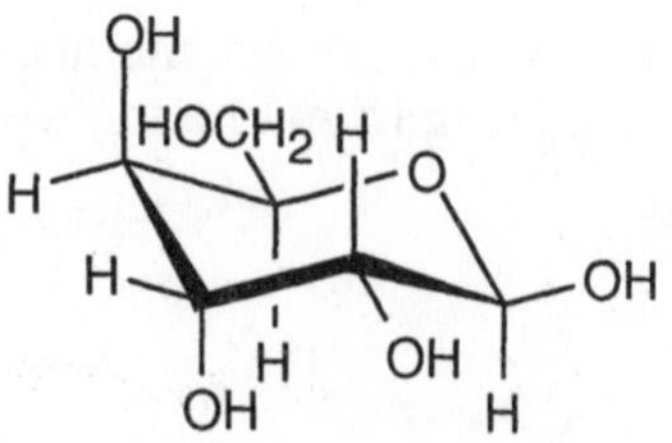

β-D-Gulopyranose

14.35

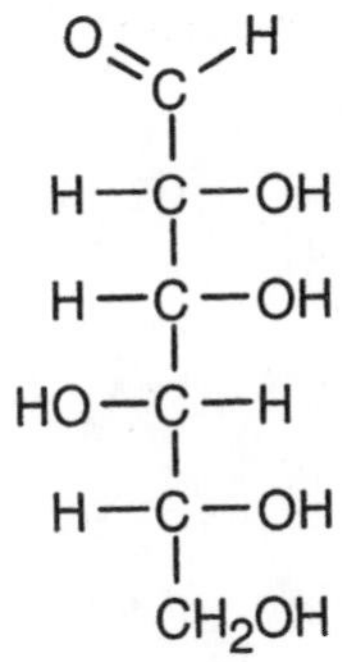

```
O=C—H
  |
H—C—OH
  |
H—C—OH
  |
HO—C—H
  |
H—C—OH
  |
 CH2OH
```

D-Gulose

14.36

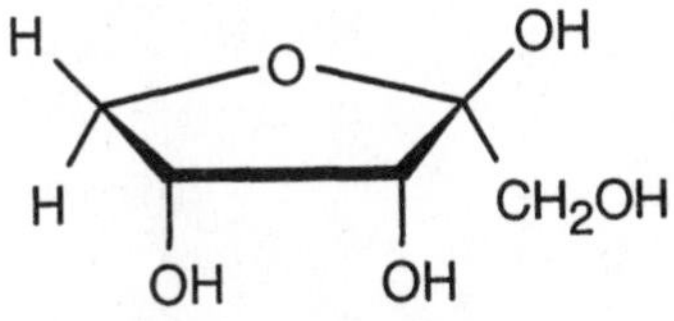

β-D-Ribulofuranose

14.37 The hemiacetal group of maltose is in equilibrium with the aldehyde, which can be reduced by $NaBH_4$. Sucrose has no hemiacetal group, and thus no aldehyde is available to be reduced by $NaBH_4$.

Maltose:

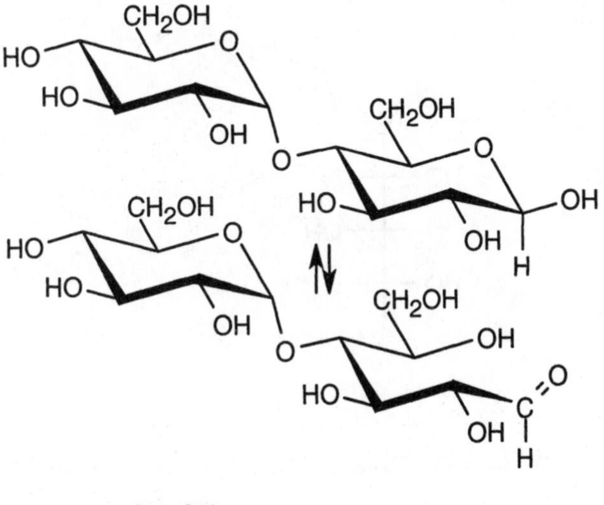

Sucrose:

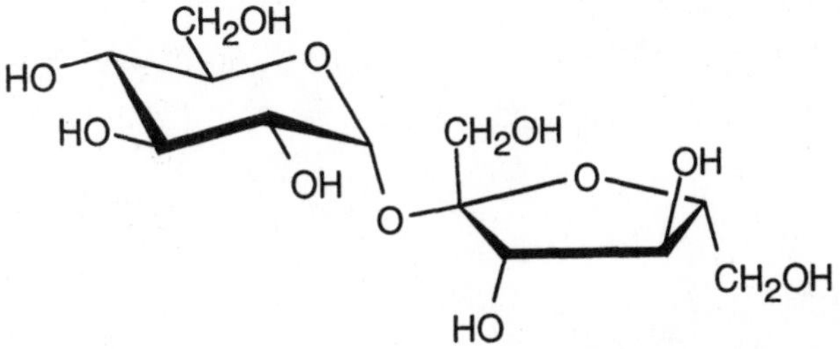

14.38

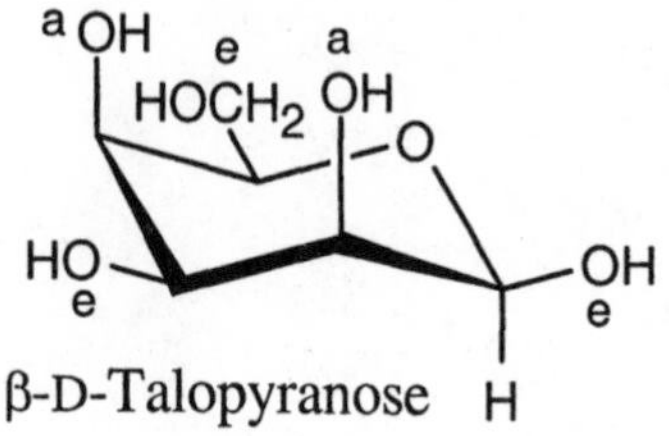

β-D-Talopyranose

14.39

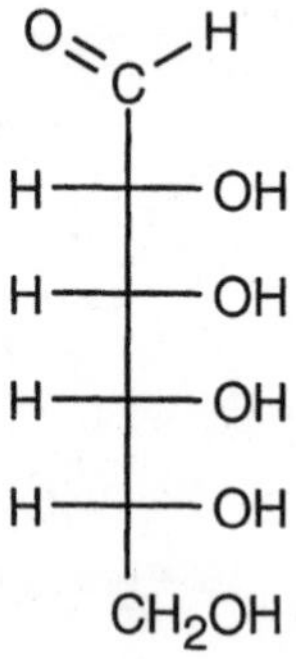

D-Allose

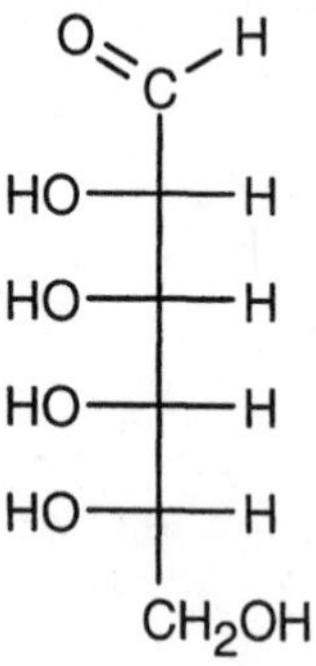

L-Allose

D-Allose and L-allose are enantiomers. Their physical properties, such as melting point, solubility in water, and density, are identical. Their specific rotations are equal in degree but opposite in sign.

14.40

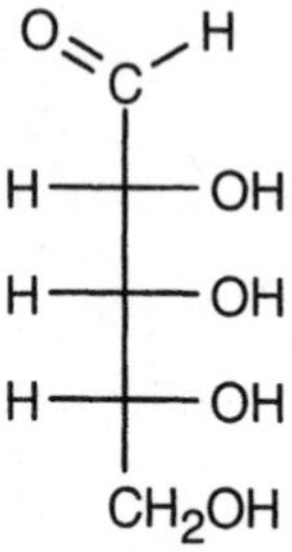

D-Ribose

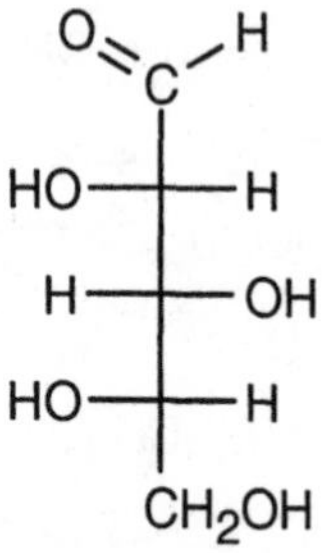

L-Xylose

D-Ribose and L-xylose are diastereomers and differ in all physical properties.

14.41 There are four D-2-ketohexoses.

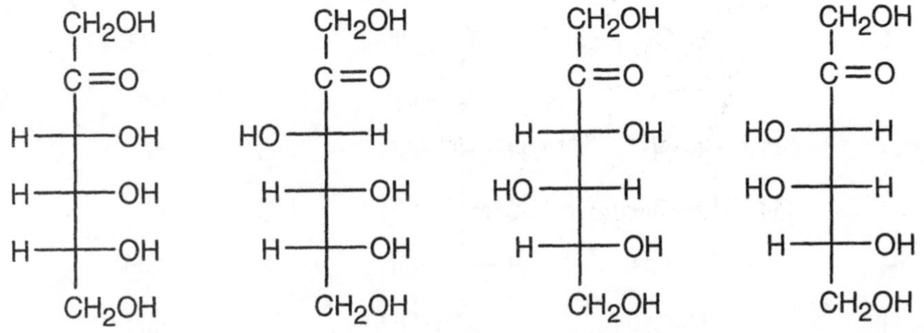
CH2OH
C=O
H — OH
HO — H
CH2OH

D-Psicose D-Fructose D-Sorbose D-Tagatose

Reactions

14.42

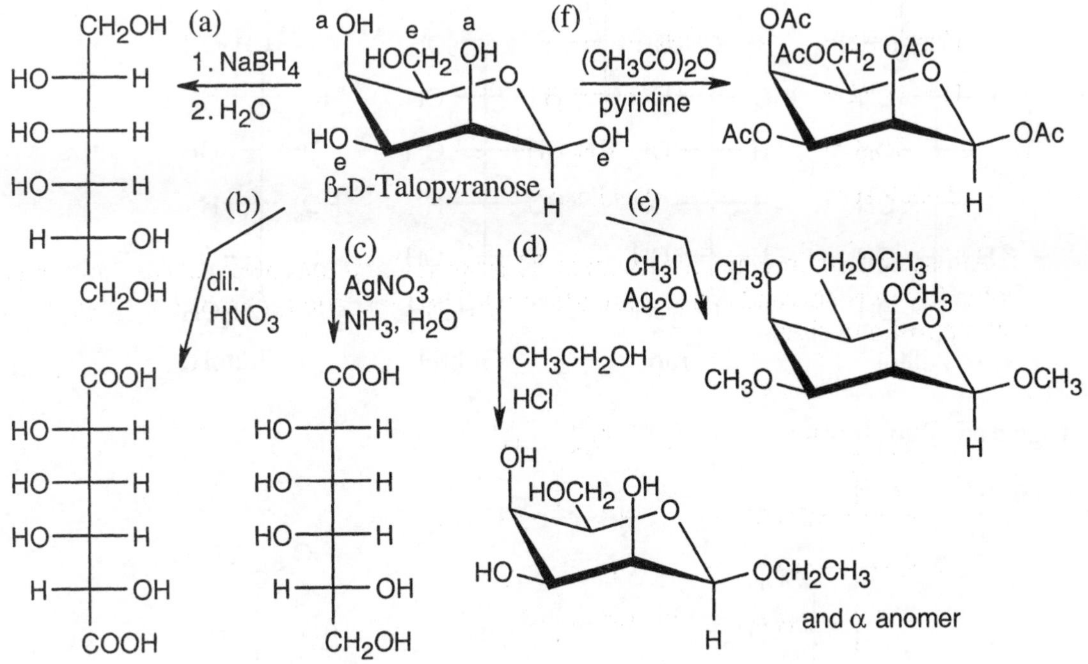
(a)
1. NaBH4
2. H2O
β-D-Talopyranose
(b)
dil. HNO3
(c)
AgNO3
NH3, H2O
(d)
CH3CH2OH
HCl
(e)
CH3I
Ag2O
(f)
(CH3CO)2O
pyridine
COOH
CH2OH
OCH2CH3
and α anomer
OAc
AcOCH2
AcO
CH3O
CH2OCH3
OCH3

14.43 - 14.44

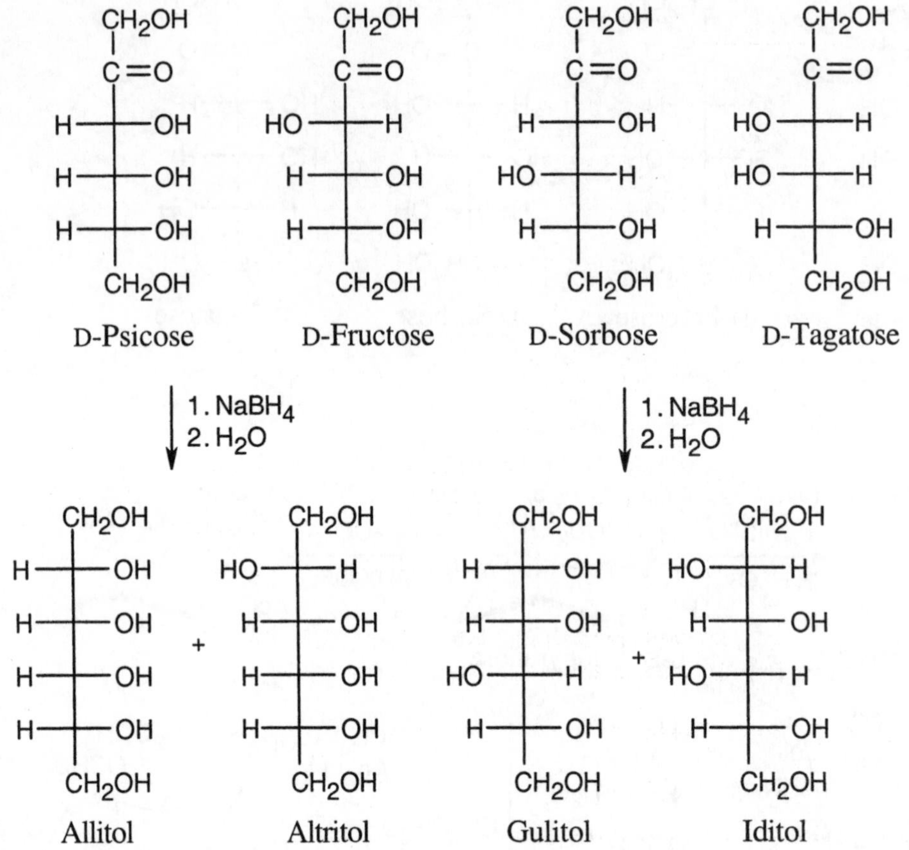

D-Psicose: CH_2OH, $C{=}O$, H–C–OH, H–C–OH, H–C–OH, CH_2OH

D-Fructose: CH_2OH, $C{=}O$, HO–C–H, H–C–OH, H–C–OH, CH_2OH

D-Sorbose: CH_2OH, $C{=}O$, H–C–OH, HO–C–H, H–C–OH, CH_2OH

D-Tagatose: CH_2OH, $C{=}O$, HO–C–H, HO–C–H, H–C–OH, CH_2OH

1. $NaBH_4$
2. H_2O

Allitol: CH_2OH, H–C–OH, H–C–OH, H–C–OH, H–C–OH, CH_2OH
+
Altritol: CH_2OH, HO–C–H, H–C–OH, H–C–OH, H–C–OH, CH_2OH

Gulitol: CH_2OH, H–C–OH, H–C–OH, HO–C–H, H–C–OH, CH_2OH
+
Iditol: CH_2OH, HO–C–H, H–C–OH, HO–C–H, H–C–OH, CH_2OH

Integrated Problems

14.45

(a) COOH; H–C–CH_3; CH_2CH_3

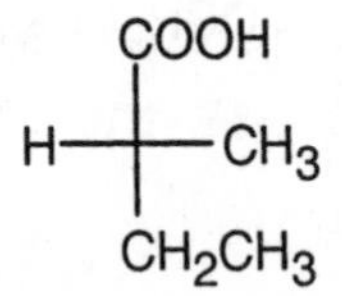

(*R*)-2-Methylbutanoic acid

(b) $O{=}C{-}CH_3$; H_3C–C–H; CH_2CH_3

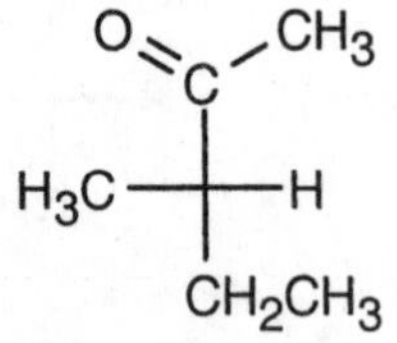

(*S*)-3-Methylpentan-2-one

14.46

(a) Br; H–C–OCH_3; CH_3 = Br; H_3C--C–OCH_3; H

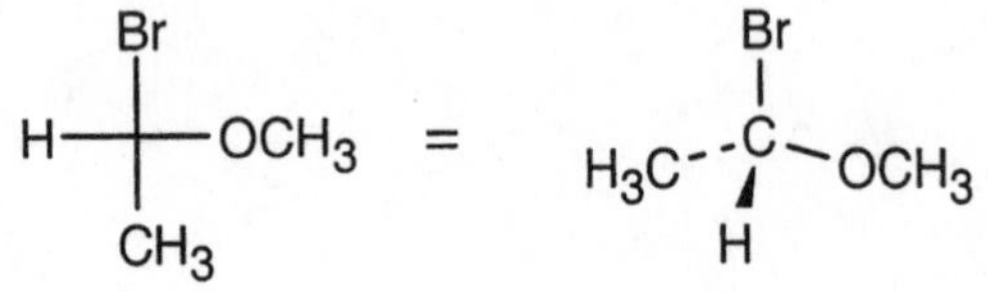

(b) CH_3; H–C–NH_2; CH_2CH_3 = CH_3; CH_3CH_2--C–NH_2; H

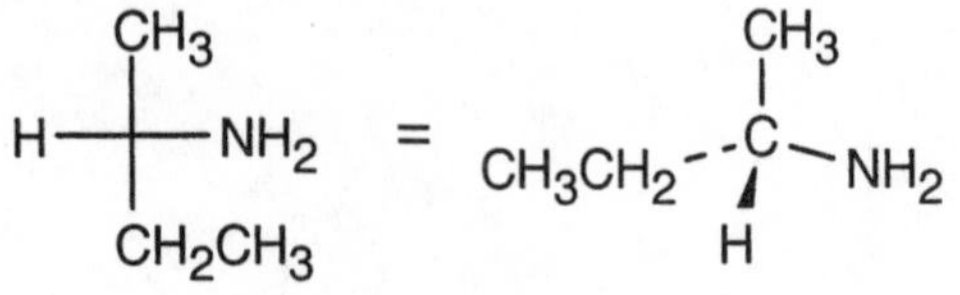

14.47

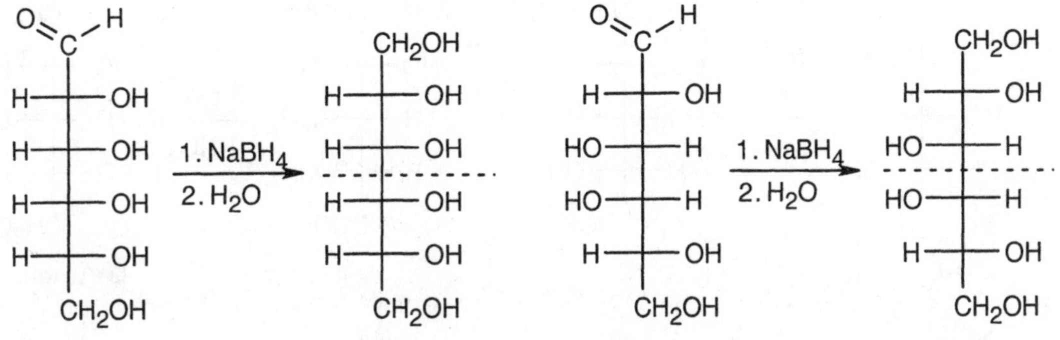

D-Allose $\xrightarrow{\text{1. }NaBH_4\text{; 2. }H_2O}$ Allitol

D-Galactose $\xrightarrow{\text{1. }NaBH_4\text{; 2. }H_2O}$ Galactitol

All of the other D-hexoses yield optically active alditols on reduction with $NaBH_4$.

14.48

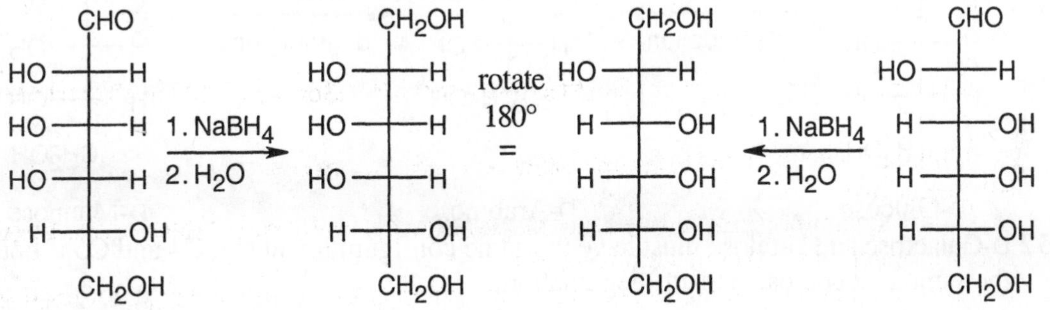

D-Talose $\xrightarrow{\text{1. }NaBH_4\text{; 2. }H_2O}$ alditol; rotate 180° = alditol $\xleftarrow{\text{1. }NaBH_4\text{; 2. }H_2O}$ D-Altrose

14.49

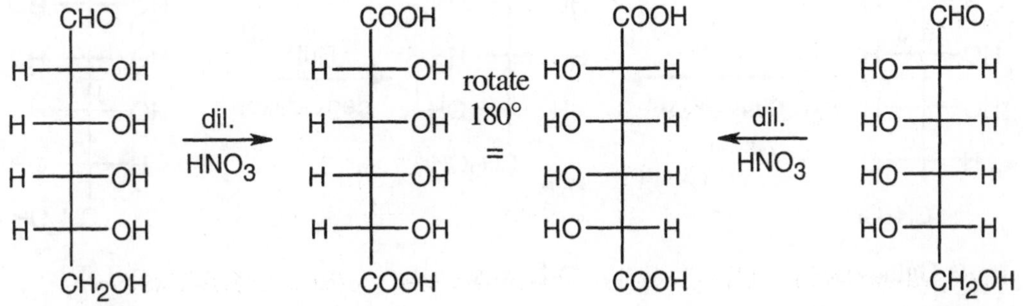

D-Allose $\xrightarrow{\text{dil. }HNO_3}$ aldaric acid; rotate 180° = aldaric acid $\xleftarrow{\text{dil. }HNO_3}$ L-Allose

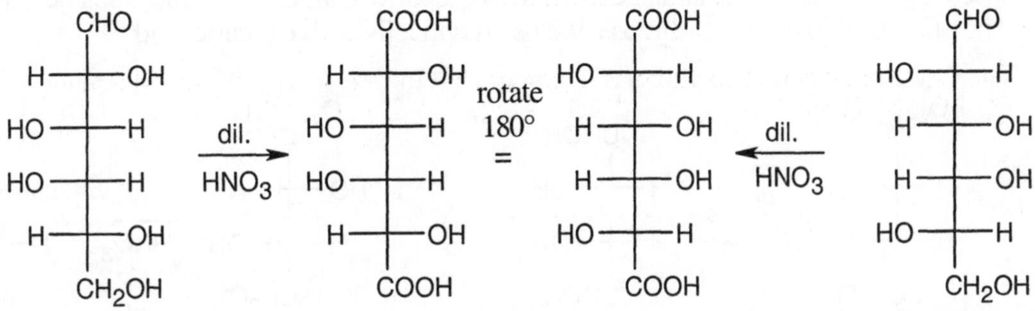

D-Galactose $\xrightarrow{\text{dil. }HNO_3}$ aldaric acid; rotate 180° = aldaric acid $\xleftarrow{\text{dil. }HNO_3}$ L-Galactose

14.50

D-Lyxose (CHO; HO–H; HO–H; H–OH; CH_2OH) $\xrightarrow[HNO_3]{dil.}$ (COOH; HO–H; HO–H; H–OH; COOH) rotate 180° = (COOH; HO–H; H–OH; H–OH; COOH) $\xleftarrow[HNO_3]{dil.}$ D-Arabinose (CHO; HO–H; H–OH; H–OH; CH_2OH)

14.51

D-Glucose (CHO; H–OH; HO–H; H–OH; H–OH; CH_2OH) $\xrightarrow[degradation]{Ruff}$ D-Arabinose (CHO; HO–H; H–OH; H–OH; CH_2OH) $\xleftarrow[degradation]{Ruff}$ D-Mannose (CHO; HO–H; HO–H; H–OH; H–OH; CH_2OH)

14.52 D-Galactose and D-talose must have the same configuration at C3, C4 and C5 if both yield the same aldopentose on Ruff degradation.

D-Galactose (CHO; H–OH; HO–H; HO–H; H–OH; CH_2OH) $\xrightarrow[degradation]{Ruff}$ D-Lyxose (CHO; HO–H; HO–H; H–OH; CH_2OH) $\xleftarrow[degradation]{Ruff}$ D-Talose (CHO; HO–H; HO–H; HO–H; H–OH; CH_2OH)

14.53 Two D-aldotetroses are possible. The one with both hydroxyl groups on the right in a Fischer projection yields an optically inactive aldaric acid, and thus must be D-erythrose. The other D-aldotetrose, D-threose, yields an optically active aldaric acid.

D-Erythrose (CHO; H–OH; H–OH; CH_2OH) $\xrightarrow[HNO_3]{dil.}$ (COOH; H–OH; H–OH; COOH)

D-Threose (CHO; HO–H; H–OH; CH_2OH) $\xrightarrow[HNO_3]{dil.}$ (COOH; HO–H; H–OH; COOH)

14.54

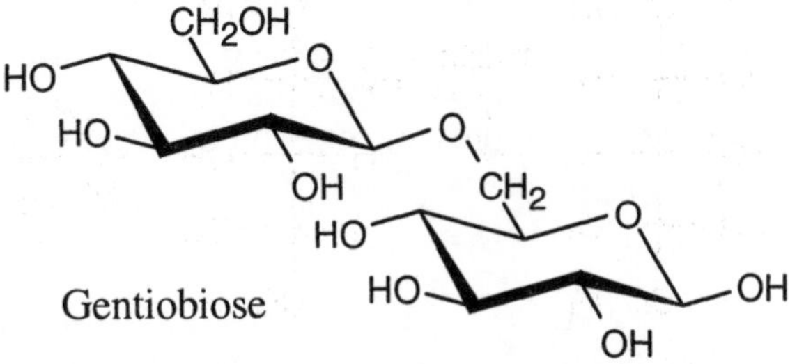

Gentiobiose

6-O-(β-D-Glucopyranosyl)-β-D-glucopyranose

14.55

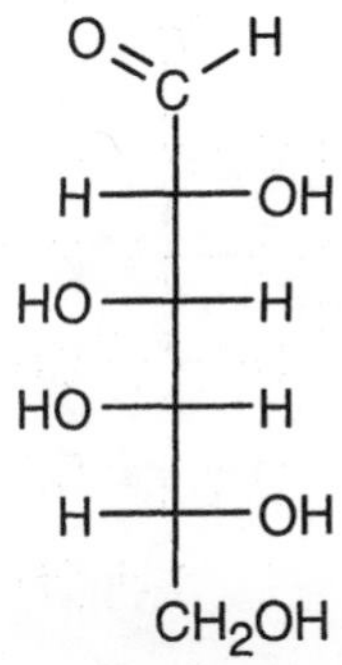

D-Galactose

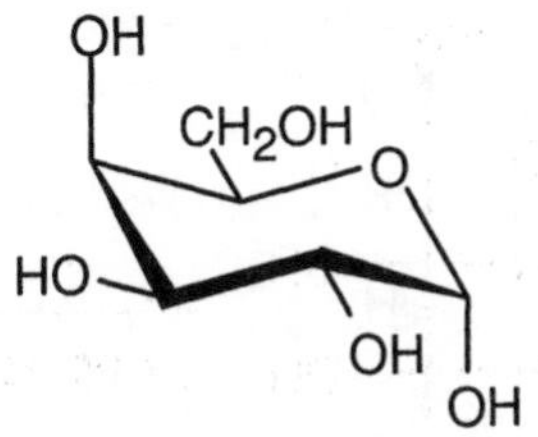

α-D-Galactopyranose
$[\alpha]_D = +150.7°$

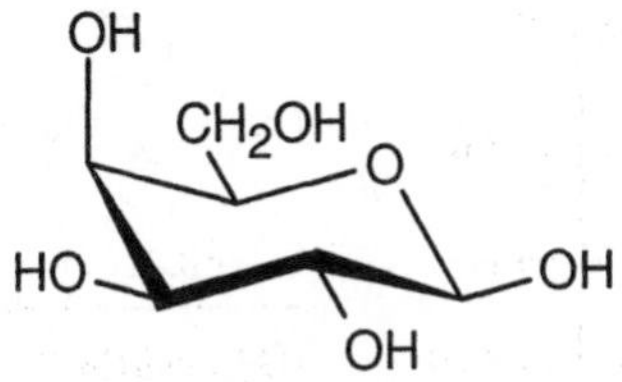

β-D-Galactopyranose
$[\alpha]_D = +52.8°$

Let x be the fraction of D-galactopyranose present as the α anomer and y be the fraction of D-galactopyranose present as the β anomer.

$$150.7°x + 52.8°y = 80.2° \qquad x + y = 1;\ y = 1 - x$$
$$150.7°x + 52.8°(1 - x) = 80.2°$$
$$97.9°x = 27.4°$$
$$x = 0.280$$
$$y = 0.720$$

28.0% of D-galactopyranose is present as the α anomer, and 72.0% is present as the β anomer.

14.56

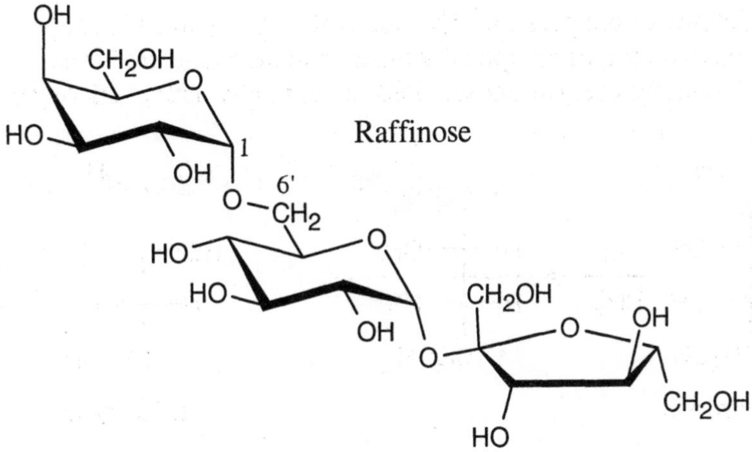

Raffinose

14.57 Raffinose is a nonreducing sugar because it contains no hemiacetal groups.

14.58 As is true for other carbonyl compounds, the hydrogens α to the carbonyl groups of glucose and fructose are acidic, and each hexose can undergo base-catalyzed enolization. In this example, glucose and fructose yield the same enol on treatment with base. This *enediol* can tautomerize to either glucose or fructose.

D-Glucose ⇌ (NaOH) enediol ⇌ (NaOH) D-Fructose

In the Medicine Cabinet

14.59

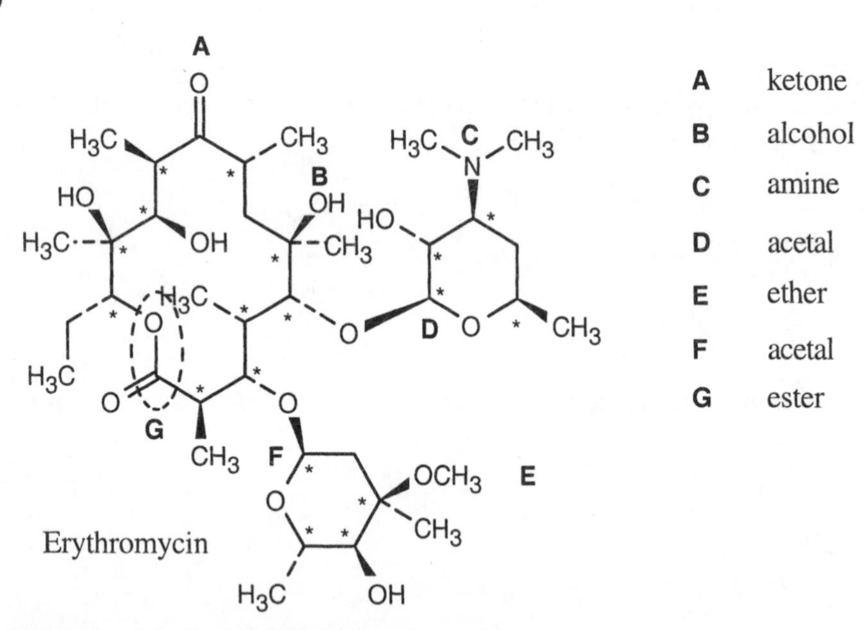

Erythromycin

Erythromycin has eighteen chirality centers.

14.60 It is likely that the circled bond is formed by cyclization of a carboxylic acid group with a hydroxyl group at the other end of the molecule, producing a lactone.

14.61

(a) This alcohol is reduced

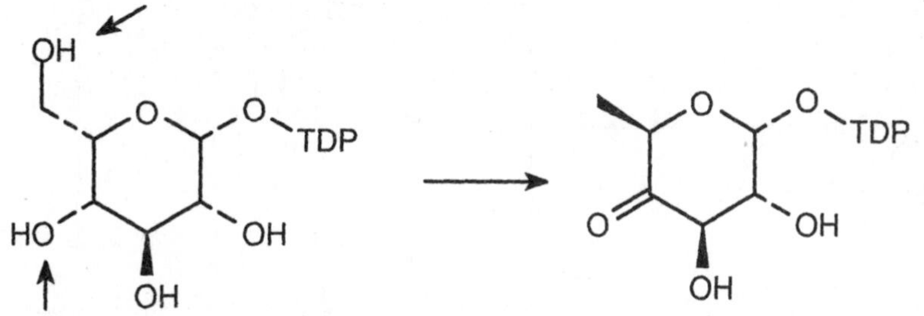

This alcohol is oxidized to a ketone

(b)

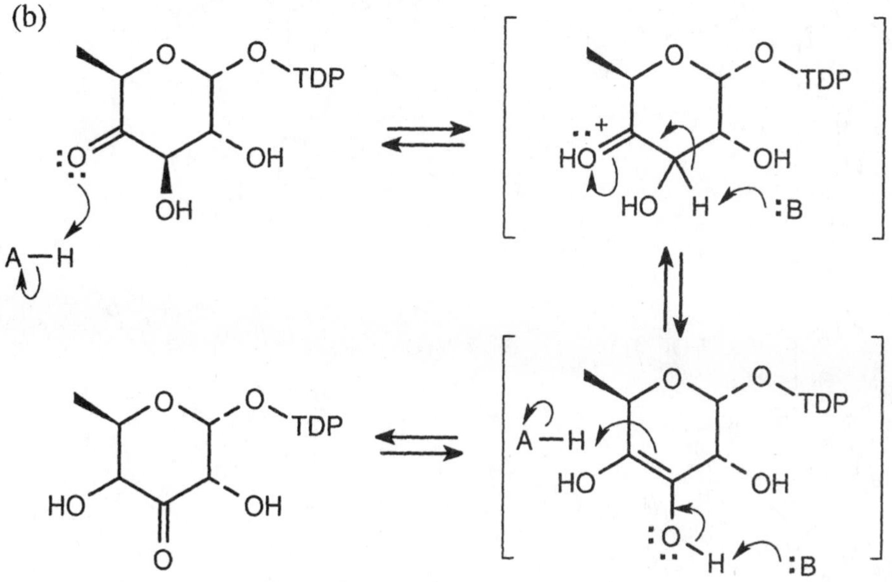

(c)

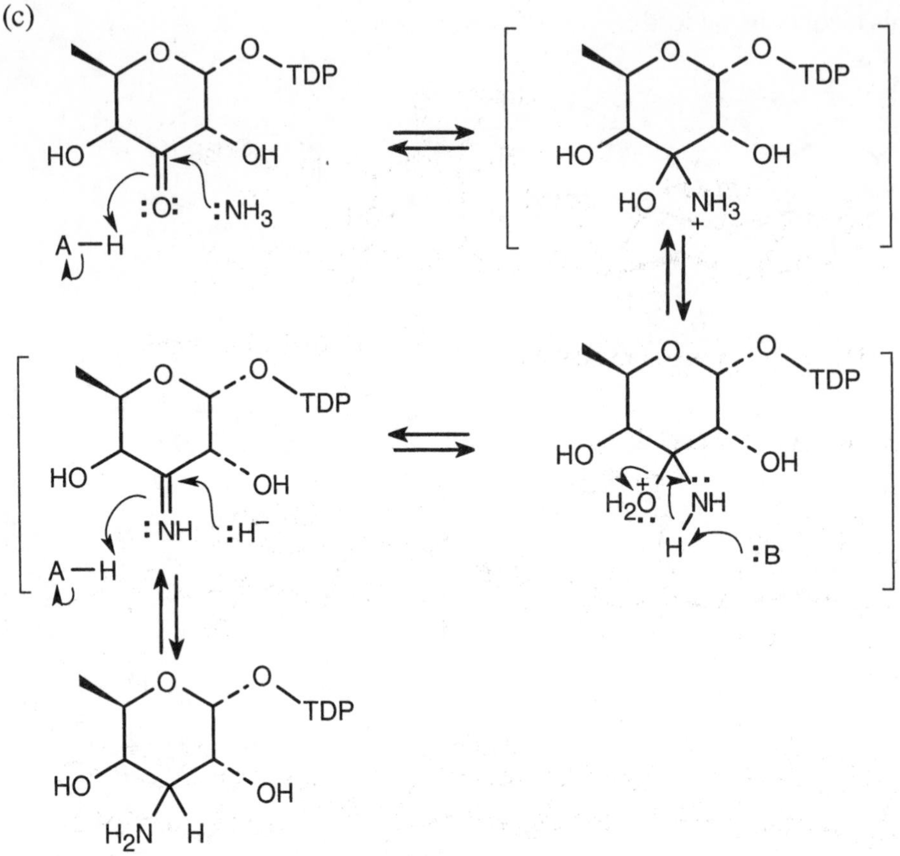

TDP HO OH :O: :NH3 A—H

TDP HO OH HO NH3 +

TDP HO OH :NH A—H :H⁻

TDP HO OH H2O NH H :B

TDP HO OH H2N H

(d)

TDP HO OH H2N: CH3—S:

TDP HO OH H—N—CH3 + :S: B: H

TDP HO OH H—N—CH3

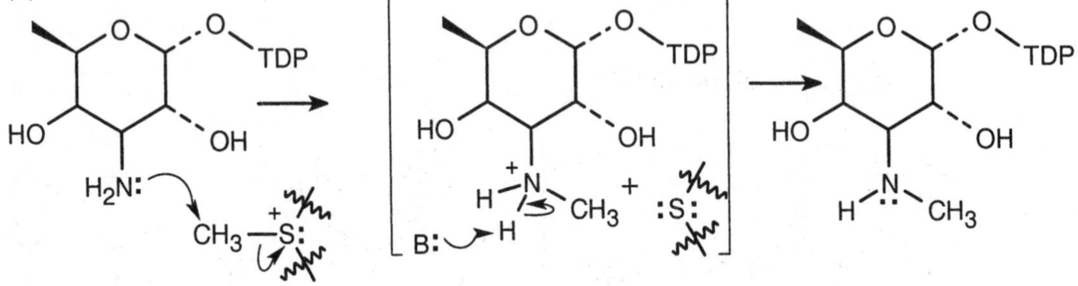

14.62

(a)

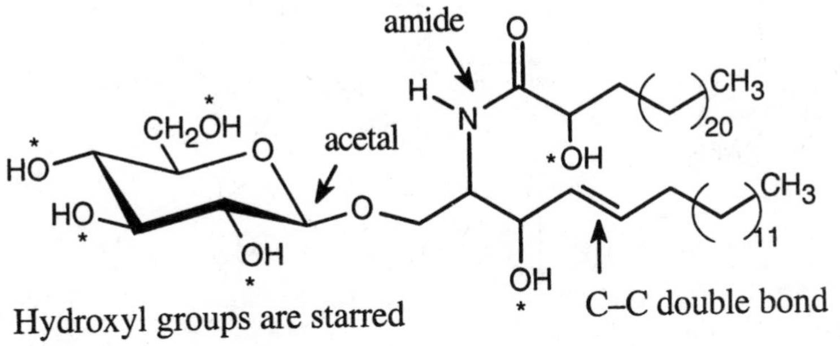

amide

acetal

C–C double bond

Hydroxyl groups are starred

(b) Hydrolysis of the acetal:

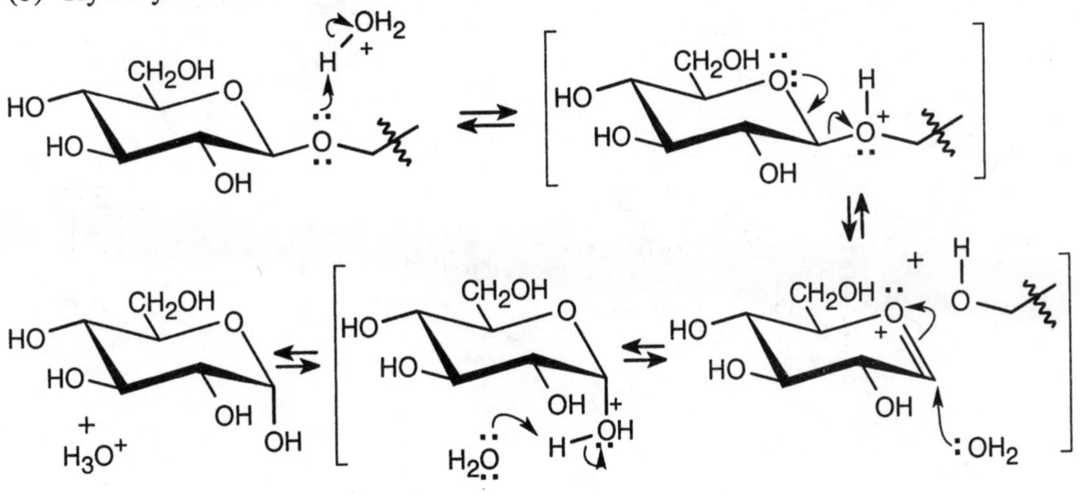

Hydrolysis of the amide:

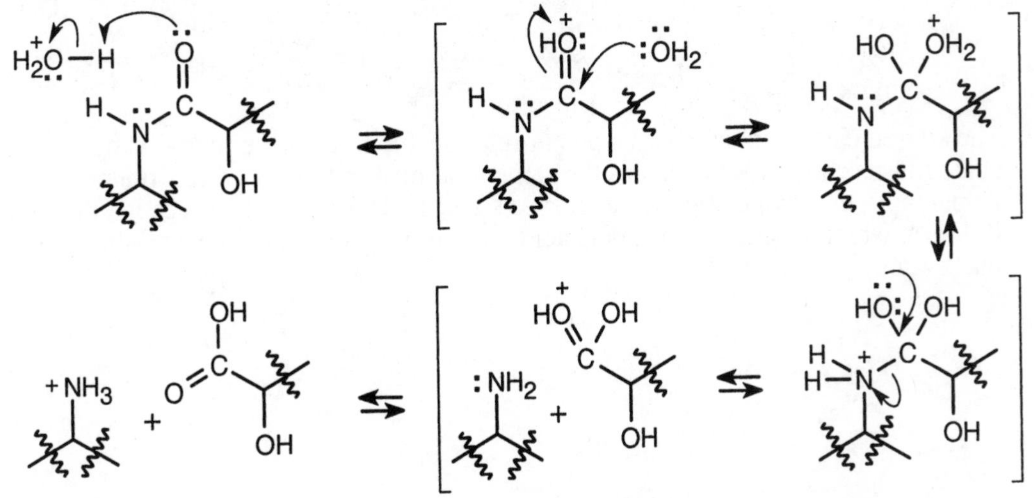

(c)

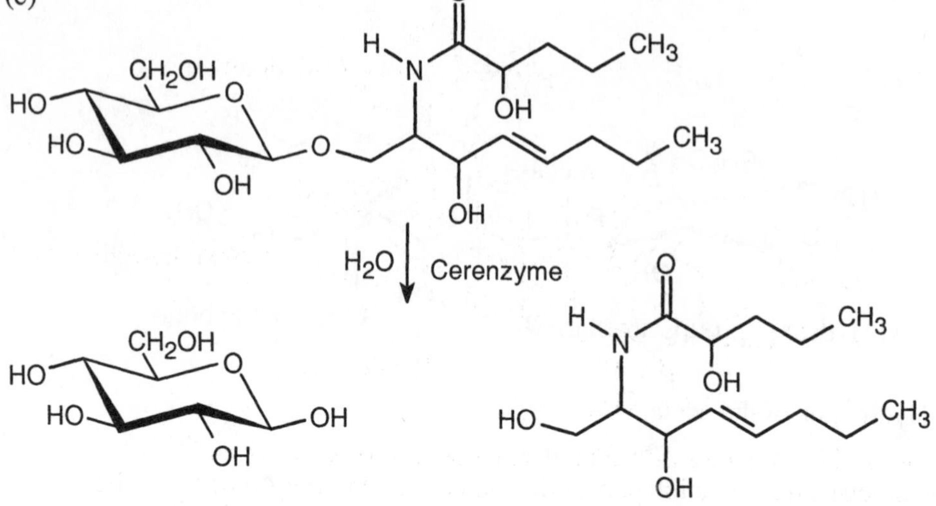

14.63

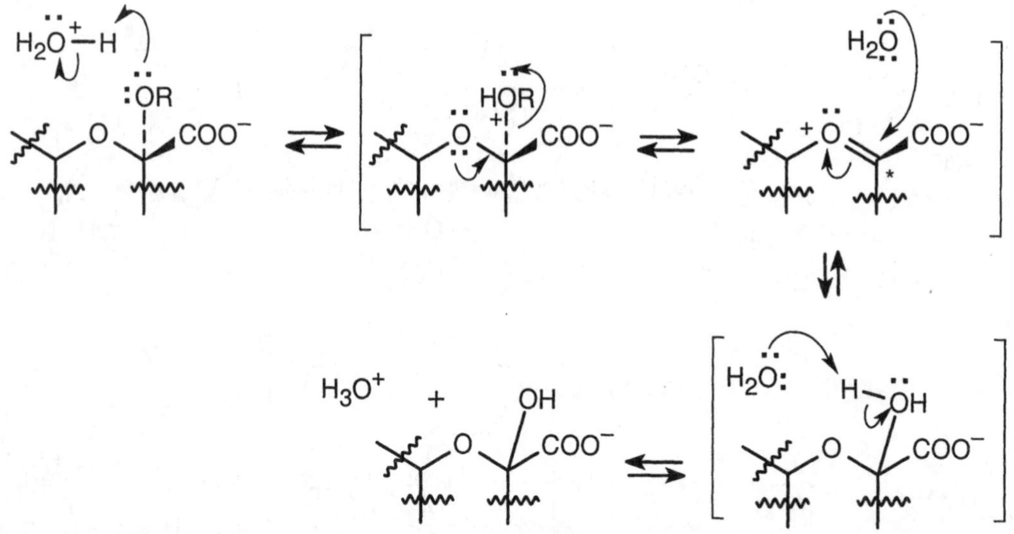

The starred carbon (above) has a planar relationship to the other carbons bonded to it. The analogous carbon of Relenza has the same relationship to the carbons bonded to it. Apparently, the viral enzyme that cleaves sialic acid from a cell surface is inhibited by Relenza, which resembles the sialic acid oxonium ion intermediate, and the virus is inactivated.

14.64

(a) – (c)

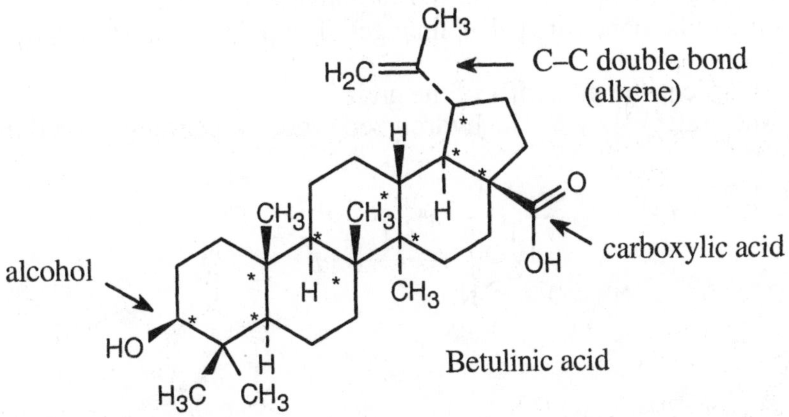

CH_3
H_2C ← C–C double bond (alkene)
H
H
CH_3 CH_3 H
O
OH ← carboxylic acid
alcohol →
H CH_3
HO
H
H_3C CH_3

Betulinic acid

Betulinic acid has ten chirality centers and could have as many as $2^{10} = 1024$ stereoisomers. The "-ic acid" part of the name is given because of the carboxylic acid group.

(d) Let betulinic acid be represented as HO–R.

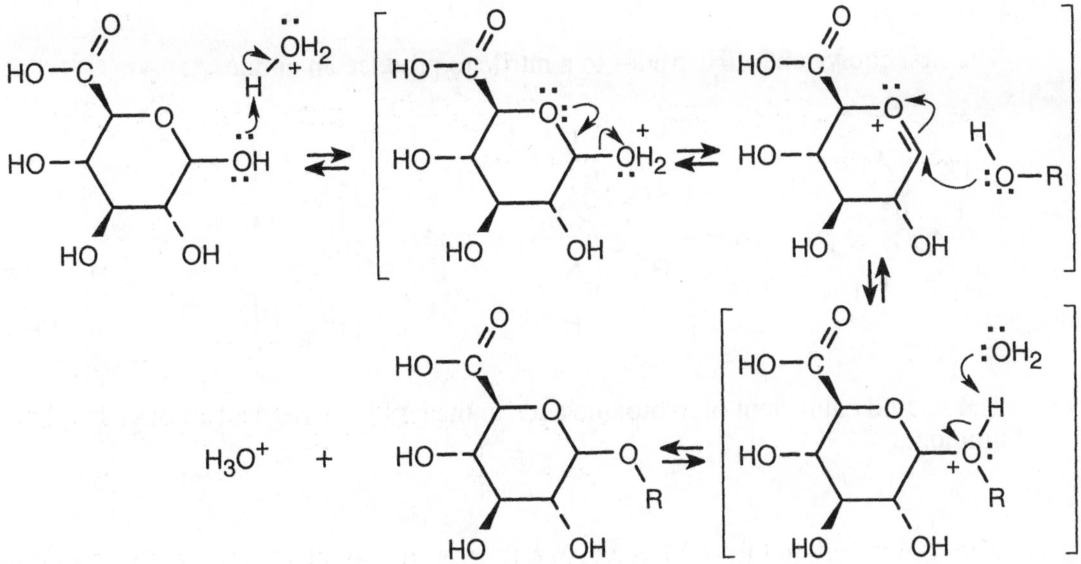

O
HO–C
O
HO–
OH
HO OH
$\ddot{O}H_2^{+}$
H
$\ddot{O}H$
⇄
[HO–C(=O), $\ddot{O}$:, HO–, $\overset{+}{O}H_2$, HO, OH]
⇄
[HO–C(=O), $\overset{+}{O}$, HO–, H, :$\ddot{O}$–R, HO, OH]
⇅
[HO–C(=O), O, HO–, :$\ddot{O}H_2$, H, $\overset{+}{O}$:, R, HO, OH]
⇆
H_3O^+ + HO–C(=O), O, HO–, O–R, HO, OH

14.65

(a) Glucose is the monomer unit of the carbohydrate chain.
(b) The glycosidic bonds are β-1-4' linkages. The polysaccharide chain resembles cellulose.
(c) Acrylonitrile is the monomer of the graft.
(d) Only the graft is shown. The hydrolyzed groups are arranged randomly.

$$\text{-CH-CH}_2\text{-CH-CH}_2\text{-CH-CH}_2\text{-}$$
(substituents, left to right: $C(=O)O^-$, $C(=O)NH_2$, $C(=O)NH_2$)

(e)

$$R-C\equiv N: \rightleftharpoons \left[R-C(OH)=N^- \; + \; H-OH \rightleftharpoons R-C(OH)=N-H \; + \; ^-OH \right] \rightleftharpoons R-C(=O)-NH_2 \text{ (amide)}$$

The first equivalent of HO⁻ adds to a nitrile to produce an amide.

$$HO^- + R-C(=O)-NH_2 \rightleftharpoons \left[R-C(HO)(O^-)-NH_2 \rightleftharpoons R-C(=O)OH \; + \; :NH_2^- \right] \rightleftharpoons R-C(=O)O^- \; + \; NH_3$$

The second equivalent of hydroxide adds to the amide to yield a carboxylate, plus ammonia.

Chapter 15 – Biomolecules: Amino Acids, Peptides, and Proteins

Chapter Outline

I. Amino acids (Sections 15.1 – 15.2).
- A. Structure of amino acids (Section 15.1).
  1. Amino acids exist in solution as zwitterions.
     - a. Zwitterions are internal salts and have many of the properties associated with salts.
       - i. They have large dipole moments.
       - ii. They are soluble in water.
       - iii. They are crystalline and high-melting.
     - b. Zwitterions can act either as acids or as bases.
       - i. The $-COO^-$ group acts as a base.
       - ii. The ammonium group acts an acid.
  2. All natural amino acids are α-amino acids: the amino group and the carboxylic acid group are bonded to the same carbon.
  3. All but one (proline) of the 20 common amino acids are primary amines.
  4. All of the amino acids are represented by both a three-letter code and a one-letter code.
  5. All amino acids except glycine are chiral.
     - a. Only one enantiomer (L) of each pair is naturally-occurring.
     - b. In Fischer projections, the carboxylic acid is at the top, and the amino group points to the left.
     - c. α-Amino acids are referred to as L-amino acids.
  6. Side chains can be acidic or basic.
     - a. Fifteen of the amino acids are neutral.
     - b. Two (aspartic acid and glutamic acid) are acidic.
       At pH = 7.3, their side chains exist as carboxylate ions.
     - c. Three (lysine, arginine and histidine) are basic.
       - i. At pH = 7.3, the side chains of lysine and arginine exist as ammonium ions.
       - ii. Histidine is not quite basic enough to be protonated at pH = 7.3.
       - iii. The double-bonded nitrogen in the histidine ring is basic.
     - d. Cysteine and tyrosine are weakly acidic.
- B. Isoelectric points (Section 15.2).
  1. The isoelectric point (p*I*) is the pH at which an amino acid exists as a neutral, dipolar zwitterion.
     - a. p*I* is related to side chain structure.
       - i. The 15 amino acids that are neutral have p*I* near neutrality.
       - ii. The two acidic amino acids have p*I* at a lower pH.
       - iii. The 3 basic amino acids have p*I* at a higher pH.
     - b. For neutral amino acids, p*I* is the average of the two pK_a values.
       - i. For acidic amino acids, p*I* is the average of the two lowest pK_a values.
       - ii. For basic amino acids, p*I* is the average of the two highest pK_a values.
       - iii. Proteins also have an overall pI.
  2. Electrophoresis allows the separation of amino acids by differences in their p*I*.
     - a. A buffered solution of amino acids is placed on a paper or gel.
     - b. Electrodes are connected to the solution, and current is applied.
     - c. Negatively charged amino acids migrate to the positive electrode, and positively charged amino acids migrate to the negative electrode.
     - d. Amino acids can be separated because the extent of migration depends on p*I*.

II. Peptides (Sections 15.3 – 15.7).
  A. Peptide structure (Sections 15.3 , 15.5).
    1. Peptide bonds (Section 15.3).
      a. A peptide is an amino acid polymer in which the amine group of one amino acid forms an amide bond with the carboxylic acid group of a second amino acid.
      b. This amino acid sequence is known as the backbone of the peptide or protein.
      c. Rotation about the amide bond is restricted.
    2. The N-terminal amino acid of the polypeptide is always drawn on the left.
    3. The C-terminal amino acid of the polypeptide is always drawn on the right.
    4. Peptide structure is described by using the three-letter codes for the individual amino acids, starting with the N-terminal amino acid.
    5. Disulfide bonds (Section 15.5).
      a. Two cysteines can form a disulfide bond (–S–S–).
      b. Disulfide bonds can link two polypeptides or introduce a loop in a polypeptide chain.
  B. Peptide synthesis (Sections 15.3 - 15.4).
    1. Laboratory synthesis (Section 15.3)
      a. Groups that are not involved in peptide bond formation are protected.
        i. Carboxyl groups are often protected as methyl or benzyl esters.
        ii. Amino groups are protected as BOC or FMOC derivatives.
      b. The peptide bond is formed by coupling with DCC.
      c. The protecting group are removed.
        i. BOC groups are removed by brief treatment with trifluoroacetic acid.
        ii. FMOC groups are removed by treatment with base.
        iii. Esters are removed by mild hydrolysis or by hydrogenolysis (benzyl).
      d. In solid phase synthesis, the C-terminal amino acid is bound to a support, and the above steps are followed.
    2. Synthesis in nature (Section 15.4).
      a. Most peptides are synthesized in ribosomes.
      b. Nonribosomal peptide synthesis is brought about by assemblies of multiple enzymes.
  C. Structure determination of peptides (Sections 15.6 – 15.7).
    1. Amino acid analysis (Section 15.6).
      a. Amino acid analysis provides the amount of each amino acid present in a protein or peptide.
      b. First, all disulfide bonds are broken and all peptide bonds are hydrolyzed.
      c. The mixture is placed on a chromatography column, and the residues are eluted.
      d. As each amino acid elutes, it undergoes reaction with ninhydrin, which produces a purple color that is detected and measured spectrophotometrically.
      e. Amino acid analysis is reproducible on properly maintained equipment; residues always elute at the same time, and only small sample sizes are needed.
    2. The Edman degradation (Section 15.7).
      a. The Edman degradation removes one amino acid at a time from the $-NH_2$ end of a peptide.
        i. The peptide is treated with phenylisothiocyanate, which reacts with the amino-terminal residue.
        ii. The PITC derivative is split from the peptide.
        iii. The residue undergoes rearrangement to a PTH, which is identified chromatographically.
        iv. The shortened chain undergoes another round of Edman degradation.

b. Since the Edman degradation can only be used on peptides containing fewer than 50 amino acids, a protein must be cleaved into smaller fragments.
   i. Partial acid hydrolysis is unselective.
   ii. The enzyme trypsin cleaves proteins at the carboxyl side of Arg and Lys residues.
   iii. The enzyme chymotrypsin cleaves proteins at the carboxyl side of Phe, Tyr and Trp residues.
c. The complete sequence of a protein results from determining the individual sequences of peptides and overlapping them.

III. Proteins (Sections 15.8 – 15.11).
  A. Classification of proteins (Section 15.8).
    1. Proteins can be classified by structure.
      a. Simple proteins yield only amino acids on hydrolysis.
      b. Conjugated proteins yield other non-protein components on hydrolysis.
    2. Proteins can be classified by shape.
      a. Fibrous proteins consist of long, filamentous polypeptide chains.
      b. Globular proteins are compact and roughly spherical.
    3. Proteins can be classified by function.
  B. Protein structure (Section 15.9).
    1. Levels of protein structure.
      a. Primary structure refers to the amino acid sequence of a protein.
      b. Secondary structure refers to the organization of segments of the peptide backbone into a regular pattern, such as a helix or sheet.
      c. Tertiary structure describes the overall three-dimensional shape of a protein.
      d. Quaternary structure describes how polypeptide subunits aggregate into a larger structure.
    2. Examples of specific proteins.
      a. α-Keratin.
        i. α-Keratin is a fibrous protein found in wool, hair and nails.
        ii. Segments of α-keratin are coiled into a right-handed helix (α-helix).
      b. Fibroin.
        i. Fibroin is found in silk.
        ii. Fibroin has a β-pleated sheet secondary structure.
      c. Myoglobin.
        i. Myoglobin is a small globular protein.
        ii. The nonpolar amino acid side chains congregate in the center of myoglobin to avoid water.
        iii. The polar side chain residues are on the surface, where they can take part in hydrogen bonding and salt bridge formation.
        iv. Myoglobin also contains a covalently-bonded heme group.
  C. Enzymes (Sections 15.10 – 15.11).
    1. Description of enzymes (Section 15.10).
      a. An enzyme is a substance (usually protein) that catalyzes a biochemical reaction.
      b. An enzyme is specific and usually catalyzes the reaction of only one substrate. Some enzymes, such as papain, can operate on a range of substrates.
      c. Most enzymes are globular proteins, and many consist of a protein portion (apoenzyme) and a cofactor.
        i. Cofactors may be small organic molecules (coenzymes) or inorganic ions.
        ii. Many coenzymes are vitamins.

d. Enzymes are grouped into 6 classes according to the reactions they catalyze.
   i. Hydrolases catalyze hydrolysis reactions.
   ii. Isomerases catalyze isomerizations.
   iii. Ligases catalyze bond formation between two molecules.
   iv. Lyases catalyze the loss of a small molecule from a substrate.
   v. Oxidoreductases catalyze oxidations and reductions.
   vi. Transferases catalyze the transfer of a group from one substrate to another.
e. The name of an enzyme has two parts, ending with *-ase*.
   i. The first part identifies the substrate.
   ii. The second part identifies the enzyme's class.

2. How enzymes work – citrate synthase (Section 15.11).
   Citrate synthase catalyzes the aldol-like addition of acetyl CoA to oxaloacetate to produce citrate.

Solutions to Problems

15.1 Amino acids with aromatic rings:

$C_6H_5-CH_2CH(NH_3^+)CO_2^-$

Phenylalanine (Phe) (F)

$HO-C_6H_4-CH_2CH(NH_3^+)CO_2^-$

Tyrosine (Tyr) (Y)

Indol-3-yl$-CH_2CH(NH_3^+)CO_2^-$ (N–H)

Tryptophan (Trp) (W)

Imidazol-4-yl$-CH_2CH(NH_3^+)CO_2^-$ (N–H)

Histidine (His) (H)

Amino acids containing sulfur:

$HSCH_2CH(NH_3^+)CO_2^-$

Cysteine (Cys) (C)

$CH_3SCH_2CH_2CH(NH_3^+)CO_2^-$

Methionine (Met) (M)

Amino acids that are alcohols:

$HOCH_2CH(NH_3^+)CO_2^-$

Serine (Ser) (S)

$CH_3CH(OH)CH(NH_3^+)CO_2^-$

Threonine (Thr) (T)

$HO-C_6H_4-CH_2CH(NH_3^+)CO_2^-$

Tyrosine (Tyr) (Y)
a phenol

Amino acids having hydrocarbon side chains:

$CH_3CH(NH_3^+)CO_2^-$
Alanine (Ala) (A)

$CH_3CH_2CH(CH_3)CH(NH_3^+)CO_2^-$
Isoleucine (Ile) (I)

$(CH_3)_2CHCH_2CH(NH_3^+)CO_2^-$
Leucine (Leu) (L)

$(CH_3)_2CHCH(NH_3^+)CO_2^-$
Valine (Val) (V)

$CHCO_2^-$ (ring, NH_2^+)
Proline (Pro) (P)

15.2 A 3-dimensional drawing of the α carbon of an L-amino acid is pictured below.

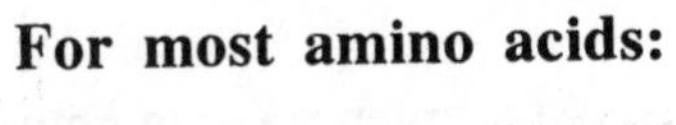

For most amino acids:

Group	*Priority*
$-NH_2$	1
$-COOH$	2
$-G$	3
$-H$	4

For cysteine:

Group	*Priority*
$-NH_2$	1
$-CH_2SH$	2
$-COOH$	3
$-H$	4

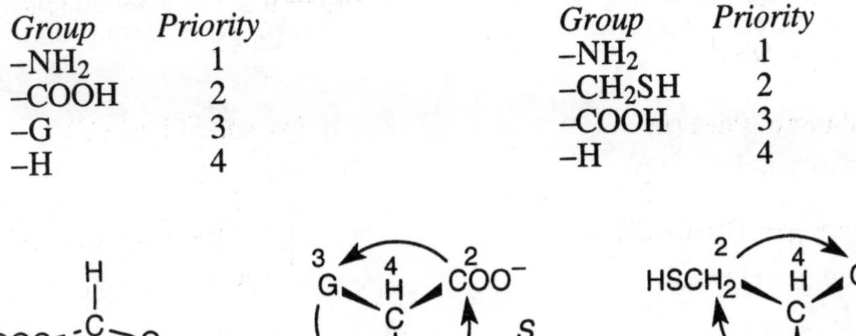

15.3

$^-OOC-C(H)(CH_3)(H_3N^+)$

L-Alanine

15.4

(a)

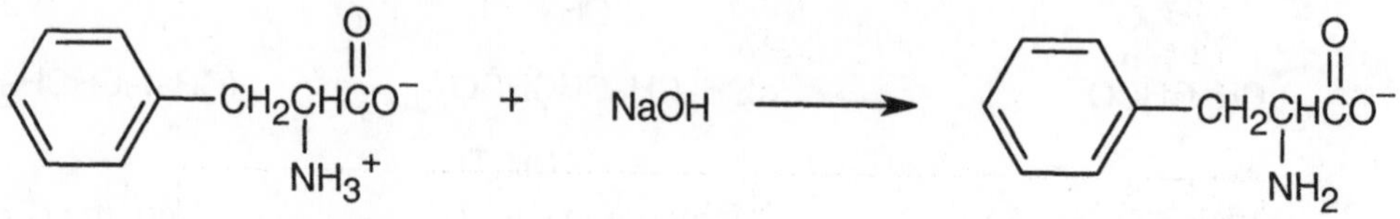

(b)

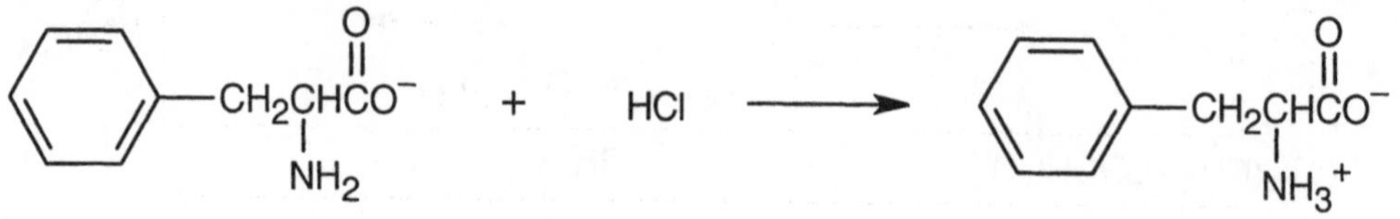

(c)

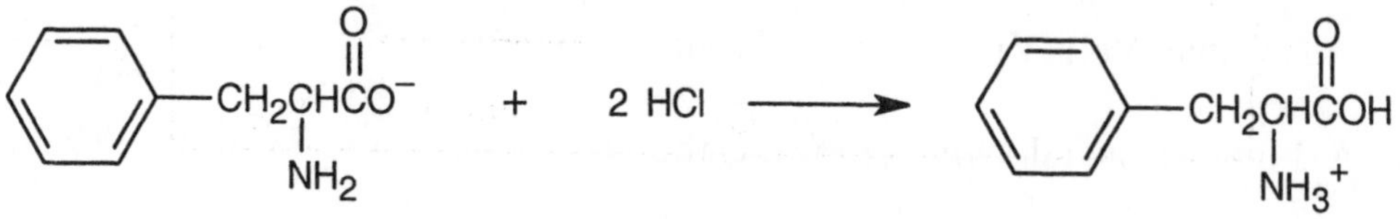

15.5 Use Table 15.1 to find pK_a values for each amino acid. At a given pK_a, an amino acid exists as a 1:1 mixture of two forms. On either side of the pK_a one form predominates.

(a)

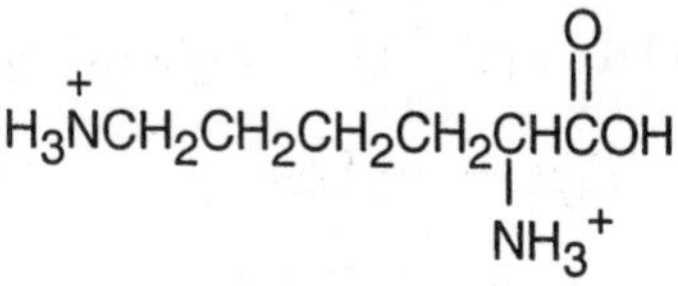

Lysine at pH = 2.0

(b)

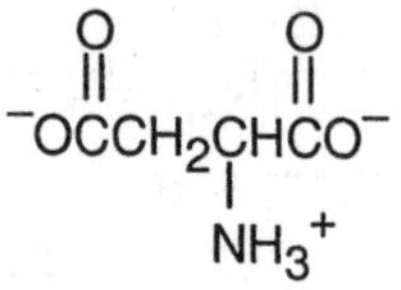

Aspartic acid at pH = 6.0

(c)

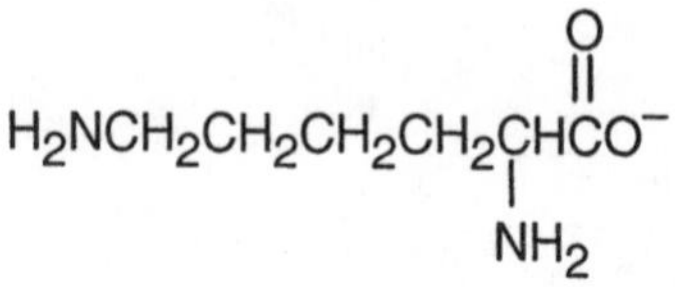

Lysine at pH = 11.0

(d)

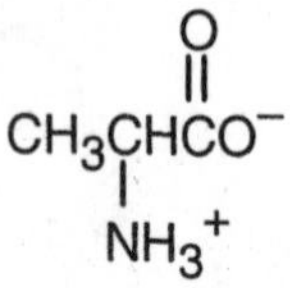

Alanine at pH = 3.0

15.6 Remember that an amino acid is positively charged at a pH lower than its pI and migrates to the negative electrode. At a pH higher than pI, the amino acid is negatively charged and migrates to the positive electrode.

(a)

Amino acid	*Isoelectric point*
Val	5.96
Glu	3.22
His	7.59

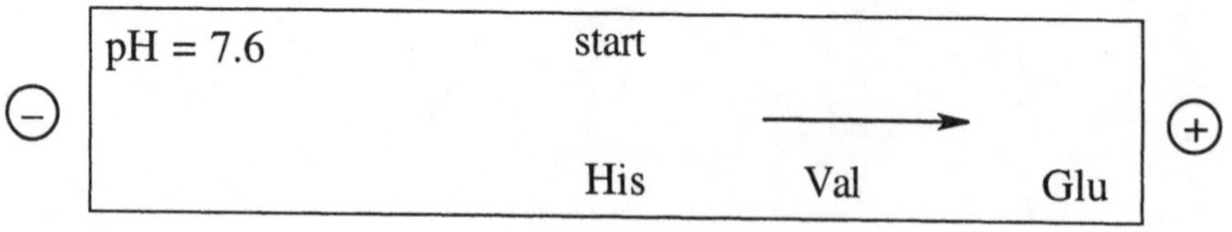

(b)

Amino acid	*Isoelectric point*
Gly	5.97
Phe	5.48
Ser	5.68

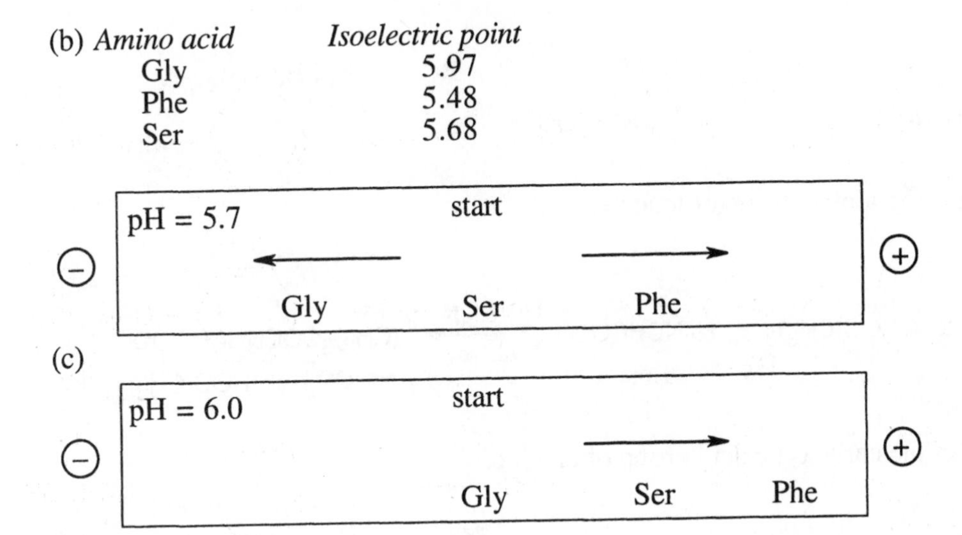

15.7

$(CH_3)_2CHCH_2$ … $H_3\overset{+}{N}$ … CH_2SH … O^-

Leu——Cys

$HSCH_2$ … $H_3\overset{+}{N}$ … $CH_2CH(CH_3)_2$ … O^-

Cys——Leu

15.8 Val–Tyr–Gly (V–Y–G) Tyr–Gly–Val (Y–G–V) Gly–Val–Tyr (G–V–Y)
Val–Gly–Tyr (V–G–Y) Tyr–Val–Gly (Y–V–G) Gly–Tyr–Val (G–Y–V)

15.9

H_2C–CH_2 … $H_3\overset{+}{N}$ … $CH_3SCH_2CH_2$ … $CH(CH_3)_2$ … COO^-

---- amide bonds

Met —Pro——Val ——Gly
M——P———V———G

15.10 Refer to Section 15.3.

$$\text{Leu} = \overset{+}{H_3N}\underset{}{CH}(CH_2CH(CH_3)_2)COO^- = \overset{+}{H_3N}CH(R)COO^- \qquad R = CH_2CH(CH_3)_2$$

1. Protect the amino group of leucine.

$$(CH_3)_3COC(=O)OC(=O)OC(CH_3)_3 + \underset{\text{Leu}}{H_2NCH(R)COO^-} \xrightarrow{Et_3N} (CH_3)_3COC(=O)NHCH(R)COO^- + CO_2 + HOC(CH_3)_3$$

2. Protect the carboxylic acid group of alanine.

$$\underset{\text{Ala}}{H_2NCH(CH_3)COOH} + CH_3OH \xrightarrow[\text{catalyst}]{H^+} H_2NCH(CH_3)COOCH_3$$

3. Couple the protected amino acids with DCC.

$$(CH_3)_3COC(=O)NHCH(R)COO^- + H_2NCH(CH_3)C(=O)OCH_3 + C_6H_{11}N{=}C{=}NC_6H_{11}$$

$$\downarrow$$

$$(CH_3)_3COC(=O)NHCH(R)C(=O){-}NHCH(CH_3)C(=O)OCH_3 + C_6H_{11}NH{-}C(=O){-}NHC_6H_{11}$$

4. Remove the leucine protecting group.

$$(CH_3)_3COC(=O)NHCH(R)C(=O){-}NHCH(CH_3)C(=O)OCH_3 \xrightarrow{CF_3COOH} \overset{+}{H_3N}CH(R)C(=O){-}NHCH(CH_3)C(=O)OCH_3 + (CH_3)_2C{=}CH_2 + CO_2$$

5. Remove the alanine protecting group.

$$\overset{+}{H_3N}CH(R)C(=O){-}NHCH(CH_3)C(=O)OCH_3 \xrightarrow[2.\ H_3O^+]{1.\ NaOH,\ H_2O} \overset{+}{H_3N}CH(CH_2CH(CH_3)_2)C(=O){-}NHCH(CH_3)C(=O)OH + CH_3OH$$

Leu — Ala

15.11

Step 1: Glycine is protected as its BOC derivative.

$$(CH_3)_3COC(=O)OC(=O)OC(CH_3)_3 + Gly \xrightarrow{Et_3N} (CH_3)_3COC(=O)-Gly-OH$$

(BOC—Gly—OH)

Step 2: BOC–Gly bonds to the polymer.

$$BOC-Gly-OH + XCH_2-\text{(Polymer)} \xrightarrow{Base} BOC-Gly-OCH_2-\text{(Polymer)}$$

Step 3: The polymer is first washed, then is treated with CF_3COOH to cleave the BOC group.

$$BOC-Gly-OCH_2-\text{(Polymer)} \xrightarrow[2.\ CF_3COOH]{1.\ wash} Gly-OCH_2-\text{(Polymer)}$$

Step 4: A BOC-protected phe is coupled to the polymer-bound glycine by reaction with DCC. The polymer is washed.

$$BOC-Phe + Gly-OCH_2-\text{(Polymer)} \xrightarrow[2.\ wash]{1.\ DCC} BOC-Phe-Gly-OCH_2-\text{(Polymer)}$$

Step 5: The polymer is treated with CF_3COOH to remove BOC.

$$BOC-Phe-Gly-OCH_2-\text{(Polymer)} \xrightarrow{CF_3COOH} Phe-Gly-OCH_2-\text{(Polymer)}$$

Step 6: A BOC-protected valine is coupled to the polymer by reaction with DCC. The polymer is washed.

$$BOC-Val + Phe-Gly-OCH_2-\text{(Polymer)} \xrightarrow[2.\ wash]{1.\ DCC} BOC-Val-Phe-Gly-OCH_2-\text{(Polymer)}$$

Step 7: Treatment with anhydrous HF removes the BOC group and cleaves the ester bond between the peptide and the polymer.

$$BOC-Val-Phe-Gly-OCH_2-\text{(Polymer)} \xrightarrow{HF}$$

$$Val-Phe-Gly + (CH_3)_2C{=}CH_2 + CO_2 + HOCH_2-\text{(Polymer)}$$

15.12 One product of the reaction of an amino acid with ninhydrin is the extensively conjugated purple ninhydrin product. The other major product is the aldehyde that results from reaction of the amino acid with ninhydrin. When valine reacts, the resulting aldehyde is 2-methylpropanal. The other products are carbon dioxide and water.

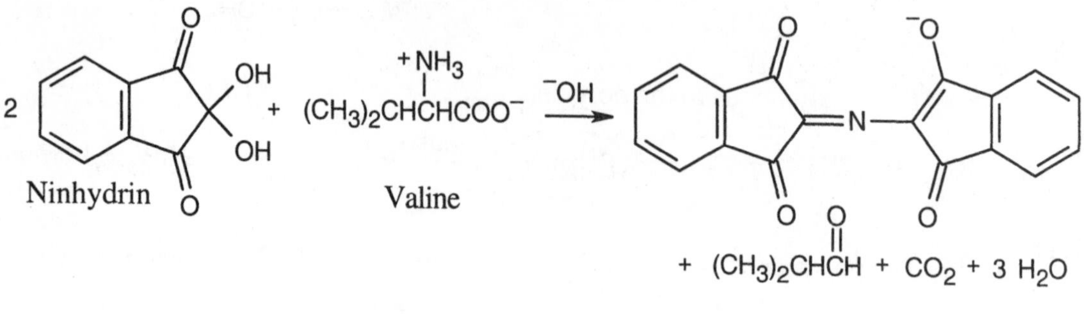

$$2\ \text{Ninhydrin} + (CH_3)_2CH\overset{+NH_3}{\overset{|}{C}}HCOO^- \xrightarrow{^-OH} \text{(purple ninhydrin product)} + (CH_3)_2CH\overset{O}{\overset{||}{C}}H + CO_2 + 3\ H_2O$$

Ninhydrin Valine

15.13 Trypsin cleaves peptide bonds at the carboxyl side of *lysine* and *arginine*. Chymotrypsin cleaves peptide bonds at the carboxyl side of *phenylalanine*, *tyrosine* and *tryptophan*.

Asp–Arg–Val–Tyr–Ile–His–Pro–Phe

Trypsin → Asp–Arg + Val–Tyr–Ile–His–Pro–Phe

Chymotrypsin → Asp–Arg–Val–Tyr + Ile–His–Pro–Phe

15.14 Arg–Pro
Pro–Leu–Gly
Gly–Ile–Val

The complete sequence: Arg–Pro–Leu–Gly–Ile–Val

15.15 The part of the PTH derivative that lies to the right of the dotted lines comes from the N-terminal residue. Complete the structure to identify the amino acid, which in this problem is methionine.

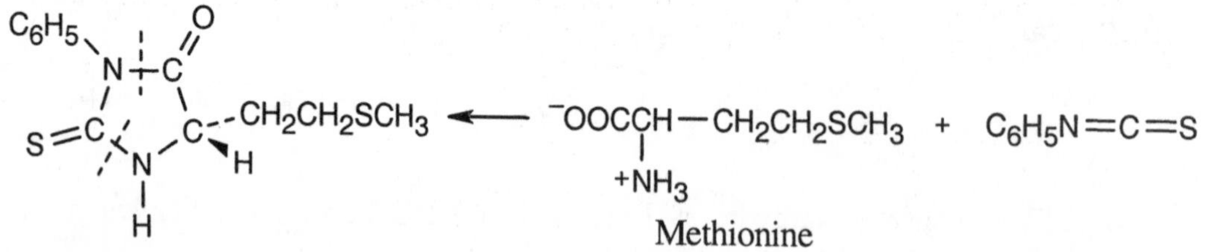

$$\text{PTH derivative (}C_6H_5\text{, }S{=}C\text{, N–H, C=O, }CH_2CH_2SCH_3\text{, H)} \longleftarrow {}^-OOC\overset{}{C}H{-}CH_2CH_2SCH_3\ (+NH_3) + C_6H_5N{=}C{=}S$$

Methionine

15.16 (a) Pyruvate decarboxylase is a lyase.
(b) Chymotrypsin is a hydrolase.
(c) Alcohol dehydrogenase is an oxidoreductase.

Visualizing Chemistry

15.17

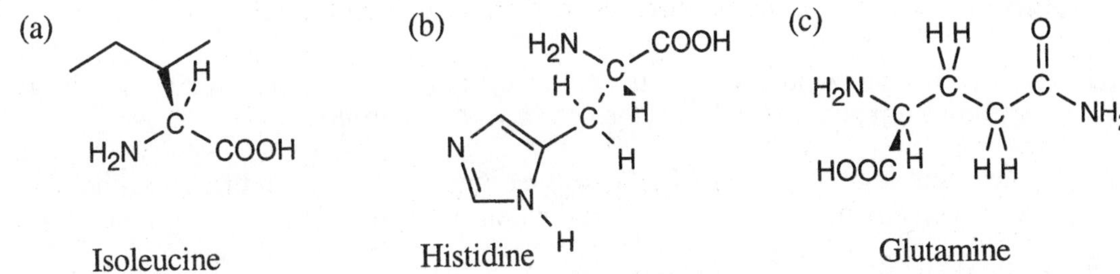

Isoleucine   Histidine   Glutamine

15.18

Cys——Lys——Ala——Asp

15.19

Isoleucine (Ile) (I)

15.20 It's possible to identify this representation of valine as the D enantiomer by drawing its Fischer projection. The amino group is on the right, just as a hydroxyl group is on the right in the Fischer projection of D-glyceraldehyde.

D-Valine
(*R*)-Valine

Additional Problems

Nomenclature

15.21 When referring to an amino acid, the prefix "α" indicates that the amino group is bonded to the carbon atom next to (alpha to) the carboxylic acid group.

15.22 (a) Ser = serine (b) Thr = threonine (c) Pro = proline
(d) F = phenylalanine (e) Q = glutamine (f) D = aspartic acid

15.23 (a) Nucleoproteins contain RNA and protein.
(b) Glycoproteins contain carbohydrate and protein.
(c) Lipoproteins contain lipids and protein.

15.24 The disulfide bridges that cysteine forms help to stabilize a protein's tertiary structure.

15.25

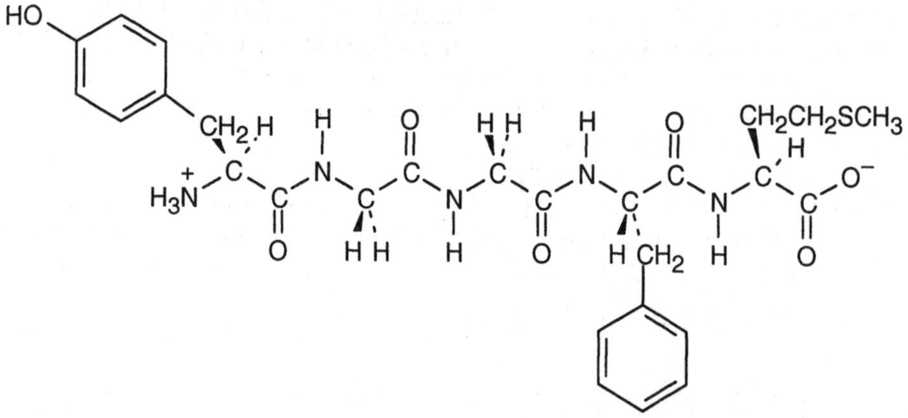

HO
CH_2 H H O H H H O $CH_2CH_2SCH_3$
H
$H_3\overset{+}{N}$ C C N C C N C C N C C N C C O^-
O H H H O H CH_2 H O

Tyr——Gly——Gly——Phe——Met
Y——G——G——F——M

Depictions and Chirality

15.26

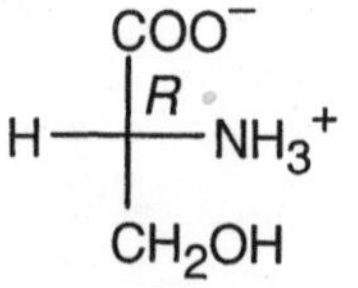

COO^-
R
H — NH_3^+
CH_2OH

(*R*)-Serine

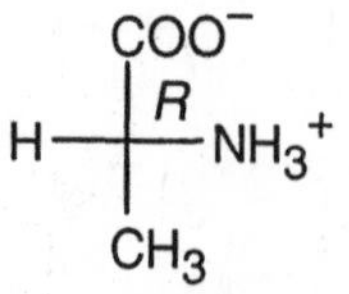

COO^-
R
H — NH_3^+
CH_3

(*R*)-Alanine

Both (*R*)-serine and (*R*)-alanine are D-amino acids.

15.27

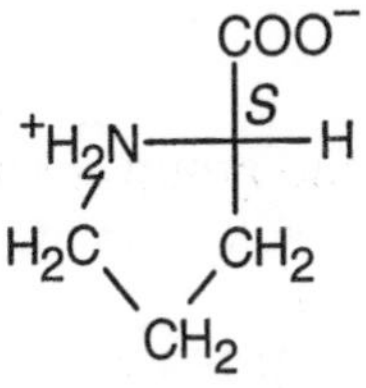

$^{+}H_2N$ — C(COO$^-$)(H), S; ring H_2C—CH_2—CH_2

(*S*)-Proline

15.28 (a) Val–Leu–Ser (V–L–S)
Val–Ser–Leu (V–S–L)
Ser–Leu–Val (S–L–V)
Ser–Val–Leu (S–V–L)
Leu–Val–Ser (L–V–S)
Leu–Ser–Val (L–S–V)

(b) Ser–Leu–Leu–Pro (S–L–L–P)
Ser–Leu–Pro–Leu (S–L–P–L)
Ser–Pro–Leu–Leu (S–P–L–L)
Pro–Leu–Leu–Ser (P–L–L–S)
Pro–Leu–Ser–Leu (P–L–S–L)
Pro–Ser–Leu–Leu (P–S–L–L)
Leu–Leu–Ser–Pro (L–L–S–P)
Leu–Leu–Pro–Ser (L–L–P–S)
Leu–Ser–Leu–Pro (L–S–L–P)
Leu–Ser–Pro–Leu (L–S–P–L)
Leu–Pro–Leu–Ser (L–P–L–S)
Leu–Pro–Ser–Leu (L–P–S–L)

15.29

(a)

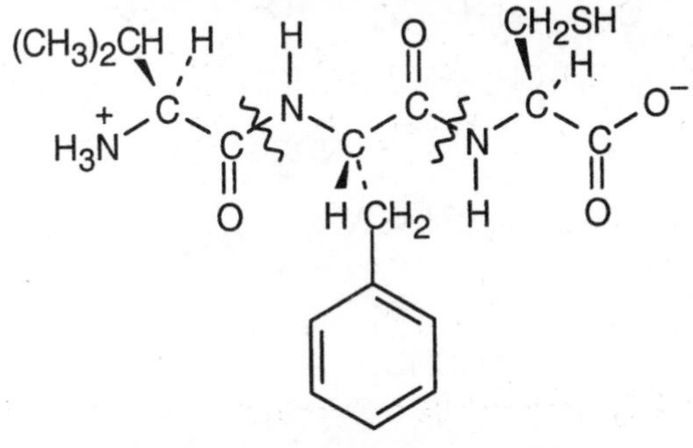

$(CH_3)_2CH$ H H O CH_2SH H $H_3\overset{+}{N}$ C C N C C N C C O$^-$ O H CH_2 H O

Val —— Phe —— Cys

V–F–C

〰 amide bonds

(b)

O $^-OCCH_2CH_2$ H $H_3\overset{+}{N}$ C C O N H H C O N C O H $CHCH_2CH_3$ CH_3 C N H H CH_2CHCH_3 CH_3 C O$^-$ O

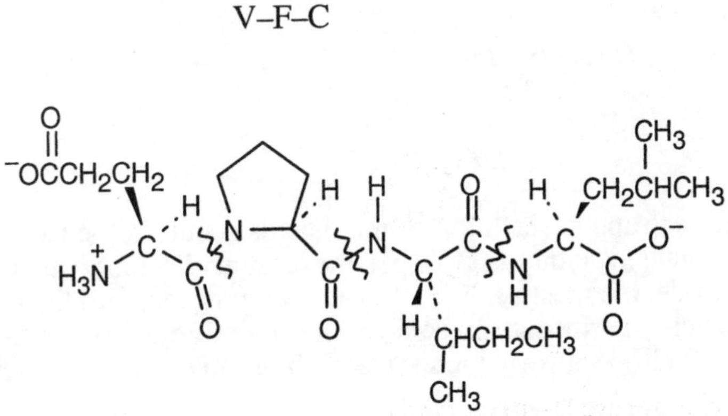

Glu —— Pro —— Ile —— Leu

E–P–I–L

15.30–15.31

D-Threose   L-Threonine   Diastereomers of L-Threonine

Acid/Base Chemistry of Amino Acids

15.32

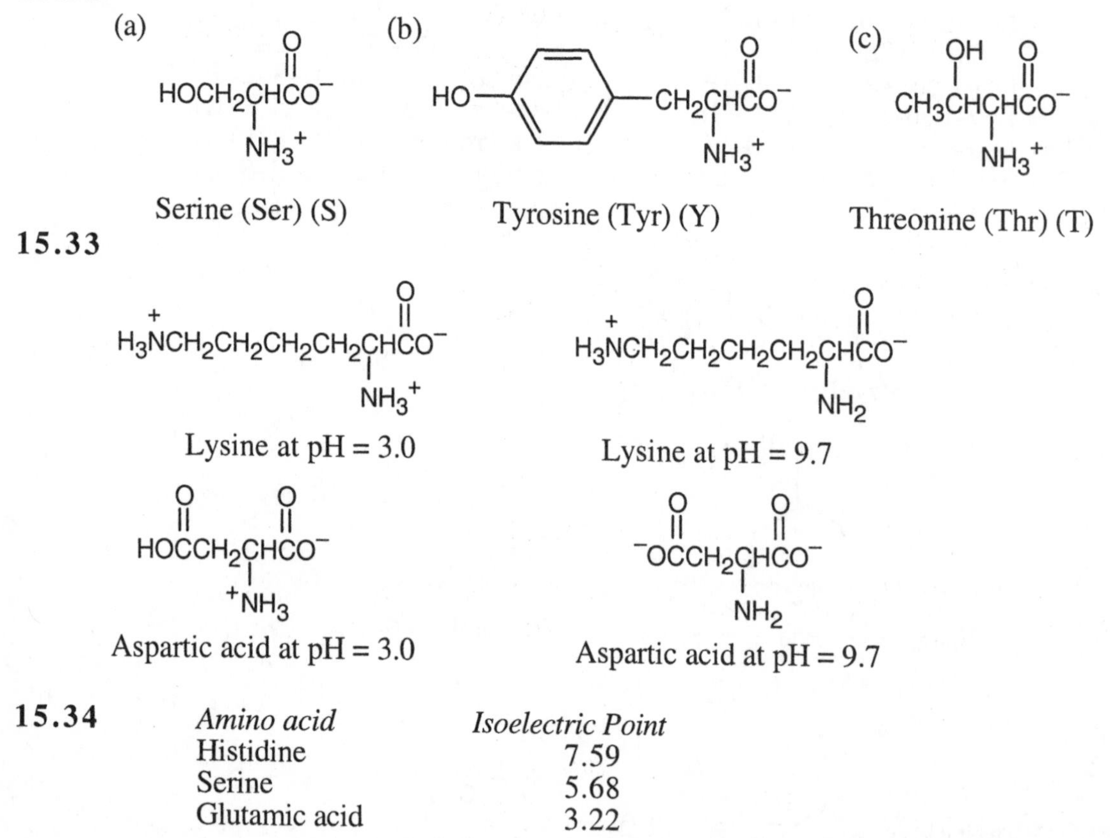

Serine (Ser) (S)   Tyrosine (Tyr) (Y)   Threonine (Thr) (T)

15.33

Lysine at pH = 3.0   Lysine at pH = 9.7

Aspartic acid at pH = 3.0   Aspartic acid at pH = 9.7

15.34

Amino acid	*Isoelectric Point*
Histidine	7.59
Serine	5.68
Glutamic acid	3.22

The optimum pH for the electrophoresis of three amino acids occurs at the isoelectric point of the amino acid intermediate in acidity. At this pH, the least acidic amino acid migrates toward the negative electrode, the most acidic amino acid migrates toward the positive electrode, and the amino acid intermediate in acidity does not migrate. In this example, electrophoresis at pH = 5.7 allows a very good separation of the three amino acids.

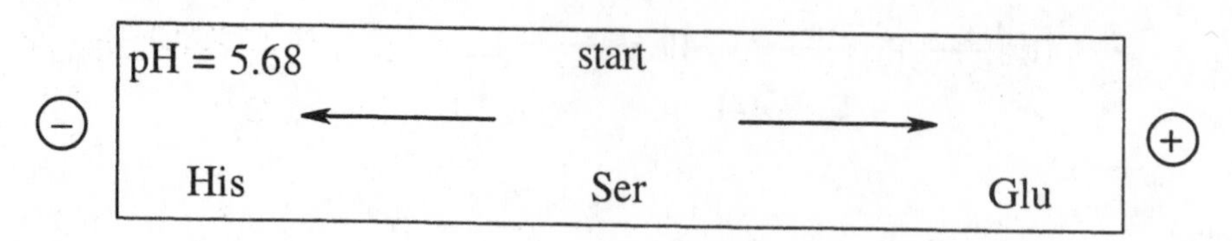

Reagents and Synthesis

15.35 Only primary amines can form the extensively conjugated purple ninhydrin product. A secondary amine such as proline yields a product containing a shorter system of conjugated bonds, which absorbs at a shorter wavelength (440 nm *vs.* 570 nm).

15.36 Treatment of a peptide with the reagent phenyl isothiocyanate (PITC) yields a *N*-phenylthiohydantoin derivative plus a peptide shortened by one amino acid at the *N*-terminal end.

Peptide $\xrightarrow{\text{PITC}}$ *Phenylthiohydantoin* + *Shortened Peptide*

(a)

Val–Leu–Gly → [phenylthiohydantoin: C_6H_5–N ring, C=O, C=S, N–H, C(H)–$CH(CH_3)CH_3$] + Leu–Gly

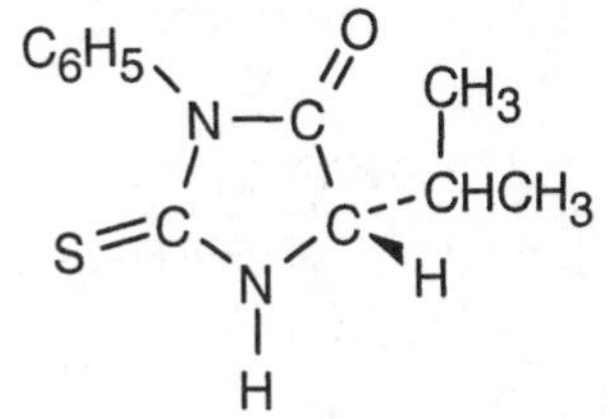

(b)

Ala–Pro–Phe → [phenylthiohydantoin: C_6H_5–N ring, C=O, C=S, N–H, C(H)–CH_3] + Pro–Phe

15.37

(a)

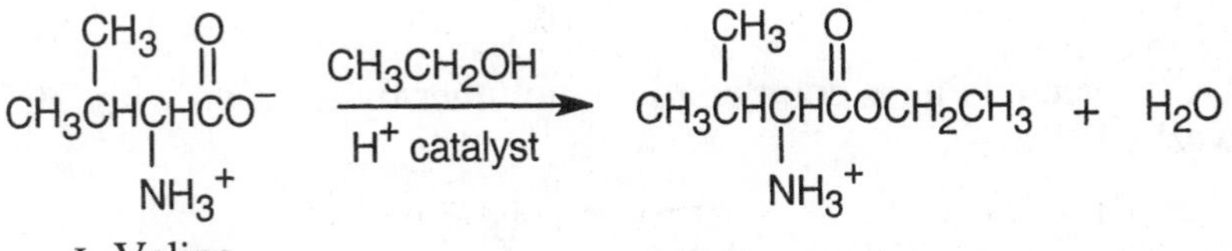

$$CH_3CH(CH_3)CH(NH_3^+)CO_2^- \xrightarrow[\text{H}^+\text{ catalyst}]{CH_3CH_2OH} CH_3CH(CH_3)CH(NH_3^+)COOCH_2CH_3 + H_2O$$

L-Valine

(b)

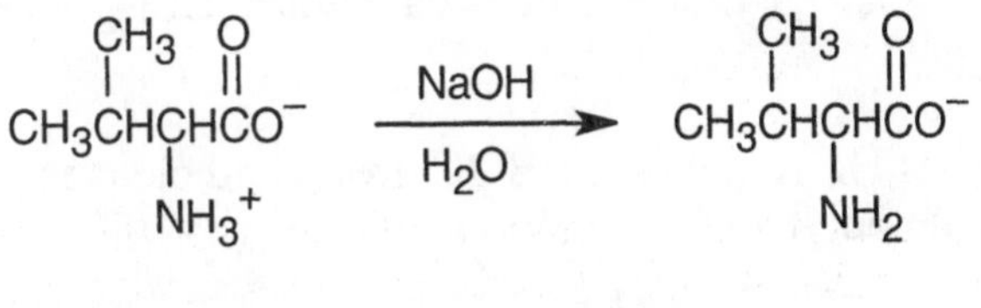

$$CH_3CH(CH_3)CH(NH_3^+)CO_2^- \xrightarrow[H_2O]{\text{NaOH}} CH_3CH(CH_3)CH(NH_2)CO_2^-$$

(c)

$$CH_3CH(CH_3)CH(NH_3^+)CO_2^- \xrightarrow{(CH_3)_3COC(=O)OC(=O)OC(CH_3)_3} CH_3CH(CH_3)CH(NHCOOC(CH_3)_3)CO_2^- + (CH_3)_3COH + CO_2$$

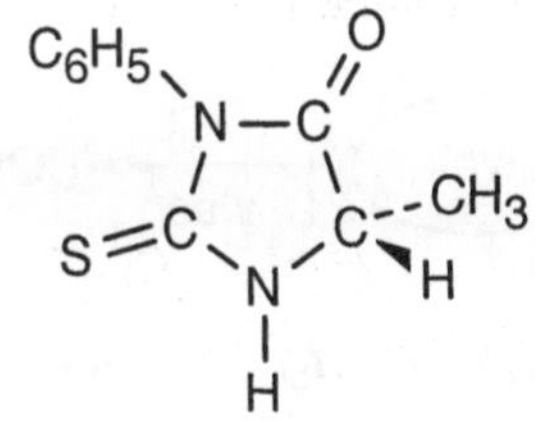

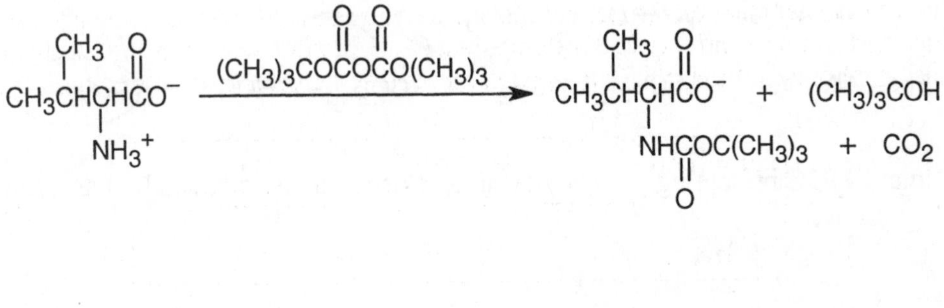

15.38

Step 1: Valine is protected as its BOC derivative.

$$(CH_3)_3COC(=O)OC(=O)OC(CH_3)_3 + Val \xrightarrow{Et_3N} (CH_3)_3COC(=O){-}Val{-}OH$$

$$(BOC{-}Val{-}OH)$$

Step 2: BOC–Val bonds to the polymer.

$$BOC{-}Val{-}OH + XCH_2{-}\text{(Polymer)} \xrightarrow{Base} BOC{-}Val{-}OCH_2{-}\text{(Polymer)}$$

Step 3: The polymer is first washed, then is treated with CF_3COOH to cleave the BOC group.

$$BOC{-}Val{-}OCH_2{-}\text{(Polymer)} \xrightarrow[\text{2. } CF_3COOH]{\text{1. wash}} Val{-}OCH_2{-}\text{(Polymer)}$$

Step 4: A BOC-protected Ala is coupled to the polymer-bound valine by reaction with DCC. The polymer is washed.

$$BOC{-}Ala + Val{-}OCH_2{-}\text{(Polymer)} \xrightarrow[\text{2. wash}]{\text{1. DCC}} BOC{-}Ala{-}Val{-}OCH_2{-}\text{(Polymer)}$$

Step 5: The polymer is treated with CF_3COOH to remove BOC.

$$BOC{-}Ala{-}Val{-}OCH_2{-}\text{(Polymer)} \xrightarrow{CF_3COOH} Ala{-}Val{-}OCH_2{-}\text{(Polymer)}$$

Step 6: A BOC-protected Phe is coupled to the polymer by reaction with DCC. The polymer is washed.

$$BOC{-}Phe + Ala{-}Val{-}OCH_2{-}\text{(Polymer)} \xrightarrow[\text{2. wash}]{\text{1. DCC}} BOC{-}Phe{-}Ala{-}Val{-}OCH_2{-}\text{(Polymer)}$$

Step 7: Treatment with anhydrous HF removes the BOC group and cleaves the ester bond between the peptide and the polymer.

$$BOC{-}Phe{-}Ala{-}Val{-}OCH_2{-}\text{(Polymer)} \xrightarrow{HF}$$

$$Phe{-}Ala{-}Val + (CH_3)_2C{=}CH_2 + CO_2 + HOCH_2{-}\text{(Polymer)}$$

Note: FMOC protecting groups may also be used but are cleaved by base, rather than acid.

15.39

(a)

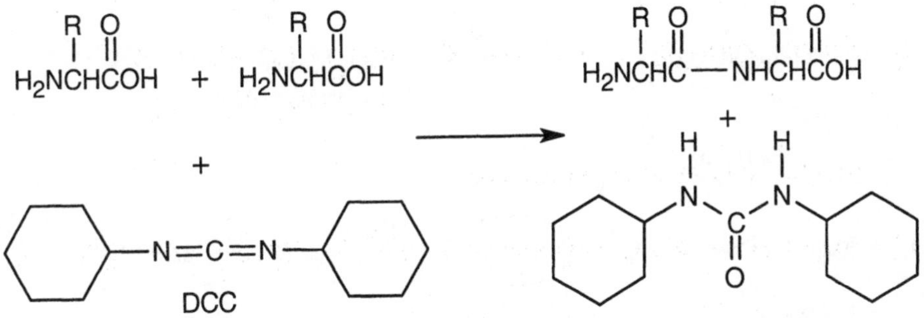

$$H_2NCH(R)COOH + H_2NCH(R)COOH + \text{DCC} \longrightarrow H_2NCH(R)C(O)\text{—}NHCH(R)COOH + \text{dicyclohexylurea}$$

DCC

In this step, a dipeptide is formed from two amino acid residues.

(b)

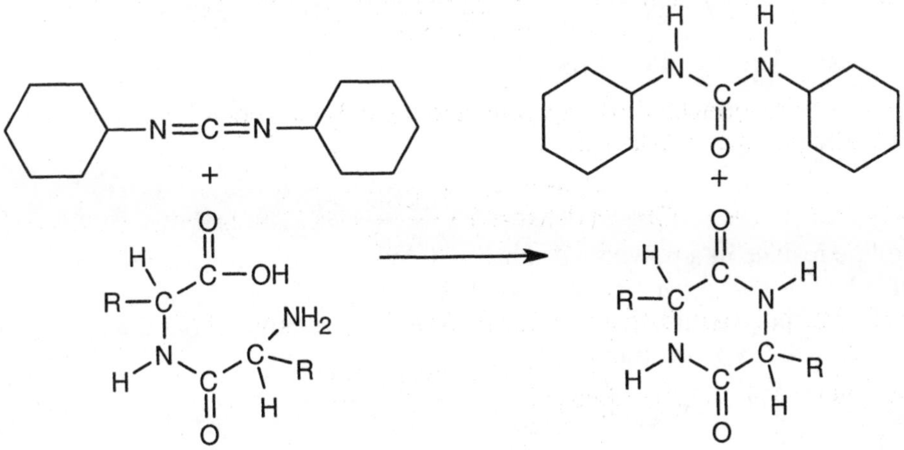

DCC + dipeptide ⟶ dicyclohexylurea + 2,5-diketopiperazine

DCC couples the carboxylic acid end of the dipeptide to the amino end to yield the 2,5-diketopiperazine.

Enzymes

15.40 (a) *Hydrolases* catalyze the hydrolysis of substrates.
(b) *Lyases* catalyze the removal of a small molecule from a substrate.
(c) *Transferases* catalyze the transfer of a group from one substrate to another.

15.41 (a) A *protease* catalyzes the hydrolysis of an amide group.
(b) A *kinase* catalyzes the transfer of a phosphate group.
(c) A *carboxylase* catalyzes the addition of CO_2 to a substrate.

15.42

Phe ~ Leu—Met—Lys ¦ Tyr ~ Asp—Gly—Gly—Arg ¦ Val—Ile—Pro—Tyr

cleaved by trypsin = ------

cleaved by chymotrypsin = ~~~~~

15.43 Gly–Asp–Phe–Pro
Phe–Pro–Val
Val–Pro–Leu

The complete sequence: Gly–Asp–Phe–Pro–Val–Pro–Leu

15.44 (a) Arg–Pro
Pro–Leu–Gly
Gly–Ile–Val

The complete sequence: Arg–Pro–Leu–Gly–Ile–Val

(b) Val–Met–Trp
Trp–Asp–Val
Val–Leu

The complete sequence: Val–Met–Trp–Asp–Val–Leu

15.45 Ser–Ile–Arg–Val–Val–Pro–Tyr–Leu–Arg

Integrated Problems

15.46 A proline residue in a polypeptide chain interrupts α-helix formation. The amide nitrogen of proline has no hydrogen that can contribute to the hydrogen-bonded structure of an α-helix. In addition, the pyrrolidine ring of proline restricts rotation about the C–N bond and reduces flexibility in the polypeptide chain.

15.47 Amino acids with hydrocarbon side chains (valine and isoleucine) are more likely to be found on the inside of a globular protein, where they can avoid water. Amino acids with charged side chains (aspartic acid and lysine) are more likely to be found on the outside of a globular protein, where they can interact with water.

15.48

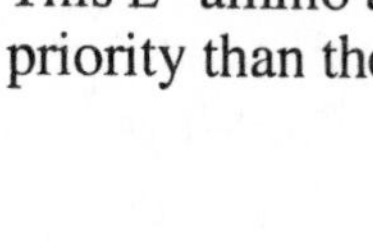

COO^-
$H_3\overset{+}{N}$ — H
CH_2Br

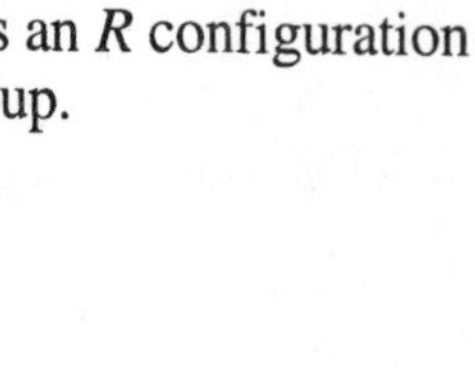

L-Bromoalanine
(*R*)-Bromoalanine

3 ^-OOC, 4 H, 1 $^+NH_3$, C, 2 CH_2Br, *R*

This L "amino acid" also has an *R* configuration because the $–CH_2Br$ side chain is higher in priority than the $–COO^-$ group.

15.49

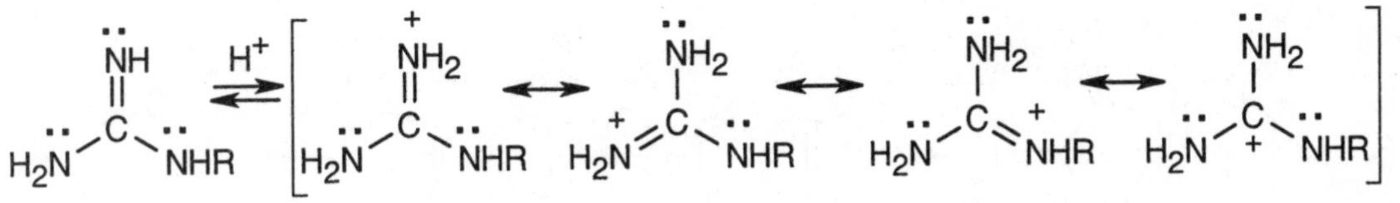

The protonated guanidino group can be stabilized by resonance.

15.50

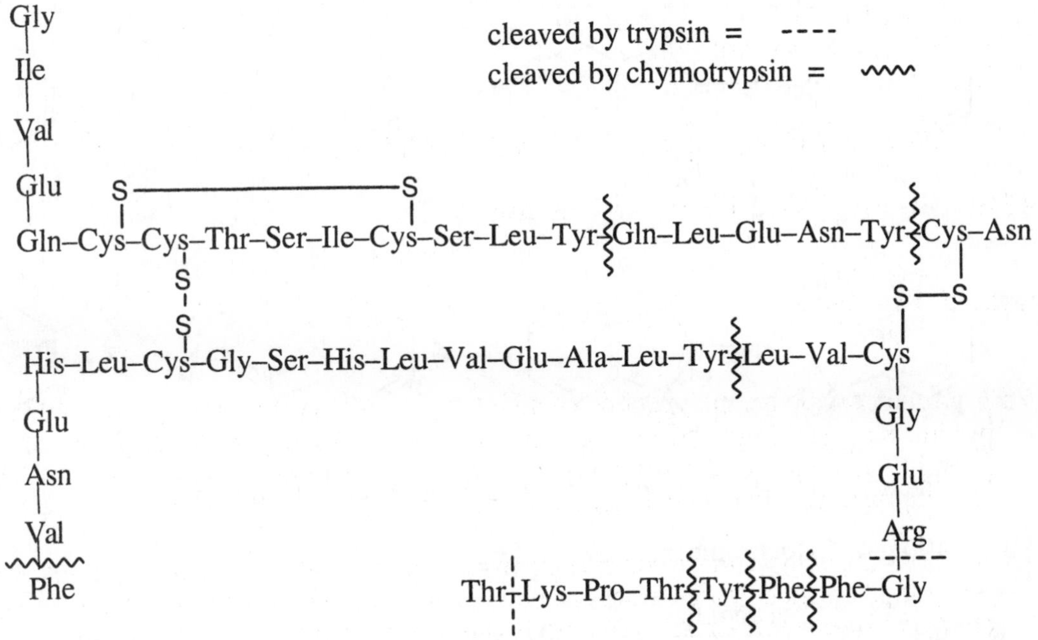

cleaved by trypsin = ----
cleaved by chymotrypsin = ∿∿∿

Gly–Ile–Val–Glu–Gln–Cys–Cys–Thr–Ser–Ile–Cys–Ser–Leu–Tyr–Gln–Leu–Glu–Asn–Tyr–Cys–Asn

His–Leu–Cys–Gly–Ser–His–Leu–Val–Glu–Ala–Leu–Tyr–Leu–Val–Cys

Glu–Asn–Val–Phe

Gly–Glu–Arg

Thr–Lys–Pro–Thr–Tyr–Phe–Phe–Gly

15.51

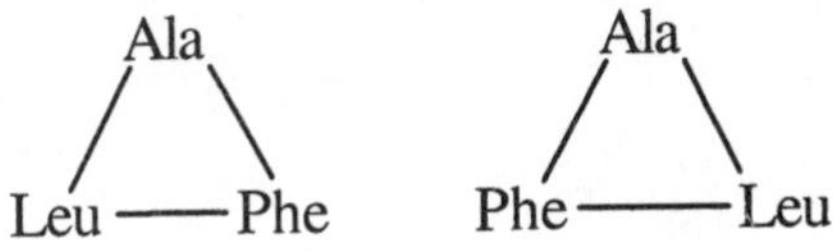

Ala, Leu, Phe   Ala, Phe, Leu

The tripeptide is cyclic.

15.52

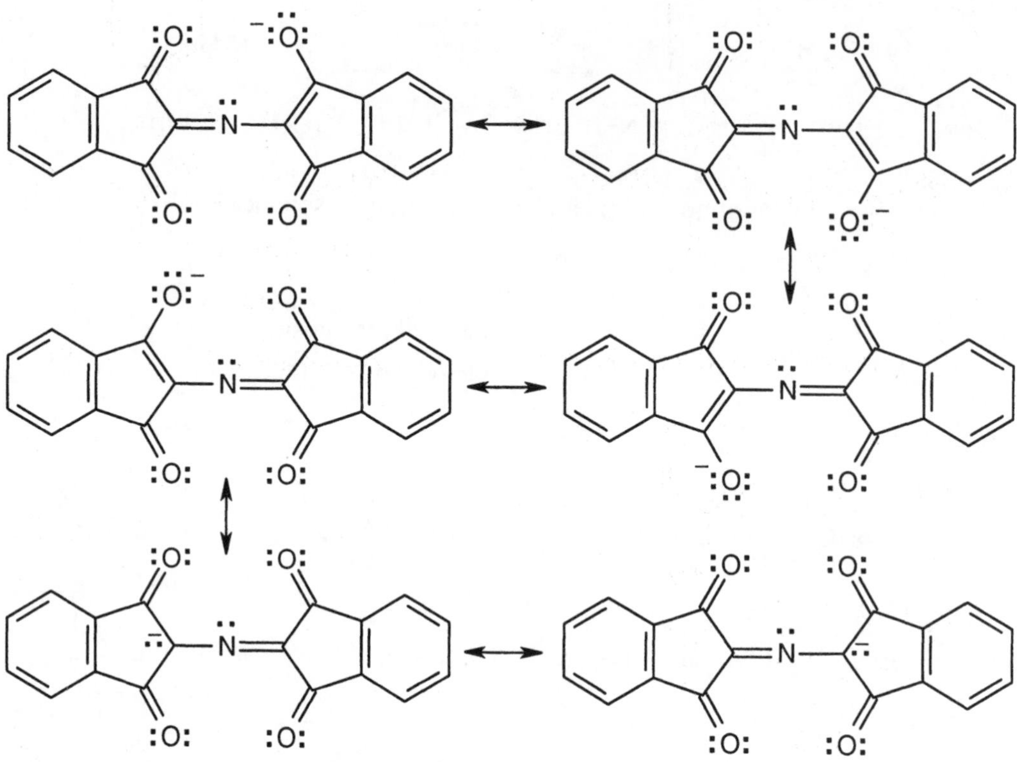

Not included are the resonance forms that involve the aromatic rings.

15.53 100 g of Cytochrome *c* contains 0.43 g iron.

$$0.43 \text{ g Fe} \times \frac{1 \text{ mol Fe}}{55.8 \text{ g Fe}} = 0.0077 \text{ mol Fe}$$

$$\frac{100 \text{ g Cytochrome } c}{0.0077 \text{ mol Fe}} = \frac{13{,}000 \text{ g Cytochrome } c}{1 \text{ mol Fe}}$$

Cytochrome *c* has a minimum molecular weight of 13,000.

15.54 Pro–Leu–Gly
Gly–Pro–Arg
Arg–Pro

The complete sequence: Pro–Leu–Gly–Pro–Arg–Pro

15.55

(a)

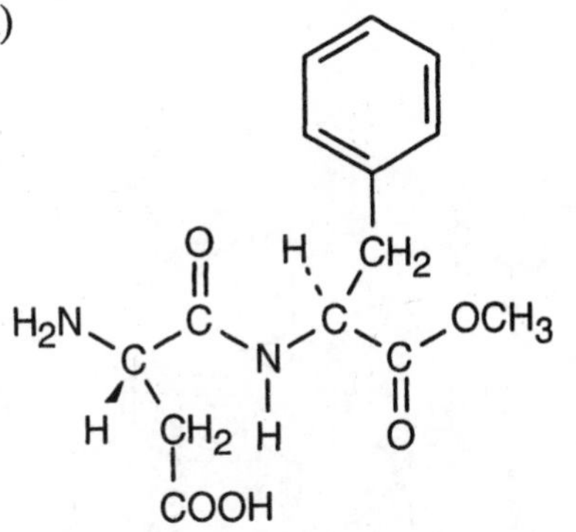

Aspartame (nonzwitterionic form)

(b)

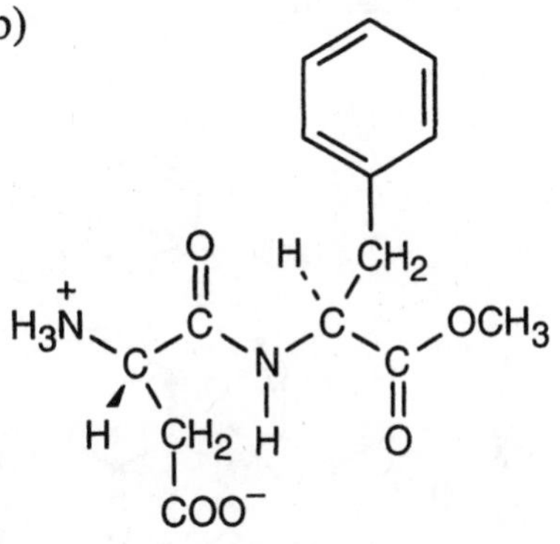

Aspartame at IP (pH = 5.9)

(c)

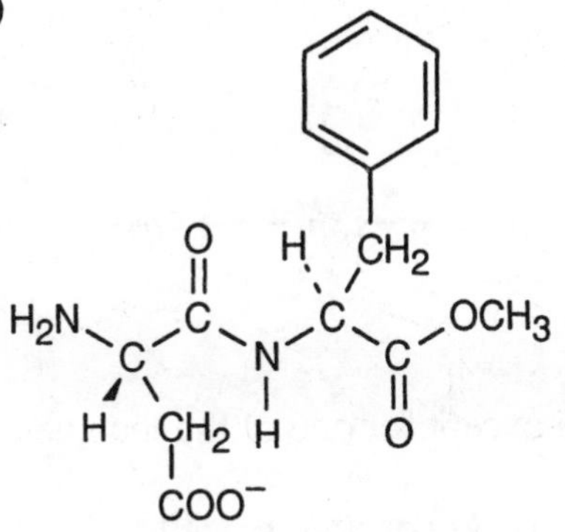

Aspartame at pH = 7.6

(d)

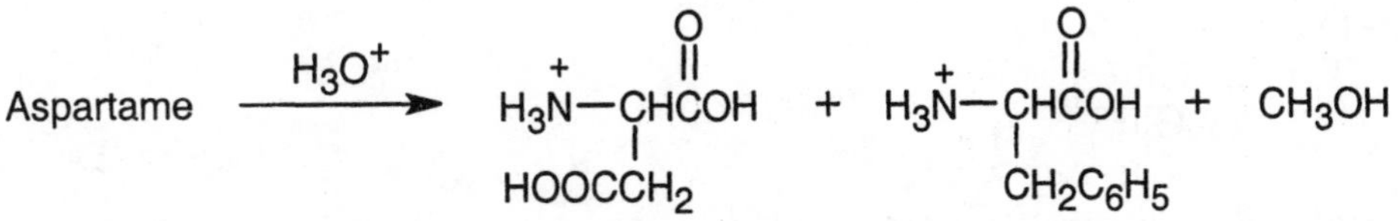

$$\text{Aspartame} \xrightarrow{H_3O^+} H_3\overset{+}{N}{-}\underset{\underset{HOOCCH_2}{|}}{CH}COOH + H_3\overset{+}{N}{-}\underset{\underset{CH_2C_6H_5}{|}}{CH}COOH + CH_3OH$$

In the Medicine Cabinet

15.56

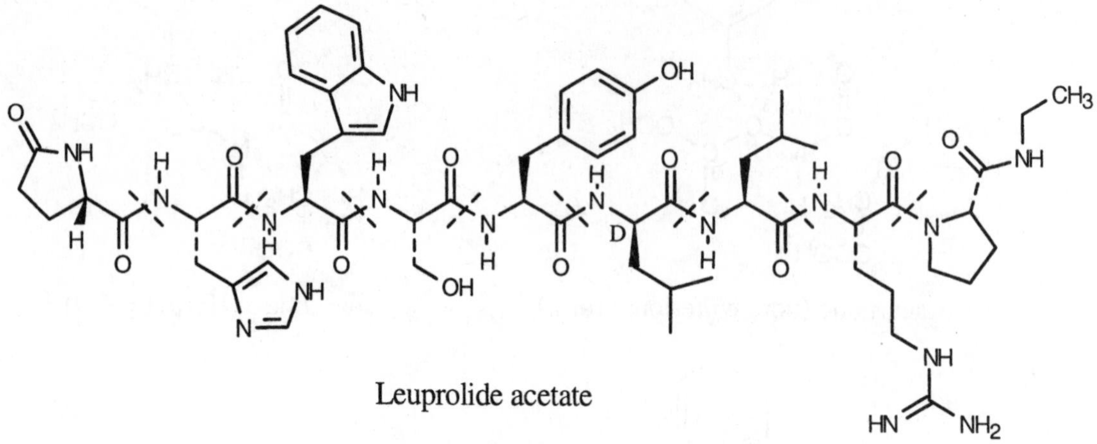

Leuprolide acetate

Glu—His—Trp—Ser—Tyr—Leu—Leu—Arg—Pro—NHEt

(a) Glutamic acid cyclizes to form a lactam.
(b) See above.
(c) All of the amino acids have an L configuration except for one of the leucines (indicated above), which has a D configuration.
(d) The charge on a peptide is due to the side chains. According to Table 15.1, the only side chain that is charged at neutral pH is arginine. Thus, leuprolide has a charge of +1 at neutral pH.

15.57 Cys–Tyr
Tyr–Ile–Glu
Ile–Glu
Glu–Asp–Cys
Asp–Cys
Cys–Pro–Leu
Leu–Gly

Reduced oxytocin: Cys–Tyr–Ile–Gln–Asn–Cys–Pro–Leu–Gly–NH_2

Oxidized oxytocin: Cys–Tyr–Ile–Gln–Asn–Cys–Pro–Leu–Gly–NH_2 (S—S bond between the two Cys)

The C–terminal end of oxytocin is an amide, but this can't be determined from the information given.

In the Field with Agrochemicals

15.58

(a)

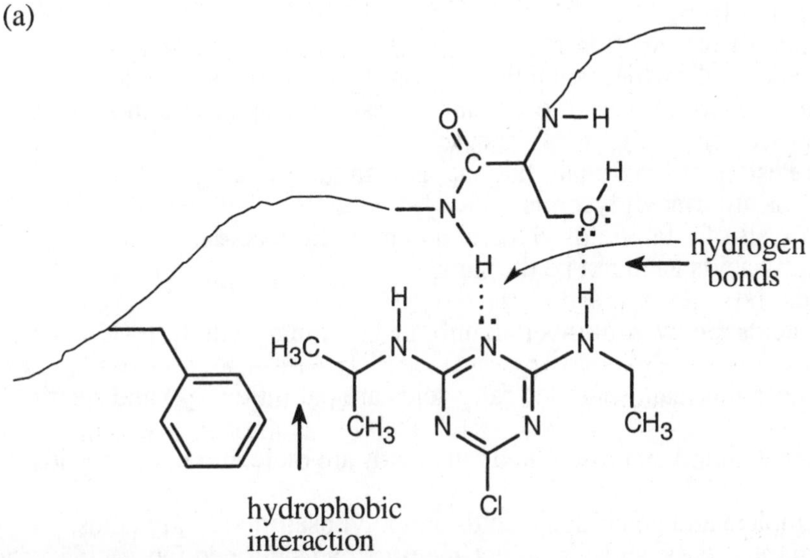

hydrogen bonds

hydrophobic interaction

(b) When glycine replaces serine, one of the two hydrogen bonds can't form. Apparently, this hydrogen bond is critical to the effectiveness of atrazine. Since substitution of phenylalanine does not change atrazine's effectiveness, a hydrogen bond must be stronger than a hydrophobic interaction.

15.59

The atrazine derivative is able to form a hydrogen bond similar to the bond lost when glycine replaced serine in the enzyme active site.

Chapter 16 – Biomolecules: Lipids and Nucleic Acids

Chapter Outline

I. Lipids (Sections 16.1 – 16.6).
  A. General characteristics (Section 16.1).
    1. Lipids are defined by their solubility in nonpolar organic solvents.
    2. Lipids fall into two classes: those that have ester linkages, and those that don't.
  B. Fats and oils (Section 16.2).
    1. Waxes are esters of long-chain fatty acids with long-chain alcohols.
    2. Fats and oils are triacylglycerols.
      a. Hydrolysis of a fat yields glycerol and three fatty acids.
      b. The fatty acids need not be the same.
    3. Fatty acids.
      a. Fatty acids are even-numbered, unbranched long-chain (C_{12}–C_{20}) carboxylic acids.
      b. The most abundant saturated fatty acids are palmitic (C_{16}) and stearic (C_{18}) acids.
      c. The most abundant unsaturated fatty acids are oleic and linoleic acids (both C_{18}).
        Linoleic and arachidonic acids are polyunsaturated fatty acids.
      d. Unsaturated fatty acids are lower-melting than saturated fatty acids because the double bonds keep molecules from packing closely.
      e. The C=C bonds can be hydrogenated to produce higher-melting fats.
  C. Soaps (Section 16.3).
    1. Soap is a mixture of the sodium and potassium salts of fatty acids produced by hydrolysis of animal fat.
    2. Soap acts as a cleanser because the two ends of a soap molecule are different.
      a. The hydrophilic carboxylate end dissolves in water.
      b. The hydrophobic hydrocarbon tails solubilize greasy dirt.
      c. In water, the hydrocarbon tails aggregate into micelles, where greasy dirt can accumulate.
    3. Soaps can form scum when they encounter Mg^{2+} and Ca^{2+} salts.
      This problem is circumvented by detergents, which don't form insoluble metal salts.
  D. Phospholipids (Section 16.4).
    1. Phosphoglycerides.
      a. Phosphoglycerides consist of glycerol, two fatty acids (at C1 and C2 of glycerol), and a phosphate group bonded to an amino alcohol at C3 of glycerol.
      b. The most important phospholipids are lecithins and cephalins.
      c. Phosphoglycerides comprise the major lipids in cell membranes.
        The phospholipid molecules are organized into a lipid bilayer, which has polar groups on the inside and outside, and nonpolar tails in the middle.
    2. Sphingolipids.
      a. Sphingolipids have sphingosine as their backbone.
      b. They are abundant in brain and nerve tissue as sphingomyelins.
  E. Steroids (Section 16.5).
    1. Facts about steroids.
      a. Steroids are found in the lipid extracts of plants and animals.

b. Steroids have a tetracyclic ring structure.
The rings adopt chair conformations but are unable to undergo ring-flips.
c. Steroids function as hormones in humans.
2. Types of steroid hormones.
a. Sex hormones.
i. Androgens (testosterone, androsterone) are male sex hormones.
ii. Estrogens (estrone, estradiol) and progestins are female sex hormones.
b. Adrenocortical hormones.
i. Mineralocorticoids (aldosterone) regulate cellular Na^+ and K^+ balance.
ii. Glucocorticoids (hydrocortisone) regulate glucose metabolism and control inflammation.
c. Synthetic steroids.
Oral contraceptives and anabolic steroids are examples of synthetic steroids.
3. Chemistry of statin drugs (Section 16.6).
a. Statin drugs inhibit the enzyme HMG-CoA reductase.
b. Part of the drug resembles the substrate HMG-CoA, and the other part is hydrophobic.

II. Nucleic acids (Sections 16.7 – 16.9).
A. Nucleic acids and nucleotides (Section 16.7).
1. Nucleosides are composed of a heterocyclic purine or pyrimidine base plus an aldopentose.
a. In RNA, the purines are adenine and guanine, the pyrimidines are uracil and cytosine, and the sugar is ribose.
b. In DNA, thymine replaces uracil, and the sugar is 2'-deoxyribose.
2. Positions on the base receive non-prime superscripts, and positions on the sugar receive prime superscripts.
3. The heterocyclic base is bonded to C1' of the sugar.
4. In nucleotides, a phosphate group is bonded to C5' of the sugar.
5. DNA is vastly larger than RNA and is found in the cell nucleus.
B. DNA (Sections 16.8 – 16.9).
1. Structure of DNA (Section 16.8).
a. DNA is composed of nucleotides connected by a phosphate ester bond between a 5' phosphate and the 3' hydroxyl group of a second nucleotide.
i. One end of the nucleic acid polymer has a free hydroxyl group and is called the 3' end.
ii. The other end has a free phosphate group and is called the 5' end.
b. The structure of DNA depends on the order of bases.
c. The sequence of bases is described by starting at the 5' end and listing the bases by their one-letter abbreviations.
2. Base-pairing in DNA (Section 16.9).
a. DNA consists of two polynucleotide strands coiled in a double helix.
Adenine and thymine hydrogen-bond with each other, and cytosine and guanine hydrogen-bond with each other.
b. Because the two DNA strands are complementary, the amount of A equals the amount of T, and the amount of C equals the amount of G.
c. The double helix is 2.0 nm wide, there are 10 bases in each turn, and each turn is 3.4 nm in height.
d. The double helix has a major groove and a minor groove into which polycyclic aromatic molecules can intercalate.

III. The transfer of genetic information (Sections 16.10 –16.13).
A. The "central dogma" of molecular genetics (Section 16.10).
1. The function of DNA is to store genetic information and to pass it on to RNA, which uses it to make proteins.

2. Replication, transcription and translation are the three processes that are responsible for carrying out the central dogma.

B. Replication of DNA (Section 16.11).
1. Replication is the enzyme-catalyzed process whereby DNA makes a copy of itself.
2. Replication is semiconservative: each new strand of DNA consists of one old strand and one newly synthesized strand.
3. How replication occurs:
   a. The DNA helix partially unwinds.
   b. New nucleotides form base-pairs with their complementary partners.
   c. Formation of new bonds is catalyzed by DNA polymerase and takes place in the 5' $\rightarrow$ 3' direction.
      Bond formation occurs by attack of the 3' hydroxyl group on the 5' triphosphate, with loss of a diphosphate leaving group.

C. Transcription (Section 16.12).
1. There are 3 types of RNA:
   a. Messenger RNA (mRNA) carries genetic information to ribosomes when protein synthesis takes place.
   b. Ribosomal RNA (rRNA), complexed with protein, comprises the physical makeup of the ribosomes.
   c. Transfer RNA (tRNA) brings amino acids to the ribosomes, where they are joined to make proteins.
2. mRNA is synthesized in the nucleus by transcription of DNA.
   a. The DNA partially unwinds, forming a "bubble".
   b. Ribonucleotides form base pairs with their complementary DNA bases.
   c. Bond formation occurs in the 5' $\rightarrow$ 3' direction and is catalyzed by RNA polymerase.
   d. Only one strand of DNA (the template strand) is transcribed.
   e. Thus, the synthesized mRNA is a copy of the coding strand (with U replacing T).
3. DNA contains "promoter sites", which indicate where mRNA synthesis is to begin, and base sequences that indicate where mRNA synthesis stops.

D. Translation (Section 16.13).
1. Translation is the process in which proteins are synthesized at the ribosomes by using mRNA as a template.
2. The message delivered by mRNA is contained in "codons" – 3-base groupings that are specific for an amino acid.
   a. Amino acids are coded by 61 of the possible 64 codons.
   b. The other 3 codons are "stop" codons.
3. Each tRNA is responsible for bringing an amino acid to the growing protein chain.
   a. A tRNA has a cloverleaf-shaped secondary structure and consists of 70–100 ribonucleotides.
   b. Each tRNA contains an anticodon complementary to the mRNA codon.
4. The protein chain is synthesized by enzyme-catalyzed peptide bond formation.
5. A 3-base "stop" codon on mRNA signals when synthesis is complete.

IV. DNA technology (Sections 16.14 –16.16).

A. Sanger DNA sequencing (Section 16.14).
1. The following mixture is assembled:
   a. The restriction fragment to be sequenced.
   b. A primer (a small piece of DNA whose sequence is complementary to that on the 3' end of the fragment).
   c. The 4 DNA nucleotide triphosphates.
   d. Small amounts of the four dideoxynucleotide triphosphates, each of which is labeled with a different fluorescent dye.

   2. DNA polymerase is added to the mixture, and a strand begins to grow from the end of the primer.
   3. Whenever a dideoxynucleotide is incorporated, chain growth stops.
   4. When reaction is complete, the fragments are separated by gel electrophoresis.
   5. Because fragments of all possible lengths are represented, the sequence can be read by noting the color of fluorescence of each fragment.

B. Polymerase chain reaction (Section 16.15).
   1. The polymerase chain reaction (PCR) can produce vast quantities of a DNA fragment.
   2. The key to PCR is *Taq* DNA polymerase, a heat-stable enzyme.
   3. Steps in PCR:
      a. The following mixture is heated to 95°C (a temperature at which DNA becomes single-stranded);
         i. *Taq* polymerase.
         ii. Mg^{2+} ion.
         iii. The 4 deoxynucleotide triphosphates.
         iv. A large excess of two oligonucleotide primers, each of which is complementary to the ends of the fragment to be synthesized.
         v. Double-strand DNA to be amplified.
      b. The temperature is lowered to 37°C – 50°C, causing the primers to hydrogen-bond to the single-stranded DNA.
      c. After raising the temperature to 72°C, *Taq* catalyzes the addition of further nucleotides, yielding two copies of the original DNA.
      d. The process is repeated until the desired quantity of DNA is produced.

C. New information about RNA (Section 16.16).
   1. RNA can store genetic information.
   2. RNA can catalyze reactions.

Solutions to Problems

16.1

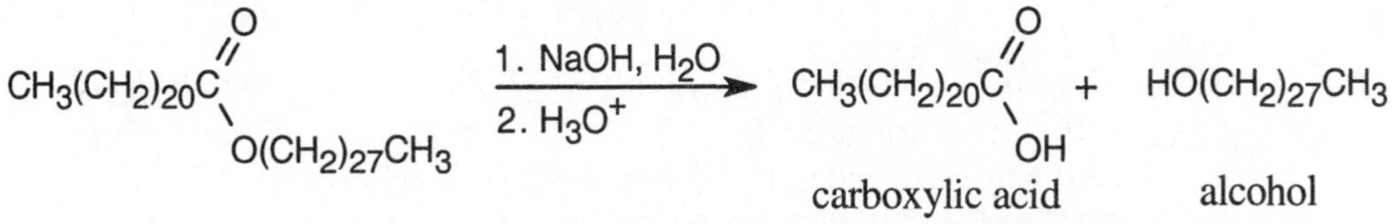

$$CH_3(CH_2)_{20}C(=O)O(CH_2)_{27}CH_3 \xrightarrow[2.\ H_3O^+]{1.\ NaOH,\ H_2O} CH_3(CH_2)_{20}C(=O)OH + HO(CH_2)_{27}CH_3$$

carboxylic acid — alcohol

16.2

(a)

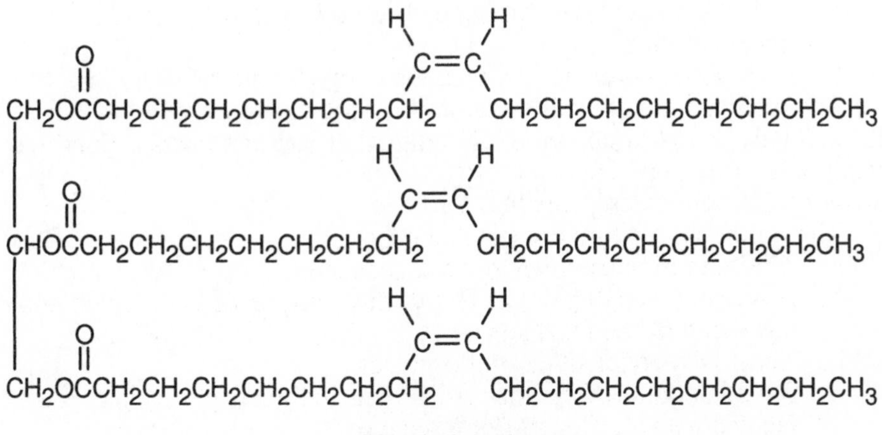

$$CH_2OC(=O)CH_2CH_2CH_2CH_2CH_2CH_2CH_2CH{=}CHCH_2CH_2CH_2CH_2CH_2CH_2CH_2CH_3$$
$$CHOC(=O)CH_2CH_2CH_2CH_2CH_2CH_2CH_2CH{=}CHCH_2CH_2CH_2CH_2CH_2CH_2CH_2CH_3$$
$$CH_2OC(=O)CH_2CH_2CH_2CH_2CH_2CH_2CH_2CH{=}CHCH_2CH_2CH_2CH_2CH_2CH_2CH_2CH_3$$

(each C=C is cis, with both H on the same side)

Glyceryl trioleate

(b)

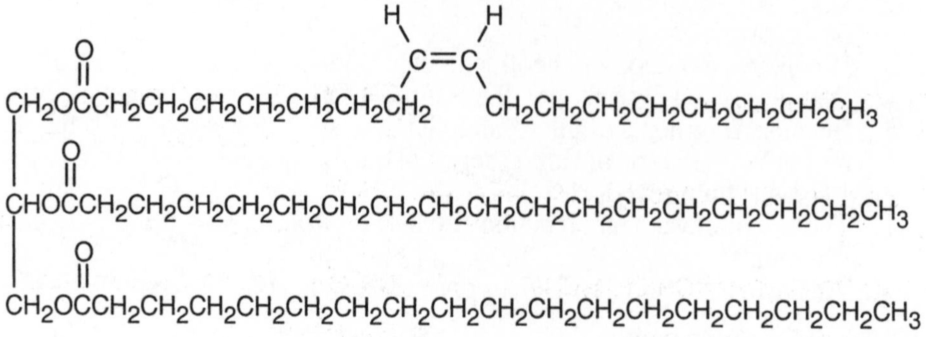

$$\begin{array}{l} CH_2OCCH_2CH_2CH_2CH_2CH_2CH_2CH_2CH{=}CHCH_2CH_2CH_2CH_2CH_2CH_2CH_2CH_3 \\ | \\ CHOCCH_2CH_2CH_2CH_2CH_2CH_2CH_2CH_2CH_2CH_2CH_2CH_2CH_2CH_2CH_2CH_2CH_3 \\ | \\ CH_2OCCH_2CH_2CH_2CH_2CH_2CH_2CH_2CH_2CH_2CH_2CH_2CH_2CH_2CH_2CH_2CH_2CH_3 \end{array}$$

(each OC is a C=O; the CH=CH double bond has both H on the same side)

Glyceryl monooleate distearate (one isomer)

Glyceryl monooleate distearate is higher melting because it contains only one unsaturated fatty acid; glyceryl trioleate contains three. Notice that two stereoisomers of fat (b) can be drawn, one with oleic acid bonded to a terminal oxygen of glycerol and one with it bonded to the central oxygen.

16.3

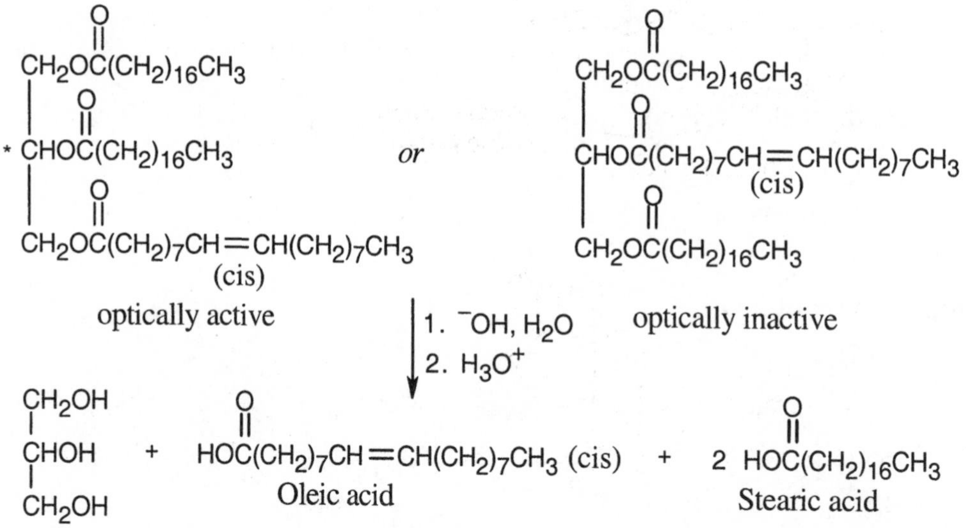

$$\begin{array}{l} CH_2OC(O)(CH_2)_{16}CH_3 \\ | \\ {}^{*}CHOC(O)(CH_2)_{16}CH_3 \\ | \\ CH_2OC(O)(CH_2)_7CH{=}CH(CH_2)_7CH_3 \\ \quad\quad\text{(cis)} \end{array} \quad \textit{or} \quad \begin{array}{l} CH_2OC(O)(CH_2)_{16}CH_3 \\ | \\ CHOC(O)(CH_2)_7CH{=}CH(CH_2)_7CH_3 \\ \quad\quad\text{(cis)} \\ | \\ CH_2OC(O)(CH_2)_{16}CH_3 \end{array}$$

optically active — optically inactive

1. ^{-}OH, H_2O
2. H_3O^+

$$\begin{array}{l} CH_2OH \\ | \\ CHOH \\ | \\ CH_2OH \end{array} \; + \; HOC(O)(CH_2)_7CH{=}CH(CH_2)_7CH_3 \text{ (cis)} \; + \; 2\; HOC(O)(CH_2)_{16}CH_3$$

Oleic acid — Stearic acid

Four different groups are bonded to the central glycerol carbon atom in the optically active fat.

16.4

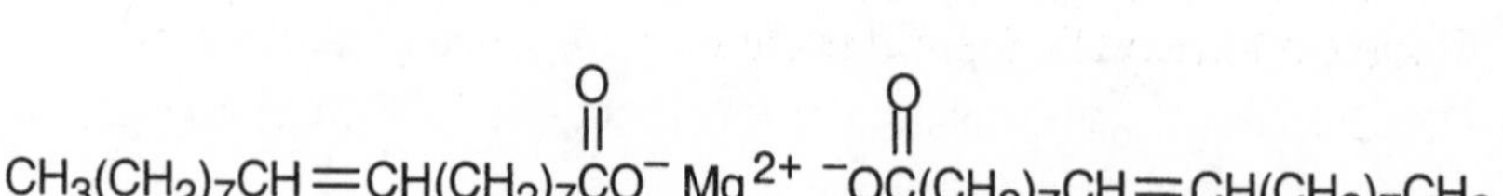

$$CH_3(CH_2)_7CH{=}CH(CH_2)_7C(O)O^- \; Mg^{2+} \; {}^{-}OC(O)(CH_2)_7CH{=}CH(CH_2)_7CH_3$$

Magnesium oleate

The double bonds are cis.

16.5

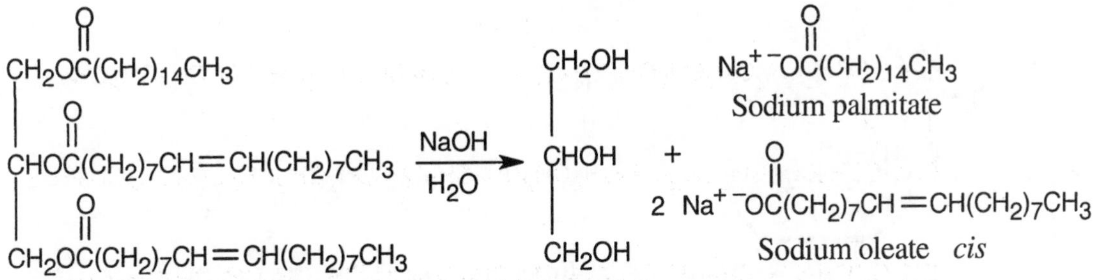

$$\begin{array}{l} CH_2OC(O)(CH_2)_{14}CH_3 \\ CHOC(O)(CH_2)_7CH{=}CH(CH_2)_7CH_3 \\ CH_2OC(O)(CH_2)_7CH{=}CH(CH_2)_7CH_3 \end{array} \xrightarrow[H_2O]{NaOH} \begin{array}{l} CH_2OH \\ CHOH \\ CH_2OH \end{array} + \begin{array}{l} Na^{+}\,{}^{-}OC(O)(CH_2)_{14}CH_3 \\ 2\ Na^{+}\,{}^{-}OC(O)(CH_2)_7CH{=}CH(CH_2)_7CH_3 \end{array}$$

Glyceryl dioleate monopalmitate (*cis* double bonds)

Glycerol

Sodium palmitate

Sodium oleate *cis*

16.6

ketone

ketone

carbon-carbon double bond

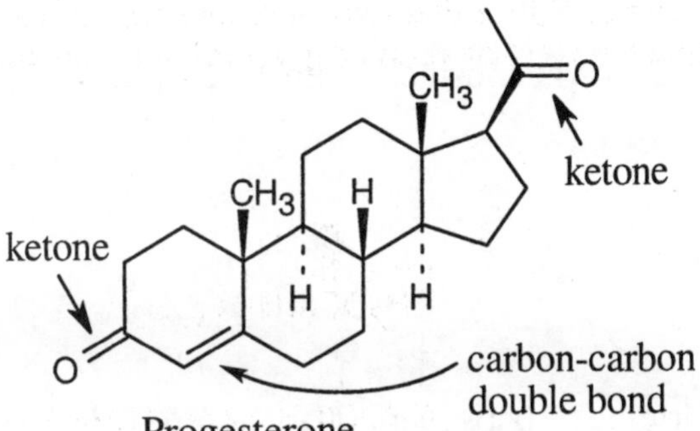

Progesterone

16.7

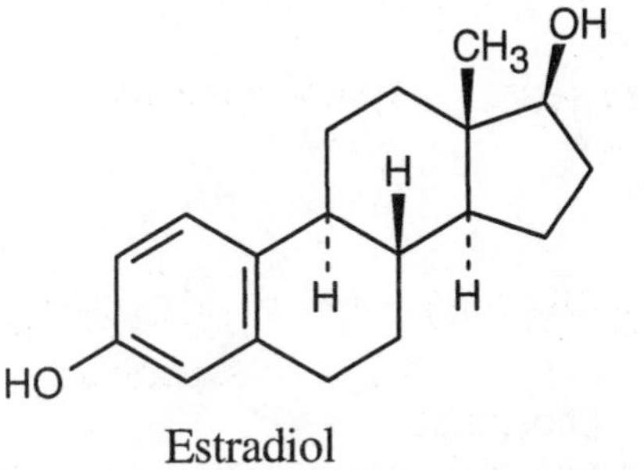

Estradiol

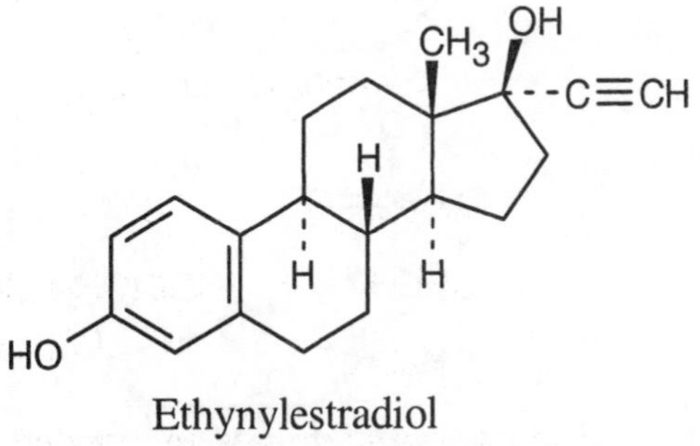

Ethynylestradiol

There is only one difference between estradiol and ethynylestradiol: ethynylestradiol has a –C≡C–H group at C17 that is not present in estradiol. Both compounds are estrogens because they both have a tetracyclic steroid skeleton with an aromatic **A** ring.

16.8

5' end

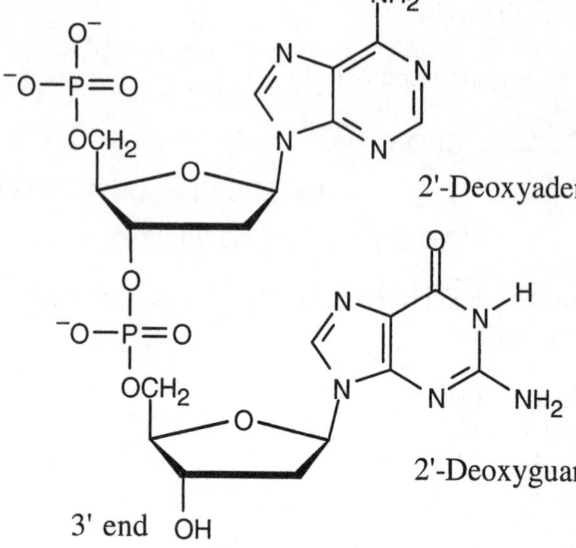

3' end

2'-Deoxyadenosine 5'-phosphate (A)

2'-Deoxyguanosine 5'-phosphate (G)

16.9

5' end

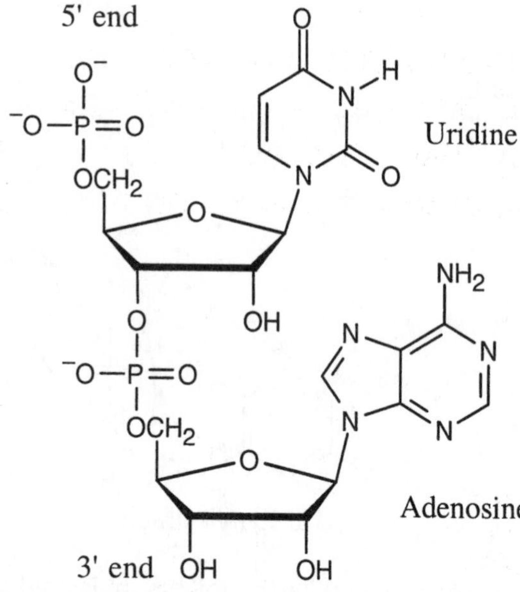

3' end

Uridine 5'-phosphate (U)

Adenosine 5'-phosphate (A)

16.10 Original DNA: 5' GGCTAATCCGT 3'
Complement: 3' CCGATTAGGCA 5'

16.11

Uracil

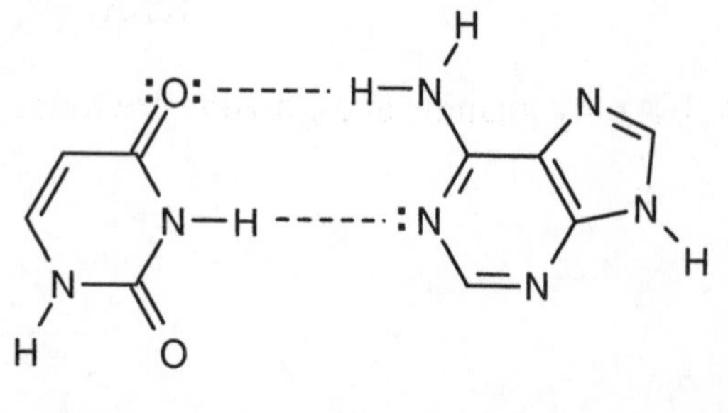

Adenine

16.12 An mRNA strand is written with the 5' end on the left and the 3' end on the right. It is transcribed from, and is complementary to, the template strand of DNA, which is written with the 3' on the left and the 5' end on the right.

DNA:	3'	GATTACCGTA	5'	is complementary to:
RNA:	5'	CUAAUGGCAU	3'	

16.13

RNA:	5'	UUCGCAGAGU	3'
DNA:	3'	AAGCGTCTCA	5'

16.14 Several different codons can code for the same amino acid. The codons are written with the 5' end on the left and the 3' end on the right.

Amino acid:	Ala	Phe	Leu	Tyr
Codon sequence:	GCU	UUU	UUA	UAU
	GCC	UUC	UUG	UAC
	GCA		CUU	
	GCG		CUC	
			CUA	
			CUG	

16.15–16.17

The mRNA base sequence:	5' CUU– AUG– GCU– UGG – CCC– UAA 3'
The amino acid sequence:	Leu – Met – Ala – Trp – Pro – (stop)
The tRNA sequences:	3' GAA UAC CGA ACC GGG 5'
The DNA sequence:	3' GAA – TAC – CGA –ACC – GGG –ATT 5'

Visualizing Chemistry

16.18

(a) Guanine (G) DNA RNA

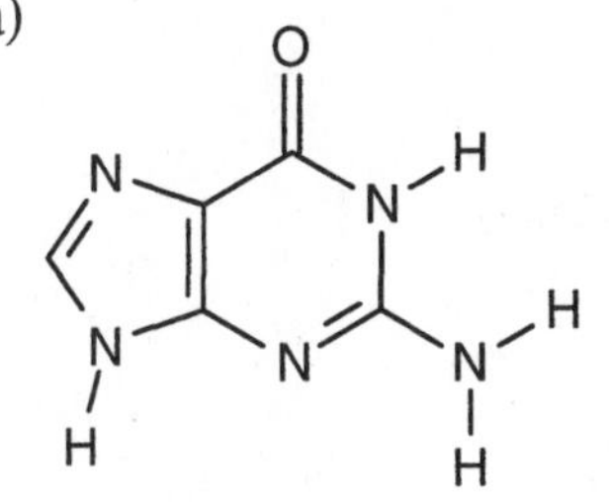

(b) Uracil (U) RNA

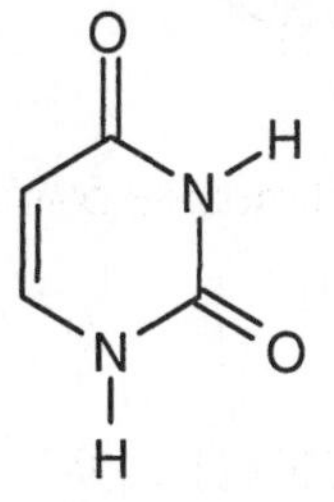

(c) Cytosine (C) DNA RNA

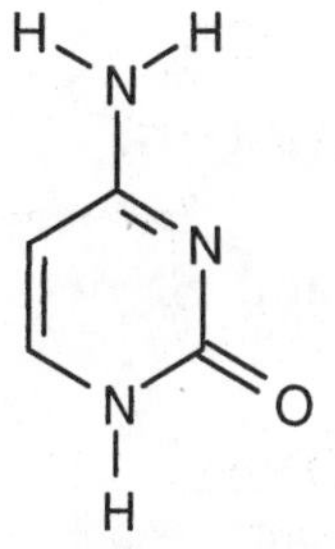

All three bases are found in RNA, but only guanine and cytosine are found in DNA.

16.19

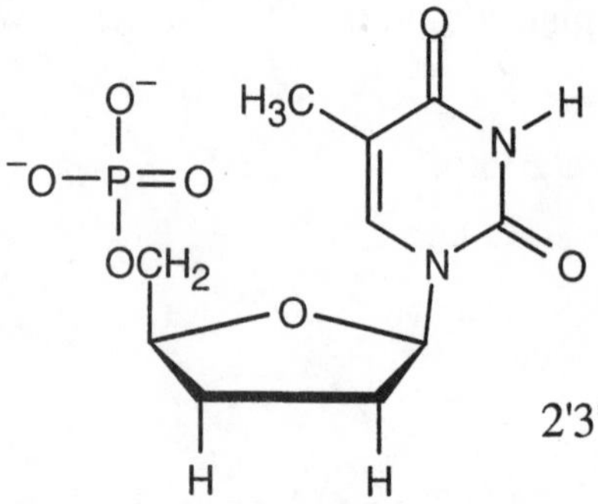

2'3'-Dideoxythymidine 5'-phosphate

The triphosphate made from 2'3'-dideoxythymidine 5'-phosphate is labeled with a fluorescent dye and used in the Sanger method of DNA sequencing. Along with the restriction fragment to be sequenced, a DNA primer, and a mixture of the four dNTPs (2'-deoxyribonucleoside triphosphates), small quantities of the four labeled dideoxyribonucleoside triphosphates are mixed together. DNA polymerase is added, and a strand of DNA complementary to the restriction fragment is synthesized. Whenever a dideoxyribonucleoside is incorporated into the DNA chain, chain growth stops. The fragments are separated by electrophoresis, and each terminal dideoxynucleotide can be identified by the color of its fluorescence. By identifying these terminal dideoxynucleotides, the sequence of the restriction fragment can be read.

16.20 The hydroxyl group in cholesterol is equatorial.

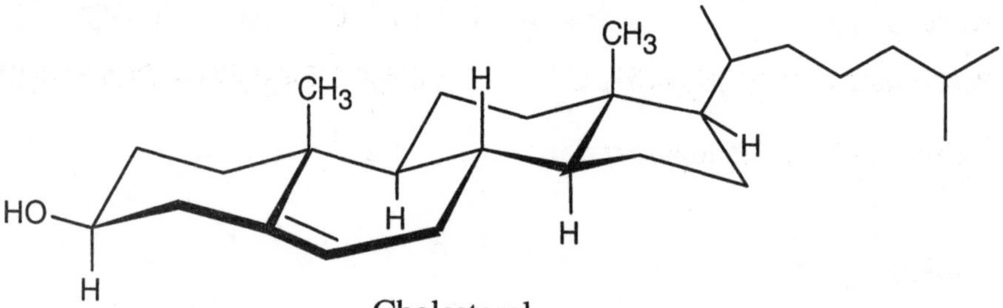

Cholesterol

Additional Problems

Nomenclature and Structure of Lipids

16.21

(a)

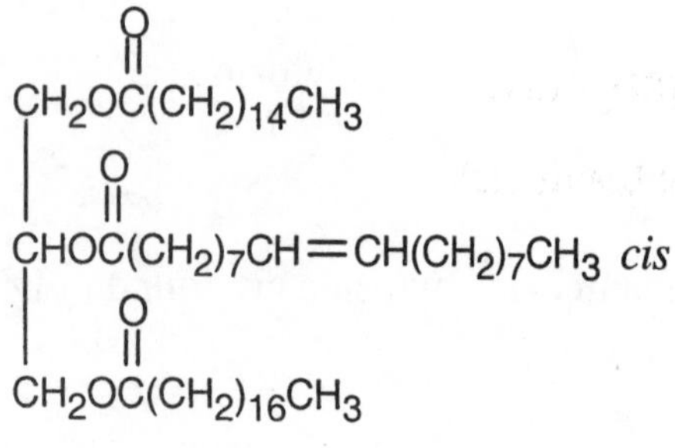

$CH_2OC(O)(CH_2)_{14}CH_3$

$CHOC(O)(CH_2)_7CH{=}CH(CH_2)_7CH_3$ *cis*

$CH_2OC(O)(CH_2)_{16}CH_3$

a fat

(b)

$CH_2OC(O)(CH_2)_7CH{=}CH(CH_2)_7CH_3$

$CHOC(O)(CH_2)_7CH{=}CH(CH_2)_7CH_3$

$CH_2OC(O)(CH_2)_7CH{=}CH(CH_2)_7CH_3$

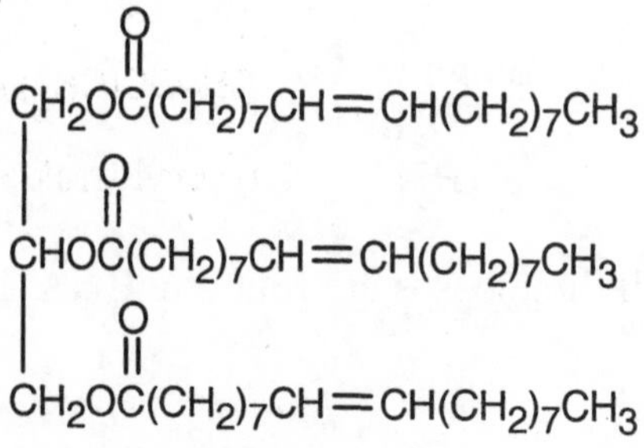

a vegetable oil
(all double bonds cis)

(c)

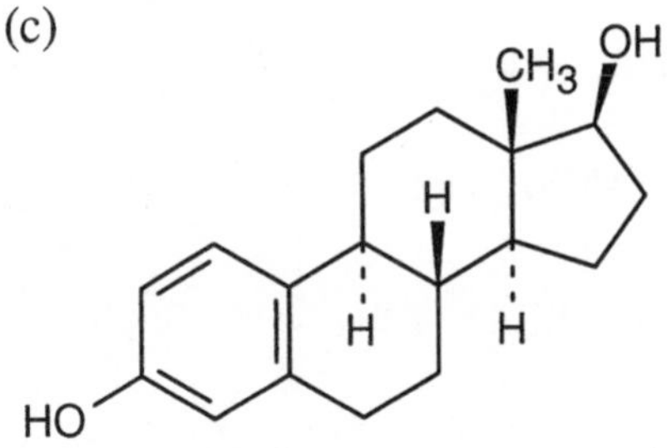

HO, CH_3, OH, H, H, H

a steroid (Estradiol)

16.22

(a)

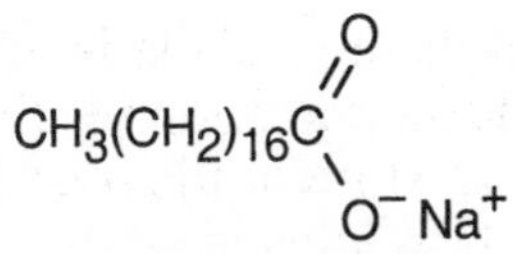

$CH_3(CH_2)_{16}C(=O)O^- Na^+$

Sodium stearate

(b)

$CH_3(CH_2)_4CH=CHCH_2CH=CH(CH_2)_7C(=O)OCH_2CH_3$

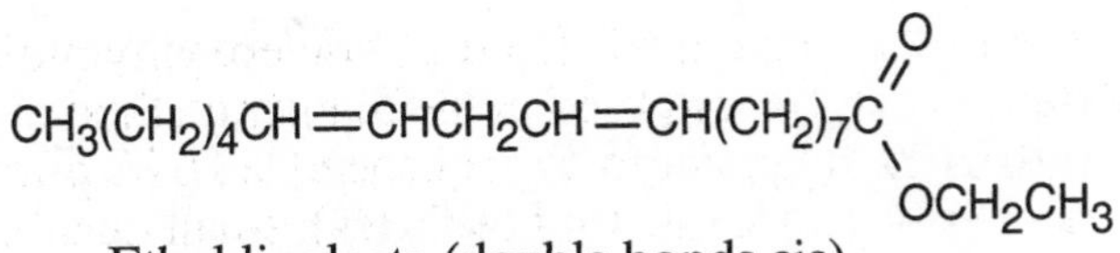

Ethyl linoleate (double bonds cis)

(c)

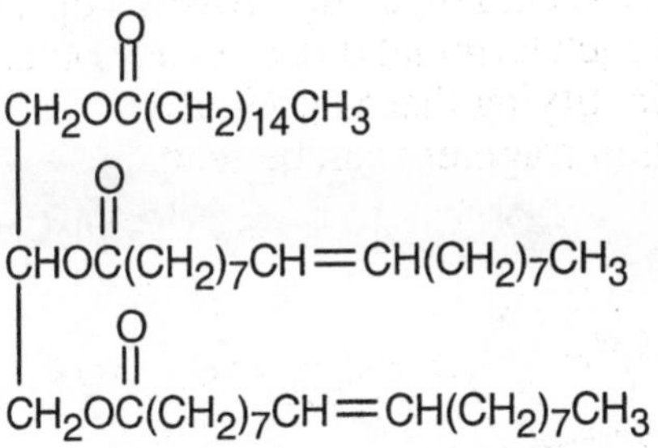

$CH_2OC(=O)(CH_2)_{14}CH_3$
$CHOC(=O)(CH_2)_7CH=CH(CH_2)_7CH_3$
$CH_2OC(=O)(CH_2)_7CH=CH(CH_2)_7CH_3$

or

$CH_2OC(=O)(CH_2)_7CH=CH(CH_2)_7CH_3$
$CHOC(=O)(CH_2)_{14}CH_3$
$CH_2OC(=O)(CH_2)_7CH=CH(CH_2)_7CH_3$

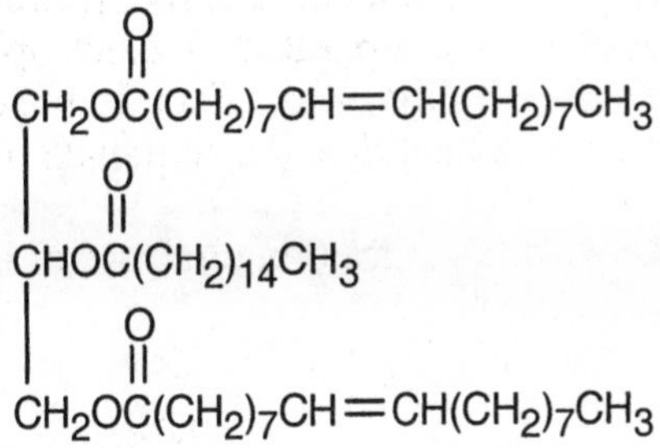

Glyceryl dioleate monopalmitate

Reactions of Lipids

16.23

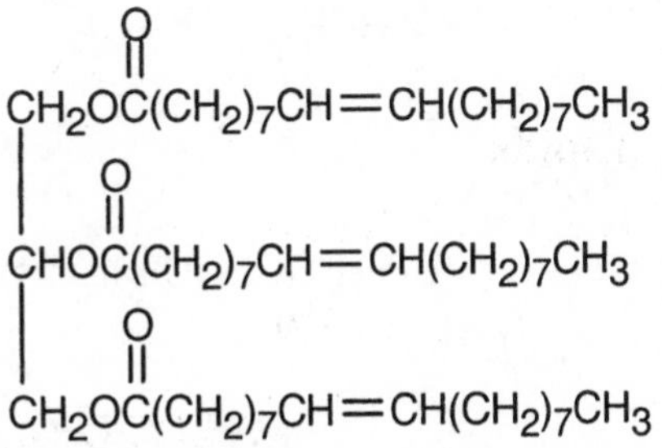

$CH_2OC(=O)(CH_2)_7CH=CH(CH_2)_7CH_3$
$CHOC(=O)(CH_2)_7CH=CH(CH_2)_7CH_3$
$CH_2OC(=O)(CH_2)_7CH=CH(CH_2)_7CH_3$

Glyceryl trioleate (double bonds cis)

(a)

$$\text{Glyceryl trioleate} \xrightarrow[CCl_4]{Br_2} \begin{array}{l} CH_2OC(=O)(CH_2)_7CH(Br)CH(Br)(CH_2)_7CH_3 \\ | \\ CHOC(=O)(CH_2)_7CH(Br)CH(Br)(CH_2)_7CH_3 \\ | \\ CH_2OC(=O)(CH_2)_7CH(Br)CH(Br)(CH_2)_7CH_3 \end{array}$$

(b)

$$\text{Glyceryl trioleate} \xrightarrow{H_2/Pd} \begin{array}{l} CH_2OC(=O)(CH_2)_{16}CH_3 \\ | \\ CHOC(=O)(CH_2)_{16}CH_3 \\ | \\ CH_2OC(=O)(CH_2)_{16}CH_3 \end{array}$$

(c)

$$\text{Glyceryl trioleate} \xrightarrow[H_2O]{NaOH} \begin{array}{l} CH_2OH \\ | \\ CHOH \\ | \\ CH_2OH \end{array} + 3\ Na^+\ {}^-OOC(CH_2)_7CH{=}CH(CH_2)_7CH_3$$

(d)

$$\text{Glyceryl trioleate} \xrightarrow[H_3O^+]{KMnO_4} \begin{array}{l} CH_2OC(=O)(CH_2)_7C(=O)OH \\ | \\ CHOC(=O)(CH_2)_7C(=O)OH \\ | \\ CH_2OC(=O)(CH_2)_7C(=O)OH \end{array} + 3\ HOC(=O)(CH_2)_7CH_3$$

(e)

$$\text{Glyceryl trioleate} \xrightarrow[2.\ H_3O^+]{1.\ LiAlH_4} \begin{array}{l} CH_2OH \\ | \\ CHOH \\ | \\ CH_2OH \end{array} + 3\ HOCH_2(CH_2)_7CH{=}CH(CH_2)_7CH_3$$

16.24

$$CH_3(CH_2)_7CH{=}CH(CH_2)_7COOH \quad \text{(cis)}$$
Oleic acid

(a)

$$\text{Oleic acid} \xrightarrow[\text{H}^+ \text{ catalyst}]{CH_3OH} CH_3(CH_2)_7CH{=}CH(CH_2)_7COOCH_3$$
Methyl oleate

(b)

$$\text{Methyl oleate from (a)} \xrightarrow[\text{Pd catalyst}]{H_2} CH_3(CH_2)_{16}COOCH_3$$
Methyl stearate

(c)

$$\text{Oleic acid} \xrightarrow[H_3O^+]{KMnO_4} CH_3(CH_2)_7\overset{O}{\overset{\|}{C}}OH + HO\overset{O}{\overset{\|}{C}}(CH_2)_7\overset{O}{\overset{\|}{C}}OH$$
Nonanoic acid + Nonanedioic acid

16.25

$$CH_3CH_2CH_2CH_2CH{=}CH{-}CH{=}CH{-}CH{=}CH{-}CH_2(CH_2)_6\overset{O}{\overset{\|}{C}}OH$$

(9Z,11E,13E)-Octadeca-9,11,13-trienoic acid
(Eleostearic acid)

$$\downarrow KMnO_4,\ H_3O^+$$

$$CH_3CH_2CH_2CH_2\overset{O}{\overset{\|}{C}}OH + HO\overset{O}{\overset{\|}{C}}{-}\overset{O}{\overset{\|}{C}}OH + HO\overset{O}{\overset{\|}{C}}{-}\overset{O}{\overset{\|}{C}}OH + HO\overset{O}{\overset{\|}{C}}(CH_2)_7\overset{O}{\overset{\|}{C}}OH$$

The stereochemistry of the double bonds can't be determined from the information given.

16.26

$$CH_3(CH_2)_7C{\equiv}C(CH_2)_7\overset{O}{\overset{\|}{C}}OH \xrightarrow[\text{Lindlar catalyst}]{H_2} CH_3(CH_2)_7CH{=}CH(CH_2)_7\overset{O}{\overset{\|}{C}}OH \ \textit{cis}$$
Stearolic acid → Oleic acid

16.27

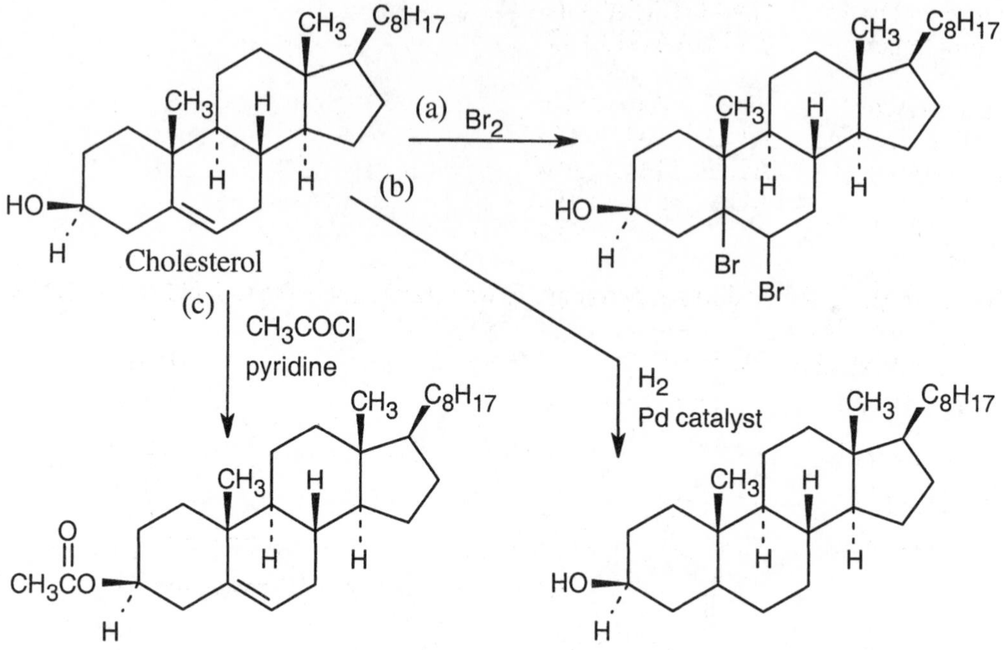

Cholesterol

(a) Br_2 →

(b) H_2, Pd catalyst →

(c) CH_3COCl, pyridine →

16.28

$$\begin{array}{l} CH_2OCOR \\ | \\ CHOCOR' \\ | \\ CH_2OCOR'' \end{array} + 3\ NaOH \xrightarrow{H_2O} \begin{array}{l} CH_2OH \\ | \\ CHOH \\ | \\ CH_2OH \end{array} + \begin{array}{l} RCO^-\ Na^+ \\ R'CO^-\ Na^+ \\ R''CO^-\ Na^+ \end{array}$$

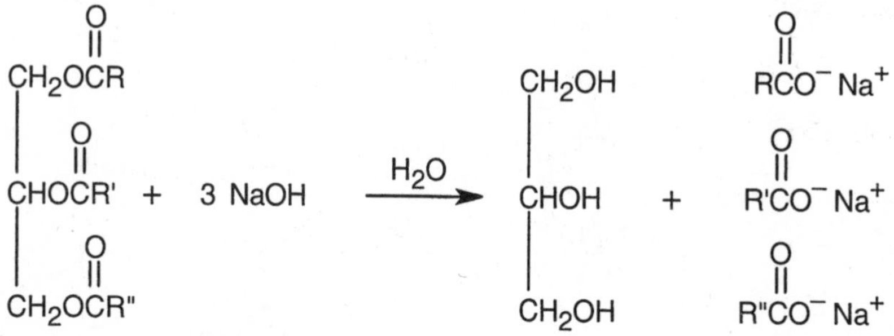

Molecular weight: 1500 g   Molecular weight: 40.0 g

$$5.00\text{ g oil} \times \frac{1\text{ mol oil}}{1500\text{ g oil}} \times \frac{3\text{ mol NaOH}}{1\text{ mol oil}} \times \frac{40.0\text{ g NaOH}}{1\text{ mol NaOH}} = 0.400\text{ g NaOH}$$

Thus, 0.400 g of NaOH is needed to saponify 5.00 g of soy oil.

Transcription and Translation

16.29 The percent of A always equals the percent of T, since A and T are complementary. A similar relationship is also true for G and C. Thus, sea urchin DNA contains about 32% each of A and T, and 18% each of G and C.

16.30

Original DNA:	GAAGTTCATGC
Complement:	CTTCAAGTACG

16.31

Amino Acid:	Ile	Asp	Thr
Codon Sequence:	AUU	GAU	ACU
(5' —> 3')	AUC	GAC	ACC
	AUA		ACA
			ACG

16.32–16.33 UAC is the codon for tyrosine. It was transcribed from ATG of the DNA chain.

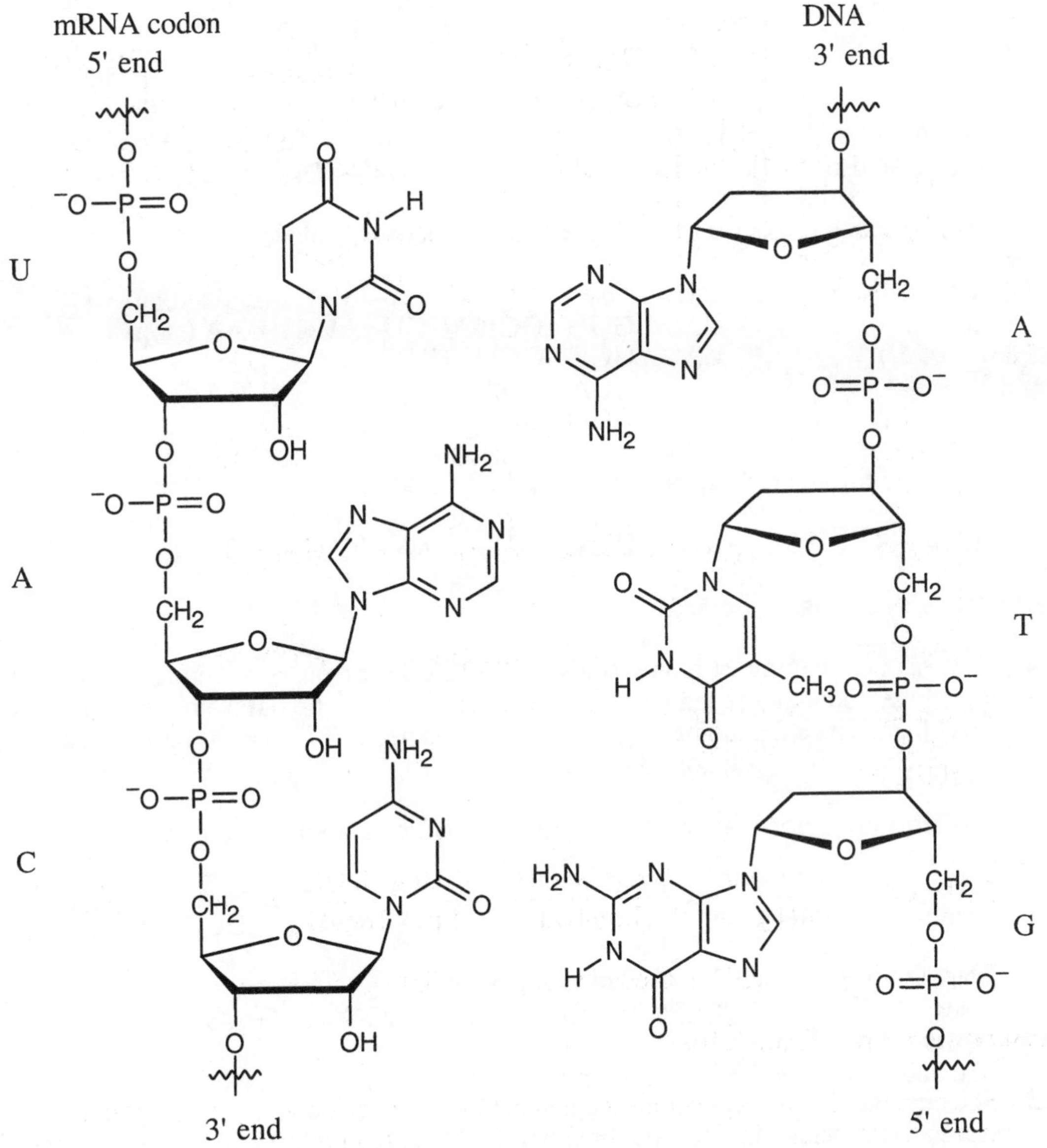

mRNA codon 5' end — U, A, C — 3' end

DNA 3' end — A, T, G — 5' end

16.34–16.36
The mRNA codons are written 5' –> 3', and the DNA sequences and tRNA anticodons are written 3' –> 5'.

	a)	b)	c)	d)	e)
mRNA codon:	AAU	GAG	UCC	CAU	ACC
Amino acid	Asn	Glu	Ser	His	Thr
DNA sequence:	TTA	CTC	AGG	GTA	TGG
tRNA anticodon:	UUA	CUC	AGG	GUA	UGG

16.37 Even though the stretch of DNA shown contains UAA in sequence, protein synthesis doesn't stop. The codons are read as 3-base individual units from start to end, and the unit UAA is read as part of two codons, not as a single codon.

16.38

	Normal	*Mutated*
DNA:	3' TAA-CCG-GAT 5'	3' TGA-CCG-GAT 5'
mRNA:	5' AUU-GGC-CUA 3'	5' ACU-GGC-CUA 3'
Amino Acids:	Ile — Gly — Leu	Thr — Gly — Leu

In the mutated protein, a threonine residue replaces an isoleucine residue.

16.39

Metenkephalin:	Tyr —	Gly —	Gly —	Phe —	Met	is coded by:
mRNA:	UAU	GGU	GGU	UUU	AUG	UAA (stop)
(5' –> 3')	UAC	GGC	GGC	UUC		UAG
		GGA	GGA			UGA
		GGG	GGG			

16.40 Using the first set of base pairs to solve this problem:

DNA: (3' –> 5') ATA – CCA – CCA – AAA – TAC – ATT

Integrated Problems

16.41 The chain containing 21 amino acids needs 21 x 3 = 63 bases to code for it. In addition, a three base codon is needed to terminate the chain. Thus, 66 bases are needed to code for the first chain. The chain containing 30 amino acids needs 30 x 3 = 90 bases, plus a three base "stop" codon, for a total of 93 bases.

16.42 Position 9: Horse amino acid = Gly Human amino acid = Ser

mRNA codons: GGU GGC GGA GGG UCU UCC UCA UCG AGU AGC

DNA bases: CCA CCG CCT CCC AGA AGG AGT AGC TCA TCG

The underlined horse DNA base triplets differ from their human counterparts by only one base.

Position 30:	Horse amino acid = Ala	Human amino acid = Thr
mRNA codons	GCU GCC GCA GCG	ACU ACC ACA ACG
DNA bases:	CGA CGG CGT CGC	TGA TGG TGT TGC

Each of the above groups of three DNA bases from horse insulin has a counterpart in human insulin that differs from it by only one base. It is possible that horse insulin DNA differs from human insulin DNA by only two bases out of 159!

16.43–16.44

mRNA sequence (5'–>3'):	CUA—GAC—CGU—UCC—AAG—UGA
Amino Acid:	Leu——Asp——Arg——Ser——Lys (stop)
tRNA anticodons (3'–>5'):	GAU CUG GCA AGG UUC
DNA sequence (3'–>5'):	GAT—CTG—GCA—AGG— TTC— ACT
DNA complement (5'–>3'):	CTA—GAC—CGT— TCC— AAG—TGA

16.45

Angiotensin II:	Asp	Arg	Val	Tyr	Ile	His	Pro	Phe	(stop)
mRNA sequence: (5'–3')	GAU	CGU	GUU	UAU	AUU	CAU	CCU	UUU	UAA
	GAC	CGC	GUC	UAC	AUC	CAC	CCC	UUC	UAG
		CGA	GUA		AUA		CCA		UGA
		CGG	GUG				CCG		
		AGA							
		AGG							

In the Medicine Cabinet

16.46–16.47

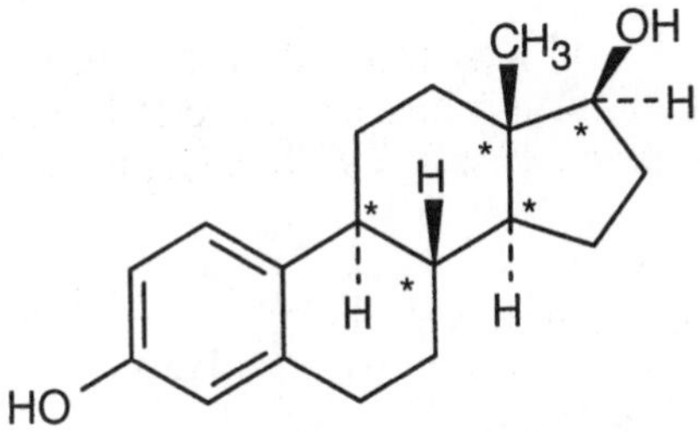

Estradiol

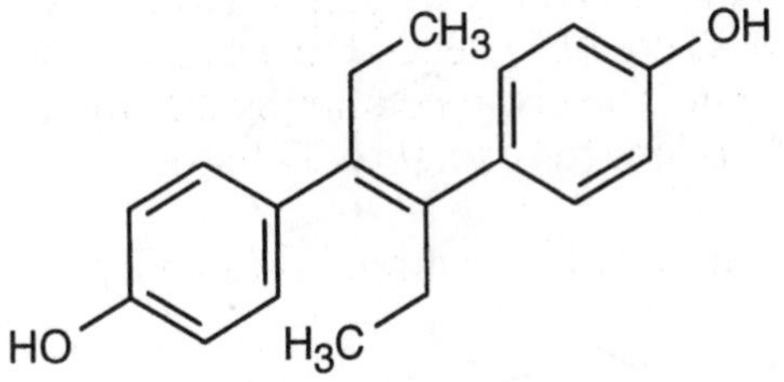

Estradiol has five stereocenters.

Diethylstilbestrol

16.48

(a)

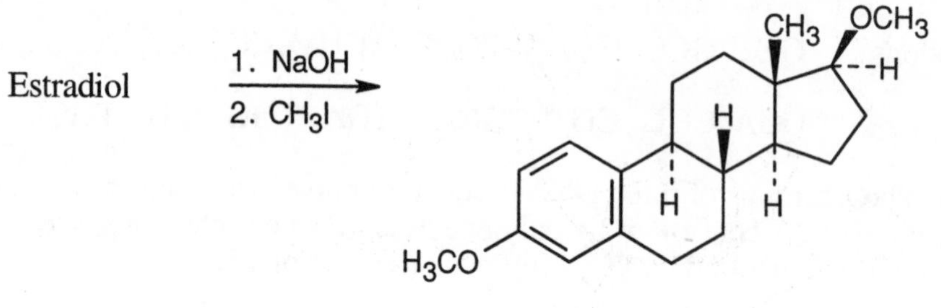

Estradiol $\xrightarrow[\text{2. } CH_3I]{\text{1. NaOH}}$

CH$_3$ OCH$_3$ H H H H$_3$CO

(b)

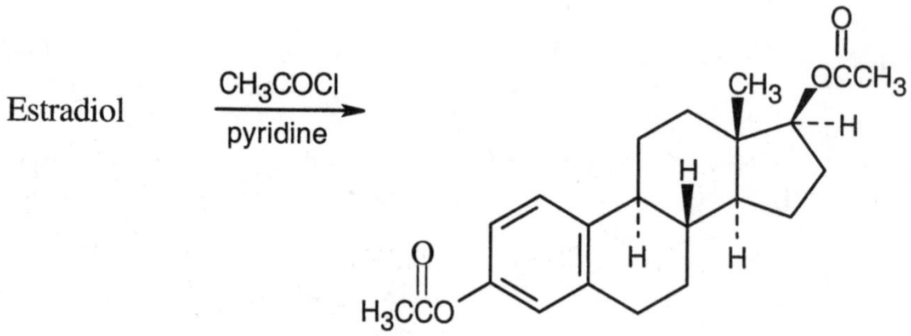

Estradiol $\xrightarrow[\text{pyridine}]{CH_3COCl}$

O OCCH$_3$ CH$_3$ H H H O H$_3$CCO

(c)

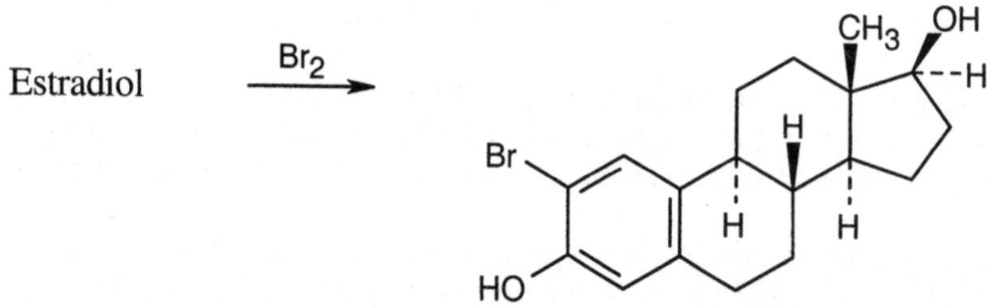

Estradiol $\xrightarrow{Br_2}$

CH$_3$ OH H Br H H HO

16.49

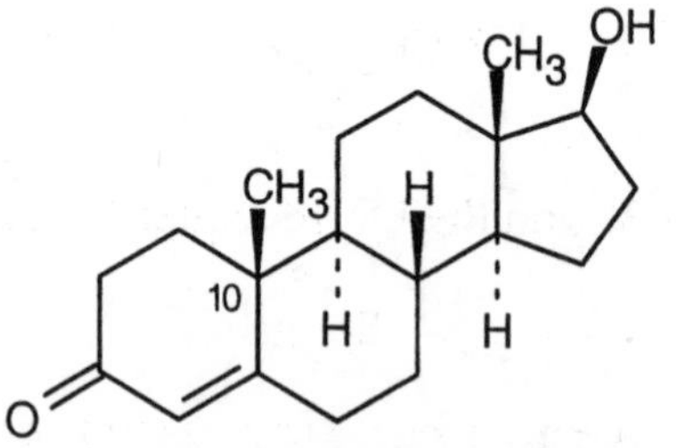

CH$_3$ OH CH$_3$ H 10 H H O

Testosterone

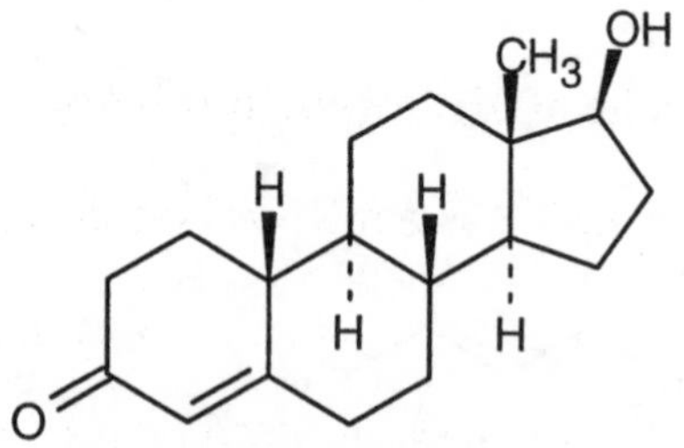

CH$_3$ OH H H H H O

Nandrolone

Testosterone and nandrolone are identical, except for the methyl group at C10 of testosterone, which is replaced by a –H in nandrolone. Both steroids have the same carbon skeleton, both have the same enone group in the A ring, and both have a hydroxyl group at C17 of the D ring.

16.50

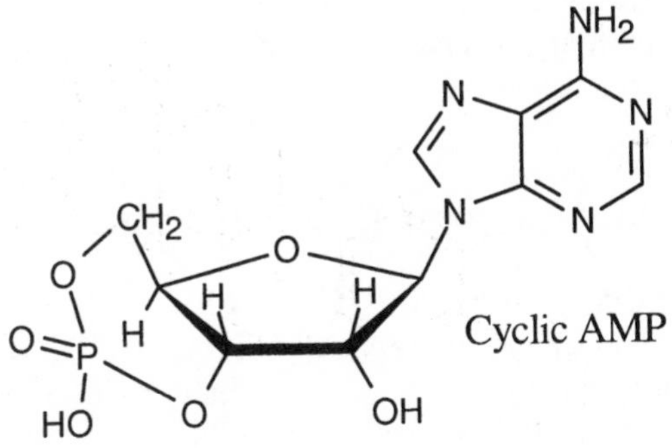

Cyclic AMP

16.51

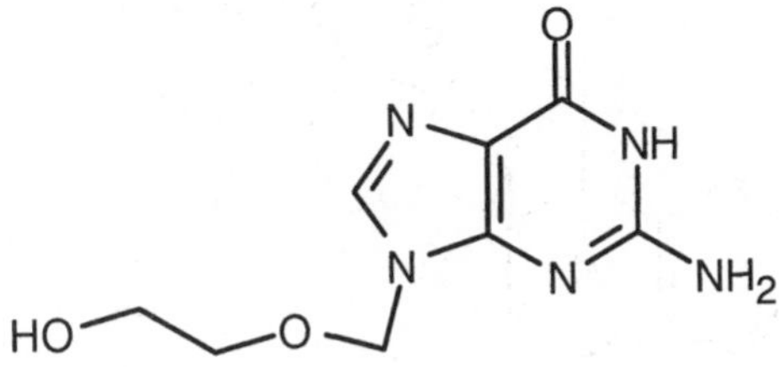

Acyclovir

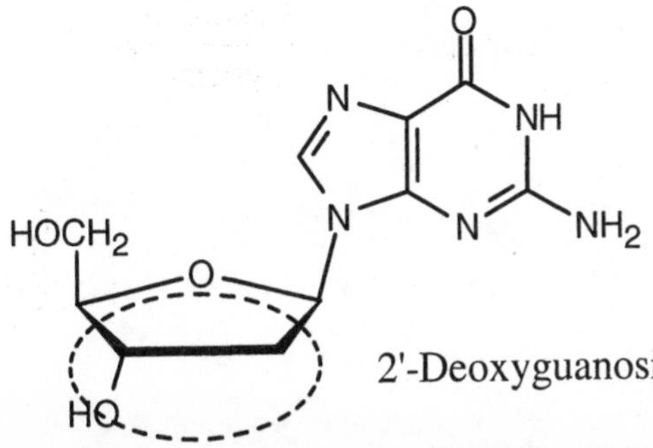

2'-Deoxyguanosine

(a) The circled atoms in 2'-deoxyguanosine, which are missing in acyclovir, are part of the deoxyribose ring. Without these atoms, a DNA chain can no longer elongate because the 3' hydroxyl group of the ring is the site of attachment of the next nucleotide in the DNA chain.

(b) Acyclovir can interfere with DNA synthesis in at least two ways. First, as mentioned in part (a), Acyclovir stops chain growth when it is incorporated into a DNA molecule. Acyclovir may also serve as an enzyme inhibitor by binding to the active site of the DNA polymerase enzyme.

In the Field with Agrochemicals

16.52 - 16.53 The mRNA codons are written 5' –> 3', and the DNA sequences are written 3' –> 5'.

	Glycine	Serine
mRNA codons:	GGU GGC GGA GGG	UCU UCC UCA UCG AGU AGC
DNA bases:	<u>CCA</u> <u>CCG</u> CCT CCC	AGA AGG ACT AGC <u>TCA</u> <u>TCG</u>

16.54 (a) The underlined three-base codons for glycine differ from the underlined codons for serine by only one base. Thus, a mutation which converts thymine to cytosine can change an amino acid from serine to glycine.

(b) Weeds with the last two serine codons (TCA, TCG) would derive resistance first because these codons are more able to mutate to glycine than the other four codons.

(c) The bases that code for the codons of alanine (CGA, CGG, CGT, CGC) also differ from their counterparts in the bases for serine (AGA, AGG, ACT, AGC) by one base. Replacing a serine by alanine might produce a partially resistant weed because the effect of having an alanine in that location is probably similar to the effect of having a glycine in that position.

Chapter 17 – The Organic Chemistry of Metabolic Pathways

Chapter Outline

I. Overview of metabolism and biochemical energy (Section 17.1).
   A. Metabolism.
      1. The reactions that take place in the cells of organisms are collectively called metabolism.
         a. The reactions that produce smaller molecules from larger molecules are called catabolism.
         b. The reactions that build larger molecules from smaller molecules are called anabolism.
      2. Catabolism can be divided into four stages:
         a. In digestion, bonds in food are hydrolyzed to yield sugars, fatty acids, and amino acids.
         b. These small molecules are degraded to acetyl CoA.
         c. In the citric acid cycle, acetyl CoA is catabolized to CO_2, and energy is produced.
         d. Energy from the citric acid cycle enters the electron transport chain, where ATP is synthesized.
   B. Biochemical energy.
      1. ATP, a phosphoric acid anhydride, is the storehouse for biochemical energy.
      2. The breaking of a P–O bond of ATP can be coupled with an energetically unfavorable reaction, so that the overall energy change is favorable.
      3. The resulting phosphates are much more reactive than the original compounds.

II. Catabolism (Sections 17.2 – 17.5).
   A. Catabolism of fats (Section 17.2).
      1. Triacylglycerols are first hydrolyzed in the stomach and small intestine to yield glycerol plus fatty acids.
         a. Glycerol is phosphorylated and enters glycolysis.
         b. Fatty acids are degraded by β-oxidation, a 4-step spiral that results in the cleavage of an *n*-carbon fatty acid into *n*/2 molecules of acetyl CoA.
         c. Before entering β-oxidation, a fatty acid is first converted to its fatty-acyl CoA.
      2. Steps of β-oxidation.
         a. Introduction of a double bond conjugated with the carbonyl group.
            i. The reaction is catalyzed by acyl CoA dehydrogenase.
            ii. The enzyme cofactor FAD is also involved.
         b. Conjugate addition of water to form an alcohol.
            The reaction is catalyzed by enoyl CoA hydratase.
         c. Alcohol oxidation.
            i. The reaction is catalyzed by L-3-hydroxyacyl CoA dehydrogenase.
            ii. The cofactor NAD^+ is reduced to $NADH/H^+$ at the same time.
            iii. The mechanism of the reaction resembles a Cannizzaro reaction, followed by conjugate addition of hydride to NAD^+.
         d. Cleavage of acetyl CoA from the chain.
            i. The reaction, which is catalyzed by β-keto thiolase, is a retro-Claisen reaction.
            ii. Nucleophilic addition of coenzyme A to the keto group is followed by loss of acetyl CoA enolate, leaving behind a chain-shortened fatty-acyl CoA.

3. An *n*-carbon fatty acid yields *n*/2 molecules of acetyl CoA after (*n*/2 – 1) passages of β-oxidation.
   a. Since most fatty acids have an even number of carbons, no carbons are left over after β-oxidation.
   b. Those with an odd number of carbons require further steps for degradation.

B. Catabolism of carbohydrates (Sections 17.3 – 17.4).
1. Glycolysis (Section 17.3).
   a. Glycolysis is a 10-step series of reactions that converts glucose to pyruvate.
   b. Steps 1–3: Phosphorylation and isomerization.
      i. Glucose is phosphorylated at the 6-position by reaction with ATP.
         The enzyme hexokinase is involved.
      ii. Glucose 6-P is isomerized to fructose 6-P by glucose-6-P isomerase.
      iii. Fructose 6-P is phosphorylated to yield fructose 1,6-bisphosphate.
         ATP and phosphofructokinase are involved.
   c. Steps 4–5: Cleavage and isomerization.
      i. Fructose 1,6-bisphosphate is cleaved to glyceraldehyde 3-phosphate and dihydroxyacetone phosphate.
         The reaction is a reverse aldol reaction catalyzed by aldolase.
      ii. Dihydroxyacetone phosphate is isomerized to glyceraldehyde 3-phosphate.
      iii. The net result is production of two glyceraldehyde 3-phosphates, both of which pass through the rest of the pathway.
   d. Steps 6–8: Oxidation and phosphorylation.
      i. Glyceraldehyde 3-phosphate is both oxidized and phosphorylated to give 3-phosphoglyceroyl phosphate.
         Oxidation occurs via a hemithioacetal to yield a product that forms the mixed anhydride.
      ii. The mixed anhydride reacts with ADP to form ATP and 3-phosphoglycerate
         The enzyme phosphoglycerate kinase is involved.
      iii. 3-Phosphoglycerate is isomerized to 2-phosphoglycerate by phosphoglycerate mutase.
   e. Steps 9–10: Dehydration and dephosphorylation.
      i. 2-Phosphoglycerate is dehydrated by enolase to give PEP.
      ii. Pyruvate kinase catalyzes the transfer of a phosphate group to ADP, with formation of pyruvate.
   f. Pyruvate can be converted to acetyl CoA, or may enter other pathways.
2. The citric acid cycle (Section 17.4).
   a. Characteristics of the citric acid cycle.
      i. The citric acid cycle is a closed loop.
      ii. The intermediates are constantly regenerated.
      iii. The cycle operates as long as NAD^+ and FAD are available, which means that oxygen must also be available.
   b. Steps 1–2: Addition to oxaloacetate.
      i. Acetyl CoA adds to oxaloacetate to form citryl CoA, which is hydrolyzed to citrate.
         The reaction is catalyzed by citrate synthase.
      ii. Citrate is isomerized to isocitrate by aconitase.
         The reaction is an E2 dehydration, followed by conjugate addition of water.
   c. Steps 3–4: Oxidative decarboxylations.
      i. Isocitrate is oxidized by isocitrate dehydrogenase to give a ketone that loses CO_2 to give α-ketoglutarate.
      ii. α-Ketoglutarate is transformed to succinyl CoA in a reaction catalyzed by an enzyme complex.

   d. Steps 5–6: Hydrolysis and dehydrogenation of succinyl CoA.
      i. Succinyl CoA is converted to an acyl phosphate, which transfers a phosphate group to GDP in a reaction catalyzed by succinyl CoA synthase.
      ii. Succinate is dehydrogenated by FAD and succinate dehydrogenase to give fumarate.
   e. Steps 7–8: Regeneration of oxaloacetate.
      i. Fumarase catalyzes the addition of water to fumarate to produce L-malate.
      ii. L-Malate is oxidized by NAD^+ and malate dehydrogenase to complete the cycle.

C. Catabolism of proteins: Transamination (Section 17.5).
   1. The pathway to amino acid catabolism:
      a. The amino group is removed by transamination.
      b. What remains is converted to a compound that enters the citric acid cycle.
   2. Transamination.
      a. The $-NH_2$ group of an amino acid adds to the aldehyde group of pyridoxal phosphate to form an imine.
      b. The imine rearranges to a different imine.
      c. The second imine is hydrolyzed to give an α-keto acid and an amino derivative of pyridoxal phosphate.
      d. The pyridoxal derivative transfers its amino group to α-ketoglutarate, to regenerate pyridoxal phosphate and form glutamate.
   3. Deamination.
      The glutamate from transamination undergoes oxidative deamination to yield ammonium ion and α-ketoglutarate.

III. A summary (Section 17.6).
   The mechanisms of biochemical reactions are almost identical to the mechanisms of laboratory reactions.

Solutions to Problems

17.1

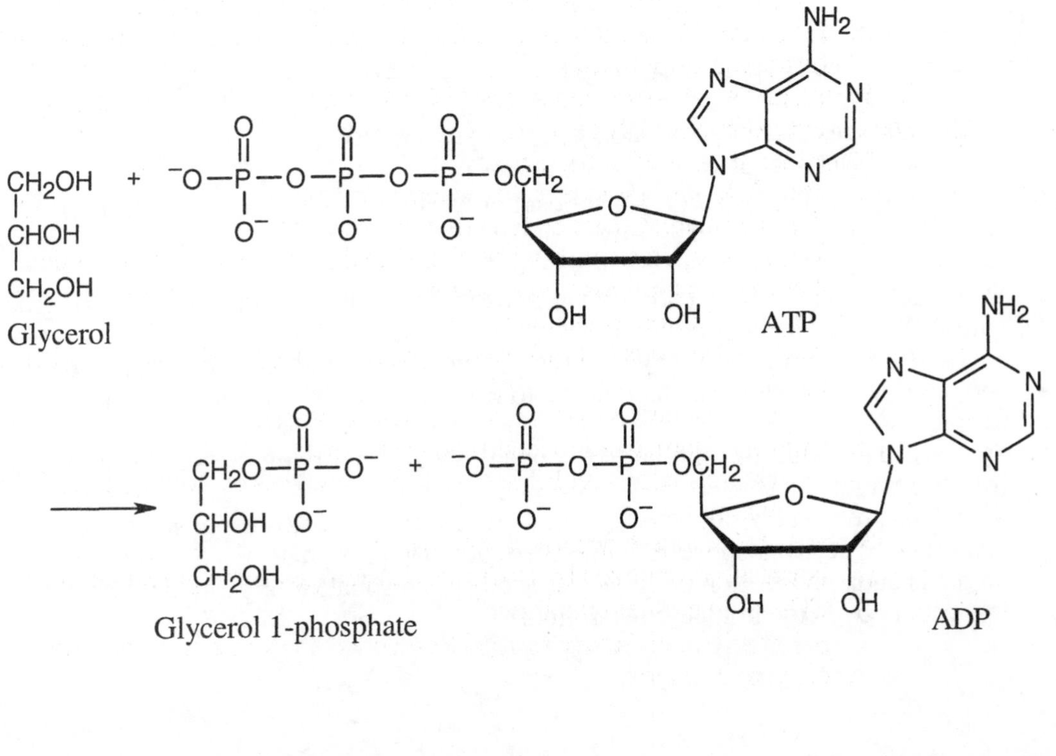

CH_2OH–CHOH–CH_2OH (Glycerol) + ATP → CH_2O–PO_3^{2-}–CHOH–CH_2OH (Glycerol 1-phosphate) + ADP

17.2

$$CH_3CH_2-CH_2CH_2-CH_2CH_2-CH_2\overset{O}{\overset{\|}{C}}SCoA$$

Caprylyl CoA

β-Oxidation (passage 4)

$$CH_3CH_2-CH_2CH_2-CH_2\overset{O}{\overset{\|}{C}}SCoA \quad + \quad CH_3\overset{O}{\overset{\|}{C}}SCoA$$

Hexanoyl CoA

β-Oxidation (passage 5)

$$CH_3CH_2-CH_2\overset{O}{\overset{\|}{C}}SCoA \quad + \quad CH_3\overset{O}{\overset{\|}{C}}SCoA$$

Butanoyl CoA

β-Oxidation (passage 6)

$$CH_3\overset{O}{\overset{\|}{C}}SCoA \quad + \quad CH_3\overset{O}{\overset{\|}{C}}SCoA$$

17.3

(a)

$$CH_3CH_2\downarrow CH_2CH_2\downarrow CH_2CH_2\downarrow CH_2CH_2\downarrow CH_2CH_2\downarrow CH_2CH_2\downarrow CH_2CH_2\downarrow CH_2COOH$$

β-oxidation

$$8\ CH_3\overset{O}{\overset{\|}{C}}SCoA$$

Seven passages of the β-oxidation pathway are needed.

(b)

$$CH_3CH_2-(CH_2CH_2)_8-CH_2COOH \xrightarrow{\beta\text{-oxidation}} 10\ CH_3\overset{O}{\overset{\|}{C}}SCoA$$

Nine passages of the β-oxidation pathway are needed.

17.4 ATP is produced in step 7 (1,3-bisphosphoglycerate —> 3-phosphoglycerate) and in step 10 (phosphoenolpyruvate —> pyruvate).

17.5 *Step 1* is a nucleophilic substitution at phosphorus by the –OH group at C6 of glucose, with ADP as the leaving group.
Step 2 is an isomerization, in which the pyranose ring of glucose 6-phosphate opens, keto-enol tautomerism causes isomerization to fructose 6-phosphate, and a furanose ring is formed.
Step 3 is a substitution, similar to the one in step 1, involving the –OH group at C1 of fructose 6-phosphate.
Step 4 is a retro-aldol condensation that cleaves fructose 1,6-bisphosphate to glyceraldehyde 3-phosphate and dihydroxyacetone phosphate.
Step 5 is an isomerization of dihydroxyacetone phosphate to glyceraldehyde 3-phosphate that occurs by keto-enol tautomerization.

In *Step 6*, the aldehyde group of glyceraldehyde 3-phosphate is oxidized by NAD^+, followed by a nucleophilic acyl substitution by phosphate, to yield the product 3-phosphoglyceroyl phosphate.
Step 7 is a nucleophilic acyl substitution reaction at phosphorus, yielding ATP and 3-phosphoglycerate.
Step 8 is an isomerization of 3-phosphoglycerate to 2-phosphoglycerate.
Step 9 is an E2 elimination of H_2O to form phosphoenolpyruvate.
Step 10 is a substitution reaction at phosphorus that forms ATP and displaces pyruvate enolate, which tautomerizes to pyruvate.

17.6 Citrate and isocitrate are tricarboxylic acids.

17.7

A–H; CO_2^-–CH_2–C(OH)(CO_2^-)–CH(H)–CO_2^- (Citrate); B–Enz; $\xrightarrow[1.]{H_2O}$ [CO_2^-–CH_2–C(=CH–CO_2^-)–CO_2^- (Aconitate); $H_2\ddot{O}$: $\xrightarrow{2.}$ CO_2^-–CH_2–C(=C(O^-)O^-)–CH($\overset{+}{O}H_2$)–CO_2^-; B:] $\longrightarrow$ CO_2^-–CH_2–HC(CO_2^-)–CH(OH)–CO_2^- (Isocitrate)

Citrate | Aconitate | Isocitrate

Enzyme-catalyzed E2 elimination of H_2O (1) is followed by nucleophilic conjugate addition of water (2) to produce an adduct that isomerizes to isocitrate.

17.8 Position leucine and α-ketoglutarate so that the groups to be exchanged are aligned. This arrangement makes it easy to predict the products of transamination reactions.

Leucine		α-Ketoglutarate		keto acid		Glutamate
CH_3		COO^-		CH_3		COO^-
$CHCH_3$		CH_2		$CHCH_3$		CH_2
CH_2	+	CH_2	$\longrightarrow$	CH_2	+	CH_2
$HC-NH_3^+$		$C=O$		$C=O$		$HC-NH_3^+$
COO^-		COO^-		COO^-		COO^-

17.9 Look for the keto group that replaced the amino group. Then try to identify the original amino acid by recognizing the side chain. The α-keto acid was formed from asparagine.

Asparagine		α-Ketoglutarate		keto acid		Glutamate
NH_2		COO^-		NH_2		COO^-
$C=O$		CH_2		$C=O$		CH_2
CH_2	+	CH_2	$\longrightarrow$	CH_2	+	CH_2
$HC-NH_3^+$		$C=O$		$C=O$		$HC-NH_3^+$
COO^-		COO^-		COO^-		COO^-

Visualizing Chemistry

17.10 The amino acid precursors are valine (a) and methionine (b).

(a)

CH_3–$CHCH_3$–HC–NH_3^+–COO^- + COO^-–CH_2–CH_2–$C{=}O$–COO^- ⟶ CH_3–$CHCH_3$–$C{=}O$–COO^- + COO^-–CH_2–CH_2–HC–NH_3^+–COO^-

Valine α-Ketoglutarate Glutamate

(b)

SCH_3–CH_2–CH_2–HC–NH_3^+–COO^- + COO^-–CH_2–CH_2–$C{=}O$–COO^- ⟶ SCH_3–CH_2–CH_2–$C{=}O$–COO^- + COO^-–CH_2–CH_2–HC–NH_3^+–COO^-

Methionine α-Ketoglutarate Glutamate

17.11 The intermediate is *S*-malate.

^-O–$C({=}O)$–$C(H)(OH)$–$C(H)(H)$–$C({=}O)$–O^- *S*-Malate

Additional Problems

General Metabolism

17.12 Digestion is the breakdown of bulk food in the stomach and small intestine. In digestion:
(a) The ester bonds of fats are hydrolyzed to give fatty acids and glycerol.
(b) Acetal bonds of complex carbohydrates are cleaved to yield simple sugars.
(c) Amide bonds of proteins are hydrolyzed to produce amino acids.

17.13 Metabolism refers to all reactions that take place inside cells; digestion is a part of metabolism in which food is broken down into small organic molecules. Metabolic processes that break down large food molecules are known as catabolism, whereas metabolic processes that assemble larger biomolecules from smaller ones are known as anabolism.

17.14

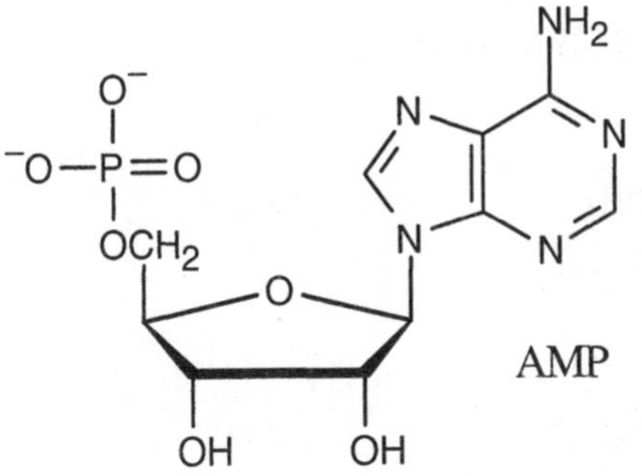

AMP

17.15 ATP is involved with phosphate transfer reactions. ATP is synthesized in catabolic reactions, and transfers a phosphate group to another molecule in anabolic reactions.

17.16 NAD^+ is a biochemical oxidizing agent that converts alcohols to aldehydes or ketones, yielding NADH as a byproduct.

17.17 FAD is an oxidizing agent that introduces a conjugated carbon-carbon double bond into a biomolecule, yielding $FADH_2$ as a byproduct.

17.18

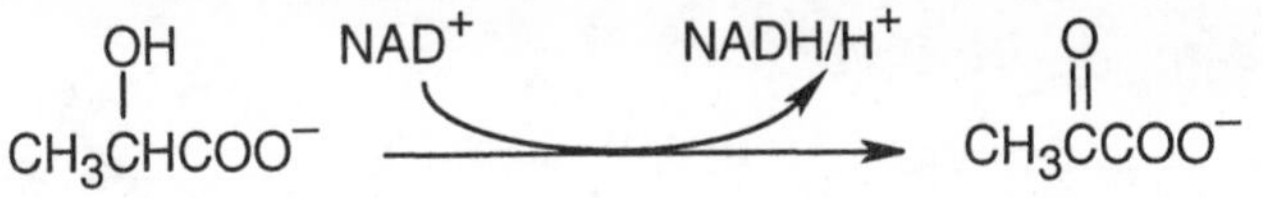

$$\underset{\text{Lactate}}{CH_3CH(OH)COO^-} \xrightarrow{NAD^+ \;\to\; NADH/H^+} \underset{\text{Pyruvate}}{CH_3C(=O)COO^-}$$

NAD^+ is needed to convert lactate to pyruvate because the reaction involves the oxidation of an alcohol.

17.19 Oxaloacetate is the starting point for the citric acid cycle.

Specific Metabolic Pathways

17.20 (a) One mole of glucose is catabolized to two moles of pyruvate, each of which yields one mole of acetyl CoA. Thus,

1.0 mol glucose —> 2.0 mol acetyl CoA

(b) A fatty acid with *n* carbons yields *n*/2 moles of acetyl CoA per mole of fatty acid. For palmitic acid ($C_{15}H_{31}COOH$),

$$1.0 \text{ mol palmitic acid} \times \frac{8 \text{ mol acetyl CoA}}{1 \text{ mol palmitic acid}} \longrightarrow 8.0 \text{ mol acetyl CoA}$$

(c) Maltose is a disaccharide that yields two moles of glucose on hydrolysis. Since each mole of glucose yields two moles of acetyl CoA,

1.0 mol maltose —> 2.0 mol glucose —> 4.0 mol acetyl CoA

17.21

	(a) Glucose	(b) Palmitic acid	(c) Maltose
Molecular weight	180.2 amu	256.4 amu	342.3 amu
Moles in 100.0 g	0.5549 mol	0.3900 mol	0.2921 mol
Moles of acetyl CoA produced	2 x 0.5549 mol =1.110 mol	8 x 0.3900 mol = 3.120 mol	4 x 0.2921 mol = 1.168 mol
Grams acetyl CoA produced	898.6 grams	2526 grams	945.6 grams

17.22 Palmitic acid is the most efficient precursor of acetyl CoA on a weight basis.

17.23

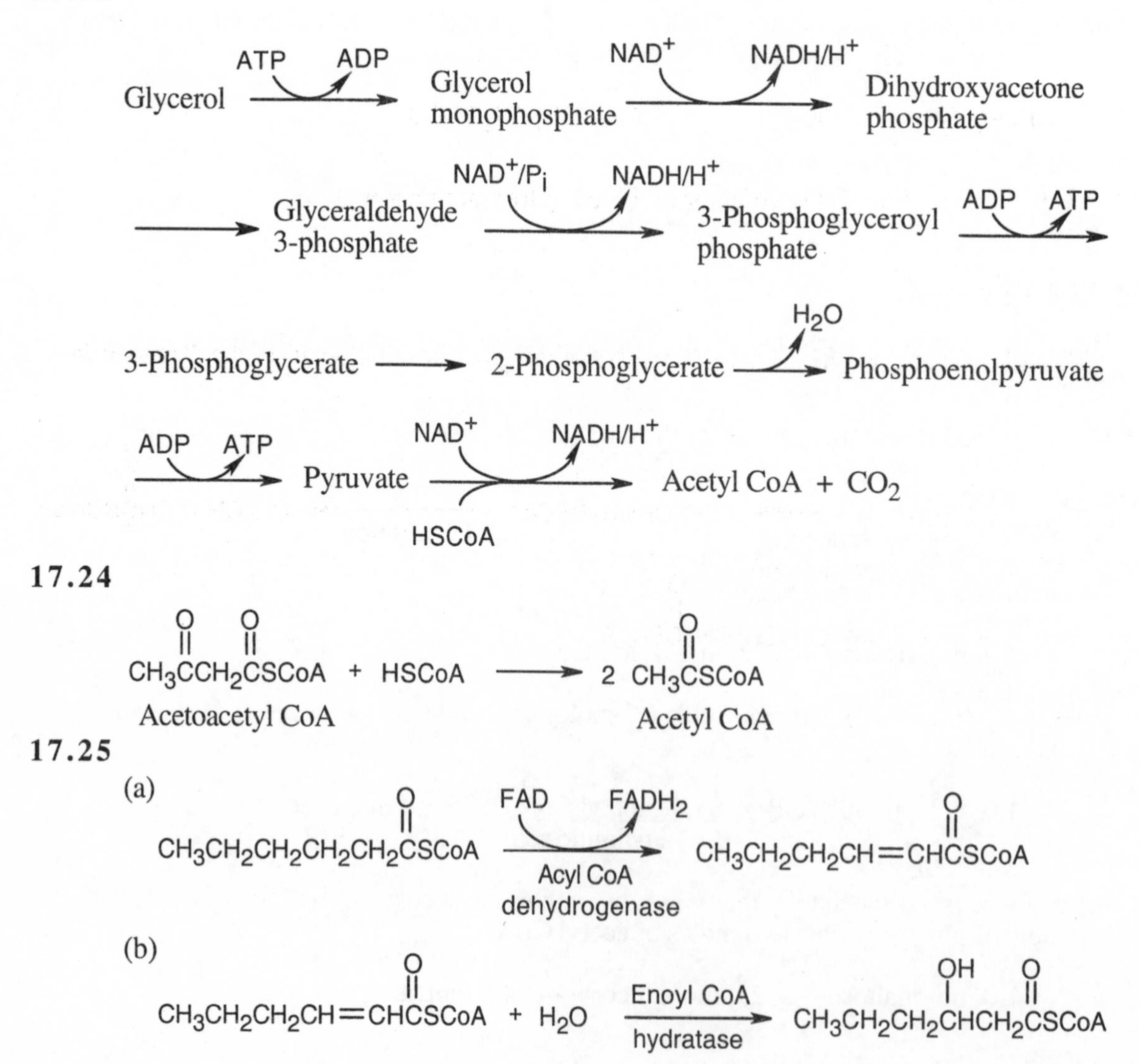

(c)

$$CH_3CH_2CH_2CH(OH)CH_2C(=O)SCoA \xrightarrow[\text{L-3-Hydroxyacyl CoA dehydrogenase}]{NAD^+ \quad NADH/H^+} CH_3CH_2CH_2C(=O)CH_2C(=O)SCoA$$

17.26

	Amino acid	*α-Keto acid*
(a)	$CH_3CH(CH_3)CH(NH_3^+)COO^-$	$CH_3CH(CH_3)C(=O)COO^-$
(b)	C_6H_5–$CH_2CH(NH_3^+)COO^-$	C_6H_5–$CH_2C(=O)COO^-$
(c)	$CH_3SCH_2CH_2CH(NH_3^+)COO^-$	$CH_3SCH_2CH_2C(=O)COO^-$

17.27 Pyridoxal phosphate is the cofactor involved in transamination.

Integrated Problems

17.28 The reactions of lipogenesis are the reverse of the reactions that occur in the β-oxidation pathway.

$$CH_3C(=O)CH_2C(=O)SCoA \xrightarrow[\text{reduction of ketone}]{} CH_3CH(OH)CH_2C(=O)SCoA \xrightarrow[\text{dehydration}]{} CH_3CH{=}CHC(=O)SCoA$$

$$\xrightarrow[\text{reduction of double bond}]{} CH_3CH_2CH_2C(=O)SCoA$$

17.29

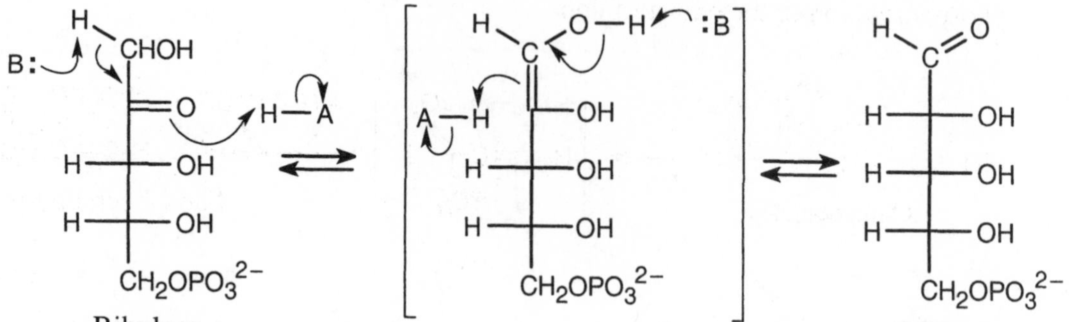

Ribulose 5-phosphate ⇌ [enolate intermediate] ⇌ Ribose 5-phosphate

The isomerization of ribulose 5-phosphate to ribose 5-phosphate occurs by way of an intermediate enolate.

17.30 This is an enzyme-catalyzed reverse aldol reaction, similar to step 4 of glycolysis.

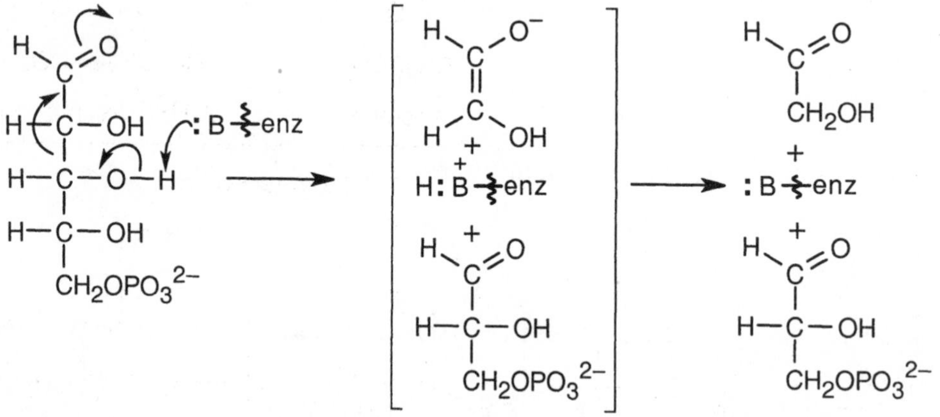

Glyceraldehyde 3-phosphate

17.31 The conversion of pyruvate to oxaloacetate is a mixed aldol condensation reaction between CO_2 and pyruvate.

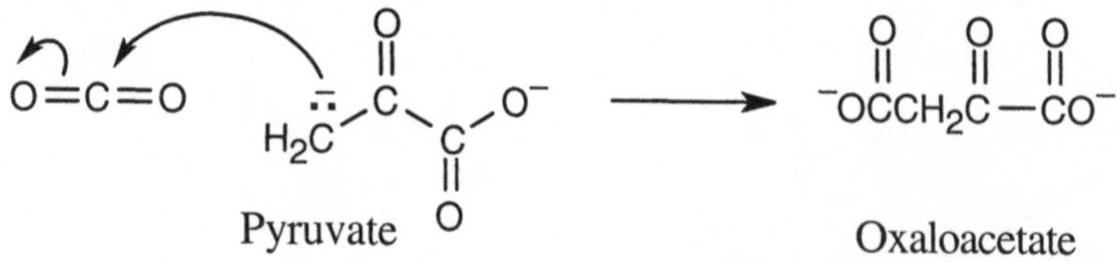

$$O{=}C{=}O + \text{Pyruvate} \longrightarrow {}^{-}OCCH_2C{-}CO^{-} \ (\text{Oxaloacetate})$$

Pyruvate

Oxaloacetate

17.32 Formation of phosphoenolpyruvate occurs by way of a reverse aldol reaction, followed by phosphorylation of the enolate anion.

$$^{-}O-\overset{O}{\overset{\|}{C}}-CH_2-\overset{O}{\overset{\|}{C}}-\overset{O}{\overset{\|}{C}}O^{-} \longrightarrow \left[H_2C=\overset{O^{-}}{\overset{|}{C}}-\overset{O}{\overset{\|}{C}}O^{-} + CO_2 \right] \xrightarrow{ATP \quad ADP} H_2C=\overset{^{2-}O_3PO}{\overset{|}{C}}-\overset{O}{\overset{\|}{C}}O^{-}$$

Oxaloacetate → Phosphoenolpyruvate

17.33

$$2\ CH_3\overset{O}{\overset{\|}{C}}SCoA \xrightarrow{HSCoA} CH_3\overset{O}{\overset{\|}{C}}CH_2\overset{O}{\overset{\|}{C}}SCoA \xrightarrow{H_2O \quad HSCoA} CH_3\overset{O}{\overset{\|}{C}}CH_2\overset{O}{\overset{\|}{C}}O^{-}$$

Acetyl CoA → Acetoacetyl CoA → Acetoacetate

Acetoacetate $\xrightarrow{-CO_2}$ $CH_3\overset{O}{\overset{\|}{C}}CH_3$ (Acetone)

Acetoacetate $\xrightarrow{NADH/H^+ \quad NAD^+}$ $CH_3\overset{OH}{\overset{|}{C}}HCH_2\overset{O}{\overset{\|}{C}}O^{-}$ (3-Hydroxybutyrate)

Acetone and 3-Hydroxybutyrate: ketone bodies

17.34

$$H_3C-\overset{:O:}{\overset{\|}{C}}-SCoA + H_2\ddot{C}^{-}-\overset{O}{\overset{\|}{C}}-SCoA \longrightarrow \left[H_3C-\overset{:\ddot{O}:^{-}}{\overset{|}{C}}(SCoA)-CH_2\overset{O}{\overset{\|}{C}}SCoA \right] \longrightarrow CH_3\overset{O}{\overset{\|}{C}}CH_2\overset{O}{\overset{\|}{C}}SCoA + {}^{-}SCoA$$

The reaction is a Claisen condensation.

17.35

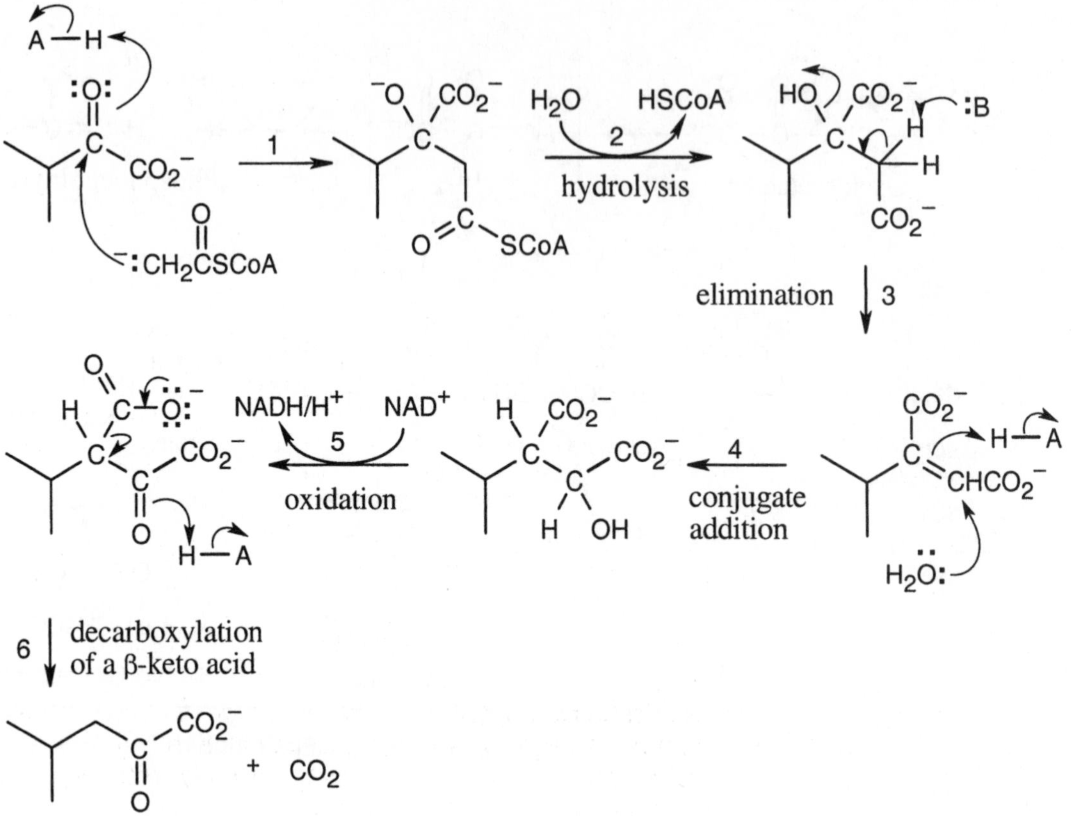

A—H :O: C CO_2^- O $^-$:CH_2CSCoA

1

$^-$O CO_2^- C C O SCoA

H_2O HSCoA 2 hydrolysis

HO CO_2^- :B C H H CO_2^-

elimination 3

O H C—O: NADH/H$^+$ NAD$^+$ 5 oxidation C CO_2^- C O H—A

H CO_2^- C CO_2^- C H OH

4 conjugate addition

CO_2^- C H—A $CHCO_2^-$ $H_2\ddot{O}$:

6 decarboxylation of a β-keto acid

CO_2^- C O + CO_2

17.36

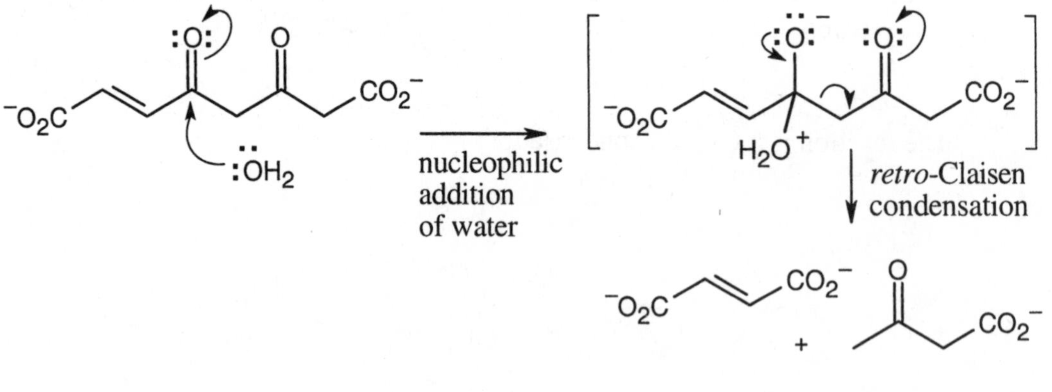

:O: O CO_2^- ^-O_2C :$\ddot{O}H_2$

nucleophilic addition of water

:O: :O: CO_2^- ^-O_2C H_2O^+

retro-Claisen condensation

^-O_2C CO_2^- + O CO_2^-

17.37 The first step in the conversion acetoacetate —> acetyl CoA is the formation of acetoacetyl CoA. This reaction also occurs as the first step in fatty acid degradation.

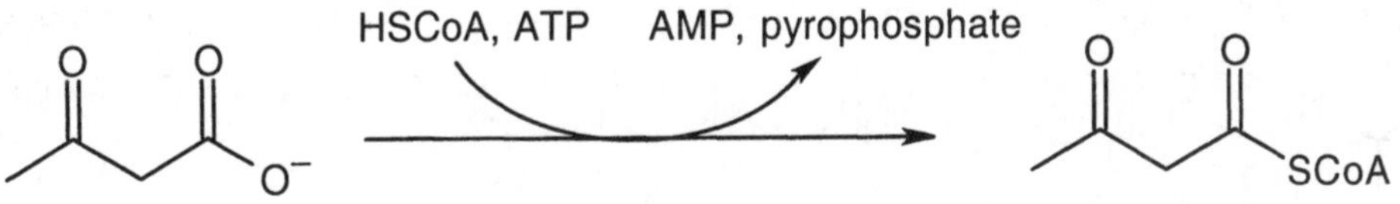

HSCoA, ATP → AMP, pyrophosphate

The final step is a retro-Claisen reaction, whose mechanism is pictured in Section 17.2 as Step 4 of β-oxidation of fatty acids.

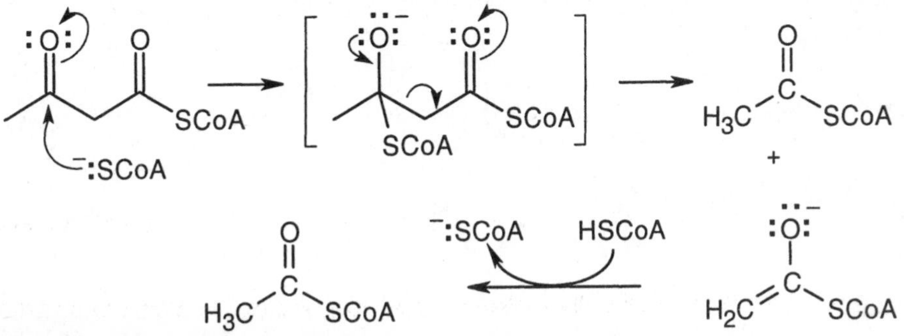

SCoA SCoA SCoA H_3C SCoA + :SCoA

:SCoA HSCoA H_3C SCoA H_2C SCoA

In the Medicine Cabinet

17.38 FAD is the enzyme cofactor associated with oxidation and reduction of conjugated double bonds. In this problem, $FADH_2$ is oxidized when folic acid is reduced to dihydrofolate, and then to tetrahydrofolate.

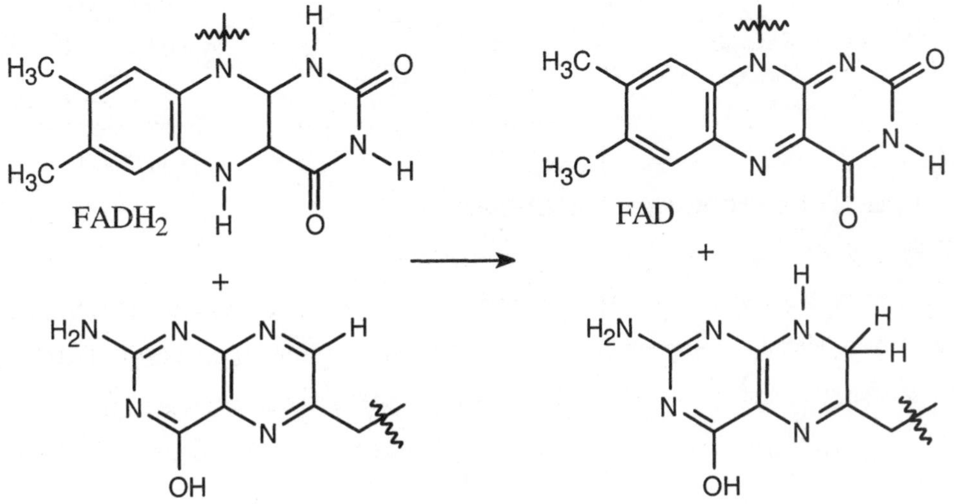

$FADH_2$ + → FAD +

17.39 The S_N2 reaction, in which a methyl group is transferred from the cofactor to homocysteine, produces two charged species that are neutralized in an acid-base reaction.

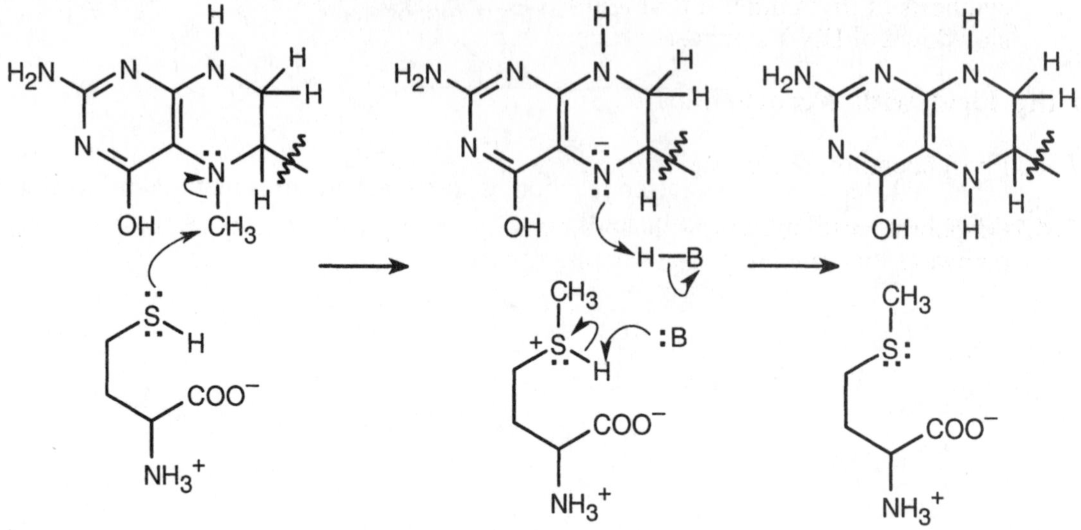

17.40

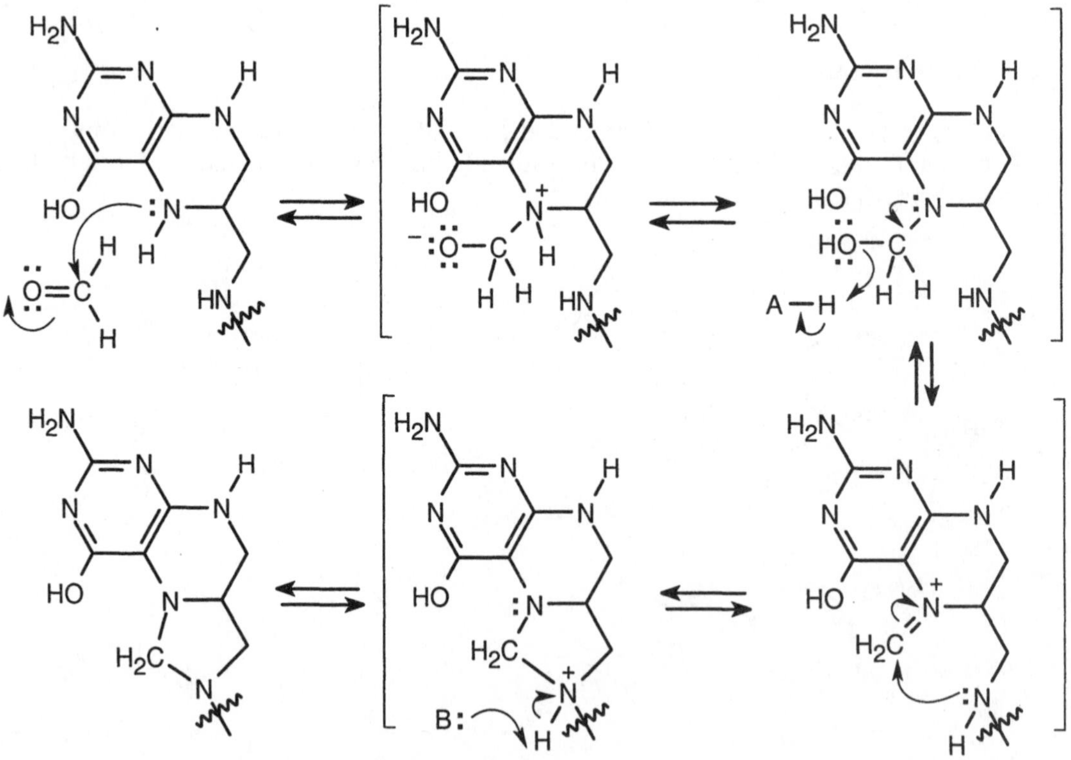

17.41 Methotrexate interferes with the reduction of folate to tetrahydrofolate. A lack of tetrahydrofolate, in turn, makes it difficult for formaldehyde groups to be used in the synthesis of thymidine. If thymidine is not available, DNA synthesis can't occur. The slowdown of DNA synthesis is a goal of anticancer drugs.

In the Field with Agrochemicals

17.42 Phosphoenolpyruvate is an intermediate in glycolysis.

17.43 Glyphosate resembles phosphoenolpyruvate and acts as an enzyme inhibitor in metabolic pathways that involve phosphoenolpyruvate.

Review Quizzes

Chapters 1, 2

Multiple Choice:

1. The compound BF_3 is a:
   (a) Brønsted–Lowry acid (b) Brønsted–Lowry base (c) Lewis acid (d) Lewis base

2. Which of the following functional groups does not contain a carbonyl group?
   (a) ester (b) ether (c) ketone (d) amide

3. Which of the following elements is more electronegative than carbon?
   (a) Al (b) S (c) N (d) Si

4.

```
                 CH2CH2CH3
                 |
CH3CHCH2CCH2CH3
       |         |
CH3CH2     CH3
```

   The above compound is named as a substituted:
   (a) hexane (b) heptane (c) octane (d) nonane

5. The compound in the previous problem contains how many secondary carbons?
   (a) 4 (b) 5 (c) 6 (d) 7

6. The Lewis electron-dot structure for CS_2 contains how many electrons?
   (a) 40 (b) 16 (c) 12 (d) 8

7. Which of the following elements has no unpaired electrons in its outer shell?
   (a) Chlorine (b) Boron (c) Calcium (d) Sulfur

8. The compound H–X has a pK_a of 4, and the compound H–Y has a pK_a of 6. All the following statements are true except:
   (a) $[H^+]$ of a 1.0 M solution of HX is approximately 2.
   (b) X^- is a stronger base than Y^-.
   (c) In a solution of 1.0 M HY and 1.0 M NaY the pH is 6.
   (d) H–X and H–Y are probably organic acids

9. The formula C_6H_{14} can represent a:
   (a) hexane (b) pentane (c) butane (d) all of these

10. In which of the following bonds does carbon have a partial negative charge?
    (a) CH_3–Na (b) CH_3–SH (c) CH_3–OCH_3 (d) CH_3–I

Tell whether the following statements are true or false:

1. If HCN has a pK_a of 9.31 and water has a pK_a of 15.74, hydroxide anion is a stronger base than cyanide anion.

2 Both aldehydes and amides can act as Lewis bases.

3. In a graph of energy vs. bond rotation, energy maxima occur when groups are in the staggered conformation.

4. A bond between carbon and sulfur is a polar covalent bond.

5. When a chemical bond is formed, energy is released.

6. A compound with four electrons in its outermost *p* subshell has no unpaired electrons.

7. Pentane is the lowest molecular weight hydrocarbon that is a liquid at room temperature.

8. A compound can be both a Lewis base and a Brønsted base.

9. One isomer of the formula C_7H_{16} is named as a substituted butane.

10. In a bond between carbon and a reactive metal, carbon bears a partial positive charge.

11. A reaction between HCN (pK_a = 9.31) and fluoride (pK_a of HF = 3.45) proceeds as written:

$$HCN + F^- \longrightarrow HF + {}^-CN$$

12. *Trans*–1,4–dimethylcyclohexane is more stable than its cis isomer.

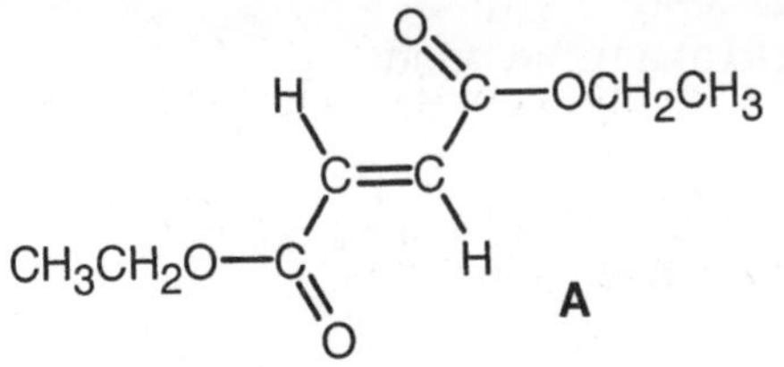

H, C=C, H, C—OCH_2CH_3 (=O), CH_3CH_2O—C (=O)

A

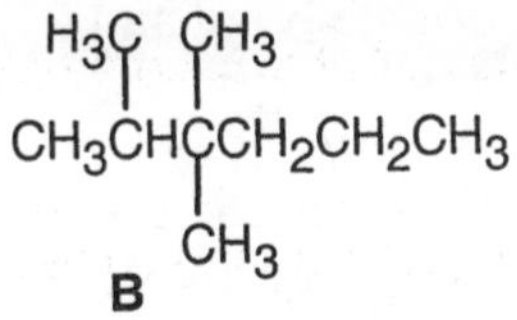

H_3C CH_3

$CH_3CHCCH_2CH_2CH_3$

CH_3

B

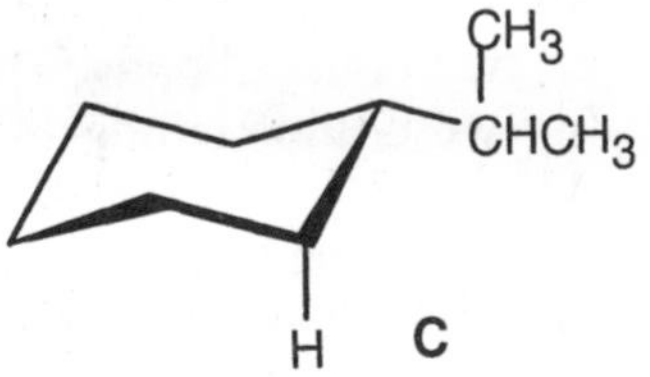

CH_3

$CHCH_3$

H **C**

What functional groups does compound **A** contain? What is its molecular formula? How many sp^2-hybridized carbons does it contain? How many π bonds? How many pairs of non-bonding electrons? Which bonds in **A** are the most polar?

What is the IUPAC name of **B**? How many of the following does **B** contain: primary carbons, secondary carbons, tertiary carbons, quaternary carbons, methyl groups, ethyl groups.

Provide a name for compound **C**. In what conformation is the alkyl group? Is this the more or less stable conformation? How might the two conformations be interconverted? What is the relationship of the indicated hydrogen atom to the alkyl group?

Chapters 3, 4, 5

Multiple Choice:

1. Which of the following compounds yields only two identical products on oxidation with acidic $KMnO_4$?
(a) 2,3-Dimethyl-2-butene (b) 1,3-Butadiene (c) 1,3-Cyclohexadiene (d) Cyclohexene

2. The correct priority ranking (from lowest to highest) of the following substituents is:
(a) $–OCH_3$, $–COOCH_3$, –SH, –Cl (b) $–COOCH_3$, $–OCH_3$, –SH, –Cl
(c) –Cl, –SH, $–OCH_3$, $–COOCH_3$ (d) $–COOCH_3$, $–OCH_3$, –Cl, –SH

3.

$$NaSH + CH_3CH_2CH_2Br \longrightarrow NaBr + CH_3CH_2CH_2SH$$

The preceding reaction is an:
(a) addition reaction (b) elimination reaction (c) substitution reaction
(d) rearrangement reaction

4. All of the following can be oxidized by reaction with acidic or neutral permanganate except:
(a) Toluene (b) 1,3-Pentadiene (c) 1-Hexene (d) Benzene

5. The compound C_5H_8 can have all of the following except:
(a) two double bonds (b) a double bond and a ring (c) a triple bond
(d) two double bonds and a ring

6. Which of the following yields one product on bromination with Br_2 and $FeBr_3$?
(a) Toluene (b) *o*– Xylene (c) *p*–Xylene (d) *p*–Ethyltoluene

7. A substituent on an aromatic ring can be all of the following except:
(a) an ortho- and para-directing activator (b) an ortho- and para-directing deactivator
(c) a meta-directing activator (d) a meta-directing deactivator

8. In the bromination of anisole (methoxybenzene), the most stable resonance form of the reaction intermediate puts the positive charge:
(a) on the methoxyl oxygen (b) on the carbon ortho to the methoxyl group
(c) on the carbon meta to the methoxyl group (d) on the carbon para to the methoxyl group

9. Which of the following statements about the bromination of benzene is false?
(a) The reaction is two-step. (b) The energy of the products is higher than the energy of the reactants. (c) The second step is faster than the first step. (d) The intermediate can be represented by three resonance forms.

10. Which of the following reactions does not introduce a carbonyl (C=O) group into a hydrocarbon?
(a) Reaction of benzene with acetyl chloride and $FeCl_3$ (b) Treatment of cyclohexene with basic permanganate (c) Hydration of 1–butyne (d) Reaction of toluene with neutral $KMnO_4$

Tell whether the following statements are true or false:

1. In ranking a substituent group attached to a double bond, –OH is of higher priority than $-CH_2OCH_3$.

2. Heterolytic reactions usually occur when the atoms in the bond to be broken differ in electronegativity.

3. A reaction that releases energy also has a negative E_{act}.

4. Addition of HBr to 2-methyl-2-pentene yields a single product.

5. A reaction can have an intermediate without having a transition state.

6. Acidic permanganate is used to hydroxylate a double bond.

7. In order for a radical chain reaction to propagate, an odd-electron species must react with an even-electron species.

8. Treatment of 2-pentyne with sodium amide produces an acetylide anion.

9. The substituent $-N(CH_3)_2$ is an ortho, para-directing activator.

10. A benzene ring is reduced with H_2 and a Pd catalyst.

11. It is easier to prepare *p*-nitrotoluene than *m* -nitrotoluene.

12 A group that is a meta-director deactivates all positions on the aromatic ring.

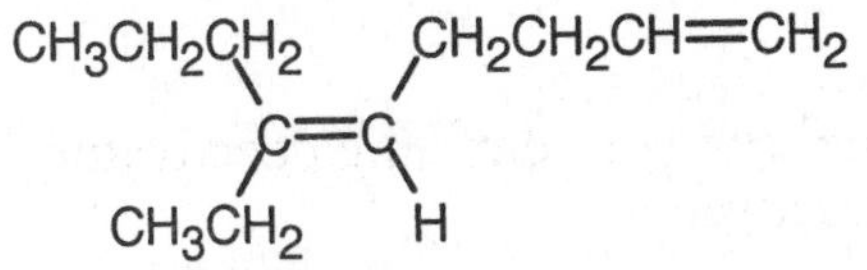

$CH_3CH_2CH_2$ $CH_2CH_2CH{=}CH_2$
C=C
CH_3CH_2 H

A

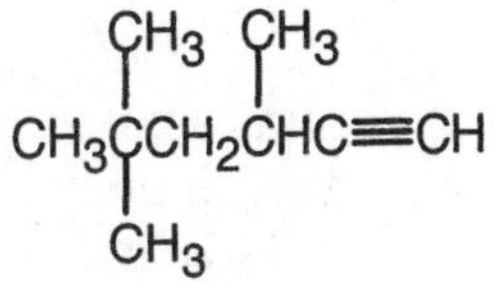

CH_3 CH_3
$CH_3CCH_2CHC{\equiv}CH$
CH_3

B

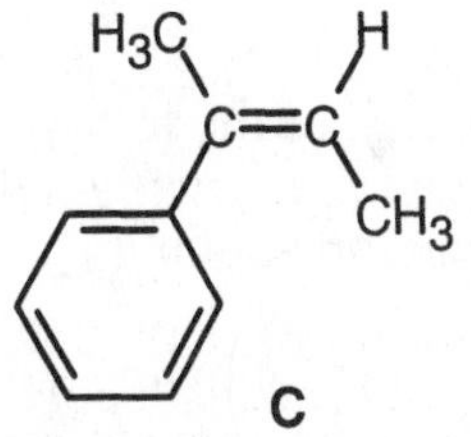

H_3C H
C=C
CH_3

C

Name compounds **A** and **B**.Which of the above compounds show *E,Z* isomerism? Assign *E* or *Z* configurations to the isomeric double bonds.Which of the above compounds have conjugated bond systems? Which of the above compounds react with the following reagents? Show the products: (a) Br_2, CH_2Cl_2 (b) H_2, Pd catalyst (c) Br_2, $FeBr_3$ (d) $NaNH_2$ (e) HBr, ether

Show the products when compounds **A** and **C** react with : (a) acidic $KMnO_4$ (b) H_2O, H^+ catalyst. What products result from treatment of **C** with HNO_3, H_2SO_4? If **C** were to react with acidic $KMnO_4$ and the resulting compound was treated with HNO_3, H_2SO_4, what product would be formed?

Chapters 6, 7, 8

Multiple Choice:

1. If you were to draw all products of radical bromination of ethane, how many structures would you draw?
(a) 2 (b) 6 (c) 9 (d) 11

2. Which of the following experiments is not useful in helping you decide if a substitution reaction proceeds by an S_N1 mechanism or an S_N2 mechanism?
(a) changing the nucleophile (b) varying the leaving group (c) varying the concentration of nucleophile (d) measuring the sign of rotation if the product has a stereocenter.

3. A reaction of a tertiary alkyl halide might occur by all of the following routes except:
(a) E1 (b) E2 (c) S_N1 (d) S_N2

4.

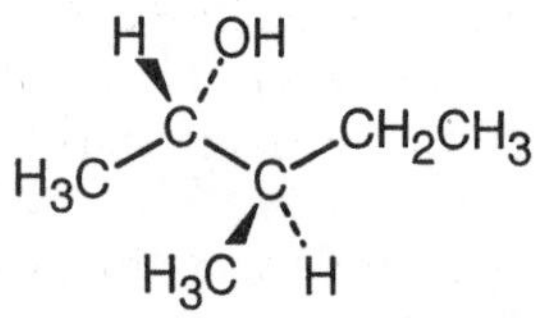

H OH
C CH_2CH_3
H_3C C
H_3C H

What are the configurations at the carbon stereocenters of the above isomer?
(a) 2*R*,3*R* (b) 2*R*, 3*S* (c) 2*S*,3*R* (d) 2*S*,3*S*

5. Which of the following is both a good nucleophile and a good leaving group?
(a) I^- (b) HO^- (c) Cl^- (d) ^-CN

6. Although all of the following compounds are of similar molecular weight, one of them is much higher boiling than the others. Which one?
(a) Tetrahydrofuran (b) 1-Butanol (c) Methyl propyl ether (d) 1,2-Propanediol

7. Which of the following substances is more acidic than phenol?
(a) Sodium bicarbonate (b) Water (c) Sodium carbonate (d) *p*–Methylphenol

8. Ethers are unreactive to all of the following except:
(a) nucleophiles (b) halogens (c) strong bases (d) strong acids

9. How many of the alcohol and ether stereoisomers of the formula $C_5H_{12}O$ have stereocenters?
(a) 2 (b) 3 (c) 4 (d) 5

10 Which of the following statements about *p*-nitrophenol is false?
(a) It is more acidic than phenol (b) It is a solid (c) Reaction with Br_2, $FeBr_3$ yields 3-bromo-4-nitrophenol (d) It reacts with potassium carbonate to form a phenoxide anion.

Tell whether the following statements are true or false:

1. *Cis*-2-pentene and *trans*-2-pentene are diastereomers.
2. The E1 reaction of a tertiary bromide occurs when strong base is present.
3. The S_N2 reaction of an *S* enantiomer with a nucleophile yields an *R* enantiomer.
4. A Grignard reagent is formed by reaction of an organohalide with Mg.
5. Two compounds that are diastereomers of the same third compound are enantiomers.
6. Acid-catalyzed epoxide ring-opening occurs by a S_N2 mechanism.
7. The reaction of a thiolate anion with an alkyl halide yields a disulfide.
8. A peroxyacid is used to form an epoxide from an alkene.
9. The physical properties of a racemic mixture are the same as those of each enantiomer.
10. In a primary alcohol, the hydroxyl group is bonded to a primary carbon.
11. In an S_N1 displacement reaction, water is a better leaving group than hydroxide ion.
12. Four different products can result from the cleavage of an unsymmetrical ether with HI.

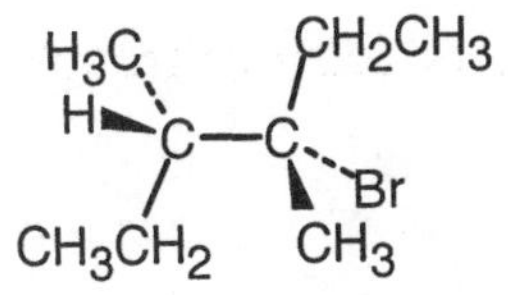

A 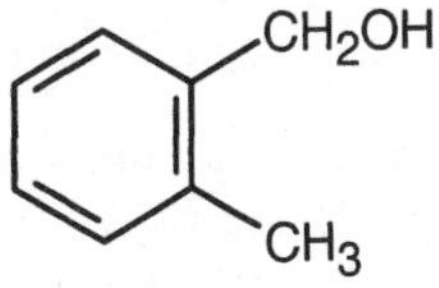 **B**

Give the name of compound **A**. What are the *R,S* configurations at the stereocenters? How many enantiomers of **A** are there? How many diastereomers? What are the configurations of the enantiomers and diastereomers? Is **A** a meso compound? Is **A** a primary, secondary or a tertiary bromide?

A mixture of two isomers of **A** can be prepared from an alcohol and a second reagent. What are the reagents? Why are two isomers formed? What are the configurations of the isomer products at the carbon stereocenters? By what mechanism is the bromide formed? What is the effect of changing the following reaction conditions on the rate of reaction: (a) doubling the concentration of the alcohol (b) tripling the concentration of the other reagent.

Name at least one other product formed as a byproduct of the above reaction to from compound **A**. By what mechanism is it formed? What is the configuration of the carbon atoms in this byproduct?

Give the name of compound **B**. Is **B** an alcohol or a phenol? Show a pathway by which **B** might be synthesized. Is **B** more or less acidic than phenol? What products result from treatment of **B** with the following reagents: (a) CrO_3, H_3O^+ (b) PBr_3, then $NaSCH_3$ (c) 2 Na (d) product of (c) + CH_3CH_2I (e) product of (d) + HI

Chapters 9, 10, 11

Multiple Choice:

1. Both aldehydes and ketones react with all of the following except:
(a) Tollens reagent (b) $NaBH_4$ (c) CH_3MgBr (d) NH_2OH

2. Which type of compound is most reactive?
(a) a nitrile (b) an amide (c) an acid chloride (d) an ester

3. Which of the following undergoes aldol self-condensation?
(a) Formaldehyde (b) Acetaldehyde (c) Benzaldehyde (d) 2,2-Dimethylbenzaldehyde

4. What reagent should be used to introduce bromine α to a carbonyl group?
(a) Br_2, $FeBr_3$ (b) Br_2, CH_3COOH (c) PBr_3 (d) HBr

5. Which of the following is most acidic?
(a) water (b) a ketone (c) a 1,3-keto ester (d) a 1,3-diketone

6. Which of the following combinations can't be used to produce 2-phenyl-2-butanol?
(a) 2-butanone and phenylmagnesium bromide (b) acetophenone and ethylmagnesium bromide (c) $C_6H_5COCH_2CH_3$ and methylmagnesium bromide (d) butanal and phenylmagnesium bromide

7. What reaction is used to produce β-keto esters?
(a) Grignard reaction (b) Claisen condensation (c) aldol condensation (d) malonic ester synthesis

8. Which of the following reactions is not a nucleophilic acyl substitution?
(a) reduction of a ketone with $NaBH_4$ (b) Fischer esterification (c) reaction of an acid chloride with an amine (d) Claisen condensation

9. Which of the following has four acidic hydrogens?
(a) 2,4-Pentanedione (b) Cyclohexanone (c) Propanoic acid (d) Acetophenone

10. A Grignard reaction can be used for all of the following except:
(a) converting a halide to an alkane (b) converting an ester to an alcohol
(c) converting a nitrile to a ketone (d) converting an amide to an alcohol

Tell whether the following statements are true or false:

1. An enolate ion is more reactive than an enol.

2. The Claisen condensation can be viewed as both an α-substitution and a nucleophilic acyl substitution.

3. An acetal is stable in aqueous solution.

4. Benzoic acid is a stronger acid than acetic acid.

5. 4-Methyl-2-hexanone is racemized by treatment with dilute acid or base.

6. In aqueous solution, some ketones exist as gem diols.

7. A nitrile can react with a Grignard reagent to yield a ketone.

8. Phenylacetic acid can be synthesized via a malonic ester synthesis.

9. The product of a Claisen condensation is an enone.

10 An amine can be formed from hydrogenation of either an amide or a nitrile.

11. Aldehydes are more reactive than ketones for steric reasons.

12. Enolate formation can be catalyzed by both acid and base.

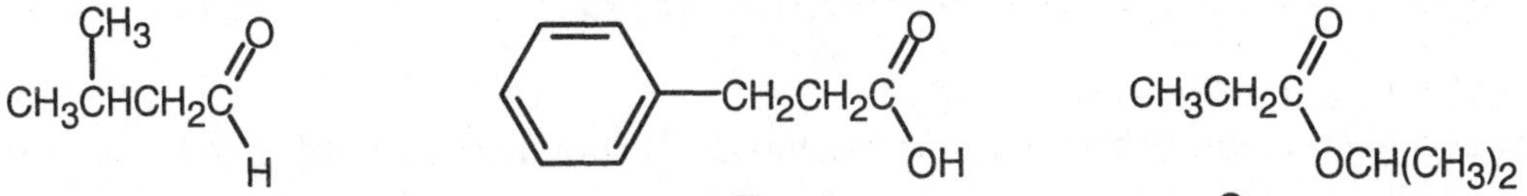

CH_3 (on C2) $CH_3CHCH_2C(=O)H$ — **A**

C_6H_5–$CH_2CH_2C(=O)OH$ — **B**

$CH_3CH_2C(=O)OCH(CH_3)_2$ — **C**

What is the name of compound **A**? How might **A** be prepared from the corresponding alcohol? What products might be expected when **A** reacts with (a) CH_3OH, H^+ catalyst (b) phenylmagnesium bromide (c) $LiAlH_4$ (d) hydroxylamine.

How many acidic hydrogens does **A** have? What products are formed when **A** is treated with: (a) Br_2, CH_3COOH (b) sodium ethoxide in ethanol (c) product of (b) + heat, acid catalyst.

Give the name of compound **B**. How can **B** be prepared from: (a) $C_6H_5CH_2CH_2Br$ (b) $C_6H_5CH_2CH_2CH_2OH$ (c) $C_6H_5CH_2Br$ (hint: malonic ester synthesis). What is the effect on carboxylic acid acidity if the aromatic ring has a nitro group in the para position?

Predict the products when **B** is treated with: (a) $SOCl_2$ (b) product of (a) + CH_3COO^- (c) CH_3OH, HCl (d) product of (a) + CH_3CH_2OH, pyridine (e) product of (a) + CH_3NH_2.

Name Compound **C**. Synthesize **C** from: (a) CH_3CH_2COOH (b) CH_3CH_2COCl (c) $CH_3CH_2C{\equiv}N$.

What products are formed when **C** reacts with: (a) H_2O, NaOH (b) $LiAlH_4$ (c) NH_3, ether (d) CH_3MgBr, ether.

How many acidic hydrogens does **C** have? What product is formed when **C** is treated with sodium ethoxide, followed by acidification?

Chapters 12, 13

Multiple Choice:

1. Which of the following compounds is the most basic?
(a) 2,4,6–Trimethylaniline (b) Aniline (c) Trimethylamine (d) Acetamide

2. A compound that shows an IR absorption at 1735 cm^{-1} is a:
(a) conjugated diene (b) ester (c) alcohol (d) amine

3. Which of the following is a fused-ring nitrogen heterocycle?
(a) Pyridine (b) Indole (c) Imidazole (d) Pyrrole

4. Which compound shows a 5-peak multiplet in its ^1H NMR spectrum?
(a) Toluene (b) Propanoic acid (c) 1,3-Dibromopropane (d) 3-Methyl-2-butanone

5. ^{13}C NMR is most useful for :
(a) identifying the functional groups in a compound (b) detecting conjugated π systems (c) counting the number of carbon atoms in a molecule (d) calculating the number of hydrogens bonded to an adjacent carbon

6. According to the "morphine rule", a biologically active opiate must have all of the following structural elements except:
(a) an aromatic ring (b) a tertiary amine (c) a methoxyl group (d) a quaternary carbon

7. Which spectroscopic technique is most useful for distinguishing between 1,3,5-octatriene and 1,4,7-octatriene?
(a) UV (b) IR (c) ^{13}C NMR (d) ^1H NMR

8. The number of peaks seen in the ^{13}C NMR spectrum of *cis*-3,4-dimethyl-3-hexene is:
(a) 3 (b) 4 (c) 5 (d) 6

9.

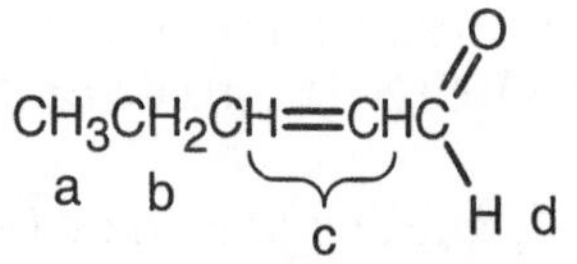

$CH_3CH_2CH{=}CHC(=O)H$

a b c H d

Arrange the ^1H NMR absorptions of the indicated hydrogens from downfield to upfield.
(a) a, b, c, d (b) d, c, b, a (c) d, b, c, a (d) b, a, c, d

10. Magnetic field strength is measured in units of:
(a) megahertz (MHz) (b) reciprocal centimeters (cm^{-1}) (c) joule (J) (d) tesla (T)

Tell whether the following statements are true or false:

1. Quaternary ammonium salts may have nitrogen stereocenters.

2. ^1H NMR is a useful technique because each unique hydrogen nucleus in a molecule experiences a slightly different magnetic field on irradiation.

3. A compound with a small value of pK_b gives up a proton more readily than a compound with a large value of pK_b.

4. The chemical shift of a carboxylic acid proton occurs at a larger δ value than does the chemical shift of any other commonly encountered proton.

5. It is difficult to synthesize butylamine from ammonia and 1-bromobutane without also forming dibutylamine and tributylamine.

6. The lone electron pair of the nitrogen atom of pyridine is part of the aromatic ring π system.

7. The larger the wave number, the higher the energy of electromagnetic radiation.

8. In the reaction of a primary alkylamine with ammonia, using a large excess of ammonia ensures that only monoalkylation product will be formed.

9. When the NMR spectrometer operating frequency changes from 60 MHz to 200 MHz the value of δ also changes.

10. Five peaks are visible in the ^{13}C NMR spectrum of *p*– nitroethylbenzene.

11. If compound **P** absorbs at a longer UV wavelength than Compound **Q**, the energy required for excitation is more for **P** than for **Q**.

12. ^{13}C NMR spectra provide information about the kinds of carbons adjacent to a given carbon.

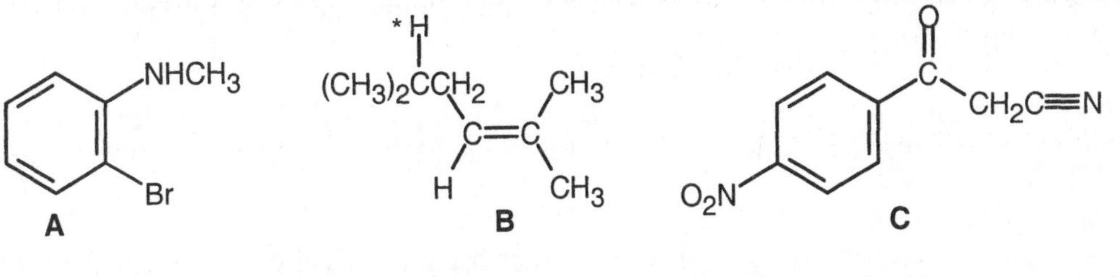

A **B** **C**

What is the name of **A**? Is **A** a primary,secondary or tertiary amine? Is **A** more or less basic than methylamine? Does the bromine on the ring make **A** more or less basic than the unbrominated compound? Design a synthetic route to **A** from benzene.

If the $-NHCH_3$ substituent on **A** is changed to $-CH_2NH_2$, what is the name of the new compound(**A´**)? Is **A´** a primary, secondary or tertiary amine? Is it more or less basic than **A**? Synthesize **A´** from benzene. Synthesize **A´** from *o*-bromobenzoic acid.

Name compound **B**. Do either UV or IR spectroscopy provide useful information about the structure of **B**? In the ^{13}C NMR spectrum of **B**, how many peaks are visible? What are the approximate chemical shifts of the peaks in the ^{13}C NMR spectrum.

How many peaks are visible in the ^1H NMR spectrum of **B**? Describe the ^1H NMR spectrum of **B**, including approximate chemical shifts and splitting patterns. If the starred atom in **B** is replaced with –Br, describe the resulting changes in the spectrum.

Does Compound **C** show a UV absorption in the range 250-400 nm? Would you expect the nitro group to increase the value of λ_{max} relative to the same compound without a nitro group? What significant peaks are visible in the IR spectrum of **C**? Does ^1H NMR provide much useful information in this problem?

Chapters 14, 15, 16, 17

Multiple Choice:

1. Which of the following amino acids doesn't contain an aromatic ring?
(a) Tyrosine (b) Proline (c) Histidine (d) Tryptophan

2. Secondary structure of a protein refers to:
(a) the sequence of covalent bonds between amino acids (b) the way that several proteins aggregate (c) the regular orientation of segments of the peptide backbone (d) the coiling of the entire molecule into a three-dimensional shape

3. All of the following types of reactions occur during β-oxidation of fatty acids except:
(a) dehydrogenation (b) retro-Claisen reaction (c) oxidation of an alcohol (d) aldol condensation

4. Which of the following is a monounsaturated fatty acid?
(a) Lauric acid (b) Linoleic acid (c) Oleic acid (d) Stearic acid

5. Which of the following monosaccharides has three carbon stereocenters?
(a) an aldohexose (b) a 2-deoxyaldopentose (c) a ketohexose (d) a ketopentose

6. If a strand of a nucleic acid has the sequence –A–C–A–T–G–G–C–T– (strand A) and a strand of a second nucleic acid has the sequence –A–C–A–U–G–G–C–U– (strand B), the relationship of strand A to strand B is:
(a) DNA coding strand and mRNA (b) DNA coding strand and DNA template strand
(c) DNA template strand and mRNA (d) DNA template strand and transfer RNA

7. Which of the following reagents is not used for oxidation of a monosaccharide?
(a) Benedict's reagent (b) Tollens reagent (c) dilute HNO_3 (d) CrO_3

8. A steroid is a lipid because:
(a) it can be extracted from tissues with a nonpolar organic solvent (b) it contains ester linkages (c) it is unsaturated (d) it has a characteristic tetracyclic ring system

9. Which amino acid has only one codon?
(a) Cysteine (b) Tyrosine (c) Lysine (d) Tryptophan

10. Which of the following is characteristic of tertiary structure?
(a) β-pleated sheet (b) hydrophobic interactions (c) superhelix (d) α-helix

Tell whether the following statements are true or false:

1. All aldoses are reducing sugars.

2. A sugar can be converted to an acetal by treatment with CH_3I and Ag_2O.

3. The purpose of the electron-transport chain is to convert acetyl groups to CO_2 and reduced coenzymes.

4. A dehydrase removes H_2 from a substrate.

5. Thymine differs from uracil in having a methyl group bonded to the pyrimidine ring.

6. The Fischer projection shown below represents an *R* enantiomer

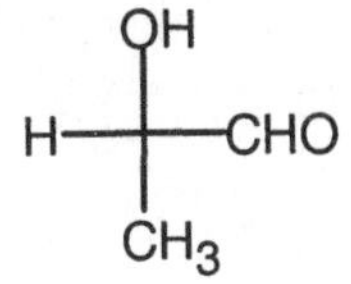

OH
H—+—CHO
CH$_3$

7. In gel electrophoresis of the products of cleavage of a DNA fragment, smaller products migrate farther from the origin than larger products.

8. An enzyme can change the equilibrium constant of a biochemical reaction.

9. Dicyclohexylcarbodiimide is used to protect amino groups during protein synthesis.

10. In the β anomer of an aldohexose, the anomeric hydroxyl group is cis to the hydroxyl group at carbon 6.

11. The hydrolysis of ATP can be coupled with an energetically unfavorable reaction so that the overall reaction has a favorable energy change.

12. Amylose contains 1,4'- and 1,6'-α-glycosidic bonds.

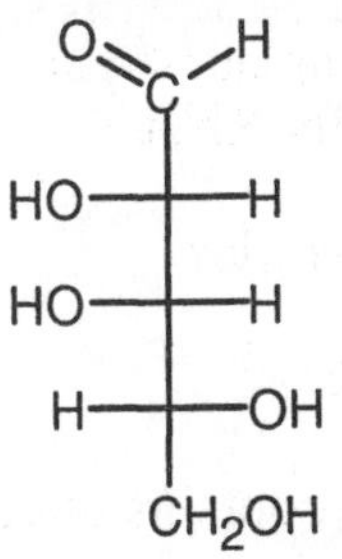

O=C—H; HO—H; HO—H; H—OH; CH$_2$OH **A**

CH$_2$CHC(=O)OH, NH$_2$ (imidazole ring, N, N—H) **B**

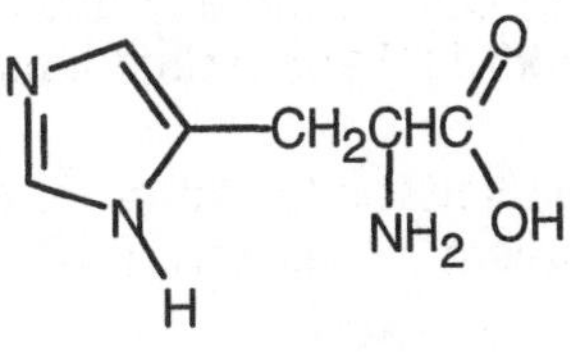

O=C—H; H—C—OH; CH$_2$OPO$_3^{2-}$ **C**

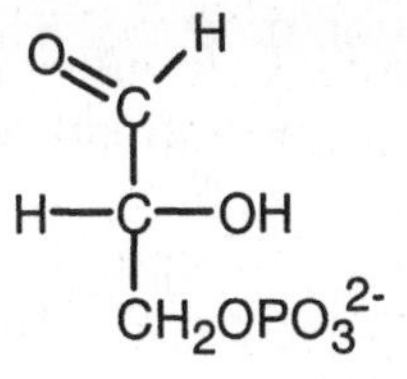

How is compound **A** classified? Is **A** a D or an L sugar? What is the *R,S* configuration at the carbon next to the carbonyl group? Is **A** a reducing sugar? Draw the β anomer of **A**, and show the products that result from reaction of **A** with: (a) $NaBH_4$ (b) CH_3OH, HCl (c) acetic anhydride (d) Tollens reagent (e) dilute HNO_3.

Which amino acid is represented by compound **B**? Is **B** acidic, neutral or basic? Draw **B** in its zwitterionic form. Is Compound **B**'s ring aromatic? Give a three-base codon for **B**, and give a three-base tRNA anticodon. What three-base sequence on DNA was used as the template?

Name compound **C**. **C** is an intermediate in which important biochemical pathway? Name the 6-carbon precursor to **C** and the type of reaction that produced **C**. Compound **C** is oxidized and phosphorylated to produce which compound? Which coenzyme is involved in this reaction? What is the end product of this biochemical pathway?

Summary of Reaction Mechanisms

The following table summarizes the most important types of polar reactions in organic chemistry. Much of organic chemistry is accounted for by these eleven mechanisms.

1. Electrophilic Additions to Alkenes (Sections 3.7–3.10)

Alkenes reacts with electrophiles such as HBr to yield saturated addition products. The reaction occurs in two steps: The electrophile first reacts with the alkene double bond to yield a carbocation intermediate that reacts further to yield the addition product.

$$\mathrm{C{=}C} \xrightarrow{\mathrm{HBr}} \left[\mathrm{H{-}C{-}C^{+}}\right] + \mathrm{Br^-} \longrightarrow \mathrm{H{-}C{-}C{-}Br}$$

Alkene — Carbocation — Addition product

2. Electrophilic Aromatic Substitutions (Sections 5.4–5.8)

Aromatic compounds react with electrophiles such as Br_2 to yield substitution products rather than addition products. The reaction occurs in two steps: The electrophile first reacts with the aromatic ring to yield a carbocation intermediate, which then loses H^+ to form a substituted aromatic product.

$$\text{Benzene} \xrightarrow[(\mathrm{Br_2} + \mathrm{FeBr_3})]{\mathrm{Br^+}} [\text{Carbocation}] \longrightarrow \text{Bromobenzene} + \mathrm{HBr}$$

Benzene — Carbocation — Bromobenzene

3. Nucleophilic Substitutions of Alkyl Halides

A. S_N2 Reaction (Section 7.5)

Alkyl halides undergo substitution when treated with nucleophiles. Primary and secondary alkyl halides react by the S_N2 mechanism, which occurs in a single step when the incoming nucleophile attacks the substrate from a direction 180° away from the leaving group. This results in an umbrella-like inversion of stereochemistry.

$$\mathrm{Br{-}C} \xrightarrow{:\ddot{O}H^-} \mathrm{C{-}OH} + \mathrm{Br^-}$$

Alkyl halide — Substitution product

B. S_N1 Reaction (Section 7.6)

Tertiary alkyl halides undergo nucleophilic substitution by the S_N1 mechanism, which occurs in two steps. Spontaneous dissociation of the alkyl halide into an anion and a carbocation intermediate takes place, followed by reaction of the carbocation with a nucleophile. The dissociation step is the slower of the two.

$(CH_3)_3C{-}Br \longrightarrow [(CH_3)_3C^+ + Br^-] \xrightarrow{H_2O} (CH_3)_3C{-}OH + HBr$

Alkyl halide — Carbocation — Substitution product

4. Elimination Reactions of Alkyl Halides

A. E2 Reaction (Section 7.7)

Alkyl halides undergo elimination of HX to yield alkenes on treatment with base. When a strong base such as hydroxide ion (HO^-) or alkoxide ion (RO^-) is used, alkyl halides react by the E2 mechanism. The E2 reaction occurs in a single step in which base removes a neighboring hydrogen at the same time that the halide ion leaves.

HO^- + H–C–C–Br $\xrightarrow{Na^+\ OH^-}$ C=C + H_2O + NaBr

Alkyl halide — Alkene

B. E1 Reaction (Section 7.8)

Tertiary alkyl halides undergo elimination by the E1 mechanism in competition with S_N1 substitution when a nonbasic nucleophile is used in a protic solvent. The reaction takes place in two steps: Spontaneous dissociation of the alkyl halide leads to a carbocation intermediate, which then loses H^+.

$(CH_3)_3C{-}Br \longrightarrow [(CH_3)_3C^+ + Br^-] \longrightarrow (H_3C)_2C{=}CH_2 + HBr$

Alkyl halide — Carbocation — Alkene

5. Nucleophilic Additions to Aldehydes and Ketones

A. Direct (1,2) Addition (Sections 9.6–9.11)

Aldehydes and ketones react with nucleophiles to yield addition products. The reaction occurs by addition of the nucleophile to the carbonyl group, producing a tetrahedrally hybridized alkoxide-ion intermediate, which is protonated to yield an alcohol.

$CH_3^-\ ^+MgBr$ → H_2O →

Ketone — Tetrahedral alkoxide-ion intermediate — Alcohol

B. Conjugate (1,4) Addition (Section 9.12)

α,β-Unsaturated aldehydes and ketones react with nucleophiles to yield saturated addition products. The reaction occurs by addition of the nucleophile to the β carbon of the double bond to give an enolate ion, followed by protonation on the α carbon.

CH_3NH_2 →

α,β-Unsaturated aldehyde/ketone — Saturated product

6. Nucleophilic Acyl Substitutions (Sections 10.7–10.12)

Carboxylic acid derivatives (acid chlorides, acid anhydrides, esters, amides) undergo a substitution reaction on treatment with nucleophiles. A nucleophilic acyl substitution reaction takes place in two steps: Addition of the nucleophile to the carbonyl group produces a tetrahedrally hybridized alkoxide-ion intermediate, which expels a leaving group to generate a new carbonyl compound.

$Na^+\ ^-OCH_3$ → → + NaCl

Acid chloride — Tetrahedral alkoxide-ion intermediate — Ester

7. Carbonyl Alpha-Substitution Reactions (Sections 11.2–11.6)

Carbonyl compounds having alpha hydrogens (hydrogens on carbon next to the carbonyl group) are in equilibrium with their enol tautomers. An alpha-substitution reaction takes place when an enol or enolate ion reacts with an electrophile to produce a substitution product.

$$H_3C{-}C(=O){-}CH_3 \underset{}{\overset{H^+}{\rightleftharpoons}} \left[H_3C{-}C(OH){=}CH_2\right] \xrightarrow{Br_2} H_3C{-}C(=O){-}CH_2Br + HBr$$

Ketone — Enol intermediate — α–substitution product

8. Carbonyl Condensation Reactions

A. Aldol condensation of aldehydes and ketones (Sections 11.8–11.10)

Aldehydes and ketones with alpha hydrogens undergo a base-catalyzed dimerization reaction leading to formation of β-hydroxy ketone/aldehyde products. The reaction occurs in two step when one molecule of ketone or aldehyde is converted into its enolate ion, which then does a nucleophilic addition reaction to the carbonyl group of a second molecule.

$$H_3C{-}C(=O){-}H \xrightarrow{:\bar{C}H_2{-}C(=O){-}H} \left[H_3C{-}C(O^-)(H){-}CH_2{-}C(=O){-}H\right] \xrightarrow{H_2O} CH_3{-}CH(OH){-}CH_2{-}C(=O){-}H$$

Aldehyde — Tetrahedral alkoxide-ion intermediate — β-Hydroxy aldehyde

B. Claisen condensation of esters (Section 11.11)

Esters with alpha hydrogens undergo a base-catalyzed dimerization reaction leading to formation of a β-keto ester product. The reaction occurs in two steps when one molecule of ester is converted into its enolate ion, which does a nucleophilic acyl substitution reaction on a second ester molecule.

$$H_3C{-}C({=}O){-}OCH_3 \xrightarrow{:\bar{C}H_2{-}C({=}O){-}OCH_3} \left[H_3C{-}C(O^-)(OCH_3){-}CH_2{-}C({=}O){-}OCH_3 \right]$$

Ester

Tetrahedral alkoxide-ion intermediate

$$\longrightarrow CH_3{-}C({=}O){-}CH_2{-}C({=}O){-}OCH_3 + CH_3O^-$$

β-Keto ester

Summary of Functional Group Preparations

The following table summarizes the synthetic methods used to prepare important functional groups. The functional groups are listed alphabetically, followed by reference to the appropriate text section and a brief description of each synthetic method.

Acetals

(Sec. 9.8)	from ketones and aldehydes by acid-catalyzed reaction with alcohols

Acid anhydrides

(Sec. 10.10)	from acid chlorides by reaction with carboxylate ions

Acid chlorides

(Sec. 10.9)	from carboxylic acids by reaction with $SOCl_2$

Alcohols

(Sec. 4.7)	from alkenes by hydroxylation with basic $KMnO_4$
(Sec. 4.4)	from alkenes by hydration with aqueous acid
(Sec. 7.5)	from alkyl halides by S_N2 reaction with hydroxide ion
(Sec. 8.7)	from ethers by acid-induced cleavage
(Sec. 8.8)	from epoxides by acid-catalyzed ring opening with H_2O
(Sec. 8.4)	from ketones and aldehydes by reduction with $NaBH_4$ or $LiAlH_4$
(Sec. 9.11)	from ketones and aldehydes by addition of Grignard reagents
(Sec. 8.4)	from carboxylic acids by reduction with $LiAlH_4$
(Secs. 8.4, 10.11)	from esters by reduction with $LiAlH_4$
(Sec. 10.11)	from esters by reaction with Grignard reagents

Aldehydes

(Sec. 4.7)	from disubstituted alkenes by reaction with acidic $KMnO_4$
(Secs. 8.5, 9.3)	from primary alcohols by oxidation

Alkanes

(Sec. 4.6)	from alkenes by catalytic hydrogenation
(Sec. 7.3)	from alkyl halides by protonolysis of Grignard reagents

Alkenes

(Sec. 7.7)	from alkyl halides by treatment with strong base (E2 reaction)
(Sec. 8.5)	from alcohols by dehydration
(Sec. 4.12)	from alkynes by catalytic hydrogenation using the Lindlar catalyst
(Sec. 11.3)	from α-bromo ketones by heating with pyridine

Amides

(Sec. 10.9)	from acid chlorides by treatment with an amine or ammonia
(Sec. 10.10)	from acid anhydrides by treatment with an amine or ammonia
(Sec. 10.11)	from esters by treatment with an amine or ammonia

Amines

(Secs. 7.5, 12.4)	from primary alkyl halides by treatment with ammonia
(Secs. 10.12, 12.4)	from amides by reduction with $LiAlH_4$
(Secs. 10.13, 12.4)	from nitriles by reduction with $LiAlH_4$
(Sec. 12.4)	From aldehydes and ketones by reductive amination

Aromatic hydrocarbons (Arenes)

(Sec. 5.6)	from arenes by Friedel-Crafts alkylation with a primary alkyl halide

Aromatic sulfonic acids

(Sec. 5.5)	from arenes by electrophilic aromatic substitution with SO_3/H_2SO_4

Arylamines

(Sec. 12.4)	from nitroarenes by reduction with either Fe, $SnCl_2$, or H_2/Pd

Carboxylic acids

(Sec. 4.7)	from alkenes by reaction with acidic $KMnO_4$
(Sec. 5.9)	from arenes by side-chain oxidation with $Na_2Cr_2O_7$ or $KMnO_4$
(Sec. 9.4)	from aldehydes by oxidation
(Sec. 8.5)	from alcohols by oxidation
(Secs. 10.4, 10.13)	from nitriles by acid or base hydrolysis
(Sec. 10.9)	from acid chlorides by reaction with aqueous base
(Sec. 10.10)	from acid anhydrides by reaction with aqueous base
(Sec. 10.11)	from esters by hydrolysis with aqueous base
(Sec. 10.12)	from amides by hydrolysis with aqueous base

Epoxides

(Sec. 8.8)	from alkenes by treatment with a peroxyacid

Esters

(Sec. 10.8)	from carboxylic acids by acid-catalyzed reaction with an alcohol (Fischer esterification)
(Sec. 10.9)	from acid chlorides by reaction with an alcohol
(Sec. 10.10)	from acid anhydrides by reaction with an alcohol
(Sec. 11.6)	from alkyl halides by alkylation with diethyl malonate

Ethers

(Sec. 8.7)	from primary alkyl halides by S_N2 reaction with alkoxide ions (Williamson ether synthesis)
(Sec. 8.8)	from alkenes by epoxidation with peroxyacids
(Sec. 8.6)	from phenols by reaction of phenoxide ions with primary alkyl halides

Halides, alkyl

(Secs. 4.1–4.2)	from alkenes by electrophilic addition of HX
(Sec. 4.5)	from alkenes by addition of halogen
(Sec. 4.12)	from alkynes by addition of halogen or HX
(Secs. 7.2, 8.4)	from alcohols by reaction with HX
(Sec. 7.2)	from alcohols by reaction with $SOCl_2$
(Sec. 7.2)	from alcohols by reaction with PBr_3
(Sec. 8.7)	from ethers by cleavage with HX
(Sec. 11.3)	from ketones by alpha-halogenation with Br_2

Halides, aryl

(Secs. 5.4, 5.5)	from arenes by electrophilic aromatic substitution with halogens

Imines

(Sec. 9.10)	from ketones or aldehydes by reaction with primary amines

Ketones

(Sec. 4.7)	from alkenes by reaction with acidic $KMnO_4$
(Sec. 4.13)	from alkynes by acid-catalyzed hydration
(Sec. 5.8)	from arenes by Lewis-acid-catalyzed reaction with an acid chloride (Friedel-Crafts acylation)
(Secs. 8.5, 9.3)	from secondary alcohols by oxidation
(Sec. 10.13)	from nitriles by reaction with Grignard reagents

Nitriles

(Secs. 7.5, 10.13)	from primary alkyl halides by S_N2 reaction with CN^-

Nitroarenes

(Sec. 5.5)	from arenes by electrophilic aromatic substitution with HNO_3/H_2SO_4

Organometallic compounds

(Sec. 7.3)	formation of Grignard reagents from organohalides by reaction with magnesium

Phenols

(Sec. 8.6) from arenesulfonic acids by heating with KOH

Quinones

(Sec. 8.6) from phenols by oxidation with $Na_2Cr_2O_7$

Sulfides

(Sec. 8.9) from thiols by S_N2 reaction of thiolate ions with primary alkyl halides

Thiols

(Sec. 8.9) from primary alkyl halides by S_N2 reaction with HS^-

Reagents Used in Organic Synthesis

The following table summarizes the uses of some important reagents in organic chemistry. The reagents are listed alphabetically, followed by a brief description of the uses of each and references to the appropriate text sections.

Acetic anhydride: Reacts with alcohols to yield acetate esters (Section 10.10).

Aluminum chloride: Catalyzes the Friedel-Crafts alkylation and acylation reactions of aromatic compounds (Section 5.6).

Bromine: Adds to alkenes yielding 1,2-dibromides (Section 4.5).

- Adds to alkynes yielding either 1,2-dibromoalkenes or 1,1,2,2-tetrabromoalkanes (Section 4.12).
- Reacts with aromatic compounds in the presence of $FeBr_3$ catalyst to yield bromoarenes (Section 5.4).
- Reacts with ketones in acetic acid solvent to yield α-bromo ketones (Section 11.3).

Di-*tert*-butoxy dicarbonate: Reacts with amino acids to give BOC protected amino acids for use in peptide synthesis (Section 15.3).

Chlorine: Adds to alkenes to yield 1,2-dichlorides (Section 4.5).

- Reacts with alkanes in the presence of light to yield chloroalkanes by a radical chain reaction pathway (Section 7.2).
- Reacts with aromatic compounds in the presence of $FeCl_3$ catalyst to yield chloroarenes (Section 5.5).

Chromium trioxide: Oxidizes alcohols in aqueous sulfuric acid to yield carbonyl-containing products. Primary alcohols yield carboxylic acids and secondary alcohols yield ketones (Sections 8.5, 9.3, 9.4).

Dicyclohexylcarbodiimide (DCC): Couples an amine with a carboxylic acid to yield an amide. DCC is often used in peptide synthesis (Section 15.3).

Ethylene glycol: Reacts with ketones or aldehydes in the presence of an acid catalyst to yield acetals, which serve as useful carbonyl protecting groups (Section 9.7).

Ferric bromide: Catalyzes the reaction of aromatic compounds with Br_2 to yield bromoarenes (Section 5.4).

Ferric chloride: Catalyzes the reaction of aromatic compounds with Cl_2 to yield chloroarenes (Section 5.5).

Grignard reagent: Adds to carbonyl-containing compounds (ketones, aldehydes, esters) to yield alcohols (Sections 9.11, 10.11).

Hydrogen bromide: Adds to alkenes to yield alkyl bromides. Markovnikov regiochemistry is observed (Sections 4.1, 4.2).
- Adds to alkynes to yield either bromoalkenes or 1,1-dibromoalkanes (Section 4.12).
- Reacts with alcohols to yield alkyl bromides (Section 7.2).

Hydrogen chloride: Adds to alkenes to yield alkyl chlorides. Markovnikov regiochemistry is observed (Section 4.2).
- Adds to alkynes to yield either chloroalkenes or 1,1-dichloroalkanes (Section 4.12).
- Reacts with alcohols to yield alkyl chlorides (Section 7.2).

Iodomethane: Reacts with alkoxide anions to yield methyl ethers (Section 8.5).
- Reacts with amines to yield methylated amines (Section 12.4).

Iron: Reduces nitroarenes in the presence of acid to yield anilines (Section 12.4).

Lindlar catalyst: Catalyzes the hydrogenation of alkynes to yield cis alkenes (Section 4.12).

Lithium aluminum hydride: Reduces ketones, aldehydes, esters, and carboxylic acids to yield alcohols (Section 8.4, 10.11).
- Reduces amides to yield amines (Section 10.12).
- Reduces nitriles to yield amines (Section 10.13).

Magnesium: Reacts with organohalides to yield Grignard reagents (Section 7.3)

Mercuric sulfate: Catalyzes the addition of water to alkynes in the presence of aqueous H_2SO_4, yielding ketones (Section 4.12).

Nitric acid: Reacts with aromatic compounds in the presence of sulfuric acid to yield nitroarenes (Section 5.5).
- Oxidizes aldoses to yield aldaric acids (Section 14.7).

Palladium on carbon: Catalyzes the hydrogenation of carbon–carbon multiple bonds. Alkenes and alkynes are reduced to yield alkanes (Sections 4.6, 4.12).
- Catalyzes the hydrogenation of nitroarenes to yield anilines (Section 12.4).

Phenyl isothiocyanate: Used in the Edman degradation of peptides to identify *N*-terminal amino acids (Section 15.7).

Phosphorus tribromide: Reacts with alcohols to yield alkyl bromides (Section 7.2).

Platinum oxide: Catalyzes the hydrogenation of alkenes and alkynes to yield alkanes (Sections 4.6, 4.12).

Potassium hydroxide: Reacts with alkyl halides to yield alkenes by an E2 reaction (Section 7.7).

Potassium permanganate: Oxidizes alkenes under basic conditions to yield 1,2-diols (Section 4.7).
- Oxidizes alkenes under neutral or acidic conditions to give carboxylic acids (Sections 4.7).
- Oxidizes aromatic compounds to yield benzoic acids (Section 5.9).

Pyridine: Reacts with α-bromo ketones to yield α,β-unsaturated ketones (Section 11.3).
- Catalyzes the reaction of alcohols with acid chlorides to yield esters (Section 10.9).
- Catalyzes the reaction of alcohols with acetic anhydride to yield acetate esters (Section 10.10).

Pyridinium chlorochromate: Oxidizes primary alcohols to yield aldehydes and secondary alcohols to yield ketones (Sections 8.5, 9.3).

Silver oxide: Oxidizes primary alcohols in aqueous ammonia solution to yield aldehydes (Tollens oxidation; Section 9.4).
- Catalyzes reaction of carbohydrates with iodomethane to yield ethers (Section 14.7).

Sodium borohydride: Reduces ketones and aldehydes to yield alcohols (Section 8.4).
- Reduces quinones to yield hydroquinones (Section 8.6).

Sodium cyanide: Reacts with alkyl halides to yield nitriles (Sections 7.5, 10.13).

Sodium dichromate: Oxidizes primary alcohols to yield carboxylic acids and secondary alcohols to yield ketones (Sections 8.5, 9.3).
- Oxidizes alkylbenzenes to yield benzoic acids (Section 5.9).

Sodium hydroxide: Reacts with arenesulfonic acids at high temperature to yield phenols (Section 8.6).

Stannous chloride: Reduces nitroarenes to yield arylamines (Section 12.4).
- Reduces quinones to yield hydroquinones (Section 8.6).

Sulfur trioxide: Reacts with aromatic compounds in H_2SO_4 to yield arenesulfonic acids (Section 5.5).

Sulfuric acid: Dehydrates alcohols to yield alkenes (Section 8.5).
- Reacts with alkynes in the presence of water and mercuric acetate to yield ketones (Section 4.12).

Thionyl chloride: Reacts with primary and secondary alcohols to yield alkyl chlorides (Section 7.2).

- Reacts with carboxylic acids to yield acid chlorides (Section 10.8).

Trifluoroacetic acid: Catalyzes cleavage of the BOC protecting group from amino acids in peptide synthesis (Section 15.3).

Organic Name Reactions

Aldol condensation reaction (Section 11.9): The nucleophilic addition of an enol or enolate ion to a ketone or aldehyde, yielding a β-hydroxy ketone.

$$2\ R\text{-}\overset{O}{\overset{\|}{C}}\text{-}\underset{|}{\overset{|}{C}}\text{-}H \xrightarrow{NaOH} R\text{-}\overset{O}{\overset{\|}{C}}\text{-}\underset{|}{\overset{|}{C}}\text{-}\underset{R}{\overset{OH}{C}}\text{-}\underset{|}{\overset{|}{C}}\text{-}H$$

Claisen condensation reaction (Section 11.11): A nucleophilic acyl substitution reaction that occurs when an ester enolate ion attacks the carbonyl group of a second ester molecule. The product is a β-keto ester.

$$2\ R\text{-}CH_2\text{-}\overset{O}{\overset{\|}{C}}\text{-}OCH_3 \xrightarrow[2.\ H_3O^+]{1.\ Na^+\ ^-OCH_3} R\text{-}CH_2\text{-}\overset{O}{\overset{\|}{C}}\text{-}\underset{R}{CH}\text{-}\overset{O}{\overset{\|}{C}}\text{-}OCH_3 + CH_3OH$$

Edman degradation (Section 15.7): A method for cleaving the *N*-terminal amino acid from a peptide by treatment of the peptide with *N*-phenylisothiocyanate.

$$Ph\text{-}N{=}C{=}S + H_2N\text{-}\underset{R}{CH}\text{-}\overset{O}{\overset{\|}{C}}\text{-}NH\text{-}\ \longrightarrow\ \text{(Ph–N ring with C=S, N–H, C(H)(R), C=O)} + H_2N\text{-}$$

Fehling's test (Section 14.7): A chemical test for aldehydes, involving treatment with cupric ion.

Fischer esterification reaction (Section 10.8): The acid-catalyzed reaction between a carboxylic acid and an alcohol to yield an ester.

$$R\text{-}\overset{O}{\overset{\|}{C}}\text{-}OH + R'\text{-}OH \xrightarrow{H^+,\ heat} R\text{-}\overset{O}{\overset{\|}{C}}\text{-}OR' + H_2O$$

Friedel-Crafts reaction (Section 5.6): The alkylation or acylation of an aromatic ring by treatment with an alkyl- or acyl chloride in the presence of a Lewis-acid catalyst.

Benzene $\xrightarrow[AlCl_3]{R{-}Cl}$ alkylbenzene (Ph–R) — Alkylation

Benzene $\xrightarrow[AlCl_3]{R{-}C(=O){-}Cl}$ Ph–C(=O)–R — Acylation

Grignard reaction (Section 9.11): The nucleophilic addition reaction of an alkylmagnesium halide to a ketone, aldehyde, or ester carbonyl group.

$$R{-}Mg{-}X + C{=}O \xrightarrow[2.\ H_3O^+]{1.\ mix} C(OH){-}R$$

Grignard reagent (Section 7.3): An organomagnesium halide, RMgX, prepared by reaction between an organohalide and magnesium metal.

Malonic ester synthesis (Section 11.6): A multi-step sequence for converting an alkyl halide into a carboxylic acid with the addition of two carbon atoms to the chain.

$$R{-}CH_2{-}X + Na^{+}\ {}^{-}{:}CH(COOCH_3){-}C(=O){-}OCH_3 \xrightarrow[2.\ H_3O^+,\ heat]{1.\ heat} RCH_2{-}CH_2COCH_3 + CO_2 + CH_3OH$$

Sanger dideoxy DNA sequencing (Section 16.14): A rapid and efficient method for sequencing long restriction fragments of DNA by using dye-labeled dideoxynucleotide triphosphates and DNA polymerase.

Tollens test (Section 9.4): A chemical test for detecting aldehydes by treatment with silver nitrate in aqueous ammonia. A positive test is signaled by formation of a silver mirror on the walls of the reaction vessel.

Walden inversion (Sections 7.4, 7.5): The inversion of stereochemistry at a stereocenter that occurs during S_N2 reactions.

$$Nu:^- + \text{C–X} \longrightarrow \text{Nu–C} + :X^-$$

Williamson ether synthesis (Section 8.5): A method for preparing ethers by treatment of a primary alkyl halide with an alkoxide ion.

$$R\text{–}O^- \ Na^+ + R'CH_2Br \longrightarrow R\text{–}O\text{–}CH_2R' + NaBr$$

Abbreviations

A	symbol for Angstrom unit (10^{-8} cm)
Ac–	acetyl group, $CH_3C(=O)$—
Ar–	aryl group
$[\alpha]_D$	specific rotation
BOC	*tert*-butoxycarbonyl group, $(CH_3)_3COC(=O)$—
bp	boiling point
sec-Bu	*sec*-butyl group, $CH_3CH_2CH(CH_3)$–
t -Bu	*tert*-butyl group, $(CH_3)_3C$–
cm^{-1}	wavenumber or reciprocal centimeter
D	stereochemical designation of carbohydrates and amino acids
DCC	dicyclohexylcarbodiimide, C_6H_{11}–N=C=N–C_6H_{11}
δ	chemical shift in ppm downfield from TMS
Δ	symbol for heat; also symbol for change
ΔH	heat of reaction
DNA	deoxyribonucleic acid
(*E*)	entgegen, stereochemical designation of double bond geometry
E_{act}	activation energy
E1	unimolecular elimination reaction
E2	bimolecular elimination reaction
Et	ethyl group, CH_3CH_2–
hν	symbol for light
Hz	Hertz
i-	iso
IR	infrared
J	Joule
J	symbol for coupling constant
K	Kelvin temperature
K_a	symbol for acidity constant
K_b	symbol for basicity constant
kcal	kilocalories
L	stereochemical designation of carbohydrates and amino acids
Me	methyl group, CH_3–
MHz	megahertz (10^6 cycles per second)
mol wt.	molecular weight
mp	melting point

NMR	nuclear magnetic resonance
–OAc	acetate group, $-OC(=O)CH_3$
PCC	pyridinium chlorochromate
Ph	phenyl group, $-C_6H_5$
pH	measure of acidity of aqueous solution (= $-\log [H_3O^+]$)
pK_a	measure of acid strength (= $-\log K_a$)
pK_b	measure of base strength (= $-\log K_b$)
ppm	parts per million
i-Pr	isopropyl group, $(CH_3)_2CH-$
R–	symbol for a generalized organic group
(*R*)	rectus, stereochemical designation of stereocenters
RNA	ribonucleic acid
(*S*)	sinister, stereochemical designation of stereocenters
sec-	secondary
S_N1	unimolecular substitution reaction
S_N2	bimolecular substitution reaction
tert-	tertiary
THF	tetrahydrofuran
TMS	tetramethylsilane, $(CH_3)_4Si$
UV	ultraviolet
(*Z*)	zusammen, stereochemical designation of double bond geometry
$\longrightarrow$	chemical reaction in direction indicated
$\rightleftharpoons$	reversible chemical reaction
$\longleftrightarrow$	resonance symbol
$\curvearrowright$	curved arrow indicating direction of electron flow
$\equiv$	is equivalent to
$>$	greater than
$<$	less than
$\approx$	approximately equal to
R-ξ	indicates that the organic fragment shown is a part of a larger molecule
◀ (wedge)	single bond coming out of the plane of the paper
-----	single bond receding into the plane of the paper
......	partial bond
$\delta+$, $\delta-$	partial charge
*	isotopically labeled atom
‡	denoting the transition state

Proton NMR Chemical Shifts

Type of Proton		Chemical Shift (δ)
Alkyl, primary	$R{-}CH_3$	0.7 – 1.3
Alkyl, secondary	$R{-}CH_2{-}R$	1.2 – 1.4
Alkyl, tertiary	$R_3C{-}H$	1.4 – 1.7
Allylic	$-C{=}C{-}C{-}H$	1.6 – 1.9
Aliphatic carbonyl	$-C(=O){-}C{-}H$	2.0 – 2.3
Benzylic	$Ar{-}C{-}H$	2.3 – 3.0
Acetylenic	$R{-}C{\equiv}C{-}H$	2.5 – 2.7
Alkyl chloride	$Cl{-}C{-}H$	3.0 – 4.0
Alkyl bromide	$Br{-}C{-}H$	2.5 – 4.0
Alkyl iodide	$I{-}C{-}H$	2.0 – 4.0
Amine	$N{-}C{-}H$	2.2 – 2.6
Epoxide	C—C(O)–H	2.5 – 3.5
Alcohol	$HO{-}C{-}H$	3.5 – 4.5
Ether	$RO{-}C{-}H$	3.5 – 4.5
Vinylic	$-C{=}C{-}H$	5.0 – 6.5
Aromatic	$Ar{-}H$	6.5 – 8.0
Aldehyde	$R{-}C(=O){-}H$	9.7 – 10.0
Carboxylic acid	$R{-}C(=O){-}O{-}H$	11.0 – 12.0
Alcohol	$R{-}O{-}H$	3.5 – 4.5
Phenol	$Ar{-}O{-}H$	2.5 – 6.0

Infrared Absorption Frequencies

Functional Group Class		Frequency (cm^{-1})
Alcohol	–O–H	3300–3600 (s)
	$>C-O-$	1050 (s)
Aldehyde	–CO–H	2720, 2820 (m)
aliphatic	$>C=O$	1725 (s)
aromatic	$>C=O$	1705 (s)
Alkane	$>C-H$	2850–2960 (s)
	$>C-C<$	800–1300 (m)
Alkene	$=C(H)-$	3020–3100 (m)
	$>C=C<$	1650–1670 (m)
	$RCH=CH_2$	910, 990 (m)
	$R_2C=CH_2$	890 (m)
Alkyne	$\equiv C-H$	3300 (s)
	$-C\equiv C-$	2100–2260 (m)
Alkyl bromide	$>C-Br$	500–600 (s)
Alkyl chloride	$>C-Cl$	600–800 (s)

Amine, *primary*	$-N(H)_2$	3400, 3500 (s)
secondary	$>N-H$	3350 (s)
Ammonium salt	$-N^{+}-H$	2200–3000 (broad)
Aromatic ring	Ar–H	3030 (m)
monosubstituted	Ar–R	690–710 (s)
		730–770 (s)
o-disubstituted		735–770 (s)
m-disubstituted		690–710 (s)
		810–850 (s)
p-disubstituted		810–840 (s)
Carboxylic acid	–O–H	2500–3300 (broad)
associated	$>C=O$	1710 (s)
free	$>C=O$	1760 (s)
Acid anhydride	$>C=O$	1820, 1760 (s)
Acid chloride		
aliphatic	$>C=O$	1810 (s)
aromatic	$>C=O$	1770 (s)

Amide, *aliphatic*	$\diagdown C{=}O$ /	1690 (s)
aromatic	$\diagdown C{=}O$ /	1675 (s)
N-substituted	$\diagdown C{=}O$ /	1680 (s)
N,N-disubstituted	$\diagdown C{=}O$ /	1650 (s)
Ester, *aliphatic*	$\diagdown C{=}O$ /	1735 (s)
aromatic	$\diagdown C{=}O$ /	1720 (s)
Ether	$-O-C\lt$ (—O—C with three bonds)	1050–1150 (s)
Ketone, *aliphatic*	$\diagdown C{=}O$ /	1715 (s)
aromatic	$\diagdown C{=}O$ /	1690 (s)
6-memb. ring	$\diagdown C{=}O$ /	1715 (s)
5-memb. ring	$\diagdown C{=}O$ /	1750 (s)
Nitrile, *aliphatic*	$-C\equiv N$	2250 (m)
aromatic	$-C\equiv N$	2230 (m)
Phenol	$-O-H$	3500 (s)

(s) = strong; (m) = medium intensity

Nobel Prizes in Chemistry

1901 **Jacobus H. van't Hoff** (The Netherlands):
"for the discovery of laws of chemical dynamics and of osmotic pressure"

1902 **Emil Fischer** (Germany):
"for syntheses in the groups of sugars and purines"

1903 **Svante A. Arrhenius** (Sweden):
"for his theory of electrolytic dissociation"

1904 **Sir William Ramsey** (Britain):
"for the discovery of gases in different elements in the air and for the determination of their place in the periodic system"

1905 **Adolf von Baeyer** (Germany):
"for his researches on organic dyestuffs and hydroaromatic compounds"

1906 **Henri Moissan** (France):
"for his research on the isolation of the element fluorine and for placing at the service of science the electric furnace that bears his name"

1907 **Eduard Buchner** (Germany):
"for his biochemical researches and his discovery of cell-less formation"

1908 **Ernest Rutherford** (Britain):
"for his investigation into the disintegration of the elements and the chemistry of radioactive substances"

1909 **Wilhelm Ostwald** (Germany):
"for his work on catalysis and on the conditions of chemical equilibrium and velocities of chemical reactions"

1910 **Otto Wallach** (Germany):
"for his services to organic chemistry and the chemical industry by his pioneer work in the field of alicyclic substances"

1911 **Marie Curie** (France):
"for her services to the advancement of chemistry by the discovery of the elements radium and polonium"

1912 **Victor Grignard** (France):
"for the discovery of the so-called Grignard reagent, which has greatly helped in the development of organic chemistry"

Paul Sabatier (France):
"for his method of hydrogenating organic compounds in the presence of finely divided metals"

1913 **Alfred Werner** (Switzerland):
"for his work on the linkage of atoms in molecules by which he has thrown new light on earlier investigations and opened up new fields of research especially in inorganic chemistry"

1914 **Theodore W. Richards** (U.S.):
"for his accurate determinations of the atomic weights of a great number of chemical elements"

1915 **Richard M. Willstätter** (Germany):
"for his research on plant pigments, principally on chlorophyll"

1916 No award

1917 No award

1918 **Fritz Haber** (Germany):
"for the synthesis of ammonia from its elements, nitrogen and hydrogen"

1919 No award

1920 **Walther H. Nernst** (Germany):
"for his thermochemical work"

1921 **Frederick Soddy** (Britain):
"for his contributions to the chemistry of radioactive substances and his investigations into the origin and nature of isotopes"

1922 **Francis W. Aston** (Britain):
"for his discovery, by means of his mass spectrograph, of the isotopes of a large number of nonradioactive elements, as well as for his discovery of the whole-number rule"

1923 **Fritz Pregl** (Austria):
"for his invention of the method of microanalysis of organic substances"

1924 No award

1925 **Richard A. Zsigmondy** (Germany):
for his demonstration of the heterogeneous nature of colloid solutions, and for the methods he used, which have since become fundamental in modern colloid chemistry"

1926 **Theodor Svedberg** (Sweden):
"for his work on disperse systems"

1927 **Heinrich O. Wieland** (Germany):
"for his research on bile acids and related substances"

1928 **Adolf O. R. Windaus** (Germany):
"for his studies on the constitution of the sterols and their connection with the vitamins"

1929 **Arthur Harden** (Britain):
Hans von Euler-Chelpin (Sweden):
"for their investigation on the fermentation of sugar and of fermentative enzymes"

1930 **Hans Fischer** (Germany):
"for his researches into the constitution of hemin and chlorophyll, and especially for his synthesis of hemin"

1931 **Frederich Bergius** (Germany):
Carl Bosch (Germany):
"for their contributions to the invention and development of chemical high-pressure methods"

1932 **Irving Langmuir** (U.S.):
"for his discoveries and investigations in surface chemistry"

1933 No award

1934 **Harold C. Urey** (U.S.):
"for his discovery of heavy hydrogen"

1935 **Frederic Joliot** (France):
Irene Joliot-Curie (France):
"for their synthesis of new radioactive elements"

1936 **Peter J. W. Debye** (Netherlands/U.S.):
"for his contributions our knowledge of molecular structure through his investigations on dipole moments and on the diffraction of X rays and electrons in gases"

1937 **Walter N. Haworth** (Britain):
"for his researches into the constitution of carbohydrates and vitamin C"

Paul Karrer (Switzerland):
"for his researches into the constitution of carotenoids, flavins, and vitamins A and B"

1938 **Richard Kuhn** (Germany):
"for his work on carotenoids and vitamins"

1939 **Adolf F. J. Butenandt** (Germany):
"for his work on sex hormones"

Leopold Ruzicka (Switzerland):
"for his work on polymethylenes and higher terpenes"

1940 No award

1941 No award

1942 No award

1943 **Georg de Hevesy** (Hungary):
"for his work on the use of isotopes as tracer elements in researches on chemical processes"

1944 **Otto Hahn** (Germany):
"for his discovery of the fission of heavy nuclei"

1945 **Artturi I. Virtanen** (Finland):
"for his researches and inventions in agricultural and nutritive chemistry, especially for his fodder preservation method"

1946 **James B. Sumner** (U.S.):
"for his discovery that enzymes can be crystallized"

John H. Northrop (U.S.):
Wendell M. Stanley (U.S.):
for their preparation of enzymes and virus proteins in a pure form"

1947 **Sir Robert Robinson** (Britain):
"for his investigations on plant products of biological importance, particularly the alkaloids"

1948 **Arne W. K. Tiselius** (Sweden):
"for his researches on electrophoresis and adsorption analysis, especially for his discoveries concerning the complex nature of the serum proteins"

1949 **William F. Giauque** (U.S.):
"for his contributions in the field of chemical thermodynamics, particularly concerning the behavior of substances at extremely low temperatures"

1950 **Kurt Alder** (Germany):
Otto P. H. Diels (Germany):
"for their discovery and development of the diene synthesis"

1951 **Edwin M. McMillan** (U.S.):
Glenn T. Seaborg (U.S.):
"for their discoveries in the chemistry of the transuranium elements"

1952 **Archer J. P. Martin** (Britain):
Richard L. M. Synge (Britain):
"for their development of partition chromatography"

1953 **Hermann Staudinger** (Germany):
"for his discoveries in the field of macromolecular chemistry"

1954 **Linus C. Pauling** (U.S.):
"for his research into the nature of the chemical bond and its application to the elucidation of the structure of complex substances"

1955 **Vincent du Vigneaud** (U.S.):
"for his work on biochemically important sulfur compounds, especially for the first synthesis of a polypeptide hormone"

1956 **Sir Cyril N. Hinshelwood** (Britain):
Nikolai N. Semenov (U.S.S.R.):
"for their research in clarifying the mechanisms of chemical reactions in gases"

1957 **Sir Alexander R. Todd** (Britain):
"for his work on nucleotides and nucleotide coenzymes"

1958 **Frederick Sanger** (Britain):
"for his work on the structure of proteins, particularly insulin"

1959 **Jaroslav Heyrovsky** (Czechoslovakia):
"for his discovery and development of the polarographic method of analysis"

1960 **Willard F. Libby** (U.S.):
"for his method to use carbon-14 for age determination in archaeology, geology, geophysics, and other branches of science"

1961 **Melvin Calvin** (U.S.):
"for his research on the carbon dioxide assimilation in plants"

1962 **John C. Kendrew** (Britain):
Max F. Perutz (Britain):
"for their studies of the structures of globular proteins"

1963 **Giulio Natta** (Italy):
Karl Ziegler (Germany):
"for their work in the controlled polymerization of hydrocarbons through the use of organometallic catalysts"

1964 **Dorothy C. Hodgkin** (Britain):
"for her determinations by X-ray techniques of the structures of important biochemical substances, particularly vitamin B_{12} and penicillin"

1965 **Robert B. Woodward** (U.S.):
"for his outstanding achievements in the 'art' of organic synthesis"

1966 **Robert S. Mulliken** (U.S.):
"for his fundamental work concerning chemical bonds and the electronic structure of molecules by the molecular orbital method"

1967 **Manfred Eigen** (Germany):
Ronald G. W. Norrish (Britain):
George Porter (Britain):
"for their studies of extremely fast chemical reactions, effected by disturbing the equilibrium with very short pulses of energy"

1968 **Lars Onsager** (U.S.):
"for his discovery of the reciprocal relations bearing his name, which are fundamental for the thermodynamics of irreversible processes"

1969 **Sir Derek H. R. Barton** (Britain):
Odd Hassel (Norway):
"for their contributions to the development of the concept of conformation and its application in chemistry"

1970 **Luis F. Leloir** (Argentina):
"for his discovery of sugar nucleotides and their role in the biosynthesis of carbohydrates"

1971 **Gerhard Herzberg** (Canada):
"for his contributions to the knowledge of electronic structure and geometry of molecules, particularly free radicals"

1972 **Christian B. Anfinsen** (U.S.):
"for his work on ribonuclease, especially concerning the connection between the amino acid sequence and the biologically active conformation"

Stanford Moore (U.S.):
William H. Stein (U.S.):
"for their contribution to the understanding of the connection between chemical structure and catalytic activity of the active center of the ribonuclease molecule"

1973 **Ernst Otto Fischer** (Germany):
Geoffrey Wilkinson (Britain):
"for their pioneering work, performed independently, on the chemistry of the organometallic sandwich compounds"

1974 **Paul J. Flory** (U.S.):
"for his fundamental achievements, both theoretical and experimental, in the physical chemistry of macromolecules"

1975 **John Cornforth** (Australia/Britain):
"for his work on the stereochemistry of enzyme-catalyzed reactions"

Vladimir Prelog (Yugoslavia/Switzerland):
"for his work on the stereochemistry of organic molecules and reactions"

1976 **William N. Lipscomb** (U.S.):
"for his studies on the structures of boranes illuminating problems of chemical bonding"

1977 **Ilya Pregogine** (Belgium):
"for his contributions to nonequilibrium thermodynamics, particularly the theory of dissipative structures"

1978 **Peter Mitchell** (Britain):
"for his contribution to the understanding of biological energy transfer through the formulation of the chemiosmotic theory"

1979 **Herbert C. Brown** (U.S.):
"for his application of boron compounds to synthetic organic chemistry"

Georg Wittig (Germany):
"for developing phosphorus reagents, presently bearing his name"

1980 **Paul Berg** (U.S.):
"for his fundamental studies of the biochemistry of nucleic acids, with particular regard to recombinant DNA"

Walter Gilbert (U.S.)
Frederick Sanger (Britain):
"for their contributions concerning the determination of base sequences in nucleic acids"

1981 **Kenichi Fukui** (Japan)
Roald Hoffmann (U.S.):
for their theories, developed independently, concerning the course of chemical reactions"

1982 **Aaron Klug** (Britain):
"for his development of crystallographic electron microscopy and his structural elucidation of biologically important nucleic acid – protein complexes"

1983 **Henry Taube** (U.S.):
"for his work on the mechanisms of electron transfer reactions, especially in metal complexes"

1984 **R. Bruce Merrifield** (U.S.):
"for his development of methodology for chemical synthesis on a solid matrix"

1985 **Herbert A. Hauptman** (U.S.):
Jerome Karle (U.S.):
"for their outstanding achievements in the development of direct methods for the determination of crystal structures"

1986 **John C. Polanyi** (Canada):
"for his pioneering work in the use of infrared chemiluminescence in studying the dynamics of chemical reactions"

Dudley R. Herschbach (U.S.):
Yuan T. Lee (U.S.):
"for their contributions concerning the dynamics of chemical elementary processes"

1987 **Donald J. Cram** (U.S.):
Jean-Marie Lehn (France):
Charles J. Pedersen (U.S.):
"for their development and use of molecules with structure-specific interactions of high selectivity"

1988 **Johann Deisenhofer** (Germany):
Robert Huber (Germany):
Hartmut Michel (Germany):
"for their determination of the structure of the photosynthetic reaction center of bacteria"

1989 **Sidney Altman** (U.S.):
Thomas R. Cech (U.S.):
"for their discovery of catalytic properties of RNA"

1990 **Elias J. Corey** (U.S.):
"for his development of the theory and methodology of organic synthesis"

1991 **Richard R. Ernst** (Switzerland):
"for his contributions to the development of the methodology of high resolution NMR spectroscopy"

1992 **Rudolph A. Marcus** (U.S.):
"for his contributions to the theory of electron-transfer reactions in chemical systems"

1993 **Kary B. Mullis** (U.S.):
"for his development of the polymerase chain reaction"

Michael Smith (Canada):
"for his fundamental contributions to the establishment of oligonucleotide-based site-directed mutagenesis and its development for protein studies"

1994 **George A Olah** (U.S.):
"for pioneering research on carbocations and their role in the chemical reactions of hydrocarbons"

1995 **F. Sherwood Rowland** (U.S.)
Mario Molina (U.S.)
Paul Crutzen (Germany)
"for their work in atmospheric chemistry, particularly concerning the formation and decomposition of ozone"

1996 **Robert F. Curl, Jr.** (U.S.)
Harold W. Kroto (U.K.)
Richard E. Smalley (U.S.)
"for their discovery of carbon atoms bound in the form of a ball (fullerenes)."

1997 **Paul D. Boyer** (U.S.)
John E. Walker (U.K.)
"for having elucidated the mechanism by which ATP synthase catalyzes the synthesis of adenosine triphosphate, the energy currency of living cells"

Jens C. Skou (Denmark)
"for his discovery of the ion-transporting enzyme Na^+–K^+ ATPase, the first molecular pump"

1998 **Walter Kohn** (U.S.)
John A. Pople (U.S.)
"to Walter Kohn for his development of the density-functional theory and to John Pople for his development of computational methods in quantum chemistry"

1999 **Ahmed H. Zewail** (Egypt, U.S.)
"for his studies of the transition states of chemical reactions using femtosecond spectroscopy."

2000 **Alan J. Heeger** (U.S.)
Alan G. MacDiarmid (U.S.)
Hideki Shirakawa (Japan)
"for opening and developing the important new field of electrically conductive polymers"

2001 **William S. Knowles** (U.S.)
Ryoji Noyori (Japan)
K. Barry Sharpless (U.S.)
"for their work on chirally catalysed hydrogenation and oxidation reactions"

2002 **John B. Fenn** (U.S.)
Koichi Tanaka (Japan)
"for their development of soft desorption ionisation methods for mass spectrometric analyses of biological macromolecules"

Kurt Wüthrich (Switzerland)
"for his development of nuclear magnetic resonance spectroscopy for determining the three-dimensional structure of biological macromolecules in solution"

2003 **Peter Agre** (U.S.)
"for the discovery of water channels in cell membranes"
Roderick MacKinnon (U.S.)
"for structural and mechanistic studies of ion channels in cell membranes"

2004 **Aaron Ciechanover** (Israel)
Avram Hershko (Israel)
Irwin Rose (U.S.)
"for the discovery of ubiquitin-mediated protein degradation"

2005 **Yves Chauvin** (France)
Robert H. Grubbs (U.S.)
Schrock, Richard R. (U.S.)
"for the development of the metathesis method in organic synthesis"